Phylogenetics, Genetics, and Breeding of Medicinal Plants

Phylogenetics, Genetics, and Breeding of Medicinal Plants

Editors

Wajid Zaman
Hakim Manghwar

MDPI

Basel • Beijing • Wuhan • Barcelona • Belgrade • Novi Sad • Cluj • Manchester

Editors
Wajid Zaman
Life Sciences
Yeungnam University
Gyeongsan
Republic of Korea

Hakim Manghwar
Scientific Research Center of
Xixia Agricultural Park
Chinese Academy of Sciences
Nanchang
China

Editorial Office
MDPI
St. Alban-Anlage 66
4052 Basel, Switzerland

This is a reprint of articles from the Special Issue published online in the open access journal *Genes* (ISSN 2073-4425) (available at: https://www.mdpi.com/journal/genes/special_issues/407QG96M2Z).

For citation purposes, cite each article independently as indicated on the article page online and as indicated below:

Lastname, A.A.; Lastname, B.B. Article Title. *Journal Name* **Year**, *Volume Number*, Page Range.

ISBN 978-3-7258-0464-1 (Hbk)
ISBN 978-3-7258-0463-4 (PDF)
doi.org/10.3390/books978-3-7258-0463-4

Contents

About the Editors

Wajid Zaman

Dr. Wajid Zaman completed his PhD from the Institute of Botany, Chinese Academy of Sciences, in June 2021. Currently, he is working as an international research professor at Yeungnam University, South Korea. He is interested in developing a career that combines teaching and research while maintaining his interest in plant sciences and advanced technologies.

Hakim Manghwar

Dr. Hakim Manghwar works at the Chinese Academy of Sciences, Nanchang, China. His current research involves the functional characterization and mechanism of genes against different stresses in Arabidopsis thaliana. Dr. Hakim has published peer-reviewed papers in journals such as *TIPS*, *Adv. Sci.*, *PBJ*, *Plant Physiol. Biochem*, *IJMS*, etc. Moreover, he is an editorial board member for well-reputed international journals.

genes

MDPI

Article

Assessing the Potential for Genome-Assisted Breeding in Red Perilla Using Quantitative Trait Locus Analysis and Genomic Prediction

Sei Kinoshita [1], Kengo Sakurai [1], Kosuke Hamazaki [2], Takahiro Tsusaka [3], Miki Sakurai [3], Terue Kurosawa [3], Youichi Aoki [3], Kenta Shirasawa [4], Sachiko Isobe [4] and Hiroyoshi Iwata [1,*]

1 Graduate School of Agricultural and Life Sciences, The University of Tokyo, Bunkyo, Tokyo 113-8657, Japan; kinoshita@ut-biomet.org (S.K.); sakurai@ut-biomet.org (K.S.)
2 RIKEN Center for Advanced Intelligence Project, Kashiwa, Chiba 227-0871, Japan; kosuke.hamazaki@riken.jp
3 TSUMURA & CO., Ami, Ibaraki 300-1155, Japan; tsusaka_takahiro@mail.tsumura.co.jp (T.T.); sakurai_miki@mail.tsumura.co.jp (M.S.); kurosawa_terue@mail.tsumura.co.jp (T.K.); aoki_youichi@mail.tsumura.co.jp (Y.A.)
4 Kazusa DNA Research Institute, Kisarazu, Chiba 292-0818, Japan; shirasaw@kazusa.or.jp (K.S.); sisobe@kazusa.or.jp (S.I.)
* Correspondence: hiroiwata@g.ecc.u-tokyo.ac.jp

Abstract: Red perilla is an important medicinal plant used in Kampo medicine. The development of elite varieties of this species is urgently required. Medicinal compounds are generally considered target traits in medicinal plant breeding; however, selection based on compound phenotypes (i.e., conventional selection) is expensive and time consuming. Here, we propose genomic selection (GS) and marker-assisted selection (MAS), which use marker information for selection, as suitable selection methods for medicinal plants, and we evaluate the effectiveness of these methods in perilla breeding. Three breeding populations generated from crosses between one red and three green perilla genotypes were used to elucidate the genetic mechanisms underlying the production of major medicinal compounds using quantitative trait locus analysis and evaluating the accuracy of genomic prediction (GP). We found that GP had a sufficiently high accuracy for all traits, confirming that GS is an effective method for perilla breeding. Moreover, the three populations showed varying degrees of segregation, suggesting that using these populations in breeding may simultaneously enhance multiple target traits. This study contributes to research on the genetic mechanisms of the major medicinal compounds of red perilla, as well as the breeding efficiency of this medicinal plant.

Keywords: breeding; genomic selection; medicinal plants; quantitative trait locus analysis; red perilla

Citation: Kinoshita, S.; Sakurai, K.; Hamazaki, K.; Tsusaka, T.; Sakurai, M.; Kurosawa, T.; Aoki, Y.; Shirasawa, K.; Isobe, S.; Iwata, H. Assessing the Potential for Genome-Assisted Breeding in Red Perilla Using Quantitative Trait Locus Analysis and Genomic Prediction. *Genes* **2023**, *14*, 2137. https://doi.org/10.3390/genes14122137

Academic Editors: Wajid Zaman and Hakim Manghwar

Received: 17 October 2023
Revised: 24 November 2023
Accepted: 25 November 2023
Published: 27 November 2023

1. Introduction

Perilla frutescens (perilla), an annual herb belonging to the Labiatae family native to China, is widely cultivated in China, Japan, Korea, and other Asian countries [1]. Perilla has long been used as an ingredient in food and traditional (Kampo) medicine. Perilla can be categorized into two types based on morphological differences in its leaves: red perilla ('aka-jiso'), which has dark red or purple leaves and stems due to the presence of anthocyanin, and green perilla ('ao-jiso'), which only contains a small amount of anthocyanin and has green leaves [2,3]. Red perilla is used as an ingredient in Kampo medicines, often to ease the stomach, cold, and anxiety, and is blended into several formulations, such as Hangekobokuto and Kososan. The main medicinal compounds in perilla are perillaldehyde (PA), rosmarinic acid (RA), and anthocyanins (ANT) [4]. PA is an odorant with antidepressant, anticancer, and antibacterial properties, whereas RA is a phenol with anti-inflammatory and antioxidant properties and is more abundant in green perilla, and ANT is a red pigment with antioxidant properties [2,3,5]. Owing to its benefits, red perilla has recently attracted attention as an ingredient in Kampo medicine, both in Asian countries and globally [6]. Therefore, the development of varieties with stable medicinal compounds is necessary to meet the increasing demand for red perilla. Although extensive research has been conducted on the bioactivity, metabolic pathways, and efficacy of red perilla medicinal compounds, no studies have investigated the enhancement of these medicinal compounds via selective breeding.

The quality of perilla plants used in Kampo medicine is determined by the concentrations of the major medicinal compounds: PA, RA, and ANT. Therefore, these compounds are considered the target traits for breeding red perilla as a medicinal plant. Conventional phenotype-selection-based breeding is not an efficient breeding method for medicinal plants because it requires extensive quantification of medicinal compounds in every individual of each generation, which is both expensive and time-consuming. Compound analysis involves plant sampling, drying, sample grinding, reagent preparation, compound extraction, quantification using ultra-high-performance liquid chromatography (UPLC), and data organization. These elaborate protocols take up to 1 month for even experienced individuals, therefore halting other experiments. In addition, there are economic costs involved in field management and labor. Genomic selection (GS) [7], an alternative approach in which genomic prediction (GP), is used to predict the phenotype from genome-wide marker information, which is suitable for improving medicinal traits. In GS, once a GP model is built, phenotypes can be predicted based solely on genomic information without the need to quantify medicinal compounds for every generation. If a target trait is controlled by a small number of quantitative trait loci (QTLs), marker-assisted selection (MAS), a selection method using markers that are strongly linked to the QTLs that control the target trait, can also be used for selection using only the genotypes of the linked markers. Genomic information can be obtained during the early growth stages of a plant, eliminating the need for field cultivation and accelerating the breeding cycle. With the development of next-generation sequencing (NGS) technology, obtaining genomic information has become easier and more economical. Therefore, GS and MAS are better suited than phenotypic selection for breeding red perilla since they minimize field cultivation and compound quantification.

GS was first proposed by Meuwissen et al. in 2001 and has since been successfully implemented in dairy cattle breeding [7,8]. In plant breeding, GS has been applied in major crops such as maize and rice, as well as other plants [9–11]. Most previous studies using GS have focused on yield or weight as target traits; however, some studies have focused on nutrients and other active ingredients, such as vitamin E in sweet corn, carotenoids in maize, or various metabolites in oats [12–14]. GS, focusing on active ingredients, is a crucial method in plant breeding. However, no studies have investigated GS in medicinal plants using medicinal compounds as target traits. Furthermore, in breeding to improve medicinal compounds, it is essential to target multiple compounds simultaneously. However, few such studies have been conducted. When several traits are improved simultaneously, it is important to consider both the correlation among the traits and the correlation between medicinal compounds and other important agronomic traits, such as yield and flowering. Therefore, it is insufficient to focus only on the phenotypic correlations among these traits. Genetic architecture and correlations estimated by genetic relationship matrices should also be utilized to assess whether each trait could be improved independently.

As few elite cultivars of red perilla exist for medicinal use, it is difficult to improve multiple medicinal compounds using only one population generated through a single cross combination. In this study, we used three populations generated by crossing red and green perilla to simultaneously enhance multiple medicinal compounds. When multiple populations with different genetic backgrounds are used, genetic correlations among traits, heritability, and accuracy of GPs are likely to differ among populations. Therefore, it is necessary to evaluate the extent of variation in these values across populations and whether the results differ from those obtained when predictions are made using all three populations together.

The present study had two main objectives: to evaluate the possibility of improving major medicinal compounds and other agronomic traits simultaneously in red perilla through genomic breeding and to identify the genetic characteristics of each population arising from differences in the effects of alleles derived from different parents and different levels of segregation in consequent generations. To achieve these objectives, we determined the genetic correlations among the traits mentioned above and conducted QTL and GP analyses using three populations with different genetic backgrounds. Based on the results of these analyses, the genetic architecture of each trait and the potential for breeding using GS and MAS are discussed. The results of this study provide valuable insights into the genetic mechanisms underlying the production of the major medicinal compounds of red perilla and will contribute substantially to red perilla breeding.

2. Materials and Methods

2.1. Plant Materials

The plant materials used in this study were the breeding populations of the F_3 and F_4 generations of two-way crosses between red and green perilla. The breeding population for each generation consisted of three populations. The cross-parents of these populations were 'SekihoS8', st27, st40, and st44. 'SekihoS8' is a representative variety of red perilla (*P. frutescens* var. *crispa f. purpurea*) owned by TSUMURA & CO. (Ami, Ibaraki, Japan), while st27, st40, and st44 were green perilla varieties (*P. frutescens Britton* var. *crispa Decne*) owned by the National Agricultural and Food Research Organization. The cross combinations were 'SekihoS8' × st27 (S827), 'SekihoS8' × st40 (S840), and 'SekihoS8' × st44 (S844), hereafter referred to as S827, S840, and S844, respectively. The plants were grown in the Ibaraki field, TSUMURA & CO, in 2021 and 2022.

The selection and breeding schemes used to derive the breeding populations are shown in Figure 1. The breeding scheme was advanced through repeated self-pollination after an initial cross with the parents mentioned above. In the F_2 generation, 200 lines were grown from each of the three crosses. From these, 100 lines from each population were selected for self-pollination, and three seeds per line were collected and cultivated for the F_3 generation. To maintain the genetic diversity of the population, in the F_3 generation, 50 lines were selected from those grown in the field, and 50 lines were selected from those not grown in the field for each population. Three individuals were cultivated in the next generation, F_4, from the seeds collected from each line.

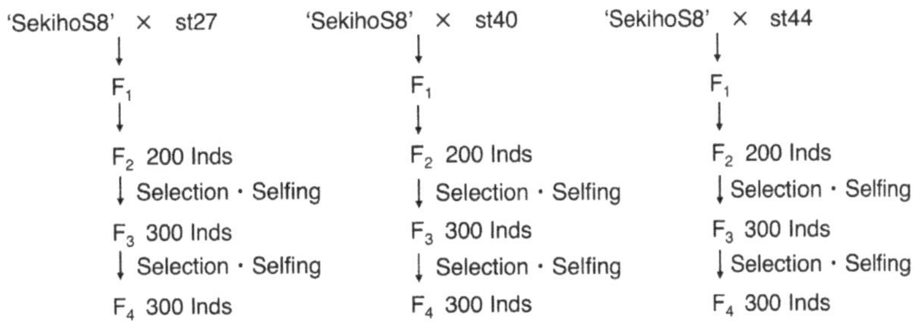

Figure 1. Selection and breeding scheme for the breeding populations used as plant parent materials in this study.

2.2. Phenotype Data

In the F_3 population, perillaldehyde (PA), rosmarinic acid (RA), and anthocyanin (ANT) content were measured in plants grown in the experimental field. In the F_4 population, PA, RA, yield, and flowering date were measured for plants grown in the experimental field.

PA and RA were quantified as described previously [15,16]. Briefly, to determine PA content, 0.1 g of each plant was weighed, placed in 25 mL of methanol, shaken, and centrifuged. The supernatant was separated and analyzed using UPLC at a wavelength of 234 nm using a UV detector. To determine the RA content, 0.05 g of each plant was weighed, placed in a 25 mL methanol/water (3:1) solution, shaken, and centrifuged, after which the supernatant was separated and analyzed by UPLC at a wavelength of 330 nm using a UV detector. The ANT content was measured thrice on 14 July, 29 July, and 12 August 2021. For each measurement, the ANT content was measured using three fully expanded leaves per plant with an anthocyanin content meter, and the average was recorded. The yield was measured thrice on 19 July, 29 July, and 8 August 2022, and all three measurements were added for subsequent analysis. For each yield measurement, the plants were harvested at a height of 45 cm using a machine, and the weight of the harvested plants was measured. The survey on flowering dates was conducted from 31 August 2022 to 30 September 2022. The number of days from 25 April, the date of sowing in the cell trays, to the flowering date was used for data analysis.

The phenotypic data for the four cross-parents collected in 2022 are shown in Table S1. SekihoS8 has a red leaf color, which maintains the content of ANT, whereas st27, st40, and st44 are superior in PA, RA, and YI to SekihoS8.

2.3. Genotype Data

Genomic DNA was extracted from all the individuals in the F_3 and F_4 generations, as well as from the four parental lines, 'SekihoS8', st27, st40, and st44. Double-digest restriction-associated DNA sequencing [17] was performed using a DNBSeq-G400RS (MGI Tech Co., Ltd. Shenzhen, China) to examine the genome-wide single nucleotide polymorphisms (SNPs). The 100 paired-end reads obtained were mapped to the reference sequence Hoko-3 (*P. frutescens cv.*), as published by Tamura et al. [18] using Bowtie2 [19]. Variant calling (max-missing 0.9) was conducted, and low-quality SNPs were filtered out using the VCFtools version 0.1.16 [20], after which 2282 SNPs were identified. SNPs with minor allele frequencies (MAF < 0.025) were excluded, and missing data were included using Beagle version 5.1 [21], after which they were filtered again using MAF < 0.025. Finally, 2063 SNPs were identified. These SNPs were filtered with MAF < 0.01 for both each population and the total three populations. In the F_3 generation, 1964 SNPs remained in the entire population, including 727, 1432, and 579 each in S827, S840, and S844, respectively. In the F_4 generation, 1964 SNPs remained in the entire population, including 862, 1432, and 579 in S827, S840, and S844, respectively. For further analysis, the score for SNPs of the same genotype as 'SekihoS8' was set to 0, 1 for heterozygous SNPs, and 2 for homozygous SNPs in the other parent.

2.4. Linkage Map Construction and QTL Analysis

The perilla used in this study was an allotetraploid (2n = 4x = 40) species with a genome size of about 1.24 Gb [18]. The F_3 generation was used to create a linkage map, and the 1964 SNPs described in the previous subsection were used as marker genotypes. The order of the markers was the same as that described in Section 2.3. Based on the expected ratio of 3:2:3 for the marker classes at each locus, the Maximum Likelihood Method was used to estimate the recombination rates between markers for each population. Markers with recombination rates greater than 0.499 were excluded and the remaining markers were transformed into map distances using the Kosambi function [22]. Subsequently, to obtain a common linkage map among the populations, smoothing was performed among the three populations using the 'loess' function in the 'stats' package version 4.1.2 in R. A common linkage map is shown in Figure S1.

Based on the common linkage map described above, QTL analysis was performed for each population using the 'cim' function in the 'qtl' package version 1.60 in R [23]. The window size was set to 10 cM, and the covariate markers were set to five. Logarithm of odds (LOD) scores were determined using 10,000 permutations. The targeted traits in the QTL analysis were PA, RA, and ANT in the F_3 generation and PA, RA, yield, and flowering date in the F_4 generation.

2.5. Genomic Heritability and GP Model

Genomic heritability was estimated for all the traits described in Section 2.2, and single- and multi-trait GP models were constructed for each generation. GPs were generated for each population and the three combined populations.

To evaluate the GP accuracy, a 10-fold cross validation was performed and repeated 10 times. For each cross validation, the Pearson correlation between the observed and predicted values was calculated, and the average of these correlations was used as the prediction accuracy.

2.5.1. Single Trait GP Model

The GBLUP [24] and BayesB [7] models were used for single-trait GP. The GBLUP model can be expressed as:

$$\mathbf{y} = \mathbf{X}\boldsymbol{\beta} + \mathbf{Z}\mathbf{u} + \varepsilon, \tag{1}$$

where n is the number of individuals, g is the number of genotypes, $\mathbf{y}$ is an n $\times$ 1 vector representing phenotypic values of a target trait, $\mathbf{X}\boldsymbol{\beta}$ is an n $\times$ 1 vector corresponding to fixed effects, $\mathbf{Zu}$ is an n $\times$ 1 vector corresponding the term of random effects. In the prediction model for each population, $\mathbf{X}\boldsymbol{\beta}$ represents an intercept, whereas in the model for all three populations, $\mathbf{X}$ represents an n $\times$ 3 design matrix, $\boldsymbol{\beta}$ is a 3 $\times$ 1 vector, and $\mathbf{X}\boldsymbol{\beta}$ is a term correcting for the average effect of each population. The random effect $\mathbf{u}$ is a g $\times$ 1 vector corresponding to the genotypic values that follow a multivariate normal distribution.

$$\mathbf{u} \sim \mathrm{MVN}(\mathbf{0}, \mathbf{G}\sigma^2{}_u), \tag{2}$$

where $\mathbf{G}$ is a g $\times$ g additive genomic relationship matrix, and $\sigma^2{}_u$ is a genetic variance. In this study, the additive genomic relationship matrix $\mathbf{G}$ was computed based on the marker genotype data using the 'calcGRM' function in the 'RAINBOWR' package version 0.1.29 in R [25]. The additive relationship matrix G was computed as a linear kernel of the marker genotype scaled by the allele frequency divided by the normalization constant, as described previously [26]. ε is an n $\times$ 1 error vector that follows a multivariate normal distribution;

$$\varepsilon \sim \mathrm{MVN}(\mathbf{0}, \mathbf{I_n}\sigma^2{}_e), \tag{3}$$

where $\mathbf{I_n}$ is an n $\times$ n identity matrix, and $\sigma^2{}_e$ is an error variance. We estimated the genetic variance $\sigma^2{}_u$ and the error variance $\sigma^2{}_e$ using the 'EMM.cpp' function in the 'RAINBOWR' package version 0.1.29 in R [24]. Based on the estimated genetic and error variances, the genomic heritability for each trait was calculated as $h^2 = \sigma^2{}_u/(\sigma^2{}_u + \sigma^2{}_e)$.

The BayesB model can be written as:

$$\mathbf{y} = \mathbf{X}\boldsymbol{\beta} + \mathbf{W}\mathbf{a} + \varepsilon, \tag{4}$$

where m is the number of markers, $\mathbf{W}$ is an n $\times$ m matrix of the marker genotype, and $\mathbf{a}$ is an m $\times$ 1 vector representing marker effects. The model was implemented using the 'BGLR' function in the 'BGLR' package version 1.1.0 in R [27].

2.5.2. Multi Trait GP Model

GBLUP [28] and BayesCπ [29] models were used for multi-trait GP. The GBLUP model is similar to Equation (1), but with the vector of phenotypic values $\mathbf{y}$ extended to an nt $\times$ 1 vector through the vertical alignment of the phenotypic values of 't' traits. Additionally, the random effect $\mathbf{u}$ now follows a multivariate normal distribution $\mathbf{u} \sim \mathrm{MVN}(\mathbf{0}, \mathbf{K} \otimes \mathbf{G})$, where $\mathbf{K}$ is a t $\times$ t genetic variance–covariance matrix between the t traits, and $\mathbf{G}$ is the g $\times$ g additive genetic matrix. The non-diagonal elements of $\mathbf{K}$ represent covariances between traits. The error ε follows a multivariate normal distribution $\varepsilon \sim \mathrm{MVN}(\mathbf{0}, \mathbf{R} \otimes \mathbf{I_n})$, where $\mathbf{R}$ is the t $\times$ t residual variance–covariance matrix of the traits, and $\mathbf{I_n}$ is the n $\times$ n identity matrix. Here, $\mathbf{A} \otimes \mathbf{B}$ represents the Kronecker product between the two matrices, A and B. The BayesCπ model is similar to Equation (4). The model was also extended to multi-trait GP by considering the covariances between the different 't' traits at each marker. Furthermore, in BayesB for the single-trait model, the prior distribution of each marker effect is a scaled-t distribution, whereas, in the BayesCπ for the multi-trait model, a $t \times 1$ vector of each marker effect for t traits follows a normal prior distribution. The multi-trait GP models were implemented using the 'Multitrait' function in the 'BGLR' package version 1.1.0 in R [27].

3. Results

3.1. QTL Analysis

We performed QTL analysis in each population and each generation. In the F_3 generation, the three target traits (PA, RA, and ANT) were investigated. In the F_4 generation, PA, RA, yield, and flowering date were investigated. Following composite interval mapping, 136 QTLs in the F_3 generation and 268 QTLs in the F_4 generation were detected above the threshold LOD score. The number of QTLs consistent across multiple populations or two generations was 48, 0, 2, 2, and 11 for PA, RA (Figure S2), ANT (Figure S3), yield (Figure S4), and flowering date (Figure S5), respectively. Among these QTLs, two each for ANT and yield were detected at identical locations. These two QTLs may represent a single pleiotropic QTL. As shown in Figure 2a, when the genotype at the detected QTL was homozygous for 'SekihoS8' (red perilla), the phenotypic value of ANT content was also higher. In S844, no QTLs were detected, and all individuals were homozygous for 'SekihoS8' at the corresponding markers. However, when the genotype at the QTLs detected for yield was homozygous for 'SekihoS8', the phenotypic value of yield was lower than that of the other genotypes (Figure 2b).

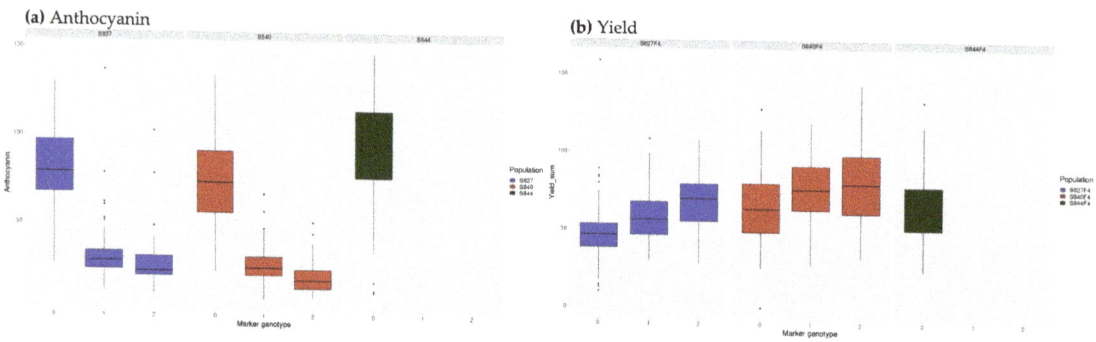

Figure 2. Genotypes at the detected quantitative trait loci (QTLs) identical to anthocyanin and yield. Genotype scores 0, 1, and 2 represent homozygous for 'SekihoS8', heterozygous, and homozygous for the other cross parents, respectively. Blue: S827; Red: S840; Green: S844. (**a**) The genotypes at the detected QTL on chr8 for anthocyanin (1st measurement); (**b**) genotypes at the detected QTL on chr8 for yield.

Eleven QTLs for flowering date were detected across multiple populations, of which 1 QTL was common to S827 and S844, and 10 QTLs were common to S840 and S844. We found that when the genotype at the detected QTLs was homozygous for 'SekihoS8', the plant showed early flowering (Figure S6).

Among the QTLs detected in PA, 16 were common to S827 and S844 on chr5. Twenty-eight QTLs were detected in S840 on chr5 and chr7, which were common to the F_3 and F_4 generations and were at different positions from the QTLs detected in the other two populations (Figure 3). For QTLs commonly detected in S827 and S844, plants with genotypes homozygous for 'SekihoS8' had lower PA content (Figure S7a). In contrast, for QTLs detected only in S840, plants with genotypes homozygous for 'SekihoS8' had high PA contents, whereas those with genotypes homozygous for 'st40' had a PA content close to zero (Figure S7b).

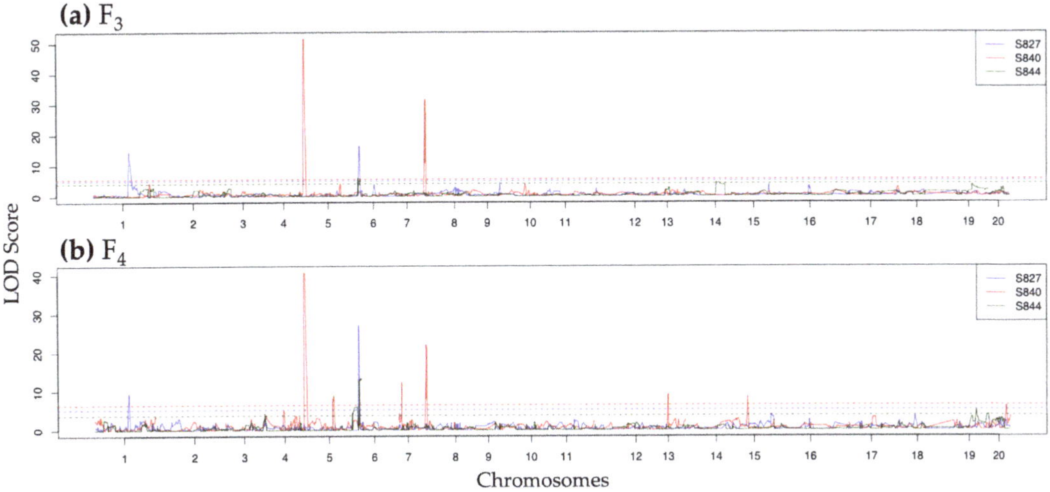

Figure 3. Quantitative trait loci detected for perillaldehyde in S827 (blue), S840 (red), and S844 (green): (**a**) QTLs detected in the F_3 population; (**b**) QTLs detected in the F_4 population. Dashed lines represent the logarithm of odds (LOD) threshold estimated by 10,000 permutations for each population.

3.2. Genomic Heritability

The genomic heritability of all the traits mentioned in Section 2.2, estimated using each population and the three populations combined, is shown in Tables 1 and 2. Genomic heritability varied substantially among traits and populations. Among all traits, RA had the lowest genomic heritability, with values ranging from 0.151 to 0.484. In contrast, PA, ANT, yield, and flowering date showed considerably higher genomic heritability. Among the three populations, the genomic heritability estimated using S844 was consistently the lowest for almost all traits except yield. When genomic heritability was estimated using the three populations combined, it was always greater than the average genomic heritability estimated using each population singularly. For the flowering date, the genomic heritability estimated using the three populations combined was 0.879, which was the highest among all estimates made using each population.

Table 1. Estimated genomic heritability in F_3 generation using each population and three populations combined.

Population	PA	RA	ANT		
			Day 1	Day 2	Day 3
S827	0.626	0.227	0.718	0.707	0.721
S840	0.505	0.484	0.803	0.726	0.760
S844	0.190	0.186	0.114	0.241	0.355
All	0.548	0.401	0.661	0.644	0.639

PA, perillaldehyde; RA, rosmarinic acid; ANT, day n; n^{th} measurement of anthocyanins.

Table 2. Estimated genomic heritability in F_4 generation using each population and three populations combined.

Population	PA	RA	YI	FD
S827	0.766	0.394	0.634	0.869
S840	0.796	0.389	0.605	0.848
S844	0.449	0.151	0.715	0.691
All	0.735	0.350	0.702	0.879

PA, perillaldehyde; RA, rosmarinic acid; YI, yield; FD, flowering date.

3.3. Genetic Correlation

The multi-trait GBLUP model was used to estimate the genetic correlations among all traits. As shown in Figure 4, variations were observed in the positivity, negativity, and magnitude of genetic correlations among different generations and populations. However, for ANT, a strong positive correlation was observed across the three measurements in all populations. For genetic correlations estimated using three populations combined, two prime medicinal compounds, PA and RA, showed weak negative correlations with ANT (−0.2 to −0.12), while PA and yield showed a moderate positive correlation. The genetic correlation between these two traits became more negative from the F_3 to the F_4 generation, except in S840. When genetic correlations were estimated for each population, S840 had the lowest absolute values for almost all traits.

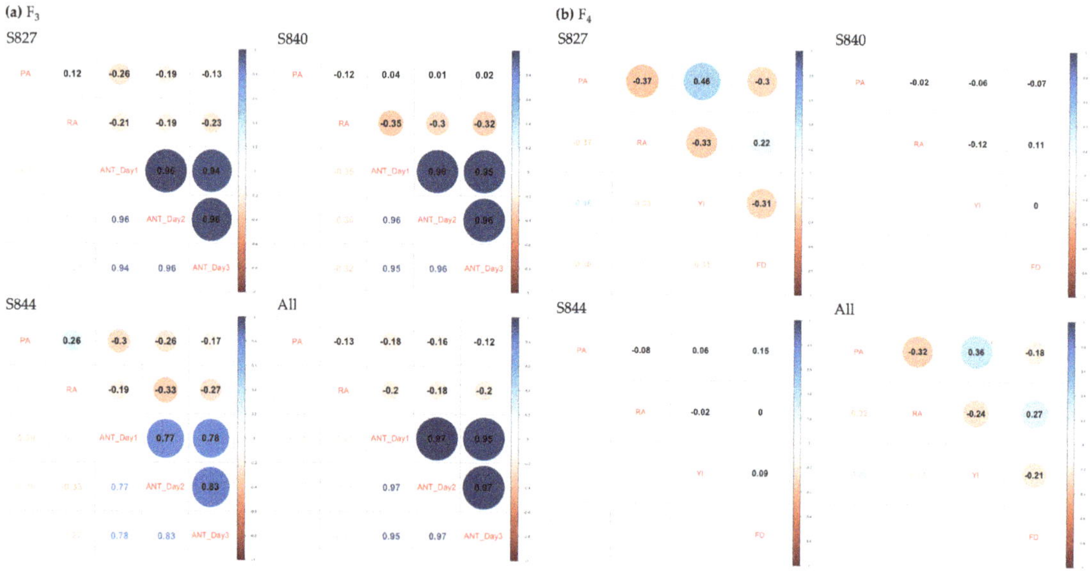

Figure 4. Genetic correlation among traits using each population and the three populations combined: (**a**) Genetic correlations using the F_3 population; (**b**) genetic correlations using the F_4 population. PA: perillaldehyde, RA: rosmarinic acid, ANT_Day n: n^{th} measurement of anthocyanin, YI: yield, FD: flowering date.

3.4. Accuracy of Genomic Prediction

GP models were constructed for each F_3 and F_4 generation using each population and the three populations combined. The prediction accuracy of the two single-trait models (GBLUP and Bayes B) is shown in Figure 5. Traits and populations with low genomic heritability tended to have a lower GP accuracy. When the models were constructed using each population, regardless of the GP model, S844 had the lowest accuracy for almost all traits, while S827 and S840 had the highest accuracy for yield and flowering dates and PA, RA, and ANT, respectively. The models constructed using the three populations combined showed improved accuracy compared with the models constructed using each population for PA and flowering date. When comparing the accuracies of the GBLUP and BayesB models, the BayesB model had better prediction accuracies for most traits, with values ranging from 0.0466% to 41.7%, higher than those of GBLUP. However, the GBLUP model was superior for some traits and populations, particularly RA, with a 0.0547–4.25% higher accuracy than the BayesB model. Furthermore, the same GP model showed higher prediction accuracy using the F_4 population than when using the F_3 population. When the multi-trait models (GBLUP and BayesCπ) were used, the prediction accuracies obtained were similar to those obtained using the single-trait model (Figure S8).

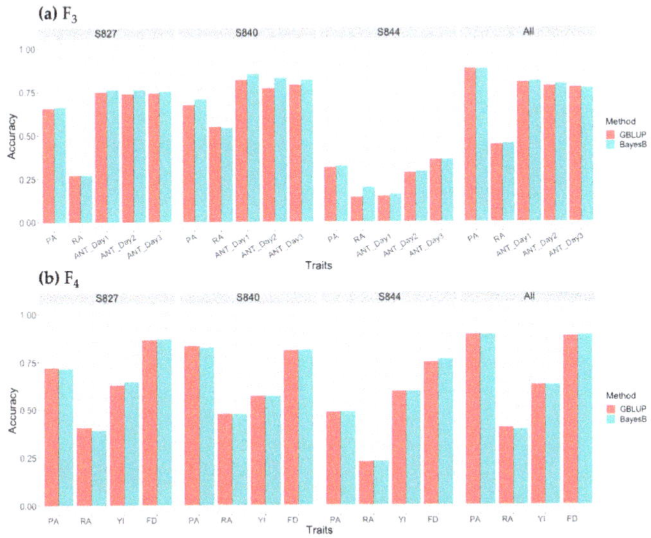

Figure 5. Prediction accuracy of single-trait genomic prediction (GP) by the GBLUP (red) and BayesB (cyan) models using each population and the three populations combined: (**a**) prediction accuracy using the F_3 population; (**b**) prediction accuracy using the F_4 population. PA: perillaldehyde, RA: rosmarinic acid, ANT_Day n: n^{th} measurement of anthocyanin, YI: yield, FD: flowering date.

4. Discussion

In this study, we conducted QTL and GP analyses for the major medicinal compounds and other agronomic traits of *P. frutescens* to reveal the genetic mechanisms underlying these traits and to evaluate the efficiency of GS and MAS. Genetic correlations, heritability, and GP accuracy revealed that each population had varying characteristics; however, reasonable values were obtained when the three populations were used together. The results of the QTL analysis for each trait showed that traits such as PA, ANT, and flowering date had significant QTLs with high LOD scores; however, no significant QTLs were detected for RA. One possible reason for this could be that no significant differences in RA content among the four cross-parents were observed, as shown in Table S1, and all parents already had genes that had a large effect on RA. These results revealed the genetic characteristics of each trait. Examination of the association between marker genotypes at the QTL and phenotypes of ANT revealed that plants with the 'SekihoS8' homozygous genotype had a significantly higher ANT content. 'SekihoS8' is a red perilla variety, which has been reported to accumulate more ANT in the leaves [2,3]. In contrast, the green perilla line, another cross-parent, has been reported to accumulate minimal amounts of ANT in its leaves. Therefore, the detected QTLs were considered to be important gene loci associated with the ANT metabolic pathway. Zhang et al. performed a genome-wide association study for ANT and detected a strong signal on chr8, which was assumed to be the MYB transcription factor regulating ANT biosynthesis in vegetative plant tissues [30]. The two QTLs for ANT detected in our study were also located on chr8 and had extremely high LOD scores, consistent with those reported by Zhang et al. [30]. QTLs for yield were identical to those for ANT, but yield tended to be lower when the genotype at these QTLs was homozygous for the 'SekihoS8′ allele. The QTLs for ANT had a contradictory effect on yield; however, simultaneous enhancement of both traits during breeding was not difficult, as the yield was also substantially affected by other markers. Therefore, individuals with high ANT content can be selected by MAS using these markers, and an improvement in yield can be achieved by combining the effects of other markers with GS. For PA, the QTLs detected in S840 differed from those detected in S827 and S844. When the marker genotypes of the QTLs detected only in S840 were homozygous for st40, PA content was nearly zero. PA is an essential oil that produces a strong odor. During cultivation, st40 differed considerably from st27 and st44. These results suggest that plant PA production is controlled by QTLs located on chr5 and chr7 of S840. Furthermore, among the perilla plants that produce PA, QTLs that determine the amount of accumulated PA may be located at different positions on chr5, which were detected in S827 and S844. Among the detected QTLs, 1 QTL was common to both S827 and S844, and 10 QTLs were common to both S840 and S844. Kang et al. performed QTL analysis on traits related to flowering time and detected six QTLs that were identified as perilla orthologs regulating flowering time [31]. However, the linkage groups in their study did not converge to 20 (the number of chromosomes in perilla). Therefore, we cannot confirm whether this is consistent with our study.

According to the genomic heritability and GP estimation, all traits except RA had considerably high genomic heritability, indicating that improvement through GS would be effective for these traits. Even for RA, which had the lowest genomic heritability among all traits, the genomic heritability was approximately 0.4 in S827 and S840. Since we evaluated the genomic heritability of each trait based only on additive genetic effects, selection accuracy could be further improved by considering non-additive genetic effects, such as dominance and epistatic effects, in the GP model, even for traits with low genomic heritability, such as RA [32].

Regarding the comparison of the prediction accuracy between different GP models, the BayesB and GBLUP models differed depending on the traits. This is because the fitness of either model depends on whether each trait is controlled by a small number of QTLs with large effects or by an accumulation of polygenes with small effects. For most traits, the Bayes B model was more accurate than the GBLUP model; however, for RA, the GBLUP model was 4.2% more accurate than the Bayes B model. This indicates that RA is regulated by numerous genes. In contrast, PA, ANT, and flowering date were significantly affected by only a few genes. Such differences in the genetic mechanisms of the traits were also evident in the QTL analysis.

The accuracy of genomic selection depends not only on genomic heritability but also on factors such as the strength of linkage disequilibrium (LD), number of QTLs, and population size [33]. For traits with low heritability, marker density may affect the accuracy of GP, with some studies showing that the accuracy improved when LD between adjacent markers was >0.2 [34]. Considering the relationship between marker density and GP prediction accuracy, of the three populations, S840 showed the highest prediction accuracy for most traits, probably because 1432 SNPs were used for prediction. This was approximately 2.5 times higher than that of the 579 SNPs used in S844, which had the lowest prediction accuracy among the three populations. Therefore, increasing the population size or number of markers may further improve the prediction accuracy for S844 and traits such as RA.

Furthermore, evaluation of the multi-trait GP model showed that the prediction accuracy was similar to that of the single-trait GP model. One of the reasons for this may be that the genetic correlation among the traits estimated in Section 3.3 was not strong. Another notable result was that the genetic correlations among the traits differed for each population. During breeding, the absence of strong negative correlations between target traits is highly desirable. However, the results revealed that some populations showed slightly negative genetic correlations for some traits. These negative genetic correlations can be reduced by simultaneously using multiple populations with different genetic backgrounds during breeding.

Finally, we examined the possibility of the genomic breeding of red perilla as a medicinal plant and discussed the different genetic mechanisms underlying the regulation of medicinal and other agronomic traits. Therefore, it is important to use various GP models and selection methods suitable for the genetic characteristics of each trait. For traits such as PA and ANT, where significant QTLs have been detected, selection can be performed using MAS. Even for traits such as RA, where no highly effective QTL is present in a breeding population to be used, selection can be made based on GP. Even for traits that are controlled by a QTL with a large effect, further improvement can be expected by using GS to accumulate genes with a small effect. Because the populations used in this study had only four cross parents and because of the strong linkage disequilibrium at the chromosome level in the F_3 and F_4 populations, the number of markers and population size on a small scale was sufficient for prediction accuracy. To breed medicinal plants, it is more realistic to create relatively small breeding populations for the application of MAS and GS. However, if one wants to make the GP model more general and more accurate for application to other populations, the model should be built using a large number of populations containing more genetic diversity [35]. Moreover, the results of this study indicate that the simultaneous use of multiple populations with different characteristics, such as different genetic correlations among target traits and different QTL positions, for breeding can result in multiple trait improvements when the size of each population is sufficient to build the prediction model. However, there are limitations in applying the GP models used in this study to multiple populations with varying genetic backgrounds in a general case. In the GP models using the three populations combined in this study, the effect of each population was included in the model as differences in the population mean (intercept), which may not be sufficient to account for differences in genetic background between populations. A possible way to model the genetic background between populations is to use a GP model that assumes that different populations have different genetic variances for the same QTL [36,37].

5. Conclusions

In this study, based on the results of the QTL analysis and the prediction accuracy of GP, the use of GS and MAS instead of conventional phenotypic selection proved to be effective in red perilla breeding. Because each trait has a different genetic mechanism, MAS should be used for traits in which large QTLs are detected, whereas GS should be used for traits controlled by polygenes. In addition, using multiple populations with different genetic backgrounds for breeding at the same time may lead to the simultaneous improvement of multiple traits. However, further research is required for its generalization. Our research contributes to the study of the genetic mechanisms underlying the production of medicinal compounds in perilla, as well as to the acceleration of medicinal plant breeding using MAS and GS.

Supplementary Materials: The following supporting information can be downloaded from https://www.mdpi.com/article/10.3390/genes14122137/s1: Table S1: Phenotype data of four cross-parents collected in 2022. Figure S1. Common genetic maps constructed using 20 chromosomes defined from three populations. Figure S2: QTLs detected for rosmarinic acid in S827, S840, and S844. Figure S3: QTLs detected for anthocyanins in S827, S840, and S844. Figure S4: QTLs detected for yield in S827, S840, and S844. Figure S5: QTLs detected for flowering dates in S827, S840, and

S844. Figure S6: Genotypes of QTL detected on chr19 for flowering date. Figure S7: Genotypes of QTLs detected for perillaldehyde. Figure S8: Prediction accuracy of multi-trait GP by the GBLUP and BayesCπ models using each population and the three populations combined.

Author Contributions: Data curation, formal analysis, software, visualization, and writing—original draft preparation, S.K.; conceptualization, K.S. (Kengo Sakurai), K.H., and H.I.; formal analysis, K.H.; funding acquisition, T.T. and H.I.; investigation, T.T. and Y.A.; resources, T.T., M.S., T.K., K.S. (Kenta Shirasawa), and S.I.; writing—review, editing, supervision, project administration, H.I. All authors have read and agreed to the published version of the manuscript.

Funding: This research was funded by Program on Open Innovation Platform with Enterprises, Research Institute and Academia, Japan Science and Technology Agency (JST, OPERA, JPMJOP1851).

Institutional Review Board Statement: Not applicable.

Informed Consent Statement: Not applicable.

Data Availability Statement: All other relevant data are available upon reasonable request.

Acknowledgments: The seeds of the cross-parents of the three populations used in this study were provided by the Genebank Project for Agricultural Biological Resources of the National Institute of Agrobiological Sciences (NIAS). We are also grateful to the technical staff of TSUMURA & CO., Kazusa DNA Research Institute, and Motoyuki Ishimori, who played important roles in selecting the cross-parents of the populations used in this study.

Conflicts of Interest: The funders had no role in the design of the study; in the collection, analyses, or interpretation of data; in the writing of the manuscript; or in the decision to publish the results.

References

1. Yu, H.-C.; Kosuna, K.; Haga, M. (Eds.) *Perilla: The Genus Perilla; Medicinal and Aromatic Plants—Industrial Profiles*; Harwood Academic Publishers: Amsterdam, The Netherlands, 1997; ISBN 978-90-5702-171-8.
2. Wu, X.; Dong, S.; Chen, H.; Guo, M.; Sun, Z.; Luo, H. *Perilla frutescens*: A traditional medicine and food homologous plant. *Chin. Herb. Med.* **2023**, *15*, 369–375. [CrossRef] [PubMed]
3. Nishimura, T.; Ohyama, K.; Goto, E.; Inagaki, N. Concentrations of perillaldehyde, limonene, and anthocyanin of Perilla Plants as affected by light quality under controlled environments. *Sci. Hortic.* **2009**, *122*, 134–137. [CrossRef]
4. Ogawa, E.; Hikosaka, S.; Goto, E. Effects of nutrient solution temperature on the concentration of major bioactive compounds in red perilla. *J. Agric. Meteorol.* **2018**, *74*, 71–78. [CrossRef]
5. Deguchi, Y.; Ito, M. Rosmarinic acid in *Perilla frutescens* and Perilla Herb analyzed by HPLC. *J. Nat. Med.* **2020**, *74*, 341–352. [CrossRef] [PubMed]
6. Oka, T. The role of Kampo (Japanese traditional herbal) medicine in psychosomatic medicine practice in Japan. *Int. Congr. Ser.* **2006**, *1287*, 304–308. [CrossRef]
7. Meuwissen, T.H.E.; Hayes, B.J.; Goddard, M.E. Prediction of total genetic value using genome-wide dense marker maps. *Genetics* **2001**, *157*, 1819–1829. [CrossRef]
8. Hayes, B.J.; Bowman, P.J.; Chamberlain, A.J.; Goddard, M.E. Invited review: Genomic selection in dairy cattle: Progress and challenges. *J. Dairy Sci.* **2009**, *92*, 433–443. [CrossRef]
9. Bernardo, R.; Yu, J. Prospects for genomewide selection for quantitative traits in maize. *Crop Sci.* **2007**, *47*, 1082–1090. [CrossRef]
10. Spindel, J.; Begum, H.; Akdemir, D.; Virk, P.; Collard, B.; Redoña, E.; Atlin, G.; Jannink, J.L.; McCouch, S.R. Genomic selection and association mapping in rice (*Oryza sativa*): Effect of trait genetic architecture, training population composition, marker number and statistical model on accuracy of rice genomic selection in elite, tropical rice breeding lines. *PLoS Genet.* **2015**, *11*, e1004982. [CrossRef]
11. Yamamoto, E.; Matsunaga, H.; Onogi, A.; Kajiya-Kanegae, H.; Minamikawa, M.; Suzuki, A.; Shirasawa, K.; Hirakawa, H.; Nunome, T.; Yamaguchi, H.; et al. A simulation-based breeding design that uses whole-genome prediction in tomato. *Sci. Rep.* **2016**, *6*, 19454. [CrossRef]
12. Baseggio, M.; Murray, M.; Magallanes-Lundback, M.; Kaczmar, N.; Chamness, J.; Buckler, E.S.; Smith, M.E.; DellaPenna, D.; Tracy, W.F.; Gore, M.A. Genome-wide association and genomic prediction models of tocochromanols in fresh sweet corn kernels. *Plant Genome* **2019**, *12*, 180038. [CrossRef] [PubMed]
13. Brzozowski, L.J.; Campbell, M.T.; Hu, H.; Caffe, M.; Gutiérrez, L.A.; Smith, K.P.; Sorrells, M.E.; Gore, M.A.; Jannink, J.L. Generalizable approaches for genomic prediction of metabolites in plants. *Plant Genome* **2022**, *15*, e20205. [CrossRef] [PubMed]
14. Owens, B.F.; Lipka, A.E.; Magallanes-Lundback, M.; Tiede, T.; Diepenbrock, C.H.; Kandianis, C.B.; Kim, E.; Cepela, J.; Mateos-Hernandez, M.; Buell, C.R.; et al. A foundation for provitamin A biofortification of maize: Genome-wide association and genomic prediction models of carotenoid levels. *Genetics* **2014**, *198*, 1699–1716. [CrossRef] [PubMed]

15. Masumoto, N.; Nishizaki, Y.; Maruyama, T.; Igarashi, Y.; Nakajima, K.; Yamazaki, T.; Kuroe, M.; Numata, M.; Ihara, T.; Sugimoto, N.; et al. Determination of perillaldehyde in perilla herbs using relative molar sensitivity to single-reference diphenyl sulfone. *J. Nat. Med.* **2019**, *73*, 566–576. [CrossRef]
16. Liu, J.; Wan, Y.; Zhao, Z.; Chen, H. Determination of the content of rosmarinic acid by HPLC and analytical comparison of volatile constituents by GC-MS in different parts of *Perilla frutescens* (L.) Britt. *Chem. Cent. J.* **2013**, *7*, 61. [CrossRef]
17. Peterson, B.K.; Weber, J.N.; Kay, E.H.; Fisher, H.S.; Hoekstra, H.E. Double digest RADseq: An inexpensive method for de novo SNP discovery and genotyping in model and non-model species. *PLoS ONE* **2012**, *7*, e37135. [CrossRef]
18. Tamura, K.; Sakamoto, M.; Tanizawa, Y.; Mochizuki, T.; Matsushita, S.; Kato, Y.; Ishikawa, T.; Okuhara, K.; Nakamura, Y.; Bono, H. A highly contiguous genome assembly of red perilla (*Perilla frutescens*) domesticated in Japan. *DNA Res.* **2023**, *30*, dsac044. [CrossRef]
19. Langmead, B.; Salzberg, S.L. Fast gapped-read alignment with bowtie 2. *Nat. Methods* **2012**, *9*, 357–359. [CrossRef]
20. Danecek, P.; Auton, A.; Abecasis, G.; Albers, C.A.; Banks, E.; DePristo, M.A.; Handsaker, R.E.; Lunter, G.; Marth, G.T.; Sherry, S.T.; et al. The variant call format and VCFtools. *Bioinformatics* **2011**, *27*, 2156–2158. [CrossRef]
21. Browning, B.L.; Zhou, Y.; Browning, S.R. A one-penny imputed genome from next-generation reference panels. *Am. J. Hum. Genet.* **2018**, *103*, 338–348. [CrossRef]
22. Kosambi, D.D. The estimation of map distances from recombination values. *Ann. Eugen.* **1943**, *12*, 172–175. [CrossRef]
23. Broman, K.W.; Wu, H.; Sen, S.; Churchill, G.A. R/qtl: QTL mapping in experimental crosses. *Bioinformatics* **2003**, *19*, 889–890. [CrossRef] [PubMed]
24. VanRaden, P.M. Efficient methods to compute genomic predictions. *J. Dairy Sci.* **2008**, *91*, 4414–4423. [CrossRef] [PubMed]
25. Hamazaki, K.; Iwata, H. RAINBOW: Haplotype-based genome-wide association study using a novel SNP-set method. *PLoS Comput. Biol.* **2020**, *16*, e1007663. [CrossRef] [PubMed]
26. Endelman, J.B.; Jannink, J.L. Shrinkage estimation of the realized relationship matrix. *Genes Genomes Genet.* **2012**, *2*, 1405–1413. [CrossRef]
27. Pérez, P.; De Los Campos, G. Genome-wide regression and prediction with the BGLR statistical package. *Genetics* **2014**, *198*, 483–495. [CrossRef]
28. Guo, G.; Zhao, F.; Wang, Y.; Zhang, Y.; Du, L.; Su, G. Comparison of single-trait and multiple-trait genomic prediction models. *BMC Genet.* **2014**, *15*, 30. [CrossRef]
29. Habier, D.; Fernando, R.L.; Kizilkaya, K.; Garrick, D.J. Extension of the Bayesian alphabet for genomic selection. *BMC Bioinform.* **2011**, *12*, 186. [CrossRef]
30. Zhang, Y.; Shen, Q.; Leng, L.; Zhang, D.; Chen, S.; Shi, Y.; Ning, Z.; Chen, S. Incipient diploidization of the medicinal plant perilla within 10,000 years. *Nat. Commun.* **2021**, *12*, 5508. [CrossRef]
31. Kang, Y.J.; Lee, B.M.; Nam, M.; Oh, K.W.; Lee, M.H.; Kim, T.H.; Jo, S.H.; Lee, J.H. Identification of quantitative trait loci associated with flowering time in perilla using genotyping-by-sequencing. *Mol. Biol. Rep.* **2019**, *46*, 4397–4407. [CrossRef]
32. Manolio, T.A.; Collins, F.S.; Cox, N.J.; Goldstein, D.B.; Hindorff, L.A.; Hunter, D.J.; McCarthy, M.I.; Ramos, E.M.; Cardon, L.R.; Chakravarti, A.; et al. Finding the missing heritability of complex diseases. *Nature* **2009**, *461*, 747–753. [CrossRef] [PubMed]
33. Zhong, S.; Dekkers, J.C.M.; Fernando, R.L.; Jannink, J.L. Factors affecting accuracy from genomic selection in populations derived from multiple inbred lines: A barley case study. *Genetics* **2009**, *182*, 355–364. [CrossRef] [PubMed]
34. Calus, M.P.L.; Veerkamp, R.F. Accuracy of breeding values when using and ignoring the polygenic effect in genomic breeding value estimation with a marker density of one SNP per CM: Including polygenic effects in genomic selection. *J. Anim. Breed. Genet.* **2007**, *124*, 362–368. [CrossRef] [PubMed]
35. Jannink, J.-L.; Lorenz, A.J.; Iwata, H. Genomic Selection in Plant Breeding: From Theory to Practice. *Brief. Funct. Genom.* **2010**, *9*, 166–177. [CrossRef]
36. Chen, L.; Li, C.; Miller, S.; Schenkel, F. Multi-Population Genomic Prediction Using a Multi-Task Bayesian Learning Model. *BMC Genet.* **2014**, *15*, 53. [CrossRef]
37. Lehermeier, C.; Schön, C.-C.; De Los Campos, G. Assessment of Genetic Heterogeneity in Structured Plant Populations Using Multivariate Whole-Genome Regression Models. *Genetics* **2015**, *201*, 323–337. [CrossRef]

genes

MDPI

Review

The Systematics, Reproductive Biology, Biochemistry, and Breeding of Sea Buckthorn—A Review

Hilde Nybom [1,*], Chengjiang Ruan [2] and Kimmo Rumpunen [3]

[1] Department of Plant Breeding–Balsgård, Swedish University of Agricultural Sciences, 29194 Kristianstad, Sweden
[2] Key Laboratory of Biotechnology and Bioresources Utilization, Ministry of Education, Institute of Plant Resources, Dalian Minzu University, Dalian 116600, China; ruan@dlnu.edu.cn
[3] Department of Plant Breeding, Swedish University of Agricultural Sciences, 23053 Alnarp, Sweden; kimmo.rumpunen@slu.se
* Correspondence: hilde.nybom@slu.se

Abstract: Both the fruit flesh and seeds of sea buckthorn have multiple uses for medicinal and culinary purposes, including the valuable market for supplementary health foods. Bioactive compounds, such as essential amino acids, vitamins B, C, and E, carotenoids, polyphenols, ursolic acid, unsaturated fatty acids, and other active substances, are now being analyzed in detail for their medicinal properties. Domestication with commercial orchards and processing plants is undertaken in many countries, but there is a large need for improved plant material with high yield, tolerance to environmental stress, diseases, and pests, suitability for efficient harvesting methods, and high contents of compounds that have medicinal and/or culinary values. Applied breeding is based mainly on directed crosses between different subspecies of *Hippophae rhamnoides*. DNA markers have been applied to analyses of systematics and population genetics as well as for the discrimination of cultivars, but very few DNA markers have as yet been developed for use in selection and breeding. Several key genes in important metabolic pathways have, however, been identified, and four genomes have recently been sequenced.

Keywords: *Hippophae*; chemical contents; cultivar development; DNA markers; genetics; medicinal plant; systematics

Citation: Nybom, H.; Ruan, C.; Rumpunen, K. The Systematics, Reproductive Biology, Biochemistry, and Breeding of Sea Buckthorn—A Review. *Genes* **2023**, *14*, 2120. https://doi.org/10.3390/genes14122120

Academic Editors: Wajid Zaman and Hakim Manghwar

Received: 15 October 2023
Revised: 11 November 2023
Accepted: 15 November 2023
Published: 24 November 2023

1. Introduction

The genus *Hippophae* L. belongs to the small plant family Elaeagnaceae. The deciduous shrubs or small trees, usually 2–5 m high (occasionally over 10 m), are wind-pollinated and dioecious (male and female flowers are produced on separate plants). The yellow to orange or sometimes reddish to red fruits contain one seed each, and these seeds are often dispersed by birds and other frugivorous animals. The most well-known species, *Hippophae rhamnoides*, commonly known as sea buckthorn, is widely distributed over China, the Indian Himalayas, Central Asia, and Russia, as well as large parts of Europe (Figure 1).

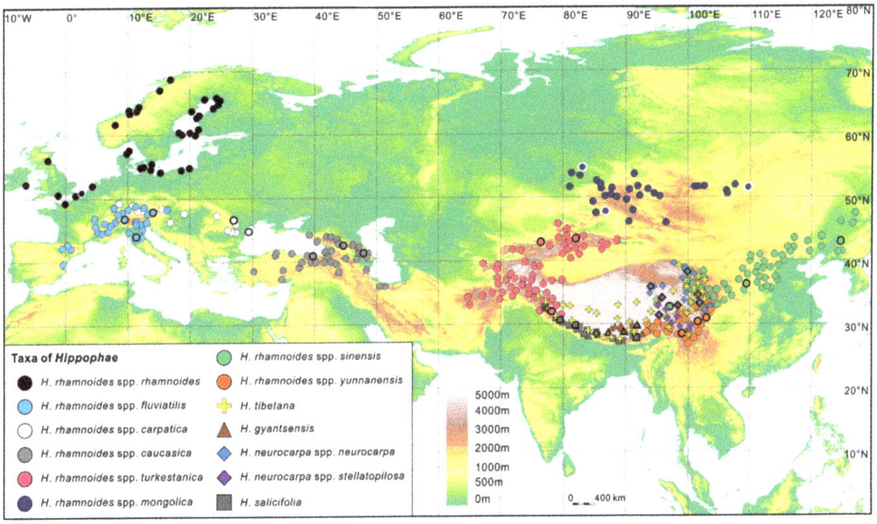

Figure 1. Geographic distribution of *Hippophae* L. (reprinted with permission from [1]), 2018, Jia and Bartish.

Both the fruit flesh and the seeds of sea buckthorn have multiple uses for medicinal and culinary purposes, including the valuable market for supplementary health foods [2]. The sea buckthorn industry has been thriving in Russia ever since the bioactive compounds started to become appreciated. By contrast, industrial use is somewhat more recent in China, despite several centuries of traditional use in Chinese medicine. Domestication with commercial orchards and/or processing plants is now undertaken for both culinary and medicinal products in, e.g., Mongolia, Nepal, Tajikistan, India, Iran, Belarus, Ukraine, Turkey, Greece, Romania, Germany, Finland, Sweden, and the Baltic countries, as well as in some other parts of Europe. Sea buckthorn has also been introduced to other countries like Canada, the USA, Bolivia, Chile, South Korea, and Japan.

The sea buckthorn is a typical pioneer plant and prefers open habitats with sandy or rocky free-draining soils, high insolation, and restricted competition from other species. Plants are spread efficiently via seeds and root suckers. Sea buckthorn plants have root nodules that contain nitrogen-fixating bacteria (the actinomycete *Frankia*), allowing them to thrive in nutritionally deficient soils. Due to its high tolerance to extreme conditions (both very low and very high temperatures, drought, salinity, and poor soils), sea buckthorn is often grown as protection against wind, for prevention of sand drift, for conservation of soil and water, and to adjust microclimate conditions, especially in China but also in India, Russia, Canada, and Bolivia [3].

In December 2020, the total sea buckthorn acreage, including both wild and cultivated plants, was estimated to amount to approximately 2.33 million ha in the world (http://www.isahome.net/news.php?id=586, accessed on 12 September 2023). Of these, about 2.07 million ha were found in China (0.77 million ha wild and 1.35 million ha planted), 16,300 ha in India, 15,000 ha in Romania, 20,000 ha in Mongolia, ~6000 ha in Russia, and 5700 ha in Pakistan.

The increasing interest in sea buckthorn is mainly centered around its properties as a medicinal and culinary plant. In addition to their long-time use in traditional Chinese medicine for, e.g., calming coughs, aiding digestion, improving blood circulation and alleviating pain, recent pharmacological studies have shown that crude extracts or compounds have anti-inflammatory, antioxidant, hepatoprotective, anticancer, hypoglycemic, hypolipidemic, neuroprotective, and antibacterial properties (review in [3,4]).

Hundreds of bioactive compounds, such as essential amino acids, vitamins B, C, and E, carotenoids, polyphenols, ursolic acid, unsaturated fatty acids, and other active substances, have been recorded in sea buckthorn, many of which are now being analyzed in detail for their medicinal properties [5,6]). Commercial products promoted as remedies for various illnesses, including cosmetics for sensitive skin treatment, are usually based on dried fruits, fruit pulp oils, and/or seed oils. These products are often manufactured as, e.g., fruit powders, oil capsules, or vitamin C tablets.

In some countries, especially in Europe and North America, sea buckthorn is highly appreciated for its tasty fruits that are used in the food industry to manufacture various culinary products like juice, wine, liquor, syrup, cakes and pies, curd, jam, candy, and ice cream [3]. The taste is tart and quite unique, often requiring considerable amounts of sweetening to produce a balanced flavor. A need for inherently sweeter fruits has therefore been addressed in some plant breeding programs.

2. Taxonomy

The genus *Hippophae* contains several diploid (2n = 24) taxa, but until now, only the most widespread and variable species, *H. rhamnoides*, has been domesticated. Nevertheless, other species, all of which are restricted to cold-temperate areas on the Qinghai–Tibet plateau and adjacent areas in Central Asia, are being investigated for valuable traits to be exploited in the future.

Despite the relatively small number of taxa, systematic treatment of the genus *Hippophae* has been quite controversial, and several treatises have been published [7]. Rousi [8] recognized three species: *H. rhamnoides*, *Hippophae salicifolia* D. Don, and *Hippophae tibetana* Schlecht., and up to nine subspecies of *H. rhamnoides*. Subsequently, several additional species and at least one subspecies have been described and published: *Hippophae goniocarpa* YS Lian, XL Chen and K Sun ex Swenson and Bartish; *Hippophae gyantsensis* (Rousi) YS Lian; *Hippophae litangensis* YS Lian and XL Chen ex Swenson and Bartish; *H. neurocarpa* SW Liu and TN He; and *Hippophae neurocarpa* subsp. *stellatopilosa* YS Lian, XL Chen and K Sun ex Swenson and Bartish.

In 2000, the first molecular marker-based treatise of the entire genus *Hippophae* (15 taxa) was presented using Random Amplified Polymorphic DNA (RAPD) markers [9]. Other RAPD-based studies have focused on a subset of the taxa, e.g., *H. rhamnoides* subsp. *sinensis* and *H. rhamnoides* subsp. *mongolica* [10] and *H. rhamnoides* subsp. *sinensis* and other Chinese taxa [11]. Subsequent investigations have been undertaken using, e.g., chloroplast DNA (cpDNA) and morphological characters [12] as well as Internal Transcribed Spacer (ITS) sequences [13]. Other molecular methods like Simple Sequence Repeats (SSR) markers [14] have been applied to investigate relationships among the different taxa of *Hippophae*, while Amplified Fragment Length Polymorphism (AFLP), Selective Amplification of Microsatellite Polymorphic Loci (SAMPL) [15], and DNA barcode markers (ITS, matK, rbcL, and rpoC1) [16] have been applied to *H. rhamnoides* subsp. *turkestanica*, *H. salicifolia*, and *H. tibetana* in northern India and the Himalayas.

Two sections, sect. *Gyantsenses* (*H. gyantsensis*, *H. tibetana*, and *H. neurocarpa*) and sect. *Hippophae* (*H. salicifolia* and *H. rhamnoides*), have been recognized based on whether the carpodermis is fused with the seed coat or not. A recent analysis based on multiple chloroplast and nuclear gene fragments has instead shown that *H. gyantsensis*, *H. neurocarpa*, and *H. salicifolia* form one clade, whereas *H. tibetana* forms a second clade together with *H. rhamnoides* [1]. In addition, three species appear to be hybridogenic: *H. gyantsensis* has probably derived through gene flow between *H. rhamnoides* subsp. *yunnanensis* and *H. neurocarpa*, whereas *H. goniocarpa* derives through gene flow between *H. rhamnoides* subsp. *sinensis* and *H. neurocarpa*, and *H. litangensis* through gene flow between *H. rhamnoides* subsp. *sinensis* and *H. neurocarpa* subsp. *stellatopilosa* [12,17,18].

The most variable and economically important species, *H. rhamnoides*, is generally divided into five subspecies in Asia: subsp. *caucasica* Rousi, subsp. *mongolica* Rousi, subsp. *turkestanica* Rousi, subsp. *sinensis* Rousi, and subsp. *yunnanensis* Rousi, and three in Europe:

subsp. *rhamnoides* L., subsp. *fluviatilis* (Soest) Rousi, and subsp. *carpatica* Rousi. Most populations are found mainly on seashores and river deltas or on valley slopes up to 5000 m above sea level, i.e., in habitats typical of early successional species. The first taxon to be cultivated was subsp. *mongolica*, which has been utilized for almost a century in Siberia, but subspecies *sinensis*, *turkestanica*, *rhamnoides*, *carpatica*, and *caucasica* have also been used in cultivation and plant breeding.

Since the phytochemical contents differ considerably among the subspecies of *H. rhamnoides*, efforts have been made to develop taxon-specific molecular markers that can detect adulteration and contamination in commercial fruits and derived products. A high-resolution melting assay based on a DNA barcoding region of ITS2 in the ribosomal DNA (rDNA) could discriminate among all seven *Hippophae* species [19]. Single Nucleotide Polymorhism (SNP) markers from the 45S rDNA region were later used to distinguish subsp. *mongolica* and subsp. *sinensis* and apparently produce reliable results even with very low DNA concentrations [20].

3. Population Genetics

Information on genetic diversity is very helpful when collecting and utilizing plant material for gene banks and plant breeding. Lately, DNA-based markers have been used to assess the amount and partitioning of genetic variability in numerous plant species. Considering the life history traits of *H. rhamnoides*, which is a perennial (wild plants generally live for 30–60 years), obligately outcrossing, wind-pollinated, early successional species with seed dispersal through birds, relatively high genetic variation is expected within populations, together with low differentiation between populations [21].

A RAPD-based study carried out on 10 populations of *H. rhamnoides* in Northern Europe showed that within-population genetic variation, estimated as expected heterozygosity H_E, was on average 0.16 [22]. This value is slightly lower than the mean value of 0.22 reported in a metastudy with 60 RAPD-based studies of different species [21]. Analyses of the impact of various life history traits showed that breeding system and successional status are highly important; outcrossing species had the highest within-population diversity (0.27 on average), while early successional species instead had the lowest (0.17 on average).

Analysis of Molecular Variation (AMOVA) allows the partitioning of genetic variation between and within populations. For sea buckthorn, only 15% of the variation occurred between populations, which is considerably lower than the mean value of 34% in 116 studies [21]. This parameter is affected by the breeding system (lowest values for outcrossing species), successional status (highest values for early and medium-early species), life form (lowest values for long-lived perennials), and seed dispersal (lowest values for species with animal-ingested seeds). Apart from the early successional status of *H. rhamnoides*, the other life history traits are in keeping with low genetic differentiation between populations.

In another RAPD-based study, 13 populations of *H. rhamnoides* subsp. *sinensis* were found to have rather low average within-population diversity = 0.17, as well as restricted population differentiation = 18 [23]. In yet another study, Inter Simple Sequence Repeat (ISSR) markers were applied to 11 sea buckthorn populations in Northeastern and Northwestern China [24]. Within-population diversity was in keeping with the other studies, ranging from 0.16 to 0.21, but population differentiation was only 7%. A second ISSR study on 15 Chinese populations yielded an average within-population diversity of 0.20 for 7 populations of subsp. *yunnanensis*, 0.22 for 7 populations of subsp. *sinensis*, and 0.14 for the single population of subsp. *gyantsensis* [25]. Overall differentiation among all 15 populations was 14.5%, and a cluster analysis showed that the subsp. *gyantsensis* population differed considerably from the others in accordance with a proposed species status as *H. gyantsensis*.

SSR markers were applied to distinguish between *H. rhamnoides* subsp. *sinensis* and *H. rhamnoides* subsp. *yunnanensis* and to estimate the genetic diversity within each taxon [26]. The 32 investigated populations could be grouped according to taxonomic status, with indications of some inter-subspecies hybridization in the zone of partly overlapping distribution. Populations of subsp. *sinensis* were more variable, with observed heterozygosity (H_o = 0.40) compared to subsp. *yunnanensis* (H_o = 0.20), whereas population differentiation was 26% and 37%, respectively, in these two subspecies. Both within-population diversity estimates are somewhat low compared to an average H_o = 0.58 in 80 SSR-based studies, whereas between-population estimates were more similar to the average of 24–26% in 51 studies [21].

Most studies have failed to find an association between genetic and geographic distances in sea buckthorn [12,23,25]. Within-population variation appears to be especially high in species and subspecies that occur close to the center of origin in Central Asia, e.g., *H. tibetana*, and in the possibly hybridogenous species *H. goniocarpa* [9]. By contrast, small and isolated populations (e.g., growing at high altitudes) often appear to be more homogenous [27,28]. For a more comprehensive review of population genetics and phylogeography in *Hippophae*, see Bartish et al. [7].

4. Sex Determination

In sea buckthorn, sex is genetically determined through an X/Y system and heteromorphic sex chromosomes [29], with the Y-chromosome being slightly longer than the X-chromosome. In commercial sea buckthorn orchards, the majority of plants are female, but with the addition of about 10% male plants to ensure adequate pollination. Plant breeding is therefore mainly directed towards the development of high-quality female plants. Male seedlings are often discarded as soon as their gender can be ascertained, but morphology-based gender determination cannot be undertaken until the plants flower for the first time, which usually takes 3–5 years. A method for early discrimination between male and female seedlings would save much time and space in plant breeding programs as well as in seedling-based plantations.

Genes related to sex have been identified in several dioecious plants, but in, e.g., papaya (*Carica papaya* L.), factors such as the environment, hormones, and genetic and epigenetic background can apparently also affect sex expression [30]. Nevertheless, molecular markers have been developed and are now being successfully used for early sex determination in papaya and other crops.

Attempts have also been made to develop gender-specific DNA markers in sea buckthorn, but these have usually not been sufficiently consistent when screened in a more diverse germplasm. The first attempt was based on RAPD analysis of offspring derived from experimental crosses in *H. rhamnoides* subsp. *rhamnoides* [31]. A male-specific DNA marker was found using Bulked Segregant Analysis (BSA), but when applied to individual plants, this marker worked only for the progeny obtained in one of the two tested crosses. Several RAPD-based studies have also been carried out on *H. rhamnoides* subsp. *turkestanica* in northern India. One male-specific marker was found but validated in only five plants of each sex [32]. In another study, two female-specific markers were found and subsequently converted into Sequence-Characterized Amplified Region (SCAR) markers [33]. These were validated in a larger material but derived from only one population. A female-specific RAPD-based marker was developed for subsp. *sinensis* and validated in more diverse material [34].

An ISSR-derived male-specific band was converted into a SCAR marker and validated in subsp. *turkestanica* plants from three geographically isolated valleys in the Ladakh region of India [35]. In a broader approach, RAPD, ISSR, SSR, and MADS box gene-specific markers were applied, with one of the RAPD primers producing two male-specific fragments [36]. These could, however, discriminate between males and females in only one of the populations when screened on material obtained from both the Ladakh region (Jammu and Kashmir) as well as Lahaul and Spiti, and Kannaur (Himachal Pradesh).

In the last two decades, transcriptomic analyses have revealed important information about gene action in many plant species. This approach has also recently been employed in sea buckthorn to search for gender-specific genes. Thus, 21 floral regulatory genes, homologous to previously established sequences in model plant species, proved to be differentially expressed across the developmental stages of male and female flowers in samples of *H. rhamnoides* [37]. Two possibly promising genes were identified: *HrCRY2* (cryptochrome receptor gene) was significantly over-expressed in female flowers, whereas *HrCO* (circadian pathway gene) was significantly over-expressed in male flowers. Further research is still needed to determine the role of these genes in the development of male and female flowers.

The observed difficulties in developing robust gender-specific markers may be connected to the low differentiation between the male and female genomes in *Hippophae*. A survey of 25 wild populations of *H. rhamnoides* subsp. *turkestanica* in northern India showed that 2–4% of the plants were polygamomonoecious (PGM) with male, female, and hermaphrodite flowers, suggesting that the transition from hermaphrodites (the original state for angiosperms) to dioecy (the derived state) happened recently in this genus [38]. Moreover, the male and female genomes appear to be very similar overall. A Representational Difference Analysis (RDA) of DNA sequences was applied to the search for gender-specific differences between male and female genomes. Some of the obtained polymorphisms looked promising initially, but none of them held up when screened on plants from more distant populations [38].

An AFLP analysis conducted on five populations of subsp. *turkestanica* was more successful and produced 4 female-specific and 2 male-specific fragments that were cloned and sequenced and subsequently used for developing SCAR primers [38]. The genomic region around each SCAR was explored through genome walking, and new primers were designed and then tested on 50 male and 50 female plants. Most of the tested regions were not informative when applied to a more diverse plant material, but with one exception: primers for the locus *HRML* (*H. rhamnoides* male locus) produced a 329 bp fragment in all male plants tested in 25 populations. This has been sequenced and further characterized, and it is now available as a 7 kb region named HRMSSR in NCBI Genbank accession KX444194 [38].

While most of the research has been carried out on *H. rhamnoides*, a similar sex determination system is probably also present in the other species. A RAPD-based female-specific marker was thus found in *H. goniocarpa* [39]. In a more recent study, 80 *H. tibetana* samples were sampled from four sites with 10 male and 10 female plants in each site and screened for male-specific markers using a comparative analysis of Restriction-Associated DNA Sequencing (RAD-seq) data [40]. A Genome-Wide Association Study (GWAS) indicated that several SNPs on Chromosome 2 are related to male sex determination. Much additional work is, however, needed to determine the location of sequences that can be used for the development of sex-specific markers.

It should also be mentioned that the occurrence of facultative apomixis, i.e., seed set without pollination, has been reported in *H. rhamnoides* based on some seed setting in bagged inflorescenses (16% fruit set as compared to 68% fruit set in unbagged inflorescenses) and on cytological evidence of both apospory and adventive (nucellar) embryony in embryo sacs and in fruits obtained after bagging, respectively [41]. The studied plant material was collected in northern India and presumably belongs to subsp. *turkestanica*. In a follow-up study on five wild sea buckthorn populations in India, only a 3% fruit set was recorded in the bagged inflorescenses, compared to a 60% fruit set in the unbagged inflorescenses [42]. This strong reduction in fruit set after bagging is not typical of apomicts in general, and the extent to which apomixis plays a role in the genus requires more research. The effects of apomixis on the genetic variability of seedling offspring after natural pollination or after the undertaking of experimental crosses have also not yet been reported to our knowledge.

5. DNA-Based Identification of Cultivars

Sea buckthorn cultivars vary widely in, e.g., fruit size, color and shape, but are still often difficult to identify properly based on only morphological traits (Figure 2). Accessions of the same cultivar can also vary to an unexpected extent when grown under different conditions that affect, e.g., fruit quality and levels of resistance. Methods are therefore needed for the unambiguous identification of germplasm (wild material, lines, or elite cultivars). Presently, this can be achieved through the application of various types of DNA markers. Since dioecious reproduction results in obligate outcrossing, all seedlings have unique genotypes even when obtained from the same seed parent. Moreover, sea buckthorn cultivars and advanced selections are propagated vegetatively and are therefore expected to consist of a single genotype that can be identified and distinguished from all other cultivars and selections with DNA fingerprinting.

Figure 2. Variation among sea buckthorn (*Hippophae rhamnoides*) genotypes.

Almost any type of molecular marker method that produces a sufficient number of data points, like RAPD, ISSR, and AFLP, will usually allow discrimination between different plant accessions [3]. For the setting up of joint international databases, SSR markers and SNP data are, however, usually preferred since results from different laboratories can be entered and compared as long as the same set of SSR primers or the same SNP arrays are used. Such databases are now developed to an increasing extent in many crops and are used as a basis for defining cultivar-specific profiles like the SNP-based Malus UNiQue genotype (MUNQ) codes in apple [43] and the SSR-based Cherry UNiQue genotype (CHUNQ) codes in cherry [44]. So far, joint international databases have not yet been developed for sea buckthorn, making it difficult to evaluate data across different studies.

The first major screening of cultivated sea buckthorn germplasm was performed on 55 accessions in a Swedish gene bank using RAPD [45]. Approximately 20 accessions derived from crosses between subsp. *mongolica* and subsp. *rhamnoides*, 15 belonged to subsp. *mongolica*, 10 belonged to subsp. *rhamnoides*, and the remainder represented either subsp. *fluviatilis*, subsp. *carpatica*, subsp. *caucasica*, or other crosses. Cluster and Principal Coordinate (PCO) analyses showed considerable grouping in accordance with the taxonomic and geographic origination.

Efficient discrimination among genotypes and grouping according to origin has been reported in other RAPD-based studies on the different subspecies of *H. rhamnoides* with wild as well as cultivated material [10,46–48]. Similar results have also been obtained when using other types of multilocus markers, like ISSR [49] and AFLP [50].

For the development of more robust databases, SSR loci are often preferred. The first report in sea buckthorn described primers for nine genomic SSR (gSSR) loci [51], and was soon followed by primers using expressed sequence tags (estSSR) [52]. Numerous analyses have since then been undertaken in different sets of sea buckthorn material using SSR loci that generally seem to be transferable across all taxa in the genus *Hippophae* [14,53–58].

Regardless of the marker system used, the grouping of the accessions has in general reflected the taxonomic classification into different subspecies as well as the geographic origin of the accessions themselves or the breeding program where they were developed. Information has been obtained about the origin of named cultivars, either as selections from local wild populations or as crosses in breeding programs. Moreover, the application of DNA markers to germplasm collections has frequently proved useful in detecting labeling problems like the presence of more than one name for the same cultivar or different cultivars sharing the same name.

6. Sea Buckthorn Breeding around the World

Sea buckthorn fruits are still harvested from wild or naturalized stands in many countries in Asia and Europe, but the increasing interest in this crop has boosted the establishment of commercial orchards. Since most orchards are managed as low-input and/or organically, i.e., without fungicides or pesticides, a healthy plant material is required. Low-cost orchards are sometimes planted with seedlings, but the utilization of genetically superior and vegetatively propagated cultivars holds more potential in the long run (Table 1). Consequently, a development in the derivation of suitable plant material can be seen, from (1) selection among wild plants or their offspring, to (2) selection among offspring after open pollination of superior genotypes, and to (3) selection among offspring derived from crosses between selected parents.

Table 1. Female sea buckthorn cultivars of special interest in cultivation history and/or in plant breeding.

Cultivar	Interesting Trait(s)	Origin	Reference
Afina	Very vigorous, large fruits, long pedicels	Russia, subsp. *mongolica*	[59]
Altaiskaya	Thornless, sweet taste	Russia, subsp. *mongolica*	[59]
Askola	Dark orange fruit, high vitamin C and E contents	Germany, subsp. *rhamnoides*	[60]
Aurelia	High yield, large fruits, long pedicels	Russia, subsp. *mongolica*, offspring of 'Avgustina'	[59]
Avgustina (Avgustinka)	Thornless, large fruits, long pedicels, early ripening	Russia, subsp. *mongolica*	[59,61]
Azhurnaya (Agurnaya)	Thornless, large fruits, long pedicels	Russia, subsp. *mongolica*	[59,61]
Baikal	Tolerant to sea buckthorn fly, sweet taste	Russia	[62]
Botanicheskaya Ljubitelskaya (Sunny)	High yield, large yellow fruits, low acidity, medium tolerant to sea buckthorn fruit fly	Russia, subsp. *mongolica* × subsp. *rhamnoides*	[62]
Chaoyang	Vigorous, late ripening, long and drooping branches	China, subsp. *mongolica* × subsp. *sinensis*	
Chechek	Thornless, compact stature, long period of maturity, high in carotenoids	Russia, subsp. *mongolica*	[59]
Chengse	Late ripening	Russia, subsp. *mongolica*	[61]
Chuyskaya (Star of Altai)	Thornless, compact stature, high yield, sweet taste, suitable for handpicking, early ripening	Russia, subsp. *mongolica*	[59,61]
Dalate	Thornless, balanced acidity/sweetness	China, subsp. *mongolica* × subsp. *sinensis*	[63]
Dar Katuni	Small fruits	Russia, subsp. *mongolica*	[59,61]
Desert maslichnyi (Maslichnaya)	Small reddish fruits, tolerant to wilt	Russia, subsp. *mongolica*	[59,62]
Dorana	Smallish stature, suitable for hand picking	Germany, subsp. *rhamnoides*	[60]
Druzhina	Compact, few branches, suitable for machine harvesting, large fruits with hard pericarps	Russia, subsp. *mongolica*, induced irradiation mutant	[64,65]
Eir	High yield, orange color, non-rancid	Sweden, subsp. *mongolica* × subsp. *rhamnoides* (BHi727102)	[9]

Table 1. *Cont.*

Cultivar	Interesting Trait(s)	Origin	Reference
Elizaveta	High yield, large fruits, long pedicels, susceptible to *Fusarium* wilt	Russia, subsp. *mongolica*, induced chem. mutant of 'Panteleevskaya'	[59,61]
Essel	High yield, large fruits, long pedicels, tolerant to *Fusarium* wilt, suited for fresh eating	Russia, subsp. *mongolica*	[59]
Etna	High yield, red fruits, very early ripening	Russia, offspring from open pollination of 'Inya'	[59]
Eva	High sucrose content, suited for a cold climate with short summers, disease resistant	Latvia, subsp. *rhamnoides* × subsp. *mongolica*	
Ezhonghuang	Almost thornless, high acidity	China, subsp. *mongolica* × subsp. *sinensis*	[63]
Ezhongxian	Thornless, balanced acidity/sweetness	China, subsp. *mongolica* × subsp. *sinensis*	[63]
Fenja	Large fruits, red-orange color, low acidity, little flavor, easy harvest	Sweden, subsp. *mongolica* × subsp. *rhamnoides* (BHi72588)	[9]
Frugana	Vigorous, high yield, early ripening	Germany, subsp. *rhamnoides*	[60]
Gaoyou 1	Vigorous, high yield, high seed oil content	China, subsp. *mongolica* × subsp. *sinensis*	
Habego (Orange Energy)	Vigorous, high yield	Germany, subsp. *rhamnoides* × subsp. *mongolica*	
Harvest Moon	Early ripening, few thorns, winter-hardy, large fruits with long pedicels	Canada, subsp. *mongolica*	[66]
Hergo	Vigorous, high yield, late-ripening, suitable for machine harvesting, small fruits	Germany, subsp. *rhamnoides*	[67]
Hongji 1	Almost thornless, red rounded fruits	China, subsp. *mongolica* × subsp. *sinensis*	[68]
Hongxia	Ornamental, densely set orange fruits, fruits stay on branches for several months	China, subsp. *sinensis*	[69]
Hongyun	Tolerant to both low and high temperatures and to draught	Russia, subsp. *mongolica*	[61]
Hualin 1	Tolerant to both low and high temperatures and to draught	China, subsp. *mongolica* × subsp. *sinensis*	[63]
Hunjin	High yield, few thorns, large fruits, long pedicels	China, offspring from subsp. *mongolica*	[70,71]
Idun	Oblong orange fruits, non-rancid	Sweden, subsp. *mongolica* × subsp. *rhamnoides* (BHi727115)	
Inya	Arching branches, large red fruits with high carotenoid content, short juvenile period	Russia, subsp. *mongolica*, induced chem. mutant of 'Panteleevskaya'	[59]
Ivushcka	Late ripening, red fruits with high carotenoid contents, can be harvested by shaking the bush	Russia, subsp. *mongolica*, induced irradiation mutant	[64,65]
Julia	Vigorous, high yield, orange fruits	Sweden, subsp. *mongolica* × subsp. *rhamnoides*	[9]
Kapriz (Caprice)	High sugar content	Russia, subsp. *mongolica*	[65]
Klavdia (Klavdija)	High yield, long pedicels, suitable for hand-picking, sweet taste	Russia, subsp. *mongolica*, offspring of 'Chuyskaya'	[59]
Krasnoplodnaya	Large fruits, high carotenoid content, sensitive to wilt	Russia	[72]
Krasny Fakel	Late ripening, red fruits high carotenoid content, can be harvested by shaking the bush	Russia, subsp. *mongolica*, induced irradiation mutant	[64,65]
Leikora	Vigorous but with restricted root suckering, winter-hardy	Germany, subsp. *rhamnoides*	[60]
Lotta	High yield, winter-hardy, long pedicels, suited for hand-picking, low acidity, non-rancid	Sweden, subsp. *mongolica* × subsp. *rhamnoides* (Bhi 31415hona)	
Lvivyanka	High yield, large fruits	Ukraine, crosses with subsp. *mongolica*	[73]

Table 1. *Cont.*

Cultivar	Interesting Trait(s)	Origin	Reference
Mariya	Low pull-out force (=fruits are easily detached)	Russia	[72]
Mary (Marija Bruvele)	Mild taste, high oil content, relatively tolerant to mycotic wilt and bud bacteriosis	Latvia, open pollination of 'Botanicheskaya Ljubitelskaya'	[54]
Medova Osin	Drought tolerance, winter hardiness, disease resistance	Ukraine, subsp. *carpatica*	[73]
Mendeleevskaya	Tolerant to wilt	Russia	[62]
Mengzhonghong	Rounded fruits, balanced sweetness/acidity	China, subsp. *mongolica* × subsp. *sinensis*	[63]
Mengzhonghuang	Almost thornless, rounded fruits, balanced sweetness/acidity	China, subsp. *mongolica* × subsp. *sinensis*	[63]
Mukshanska	High yield, large fruits	Russia, offspring from open pollination of 'Inya'	[73]
Nivelena	High yield	Russia	[72]
Novost Altaya	Small fruits, tolerant to Fusarium wilt	Russia, subsp. *mongolica*	[59,61]
Ognistaya	Late ripening, red fruits high in carotenoids, can be harvested by shaking the bush	Russia, subsp. *mongolica*, induced irradiation mutant	[64,65]
Ognivo	Thornless, large, and red fruits, long pedicels, late ripening	Russia, subsp. *mongolica*	[59]
Oranzhevaya (Orangenaya)	High yield, few thorns, small fruits	Russia, subsp. *mongolica*	[59,61,74]
Osinnia krasunia	High yield, large fruits	Ukraine, crosses with subsp. *mongolica*	[73]
Panteleevskaya	Large fruits	Russia, subsp. *mongolica*	[59]
Plamennaya	High yield, large fruits	Belarus	[61,72]
Podaruk Sadu	High yield	Russia	[72]
Podrugha	High sugar content	Russia, subsp. *mongolica*, induced irradiation mutant	[64,65]
Qiuyang	Drought tolerant	China, open pollination of subsp. *mongolica* 'Wulangmu'	[75]
Raisa	Disease resistant, high acidity	Finland, subsp. *rhamnoides* × subsp. *caucasica*	[76,77]
Rapsodiia	High yield, large fruits	Ukraine, crosses with subsp. *mongolica*	[73]
Sibirsky Rumyanets	Red fruits high in carotenoids, very earrly-ly ripening	Russia	[65]
Sirola	Earrly-ly ripening	Germany, subsp. *rhamnoides* × subsp. *mongolica*	
Syurpriz Baltiki (Sjurpriz Pribaltiki)	Tolerant to wilt, medium tolerant to sea buckthorn fruit fly	Russia	[62]
Skibes siev		Latvia, from wild-growing subsp. *rhamnoides*	[54]
Sol	High yield, vigorous, round fruits, rich flavourflavor, suitable for harvesting by cutting off branches	Sweden, subsp. *mongolica* × subsp. *rhamnoides* (BHi10726)	[9]
Sudaruschka	High yield, large fruit, high carotenoid content	Russia, subsp. *mongolica*, induced chem. mutant of 'Panteleevskaya'	[59]
Tatjana	Reddish fruits, high oil content, relatively tolerant to mycotic wilt and bud bacteriosis	Latvia, open pollination of 'Botanicheskaya Ljubitelskaya'	[54]
Terhi	Winter hardy, high in vitamin C, tolerant to stem canker	Finland, subsp. *rhamnoides*	[76,77]
Torun	High yield, large and non-rancid fruits, suitable for harvesting by cutting off branches	Sweden, subsp. *mongolica* × subsp. *rhamnoides* (BHi727137)	
Trofimovskaya	High yield, susceptible to sea buckthorn fly	Russia, subsp. *mongolica* × subsp. *rhamnoides*	[62,72]
Tytti	Winter hardy, moderate vigor, high in vitamin C, tolerant to stem canker	Finland, subsp. *rhamnoides*	[76,77]

Table 1. *Cont.*

Cultivar	Interesting Trait(s)	Origin	Reference
Vitaminnaya	Vigorous thornless bush, small fruits	Russia, subsp. *mongolica*	[59,61]
Wanhuang	High yield, late ripening yellow fruits that stay on the branches for several months	China, subsp. *mongolica* × subsp. *sinensis*	
Wanxia	High yield, late ripening orange-red fruits that stay on the branches for several months	China, subsp. *mongolica* × subsp. *sinensis*	
Wucifeng	High yield, few thorns, large fruits, long pedicels	China, offspring from subsp. *mongolica*	[70,71]
Wulanshalin	High yield, few thorns, large fruits, long pedicels	China, offspring from subsp. *mongolica*	[70,71]
Yolochka	Tolerant to wilt	Russia	[62]
Zarnitsa		Russia, subsp. *mongolica*, induced irradiation mutant	[64]
Zhemchuzhnitsa	High yield, sweet and aromatic fruits, suited for fresh consumption	Russia, subsp. *mongolica*, linebreeding of 'Chuyskaya'	[59]
Zhongji 3	Few thorns, yellow rounded fruits	China, subsp. *mongolica* × subsp. *sinensis*	[78]
Zhongji 4	Few thorns, early ripening, yellow rounded fruits	China, subsp. *mongolica* × subsp. *sinensis*	
Zlata	High yield, thornless, large fruits, late-ripening	Russia, subsp. *mongolica*	[59]
Zolotoi Kaskad	High sugar content	Russia, subsp. *mongolica*, induced irradiation mutant	[64,65]
Zolotoi klyuchik	Tolerant to wilt, medium tolerant to sea buckthorn fruit fly	Russia	[62]

Conventional breeding programs for improving adaptability, tolerance, yield, and quality of sea buckthorn have been undertaken since the early 1900s, and more than 150 cultivars have been developed for culinary and/or medicinal use [79]. Breeding was initiated in Russia in the 1930s and in China in 1985 [61,80] using *H. rhamnoides* subsp. *mongolica* from the Altai mountains, and subsp. *sinensis* in Northern China, respectively. Cultivars based on subsp. *sinensis* are usually adapted to harsh environments, grow fast, and have a very high vitamin C content, but their major drawbacks are very thorny branches, low fruit yield, small fruits, and high acidity. By contrast, cultivars based on subsp. *mongolica* often have fewer thorns, a high fruit yield (5–12 kg/plant), large fruits (30–120 g/100 fruits), high oil content, and less acidity. Unfortunately, these cultivars are less well adapted to high summer temperatures and drought and are more susceptible to fungal diseases and pests. The third major subspecies to be used in plant breeding is subsp. *rhamnoides*, which in many respects takes an intermediate position compared to subsp. *sinensis* and subsp. *mongolica*. Russia (over 70 cultivars) and China (over 60 cultivars) are still the leading countries in sea buckthorn improvement, and many cultivars from these countries are planted around the world.

In Russia, selection and breeding have been carried out in different locations, from the maritime climate in Leningrad to a more continental climate in Moscow, where selection and breeding were initiated in 1952 [81]. The domestication, however, started further east, in Barnaul in the Altai region of Siberia. Three cultivars ('Dar Katuni', 'Novost Altaya', and 'Zolotoy Pochatok') derived by selection in wild stands from Katun (Gomy Alta) were released in the 1960s by the Lisavenko Research Institute of Horticulture for Siberia [61]. Accessions were also collected and evaluated from a wider geographic area, and additional cultivars were released, like 'Maslichnaya', 'Vitaminnaya', 'Oranzhevaya', and 'Chuyskaya'. Later, well-known cultivars were also developed at other locations in Russia, from crosses between different accessions of subsp. *mongolica* as well as from crosses between subsp. *mongolica* and subsp. *rhamnoides* like, e.g., 'Botanicheskaya Ljubitelskaya' (Figure 3) and 'Trofimovskaya' developed in Moscow. Some breeding has also taken place in Mongolia, resulting in, e.g., 'Ulaangom' and 'Chandman'.

Figure 3. 'Botanicheskaya Ljubitelskaya', Russian cultivar (*Hippophae rhamnoides* subsp. *rhamnoides* × subsp. *mongolica*).

In China, breeding and selection were initially based on the indigenus subsp. *sinensis* (Figure 4). Five cultivars, including 'Hongxia' and 'Wucixiong' (male), were obtained through the selection of open-pollinated offspring from local stands [69]. Another 15 cultivars, including 'Wulanshalin', 'Hunjin', and 'Wucifeng' that have no or few thorns, large fruits and long fruit stalks, and yields of 10,000–20,000 kg/ha, were selected among offspring from introduced sea buckthorn accessions from Mongolia and Russia (subsp. *mongolica*) [70,71]. The hardy and highly drought-resistant cultivar 'Qiuyang' was selected among seedlings derived from open-pollination of the subsp. *mongolica* cultivar 'Wulangmu' [75]. This step in the domestication process was then followed by making controlled crosses between subsp. *sinensis* and subsp. *mongolica* to improve adaptability, yield, and quality [80]. About 20 cultivars, like 'Hualin 1′', 'Mengzhonghuang', 'Mengzhonghong' (Figure 5), 'Dalate', 'Ezhonghuang', 'Ezhongxian', 'Chengse', 'Hongyun', 'Hongji 1′', 'Zhongji 3′', and 'Zhongji 4′', have now been released based on these crosses [63]. Some other new cultivars, such as 'Gaoyou 1′', 'Chaoyang', 'Wanxia', and 'Wanhuang', have also been released recently in China.

Many other Asian countries have now initiated their own programs for developing sea buckthorn as a crop plant. Harvesting has initially been conducted in native stands of, e.g., subsp. *turkestanica* (India) or subsp. *caucasica* (Iran and Turkey), but these are in general unsuited for large-scale commercial plantations due to thorniness, small fruit size, and low fruit yield. In India, evaluation and selection are presently undertaken in locally grown offspring obtained from the seeds of Russian subsp. *mongolica* cultivars [82], and a similar development is also seen in other Asian countries.

In Ukraine, plant breeding has resulted in cultivars like 'Lvivyanka', 'Osinnia krasunia', 'Mukshanska', 'Rapsodiia', and 'Medova osin' [73]. The first four of these cultivars are based on crosses involving subsp. *mongolica* and produce larger fruits and higher yields compared to the fifth, which is derived from subsp. *carpatica* and instead is characterized by high winter hardiness, drought tolerance, and disease resistance. Some cultivars have also been developed in neighboring Belarus, including the productive 'Plamennaya' with fruits of 80 g/100 fruits [61]. Commercial production is also undertaken in, e.g., Romania and, recently, Greece, but is mostly based on Russian cultivars derived from subsp. *mongolica*.

Figure 4. Plant of *Hippophae rhamnoides* subsp. *sinensis*.

Figure 5. 'Mengzhonghong', Chinese cultivar (*Hippophae rhamnoides* subsp. *mongolica* × subsp. *sinensis*).

In Berlin, in Germany, cultivars like 'Askola', 'Dorana', 'Frugana', 'Hergo', and 'Leikora' were released from 1979 onwards, based on selection in wild stands of subsp. *rhamnoides* [60]. Increased focus on oil content prompted a second step in the breeding program, with renewed selection activities in subsp. *rhamnoides* as well as the undertaking of crosses with subsp. *mongolica* from the Altai region, resulting in 'Sirola' and 'Habego' ('Orange Energy').

Domestication of sea buckthorn in the Nordic and Baltic countries (mainly Sweden, Finland, Estonia, and Latvia) was influenced by both Germany and Russia. Some of the German cultivars were not sufficiently cold-hardy to ensure survival in Finland and the Baltic countries [67] and ripened too late. In Finland, 'Terhi' and 'Tytti' were obtained from native stands of subsp. *rhamnoides* [76], while 'Raisa' was derived from a cross between subsp. *rhamnoides* and subsp. *caucasica* (Figure 6). Nowadays, breeding relies mainly on crosses using subsp. *rhamnoides* or subsp. *fluviatilis* as one parent and subsp. *mongolica* as the other parent. Production of juice is a major goal, and the hybrid cultivars have larger fruits, reduced acidity, perceived sweeter juice, superior juicing capacity (80–90% compared to 50–60% in subsp. *rhamnoides*), and significantly fewer troublesome stellate hairs on the fruits. Hybrid cultivars developed in Sweden, where a breeding program started in 1986, are, e.g., 'Julia', 'Lotta', 'Sol' (Figure 7), 'Idun', 'Eir', 'Fenja', and 'Torun', and in Latvia, e.g., 'Mary' ('Marija Bruvele') and 'Tatjana' [54].

Figure 6. 'Raisa', Finnish cultivar (*Hippophae rhamnoides* subsp. *rhamnoides* × subsp. *caucasica*).

Figure 7. 'Sol', Swedish cultivar (*Hippophae rhamnoides* subsp. *rhamnoides* × subsp. *mongolica*).

In Canada, a breeding program was initiated in 1995, resulting in, e.g., 'Harvest Moon' derived from open pollination of a subsp. *mongolica* seedling obtained from Siberia. This early-maturing, relatively thornless, and winter-hardy cultivar has large reddish-orange fruits with long pedicels, which facilitate hand harvesting and is well suited for growing in the Canadian Prairies and northern Great Plains of the USA [66].

In the future, transgenesis and genome editing with Clustered Regularly Interspaced Short Palindromic Repeats-associated Protein 9 (CRISPR/Cas9) may become useful for improving commercially important traits in sea buckthorn. Still, the notion of using gene transfer or gene editing methods may prove unacceptable for a crop destined for the health food market. Since only a few protocols are available for regeneration [83] and transformation of sea buckthorn [84], a lack of efficient regeneration protocols could also hamper the application of these methods.

7. Plant Breeding Goals

In order to maximize the economic, ecological, and social benefits of sea buckthorn cultivation, breeding goals have focused on (1) enhancing the yield of fruits and seeds, (2) improving the contents and quality of oils and bioactive components, and (3) increasing the tolerance to extreme temperatures (both high and low), drought, waterlogging, diseases, and pests.

7.1. Plant Architecture and Yield

The small and inconspicuous flowers are borne in tight clusters on 2-year-old branches. It takes approximately 100–120 days from fertilization to ripe fruits. Yield is influenced by several factors like plant growth, habit and shape, internode length, fruit size, and pedicel length; phenological events like flowering period and time of fruit maturity; and physiological traits such as disease resistance and tolerance to stress.

The ideotype for sea buckthorn plant shape depends on the harvesting method, which usually consists of hand-picking in the field, direct juicing in the field, cutting off whole branches, which are then frozen and threshed, or machine harvesting in the field [85]. For hand-picking and juicing in the field, a rather short plant with as few thorns as possible is desirable. Thorniness does not matter that much if whole branches are cut off, but longer fruit stalks (5–10 mm) allow easier detachment of fruits. Plant size and ripening time can also be important depending on the type of branch cutting applied (e.g., total cut, lower cut, vertical split, and horizontal split [86]). The machine harvesting of sea buckthorns is conducted on a restricted scale, usually with either vacuum or vibrations; both types of harvesters work better on cultivars with long fruit stalks, and the latter type also requires branches that bend easily without breaking.

Late-ripening plants are generally expected to have higher yields than early-ripening plants since their period of active growth and development is longer. Very-late-ripening plants are, however, vulnerable to early periods of freezing before full ripening. In high-latitude regions, late-ripening individuals may also fail to form sufficient flower primordia for the following year [87].

The fruit of sea buckthorn is not really a berry and is better described as a "pseudo-drupe" [41]. Fruit size and shape are commercially important traits that are determined by several processes, including cell division and cell expansion. Interspecific variation in these processes has been demonstrated for three species of *Hippophae* [88], but this information is not yet immediately applicable in applied plant breeding.

Generally, cultivars derived from subsp. *mongolica* have the largest fruits, like the Siberian 'Elizaveta', 'Agurnaya', and 'Avgustina', which are reported to produce 100–120 g/100 fruits when grown in Russia [61]. In a study of 78 accessions in Northern China, cultivars of subsp. *sinensis* had much smaller fruits (average of 11 g/100 fruits) compared to subsp. *mongolica* (average of 48 g/100 fruits), while hybrids took an intermediate position (average of 31 g/100 fruits) [58]. Cultivars based on subsp. *caucasica* also tend to produce small fruits, as reported for 10 accessions in Eastern Anatolia, where weight ranged from 14 to 24 g/100 fruits [89]. One subsp. *carpatica* cultivar in Ukraine had larger fruits, 37 g/100 fruits, while four hybrid cultivars (subsp. *mongolica* × subsp. *carpatica*) varied from 63 to 71 g/100 fruits, i.e., similar to pure subsp. *mongolica* (68 g/100 g) [73].

7.2. Resistance against Abiotic Dress

Cold tolerance: Plants of *H. rhamnoides* have been reported to tolerate temperatures between −43 °C and +40 °C but grow best when the average temperature in the hottest month ranges from 15 to 25 °C [90]. There is considerable variation in cold hardiness between plants from different geographic origins. Both subsp. *mongolica* and subsp. *sinensis* are adapted to a continental climate and can tolerate very cold winters. In a cool maritime or semi-maritime climate with fluctuating winter temperatures, as in Northern Europe, imported cultivars of subsp. *mongolica* suffer considerably more winter injury compared to native plants of subsp. *rhamnoides*. Several inter-subspecies hybrids originally bred in Russia (especially Moscow) instead have satisfactory cold hardiness when grown in, e.g., Finland, Sweden, Estonia, and Latvia [54,91], and similar crosses are now undertaken in these countries. *H. tibetana*, which grows in alpine plateau regions, can tolerate very low temperatures during the growing season but has not yet been used in breeding programs.

Variation in cold hardiness among genotypes of various origins is, to a large extent, affected by their biochemical content. Cold acclimation increases the concentrations of sugars and dehydrins, resulting in elevated freezing tolerance. Levels of dehydrin mRNA as well as sugar components (especially sucrose) appear to correlate with the cold hardiness of the leaves [87]. Further analyses of the biochemistry of cold hardiness may result in the identification of key genes and the development of DNA markers for application in plant breeding. Large phenotyping projects are, however, needed to supply information about allelic variation in these genes. Thus, data on carefully evaluated parameters involved in cold hardiness must be collected for a large set of genotypes.

Drought tolerance: Although sea buckthorn plants can endure lengthy periods of drought better than many other woody shrubs, flowering and fruit set are adversely affected by inadequate soil moisture, particularly during early spring. Irrigation may therefore be required for orchards in areas with low precipitation. Since this can be difficult and costly to implement, naturally high levels of drought tolerance would be beneficial for cultivars grown in dry areas.

There is, as of yet, little information on genetic variation in parameters involved in water uptake and use. Research on drought stress has shown that leaf water potential, photosynthetic rate, and stomatal conductance decrease in genotypes of both subsp. *sinensis* and subsp. *mongolica*, whereas the content of polyphenols such as flavone, flavonol, isoflavone, and flavanone decreases only in subsp. *mongolica* [92]. By contrast, flavone and abscisic acid (ABA) contents were significantly higher in subsp. *sinensis*. In a transcriptomics analysis, numerous Differentially Expressed Genes (DEGs) were identified under drought stress in the two genotypes [92]. These DEGs were associated mainly with carotenoid biosynthesis, flavonoid biosynthesis, photosynthesis, and plant hormone signal transduction. Six hub DEGs, which play a role in the ABA-dependent signaling pathway, were identified. The mutual regulation of ABA and flavonoid signaling apparently contributes to the difference in drought resistance between the two analyzed samples. Any kind of stress is known to activate plant-specific transcription factor *TCP* genes. In sea buckthorn, *HrTCP20* was significantly up-regulated under drought stress, suggesting that this *TCP* has a role in drought tolerance, possibly by affecting the biosynthesis of the important plant hormone jasmonic acid [93].

7.3. Resistance against Fungal Diseases

Fungal diseases have become increasingly problematic in sea buckthorn, causing serious damage in both wild stands and commercial orchards all over the world [94]. Unfortunately, there is a lack of in-depth pathological studies with bioassays to quantify the symptoms caused by unambiguously identified fungi. Commonly observed symptoms in the field after natural infection are various cankers and lesions on the stems of trunks and main branches, wilting leaves and buds, followed by necrosis of both shoots and roots, and eventual death of the entire plant. In addition, several fungi have been reported to affect leaves and/or fruits, but these symptoms are seldom quite as serious as the three diseases described below.

Wilt: Drying up of sea buckthorn plants, also known as wilt, is a serious worldwide problem. Soil-borne fungi belonging to the genera *Fusarium* (several species) and *Verticillium* (*Verticillium dahliae* and *Verticillium albo-atrum*) have been identified in connection with wilt in Europe, including Russia, as well as in India and Canada. Early symptoms vary between orchards in different countries and are possibly associated with the origin of the planted material but also with environmental factors like drought and excessive soil moisture. Thus, Russian cultivars, presumably derived from subsp. *mongolica*, were reported to be more sensitive to *Verticillium* in Germany in comparison with other cultivars [67]. Cultivars with above-average tolerance to wilt in a field trial with 42 cultivars and advanced selections in Belarus include 'Desert maslichnyi', 'Yolochka', 'Mendeleevskaya', 'Syurpriz Baltiki', and 'Zolotoi klyuchik' [62].

Stem canker: Several fungi, e.g., *Hymenopleella hippophaeicola*, *Cytospora* spp., and *Stigmina* spp., have been reported to cause stem canker and leaf spot on sea buckthorn around the world. The symptoms can lead to the death of entire shoots and may eventually kill the whole plant. In Finland, Russian cultivars derived from subsp. *mongolica* appear to be especially susceptible to damage by *Stigmina*, whereas wild stands of subsp. *rhamnoides* remain healthy [95]. Tolerance to *Stigmina* seems to depend on the ability to withstand frost damage in spite of the fluctuating temperatures during autumn typical of a coastal climate.

Dried shrink disease: The concept of Dried Shrink Disease (DSD) was introduced to denote a common infection pattern caused by various fungi in China but has also been reported in, e.g., Sweden and Russia [3]. DSD has been a major factor in limiting the success of sea buckthorn plantings in China for several decades. Leaves of infected plants become increasingly chlorotic at the beginning of summer, and branches bend down, wilt, and eventually dry up. Fruits of infected plants do not develop properly, and many drop well before maturation. Possibly, wilt and stem canker, as described above, are part of the DSD concept, but other symptoms also occur. In addition to *Fusarium acuminatum*, *Fusarium oxysporum*, *Fusarium camptoceras*, *Stigmina*, and *Verticillium*, fungi that have been reported in connection with DSD include *Phomopsis* (=*Diaporthe*), *Phellinus hippophaeicola*, and *Dothidea hippophaes* (=*Plowrightia hippophaes*) [3,96]. *Hymenopleella* and *Diaporthe* were the most frequently found species in symptomatic samples in Germany [97].

7.4. Resistance against Insects

Numerous insects have been reported to infest sea buckthorn orchards around the world [3]. In China, the most dangerous pest is *Holcocerus hippophaecolus*, found in plantations in, e.g., the Liaoning province and Inner Mongolia, where more than 0.8 million ha have been damaged. As of yet, there is no information on cultivar-dependent variation in susceptibility to this pest. Outside of China, the sea buckthorn fruit fly *(Rhagoletis batava)* is potentially the single biggest threat to commercial production. Infestations can decimate fruit yield and ultimately lead to total loss of the harvest (Figure 8). This fly probably originates from Siberia and has gradually spread west from Siberia into Eastern Europe and south towards Mongolia and China. Severe damage has been reported in commercial orchards in the Baltic countries, Scandinavia, Poland, Romania, and Germany, with yearly yield losses. In China, the extent of fly damage is currently less prevalent, possibly because this insect has not yet adapted to the climate.

Susceptibility to the sea buckthorn fruit fly has been reported to vary among cultivars, with 'Baikal' being highly resistant [62]. Early-ripening cultivars may be more susceptible since their fruits are at a suitable stage for egg-laying. Fruit size, color, and shape may also play a role since female flies of this genus are known to evaluate these traits carefully before laying their eggs. Thus, the majority of relatively resistant varieties appear to have small, reddish, and late-ripening fruit [98]. Proper studies of the inheritance for this trait have, however, not yet been presented.

Figure 8. Damage caused by sea buckthorn fruit fly *Rhagoletis batava*.

Molecular mechanisms involved in insect resistance are not yet well documented, but brassinosteroid (BR) appears to be involved in the stress tolerance of many plant species. In sea buckthorn, BR content was much higher in fruits infected by the sea buckthorn fruit fly compared to uninfected fruits [99]. Moreover, since damage was lower in fruit treated with BR, this compound may enhance resistance in sea buckthorn. Several BR biosynthesis-related *HrCYPs* genes were identified in a transcriptome analysis. The most up-regulated gene in infected sea buckthorn fruits was *HrCYP90B1*, which may act as a positive regulator in resistance to the sea buckthorn fruit fly.

7.5. Chemical Contents

An increasing part of the sea buckthorn cultivation is aimed at producing raw materials for the health food, medicinal, and cosmetic industries. Research has therefore focused on medicinal properties in mainly fruits, with special emphasis on vitamins, phenols, lipids, phytosterols, carotenoids, tocopherols, and triterpenoids. Sea buckthorn is indeed rich in many bioactive phytochemicals, but the contents of the fruit pulp, seeds, and leaves vary considerably. This variation can be attributed to species, ripening time, cultivar, harvest date, influence of year, as well as their interactions, as noted for, e.g., content of acids, sugars and sugar alcohols [100], carotenoids [101], tocopherols [102]), and total phenolic compounds [103]. Also, location [104], orchard management, and sample treatment can influence the reported results, and therefore published data are not easily compared. Future analyses of chemical compounds in fruits should be conducted on samples where the maturity stage is more precisely defined using, e.g., pheophytin as a maturity marker [101].

Ideally, phenotyping data are obtained from samples of replicated plants in formal multi-year field trials, but most sea buckthorn studies are instead based on a limited number of samples without true biological replicates, and only a few studies have focused on the comparison of several species and subspecies. Furthermore, a lack of cheap and efficient biochemical analytical methods has historically restricted the study of phytochemical variation, but several advanced Liquid Chromatography–Mass Spectrometry (LC-MS) and High-Performance Liquid Chromatography–Mass Spectrometry (HPLC-MS) methods are now commonly used to provide reliable data. Fast and efficient rapid proton Nuclear Magnetic Resonance spectroscopy (^{1}H NMR) analysis has been used for the metabolic profiling and discrimination of wild sea buckthorn fruits grown at different locations in Finland (subsp. *rhamnoides*) and China (subsp. *sinensis*) [104]. The two subspecies and different growth sites could be distinguished, and variation within subsp. *rhamnoides* was shown to be associated mainly with the higher temperature, solar radiation, and humidity, as well as the lower precipitation in southern Finland, yielding higher levels of O-ethyl β-d-glucopyranoside and d-glucose and lower levels of malic, quinic, and ascorbic acids. Significant metabolic differences in genetically identical fruits were observed between latitudes 60° and 67° north in Finland. High altitudes (>2000 m) were correlated with higher levels of malic and ascorbic acids in subsp. *sinensis*.

Using NMR, effects of location were also studied on samples of subsp. *rhamnoides* cultivars grown in northern and southern Finland as well as in Canada, showing that the metabolic profile of the northernmost fruits was distinctly different from those grown in southern Finland or Canada, thus demonstrating considerable plasticity in the acclimatization to growth environments [105]. In another study, NMR metabolomics and multivariate data analysis were used to study variation in seven species and seven subspecies of *Hippophae* (90 accessions), with metabolites being quantified with quantitative NMR (qNMR) [106]. Different species were clearly discriminated against by their content of organic acids, L-quebrachitol, and carbohydrates.

NMR non-targeted metabolomics have also been applied to the leaves of sea buckthorn. In a study of two cultivars (subsp. *rhamnoides*) grown in the south and north of Finland during two consecutive growth seasons, the highest variance in the metabolic profile was linked to the growth stage, with the second highest variance attributed to location [107]. The north–south comparison identified fatty acids and sugars as discriminatory metabolites.

An NMR-based approach was used on sea buckthorn juice to reveal metabolic changes during fermentation with different strains of *Lactiplantibacillus plantarum* [108]. In total, 46 metabolites were identified from the fresh and fermented juice, including various sugars, amino acids, organic acids, ketones, nucleosides, and one alkaloid. Thus, NMR-based metabolomics could be a useful approach for simultaneous metabolic profiling, species and subspecies differentiation, and quality assessment of sea buckthorn, as well as for discriminating the effects of the geographical origin of sea buckthorn fruits and leaves.

Sugars and sugar alcohols: Fructose, glucose, and ethyl glucose are usually the most abundant sugars in sea buckthorn fruits, and L-quebrachitol is the most abundant sugar alcohol, with methyl-myo-inositol and myo-inositol being present in smaller amounts. Interestingly, changes in sugar content during maturation seem to differ between species. Levels of fructose and glucose in subsp. *sinensis* fruit juice increased during ripening, whereas corresponding contents in samples of subsp. *rhamnoides* decreased [100].

Fruits were collected from nine natural stands of subsp. *sinensis* in China during three years, and the influence of latitude and altitude on sugars and sugar alcohols was investigated [109]. Although samples were collected at optimal maturity, the contents of fructose, glucose, and L-quebrachitol in the fruit juice varied widely: 0.01–7.17, 0.05–7.85, and 0.21–1.09 g/100 mL, respectively. The contents of fructose, glucose, and total sugar correlated positively with the growth latitude but negatively with the altitude. The contents of L-quebrachitol correlated strongly and positively with latitude.

In a study of inositols and methylinositols in the juice of three subspecies, wild Chinese fruits (subsp. *sinensis*) contained higher levels of L-quebrachitol (average 615 mg/100 mL juice) and methyl-myo-inositol (average 58 mg/100 mL juice) than the Finnish (subsp. *rhamnoides*, 276 and 11 mg/100 mL juice, respectively) and Russian (subsp. *mongolica*, 228 and 16 mg/100 mL juice, respectively) fruits [110]. The content of myo-inositol was higher in the Chinese and the Russian fruits than in the Finnish (26 and 20 mg/100 mL juice vs. 8 mg/100 mL juice). In the Chinese fruits, the contents of methyl-myo-inositol and L-quebrachitol increased, whereas that of myo-inositol decreased from late September to late November. The content of L-quebrachitol in the Chinese fruits correlated negatively with the air temperature and the number of frost-free days.

In another study, sugars, ethyl β-d-glucopyranose, and methylinositol were analyzed in fruit juice from three subspecies (subsp. *sinensis, rhamnoides,* and *mongolica*) collected in China, Finland, and Russia during four years [100]. Origin and harvesting date had a significant impact on the content of sugars. During the harvesting period, sugar content developed differently in fruits of the different subspecies. Fructose (subsp. *sinensis* 1.5–11.7 g/100 mL, subsp. *rhamnoides* 0.1–0.6 g/100 mL, and subsp. *mongolica* 0.9–4.3 g/100 mL) and glucose (subsp. *sinensis* 1.6–12.5 g/100 mL, subsp. *rhamnoides* 0.8–2.9 g/100 mL, and subsp. *mongolica* 4.3–7.2 g/100 mL) were the main sugars in all three subspecies. Ethyl glucose was present in the sugar fraction of subsp. *rhamnoides* (0.1–1.9 g/100 mL) but was found only in trace amounts in the other two subspecies. In subsp. *rhamnoides*, the level of ethyl glucose increased during the harvesting period and was accompanied by a decrease in glucose content. Methylinositol was present in higher levels in subsp. *sinensis* (0.3–1.6 g/100 mL) than in the other two subspecies (0.1–0.5 and 0.2–0.3 g/100 mL for subsp. *rhamnoides* and subsp. *mongolica*, respectively). The total amount of sugars was highest in subsp. *sinensis* and lowest in subsp. *rhamnoides*.

Organic acids and vitamin C: Sea buckthorn is a rich source of organic acids, mainly malic, quinic, and ascorbic acids. Maturity has a significant influence on the content of ascorbic acid, with an average reduction of 25% from early to late maturity reported for, e.g., hybrid cultivars derived from subsp. *mongolica* [111] as well as for wild-growing subsp. *rhamnoides* [112]. The time of sampling is therefore crucial to obtaining comparable data. Ascorbic acid seems to be especially high in the fruits of *H. salicifolia*, but also *H. rhamnoides* subsp. *sinensis* and subsp. *yunnanensis* have a high content [113].

Wild fruits of subsp. *sinensis* are reported to contain 5−10 times more vitamin C in the juice fraction than fruits of subsp. *rhamnoides* and subsp. *mongolica* [112]. In subsp. *sinensis*, the content of organic acids in fruits has been shown to vary among genotypes and sites [109]. Malic acid (1.55–8.84 g/100 mL juice) and ascorbic acid (0.25–1.66 g/100 mL) thus increased as the altitude increased and as the latitude decreased, while the content of quinic acid (0.07–2.94 g/100 mL) correlated strongly and positively with the latitude.

In another study, organic acids were analyzed in the fruit juice of three subspecies over four years [100]. Similarly, as for sugars, growth location and harvesting date had a significant impact on the content of acids. The content of malic acid decreased from initial ripening to full ripening and thereafter increased slightly. Malic acid content varied in all three subspecies (subsp. *sinensis* 1.9–9.2 g/100 mL, subsp. *rhamnoides* 2.3–4.7 g/100 mL, and subsp. *mongolica* 0.8–2.7 g/100 mL), as did quinic acid (subsp. *sinensis* (0.7–7.5 g/100 mL, subsp. *rhamnoides* 0.7–4.3 g/100 mL, and subsp. *mongolica* 1.3–2.6 g/100 mL).

Tocopherols, tocotrienols, and vitamin E: Both tocopherols and tocotrienols have vitamin E activity, with α-tocopherol being the most important in the fruit flesh (pulp) of sea buckthorn. In seed oil, the content of vitamin E reported for different sea buckthorn species (*H. goniocarpa, H. gyantsensis, H. neurocarpa, H. rhamnoides, H. salicifolia,* and *H. tibetana*) and subspecies was in the range of 98–273 mg per 100 g of oil, with the highest amount for *H. neurocarpa* subsp. *stellatopilosa* and the lowest for *H. tibetana* [113]. In the same plant material, vitamin E content in pulp oil ranged from 54 to 181 mg per 100 g of oil, with the highest amount for *H. goniocarpa* subsp. *goniocarpa* and the lowest for *H. neurocarpa* subsp. *stellatopilosa* [113].

Seeds of subsp. *sinensis* contained fewer tocopherols and tocotrienols (average 130 mg/kg) compared with seeds of subsp. *rhamnoides* (average 290 mg/kg) and *mongolica* (average 250 mg/kg) in a large study with both wild and cultivated material [112]. By contrast, the fruit flesh of subsp. *sinensis* berries had 2−3 times higher contents of tocopherols and tocotrienols compared to the other two subspecies (120 mg/kg vs. 40 mg/kg in subsp. *rhamnoides* and 50 mg/kg in subsp. *mongolica*). Overall, the fresh whole fruits of subsp. *sinensis* were clearly the best source of total tocopherols and tocotrienols. In the same study, it was shown that the total content of tocopherols and tocotrienols in fruit flesh reached its maximal level around early to mid-September, whereas the content in seeds continued to increase until the end of November.

In a study of four cultivars over three years, 'Ljubitelskaya' (synonym 'Botanicheskaya Ljubitelskaya'), derived from a cross between subsp. *mongolica* and subsp. *rhamnoides*, generally had a higher content of most tocopherols and tocotrienols in the fruit flesh than the other cultivars studied [102]. The total amounts of tocopherols plus tocotrienols were two-fold higher compared with 'Eir'. The most extreme difference between the cultivars was noted for δ-tocopherol, with the highest amounts in 'Ljubitelskaya' and only trace amounts in 'Eir'. Levels of α-tocopherol were higher early in the ripening period, while at later dates, δ-tocopherol levels increased.

Carotenoids and vitamin A: In the fruit flesh of sea buckthorn, β-carotene, γ-carotene, and lycopene are the most important carotenes, while zeaxanthin, lutein, and β-cryptoxanthin are the most important xanthophylls. Vitamin A activity stems mainly from the content of β-carotene. In *H. rhamnoides*, β-carotene constitutes approximately 15–55% of the total carotenoids, with a range of 100–500 and 20–100 mg/100 g in pulp and seed oil, respectively [114].

The fruit pulp of four cultivars was analyzed for carotenoid contents during ripening in three consecutive years [101]. The different carotenoids generally increased in concentration during ripening and comprised from 120 to 1425 μg/g dw (dry weight) of total carotenoids (1.5–18.5 mg/100 g of pulp fresh weight) depending on cultivar, harvest time, and year. The analyses showed that the effect of cultivar was considerably larger than the effect of year and harvest time.

Lipids and fatty acids: Sea buckthorn fruits are rich in oils, both in pulp and seeds. The oil content is generally lower in the pulp compared to the seeds when based on pulp fresh weight, but larger when based on dry weight. In a study of 78 accessions, oil content proved to be higher in pulp (3.5–38.6% dw) than in seeds (3.9–12.8% dw), especially in subsp. *mongolica* accessions [58]. The lowest total oil content in *H. rhamnoides* has been reported for subsp. *sinensis* (2–3%, whole fruit fresh weight), whereas subsp. *turkestanica* and subsp. *mongolica* have the highest (4–14% and 2–10%, respectively) [114]. In *H. goniocarpa*, *H. gyantsensis*, *H. rhamnoides* (including all subspecies except subsp. *caucasica*, for which data are missing), *H. salicifolia*, and *H. tibetana*, pulp oil content ranges between 1.6 and 7.8% (fresh weight), but is considerably higher in *H. neurocarpa*, 10.6–19.8% [113]. Seed oil content ranges between 5.5 and 13.0% in *H. goniocarpa*, *H. gyantsensis*, *H. rhamnoides*, and *H. salicifolia* but is higher in *H. neurocarpa* and *H. tibetana* (8.6–16.2% and 15.3–19.6%, respectively).

The fatty acid profile differs between pulp oil and seed oil, but the major fatty acids are apparently the same in the commonly cultivated subspecies, i.e., subsp. *caucasica*, subsp. *mongolica*, subsp. *rhamnoides*, and subsp. *sinensis*. The content of monounsaturated fatty acids is over 50% in the pulp oil of subsp. *mongolica*, subsp. *rhamnoides*, and subsp. *sinensis* (51–59%), but lower for subsp. *caucasica* (27–43%) [58,89]. The content of polyunsaturated fatty acids is 10–20% in subsp. *mongolica*, subsp. *rhamnoides*, and subsp. *sinensis*, but higher for subsp. *caucasica* (28–43%) [89]. Total content of polyunsaturated fatty acids is much higher in seed oil (55–78%) compared to in pulp oil (27–59%), but variation among subspecies of *H. rhamnoides* seems to be as large as or larger than the variation among species when many accessions are investigated [58].

The development of cultivars with high total oil content and high content of the rare palmitoleic acid, which can increase basal and insulin-stimulated glucose uptake and reduce de novo fatty acid synthesis and activity of lipogenic enzymes, should be possible using the genetic variation within and among species and subspecies. Recently, the main enzymes involved in pulp oil accumulation in sea buckthorn were identified [6,115]. A Real-Time Quantitative Reverse Transcription-Polymerase Chain Reaction (qRT-PCR) analysis of 15 genes involved in fatty acid and triacylglycerol (TAG) biosynthesis in two sea buckthorn genotypes helped to understand the mechanism of high C16:1n7 accumulation in fruit pulp [116]. A thorough understanding of the lipid synthesis pathway and its mechanisms for regulation is important and could in the future facilitate breeding of sea buckthorn cultivars with high oil contents and a specific composition of fatty acids using gene editing tools.

Phenolic compounds: Phenolic compounds are a very large group of phytochemicals present in all parts of the sea buckthorn plant, with the major phenols being flavonols (quercetin, isorhamnetin, and kaempferol derivatives), polymeric procyanidins, flavan-3-ols, and phenolic acids. In a study of cultivars derived from subsp. *mongolica* grown in Poland, leaves had the highest content of flavonols (921.7 mg/100 g dw) and branches the lowest (41.3 mg/100 g dw), with skin, flesh, seeds, and endocarp in between [117] (Table 2). Leaves also had the highest content of phenolic acids, albeit still low. The highest content of flavan-3-ols and polymeric procyanidins was found in branches and seeds, followed by leaves. The procyanidins in sea buckthorn are of the B-type [118]. Among flavonols, leaves are especially rich in quercetin derivatives. Quercetin derivatives were also high in skin and fruit flesh, followed by isorhamnetin derivatives and kaempferol derivatives [117].

Table 2. Average content of phenols (mg/100 g dw) in different parts of sea buckthorn fruits, branches, and leaves of 7 cultivars derived from *H. rhamnoides* subsp. *mongolica*. Table compiled from data in Tkacz et al. [117]). SD = standard deviation, DP = degree of procyanidin polymerization, and nd = not detected.

Phenolic Group	Skin	SD	Flesh	SD	Endocarp	SD	Seeds	SD	Branches	SD	Leaves	SD
Qercetin derivatives	397.1	235.4	186.6	65.7	58.1	10.3	65.7	13.2	38.1	48.7	729.1	188.2
Isorhamnetin derivatives	308.0	208.0	114.1	62.8	72.7	18.7	112.8	24.0	3.3	3.1	173.4	36.6
Kaempferol derivatives	16.1	6.0	9.6	6.7	7.1	5.9	1.2	0.2	nd	nd	19.3	9.1
Total flavonols	721.2		310.2		137.9		179.8		41.3		921.7	
Phenolic acids	3.1	1.2	2.4	0.8	nd	nd	nd	nd	nd	nd	5.4	1.4
Flavan-3-ols	65.7	17.3	53.3	8.5	137.0	43.4	41.6	9.2	2001.3	413.6	124.7	36.8
Polymeric procyanidins	503.4	83.2	141.3	35.4	424.9	111.8	1266.8	203.2	7276.0	1574.5	974.5	533.0
DP	4.5	0.8	4.4	1.1	3.3	0.8	8.0	2.1	2.4	0.1	2.6	0.4
Total phenolics	1293.3	434.7	507.2	127.5	699.8	93.5	1488.2	241.1	9318.7	1909.5	2026.4	759.4

Flavonol glycosides were identified and quantified in the fruits of plants grown in Finland and Canada but derived from wild-growing subsp. *sinensis* in China and cultivated subsp. *mongolica* [119]. Twenty-six flavonol glycosides were found, with isorhamnetin and quercetin as the major aglycones. The contents of flavonol glycosides ranged from 23 to 250 mg/100 g of fresh fruits and were significantly higher in subsp. *sinensis* than in subsp. *mongolica*. Among the subsp. *mongolica* cultivars, the fruits of 'Oranzhevaya' had the lowest (23 mg/100 g) content, and those of 'Prevoshodnaya' had the highest (80 mg/100 g) flavonol glycosides. Also, samples grown in Kittilä (northern Finland) had higher levels of most flavonol glycosides than samples grown in Turku (southern Finland) and Québec. For subsp. *sinensis*, the contents of most compounds increased as the altitude of the growth site increased and as the latitude decreased. Fruits from stands originally collected in Sichuan had remarkably high contents and unique profiles of flavonol glycosides.

Differences were found between subsp. *rhamnoides* and subsp. *mongolica* accessions, as well as for procyanidins [118]. Furthermore, samples of subsp. *rhamnoides* grown in northern Finland were characterized by a high amount of total procyanidins, typically 2–3 times higher than found in samples grown in Southern Finland. In subsp. *sinensis*, altitude did not correlate with the PA (proanthocyanidins) composition [118].

The flavonol profile of sea buckthorn fruits has turned out to be useful in separating species, and a fingerprint method based on 12 flavonoids identified from HPLC chromatograms has been developed [120]. Potentially, this method could also be used for quality assurance of sea buckthorn fruits and extracts.

In a recent study of two subsp. *mongolica* accessions with similar flavonoid profiles, chalcone synthase (CHS) and flavanone-3-hydroxylase (F3H) were the main enzymes responsible for the difference in flavonoid synthesis and accumulation in sea buckthorn fruits [6], which is in agreement with previous studies [121]. A deeper understanding of the flavonoid synthesis pathway could help breeders develop strategies for future breeding.

The content of different phenolic compounds is strongly influenced by developmental stage, as shown for both leaves [122] and fruits [6]. A proper sampling strategy must therefore be applied to obtain reliable and comparable data.

Sterols: Phytosterols can lower serum cholesterol levels in humans and therefore constitute an important group of bioactive compounds in plants. In a study of pulp oil in eight subsp. *mongolica* cultivars, 14 phytosterols were identified and quantified, the major compounds being β-sitosterol, 24-methylenecykloartanol, and squalene [123]. Although β-sitosterol was always the major sterol, the total amount and profile differed widely among cultivars (Table 3). In another study, sterols were analyzed in seeds, pulp/peel fractions, and whole fruit samples derived from two major subspecies (subsp. *sinensis* and subsp. *rhamnoides*) from Finland and China [124]. Total sterol contents in the seeds, the fresh pulp/peel, and the whole fruits were 1200–1800, 240–400, and 340–520 mg/kg, respectively. The corresponding values in the extracted oils were 12–23, 10–29, and 13–33 g/kg. β-sitosterol constituted 57–76 and 61–83%, respectively, of the seed and pulp/peel sterols. The sterol content and composition showed little variation between subspecies and collection sites. Different harvesting dates showed significant effects on the levels of some sterols, both in the seeds and in the pulp/peel. The content of sterols can depend on the extraction method, and supercritical carbon dioxide has proven to be an efficient and reliable method for exhaustive extraction of sterols [125].

Table 3. Composition and content of phytosterols in different sea buckthorn cultivars (μg/100 mL of pulp oil). Numbers followed by the same lower-case letter constitute statistically homogeneous groups (Duncan test, $p \leq 0.05$), and nd = not detected. Modified from Teleszko et al. [123].

Phytosterol	Aromatnaja	Avgustinka	Botaniczeskaja	Botaniczeskaja Ljubitelskaja	Luczistaja	Moskwiczanka	Podarok Sadu	Porożrachnaja
Squalene	1638 c	1205 e	2714 a	1872 b	885 g	1110 ef	1516 d	1026 f
Kampesterol	167 b	139 c	201 a	95 e	82 e	44 f	118 d	127 cd
Stigmasterol	nd	68 a	nd	nd	nd	24 b	nd	nd
β-sitosterol	6145 a	4942 b	3801 f	4022 e	2049 g	2036 g	4216 d	4409 c
Sitostanol	217 b	172 c	217 b	110 e	97 f	96 f	254 a	150 d
Δ5-avenasterol	274 b	251 c	377 a	197 d	198 d	114 e	249 c	243 c
α-amyrin	314 a	110 f	302 a	130 e	166 d	112 f	203 c	228 b
Cycloartenol	441 b	399 c	462 a	439 b	299 e	293 e	474 a	360 d
Δ7-avenasterol	156 c	152 cd	154 c	138 e	147 d	80 f	186 b	194 a
28-methylobtusifoliol	249 a	130 d	251 a	145 c	152 b	70 f	89 e	153 b
24-methylene-cycloartanol	1554 f	4048 a	2711 c	2521 d	1994 e	1454 g	3131 b	2661 c
Erythrodiol	818 a	680 b	590 c	367 f	338 g	284 h	444 d	407 e
Citrostadienol	663 a	327 d	417 c	309 e	244 g	212 h	444 b	269 f
Friedelan-3-ol	737 a	583 c	606 b	398 d	257 f	232 g	260 f	320 e
Total	13,378	13,212	12,810	10,749	6914	6168	11,592	10,553

Triterpenoids: The pharmaceutical potential makes triterpenoids a relevant target for future breeding efforts. Recently, a detailed account of triterpenoids in the skin, flesh, endocarp, seed, leaves, and branches of cultivars derived from subsp. *mongolica* was given [117] (Table 4). The 11 identified triterpenoids were divided into two groups according to the amount detected: (1) betulin, betulinic acid, corosolic acid, maslinic acid, pomolic acid, oleanolic acid, and ursolic acid, all of which were present in significant amounts, and (2) α-boswellic acid, erythrodiol (only in branches and leaves), tormentic acid, and uvaol, all found in amounts below 1 mg/100 g dw [117]. In fruit fractions (skins, flesh, endocarp, and seed), pomolic acid dominated and constituted 62% of total triterpenoids in the fruit flesh. The highest content of maslinic acid was also found in the fruit flesh, constituting 16% of total triterpenoids. The highest content of corosolic acid and betulinic acid was found in branches. Skins had the highest amounts of oleanolic acid, ursolic acid, and betulin. Leaves had similar levels of ursolic acid as skins; in leaves, it constituted 42% of total triterpenoids. Large differences were also found among cultivars; thus, 'Botanicheskaya Ljubitelskaya' had the highest amounts of triterpenoids in skins and seeds, whereas 'Golden Rain' and 'Tatiana' had the highest triterpenoid levels in branches and the lowest in seeds. Thus, breeding strategies should consider both genotype and utilized plant parts in order to optimize the contents.

Table 4. Average content of major triterpenoids (mg/100 g dw) in different parts of sea buckthorn berries, branches, and leaves of 7 cultivars derived from *H. rhamnoides* ssp. *mongolica*. Table compiled from data in Tkacz et al. [117]. SD = standard deviation.

Triterpenoids	Skin	SD	Flesh	SD	Endocarp	SD	Seeds	SD	Branches	SD	Leaves	SD
Maslinic acid	5.15	2.08	11.06	1.89	4.50	1.01	3.22	0.47	2.43	0.58	0.51	0.12
Pomolic acid	21.45	11.7	43.08	3.06	17.81	3.64	15.19	2.79	8.95	3.13	1.98	0.47
Corosolic acid	8.99	1.71	9.70	1.27	5.21	0.79	6.68	0.43	10.93	0.97	3.39	0.34
Betulinic acid	0.66	0.43	0.87	0.31	0.29	0.12	1.77	1.21	1.93	0.28	0.56	0.21
Oleanolic acid	4.73	3.59	2.65	2.93	1.41	0.97	1.85	1.22	3.70	0.33	0.63	0.08
Ursolic acid	6.45	1.53	0.46	0.09	1.31	1.35	0.39	0.06	4.28	1.00	6.24	2.70
Betulin	13.27	2.90	0.74	0.22	0.31	0.09	0.28	0.10	0.59	0.21	1.17	0.24
Total	62.26	12.76	69.61	4.97	31.3	4.55	30.21	4.15	33.75	4.34	14.72	2.41

Sensory aspects: In many countries, sea buckthorn juice is the main or only product derived from commercial plantations. Naturally, the amount and balance of chemical compounds that affect sensory components are crucial for cultivars grown mainly for their culinary properties. Astringency, sourness, and bitterness were negatively correlated with pleasantness and favored only by a few members of a sensory panel in Finland testing sea buckthorn juice, whereas sweetness showed a strong positive correlation with pleasantness [126]. In that study, genotypes obtained from crosses between subsp. *rhamnoides* and subsp. *mongolica* gave the most acceptable juice, possibly due to a lower total acidity. In another Finnish study, the total content of sugars and the sugar/acid ratio correlated positively with sweetness and negatively with sourness and astringency in the juice of six cultivars derived from subsp. *mongolica* and one cultivar from subsp. *rhamnoides* [127]. Total acidity and titratable acidity instead correlated positively with sourness and astringency and negatively with sweetness.

The slightly astringent and bitter taste of sea buckthorn fruits has also been shown to correlate with the contents of flavonols, proanthocyanidins, and ethyl b-D-glucopyranoside [128,129]. However, the low average degree of procyanidin polymerization (number of flavan-3-ol units in polymers) ranging from 2.4 (for branches), followed by 2.6 (leaves), 4.4 (fruit flesh), to 8.0 (for seeds), and the presence of preferably dimeric and oligomeric flavan-3-ols indicate a potentially low intensity of astringency in food applications [117].

Besides these non-volatile compounds, odor-active volatiles have a crucial influence on the sensory quality of sea buckthorn fruits [130]. Fatty acid oxidation during the late ripening stages results in unattractive fragrances and off-flavors in many cultivars, but breeding and selection can help reduce rancidity, as shown in several of the recently developed cultivars in Sweden.

Fruit color: Sea buckthorn fruit color is an important commercial trait and varies across species from bright yellow, over orange and red, to brown in mature fruits. Integrative analyses of the transcriptome and targeted metabolome, including carotenoids, flavonoids, and chlorophylls, were recently performed for five accessions: subsp. *sinensis* 'FengNing' (yellow), subsp. *mongolica* 'XiangYang' (orange), and two accessions of red-colored and yellow-colored hybrid offspring, as well as a brown *H. neurocarpa* subsp. *neurocarpa* fruit [131]. A total of 209 flavonoids and 41 carotenoids were identified, but a high content of chlorophyll was found only in the brown fruits. It was shown that the quantities and relative proportions of the flavonoids, carotenoids, and chlorophyll led to the different colors of the sea buckthorn fruits, and key genes related to the carotenoids and chlorophyll metabolism were identified. The high content of chlorophyll in the brown fruit was closely related to the downregulated expression of key genes in the chlorophyll degradation pathway. In yellow fruits, carotenes dominated, while lycopene contributed significantly to the color in orange and red fruits. Future breeding efforts could therefore target key enzymes affecting the synthesis and accumulation of carotenoids and chlorophylls.

8. Breeding of Male Plants

Male sea buckthorn plants are not given as much attention from breeders as the females since their only role is usually to provide prolific amounts of pollen. Male plants are therefore selected mainly for plant vigor, persistence, and the timing of pollen dispersal, which should overlap with the flowering time of the female plants. Insufficient cold hardiness has, however, been reported as a major problem with continental male cultivars, making them unsuitable for production in Finland and the Baltic countries [67]. Another potential goal is to produce plants for commercial pollen-based medicinal products [132]. Since sea buckthorn is a dioecious wind-pollinated species, the inheritance of fruit characteristics from male plants can only be evaluated from the performance of their offspring following controlled crosses.

Two thornless male cultivars with winter-hardy flower buds were selected at the Lisavenko Research Institute in Russia [59]. 'Aley' is very vigorous and is mainly recommended for commercial plantations, whereas the compact 'Gnom' is suitable both for commercial plantings and home gardens. In China, 'Wucixiong' was selected from wild stands of subsp. *sinensis*, whereas the almost thornless 'Mengzhongxiong' derives from subsp. *mongolica* × subsp. *sinensis.*

'Pollmix' is the common name for a group of four different genotypes selected from wild stands of subsp. *rhamnoides* in Germany [60], while 'Tarmo' was selected from wild stands of subsp. *rhamnoides* in Finland, and 'Edgars' and 'Skibes vir' in Latvia. Later on, 'Lord' was selected in Latvia from a cross between a subsp. *mongolica* × subsp. *rhamnoides* hybrid and local subsp. *rhamnoides*. In Sweden, 'Romeo' was selected from a subsp. *mongolica* × subsp. *rhamnoides* population, and 'Svenne' was selected from an open-pollinated local subsp. *rhamnoides* selection grown in a breeding collection (Figure 9). More recently, 'Balsgård Hubert' was released, a vigorous male selected from an open-pollinated local selection of a hybrid between subsp. *mongolica* and subsp. *rhamnoides*.

Figure 9. 'Svenne', male Swedish cultivar (*Hippophae rhamnoides* subsp. *rhamnoides*).

Although sea buckthorn is a predominantly dioecious plant, hermaphrodite flowers sometimes develop on normal male plants [64]. Thus, a future breeding goal could be to develop truly hermaphrodite plants, which could potentially exclude the need for male pollinator plants.

9. Breeding Methods

9.1. Conventional Breeding Methods

Sea buckthorn breeding started with the mass selection of superior plants in wild populations. Mass selection is still used to ensure locally adapted plant material for use in breeding programs. However, hybridization has increasingly replaced mass selection as a breeding method.

For hybridization, flowers need to be isolated 2–3 days prior to the beginning of flowering, preferably using bags of dense fabric to cover branches on female plants. A few days later, branches with open male flowers are placed together with the isolated female flowers, and bags are shaken the following day to ensure good pollination. A different approach consists of removing the bags (which requires very calm weather with absolutely no wind) and carefully dusting the female flowers with previously collected pollen. The female flower can receive pollen for about a week, depending on the temperature. The bags must, however, be removed when the stigma is no longer receptive to allow proper ovary development, which is dependent on light [133].

Sea buckthorn pollen can be collected from forced twigs being cut 2–3 days before flowering. The pollen grains can be stored cool and dry for later use, but little information is available on the possibility of long-term storage of pollen and its viability. Germination of pollen grains is reported to occur in 3–4 h and fertilization takes 5–10 days, depending on temperature [134].

Another approach is to grow female and male plants in pots and place them together in a greenhouse chamber for hybridization. It is important to shake the male plants daily to ensure proper pollen dispersal and pollination.

As soon as the fruits have ripened, seeds can be collected, stored dry, or immediately sawn since they have a short-term physiological dormancy [135]. However, cold stratification for 15–90 days reduces dormancy and improves germination of *H. rhamnoides*. Optionally, soaking seeds in water for 7 days could replace stratification [136]. Seeds can maintain germinability for more than two years when stored at low humidity, even at room temperature [135].

Sea buckthorn seedlings enter the adult stage after 3–5 years and can then be evaluated for fruit characteristics and yield potential. Selected seedlings are propagated for comparative field trials. Sea buckthorn is easily propagated by hardwood and softwood cuttings [137], but it can also be micro-propagated [138]. Following propagation, plants usually start to yield after 3-4 years and should be evaluated for 3 years at least in observation trials before superior selections are propagated for large-scale yield trials in different locations. The time taken from initial cross to release of cultivars is thus at least 15 years.

9.2. Treatment with Mutagenics

Induced mutagenesis with a sub-lethal dose of ionizing irradiation or with chemicals can increase the genetic variability of sea buckthorn and shorten the time required for developing a new variety [139]. Using irradiation, several cultivars have been developed in Russia, e.g., 'Druzhina', 'Podrugha', 'Zolotoy Kaskad', 'Ivushcka', 'Ognistaya', 'Zarnitsa', and 'Krasny Fakel'. Some of these are reported to have large aromatic fruits, a high yield, a high content of vitamins and other biologically active substances, long pedicels, and a low number of thorns [64]. 'Elizaveta', 'Inya', and 'Sudaruschka' were instead derived as chemically induced mutants of 'Panteleevskaya' and are presently being marketed for qualities like high yield, high carotenoid contents, and a short juvenile period.

Polyploid sea buckthorn plants (2n = 24) can be found in wild populations but have also been developed using colchicine [140]. Although polyploidization frequently results in abnormal and weak plants, it could also contribute to novel and useful variation.

9.3. Molecular Marker-Assisted Breeding

Although considerable initial progress can be attained from choosing parents with above-average performance in desirable traits, DNA marker-based selection can be a valuable tool, especially for traits that are difficult to evaluate in the field. DNA markers are now being used to screen putative parents in many fruit and berry crops for traits like disease resistance and fruit texture components. The cost-effectiveness of screening the resulting seedlings depends on the percentage of seedlings that can be discarded at an early stage, which in turn depends on the number and efficiency of the DNA markers scored, as shown for apples [141].

In sea buckthorn, very few molecular markers have been developed as of yet, apart from those used for sex determination. Some attempts for disease resistance have been made. Using a disease index for resistance/susceptibility to DSD, four ISSR markers [142] and eleven Sequence-Related Amplified Polymorphism (SRAP) markers were shown to be associated with the number of disease symptoms in the investigated plant material [143]. These multilocus marker methods do, however, not lend themselves well to large-scale screenings. Also, the identity of the fungi causing DSD varies considerably between locations as well as the type of germplasm. For the identification of causative genes and the development of single-locus DNA-based markers, large amounts of phenotypic data must first be collected, preferably based on carefully conducted inoculation experiments with properly identified fungal strains. In the case of sea buckthorn, this will first require considerable research into the plant pathology of the species.

The detection of key genes for important traits in other crops has been attempted via different avenues. Traditionally, major genes and Quantitative Trait Loci (QTL) are identified based on a molecular-marker mapping population with offspring that segregate for the trait of interest. So far, really dense (marker-rich) genetic maps have, however, not been published for sea buckthorn. Another, more recent, and perhaps more promising approach is to conduct a GWAS, which exploits the linkage disequilibrium present among individuals from natural populations or germplasm collections. These are usually more diverse than segregating progenies and can be used to map QTL with high resolution. GWAS is facilitated in many crops by the recent development of high-density SNP arrays with uniform coverage of the whole genome. Genes can, however, also be located using the increasingly affordable methods for whole-genome re-sequencing (see below).

The identification of important genes and the development of molecular markers for use in genetic research and plant breeding are facilitated by the existence of a chromosome-level genome assembly obtained via in-depth sequencing. This has been lacking for the entire family Elaeagnaceae, but no less than four genomes were published in 2022, three of which were developed for *Hippophae* [140,144,145].

The first of these was *H. rhamnoides* subsp. *mongolica* cultivar 'Sunny', with a total genome length of 849.04 Mb [144]. The quality and completeness of this assembly were ascertained, and high synteny between the physical map and the genetic map was also demonstrated. A total of almost 31,000 genes were predicted with an average length of 4900 bp and an average coding-sequence length of 1307 bp. Two sequential whole-genome duplication events were proposed, the first occurring about 36–41 million years ago (Mya) and the second 24–27 Mya. These events have prompted an expansion of genes related to ascorbate and aldarate metabolism, lipid biosynthesis, and fatty acid elongation. Several key genes contributing to the high contents of polyunsaturated fatty acids and ascorbic acid were identified via transcriptome sequencing. A subsequent population genomic analysis based on whole-genome sequencing of 55 *H. rhamnoides* accessions (wild and cultivated subsp. *mongolica* and wild subsp. *sinensis*) yielded a large number of SNPs that were used for analysis of genetic relatedness. A search for candidate genomic regions under positive selection indicated that selective sweeps have contributed to the richness of fatty acids and ascorbic acid in the fruits of domesticated subsp. *mongolica* plants [144]. Numerous genes for economically important traits were identified in the analyzed accessions, which will hopefully soon result in easy-to-use molecular tools for marker-assisted plant breeding.

The second genome assembly was undertaken with a sample of *H. rhamnoides* subsp. *sinensis* [145]. The obtained genome was 730.46 Mb, and many genes of importance for, e.g., vitamin C synthesis were identified. The sequenced genome was shown to share a high number of conserved symbiotic root nodulation genes with other species (e.g., *Lotus* and *Medicago*) that also have rhizobia-induced symbiosis.

The third genome assembly was developed for a different species in the *Hippophae* genus, namely *H. tibetana*, which is an extreme high-altitude plant growing at 3000–5200 m above sea level on the Qinghai–Tibet Plateau [40]. This genome had undergone two duplications, just like the genomes assembled for *H. rhamnoides* [144,145], but was somewhat larger, 1452.75 Mb, than the other two. An adaptation to the more extreme and variable conditions at high altitudes (cold, drought, extremely low supplies of nitrogen, low carbon dioxide concentration, and high UV-B ultraviolet light stress) is indicated by the high number of genes related to adaptation to high ultraviolet and low temperature in *H. tibetana*, as well as more copies of nitrogen-fixing genes.

In many crops, the development of molecular markers has been hampered by a lack of high-quality phenotyping data. In addition to exact and informative measurements, the potential to assess a genotype–environment interaction by measuring the same trait over several years—and preferably on replicated plants of the same genotype growing in different orchards—would be very valuable. In apple, large field collections have recently been implemented for this purpose, such as the international REFPOP project with the same apple genotype being planted at six different locations [146]. A similar approach in sea buckthorn could be very useful, but except for the joint Russian–Chinese project "Transcriptional regulation mechanisms of tissue-specific expression of key genes involved in lipid biosynthesis in sea buckthorn" (2023–2025), large international co-operative projects have been few for this crop.

10. Summary and Outlook

DNA markers have been applied to quantify the amount of genetic variation between and within taxa and populations, as well as for discrimination among sea buckthorn accessions in germplasm collections. Modern analytical techniques have demonstrated that there is genetic variation in the biochemical compounds among different species, subspecies, and genotypes, but also an environmental impact from differences in harvesting time, storage conditions, and processing methods. Research involving transcriptomics, metabolomics, and proteomics has become very valuable for the identification of chemical pathways and the enzymes that are responsible for crucial steps.

Until now, there have been few studies on the inheritance of yield, chemical contents, resistance traits, or the identification of markers for desirable genes. Technical developments are, however, moving fast; large-scale genotyping-by-sequencing, together with improved data analyses such as GWAS and Genomics-Assisted Prediction (GAP), will most likely soon improve our understanding of the intricate networks that regulate economically important traits in sea buckthorn. Such information, together with the recently published reports on the sea buckthorn genome and improved phenotypic data, if these can be acquired, will be crucial in developing large DNA-based assays with carefully selected markers for key genes to be used in applied breeding programs. The need for high-quality phenotypic data must, however, be addressed, e.g., by large co-operative multi-site projects.

Funding: This review was supported by the National Natural Science Foundation of China (32261133521, 32111540255, and 32071799), the International Cooperation Plan Project of Liaoning Province (2023JH2/10700016), the Central Guidance on Local Science and Technology Development Fund of Tibet Autonomous Region (XZ202201yd0025c), and the Project of Sea buckthorn Development and Management Center of the Ministry of Water Resources (2023-zg-kj-013).

Conflicts of Interest: The authors declare no conflict of interest.

References

1. Jia, D.R.; Bartish, I.V. Climatic changes and orogeneses in the Late Miocene of Eurasia: The main triggers of an expansion at a continental scale? *Front. Plant Sci.* **2018**, *9*, 1400. [CrossRef] [PubMed]
2. Wang, Z.; Zhao, F.; Wei, P.; Chai, X.; Hou, G.; Meng, Q. Phytochemistry, health benefits, and food applications of sea buckthorn (*Hippophae rhmanoides* L.): A comprehensive review. *Front. Nutr.* **2022**, *9*, 1036295. [CrossRef] [PubMed]
3. Ruan, C.J.; Nybom, H.; Rumpunen, K. Advances in improvement of quality and resistance in a multipurpose crop: Sea buckthorn. *Crit. Rev. Biotechnol.* **2013**, *33*, 126–144. [CrossRef] [PubMed]
4. Ma, Q.G.; He, N.X.; Huang, H.L.; Fu, X.M.; Zhang, Z.L.; Shu, J.C.; Wang, Q.Y.; Chen, J.; Wu, G.; Zhu, M.N.; et al. *Hippophae rhamnoides* L.: A comprehensive review on the botany, traditional uses, phytonutrients, health benefits, quality markers, and applications. *J. Agric. Food Chem.* **2023**, *71*, 4769–4788. [CrossRef]
5. Rentsendavaa, C. Phytochemical Investigations on Sea Buckthorn Juice Enriched with Pomace. Ph. D. Thesis, Food Science Doctoral School, Hungarian University of Agricultural and Life Sciences, Gödöllö, Hungary, 2021.
6. Du, W.; Ding, J.; Lu, S.; Wen, X.; Hu, J.; Ruan, C.J. Identification of the key flavonoid and lipid synthesis proteins in the pulp of two sea buckthorn cultivas at different developmental stages. *BMC Plant Biol.* **2022**, *22*, 2999. [CrossRef]
7. Bartish, I.V. An ancient medicinal plant at the crossroads of modern horticulture and genetics: Genetic resources and biotechnology of sea buckthorn (*Hippophae* L., Elaeagnaceae). In *Gene Pool Diversity and Crop Improvement*; Springer: Berlin/Heidelberg, Germany, 2016; pp. 415–446.
8. Rousi, A. The genus *Hippophae* L. A taxonomic study. *Ann. Bot. Fenn.* **1971**, *8*, 177–227.
9. Bartish, I.V.; Jeppsson, N.; Bartish, G.I.; Lu, R.; Nybom, H. Inter- and intraspecific genetic variation in *Hippophae* (Elaeagnaceae) investigated by RAPD markers. *Plant Syst. Evol.* **2000**, *225*, 85–101. [CrossRef]
10. Ruan, C.J.; Teixeira da Silva, J.A.; Li, Q.; Li, H.; Zhang, J. Pathogenicity of dried-shrink disease and evaluation of resistance in a germplasm collection of sea buckthorn (*Hippophae* L.) from China and other countries. *Sci. Hortic.* **2010**, *127*, 70–78. [CrossRef]
11. Sheng, H.M.; An, L.Z.; Chen, T.; Xu, S.J.; Liu, G.X.; Zheng, X.L.; Pu, L.L.; Liu, Y.J.; Lian, Y.S. Analysis of the genetic diversity and relationships among and within species of *Hippophae* (Elaeagnaceae) based on RAPD markers. *Plant Syst. Evol.* **2006**, *260*, 25–37. [CrossRef]
12. Bartish, I.V.; Jeppsson, N.; Swenson, U.; Nybom, H. Phylogeny of *Hippophae* (Elaeagnaceae) inferred from parsimony analysis of chloroplast DNA and morphology. *Syst. Bot.* **2002**, *27*, 41–54.
13. Sun, K.; Chen, X.; Ma, R.; Li, C.; Wang, Q.; Ge, S. Molecular phylogenetics of *Hippophae* L. (Elaeagnaceae) based on the internal transcribed spacer (ITS) sequences of nrDNA. *Plant Syst. Evol.* **2002**, *235*, 121–134. [CrossRef]
14. Wang, L.Y.; Li, S.S.; Wang, T.Y.; He, C.Y.; Luo, H.M.; Zhang, J.G.; Zeng, Y.F. Genomic SSR and EST-SSR markers for phylogenetic and pedigree construction–a comparison in sea buckthorn. *Plant Breed.* **2021**, *140*, 167–183. [CrossRef]
15. Raina, S.N.; Jain, S.; Seghal, D.; Kumar, A.; Dar, T.H.; Bhat, V.; Pandey, V.; Vaishnavi, S.; Bhargav, A.; Singh, V.; et al. Diversity and relationships of multipurpose seabuckthorn (*Hippophae* L.) germplasm from the Indian Himalayas as assessed by AFLP and SAMPL markers. *Genet. Res. Crop Evol.* **2012**, *59*, 1033–1053. [CrossRef]
16. Malik, S.; Babbar, S.; Chaudhary, M.; Sharma, S.; Raina, S.N.; Babbar, S.B. Authentication and deciphering interrelationships of *Hippophae* species using DNA barcodes. *Nucl.-India* **2022**, *65*, 165–177. [CrossRef]
17. Jiang, Y.F.; Yan, R.; Su, X.; Wen, C.; Sun, K. Molecular evidence for bidirectional hybrid origin and *Hippophae rhamnoides* ssp. *sinensis* as the mainly maternal plant of the diploid hybrid *H. goniocarpa* (Elaeagnaceae). *Bull. Bot. Res.* **2014**, *34*, 32–36.
18. Jia, D.R.; Wang, Y.J.; Liu, T.-L.; Wu, G.-L.; Kou, Y.-X.; Cheng, K.; Liu, J.-Q. Diploid hybrid origin of *Hippophae gyantsensis* (Elaeagnaceae) in the western Qinghai-Tibet Plateau. *Biol. J. Linn. Soc.* **2016**, *117*, 658–671. [CrossRef]
19. Liu, Y.; Xiang, L.; Zhang, Y.; Lai, X.; Xiong, C.; Li, J.; Su, Y.; Sun, W.; Chen, S. DNA barcoding based identification of *Hippophae* species and authentication of commercial products by high resolution melting analysis. *Food Chem.* **2018**, *242*, 62–67. [CrossRef]
20. Piao, X.M.; Mohanan, P.; Anandhapadmanaban, G.; Ahn, J.C.; Park, J.K.; Yang, D.C.; Kwak, G.-Y.; Wang, Y.P. Authentication of *Hippophae rhamnoides* ssp. *sinensis* and ssp. *mongolica* based on single nucleotide polymorphism at ribosomal DNA and their vitamin content analysis. *Plants* **2022**, *11*, 1843. [CrossRef]
21. Nybom, H. Comparison of different nuclear DNA markers for estimating intraspecific genetic diversity in plants. *Mol. Ecol.* **2004**, *13*, 1143–1155. [CrossRef]
22. Bartish, I.V.; Jeppsson, N.; Nybom, H. Population genetic structure in the dioecious pioneer plant species *Hippophae rhamnoides* investigated by random amplified polymorphic DNA (RAPD) markers. *Mol. Ecol.* **1999**, *8*, 791–802. [CrossRef]
23. Sun, K.; Chen, W.; Ma, R.; Chen, X.; Li, A.; Ge, S. Genetic variation in *Hippophae rhamnoides* ssp. *sinensis* (Elaeagnaceae) revealed by RAPD markers. *Biochem. Genet.* **2006**, *44*, 186–197. [PubMed]
24. Tian, C.; Lei, Y.; Shi, S.; Nan, P.; Chen, J.; Zhong, Y. Genetic diversity of sea buckthorn (*Hippophae rhamnoides*) populations in northeastern and northwestern China as revealed by ISSR markers. *New For.* **2004**, *27*, 229–237. [CrossRef]
25. Tian, C.; Nan, P.; Shi, S.; Chen, J.; Zhong, Y. Molecular genetic variation in Chinese populations of three subspecies of *Hippophae rhamnoides*. *Biochem. Genet.* **2004**, *42*, 259–267. [CrossRef]
26. Wang, T.Y.; Xu, Y.; Wang, L.Y.; Zhang, J.G.; Zeng, Y.F. Genetic differentiation and genetic diversity of *Hippophae rhamnoides* subsp. *sinensis* and *H. rhamnoides* subsp. *yunnanensis*. *For. Res.* **2021**, *34*, 13–21.
27. Zhao, C.; Chen, G.; Wang, Y.; Korpelainen, H.; Li, C. Genetic variation of *Hippophae rhamnoides* populations at different altitudes in the Wolong Nature Reserve based on RAPDs. *Chin. J. Appl. Environ. Biol.* **2007**, *13*, 753–758.

28. Wang, Y.H.; Jiang, H.; Peng, S.M.; Korpelainen, H. Genetic structure in fragmented populations of *Hippophae rhamnoides* ssp. *sinensis* in China investigated by ISSR and cpSSR markers. *Plant Syst. Evol.* **2011**, *295*, 97–107.
29. Shchapov, N.S. On the karyology of *Hippophae rhamnoides* L. *Tsitologiya I Genet.* **1979**, *13*, 45–47.
30. Ávila-Hernández, J.G.; del Rosario Cárdenas-Aquino, M.; Camas-Reyes, A.; Martínez-Antonio, A. Sex determination in papaya: Current status and perspectives. *Plant Sci.* **2023**, *335*, 111814. [CrossRef]
31. Persson, H.A.; Nybom, H. Genetic sex determination and RAPD marker segregation in the dioecious species sea buckthorn (*Hippophae rhamnoides* L.). *Hereditas* **1998**, *129*, 45–51. [CrossRef]
32. Sharma, A.; Zinta, G.; Rana, S.; Shirko, P. Molecular identification of sex in *Hippophae rhamnoides* L. using isozyme and RAPD markers. *For. Stud. China* **2010**, *12*, 62–66. [CrossRef]
33. Korekar, G.; Sharma, R.K.; Kumar, R.; Meenu Bisht, N.C.; Srivastava, R.B.; Ahuja, P.S.; Stobdan, T. Identification and validation of sex-linked SCAR markers in dioecious *Hippophae rhamnoides* L. (Elaeagnaceae). *Biotechnol. Lett.* **2012**, *34*, 973–978. [CrossRef] [PubMed]
34. Zhou, W.; Wang, Y.W.; Zhang, G.; Luan, G.X.; Chen, S.S.; Meng, J.; Wang, H.L.; Hu, N.; Suo, Y.R. Molecular sex identification in dioecious *Hippophae rhamnoides* L. via RAPD and SCAR markers. *Molecules* **2018**, *23*, 1048. [CrossRef] [PubMed]
35. Das, K.; Ganie, S.H.; Mangla, Y.; Dar, T.U.; Chaudhary, M.; Thakur, R.K.; Tandon, R.; Raina, S.N.; Goel, S. ISSR markers for gender identification and genetic diagnosis of *Hippophae rhamnoides* ssp. *turkestanica* growing at high altitudes in Ladakh region (Jammu and Kashmir). *Protoplasma* **2017**, *254*, 1063–1077.
36. Jadhav, M.S.; Sharma, T.R. Identification of gender specific DNA markers in sea buckthorn (*Hippophae rhamnoides* L.). *Indian Res. J. Genet. Biotechnol.* **2014**, *6*, 464–469.
37. Chawla, A.; Stobdan, T.; Srivastava, R.B.; Jaiswal, V.; Chauhan, R.S.; Kant, A. Sex-biased temporal gene expression in male and female floral buds of seabuckthorn (*Hippophae rhamnoides*). *PLoS ONE* **2015**, *10*, e0124890. [CrossRef]
38. Mangla, Y.; Das, K.; Bali, S.; Ambreen, H.; Raina, S.N.; Tandon, R.; Goel, S. Occurrence of subdioecy and scarcity of gender-specific markers reveal an ongoing transition to dioecy in Himalayan seabuckthorn (*Hippophae rhamnoides* ssp. *turkestanica*). *Heredity* **2019**, *122*, 120–132. [CrossRef]
39. Dong, L.N.; Sun, K.; Su, X.; Zhang, L.; Ding, S.S.; Ma, R.J. RAPD markers related to sex locus in *Hippophae goniocarpa*. *Bull. Bot. Res.* **2007**, *27*, 73–76.
40. Wang, R.; Wu, B.; Jian, J.; Tang, Y.; Zhang, T.; Song, Z.; Zhang, W.; Qiong, L. How to survive in the world's third poplar: Insights from the genome of the highest altitude woody plant, *Hippophae tibetana* (Elaeagnaceae). *Front. Plant Sci.* **2022**, *13*, 1051587. [CrossRef] [PubMed]
41. Mangla, Y.; Chaudhary, M.; Gupta, H.; Thakur, R.; Goel, S.; Raina, S.N.; Tandon, R. Facultative apomixis and development of fruit in a deciduous shrub with medicinal and nutritional uses. *AoB Plants* **2015**, *7*, plv098. [CrossRef]
42. Ali, A.; Kaul, V. Incidence of apomixis in sea buckthorn (*Hippophae rhamnoides* L.) a dioecious taxon of immense economic importance. *Curr. Sci.* **2017**, *112*, 1994–1996.
43. Muranty, H.; Denancé, C.; Feugey, L.; Crépin, J.-L.; Barbier, Y.; Tartarini, S.; Ordidge, M.; Troggio, M.; Lateur, M.; Nybom, H.; et al. Using whole-genome SNP data to reconstruct a large multi-generation pedigree in apple germplasm. *BMC Plant Biol.* **2020**, *20*, 2. [CrossRef] [PubMed]
44. Barreneche, T.; Carcamo de la Concepcion, M.; Blouin-Delmas, M.; Ordidge, M.; Nybom, H.; Lacis, G.; Feldmane, D.; Sedlák, J.; Meland, M.; Kaldmäe, H.; et al. SSR-based analysis of genetic diversity and structure of sweet cherry (*Prunus avium* L.) from 19 countries in Europe. *Plants* **2021**, *10*, 1983. [CrossRef]
45. Bartish, G.I.; Jeppsson, N.; Bartish, I.V.; Nybom, H. Assessment of genetic diversity using RAPD analysis in a germplasm collection of sea buckthorn. *Agric. Food Sci. Finl.* **2001**, *9*, 279–288. [CrossRef]
46. Singh, R.; Mishra, S.N.; Dwivedi, S.; Ahmed, Z. Genetic variation in seabuckthorn (*Hippophae rhamnoides* L.) populations of cold arid Ladakh (India) using RAPD markers. *Curr. Sci.* **2006**, *91*, 1321–1322.
47. Simon-Gruita, A.; Tataru, E.; Constantin, N.; Duta Cornescu, G.; Pavlusenco, C.E.; Rati, V.; Rati, L.; Stoian, V. The assessment of the genetic diversity of sea buckthorn populations from Romania using RAPD markers. *Rom. Biotechnol. Lett.* **2012**, *17*, 7749–7756.
48. Yang, G.; Ding, J.; Wu, L.R.; Duan, Y.D.; Li, A.Y.; Shan, J.Y.; Wu, Y.X. A new strategy for complete identification of sea buckthorn cultivars by using random amplified polymorphic DNA markers. *Genet. Mol. Res.* **2015**, *14*, 1836–1845. [CrossRef]
49. Ding, J.; Ruan, C.J.; Bao, Y.H.; Guan, Y.; Shan, J.Y.; Li, H.; Ding, G.J. Analysis of genetic relationships in sea buckthorn (*Hippophae rhamnoides*) germplasm from China and other countries using ISSR markers. *J. Horticult. Sci. Biotechnol.* **2015**, *90*, 599–606. [CrossRef]
50. Ruan, C.; Li, D. AFLP fingerprinting analysis of some cultivated varieties of sea buckthorn (*Hippophae rhamnoides*). *J. Genet.* **2005**, *84*, 311–316. [CrossRef]
51. Wang, A.L.; Zhang, Q.; Wan, D.S.; Yang, Y.Z.; Liu, J.Q. Nine microsatellite DNA primers for *Hippophae rhamnoides* ssp. *sinensis* (Elaeagnaceae). *Conserv. Genet.* **2008**, *9*, 969–971. [CrossRef]
52. Jain, A.; Ghangal, R.; Grover, A.; Raghuvanshi, S.; Sharma, P.C. Development of EST-based new SSR markers in seabuckthorn. *Physiol. Mol. Biol. Plants* **2010**, *16*, 375–378. [CrossRef]
53. Jain, A.; Chaudhary, S.; Sharma, P.C. Mining of microsatellites using next generation sequencing of seabuckthorn (*Hippophae rhamnoides* L.) transcriptome. *Physiol. Molec. Biol. Plants* **2014**, *20*, 115–123. [CrossRef]

54. Lacis, G.; Kota-Dombrovska, I. Assessment of genetic diversity of Latvian sea buckthorn (*Hippophae rhamnoides* L.) germplasm using molecular markers. *Zemdirb.-Agric.* **2014**, *101*, 333–340. [CrossRef]
55. Islam, M.A.; Sinha, P.; Sharma, S.S.; Negi, M.S.; Tripathi, S.B. Isolation and characterization of novel polymorphic microsatellite loci in *Hippophae rhamnoides*. *Proc. Natl. Acad. Sci. India Sect. B Biol. Sci.* **2015**, *87*, 727–732. [CrossRef]
56. Li, H.; Ruan, C.J.; Wang, L.; Ding, J.; Tian, X.J. Development of RNA-Seq SSR markers and application to genetic relationship analysis among sea buckthorn germplasm. *J. Am. Soc. Hortic. Sci.* **2017**, *142*, 200–208. [CrossRef]
57. Li, S.S.; Zeng, Y.F.; He, C.Y.; Zhang, J.G. Development of EST-SSR markers based on sea buckthorn transcriptomic sequences. *For. Res.* **2017**, *30*, 69–74.
58. Li, H.; Ruan, C.J.; Ding, J.; Li, J.B.; Wang, L.; Tian, X. Diversity in sea buckthorn (*Hippophae rhamnoides* L.) accessions with different origins based on morphological characteristics, oil traits, and microsatellite markers. *PLoS ONE* **2020**, *15*, e0230356. [CrossRef]
59. Zubarev, Y. Seabuckthorn Varieties of the Lisavenko Institute. Available online: www.oblepiha22.ru (accessed on 11 September 2023).
60. Albrecht, H.J. Research on Seabuckthorn (*Hippophae rhamnoides* L.) in Germany. In *Seabuckthorn (Hippophae L.): A Multipurpose Wonder Plant*; Singh, V., Ed.; Indus Publishing Company: New Delhi, India, 2003; Volume 1, pp. 178–186.
61. Singh, V.; Zubarev, Y. Breeding strategies of Russian seabuckthorn (*Hippophae rhamnoides* ssp. *mongolica*) varieties and their global introduction. In *Seabuckthorn (Hippophae L.): A Multipurpose Wonder Plant. Emerging Trends in Research & Technologies*; Singh, V., Ed.; Daya Publishing House: New Delhi, India, 2014; Volume 4, pp. 71–88.
62. Shalkevich, M.S.; Koltun, N.Y.; Pleskatsevich, R.I. Sea buckthorn pests and diseases in Belarus. In Proceedings of the 3rd European Workshop on Sea Buckthorn EuroWorkS2014, Naantali, Finland, 14–16 October 2014; Natural Resources Institute Finland: Helsinki, Finland, 2015; pp. 83–87.
63. Lu, S.G.; Hu, J.Z.; Gao, Y.; Wen, X.F. Introduction of China's main seabuckthorn cultivars for plantations. In Proceedings of the 9th International Sea Buckthorn Association Conference, Thessaloniki, Greece, 22–25 May 2023; pp. 19–20.
64. Privalov, G.F.; Solonenko, L.P.; Skuridin, G.M. Mutagensis breeding of seabuckthorn (*Hippophae rhamnoides* L.). In *Seabuckthorn (Hippophae L.): A Multipurpose Wonder Plant*; Singh, V., Ed.; Indus Publishing Company: New Delhi, India, 2003; Volume 1, pp. 194–210.
65. Shcherbakov, D.N.; Galitsyn, G.Y.; Kukina, T.P.; Panteleeva, N.V.; Salnikova, O.I.; Kolosov, P.V. Prospects for comprehensive use of sea buckthorn of Novosibirsk selection. *AIP Conf. Proc.* **2022**, *2390*, 030085. [CrossRef]
66. Schroeder, W.R.; Soolanayakanahally, R.Y.; Naeem, H.A. Harvest Moon seabuckthorn. *Can. J. Plant Sci.* **2014**, *9*, 1309–1312. [CrossRef]
67. Höhne, F. Overview of cultivation technologies and their challenges. In Proceedings of the 3rd European Workshop on Sea Buckthorn EuroWorkS2014, Naantali, Finland, 14–16 October 2014; Natural Resources Institute Finland: Helsinki, Finland, 2015; pp. 31–35.
68. Duan, A.G.; Zhang, J.G.; Luo, H.M.; Li, J.X.; He, C.Y. A new red fruit *Hippophae rhamnoides* cultivar 'Hongji 1'. *Acta Horticult. Sinica* **2018**, *45*, 2865–2866.
69. Zhao, H.Z. Advanced in genetic improvement of *Hippophae* varieties in China. *Hippophae* **1996**, *9*, 1–4. (In Chinese with English Abstract)
70. Huang, Q.; Zhao, Y. Selection and breeding of Wucifeng and Shengqiuhong sea buckthorn. *Hippophae* **2004**, *17*, 7–9. (In Chinese with English Abstract)
71. Zhang, J.G.; Luo, H.M.; Huang, Q.; Shan, J.Y.; Wang, C.Y. A comparative study on berry characteristics of large berry cultivars of sea buckthorn. *Forest Res.* **2005**, *18*, 643–650. (In Chinese with English Abstract)
72. Shalkevich, M.S.; Radkevich, D.B.; Bieniek, A. Seabuckthorn (*Hippophae rhamnoides* L.) research in Belarus. In Proceedings of the International Scientific Conference Sustainable Fruit Growing: From Plant to Product, Jurmala–Dobele, Latvia, 28–31 May 2008; pp. 108–116.
73. Moskalets, T.Z.; Vovkohon, A.H.; Pelekhata, N.P.; Ovezmyradova, O.B.; Pelekhatyi, V.M. Sea buckthorn: New promising varieties and using their berries for the manufacture of functional products. *Ukrainian J. Ecol.* **2021**, *11*, 137–143.
74. Ficzek, G.; Matravolgyi, G.; Furulyas, D.; Rentsendavaa, C.; Jocsak, I.; Papp, D.; Simon, G.; Vegvari, G.; Steger-Mate, M. Analysis of bioactive compounds of three sea buckthorn cultivars (*Hippophae rhamnoides* L. 'Askola', 'Leikora', and 'Orangeveja') with HPLC and spectrophotometric methods. *Eur. J. Hortic. Sci.* **2019**, *84*, 31–38. [CrossRef]
75. Liu, H.Z.; Liu, S.Y.; Wen, L.K. Breeding of a new cultivar of *Hippophae rhamnoides*, 'Qiuyang'. *China Fruits* **2010**, *6*, 13–15. (In Chinese with an English Abstract)
76. Yao, Y.; Tigerstedt, P. Genetic diversity in *Hippophae* L. and its use in plant breeding. *Euphytica* **1994**, 165–169. [CrossRef]
77. Karhu, S.; Hellström, J.; Hietaranta, T.; Mattila, P.; Pihlava, J.-M.; Tahvonen, R.Y. MTT's berry varieties–rich in phytochemicals. *NJF Rep.* **2010**, *6*, 99–100.
78. Duan, A.G.; Zhang, J.G.; Luo, H.M.; Li, J.X.; He, C.Y. A new large fruit *Hippophae rhamnoides* cultivar 'Zhongji 3'. *Acta Horticult. Sinica* **2018**, *45*, 2871–2872.
79. Piłat, B.; Bieniek, A.; Zadernowski, R. Common sea buckthorn (*Hippophae rhamnoides* L.) as an alternative orchard plant. *Pol. J. Natur. Sci.* **2015**, *30*, 417–430.
80. Huang, Q. The research progress on sea-buckthorn breeding in China. In Proceedings of the 2nd International Sea Buckthorn Association Conference, Beijing, China, 26–29 August 2005.

81. Aksenova, N.A.; Dolgacheva, V.S. Seabuckthorn culture in the botanical garden of the Moscow State University. In *Seabuckthorn (Hippophae L.): A Multipurpose Wonder Plant*; Singh, V., Ed.; Indus Publishing Company: New Delhi, India, 2003; Volume 1, pp. 165–169.
82. Singh, V. Performance of Russian seabuckthorn varieties during 4 years in Lahaul valley, cold desert of Himachal Himalaya. In Proceedings of the 9th International Sea Buckthorn Association Conference, Thessaloniki, Greece, 22–25 May 2023.
83. Sriskandarajah, S.; Lundquist, P.-O. High frequency shoot organogenesis and somatic embryogenesis in juvenile and adult tissues of seabuckthorn (*Hippophae rhamnoides* L.). *Plant Cell Tissue Organ Cult. (PCTOC)* **2009**, *99*, 259–268. [CrossRef]
84. Sriskandarajah, S.; Clapham, D.; Lundquist, P.-O. Development of a genetic transformation method for seabuckthorn (*Hippophae rhamnoides* L.). *Am. J. Plant Sci.* **2014**, *5*, 528–534. [CrossRef]
85. Fu, L.S.; Su, H.D.; Li, R.; Cui, Y.J. Harvesting technologies for sea buckthorn fruit. *Eng. Agric. Environ. Food* **2014**, *7*, 64–69. [CrossRef]
86. Bruvelis, A. Experineces about sea buckthorn cultivation and harvesting in Latvia. In Proceedings of the 3rd European Workshop on Sea Buckthorn EuroWorkS2014, Naantali, Finland, 14–16 October 2014; Natural Resources Institute Finland: Helsinki, Finland, 2015; pp. 36–41.
87. Tang, X.; Tigerstedt, P.M.A. Inheritance of flowering, maturity, yield and hardiness of sea buckthorn (*Hippophae rhamnoides* L.). *J. Am. Soc. Hortic.* **2001**, *126*, 744–749. [CrossRef]
88. Zhao, J.; Zhang, Z.; Zhou, H.; Bai, Z.; Sun, K. The study on sea buckthorn (genus *Hippophae* L.) fruit reveals cell division and cell expansion to promote morphogenesis. *Plants* **2023**, *12*, 1005. [CrossRef]
89. Ilhan, G.; Gundogdu, M.; Karlovic, K.; Zidovec, V.; Vokurka, A.; Ercisli, S. Main agro-morphological and biochemical berry characteristics of wild-grown sea buckthorn (*Hippophae rhamnoides* L. ssp. *caucasica* Rousi) genotypes in Turkey. *Sustainability* **2021**, *13*, 1198. [CrossRef]
90. Lu, R. *Hippophae* and its general chemical compositions. In Proceedings of the 1st Congress of the International Sea Buckthorn Association, Berlin, Germany, 14–18 September 2003; pp. 20–35.
91. Univer, T.; Univer, N. The hardiness of sea buckthorn cultivars in Estonian climatic conditions. In Proceedings of the 3rd European Workshop on Sea Buckthorn EuroWorkS2014, Naantali, Finland, 14–16 October 2014; Natural Resources Institute Finland: Helsinki, Finland, 2015; pp. 95–97.
92. Gao, G.R.; Lv, Z.R.; Zhang, G.Y.; Li, J.Y.; Zhang, J.G.; He, C.Y. An ABA-flavonoid relationship contributes to the differences in drought resistance between different sea buckthorn subspecies. *Tree Physiol.* **2021**, *41*, 744–755. [CrossRef] [PubMed]
93. Yao, Y.; Dong, L.J.; Fu, X.H.; Zhao, L.; Wei, J.R.; Cao, J.F.; Sun, Y.Y.; Liu, J.F. HrTCP20 dramatically enhance drought tolerance of sea buckthorn (*Hippophae rhamnoides* L). by mediating the JA signaling pathway. *Plant Physiol. Biochem.* **2022**, *174*, 51–62. [CrossRef] [PubMed]
94. Drevinska, K.; Moroèko-Bièevska, I. Sea buckthorn diseases caused by pathogenic fungi. *Proc. Latv. Acad. Sci. Sect B* **2022**, *76*, 393–401. [CrossRef]
95. Prokkola, S. Reliability of Russian sea buckthorn cultivars in North Ostrobothnia, Finland. *Acta Hortic.* **2003**, *626*, 381–387. [CrossRef]
96. Ruan, C.J.; Qin, P.; Zheng, J.W.; He, Z.X. Genetic relationships among some cultivars of sea buckthorn from China, Russia and Mongolia based on RAPD analysis. *Sci. Hortic.* **2004**, *101*, 417–426. [CrossRef]
97. Popp, C.; Kind, S.; Balbín-Suárez, A.; Behrens, F.H.; Fischer, M.; Jelkmann, W. Investigations on fungi in context with sea buckthorn dieback in Northern Germany. In Proceedings of the 9th International Sea Buckthorn Association Conference, Thessaloniki, Greece, 22–25 May 2023.
98. Shamanskaya, L.D. Bioecology of the sea-buckthorn fly (*Rhagoletis batava obscuriosa* Kol.) and pest control treatment in Altai. In Proceedings of the 3rd European Workshop on Sea Buckthorn EuroWorkS2014, Naantali, Finland, 14–16 October 2014; Natural Resources Institute Finland: Helsinki, Finland, 2015; pp. 7–20.
99. Liu, J.F.; Wang, Z.Y.; Zhao, J.; Zhao, L.; Wang, L.; Su, Z.; Wei, J.R. HrCYP90B1 modulating brassinosteroid biosynthesis in sea buckthorn (*Hippophae rhamnoides* L.) against fruit fly (*Rhagoletis batava obseuriosa* Kol.) infection. *Tree Physiol.* **2021**, *41*, 444–459. [CrossRef] [PubMed]
100. Yang, B. Sugars, acids, ethyl β-d-glucopyranose and a methyl inositol in sea buckthorn (*Hippophaë rhamnoides*) berries. *Food Chem.* **2009**, *112*, 89–97. [CrossRef]
101. Andersson, S.C.; Olsson, M.; Johansson, E.; Rumpunen, K. Carotenoids in sea buckthorn (*Hippophae rhamnoides* L.) berries during ripening and use of pheophytin as a maturity marker. *J. Agric. Food Chem.* **2009**, *57*, 250–258. [CrossRef]
102. Andersson, S.C.; Rumpunen, K.; Johansson, E.; Olsson, M.E. Tocopherols and tocotrienols in sea buckthorn (*Hippophae rhamnoides* L.) berries during ripening. *J. Agric. Food Chem.* **2008**, *56*, 6701–6706. [CrossRef]
103. Yang, H.; Yang, S.; Chen, X.; Zhang, J.; Zhang, Y. Dynamic changes in flavonoid, phenolic, and polysaccharide contents in leaves and fruits of sea buckthorn during the growing season in southeastern Tibet plateau. *Sci. Hortic.* **2023**, *307*, 111497. [CrossRef]
104. Kortesniemi, M.; Sinkkonen, J.; Yang, B.; Kallio, H. ^{1}HNMR spectroscopy reveals the effect of genotype and growth conditions on composition of sea buckthorn (*Hippophaë rhamnoides* L.) berries. *Food Chem.* **2014**, *147*, 138–146. [CrossRef]
105. Kortesniemi, M.; Sinkkonen, J.; Yang, B.; Kallio, H. NMR metabolomics demonstrates phenotypic plasticity of sea buckthorn (*Hippophae rhamnoides*) berries with respect to growth conditions in Finland and Canada. *Food Chem.* **2017**, *219*, 139–147. [CrossRef]

106. Liu, Y.; Fan, G.; Zhang, J.; Zhang, Y.; Li, J.; Xiong, C.; Zhang, Q.; Li, X.; Lai, X. Metabolic discrimination of sea buckthorn from different *Hippophaë* species by 1H NMR based metabolomics. *Sci. Rep.* **2017**, *7*, 1585. [CrossRef]
107. Pariyani, R.; Kortesniemi, M.; Liimatainen, J.; Sinkkonen, J.; Yang, B. Untargeted metabolic fingerprinting reveals impact of growth stage and location on composition of sea buckthorn (*Hippophaë rhamnoides*) leaves. *Food Sci.* **2020**, *85*, 364–373. [CrossRef]
108. Markkinen, N.; Pariyani, R.; Jokioja, J.; Kortesniemi, M.; Laaksonen, O.; Yang, B. NMR-based metabolomics approach on optimization of malolactic fermentation of sea buckthorn juice with *Lactiplantibacillus plantarum*. *Food Chem.* **2022**, *366*, 130630. [CrossRef]
109. Zheng, J.; Kallio, H.; Linderborg, K.; Yang, B. Sugars, sugar alcohols, fruit acids, and ascorbic acid in wild Chinese sea buckthorn (*Hippophaë rhamnoides* ssp. *sinensis*) with special reference to influence of latitude and altitude. *Food Res. Int.* **2011**, *44*, 2018–2026. [CrossRef]
110. Yang, B.; Zheng, J.; Kallio, H. Influence of origin, harvesting time and weather conditions on content of inositols and methylinositols in sea buckthorn (*Hippophaë rhamnoides*) berries. *Food Chem.* **2011**, *125*, 388–396. [CrossRef]
111. Jeppsson, N.; Gao, X. Changes in the contents of kaempherol, quercetin and L-ascorbic acid in sea buckthorn berries during maturation. *Agric. Food Sci. Finland* **2000**, *9*, 17–22. [CrossRef]
112. Kallio, H.; Yang, B.; Peippo, P. Effects of different origins and harvesting time on vitamin C, tocopherols, and tocotrienols in sea buckthorn (*Hippophaë rhamnoides*) berries. *J. Agric. Food Chem.* **2002**, *50*, 6136–6142. [CrossRef] [PubMed]
113. Rongsen, L. Biochemical characteristics of seabuckthorn (*Hippophae* L.). In *Seabuckthorn (Hippophae L.) A Multipurpose Wonder Plant. Vol. 2: Biochemistry and Pharmacology*; Singh, V., Ed.; Daya Publishing House: New Delhi, India, 2006; pp. 98–107.
114. Yang, B.; Kallio, H. Composition and physiological effects of sea buckthorn (*Hippophaë*) lipids. *Trends Food Sci. Technol.* **2002**, *13*, 160–167. [CrossRef]
115. Du, W.; Xiong, C.W.; Ding, J.; Nybom, H.; Ruan, C.J.; Guo, H. Tandem mass tag based quantitative proteomics of developing sea buckthorn berries reveals candidate proteins related to lipid metabolism. *J. Proteome Res.* **2019**, *18*, 1958–1969. [CrossRef] [PubMed]
116. Ding, J.; Ruan, C.J.; Du, W.; Guan, Y. RNA-seq data reveals a coordinated regulation mechanism of multigenes involved in the high accumulation of palmitoleic acid and oil in sea buckthorn berry pulp. *BMC Plant Biol.* **2019**, *19*, 207–224. [CrossRef] [PubMed]
117. Tkacz, K.; Wojdyło, A.; Turkiewicz, I.P.; Nowicka, P. Triterpenoids, phenolic compounds, macro- and microelements in anatomical parts of sea buckthorn (*Hippophaë rhamnoides* L.) berries, branches and leaves. *J. Food Comp. Anal.* **2021**, *103*, 104107. [CrossRef]
118. Yang, W.; Laaksonen, O.; Kallio, H.; Yang, B. Proanthocyanidins in sea buckthorn (*Hippophaë rhamnoides* L.) berries of different origins with special reference to the influence of genetic background and growth location. *J. Agric. Food Chem.* **2016**, *64*, 1274–1282. [CrossRef]
119. Ma, X.; Laaksonen, O.; Zheng, J.; Yang, W.; Trépanier, M.; Kallio, H.; Yang, B. Flavonol glycosides in berries of two major subspecies of sea buckthorn (*Hippophaë rhamnoides* L.) and influence of growth sites. *Food Chem.* **2016**, *200*, 189–198. [CrossRef]
120. Chen, C.; Zhang, H.; Xiao, W.; Yong, Z.P.; Bai, N. High-performance liquid chromatographic fingerprint analysis for different origins of sea buckthorn berries. *J. Chromatogr. A* **2007**, *1154*, 250–259. [CrossRef]
121. Fatima, T.; Kesari, V.; Watt, I.; Wishart, D.; Todd, J.F.; Schroeder, W.R.; Paliyath, G.; Krishna, P. Metabolite profiling and expression analysis of flavonoid, vitamin C and tocopherol biosynthesis genes in the antioxidant-rich sea buckthorn (*Hippophae rhamnoides* L.). *Phytochemistry* **2015**, *118*, 181–191. [CrossRef] [PubMed]
122. Morgenstern, A.; Ekholm, A.; Scheewe, P.; Rumpunen, K. Changes in content of major phenolic compounds during leaf development of sea buckthorn (*Hippophae rhamnoides* L.). *Agric. Food Sci.* **2014**, *23*, 207–219. [CrossRef]
123. Teleszko, M.; Wojdyło, A.; Rudzińska, M.; Oszmiański, J.; Golis, T. Analysis of lipophilic and hydrophilic bioactive compounds content in sea buckthorn (*Hippophaë rhamnoides* L.) berries. *J. Agric. Food Chem.* **2015**, *63*, 4120–4129. [CrossRef] [PubMed]
124. Yang, B.; Karlsson, R.M.; Oksman, P.H.; Kallio, H.P. Phytosterols in sea buckthorn (*Hippophaë rhamnoides* L.) berries: Identification and effects of different origins and harvesting times. *J. Agric. Food Chem.* **2001**, *49*, 5620–5629. [CrossRef] [PubMed]
125. Li, T.S.C.; Beveridge, T.H.J.; Drover, J.C.G. Phytosterol content of sea buckthorn (*Hippophae rhamnoides* L.) seed oil: Extraction and identification. *Food Chem.* **2007**, *101*, 1633–1639. [CrossRef]
126. Tang, X.; Kälviäinen, N.; Turila, H. Sensory and hedonic characteristics of juice of sea buckthorn (*Hippophae rhamnoides* L.) origins and hybrids. *LWT–Food Sci. Technol.* **2001**, *34*, 102–110. [CrossRef]
127. Tiitinen, K.M.; Hakala, M.A.; Kallio, H.P. Quality components of sea buckthorn (*Hippophae rhamnoides*) varieties. *J. Agric. Food Chem.* **2005**, *53*, 1692–1699. [CrossRef] [PubMed]
128. Ma, X.; Laaksonen, O.; Heinonen, J.; Sainio, T.; Kallio, H.; Yang, B. Sensory profile of ethyl β-D-glucopyranoside and its contribution to quality of sea buckthorn (*Hippophaë rhamnoides* L.). *Food Chem.* **2017**, *233*, 263–272. [CrossRef]
129. Ma, X.; Yang, W.; Laaksonen, O.A.; Nylander, M.; Kallio, H.; Yang, B. Role of flavonols and proanthocyanidins in the sensory quality of sea buckthorn (*Hippophaë rhamnoides* L.) berries. *J. Agric. Food Chem.* **2017**, *65*, 9871–9879. [CrossRef]
130. Lundén, S.; Tiitinen, K.; Kallio, H. Aroma analysis of sea buckthorn berries by sensory evaluation, headspace SPME and GC-olfactometry. In *Expression of Multidisciplinary Flavour Science. Proceedings of the 12th Weurman Symposium*; Blank, I., Wüst, M., Yeretzian, C., Eds.; Institute of Chemistry and Biological Chemistry, Zürich University of Applied Sciences: Zürich, Switzerland, 2010; pp. 490–493.

131. Liang, J.; Zhang, G.; Song, Y.; He, C.; Zhang, J. Targeted metabolome and transcriptome analyses reveal the pigmentation mechanism of *Hippophae* (sea buckthorn) fruit. *Foods* **2022**, *11*, 3278. [CrossRef]
132. Panteev, A.V.; Shalkevich, M.S. The breeding of male plants of sea-buckthorn (*Hippophae rhamnoides*) for medicinal use of pollen. *Acta Hort.* **1995**, *390*, 137–140. [CrossRef]
133. Fefelov, V.A.; Selekhov, V.V. Methods of seabuckthorn breeding. In *Seabuckthorn (Hippophae L.): A multipurpose Wonder Plant*; Singh, V., Ed.; Indus Publishing Company: New Delhi, India, 2003; Volume 1, pp. 152–156.
134. Fefelov, V.A.; Selekhov, V.V. Biology of flowering, pollination and fertilization in seabuckthorn. In *Seabuckthorn (Hippophae L.): A Multipurpose Wonder Plant*; Singh, V., Ed.; Indus Publishing Company: New Delhi, India, 2003; Volume 1, pp. 99–104.
135. Fefelov, V.A.; Selekhov, V.V. Propagation of seabuckthorn (*Hippophae rhamnoides* L.) from seeds. In *Seabuckthorn (Hippophae L.): A Multipurpose Wonder Plant*; Singh, V., Ed.; Indus Publishing Company: New Delhi, India, 2003; Volume 1, pp. 334–337.
136. Garanovich, I.N. Introduction of seabuckthorn (*Hippophae rhamnoides* L.) in Belarus. In *Seabuckthorn (Hippophae L.): A Multipurpose Wonder Plant*; Singh, V., Ed.; Indus Publishing Company: New Delhi, India, 2003; Volume 1, pp. 137–151.
137. Singh, V.; Gupta, R.K. Propagation of seabuckthorn (*Hippophae rhamnoides* L.). In *Seabuckthorn (Hippophae L.): A Multipurpose Wonder Plant*; Singh, V., Ed.; Indus Publishing Company: New Delhi, India, 2003; Volume 1, pp. 315–333.
138. Grover, A.; Gupta, S.M.; Bala, M. Biotechnological approaches for seabuckthorn improvement. In *The Seabuckthorn Genome. Compendium of Plant Genomes*; Sharma, P.C., Ed.; Springer: Cham, Switzerland, 2022; pp. 173–186.
139. Zubarev, Y.A. The main directions of sea-buckthorn breeding program in Siberia. In Proceedings of the 2nd International Sea Buckthorn Association Conference, Beijing, China, 26–29 August 2005.
140. Demidova, N. Research on the introduction of seabuckthorn varieties in North Russia. In *Seabuckthorn (Hippophae L.): A Multipurpose Wonder Plant*; Singh, V., Ed.; Indus Publishing Company: New Delhi, India, 2003; Volume 1, pp. 125–135.
141. Wannemuehler, S.D.; Luby, J.J.; Yue, C.; Bedford, D.S.; Gallardo, R.K.; McCracken, V.A. A cost–benefit analysis of DNA informed apple breeding. *HortScience* **2019**, *54*, 11. [CrossRef]
142. Ruan, C.J.; Li, H.; Mopper, S. Characterization and identification of ISSR markers associated with resistance to dried-shrink disease in sea buckthorn. *Mol. Breed.* **2009**, *24*, 255–268. [CrossRef]
143. Li, H.; Ruan, C.J.; Teixeira da Silva, J.A.; Liu, B.Q. Associations of SRAP markers with dried-shrink disease resistance in a germplasm collection of sea buckthorn (*Hippophae* L.). *Genome* **2010**, *53*, 447–457. [CrossRef]
144. Yu, L.Y.; Diao, S.F.; Zhang, G.Y.; Yu, J.G.; Zhang, T.; Luo, H.M.; Duan, A.G.; Wang, J.P.; He, C.Y.; Zhang, J.G. Genome sequence and population genomics provide insights into chromosomal evolution and phytochemical innovation of *Hippophae rhamnoides*. *Plant Biotechnol. J.* **2022**, *20*, 1257–1273. [CrossRef] [PubMed]
145. Wu, Z.; Chen, H.; Pan, Y.; Feng, H.; Fang, D.; Yang, J.; Wang, Y.; Yanh, J.; Sahu, S.K.; Liu, J.; et al. Genome of *Hippophae rhamnoides* provides insights into a conserved molecular mechanism in actinorhizal and rhizobial symbioses. *New Phytol.* **2022**, *235*, 276–291. [CrossRef] [PubMed]
146. Jung, M.; Keller, B.; Roth, M.; Aranzana, M.J.; Auwerkerken, A.; Guerra, W.; Al-Rifai, M.; Lewandowski, M.; Sanin, N.; Rymenants, M.; et al. Genetic architecture and genomic predictive ability of apple quantitative traits across environments. *Horticult. Res.* **2022**, *9*, uhac028. [CrossRef] [PubMed]

genes

Article

Comparative Analysis of the Chloroplast Genome of *Sicyos angulatus* with Other Seven Species of Cucurbitaceae Family

Muniba Kousar and Joonho Park *

Department of Fine Chemistry, Seoul National University of Science and Technology, 232-Gongneung-ro, Nowon-gu, Seoul 01811, Republic of Korea
* Correspondence: jhpark21@seoultech.ac.kr

Abstract: *Sicyos angulatus* (SA) is an annual plant from the Cucurbitaceae family that is native to the eastern part of North America. This study aims to assemble and annotate the chloroplast genome of *S. angulatus*, and then compare it with plastomes of the other species representing the Cucurbitaceae family. The chloroplast genome size of *S. angulatus* is 154,986 bp, including a pair of inverted repeats (IR) of 26,276 bp, and small single-copy region (SSC) of 18,079 bp and large single-copy region (LSC) of 84,355 bp. Compared to other Cucurbitaceae species, the chloroplast genome of *S. angulatus* is almost 4222 bp smaller than the plastome *Gynostemma pentaphyllum*. All other seven species have an identical set of tRNA (37), except *Citrullus laevigata*, which contains 36 tRNA. The IRa/LSC junction in all eight species is located upstream of *rpl2* and downstream of *trnH* gene. Moreover, variation in the size of the gene and the presence of pseudogene *ycf1* has been seen because of the IR contraction and expansion. The highest number of tandem repeats was seen in *G. pentaphyllum*, and then *Corynocarpus leavigata*. The sequence divergence analysis and topology of the phylogenetic tree indicate that *S. angulatus* is more similar to genus *Citrullus* as compared to genus *Gynostemma*. These findings contribute to developing the genomic marker for the purpose of future genetic studies.

Keywords: *Sicyos angulatus*; comparative chloroplast genome analysis; Cucurbitaceae family; phylogenetic relationship

Citation: Kousar, M.; Park, J. Comparative Analysis of the Chloroplast Genome of *Sicyos angulatus* with Other Seven Species of Cucurbitaceae Family. *Genes* **2023**, *14*, 1776. https://doi.org/10.3390/genes14091776

Academic Editors: Wajid Zaman and Zhiqiang Wu

Received: 26 June 2023
Revised: 31 August 2023
Accepted: 6 September 2023
Published: 8 September 2023

1. Introduction

Sicyos Angulatus (SA), commonly called Bur Cucumber, is an annual vine plant in the gourd family, Cucurbitaceae, that is native to the eastern part of North America. The Cucurbitaceae family consists of approximately 965 species and around 95 genera [1]. *S. angulatus* is also known as invasive plant and was introduced to South Korea as a parent stock for cucumber cultivation in the 1980s [2,3]. Indeed, reports of the *Sicyos angulatus* have been found in India and other countries in Asia [4]. It is adapted to wet habitats, river floodplains, and colonizes opened habitats, like fencerows, roadsides, and woodland borders. It is found in various eastern parts of the Rocky Mountains, Canada, Mexico, and eastern Asia [5,6]. The genus *Sicyos* contains almost 40 species, and it belongs to the most diversified genera in the Cucurbitaceae family [7]. The features of chloroplast genomes (cp) of some Cucurbitaceae species have been investigated [8]. Moreover, one study revealed the potential therapeutic effect of *Sicyos Angulatus* in acute liver disorder and an atherosclerosis mouse model [9,10]. SA extracts also show an inhibitory effect of flavone glycosides on hepatic lipid accumulation [11]. In the latest study, the anti-obesity effect of *S. angulatus* in HFD-induced (High-Fat diet) obese mice by inhibiting the accumulation of fat has been reported [12]. These findings emphasize *Sicyos angulatus*'s multifaceted nature, including its ecological adaptation, agricultural significance, and medicinal potential. Future studies of this plant species and its associated molecular features, such as the chloroplast genome, can reveal important details about its evolutionary background, genetic diversity, and potential uses in a variety of fields, including plant breeding programs and research on medications.

The chloroplast (cp) arose from an endosymbiotic relationship between photosynthetic bacteria and a non-photosynthetic host [13]. Moreover, molecular studies show that it still contains the remnants of the eubacterial genomes, but during evolution, most of their genes transferred to the nucleus [14,15]. The chloroplast genome has been used widely for multiples purposes, such as source of molecular markers, barcode identification, phylogenetic analysis, and providing a wealth of other valuable feature that make it an interesting and highly desirable object of study in molecular research. The chloroplast genome has a remarkable variety of characteristics that contribute to its importance, including its highly conserved nature, compact size, ease of extraction, relatively low mutation rates, abundant presence within plant cells, and its capacity to encode crucial genes involved in photosynthesis and other crucial cellular process. These numerous characteristics further increase the value and appeal of the chloroplast genome as a flexible tool in different scientific studies [16–18]. The chloroplast is a photosynthetic organelle that plays a vital role in the synthesis of starch, amino acids, and fatty acids [19,20]. In general, the chloroplast genome size usually ranges from 120 to 160 kb; it exhibits a highly conserved circular quadripartite structure, which includes four major regions: a large single copy (LSC), small single copy (SSC), and two inverted repeat regions (IR) [21,22]. Moreover, the structural characteristics of plastome are highly common among land plants. Additionally, their usage as significant models for evolutionary studies is made possible by uniparental inheritance, low rates of substitution, and their small size as compared to that of the nuclear genome.

The number of sequenced chloroplast genomes has soared in correlation with the advancements made in second-generation high-throughput sequencing technology [23]. Based on chloroplast genomic information, phylogenetic relationships for many plant species as well as the reclassification of different plant taxa have been inferred [24]. Therefore, the exploration of the *Sicyos* cp genome may provide a new breakthrough to solve the issues of the genus' evolution and its phylogenetic relationship with other taxa. By comparing cp genomes of different *Sicyos* species and related taxa, researchers can find genetic variants, sequence similarities, and structural rearrangement that can reveal important information about their evolutionary history.

In this research, the complete chloroplast genome of *Sicyos angulatus* and the other seven species of Cucurbitaceae family was chosen as a research target. The aims of this study are (I) to characterize the structure of the newly sequenced chloroplast genome of *S. angulatus*, (II) to perform the comparative analysis of the cp genomes of *S. angulatus* and selected representatives of Cucurbitaceae family, and (III) to reconstruct phylogenetic relationships within the Cucurbitaceae family based on plastome sequences for selected species. These data will be helpful to develop a genomic marker for the purpose of future genetic studies.

2. Materials and Methods

Gene Annotation, Sequence Alignment, and Repeat Prediction

The complete chloroplast sequences of eight species were obtained from GenBank for comparative analysis with the following accession numbers: *S. angulatus* (NC_062884.1), *Corynocarpus laevigata* (NC_014807.1), *Gynostemma pentaphyllum* (NC_029484.1), *Gynostemma pentagynum* (NC_036136.1), *Gynostemma longipes* (NC_036140.1), *Citrullus naudinianus* (NC_058581.1), and *Sicyos edulis* (MT and *Citrullus ecirrhosus* (NC_058582.1).

Geneious prime software (Geneious Prime v.2023.2.1) was used to annotate the chloroplast genome, and the manual evaluation of annotation results was conducted. OGDRAW software version number 1.3.1 was used to plot the circular DNA map [25]. Pairwise sequence alignment was performed with online comparison tool mVISTA [26] using the annotation of *S. angulatus* as reference with the Shuffle-LAGAN mode and a 70% cut-off identity [27]. Tandem repeats were analyzed by using Tandem Repeat Finder version 4.09 with the following parameters, 2 for alignment parameters match, 7 for mismatch, and 7 for indels [28]. The comparison of the LSC/IRB/SSC/IRA boundaries among eight species was performed by using an online tool IRSCOPE (https://irscope.shinyapps.io/irapp/

accessed on 1 August 2023) [29]. Microsatellites (SSRs) analysis was performed using Krait software (Krait V1.3.3, Microsatellite identification and Primer Design; https://krait.biosv.com/en/latest/ accessed on 3 August 2023) with the following parameters: 10 for mono-nucleotide, 5 for di-nucleotide, 4 for tri-nucleotide, 3 for tetra-nucleotide, 3 for penta-nucleotide, and 3 for hexa-nucleotide repeats, and the motif standardization level was set to level 3 [30]. To investigate the phylogenetic relationship, CDs sequence of 25 species were extracted from whole cp genome and aligned with MAFFT v.7.450 [31]. The phylogenetic tree was constructed using the Neighbor-Joining method and the Tamura–Nei genetic distance model with geneious software; *Begonia versicolor* was selected as an outgroup. For the clade support, a 100 bootstrap value was used.

3. Results

3.1. General Characteristics of the *Sicyos angulatus* Chloroplast Genome

S. angulatus has a complete CP genome with a length of 154,986 bp and displays a quadripartite circular structure. It consists of a pair of IR regions, each 26,276 bp in length, separated by an LSC region and an SSC region with lengths of 84,355 bp and 18,079 bp, respectively. The complete genome contains a total of 129 genes, including 84 protein-coding gene (PCGs), 37 tRNA, and 8 rRNA genes. The LSC region contains 83 genes, while the SSC region consists of 14 genes. Additionally, 17 genes are duplicated in the IRs (Figure 1). The protein-coding gene accounts for 51.24% (79,419 bp) of the total genome, whereas the remaining regions are composed of rRNA, tRNA, and intron and intergenic spaces.

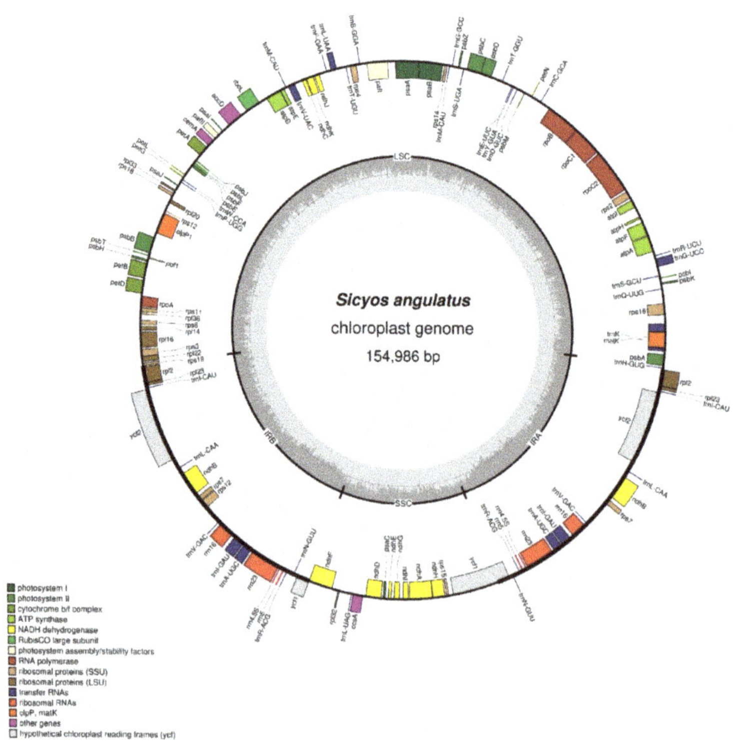

Figure 1. Map of *S. angulatus* chloroplast genome. The genes drawn inside are transcribed anticlockwise, while the outside genes are transcribed clockwise. The different colors represent genes from different functional groups.

In addition, we observed 22 intron-containing genes, out of which 19 genes contained one intron (11 protein-coding genes and 8 tRNA genes) and 2 genes contained two introns (*pafl* and *clpP1*). The total GC content of cpDNA was 37.2%, but we observed distinct differences between the three regions of cp genome when this feature was considered. The highest GC content exhibited in the IR regions (42.8%) were followed by the LSC and SSC regions (35.1% and 31.0%, respectively).

3.2. Comparative Analysis with Other Chloroplast Genomes from the Cucurbitaceae Family

The *S. angulatus* chloroplast genome was compared to seven other species of the Cucurbitaceae family.

3.2.1. Genome Size and Gene Content

The complete chloroplast genome of *S. angulatus* is compared to other seven species of the Cucurbitaceae family.

The complete Cucurbitaceae CP genome has highly conserved structures, regarding its gene content. Each chloroplast genome encodes almost 133 genes, except *C. laevigata* and *S. angulatus* which possess 128 and 129 genes, respectively. However, the chloroplast genome of *S. edulis* encodes a low total gene content, approximately 123 genes. Moreover, they have 37 tRNA (except *C. laevigata*), 8 rRNA, and two genes with double introns. A total of 19 genes in *G. longipes*, *G. pentaphyllum*, *C. naudinianus*, and *S. angulatus*, 20 genes in *C. leavigata*, *C. ecirrhosus*, and 21 genes in *G. pentagynum* contain single introns, However, two genes contain double introns (*clpP* and *ycf3*).

The CP genome size of *G. pentaphyllum* is the largest among the eight studied genomes (159,208 bp), and it is 4222 bp longer than the plastomes of *S. angulatus* (Table 1).

Table 1. Summary of the chloroplast genome features of eight species representing the Cucurbitaceae family.

Features	*S. angulatus*	*S. edulis*	*G. pentagynum*	*G. longipes*	*G. pentaphyllum*	*C. laevigata*	*C. ecirrhosus*	*C. naudinianus*
Genomic Size (bp)	154,986	154,558	157,791	157,601	159,208	159,202	157,009	156,926
LSC region (bp)	84,355	84,577	86,610	86,780	88,487	88,862	86,823	86,728
SSC region (bp)	18,079	20,504	18,869	18,647	18,385	18,992	17,890	17,908
IR region (bp)	26,276	24,758	26,156	26,087	25,982	25,674	26,148	26,145
Coding size	79,419	69,423	80,385	80,478	58,548	77,355	81,240	80,010
Total genes	129	123	133	133	132	128	133	133
No. of PCGs	84	79	87	87	79	83	89	88
tRNA (genes)	37	36	37	37	37	36	37	37
rRNA (genes)	8	8	8	8	8	8	8	8
GC content %	37.2	37.2	37.0	37.0	36.9	36.6	37.2	37.1
LSC (GC content)	35.1	35.0	34.8	34.8	34.6	34.2	35.0	34.9
SSC (GC content)	31.0	31.8	30.6	30.6	31.1	30.6	31.5	31.4
IR (GC content)	42.8	43.2	42.8	42.8	42.8	42.8	42.8	42.8
Single introns	19	19	21	19	19	20	20	19
Double introns	3	2	2	2	2	2	2	2
Duplicated genes	18	17	17	17	19	19	19	19

Only *C. leavigata* lacks *trnfM-CAU* gene, although it has one *trnl-CAU*, andone additional *trnM-CAU* gene. All rRNAs (*rrn4.5*, *rrn5*, *rrn16*, and *rrn23*) are in the IR region. In addition to this, in *C. leavigata* and *G. pentaphyllum*, some of duplicated genes have not same sequence, consisting of a total of 21 and 15 duplicated genes, respectively.

All the species share seventy-four protein coding genes, except for five protein coding genes, as shown in Table 2. Genes without any mark are single-copy genes. The *ndhA* is a single-copy gene, but it is duplicated in *C. ecirrhosus*. Moreover, there are differences in the annotation as well, like in *S. angluatus*, *ycf3* and *ycf4* are annotated as *pafl* and *pafll*, respectively.

Table 2. List of 74 protein coding genes shared among eight species of Cucurbitaceae family.

accD	*atpA*	*atpB*	*atpE*	*atpH*	*atpI*	*ccsA*
cemA	*clpP*	*matK*	*ndhA* (3)	*ndhB* (1)	*ndhC*	*ndhE*
ndhF	*ndhG*	*ndhH*	*ndhI*	*ndhJ*	*ndhK*	*petA*
petB	*petD*	*petG*	*petL*	*petN*	*psaA*	*psaC*
psaC	*psaI*	*psaJ*	*psbA*	*psbB*	*psbC*	*psbD*
psbE	*psbD*	*psbH*	*psbI*	*psbJ*	*psbK*	*psbL*
psbM	*psbN* (2)	*psbT*	*psbZ*	*rbcL*	*rpl2* (1)	*rpl14*
rpl16	*rpl20*	*rpl22*	*rpl23* (1)	*rpl32*	*rpl33*	*rpl36*
rpoA	*rpoB*	*rpoc1*	*rpoc2*	*rps2*	*rpsu3*	*rpsu4*
rps7 (1)	*rps8*	*rps11*	*rps12* (4)	*rps14*	*rps15*	*rps16*
rps18	*rps19*	*ycf3* (5)	*ycf4* (5)			

(1) duplicated genes in all species; (2) *pbf1* in *S. angulatus* genome, but annotated as *psbN* in other species; (3) single copy in most species but duplicated in *C. ecirrhosus* genome; (4) not duplicated in *S. angulatus* genome; (5) in *S. angulatus* genome, *ycf3* is annotated as *pafI* and *ycf4* is annotated as *pafII*.

3.2.2. Sequence Divergence, Tandem Repeats, and SSR Analysis

The many complete CP genome permitted us to evaluate the sequence variation among species. The divergence of the sequence in the CP genome among the eight species was plotted with a cut-off of 70% identity by using mVISTA. The genome of *S. angulatus* was set as the reference genome.

Figure 2 shows that most of the divergence was seen in both the LSC and SSC regions. The IR region and protein-coding genes are more conservative compared to the non-coding region, while some portions of coding region also contain fair observable divergences among *atpF*, *rpoC2*, *rpoC1*, *clpP*, *ndhA*, *petB*, and *psbM*. In comparison to genus *Gynostemma*, *S. angulatus* is more closely linked to *Sicyos edulis*, and *Citrullus* genus, according to the sequence divergence analysis.

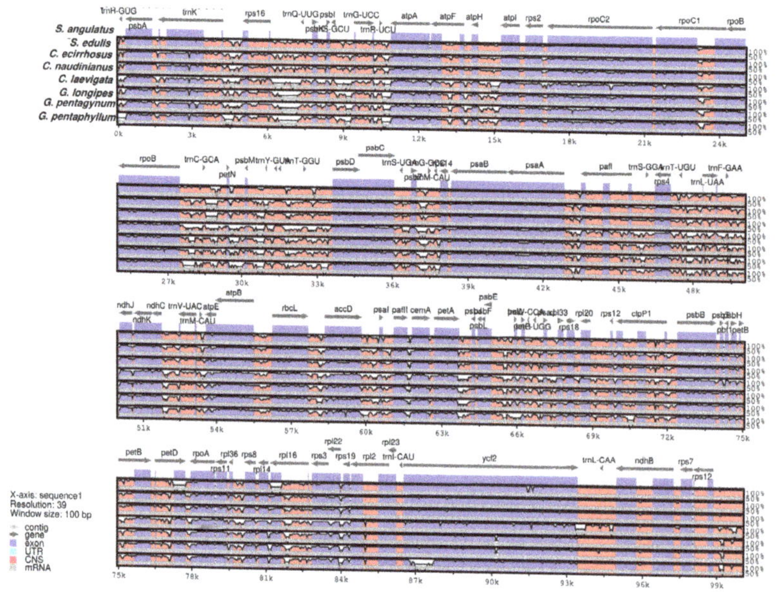

Figure 2. Sequence alignment plot for eight Cucurbitaceae species using mVISTA with *S. angulatus* genome as a reference. Gray arrows represent gene orientation and position. The *y*-axis point the percentage identity range of 50–100%. Red and blue bars indicate non-coding sequence (CNSs) and exons, respectively.

Tandem repeats among the eight species are present along the intergenic spaces and coding sequence region. The repeat sequence in *G. pentaphyllum* and *C. laevigata* is the most frequent in the intergenic spaces 74 and 60, respectively, while the repeat sequence of *C. ecirrhosus* (13) is less frequent in the intergenic space (Table 3). However, among the repeats with a range of 25–45, *C. laevigata* has the highest number of repeats, and *S. edulis* shows second lowest number of repeats, followed by *C. ecirrhosus* (Figure 3A). The overall distribution of these genes throughout the intergenic space and coding sequence region is illustrated in Figure 3B in the form of a percentage.

Table 3. Tandem repeats distribution among eight species of Cucurbitaceae family.

	Intergenic Space	**Coding Sequence**
S. angulatus	22	7
C. levigate	60	9
G. pentaphyllum	74	7
G. pentagynum	24	8
G. longupes	21	9
C. naudinaus	25	6
C. ecirrhosus	13	9
S. edulis	16	7

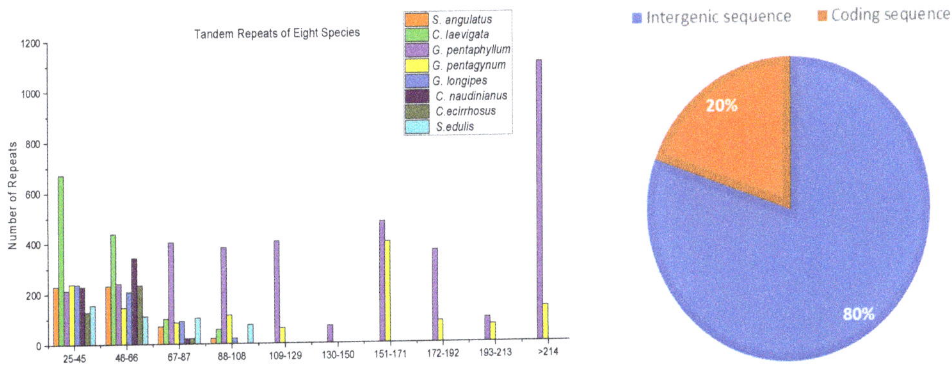

(A) (B)

Figure 3. Length and distribution of tandem repeats using tandem repeat finder among eight species from Cucurbitaceae family. The number of repeats is shown on the *y*-axis (**A**). Location distribution of overall tandem repeats in percentage (**B**).

SSR analysis revealed that among eight species, mononucleotide repeats were the most abundant, followed by the di-, tri-, tetra, penta-, and hexanucleotide repeats (Figure 4A). The highest percentage count of mononucleotide repeats was observed in *C. laevigata* (72.8%), followed by *S. edulis* (80.0%), and lowest percent count of mononucleotide repeats was in *G. pentagynum* (46.6%). In mononucleotide repeats, A was found to be abundant in *C. laevigata* (93 counts), followed by *S. edulis* (60 counts) and *S. angulatus* (49 counts) (Figure 4B). In di-nucleotide repeats, AT in tri-nucleotide AAT, in tetra-nucleotide AAAT, in penta-nucleotide AGGGG, and in hexanucleotide AAATGG is most abundant among eight species of the Cucurbitaceae family. Moreover, the SSRs length distribution for mononucleotide repeats is highest in *S. edulis* (78.2%), followed by *S. angulatus* (76.8%), *C. laevigata* (73.4%), *C. naudinianus* (71.2%), *C. ecirrhosus* (61.4%), *G. longipes* (46.5), *G. pentagynum* (42.9%), and *G. pentaphyllum* (39.7%) (Figure 4C).

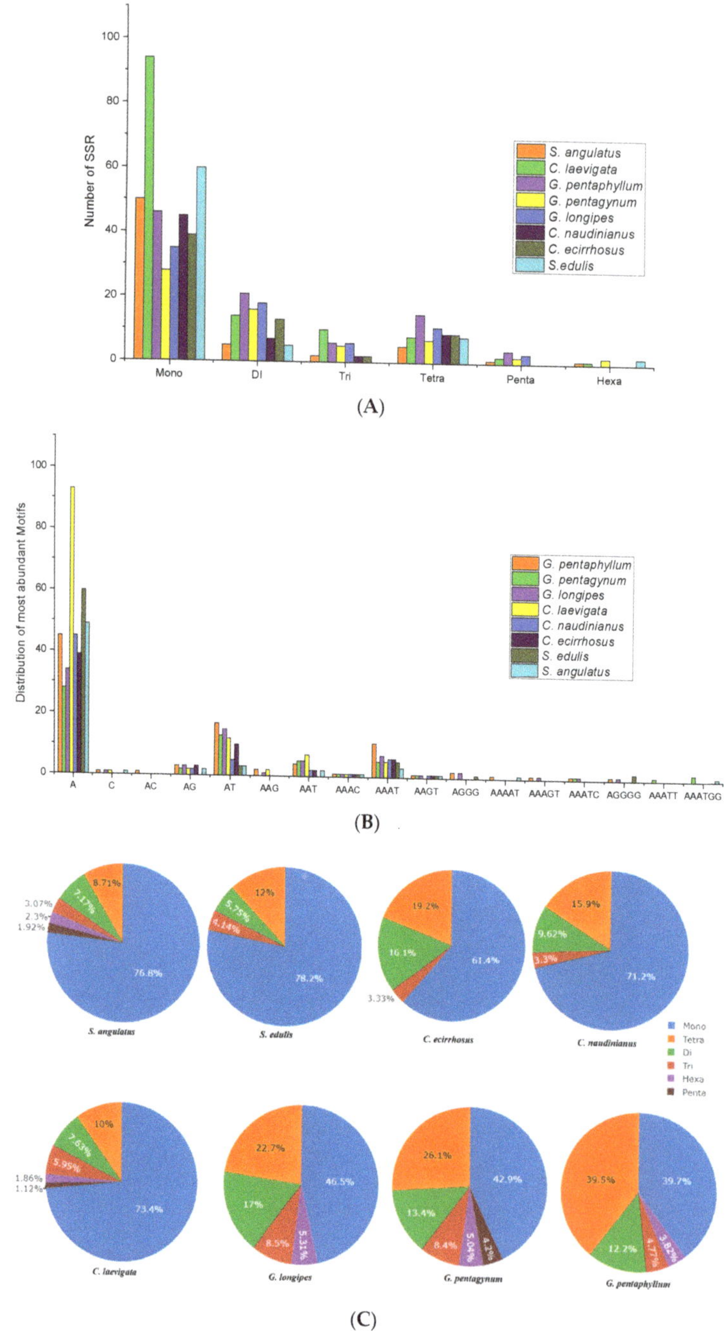

Figure 4. Types of SSRs (simple sequence repeats) in eight species of Cucurbitaceae family. (**A**) Distribution of various SSRs types. (**B**) The abundance of each standard motif category for each type. (**C**) SSR length percentage distribution for each type.

3.2.3. Contraction and Expansion of IRs

The comparison of the genes adjacent to the IR/SSC and IR/LSC boundaries of the analyzed plastomes shows a slight variation, as represented in Figure 5. In this study, the 46 bp and 19 bp of *rps19* genes span the LSC/IRb regions of *C. laevigata* and *G. pentagynum*, respectively, while the other species *rps19* has shares 2 bp with the LSC/IRb region, but in *S. edulis* species, *rps19* is completely located in the LSC region. Notable differences were observed on the junction of IRb/SSC (JSB). At this junction, *ycf1* was absent in *C. laevigata*, while the remaining six species contain *ycf1ψ* (pseudogene). Additionally, the *ndhF* gene was located at the IRb/SSC junction in *G. pentagynum* and *G. longipes*, *C. ecirhosus*, and *S. angulatus*, wherein 12 bp was located in the IRb region in *G. pentagynum* and *G. longipes*, while 7 and 6 bp were found in the IRb region in *C. ecirhosus* and *S. angulatus*, respectively. It is noteworthy to mention *ycf1*, which spans the SSC/IRa junction in all eight species. Moreover, on the JSB junction, the *ycf1* gene spans IRb/SSC border, except in *G. pentaphyllym*, wherein *ycf1* is entirely located in the IRb region. On the other hand, *ycf1* is absent in *S. edulis* species. *psbA* and *trnH* genes are located entirely in the LSC region.

Figure 5. Comparison of the LSC, SSC, and IR junctions among the eight chloroplast genome sequence examined in this study.

3.2.4. Phylogenetic Analysis

To determine the phylogenetic relationship between *S. angulatus* and other species of Cucurbitaceae family, the Tamura–Nei method was used based on the plastome of all the species by using the CDs sequence from the complete cp genome sequence. All 25 species developed two branches with dedicated support using bootstrap values of 100% (Figure 6). The phylogenetic tree indicates that *Sicyos angulatus* is clustered with *Sicyos edulis*. In addition, *Sicyos* genus is close to the *Trichosanthes* genus. Also, the *Sicyos* genus is closely related to *Citrullus ecirrhosus* and *Citrullus naudinianus* (*Citrullus* genus). However, it can be seen that the *Sicyos genus* is more closer to *Gynostemma longipes*, *Gynostemma pentagynym*, *Gynostemma pentaphyllum* (*Gynostemm* genus), as well as *Corvynocarpus laevigata* from the Cucurbitaceae family.

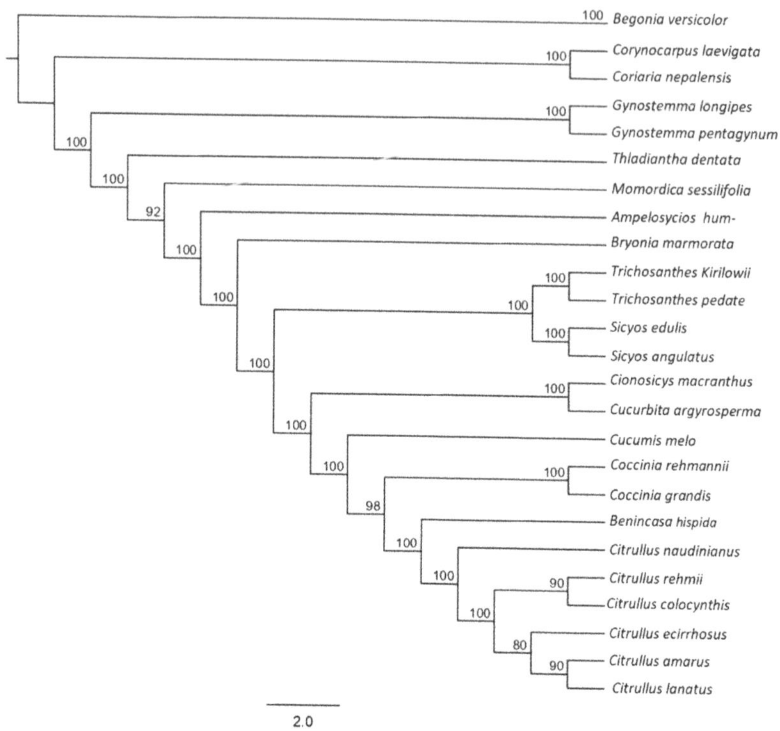

Figure 6. Reconstruction of phylogenetic relationship among selective representatives of the Cucurbitaceae family based on CDs sequence from complete chloroplast genomes.

4. Discussion

The chloroplast genome of *Sicyos angulatus* (SA) was assembled and annotated in this study, and these efforts shed light on the genetic makeup of this species and its relationship with other species in the Cucurbitaceae family [32]. We have obtained deeper knowledge of their genomic characteristics and evolutionary dynamics by contrasting the chloroplast genomes of SA with those of other species in the family. Moreover, this study offers a thorough grasp of the genomic size, gene content, and structural organization of the chloroplast genomes of the *Sicyos* and other genera belonging to the same family.

The selected eight species were chosen to represent various genera within the Cucurbitaceae family and span a variety of taxonomic diversities. Because these species exhibit interesting ecological traits and phenotype diversity, they are useful for comparative analysis. It was observed that the *G. pentaphyllum* and *C. laevigata* chloroplast genome sizes were larger than that of *S. angulatus*. *C. ecirrhosus* has the largest coding size. However, *G. pentaphyllum* and *S. edulis* have the smallest coding size among the selected species. The notable causes of this fluctuation in the chloroplast genome size are intergenic region variance, the shrinkage and expansion of inverted repeats regions, and the loss of genes/introns, which are consistent with earlier reports [33]. This size discrepancy reflects possible changes in the structural and functional components of chloroplast genomes and points to genetic diversity within the family. For an understanding of the evolutionary trends and adaptive tactics within the Cucurbitaceae family, it is essential to comprehend these variances.

The location of the boundaries between the four chloroplast sections is another crucial aspect of the chloroplast genome that is valuable for evolutionary investigation. Analyzing their contraction and expansion can provide an insight into how some taxa have evolved [34]. All eight species have similar placements of the IRa/LSC junction between the upstream and downstream regions of *rpls2* and *trnH* genes, suggesting that the family's genomic organization is conserved. IR contraction or expansion affects the genes which are placed in a close vicinity to the LSC/IR/SSC borders.

This study of eight complete chloroplast genomes (for *S. angulatus* and seven other representatives of the family Cucurbitaceae) gave us the chance to compare the sequence variance among the eight species. For visualizing the divergence in the CP genome sequence, we used the mVISTA tool. Similar to prior research, this study also found that noncoding areas were more diverged than the coding regions are [35,36]. The overall results demonstrated that *S. angulatus* is more like genus *Citrullus* than it is like the genus *Gynostemma*. This finding points to a more recent common ancestor or shared genetic inheritance between *S. angulatus* and *Citrullus* genera, indicating a closer evolutionary link between these two species.

Plant cp genomes frequently contain SSRs, which are frequently employed as molecular markers for polymorphism studies [37]. In this study, SSRs' length distribution for the mono-nucleotide type showed significantly similar repeats in *S. angulatus* and *S. edulis*. However, the SSR count distribution and each standard motif in the mononucleotide were abundant in *C. laevigata* among eight species of the Cucurbitaceae family. Overall, a higher number of mononucleotides repeats as compared to those of di-hexa SSRs were observed in this study, which is consistent with one other study [38].

Previous studies have shown that the variation in size of the chloroplast genome can be due to the expansion and contraction of its region, mainly the IRs [39]. The nucleotide sequence was seen to be highly conserved in the IR regions, which can provide a powerful means to correct unavoidable mutations [40]. In addition, ycf1 lies on the IRa/SSC junction. The *ndhF* genes in *C. laevigata*, *G. pentaphyllum*, and *C. naudinianus* is completely on the SSC region, while 12 bp in *G. pentaphyllum* and *G. longipes*, 7 bp in *C. ecirrhosus*, and 6 bp in *S. angulatus* of *ndhF* overlap in the IRb region. No variation was seen at the JSA junction. *Ycf1* is located at the JSA junction. *trnH* is completely in the LSC region. In addition, *ycf1* is completely lacking in *sicyos edulis*. This finding is consistent with previous research that indicated *ycf1* is not onlylacking in grasses but also in cranberries [41].

Phylogenetic analysis demonstrated SA's evolutionary relationship with the other examined species. Our phylogenetic tree indicated a very clear internal relationship between *Sicyos angulatus* and *Sicyos edulis*, which is similar to that in previous research [32]. Moreover, the tree suggests that the *Sicyos* genus is closely related to the *Citrullus* species: *Citrullus ecirrhosus* and *Citrullus naudinianus*. The phylogenetic differences observed in the tree are likely related to both physiological features and differences in the analyzed chloroplast genome. This study's limitations in terms of its scope and emphasis may account for the lack of a thorough examination of the connections between the results of the phylogenetic analysis and other cp genome diversities. Instead of in-depth discussion about the link with other cp genome diversities, the primary goal may have been to investigate the cp genomes of chosen species and discover the phylogenetic relationships within the Cucurbitaceae family.

In conclusion, insights into the evolutionary relationships within the Cucurbitaceae family were gained through the assembly and annotation of the *Sicyos angulatus* chloroplast genome, as well as comparative research. Our knowledge of the genetic diversity and evolution of this plant family is influenced by the structural changes, gene makeup, and sequence divergence that have been found. The cp genome for *S. angulatus* may become resource to develop genomic markers, which can further help with future genetic research and support breeding initiatives, conservation efforts, and phylogenetic analyses of the family Cucurbitaceae.

Author Contributions: Conceptualization, J.P. and M.K.; data analysis, M.K.; writing—original draft preparation, M.K.; writing—review and editing, J.P. and M.K.; Supervision, J.P. All authors have read and agreed to the published version of the manuscript.

Funding: This research was supported by the Research Program funded by SeoulTech (Seoul National University of Science and Technology, 2022-0167).

Institutional Review Board Statement: Not applicable.

Informed Consent Statement: Not applicable.

Data Availability Statement: All other relevant data are available from the corresponding author upon reasonable request.

Conflicts of Interest: The authors declare no conflict of interest.

References

1. Christenhusz, M.J.; Byng, J.W. The number of known plants species in the world and its annual increase. *Phytotaxa* **2016**, *261*, 201–217. [CrossRef]
2. Na, C.; Lee, Y.; Murai, Y.; Iwashina, T.; Kim, T.; Hong, S. Flavonol 3, 7-diglycosides from the aerial parts of *Sicyos angulatus* (Cucurbitaceae) in Korea and Japan. *Biochem. Syst. Ecol.* **2013**, *48*, 235–237. [CrossRef]
3. Lee, C.H.; Kim, D.; Cho, H.; Lee, H. The Riparian Vegetation Disturbed by Two Invasive Alien Plants, *Sicyos angulatus* and *Paspalum distichum* var. *indutum* in South Korea. *Ecol. Resilient Infrastruct.* **2015**, *2*, 255–263.
4. Thakur, A.K. *Sicyos angulatus* L. (Cucurbitaceae): A new adventive species for the flora of India. *Curr. Sci.* **2016**, *111*, 789.
5. Choi, J.S.; Park, N.J.; Lim, H.K.; Ko, Y.K.; Kim, Y.S.; Ryu, S.Y.; Hwang, I.T. Plumbagin as a new natural herbicide candidate for Sicyon angulatus control agent with the target 8-amino-7-oxononanoate synthase. *Pestic. Biochem. Physiol.* **2012**, *103*, 166–172. [CrossRef]
6. Lee, S.M.; Radhakrishnan, R.; Kang, S.M.; Kim, J.H.; Lee, I.Y.; Moon, B.K.; Yoon, B.W.; Lee, I.J. Phytotoxic mechanisms of bur cucumber seed extracts on lettuce with special reference to analysis of chloroplast proteins, phytohormones, and nutritional elements. *Ecotoxicol. Environ. Saf.* **2015**, *122*, 230–237. [CrossRef]
7. Kobayashi, H.; Kurokawa, S.; Ikeda, K. Dairyland populations of bur cucumber (*Sicyos angulatus*) as a possible seed source for riverbank populations along the Abukuma River, Japan. *Weed Biol. Manag.* **2012**, *12*, 147–155. [CrossRef]
8. Zhang, L.B.; Simmons, M.P.; Kocyan, A.; Renner, S.S. Phylogeny of the Cucurbitales based on DNA sequences of nine loci from three genomes: Implications for morphological and sexual system evolution. *Mol. Phylogenet. Evol.* **2006**, *39*, 305–322. [CrossRef]
9. Kim, Y.H.; Noh, J.R.; Hwang, J.H.; Kim, K.S.; Choi, D.H.; An, J.P.; Oh, W.K.; Lee, C.H. Sicyos angulatus ameliorates atherosclerosis through downregulation of aortic inflammatory responses in apolipoprotein E-deficient mice. *Exp. Ther. Med.* **2017**, *14*, 5863–5870. [CrossRef]
10. Kim, H.Y.; Noh, J.R.; Moon, S.J.; Choi, D.H.; Kim, Y.H.; Kim, K.S.; Yook, S.H.; An, J.P.; Oh, W.K.; Hwang, J.H.; et al. Sicyos angulatus ameliorates acute liver injury by inhibiting oxidative stress via upregulation of antioxidant enzymes. *Redox Rep.* **2018**, *23*, 206–212. [CrossRef]
11. An, J.P.; Dang, L.H.; Ha, T.K.Q.; Pham, H.T.T.; Lee, B.W.; Lee, C.H.; Oh, W.K. Flavone glycosides from Sicyos angulatus and their inhibitory effects on hepatic lipid accumulation. *Phytochemistry* **2019**, *157*, 53–63. [CrossRef] [PubMed]
12. Choi, J.H.; Noh, J.R.; Kim, Y.H.; Kim, J.H.; Kang, E.J.; Choi, D.H.; Choi, J.H.; An, J.P.; Oh, W.K.; Lee, C.H. *Sicyos angulatus* prevents high-fat diet-induced obesity and insulin resistance in mice. *Int. J. Med. Sci.* **2020**, *17*, 787. [CrossRef] [PubMed]
13. Xiang, B.; Li, X.; Qian, J.; Wang, L.; Ma, L.; Tian, X.; Wang, Y. The complete chloroplast genome sequence of the medicinal plant Swertia mussotii using the PacBio RS II platform. *Molecules* **2016**, *21*, 1029. [CrossRef]
14. Martin, W.; Stoebe, B.; Goremykin, V.; Hansmann, S.; Hasegawa, M.; Kowallik, K.V. Gene transfer to the nucleus and the evolution of chloroplasts. *Nature* **1998**, *393*, 162–165. [CrossRef]
15. Martin, W.; Rujan, T.; Richly, E.; Hansen, A.; Cornelsen, S.; Lins, T.; Leister, D.; Stoebe, B.; Hasegawa, M.; Penny, D. Evolutionary analysis of Arabidopsis, cyanobacterial, and chloroplast genomes reveals plastid phylogeny and thousands of cyanobacterial genes in the nucleus. *Proc. Natl. Acad. Sci. USA* **2002**, *99*, 12246–12251. [CrossRef]
16. Wu, M.; Li, Q.; Hu, Z.; Li, X.; Chen, S. The complete Amomum kravanh chloroplast genome sequence and phylogenetic analysis of the commelinids. *Molecules* **2017**, *22*, 1875. [CrossRef] [PubMed]
17. Provan, J.; Powell, W.; Hollingsworth, P.M. Chloroplast microsatellites: New tools for studies in plant ecology and evolution. *Trends Ecol. Evol.* **2001**, *16*, 142–147. [CrossRef]
18. Park, I.; Kim, W.J.; Yeo, S.M.; Choi, G.; Kang, Y.M.; Piao, R.; Moon, B.C. The complete chloroplast genome sequences of Fritillaria ussuriensis Maxim. and Fritillaria cirrhosa D. Don, and comparative analysis with other Fritillaria species. *Molecules* **2017**, *22*, 982. [CrossRef]
19. Neuhaus, H.E.; Emes, M.J. Nonphotosynthetic metabolism in plastids. *Annu. Rev. Plant Biol.* **2000**, *51*, 111. [CrossRef]
20. Rodríguez-Ezpeleta, N.; Brinkmann, H.; Burey, S.C.; Roure, B.; Burger, G.; Löffelhardt, W.; Bohnert, H.J.; Philippe, H.; Lang, B.F. Monophyly of primary photosynthetic eukaryotes: Green plants, red algae, and glaucophytes. *Curr. Biol.* **2005**, *15*, 1325–1330. [CrossRef]
21. Zhou, J.; Zhang, S.; Wang, J.; Shen, H.; Ai, B.; Gao, W.; Zhang, C.; Fei, Q.; Wu, Z.; Liao, X. Chloroplast genomes in Populus (Salicaceae): Comparisons from an intensively sampled genus reveal dynamic patterns of evolution. *Sci. Rep.* **2021**, *11*, 9471. [CrossRef] [PubMed]

22. Palmer, J.D. Plastid chromosomes: Structure and evolution. *Mol. Biol. Plast.* **1991**, *7*, 5–53.
23. Zhang, X.; Gu, C.; Zhang, T.; Tong, B.; Zhang, H.; Wu, Y.; Yang, C. Chloroplast (Cp) Transcriptome of P. davidiana Dode× P. bolleana Lauch provides insight into the Cp drought response and Populus Cp phylogeny. *BMC Evol. Biol.* **2020**, *20*, 51. [CrossRef] [PubMed]
24. Jansen, R.K.; Raubeson, L.A.; Boore, J.L.; Depamphilis, C.W.; Chumley, T.W.; Haberle, R.C.; Wyman, S.K.; Alverson, A.J.; Peery, R.; Herman, S.J.; et al. Methods for obtaining and analyzing whole chloroplast genome sequences. *Meth. Enzymol.* **2005**, *395*, 348–384.
25. Lohse, M.; Drechsel, O.; Bock, R. OrganellarGenomeDRAW (OGDRAW): A tool for the easy generation of high-quality custom graphical maps of plastid and mitochondrial genomes. *Curr. Genet.* **2007**, *52*, 267–274. [CrossRef] [PubMed]
26. Mayor, C.; Brudno, M.; Schwartz, J.R.; Poliakov, A.; Rubin, E.M.; Frazer, K.A.; Pachter, L.S.; Dubchak, I. VISTA: Visualizing global DNA sequence alignments of arbitrary length. *Bioinformatics* **2000**, *16*, 1046–1047. [CrossRef]
27. Frazer, K.A.; Pachter, L.; Poliakov, A.; Rubin, E.M.; Dubchak, I. VISTA: Computational tools for comparative genomics. *Nucleic Acids Res.* **2004**, *32* (Suppl. 2), W273–W279. [CrossRef]
28. Herrero, J.; Muffato, M.; Beal, K.; Fitzgerald, S.; Gordon, L.; Pignatelli, M.; Vilella, A.J.; Searle, S.M.J.; Amode, R.; Brent, S.; et al. Ensembl comparative genomics resources. *Database* **2016**, *2016*, bav096. [CrossRef]
29. Amiryousefi, A.; Hyvönen, J.; Poczai, P. IRscope: An online program to visualize the junction sites of chloroplast genomes. *Bioinformatics* **2018**, *34*, 3030–3031. [CrossRef]
30. Du, L.; Zhang, C.; Liu, Q.; Zhang, X.; Yue, B. Krait: An ultrafast tool for genome-wide survey of microsatellites and primer design. *Bioinformatics* **2018**, *34*, 681–683. [CrossRef]
31. Katoh, K.; Rozewicki, J.; Yamada, K.D. MAFFT online service: Multiple sequence alignment, interactive sequence choice and visualization. *Brief. Bioinform.* **2019**, *20*, 1160–1166. [CrossRef] [PubMed]
32. Choi, T.Y.; Kang, E.S.; Son, D.C.; Lee, S.R. The complete chloroplast genome sequences of the *Sicyos angulatus* (Cucurbitaceae). *Mitochondrial DNA Part. B* **2022**, *7*, 1243–1245. [CrossRef] [PubMed]
33. Xiao-Ming, Z.; Junrui, W.; Li, F.; Sha, L.; Hongbo, P.; Lan, Q.; Qingwen, Y. Inferring the evolutionary mechanism of the chloroplast genome size by comparing whole-chloroplast genome sequences in seed plants. *Sci. Rep.* **2017**, *7*, 1555. [CrossRef] [PubMed]
34. Nazareno, A.G.; Carlsen, M.; Lohmann, L.G. Complete chloroplast genome of *Tanaecium tetragonolobum*: The first Bignoniaceae plastome. *PLoS ONE* **2015**, *10*, e0129930. [CrossRef] [PubMed]
35. Huang, H.; Shi, C.; Liu, Y.; Mao, S.Y.; Gao, L.Z. Thirteen Camelliachloroplast genome sequences determined by high-throughput sequencing: Genome structure and phylogenetic relationships. *BMC Evol. Biol.* **2014**, *14*, 151. [CrossRef]
36. Hong, Z.; Wu, Z.; Zhao, K.; Yang, Z.; Zhang, N.; Guo, J.; Xu, D. Comparative analyses of five complete chloroplast genomes from the genus *Pterocarpus* (Fabacaeae). *Int. J. Mol. Sci.* **2020**, *21*, 3758. [CrossRef]
37. Pauwels, M.; Vekemans, X.; Godé, C.; Frérot, H.; Castric, V.; Saumitou-Laprade, P. Nuclear and chloroplast DNA phylogeography reveals vicariance among European populations of the model species for the study of metal tolerance, *Arabidopsis halleri* (Brassicaceae). *New Phytol.* **2012**, *193*, 916–928. [CrossRef]
38. George, B.; Bhatt, B.S.; Awasthi, M.; George, B.; Singh, A.K. Comparative analysis of microsatellites in chloroplast genomes of lower and higher plants. *Curr. Genet.* **2015**, *61*, 665–677. [CrossRef]
39. Wang, R.J.; Cheng, C.L.; Chang, C.C.; Wu, C.L.; Su, T.M.; Chaw, S.M. Dynamics and evolution of the inverted repeat-large single copy junctions in the chloroplast genomes of monocots. *BMC Evol. Biol.* **2008**, *8*, 36. [CrossRef]
40. Presting, G.G. Identification of conserved regions in the plastid genome: Implications for DNA barcoding and biological function. *Botany* **2006**, *84*, 1434–1443. [CrossRef]
41. De Vries, J.; Sousa, F.L.; Bölter, B.; Soll, J.; Gould, S.B. YCF1: A green TIC? *Plant Cell* **2015**, *27*, 1827–1833. [CrossRef] [PubMed]

genes

MDPI

Communication

Application of High-Resolution Melting and DNA Barcoding for Discrimination and Taxonomy Definition of Rocket Salad (*Diplotaxis* spp.) Species

Pasquale Tripodi

Research Centre for Vegetable and Ornamental Crops, Council for Agricultural Research and Economics (CREA), 84098 Pontecagnano Faiano, Italy; pasquale.tripodi@crea.gov.it; Tel.: +39-089-386-217

Abstract: Nuclear and cytoplasmic DNA barcoding regions are useful for plant identification, breeding, and phylogenesis. In this study, the genetic diversity of 17 *Diplotaxis* species, was investigated with 5 barcode markers. The allelic variation was based on the sequences of chloroplast DNA markers including the spacer between *trnL* and *trnF* and *tRNA-Phe* gene (*trnL-F*), the rubisco (*rbcl*), the maturase K (*matk*), as well as the internal transcribed spacer (ITS) region of the nuclear ribosomal DNA. A highly polymorphic marker (HRM500) derived from a comparison of cytoplasmic genome sequences in Brassicaceae, was also included. Subsequently, a real-time PCR method coupled with HRM analysis was implemented to better resolve taxonomic relationships and identify assays suitable for species identification. Integration of the five barcode regions revealed a grouping of the species according to the common chromosomal set number. Clusters including species with n = 11 (*D. duveryrieriana* or *cretacea*, *D. tenuifolia*, *D. simplex* and *D. acris*), n = 8 (*D. ibicensis*, *D. brevisiliqua* and *D. ilorcitana*), and n = 9 (*D. brachycarpa*, *D. virgata*, *D. assurgens*, and *D. berthautii*) chromosomes were identified. Both phylogenetic analysis and the genetic structure of the collection identified *D. siifolia* as the most distant species. Previous studies emphasized this species' extremely high glucosinolate content, particularly for glucobrassicin. High-resolution melting analysis showed specific curve patterns useful for the discrimination of the species, thus determining ITS1 as the best barcode for fingerprinting. Findings demonstrate that the approach used in this study is effective for taxa investigations and genetic diversity studies.

Keywords: *Diplotaxis* spp.; plastid markers; nuclear markers; DNA sequencing; structure analysis; high resolution melting; phylogenesis

Citation: Tripodi, P. Application of High-Resolution Melting and DNA Barcoding for Discrimination and Taxonomy Definition of Rocket Salad (*Diplotaxis* spp.) Species. *Genes* **2023**, *14*, 1594. https://doi.org/10.3390/genes14081594

Academic Editors: Wajid Zaman and Hakim Manghwar

Received: 13 July 2023
Revised: 4 August 2023
Accepted: 4 August 2023
Published: 6 August 2023

1. Introduction

The genus *Diplotaxis* DC. is a member of the large Brassicaceae family, which includes over 4000 species with a relevant diversity in terms of plant architecture, leaf morphology, and content of nutraceutical compounds [1,2]. The genus includes over 30 species that originated from two main areas, including the Mediterranean basin, with a high level of endemic diversity in the northwest part of Africa and the Iberian Peninsula [3], and the western and southern Asian countries (mostly Turkey, Pakistan, and India) [3,4]. All species are diploid with different gametic chromosome numbers (7, 8, 9, 10, 11, and 13), except *D. muralis* (L.) DC, which is the only tetraploid (n = 21) derived from the hybridization of *D. tenuifolia* (L.) DC (n = 11) and *D. viminea* (L.) DC (n = 10) [5]. Since antiquity, *Diplotaxis* species have been widely used as a food and non-food commodity, such as oil, deodorant, cosmetic, and for medicinal purposes due to their inflammatory and depurative effects. Among them, *D. tenuifolia* (L.) DC, known as wild or perennial rocket, is the most relevant for alimentary uses and it is consumed worldwide as a leafy vegetable in mixing ready-to-use salads. Although the other species are not recognized for economic importance, they are reported to encompass a discrete level of diversity for the content of glucosinolates [6]. So far, the relationship within the taxa has been investigated through different approaches, including morphological assessments [7,8], biochemical studies [9,10], random DNA marker assays such as inter simple sequence repeat (ISSR) [4], and random amplification of polymorphic DNA (RAPD) [11]. These studies demonstrated the existence of two major clusters that partially distribute the accessions according to chromosome number.

A first branch classifies accessions with n = 11 (*D. tenuifolia*, *D. cretacea*, and *D. simplex*), n = 10 (*D. viminea*), and the derived allopolyploid (*D. muralis*), while a second branch classifies the remaining accessions according to a diverse degree of relationship. Chloroplast and nuclear DNA markers showed a division of the genus into the *Brassica rapa/oleoracea* and *Brassica nigra* lineages [12,13]. The close relationship makes these species compatible for hybridization, thus suitable for the genetic improvement of cultivated brassicas. Although in the Brassicaceae family, the systematics of *Diplotaxis* spp. in relation to close taxa have been clearly determined, inconsistencies still occur for species relationships internal to the genus. These are probably due to the typology of markers so far used as well as the method used for polymorphism detection based on gel-electrophoresis techniques. Therefore, the phylogeny is still not well-defined and no further advances for a better definition of taxonomic relationships through genetic investigation have been reported over the past 15 years.

Amplification of highly conserved sequences in plants represents a powerful strategy for fingerprinting and taxonomy objectives [14]. DNA barcoding is a flexible and highly accurate tool for species authentication. Barcode regions mostly involve the chloroplast genome that is uniparentally inherited from the maternal side, which has a very low rate of mutation and is unaffected by hybridization events [15]. The most promising markers designed on the chloroplast genome include the maturase K gene *matK* [16], the non-coding *trnL-F* intergenic spacer, and the Rubisco ribulose biphosphate carboxylase large chain *rbcl* gene [17]. Furthermore, the internal transcribed spacer ITS of nuclear ribosomal DNA (18S-26S) is a widely employed region for taxonomic research [18]. These genes are accepted by the community as the universal standards for species identification and molecular systematics [19,20], being extensively applied in the plant kingdom thanks to the easiness of amplification and sequencing [21]. Sequencing of barcode regions ensures the identification of single nucleotide polymorphisms (SNPs) that underlie the molecular signature of species taxa [22].

Implementing high-resolution melting (HRM) methods based on quantitative polymerase chain reaction (qPCR) coupled with melting curve analysis provides a fast and highly sensitive approach to genotyping [23]. HRM relies on the detection of mutations in the target sequence of amplicons based on dissociation curves generated during DNA denaturation [24]. Several studies combine HRM technology with DNA barcodes for plant authentication with high accuracy [25–28]. Therefore, HRM offers a valid method to be applied for the detection of genetic diversity in crops toward species identification and phylogenesis aims.

Here is reported the investigation of genetic variability and phylogenetic relationships of the *Diplotaxis* genus at the intraspecific level, analyzing both chloroplast and nuclear barcode DNA regions. Hence, it provides an effective and rapid method to identify *Diplotaxis* species and detect polymorphisms through combined sequencing and HRM approaches. This study represents a first attempt to investigate *Diplotaxis* species complementing barcode marker sequencing and quantitative PCR high-resolution melting.

2. Materials and Methods

2.1. Plant Material

The plant material consisted of 17 *Diplotaxis* species retrieved from different European genebanks (Figure 1) including the Leibniz-Institut für Pflanzengenetik und Kulturpflanzenforschung, IPK (Gatersleben, Germany), the Royal Botanical Garden Kew (Richmond, United Kingdom), the Universidad Politécnica de Madrid UPM (Madrid, Spain), and the Universidad de Castilla—La Mancha UCLM (Ciudad Real, Spain).

Figure 1. Plant material. (**a**) *Diplotaxis acris* (acc. 554488); (**b**) *Diplotaxis assurgens* (Delile) Thell. (acc. S17021002); (**c**) *Diplotaxis berthautii* Braun-Blanq. and Maire (acc. S17021001); (**d**) *Diplotaxis brachycarpa* (acc. S17021009); (**e**) *Diplotaxis brevisiliqua* (Coss.) Mart.-Laborde (acc. S17021010); (**f**) *Diplotaxis duveyrieriana* Coss. (cretacea) (acc. S17021008); (**g**) *Diplotaxis eruicoides* (L.) DC. (acc. DIPLO2); (**h**) *Diplotaxis harra* (acc. 338167); (**i**) *Diplotaxis ibicensis* (Pau) Gómez-Campo (acc. S17021006); (**j**) *Diplotaxis ilorcitana* (acc. S17021007); (**k**) *Diplotaxis muralis* (L.) DC. (acc. DIPLO 5); (**l**) *Diplotaxis siifolia* Kunze (acc. S17021003); (**m**) *Diplotaxis simplex* (Viv.) Spreng. (acc. S17021004); (**n**) *Diplotaxis tenuifolia* (L.) DC. (acc. DIPLO 12); (**o**) *Diplotaxis tenuisiliqua* (acc. DIPLO 10); (**p**) *Diplotaxis viminea* (L.) DC. (acc. S17021005); (**q**) *Diplotaxis virgata* (Cav.) DC (acc. CM 0025909). The Royal Botanical Garden Kew (Richmond, United Kingdom) provided (**a**,**h**). The Universidad Politécnica de Madrid UPM (Madrid, Spain) provided (**b**–**f**,**i**,**j**,**l**,**m**,**p**). The Leibniz-Institut für Pflanzengenetik und Kulturpflanzenforschung, IPK (Gatersleben, Germany) provided (**g**,**k**,**n**,**o**). The Universidad de Castilla—La Mancha UCLM (Ciudad Real, Spain) provided (**q**). All images are on the same size scale.

Seeds were sown in pots under climate-grown chamber conditions at the Research Centre for Vegetable and Ornamental Crops (Italy). After 30 days from germination, 100 mg of fresh leaves were collected, followed by nucleic acid isolation using the DNeasy plant mini kit (Qiagen, Hilden, Germany). The DNA purity was estimated by absorbance at 280 and 260 nm, respectively, using a UV-Vis Nanodrop (Thermo Scientific, Wilmington, DE, USA), and the integrity by electrophoresis on a 1.0% agarose gel. Concentration was measured using the Qubit 3 Fluorometer (Thermofisher, Waltham, MA, USA). A volume of 10 µL of extracted DNA, as well as the standards, were diluted in 190 µL of buffer prepared using 1 ul of dsDNA BR Reagent 200 × and 199 µL of dsDNA BR buffer furnished with a Qubit® dsDNA BR assay kit (Thermofisher, Waltham, MA, USA). DNA was diluted at a working concentration (15 ng/µL) and then stored at −20 °C prior to analysis.

2.2. DNA Barcode Primer Design

Specific regions for DNA barcoding were designed on *Diplotaxis tenuifolia* sequences deposited in the nucleotide database at the NCBI (National Center for Biotechnology Information), including the 556 bp nucleotide sequence of the internal transcribed spacer 1, (ITS1) 5.8S nuclear ribosomal RNA gene (Genbank: EF601913.1), and 3 chloroplast sequences as following: (i) The 357 bp nucleotide sequence of the intergenic spacer *trnL*-

trnF and *tRNA-Phe(trnF)* gene (Genbank: DQ984109.1), (ii) The 583 bp nucleotide sequence of the partial *rbcL* (ribulose-1,5-bisphosphate carboxylase/oxygenase large subunit) gene for RuBisCo (Genbank: HE963454.1), (iii) The 708 bp nucleotide sequence of the partial *matK* gene for maturase K (Genbank: HE967405.1). Primers were designed using Primer 3.0 (https://primer3.ut.ee/), ensuring product sizes ranging from 100 to 267 base pairs. In addition, the HRM500 designed on cpDNA polymorphic sites of brassica was included [29]. Marker details are in Table 1.

Table 1. Molecular markers for DNA barcoding used in the present study.

Marker Name	Foward Primer (5′–3′)	Reverse Primer (3′–5′)	Amplicon Size
ITS1	TTAGGCCGTGCGTATAGCTT	TTGCGTTCAAAGACTCGATG	249 bp
trnL-F	AGAAATTCCCGGTCCAAAAC	GGCCGTTACCGAAGTATCATT	107 bp
rbcL	CGGAGTTCCACCTGAAGAAG	TTGTAACGGTCAAGGCTGGT	105 bp
matK	TACGCCGCTTCTGATGAATA	TCTTTAGCCAACGACCCAAT	267 bp
HRM500	GATTCGAACCGTAGACCTGCTC	CCTTAAGGTGTAGCAAGTTTCA	115 bp

2.3. Amplification and Sequencing

For each barcode marker, DNA amplification was performed in 15 µL reactions containing 30 ng of template DNA, 2.0 pmol of each forward and reverse primer, 0.2 mM of each dNTP, and 1.0 U of high-fidelity *PFU Taq* DNA polymerase (Promega, Madison, WI, USA). A C-1000 Touch™ thermal cycler (Bio-Rad, Hercules, CA, USA) was used for nucleic acid amplification. The PCR cycle was performed as follows: one cycle at 95 °C for 1 min; 35 cycles at 95 °C for 30 s, 58 °C for 45 s, and 72 °C for 1 min; one cycle of final extension of 72 °C for 3 min; and soak at 12 °C. Amplification products were visualized on 1% agarose (Lonza, USA) gels in buffer TBE (EDTA 2 mM, Tris base 89 mM, Boric acid 89 mM) and molecular size was assessed with 1 Kb Plus DNA ladder (Life TechnologiesTM, Carlsbad, CA, USA). The visualization of PCR products was performed by staining agarose gel with SYBR® safe (Life TechnologiesTM), and the fluorescence was viewed using Gel DocTM XR (Biorad). Amplicons were then purified with ExoSAP-IT™ (Thermofisher, Waltham, MA, USA). The sequencing reaction was prepared with a Big Dye Terminator v3.1 Cycle sequencing kit (Thermofisher, Foster City, CA, USA). The sequencing cycle consisted of a cycle of denaturation (96 °C, 1 min) and 25 cycles of amplification (96 °C, 10 s; 50 °C, 5 s; 60 °C, 2 min). Sequencing reactions were then purified using the X-Terminator Purification Kit. The SeqStudio™ Genetic Analyzer (Thermofisher, Waltham, MA, USA) was used for Sanger sequencing. SeqScape® v2.0 (Thermofisher, Waltham, MA, USA) was used for base calling.

2.4. High-Resolution Melting Analysis

Real-time PCR was performed in 10 µL of total reaction volume containing 37.5 nanograms of genomic DNA (2.5 µL with conc of 15 ng/µL), 1× Precision Melt Supermix (Bio-Rad, Inc., Hercules, CA, USA) (5 µL of 2× Supermix), and 0.4 µL of unlabeled forward and reverse primers (final concentration 200 nM). For each sample, analysis was performed in triplicate, and a negative control was included. The assay was performed on a CFX 96 RealTime PCR System (Bio-Rad, Inc., Hercules, CA, USA) using the following protocol: 95 °C for 2 min by 40 cycles at 95 °C for 15 s and 58 °C for 30 s. The melting curve was obtained with an initial step at 95 °C for 30 s (heteroduplex formation) and 64 °C for 3 min, then rising of temperature from 65 °C to 95 °C with increments of 0.2 °C/cycle every 10 s. Fluorescence data were analyzed with the Precision Melt Analysis™ Software (Bio-Rad, Inc., Hercules, CA, USA).

The generated melting profiles allowed the discrimination of genotypes by assignment to specific clusters according to the different temperatures of melting. A percent of confidence to estimate that a given sample was properly assigned with the cluster was calculated. Values above 95% were considered as the threshold for properly assigned samples to specific clusters.

2.5. Phylogenetic Tree Data Analysis

Single sequences were trimmed for low quality and manually edited using Chromas Lite. The sequences were aligned with the CLUSTAL W program implemented in MEGA X using the default settings [30]. Aligned sequences were trimmed to the same size by removing any sequence gaps and unaligned ends. All trimmed sequences are in Supplementary Table S1. A phylogenetic tree was drawn with the Kimura two-model implemented in the neighbor-joining method. A total of 10,000 bootstraps were considered. Sequences were then concatenated, re-aligned, and trimmed to obtain a concatenated tree following the method and model above described. Pairwise genetic distances (p-distance model) between all generated sequences were calculated between the 17 *Diplotaxis* species. For HRM profiles, melting curves were transformed into binary data prior to analyses. Analyses were conducted in MEGA X software.

2.6. Genetic Diversity and Population Structure

Analysis of genetic structure was performed with the Bayesian clustering method implemented in STRUCTURE v.2.4 [31]. The admixture model and MCMC (Markov chain Monte Carlo) method for allele frequency calculation and detection of the best number of population (K) were used. Runs were performed using 50,000 MCMC iterations and 50,000 burn-in cycles, with the number of K ranging between 1 and 15, with 5 independent runs. The optimal numbers of subpopulations were determined according to Evanno's test implemented in Structure Harvester [32]. A membership coefficient (qi) ≥ 0.50 was considered to infer individuals to a specific subpopulation. Accessions with values lower than 0.5 at each assigned K were considered as admixed. Principal component analysis was conducted in R by the function *prcomp* (package stats), and the biplot was drawn using the ggplot2 R package [33].

3. Results

3.1. Sequence Comparisons

A total of 85 sequences were produced (Supplementary Table S1). Sequence length and nucleotide percentage for the five considered barcode regions are reported in Table 2. HRM500 was the longest sequence, consisting of 322 aligned sites and 26 variant sites. Sequence length ranged from 309 (*D. viminea*) to 316 (*D. tenuisiliqua*) nucleotides. The nucleotide composition of HRM500 was A/T rich (34.27% A, 15.38% C, 14.44% G, 35.88% T). ITS1 consisted of 274 aligned sites and a larger number of polymorphic sites (163). Sequence length ranged from 214 (*D. acris, D. ilorcitana*, and *D. simplex*) to 264 (*D. siifolia*) nucleotides with a balanced A/T-G/C composition (30.37% A, 23.82% C, 24.84% G, 20.84% T). Of the 241 bp of total aligned nucleotides for *matk*, 46 were polymorphic. A minimum sequence length was observed in *D. acris* and *D. virgata* (233 bp), while the maximum length was in *D. muralis* (239 bp). The nucleotide makeup of the *matk* region was A/T rich, with 32.71% A, 12.71% C, 16.19% G, and 38.35% T. Both *rbcl* and *trnL-F* were the smallest analyzed regions with a total of 73 bp and 80 bp of aligned sequence, respectively.

The former showed a slightly higher G/C average content (20.32% A, 27.19% C, 25.02% G, 27.47% T), whereas the latter a higher A/T (30.51% A, 20.60% C, 15.53% G, 32.76% T). A total of 9 polymorphic sites were observed for *rbcl*, while *trnL-F* exhibited a high variant rate being 97.5%. After trimming, 304 nucleotides were retained for HRM500, 200 nucleotides for ITS1, and 231 nucleotides for *matk*, whereas for *rbcl* and *trnL-F*, a total of 65 and 60 nucleotides were kept, respectively. By merging, realigning, and trimming all sequences of the 5 barcode regions, a total of 857 nucleotides were obtained and used for the consensus concatenated phylogenetic tree (Figure 2). In total, 158 base substitution characters were obtained. The concatenated sequence was A/T rich, comprising on average 31.39% A, 31.97 T, 17.96 C, and 18.67 G.

Table 2. Nucleotide components of five nuclear and chloroplast regions for the 17 *Diplotaxis* species. For each marker, the total length of the aligned sequence is indicated in brackets.

	HRM500 (322)					ITS1 (274)					*matk* (241)					*rbcl* (73)					*trnL-F* (80)				
	Seq #	A	C	G	T	Seq	A	C	G	T	Seq	A	C	G	T	Seq	A	C	G	T	Seq	A	C	G	T
D. ac *	312	106	49	45	112	214	63	52	56	43	233	75	28	39	91	67	13	18	17	19	76	23	15	12	26
D. as	312	109	48	45	110	215	64	51	54	46	235	77	28	40	90	67	13	18	17	19	76	24	14	12	26
D. be	313	108	48	45	112	216	66	52	53	45	234	75	30	39	90	67	13	19	17	18	71	24	22	10	15
D. br	315	110	47	44	114	217	68	52	52	45	235	78	29	37	91	67	13	18	17	19	67	23	13	11	20
D. bv	313	106	48	45	114	216	62	53	57	44	234	76	31	37	90	69	14	18	17	20	80	23	18	12	27
D. du	313	105	49	47	112	217	66	52	55	44	235	76	30	38	91	68	14	19	17	18	76	23	15	12	26
D. er	313	108	48	45	112	218	68	51	53	46	234	76	29	38	91	68	14	18	17	19	78	23	17	12	26
D. ha	313	108	48	45	112	217	70	50	52	45	235	77	29	38	91	68	14	18	17	19	78	23	17	12	26
D. ib	313	108	48	45	112	214	67	52	52	43	236	77	31	38	90	70	14	20	17	19	78	24	16	12	26
D. il	314	108	48	45	113	213	63	51	56	43	236	77	31	38	90	68	14	18	17	19	79	23	17	12	27
D. mu	310	109	48	44	109	215	64	51	56	44	239	78	31	38	92	69	15	19	17	18	79	24	16	12	27
D. si	314	110	48	44	112	264	79	68	51	66	235	76	29	39	91	68	14	18	17	19	78	25	14	13	26
D. sm	314	104	49	47	114	214	63	51	56	44	235	76	31	38	90	68	14	19	17	18	76	23	15	12	26
D. te	312	104	49	47	112	216	65	51	56	44	236	77	29	40	90	68	14	19	17	18	77	24	15	12	26
D. tn	316	108	49	44	115	216	64	51	54	47	236	82	29	37	88	68	14	19	17	18	76	25	13	12	26
D. vi	309	109	47	43	110	217	66	52	55	44	236	77	31	39	89	68	14	18	17	19	78	24	16	12	26
D. vr	315	110	48	46	111	215	67	48	52	48	233	77	32	35	89	67	13	18	17	19	79	23	16	14	26

* Acronyms for the 17 assayed *Diplotaxis* species: *D. ac*, *D. acris*; *D. as*, *D. assurgens*; *D. be*, *D. berthautii*; *D. br*, *D. brachycarpa*; *D. bv*, *D. brevisiliqua*; *D. du*, *D. duveyrieriana*; *D. er*, *D. eruicoides*; *D. ha*, *D. harra*; *D. ib*, *D. ibicensis*; *D. il*, *D. ilorcitana*; *D. mu*, *D. muralis*; *D. si*, *D. siifolia*; *D. sm*, *D. simplex*; *D. te*, *D. tenuifolia*; *D. tn*, *D. tenuisiliqua*; *D. vi*, *D. viminea*; *D. vr*, *D. virgata*. # Seq = sequence length

The overall pairwise genetic distance was 0.034 among all *Diplotaxis* species, with values between accessions ranging from 0.006 (*D. tenuifolia-D. duveyrieriana;*) to 0.103 (*D. siifolia- D. berthautii*) Table 3. The highest values were observed between *D. siifolia* and the rest of the species (average p-distance = 0.087), whereas *D. ilorcitana* showed the lowest values (average p-distance = 0.024).

Table 3. Estimates of pairwise genetic distance (below the diagonal) and standard error estimate(s) (above the diagonal) within 17 Diplotaxis species. Species acronyms are reported in Table 2.

	D. ac	*D. as*	*D. be*	*D. br*	*D. bv*	*D. du*	*D. er*	*D. ha*	*D. ib*	*D. il*	*D. mu*	*D. si*	*D. sm*	*D. te*	*D. tn*	*D. vi*	*D. vr*
D. ac		0.005	0.007	0.005	0.005	0.004	0.005	0.005	0.004	0.003	0.004	0.009	0.004	0.004	0.005	0.004	0.006
D. as	0.025		0.006	0.004	0.006	0.006	0.005	0.006	0.005	0.005	0.006	0.009	0.006	0.006	0.006	0.006	0.005
D. be	0.049	0.032		0.007	0.008	0.007	0.007	0.008	0.007	0.007	0.008	0.010	0.008	0.008	0.008	0.008	0.007
D. br	0.026	0.018	0.042		0.006	0.006	0.005	0.006	0.006	0.005	0.006	0.009	0.006	0.006	0.006	0.006	0.005
D. bv	0.022	0.033	0.058	0.034		0.006	0.006	0.006	0.005	0.004	0.006	0.009	0.006	0.006	0.007	0.006	0.006
D. du	0.014	0.029	0.049	0.030	0.030		0.005	0.006	0.005	0.004	0.005	0.010	0.003	0.003	0.006	0.005	0.006
D. er	0.021	0.020	0.045	0.020	0.029	0.026		0.006	0.005	0.005	0.005	0.009	0.005	0.006	0.006	0.005	0.005
D. ha	0.026	0.035	0.059	0.035	0.032	0.029	0.032		0.005	0.005	0.006	0.010	0.006	0.006	0.006	0.005	0.006
D. ib	0.013	0.026	0.049	0.027	0.022	0.019	0.025	0.026		0.003	0.004	0.009	0.005	0.005	0.006	0.003	0.006
D. il	0.011	0.021	0.046	0.022	0.014	0.015	0.020	0.025	0.009		0.004	0.009	0.004	0.004	0.006	0.004	0.006
D. mu	0.014	0.027	0.052	0.028	0.028	0.019	0.023	0.030	0.018	0.013		0.010	0.004	0.004	0.005	0.003	0.006
D. si	0.083	0.080	0.103	0.083	0.084	0.088	0.084	0.091	0.078	0.078	0.085		0.010	0.010	0.010	0.009	0.010
D. sm	0.014	0.029	0.052	0.030	0.030	0.008	0.026	0.030	0.020	0.015	0.015	0.086		0.003	0.006	0.004	0.006
D. te	0.016	0.032	0.054	0.033	0.033	0.006	0.028	0.033	0.022	0.018	0.016	0.091	0.008		0.005	0.005	0.006
D. tn	0.026	0.029	0.052	0.030	0.039	0.030	0.030	0.036	0.032	0.027	0.026	0.092	0.028	0.026		0.006	0.006
D. vi	0.014	0.027	0.052	0.028	0.028	0.019	0.023	0.026	0.011	0.013	0.007	0.084	0.016	0.019	0.028		0.006
D. vr	0.030	0.020	0.045	0.022	0.035	0.035	0.025	0.036	0.032	0.027	0.028	0.087	0.033	0.035	0.032	0.028	

3.2. High-Resolution Melting Profiles

Polymorphisms among the 17 *Diplotaxis* species were detected based on the pattern derived from the normalized melt curve and derived difference curves (Figure 3). In total, 26 HRM profiles were revealed by the 5 barcode markers used, ranging from 2 (*rbcl*) to 9 (ITS1). HRM500 (Figure 3a) showed 5 melting curves: 2 grouping 10 and 4 *Diplotaxis* species, respectively, whereas 3 were specific for *D. siifolia*, *D. brachycarpa*, and *D. simplex*. A higher number of melting curves was shown by ITS1 (Figure 3b), which clearly discriminated 7 out of the 17 studied species including *D. brachycarpa*, *D. berthautii*, *D. eruicoides*, *D. siifolia*, *D. assurgens*, *D. virgata*, and *D. tenuisiliqua*. *Matk* (Figure 3c) evidenced 5 distinct melting curves that grouped from 2 to 7 species, except for *D. harra*, which showed a singular pattern. *Rbcl* (Figure 3d) instead showed two different melting profiles, distinguishing *D. duveyrieriana*, *D. simplex*, *D. tenuifolia*, and *D. tenuisiliqua* from the rest. Finally, *trnL-F*

(Figure 3e) showed 5 melting curves with a major group, including 13 species and 4 profiles discriminating *D. tenuisiliqua*, *D. eruicoides*, *D. brachycarpa*, and *D. assurgens*.

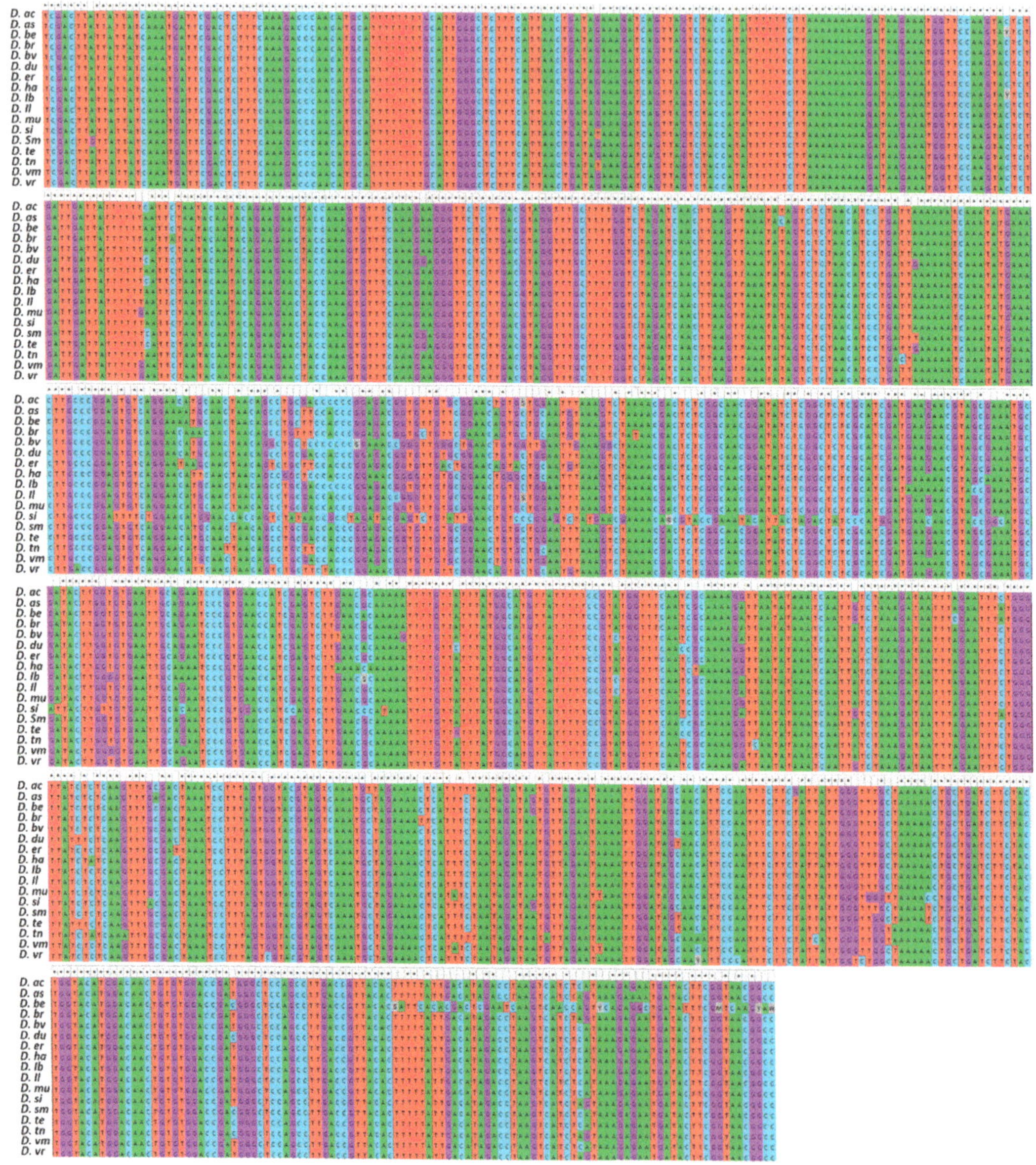

Figure 2. DNA sequence alignment analysis of concatenated ITS1, *trnL-F*, *rbcl*, *matk*, and HRM500 sequences for the *Diplotaxis* species considered in the present study. For acronyms, see Table 2 caption.

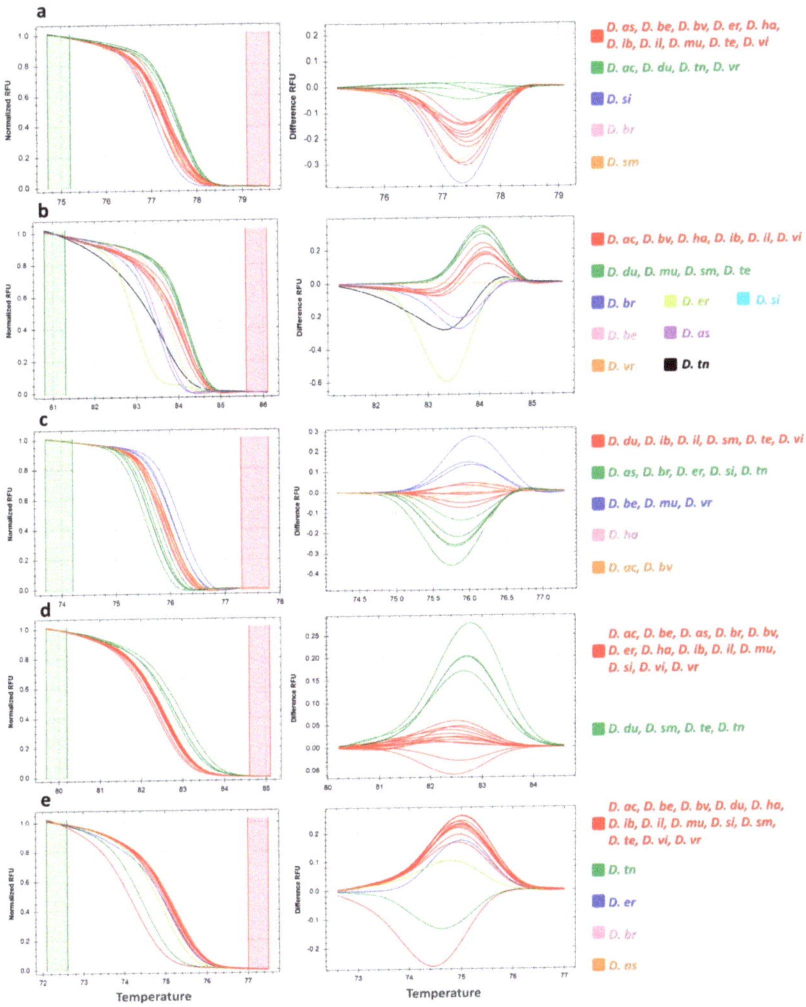

Figure 3. High-resolution melting profile showing normalized melting curves (left) and difference curves (right) for (**a**) HRM500, (**b**) ITS1, (**c**) *matk*, (**d**) *rbcl*, and (**e**) *trnL-F*. For acronyms, see Table 2 caption.

Overall, among the five barcode regions, ITS1 was especially noteworthy for the identification of *Diplotaxis* species, whereas the other markers showed less specific melting curves for the considered species. In all instances, *D. ibicensis*, *D. ilorcitana,* and *D. virgata* showed the same curve patterns, whereas *D. tenuifolia* and *D. duveyrieriana* showed the same melting profile for all markers except ITS1. Singular patterns were instead found for *D. brachycarpa* using HRM500, ITS1, and *trnL-F*.

3.3. Phylogenetic Analysis

Phylogenetic trees using separate matrices are shown in Figure 4.

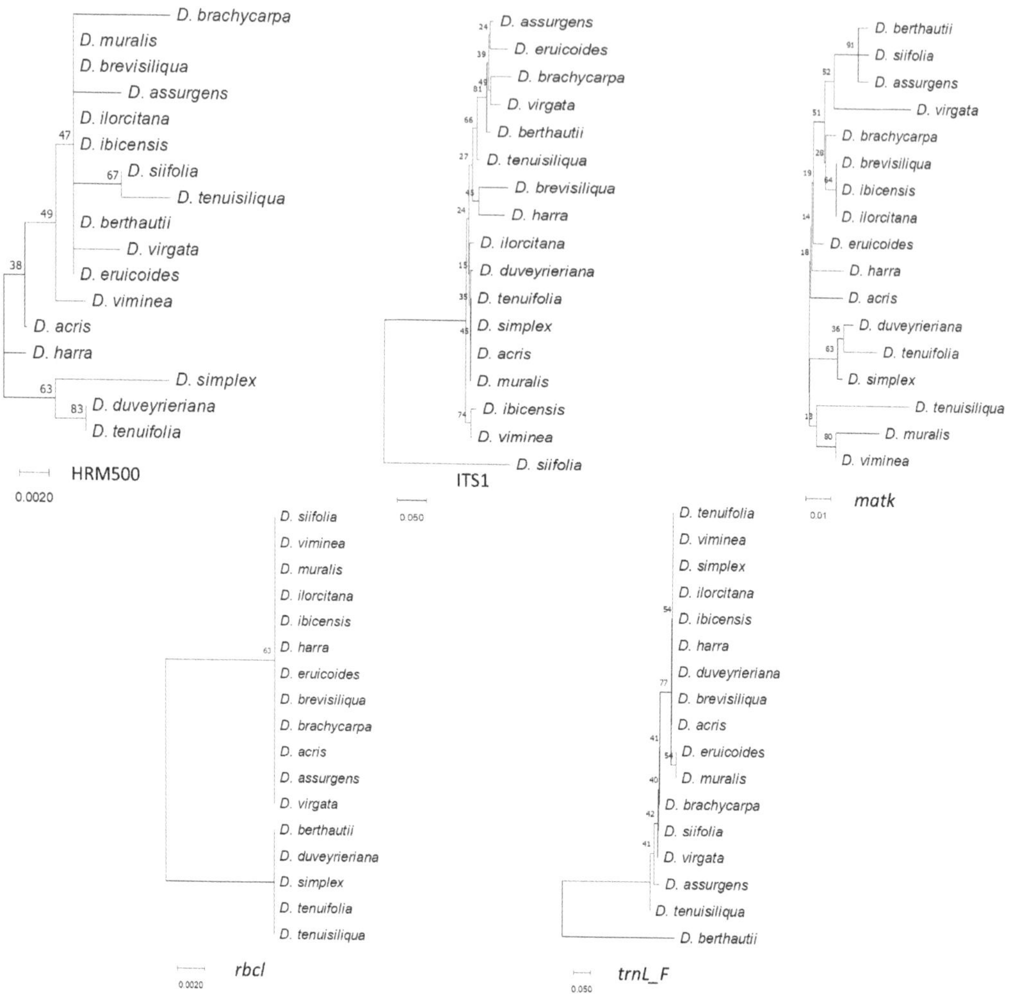

Figure 4. Phylogenetic analysis obtained from the five barcode sequences. The phylogenetic tree was inferred with 10,000 bootstraps using the neighbor-joining method. The evolutionary distances were computed using the Kimura 2-parameter model. Numbers above branches indicate the percentage of replicate trees, in which the associated taxa clustered together in the bootstrap test.

Although different relationships were resolved in each tree, clustering of *D. tenuifolia*, *D. duveyrieriana*, and *D. simplex* was observed in all instances. *D. muralis* clustered close to *D. viminea* with *matk* and *rbcl*, while HRM500, ITS1, and *trnL-F* grouped *D. muralis* with *D. brevisiliqua*, *D. acris*, and *D. eruicoides*, respectively. *D. brevisiliqua* clustered close to *D. ilorcitana* with *matk*, *rbcl*, and *trnL-F*, and close to *D. harra* with ITS1. Overall, ITS1, *rbcl*, and *trnL-F* divided the accessions into main groups, whereas HRM500 and *matk* highlighted a higher number of specific subgroups. For a better resolution of the diversity, the individual sequences of the 17 accessions for the 5 markers were combined. The consensus tree (Figure 5a) separated *D. siifolia* from the rest and clustered the remaining accessions in 2 main groups including 11 and 5 accessions, respectively, defining furthermore several subclusters. The cluster of *D. tenuifolia*, *D. duveyrieriana*, and *D. simplex* was confirmed, whereas close relationships were found between *D. muralis* and *D. viminea* as well as *D. brevisiliqua*, *D. ilorcitana*, and *D. ibicensis*. A second group comprised the accessions

D. eruicoides, *D. brachicarpa*, *D. virgata*, *D. assurgens*, and *D. berthautii*, with the latter two species showing a high level of similarity. The phylogenetic tree drawn using high-resolution melting analysis (Figure 5b) mostly confirmed the two main clusters observed by sequence alignments with few differences related to *D. virgata*, which was positioned separately in the larger group, and *D. siifolia*, which clustered within the second group, although tended to be as an outgroup accession. The high degree of similarity between *D. ibicensis*/*D ilorcitana*, *D. muralis*/*D. viminea*, *D. tenuifolia*/*D. duveyrieriana* was confirmed.

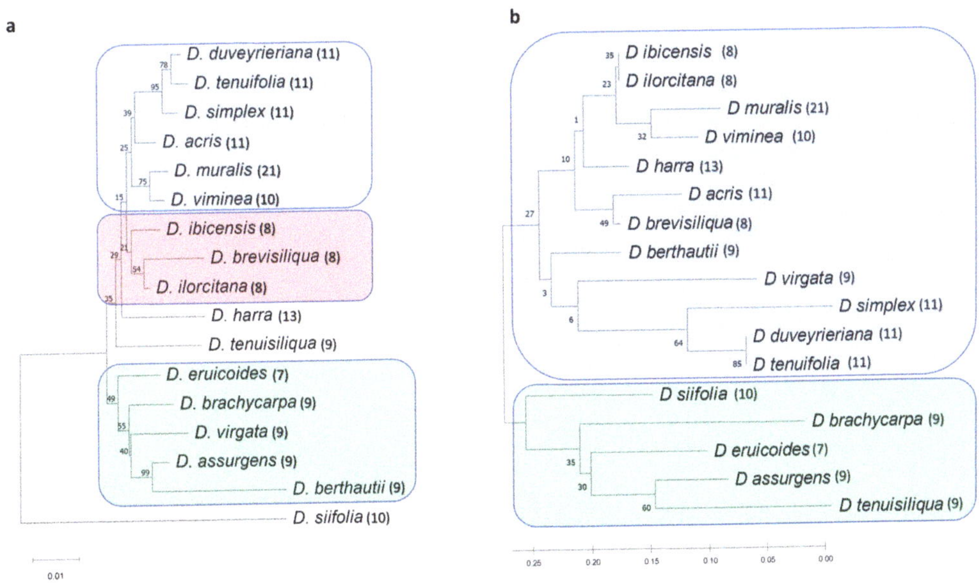

Figure 5. Phylogenetic analysis on combined polymorphisms from the five barcode regions. The phylogenetic tree was inferred by using the neighbor-joining method and the Kimura 2-parameter model. Bootstrap trees inferred from 10,000 replicates are shown: (**a**) Phylogenetic tree on concatenated barcode sequences; (**b**) Phylogenetic tree considering polymorphisms from HRM profiles. Numbers above branches indicate the percentage of replicate trees, in which the associated taxa clustered together in the bootstrap test. Phylogenetic analyses were conducted in MEGA X. In brackets are indicated the haploid chromosome numbers for each *Diplotaxis* species.

3.4. Population Structure

Polymorphisms were based on STRUCTURE analysis (Figure 6a), and the collection was divided into K = 2 as the likely number of subpopulations according to Evanno's test (Figure 6b). The main group comprised 16 accessions, of which 8 (*D. ibicensis*, *D. duveyrieriana*, *D. simplex*, *D. tenuifolia*, *D. acris*, *D. muralis*, *D ilorcitana*, and *D. viminea*) had a high coefficient of membership (qi $\geq$ 0.9). *D. berthautii* and *D. siifolia* clustered in the second subpopulation with a coefficient of membership (qi) of 0.73 and 0.97, respectively. The principal component analysis grouped the accessions in both positive and negative axis of the first (PC_1) and second (PC_2) components. Accessions were mostly separated along the first component with 2 main groups comprising 11 and 5 genotypes positioned in the negative and positive axis of PC_1, respectively. Different clusters with tightly close accessions were identified. The biggest ones included four (cluster A: *D. tenuifolia*, *D. duveyrieriana*, *D. simplex*, and *D. harra*) and five species (cluster B: *D. viminea*, *D. acris*, *D. ibicensis*, *D. muralis*, and *D. ilorcitana*), respectively. Two additional clusters, each comprising two species, were defined (cluster C: *D. tenuisiliqua* and *D. eruicoides*; cluster D: *D. virgata* and *D. brachycarpa*). *D. berthautii* and *D. siifolia* were confirmed to be highly diverse from the rest, being both located in the extreme parts of both PC_1 and PC_2.

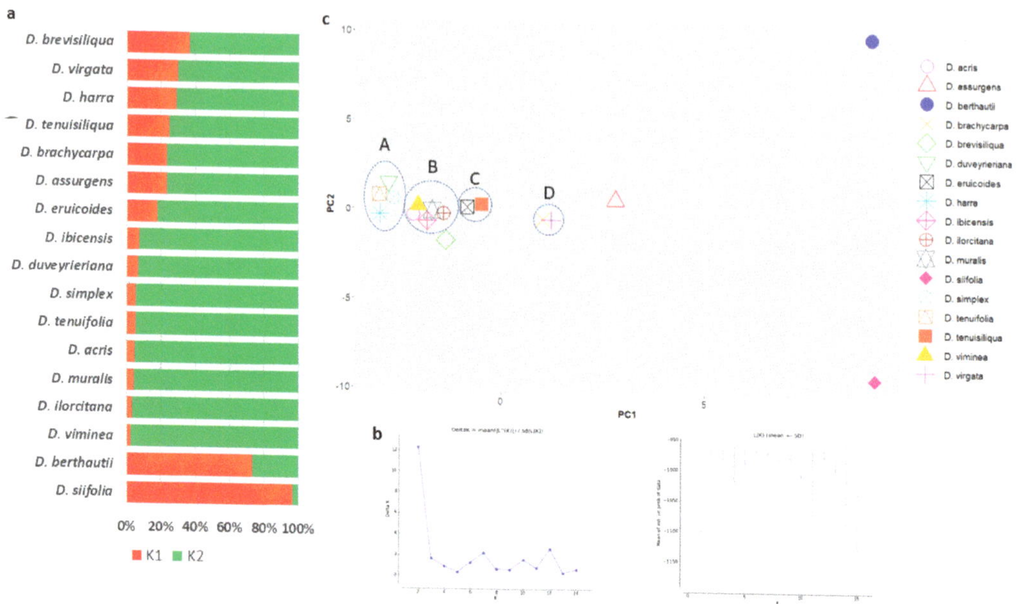

Figure 6. Population structure of 17 *Diplotaxis* spp. accessions based on 5 barcode regions. (**a**) STRUCTURE analysis considering K = 2 clusters. For each genotype, horizontal bars represent the allele frequency for each subpopulation (indicated in abscissis). (**b**) Evaluation of the best number of subpopulations within the Bayesian clustering analysis using Evanno's method: on the right is the plot generated by STRUCTURE HARVESTER for the detection of the most likely number of clusters, the highest value was at K = 2, indicating that the 17 accessions likely form 2 sub-populations; on the left is the plot of mean likelihood L(K) and variance for 5 independent runs for each value of K for K = 1–15. (**c**) Loading plot in the first two components, showing the diversity of the collection.

4. Discussion

Molecular markers have innumerable uses being effectively employed to investigate the diversity in plant species, determine their phylogenetic relationships, or serve as selection markers in plant breeding [34]. Among these, marker assays targeting barcode regions offer a promising tool for species identification given the possibility to select standard loci to analyze taxonomically different specimens, thus producing comparable data [15]. The barcodes used in this study have been demonstrated to be powerful for species discrimination in diverse organisms including plants [28,35], mammals [36], fungi [37], and bacteria [38]. Furthermore, the application of HRM analysis coupled with DNA barcode has shown its potential for crop species identification [39,40].

Based on morphological characteristics, Prantl [41] and Schulz [42] placed the *Diplotaxis* species assayed in the present study into three subsections: Anocarpum, including *D. tenuifolia*, *D. simplex*, *D. viminea*, and *D. muralis*; Rynchocarpum, including *D. virgata*, and *D. erucoides*; and Catocarpum, including *D. harra*, *D. tenuifolia*, and *D. cretacea*. Gomez-Campo and Martínez-Laborde [43] subdivided Rhyncocarpum into three subclades: (i) Rhynchocarpum, including *D. assurgens*, *D. berthautii*, *D. brachycarpa*, *D. siifolia*, *D. tenuisiliqua*, and *D. virgata*; (ii) Heterocarpum, including *D. brevisiliqua*, *D. ibicensis*, and *D. ilorcitana*; and (iii) Heteropetalum, containing *D. eruicoides*. Based on plastid DNA variation, Warwick et al. [12] placed these taxa into two different lineages of the subtribe *Brassicinae* (Brassicaceae family): the Rapa/Oleoracea combined seven species in three groups including *D. eruicoides* (Group A), *D. tenuifolia*, *D. duveryrieriana*, *D. simplex*, and *D. harra* (Group B), and *D. viminea* and *D. muralis* (Group C); the Nigra lineage included seven species in three subgroups: *D. brevisiliqua* and *D. ibicensis* (Group D), *D. brachycarpa* (Group E), and *D. tenuisiliqua*, *D. virgata*, *D. siifolia*, and *D. berthautii* (Group F).

This study showed a grouping of species according to the common chromosomal set number in agreement with previous investigations based on cytological [5], cross compatibility [44], morphological [43], and molecular [4,11] approaches. The species with 11 chromosomes (n = 11) (*D. duveryrieriana* or *cretacea*, *D. tenuifolia*, *D. simplex*, and *D. acris*) constituted a group of closely related taxa that clustered in the same group with *D. viminea* (n = 10) and the amphidiploid *D. muralis* (n = 21), derived by the cross of *D. viminea* × *D. tenuifolia*. The close relationship of *D. muralis* with *D. viminea* rather than *D. tenuifolia* confirmed previous studies. The former line has been demonstrated to be the female parent of the amphidiploid from the peptide structure of the Rubisco enzyme and analysis with chloroplast markers [11,12,45]. The significant degree of resemblance between *D. tenuifolia* and *D. duveryriana* corroborated earlier findings reporting these two species belonging to a single cluster, with *D. simplex* joining a part [4]. Furthermore, analysis with ISSR highlighted a high level of similarity between *D. viminea* and *D. muralis* compared to the other species with n = 11 chromosomes [4]. From the biochemical point of view, this homogeneous complex has been reported to have a low total glucosinolate content and profiles of main glucosinolate compounds not dominated by any specific components [9]. The other group that includes the 3 species with 8 chromosomes (*D. ibicensis*, *D. brevisiliqua*, and *D. ilorcitana*) has confirmed previous investigations using RAPD markers [11], which placed *D. brevisiliqua* and *D. ilorcitana* in the same subcluster, whereas *D. ibicensis* was on another branch. These accessions were singled out by D'Antuono et al. [9] for their medium-high overall glucosinolate content, which was particularly high in sinigrin (allyl-glucosinolate). Beyond these two groupings, past investigations did not adequately resolve the remaining species under investigation. Four species with nine chromosomes, including *D. brachycarpa*, *D. virgata*, *D. assurgens*, and *D. berthautii*, were found in a primary cluster with *D. eruicoides*, whereas *D. tenuisiliqua* was on a different branch. *D. virgata* and *D. brachycarpa* were instead separated from *D. assurgens*, which was combined with *D. tenuisiliqua* in a study using RAPD markers [11]. Additionally, *D. eruicoides* was clustered on a separate branch in disaccord with the lineage-based classification [12], which grouped this species with taxa having 11 chromosomes. Considering biochemical profiles, *D. berthautii*, *D. tenuisiliqua*, and *D. virgata* were characterized by a high content of sinigrin, with *D. virgata* being also rich in gluconapin [9]. Based on sequence variation, *D. siifolia* was the most distant species from the rest, being positioned on a separated branch. This trend was confirmed by population structure analysis. However, the high-resolution melting-based dendrogram partially agreed with both the Rhynchocarpum classification proposed by Gomez-Campo and Martínez-Laborde [8] as well as the findings of Eschmann-Grupe and collaborators [11], which included this species in a cluster with *D. tenuisiliqua* and *D. assurgens*. D'Antuono et al. [9] indeed highlighted the very high glucosinolate content of *D. siifolia* in particular for the glucobrassicin compound. In agreement with the metabolic profile and considering the leaf morphology diversity compared to the other *Diplotaxis* species, barcode DNA fingerprinting supports the high diversity of *D. siifolia* compared to the rest.

Both DNA barcode and HRM analysis highlighted the similar relationships between species with the same chromosomal number, thus providing insight into the possibility of successful interspecific crosses. Beyond taxonomical investigation, HRM provides a reliable and cost-effective method for rapid identification of *Diplotaxis* species. The discrimination is crucial when the constituent species are in processed form and/or in mixed packages where alien product detection is difficult. Barcoding through HRM enables the detection of adulterants in admixed samples at very low concentrations, since it relies on sensitive melting curve changes through the release of a saturated intercalating dye from DNA duplex denaturation following the raising of the temperature [46]. In the present study, ITS1 revealed a greater rate of percent variation compared to the other barcode regions, thus allowing for distinguishing a higher number of species. These results agree with the larger discrimination power at the low taxonomic level of the nuclear internal transcribed region compared to the plastid regions [15,47]. Raising of availability of genomic resources will benefit the discovery of novel SNP across the genome, boosting the development of suitable HRM markers for genetic fingerprinting.

5. Conclusions

This study represents the first attempt to investigate *Diplotaxis* species with newly developed barcode markers which combine DNA sequencing and high-resolution melting analysis. The phylogenetic relationships among the 17 assayed species confirmed previous inferences from morphological, biochemical, and molecular data, thus indicating the reliability of the marker tested. Results better resolved the evolutionary distance of *D. siifolia* within the *Diplotaxis* gene pool. ITS1 has been found as the barcoding sequence with the highest discriminatory power for species identification. The DNA barcode and HRM confirm the effectiveness of the strategy utilized in this work for identifying species and examining genetic relationships.

Supplementary Materials: The following supporting information can be downloaded at https://www.mdpi.com/article/10.3390/genes14081594/s1, Table S1: Raw sequences after trimming for the 17 *Diplotaxis* species with 5 universal barcode markers.

Funding: This research was funded by the Italian Ministry of University and Research (MUR), project "Conservabilità, qualità e sicurezza dei prodotti ortofrutticoli ad alto contenuto di servizio—ARS01_00640—POFACS", D.D. 1211/2020 and 1104/2021.

Institutional Review Board Statement: Not applicable.

Informed Consent Statement: Not applicable.

Data Availability Statement: All data generated in this study are available as Supplementary Materials.

Conflicts of Interest: There are no conflicts of interest to disclose. The funders had no role in the design of the study; in the collection, analyses, or interpretation of data; in the writing of the manuscript; or in the decision to publish the results.

References

1. Warwick, S.I. Brassicaceae in Agriculture. In *Genetics and Genomics of the Brassicaceae*; Warwick, S.I., Ed.; Springer: New York, NY, USA, 2010; pp. 33–65.
2. Tripodi, P.; Coelho, P.S.; Guijarro-Real, C. Breeding advances and prospects in rocket salad (*Eruca vesicaria* ssp. *sativa* Mill.) cultivation. In *Advances in Plant Breeding Strategies: Vegetable Crops*; Springer International Publishing: Cham, Switzerland, 2021; pp. 95–133.
3. Gómez-Campo, C. Taxonomy. In *Biology of Brassica Coenospecies*; Elsevier: Amsterdam, The Netherlands, 1999; pp. 3–32.
4. Martín, J.P.; Sánchez-Yélamo, M.D. Genetic relationships among species of the genus *Diplotaxis* (Brassicaceae) using inter-simple sequence repeat markers. *Theor. Appl. Genet.* **2000**, *101*, 1234–1241. [CrossRef]
5. Harberd, D.J.; McArthur, E.D. The chromosome constitution of *Diplotaxis muralis* (L.) DC. *Watsonia* **1972**, *9*, 131–135.
6. Tripodi, P.; Francese, G.; Mennella, G. Rocket Salad: Crop Description, Bioactive Compounds and Breeding Perspectives. *Adv. Hortic. Sci.* **2017**, *31*, 107–113.
7. Gómez-Campo, C. Morphology and morphotaxonomy of the Tribe Brassiceae. In *Brassica Crops and Wild Allies*; Tsunoda, S., Hinata, K., Gómez-Campo, C., Eds.; Japan Scientific Societies Press: Tokyo, Japan, 1980; pp. 3–31.
8. Gómez-Campo, C.; Martínez-Laborde, J.B. Reajustes taxonómicos y nomenclaturales en la tribu Brassiceae (Cruciferae). *An. Jardín Bot. Madrid* **1998**, *56*, 379–381.
9. D'antuono, L.F.; Elementi, S.; Neri, R. Glucosinolates in *Diplotaxis* and *Eruca* leaves: Diversity, taxonomic relations and applied aspects. *Phytochemistry* **2008**, *69*, 187–199. [CrossRef] [PubMed]
10. Sánchez-Yélamo, M.D. A chemosystematic survey of flavonoids in the *Brassicinae*: *Diplotaxis*. *Bot. J. Linn. Soc.* **1994**, *115*, 9–18. [CrossRef]
11. Eschmann-Grupe, G.; Hurka, H.; Neuffer, B. Species relationships within *Diplotaxis* (Brassicaceae) and the phylogenetic origin of *D. muralis*. *Plant. Syst. Evol.* **2004**, *243*, 13–29. [CrossRef]
12. Warwick, S.I.; Black, L.D.; Aguinagalde, I. Molecular systematics of Brassica and allied genera (subtribe *Brassicinae*, Brassiceae)—Chloroplast DNA variation in the genus *Diplotaxis*. *Theorl Appl. Genet.* **1992**, *83*, 839–850. [CrossRef]
13. Warwick, S.I.; Sauder, C.A. Phylogeny of tribe Brassiceae (Brassicaceae) based on chloroplast restriction site polymorphisms and nuclear ribosomal internal transcribed spacer and chloroplast trnL intron sequences. *Can. J. Bot.* **2005**, *83*, 467–483.

14. Yan, M.; Xiong, Y.; Liu, R.; Deng, M.; Song, J. The application and limitation of universal chloroplast markers in discriminating east Asian evergreen oaks. *Front. Plant. Sci.* **2018**, *9*, 569. [CrossRef]
15. Hollingsworth, P.M.; Graham, S.W.; Little, D.P. Choosing and Using a Plant DNA Barcode. *PLoS ONE* **2011**, *6*, e19254. [CrossRef]
16. Dunning, L.T.; Savolainen, V. Broad-scale amplification of *matK* for DNA barcoding plants, a technical note. *Bot. J. Linn. Soc.* **2010**, *164*, 1–9. [CrossRef]
17. Bafeel, S.O.; Arif, I.A.; Bakir, M.A.; Al Homaidan, A.A.; Al Farhan, A.H.; Khan, H.A. DNA barcoding of arid wild plants using rbcL gene sequences. *Genet. Mol. Res.* **2012**, *11*, 1934–1941. [CrossRef] [PubMed]
18. China Plant BOL Group; Li, D.Z.; Gao, L.M.; Li, H.T.; Wang, H.; Ge, X.J.; Liu, J.Q.; Chen, Z.D.; Zhou, S.L.; Chen, S.L.; et al. Comparative analysis of a large dataset indicates that internal transcribed spacer (ITS) should be incorporated into the core barcode for seed plants. *Proc. Natl. Acad. Sci. USA* **2011**, *108*, 19641–19646. [PubMed]
19. Hollingsworth, P.M.; Forrest, L.L.; Spouge, J.L.; Hajibabaei, M.; Ratnasingham, S.; van der Bank, M.; Chase, M.W.; Cowan, R.S.; Erickson, D.L.; Fazekas, A.J.; et al. A DNA barcode for land plants. *Proc. Natl. Acad. Sci. USA* **2009**, *106*, 12794–12797.
20. Kress, W.J. Plant DNA Barcodes: Applications Today and in the Future. *Front. Plant. Syst. Evol.* **2017**, *55*, 291–307. [CrossRef]
21. Müller, K.F.; Borsch, T.; Hilu, K.W. Phylogenetic utility of rapidly evolving DNA at high taxonomical levels: Contrasting matK, trnT-F, and rbcL in basal angiosperms. *Mol. Phylogen. Evol.* **2006**, *41*, 99–117. [CrossRef]
22. Howard, C.; Lockie-Williams, C.; Slater, A. Applied Barcoding: The Practicalities of DNA Testing for Herbals. *Plants* **2020**, *9*, 1150. [CrossRef] [PubMed]
23. Simko, I. High-resolution DNA melting analysis in plant research. *Trends Plant. Sci.* **2016**, *21*, 528–537. [CrossRef]
24. Jaakola, L.; Suokas, M.; Häggman, H. Novel Approaches Based on DNA Barcoding and High-Resolution Melting of Amplicons for Authenticity Analyses of Berry Species. *Food Chem.* **2010**, *123*, 494–500. [CrossRef]
25. Chen, W.; Chen, X.; Xu, J.; Cai, J.; Wang, X. Identification of Dendrobium officinale using DNA barcoding method combined with HRM and qPCR technology. *Food Anal. Meth.* **2022**, *15*, 1310–1320. [CrossRef]
26. Song, M.; Li, J.; Xiong, C.; Liu, H.; Liang, J. Applying high-resolution melting (HRM) technology to identify five commonly used Artemisia species. *Sci. Rep.* **2016**, *6*, 34133. [CrossRef] [PubMed]
27. Sun, W.; Li, J.; Xiong, C.; Zhao, B.; Chen, S. The Potential Power of Bar-HRM Technology in Herbal Medicine Identification. *Front. Plant. Sci.* **2016**, *7*, 367. [CrossRef] [PubMed]
28. Yu, J.; Wu, X.; Liu, C.; Newmaster, S.; Ragupathy, S.; Kress, W.J. Progress in the Use of DNA Barcodes in the Identification and Classification of Medicinal Plants. *Ecotoxicol. Environ. Saf.* **2021**, *208*, 111691. [CrossRef]
29. Qiao, J.; Zhang, X.; Chen, B.; Huang, F.; Xu, K.; Huang, Q.; Huang, Y.; Hu, Q.; Wu, X. Comparison of the cytoplastic genomes by resequencing: Insights into the genetic diversity and the phylogeny of the agriculturally important genus Brassica. *BMC Genom.* **2020**, *21*, 480. [CrossRef]
30. Kumar, S.; Stecher, G.; Li, M.; Knyaz, C.; Tamura, K. MEGA X: Molecular Evolutionary Genetics Analysis across Computing Platforms. *Mol. Biol. Evol.* **2018**, *35*, 1547–1549. [CrossRef]
31. Falush, D.; Stephens, M.; Pritchard, J.K. Inference of population structure using multilocus genotype data: Dominant markers and null alleles. *Mol. Ecol. Not.* **2007**, *7*, 574–578. [CrossRef]
32. Earl, D.A.; VonHoldt, B.M. STRUCTURE HARVESTER: A website and program for visualizing STRUCTURE output and implementing the Evanno method. *Cons. Genet. Res.* **2012**, *4*, 359–361. [CrossRef]
33. Wickham, H. Programming with Ggplot2. In *ggplot2*; Springer: Berling/Heidelberg, Germany, 2016; pp. 241–253.
34. Adhikari, S.; Saha, S.; Biswas, A.; Rana, T.S.; Bandyopadhyay, T.K.; Ghosh, P. Application of molecular markers in plant genome analysis: A review. *Nucleus* **2017**, *60*, 283–297. [CrossRef]
35. Mehmood, F.; Abdullah; Ubaid, Z.; Bao, Y.; Poczai, P.; Mirza, B. Comparative plastomics of Ashwagandha (Withania, Solanaceae) and identification of mutational hotspots for barcoding medicinal plant. *Plants* **2020**, *9*, 752. [CrossRef]
36. Hellberg, R.S.; Isaacs, R.B.; Hernandez, E.L. Identification of shark species in commercial products using DNA barcoding. *Fish. Res.* **2019**, *210*, 81–88. [CrossRef]
37. Wurzbacher, C.; Larsson, E.; Bengtsson-Palme, J.; Van den Wyngaert, S.; Svantesson, S.; Kristiansson, E.; Kagami, M.; Nilsson, R.H. Introducing ribosomal tandem repeat barcoding for fungi. *Mol. Ecol. Resour.* **2019**, *19*, 118–127. [CrossRef]
38. Choudhary, P.; Singh, B.N.; Chakdar, H.; Saxena, A.K. DNA Barcoding of Phytopathogens for Disease Diagnostics and Bio-Surveillance. *World J. Microbiol. Biotechnol.* **2021**, *37*, 54. [CrossRef]
39. Ongchai, S.; Chokchaitaweesuk, C.; Kongdang, P.; Chomdej, S.; Buddhachat, K. In vitro chondroprotective potential of *Senna alata* and *Senna tora* in porcine cartilage explants and their species differentiation by DNA barcoding-high resolution melting (Bar-HRM) analysis. *PLoS ONE* **2019**, *14*, e0215664. [CrossRef]
40. Thongkhao, K.; Tungphatthong, C.; Phadungcharoen, T.; Sukrong, S. The use of plant DNA barcoding coupled with HRM analysis to differentiate edible vegetables from poisonous plants for food safety. *Food Control* **2020**, *109*, 106896. [CrossRef]
41. Prantl, K. Diplotaxis. In *Die Naturlichen Pflanzenfamilien*; Wilhelm Engelmann: Leipzig, Germany, 1915.
42. Schulz, O.E. Diplotaxis. In *Das Pflanzenreich IV*; Engler, A., Ed.; Wilhelm Engelmann: Leipzig, Germany, 1919; Volume 105, pp. 149–179.
43. Martínez-Laborde, J.B. Notes on the taxonomy of Diplotaxis DC. (Brassiceae). *Bot. J. Linn. Soc.* **1991**, *106*, 67–71. [CrossRef]
44. Takahata, Y.; Hinata, K. Studies on cytodemes in subtribe *Brassicinae* (Cruciferae). *Tohoku J. Agric. Res.* **1983**, *33*, 111–124.

45. Mummenhoff, K.; Eschmann-Grupe, G.; Zunk, K. Subunit polypeptide composition of Rubisco indicates *Diplotaxis viminea* as maternal parent species of amphiploid *Diplotaxis muralis*. *Phytochemistry* **1993**, *34*, 429–431. [CrossRef]
46. Mader, E.E.; Ruzicka, J.; Schmiderer, C.; Novak, J. Quantitative high-resolution melting analysis for detecting adulterations. *Anal. Biochem.* **2011**, *409*, 153–155. [CrossRef]
47. Jiang, K.W.; Zhang, R.; Zhang, Z.F.; Pan, B.; Tian, B. DNA barcoding and molecular phylogeny of Dumasia (Fabaceae: Phaseoleae) reveals a cryptic lineage. *Plant. Divers.* **2020**, *42*, 376–385. [CrossRef]

genes

MDPI

Review

On the Future Perspectives of Some Medicinal Plants within *Lamiaceae* Botanic Family Regarding Their Comprehensive Properties and Resistance against Biotic and Abiotic Stresses

Dan Ioan Avasiloaiei [1], Mariana Calara [1], Petre Marian Brezeanu [1,*], Otilia Cristina Murariu [2] and Creola Brezeanu [1,*]

1 Vegetable Research and Development Station, 600388 Bacău, Romania; avasiloaiei_dan_ioan@yahoo.com (D.I.A.); calaramariana@gmail.com (M.C.)
2 Department of Food Technology, Iasi University of Life Sciences (IULS), 700490 Iasi, Romania; otiliamurariu@uaiasi.ro
* Correspondence: brezeanumarian@legumebac.ro (P.M.B.); creola.brezeanu@yahoo.com (C.B.)

Abstract: *Lamiaceae* is one of the largest botanical families, encompassing over 6000 species that include a variety of aromatic and medicinal spices. The current study is focused on three plants within this botanical family: basil (*Ocimum basilicum* L.), thyme (*Thymus vulgaris* L.), and summer savory (*Satureja hortensis* L.). These three species contain primary and secondary metabolites such as phenolic and flavonoid compounds, fatty acids, antioxidants, and essential oils and have traditionally been used for flavoring, food preservation, and medicinal purposes. The goal of this study is to provide an overview of the nutraceutical, therapeutic, antioxidant, and antibacterial key features of these three aromatics to explore new breeding challenges and opportunities for varietal development. In this context, a literature search has been performed to describe the phytochemical profile of both primary and secondary metabolites and their pharmacological uses, as well as to further explore accession availability in the medicine industry and also to emphasize their bioactive roles in plant ecology and biotic and abiotic stress adaptability. The aim of this review is to explore future perspectives on the development of new, highly valuable basil, summer savory, and thyme cultivars. The findings of the current review emphasize the importance of identifying the key compounds and genes involved in stress resistance that can also provide valuable insights for further improvement of these important medicinal plants.

Keywords: aromatics; nutrient content; essential oils; salinity stress; drought stress; breeding prospects

Citation: Avasiloaiei, D.I.; Calara, M.; Brezeanu, P.M.; Murariu, O.C.; Brezeanu, C. On the Future Perspectives of Some Medicinal Plants within *Lamiaceae* Botanic Family Regarding Their Comprehensive Properties and Resistance against Biotic and Abiotic Stresses. *Genes* **2023**, *14*, 955. https://doi.org/10.3390/genes14050955

Academic Editors: Qingyi Yu, Wajid Zaman and Hakim Manghwar

Received: 17 March 2023
Revised: 7 April 2023
Accepted: 20 April 2023
Published: 22 April 2023

1. Introduction

The economic importance of medicinal and aromatic plants within the context of agro-alimentary, pharmaceutical, natural cosmetics, and perfume development uses is of paramount significance. In addition to providing food flavoring and pleasing aromas, their secondary metabolites and antioxidants provide additional nutritional value and make them an invaluable part of the human diet, alongside cereals, fruits, and vegetables [1,2].

The *Lamiaceae* botanical family encompasses around 236 genera and over 6000 species of herbs and shrubs that have a global distribution. Within this family, basil (*O. basilicum* L.), thyme (*T. vulgaris* L.), and summer savory (*S. hortensis* L.) are species of particular importance, both for their specialized metabolites, such as essential oils and various non-volatile constituents with multiple applications in the food industry, cosmetics, and medicine, and their amazing adaptability to a whole range of biotic and abiotic stress factors. They are highly valued for their nutritional, medicinal, and industrial properties, providing flavor and fragrance to food, promoting health and well-being in traditional medicine due to the anti-inflammatory, antiseptic, and analgesic properties of their essential oils, and serving as valuable ingredients in various industries (perfumes, cosmetics, and cleaning agents) [1,2].

With regards to the main abiotic stress factors, the climate change evidence from the past decade indicates that salinity and drought will likely be the highest causes of significant concern. Salinity has emerged as a major environmental factor that has impacted over 20% of cultivated land worldwide, with the affected regions continuing to expand every year [3,4]. In fact, there is a significant risk that salinization may affect more than 50% of arable land by the mid-21st century [5], posing a serious challenge to global food security. Salinity can have a negative impact on *Lamiaceae* aromatic plants, affecting their growth, yield, and quality. High levels of salt in the soil decrease water availability and impose osmotic stress conditions on plants, leading to reduced photosynthesis and yield limitation [3]. In response to such conditions, the uptake and transport of the major essential ions (such as Ca, K, Mg, and nitrate) is transiently decreasing, and as a consequence, it impacts the amount and quality of secondary metabolites, such as essential oils. There is a significant increase in the reactive oxygen species (ROS) production under osmotic stress conditions that can exceed the scavenging ability of the plants. The accumulation of ROS has harmful cellular effects, such as DNA damage, membrane lipid peroxidation, and enzymatic activity impairment.

Drought stress is another significant environmental concern, representing a serious threat to agriculture worldwide by reducing the yield and quality for principal cash crops. It may have a negative impact on *Lamiaceae* aromatic plants as well affecting their growth, yield, and quality. However, most of these plants possess the ability to respond to drought stress at morphological, anatomical, physiological, biochemical, and molecular levels with a series of adjustments, allowing the plant to avoid the stress or to increase its tolerance [6,7].

2. Methods

Plenty of scientific literature discusses the benefits of these three aromatic species, and there is an abundance of data regarding their adaptation to key stressors and breeding perspectives. As a result, we searched SCOPUS and the Google Academic database for topics relating to "*O. basilicum*", "*T. vulgaris*", and "*S. hortensis*" and performed an extensive keyword search for "(Aromatic plant species name) composition and properties" and "(Aromatic plant species name) (a) biotic stress resistance." Our search generated around 350 documents published over a period of 35 years (1989–2022). The returned results highlighted a multitude of studies from countries with a long-standing traditional cultivation of these plants, especially from the Middle and Far East, focusing on a variety of complex topics.

This review aims to present the progress that has been made in studying the resistance potential of these three species on a series of stressors and to emphasize the need to focus on breeding and developing new cultivars with higher resistance capacity.

3. Discussion

3.1. Aromatic Plant Composition and Accessions Availability

A synthetic characterization of the main biocomponents of the three species studied, including the volatile oils, is presented in Table 1.

Table 1. Main biocomponent profile of some *Lamiaceae* family aromatic plants presented by original research papers.

Aromatic Plant Species	Chemical Composition	Fresh Leaves	References
		Volatile Oil	
O. basilicum L.		Dry matter (909.1 g kg^{-1}), cude ash (89.84 g kg^{-1}), crude protein (208.8 g kg^{-1}), ether extract (11.21 g kg^{-1}), crude fiber (45.91 g kg^{-1}), NFI (sugars readily hydrolyzed) (553.3 g kg^{-1}), Mg (79.8 μg g^{-1}), Ca (1278 μg g^{-1}), K (2135 μg g^{-1}), Na (218.5 μg g^{-1}), Fe (26.31 μg g^{-1}), Cu (1.95 μg g^{-1}), Mn (8.56 μg g^{-1}) and Zn (45.14 μg g^{-1}) Alkaloids, tannins, flavonoids, cholesterol, terpernoids, glycosides, cardiac glycosides, phenols, carbohydrates, and phlobatannins	[8,9]
		(~6.20 mg/g) Linalool (56.7–60.6%), *epi*-α-cadinol (8.6–11.4%), α-bergamotene (7.4–9.2%) and γ-cadinene (3.2–5.4%, germacrene D (1.13.3%), camphor (1.13.1%)	[10,11]
T. vulgaris L.		Oxygen terpene derivatives (1,8-cineole, linalool, followed by camphor, endo-borneol, α-terpineol and linalyl acetate), terpene hydrocarbons (α-pinene, camphene and β–pinene, trans-caryophylle, four flavonoids (two flavanones and two flavones)—sakuranetin, 6,7-dimethylcarthamidin, respectively 5-desmethylsinensetin and -hydroxy-3,7,8,2′,4′-pentamethoxy-flavone	[12]
		(12 mL/kg ≤) Thymol (~47.59%), γ-Terpinene (~30.90%), para-Cymene (~8.41%), Carene<δ-2-> (~3.76%), Caryophyllene (2.68%), α-Thujene, α-Pinene, β-Pinene, β-Myrcene, α-Phellandrene, D-Limonene, β-Phellandrene, Terpineol, Terpinen-4-ol, Cyclohexene, 1-methyl-4-(5-methyl-1-methylene-4-hexenyl)	[13]
S. hortensis L.		Moisture (72%), protein (4.2%), fat (1.65%), sugar (4.45%), fibre (8.60%), ash (2.11%) Minerals: K (1.68–3.38 mg·kg^{-1} DM),P (0.31–0.72 mg·kg^{-1} DM), Ca (1.08–2.84 mg·kg^{-1} DM), Mg (0.25–0.61 mg·kg^{-1} DM), Fe (242–726 mg·kg^{-1} DM), and Na (0.007–0.013 mg·kg^{-1} DM)	[14,15] [16]
		(≥ 5%) Carvacrol (11–67%, Thymol (0.3–28.2%), γ-terpinene (15.30–39%), p-cymene (3.5–19.6%), α-phellandrene, α- and β-pinene, Sabinene, terpineol, α-thujene	[15,17], [18–22]

3.2. *O. basilicum L. Row Plant and Essential Oil Composition and Accession Availability*

Ocimum is one of the largest genera in the *Lamiaceae* family, which consists of 65 species native to Africa, South America, and Asia [23]. Among the species, sweet basil (*O. basilicum* Linn.), which originated from the warm tropical climates of India, Africa, and southern Asia, is probably the most important crop, being cultivated as a culinary herb worldwide under various ecological circumstances [24]. Both the raw plant and essential oil of basil have marked culinary, pharmaceutical, and cosmetic purposes [23]. It is used frequently in traditional medicine, having antispasmodic, stomachic, carminative, anti-ulcerogenic, anti-inflammatory, anti-carcinogenic, analgesic, stimulant radioprotective, and febrifuge properties [25]. Basil's leafy components exhibit antimicrobial properties and antioxidant activity [26,27] that can be used for alleviating pain and otitis [28]. The essential oil extracted from European genotype basil is recognized for its superior aroma, primarily comprising linalool and methyl chavicol.

Due to their vast diversity, various plant species and cultivars exhibit unique levels of resistance and ability to withstand physiological functions and produce yields in varying environmental conditions and stressful situations.

Furthermore, at the European level, based on the EURSICO National Inventory Report Taxonomy (ipk-gatersleben.de), a total number of 834 *O. basilicum* L. accessions are available for multiplication and future breeding perspectives (Figure 1). The countries that have uploaded the largest number of accessions are Germany (268 acc.), Croatia (119), the Czech Republic (64), and Romania (38).

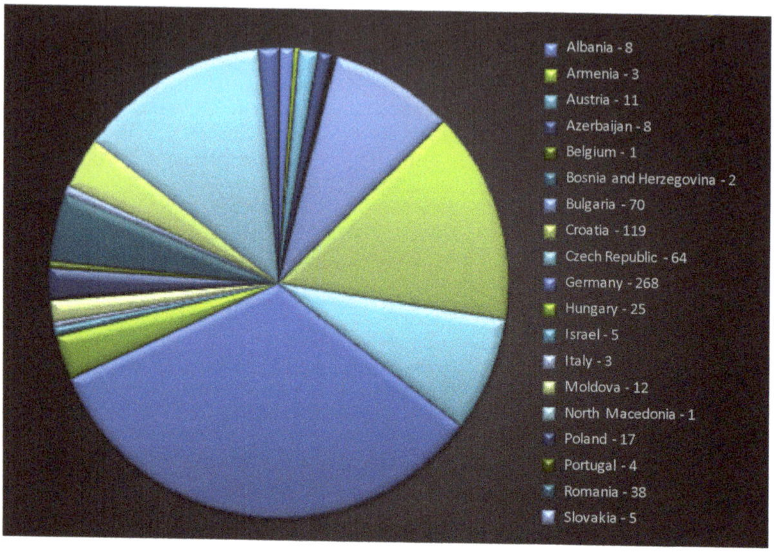

Figure 1. *O. basilicum* L. accession availability at the European level (https://eurisco.ipk-gatersleben.de/apex/eurisco_ws/r/eurisco/taxon-search-results?p26_genus=OCIMUM&p26_species=BASILICUM, accessed on 14 February 2023).

3.3. *T. vulgaris* L. Composition and Accession Availability

The *Thymus* genera of the *Lamiaceae* family is highly significant due to its large number of species [29]. Thyme (*T. vulgaris* L.), commonly known as garden thyme or common thyme, appears in many different areas worldwide, including the drier Mediterranean regions [30], and is the most commercially cultivated species in the genus *Thymus* [31] due to the dietary trends in the recent decades [32]. It has several aromatic and medicinal properties. The leaves can be used either fresh or dried as a flavoring component in various culinary preparations, containing a high ratio of minerals (K, Ca, Mg, Fe, Mn, and Se), antioxidants (flavonoids, phenolic compounds such as pigenin, naringenin, luteolin, thymonin, lutein, and zeaxanthin), and vitamins (A, B6, B9, C, E, and K) [7,33,34]. In addition, the essential oil of common thyme is extracted by the distillation of the fresh leaves and flowering tops, and it contains 20–54% thymol, which is a monoterpene known as the main active ingredient with a wide range of pharmacological properties [7,35,36]. It has demonstrated anti-pathogenic and antioxidant effects, being intensively utilized in various industries, particularly in medication, agriculture, and food production.

At the European level, the total number of *T. vulgaris* L. accessions is 202, with Spain (77), Albania (51), Poland (15), and Ukraine (14) having uploaded the highest numbers of accessions to the EURISCO database (Figure 2).

3.4. *S. hortensis* L. Composition and Accession Availability

Summer savory (*S. hortensis* L.) is an annual herbaceous plant that is among the principal *Satureja* species grown in southern Europe, as well as central and southwestern Asia [37]. It contains numerous vitamins, including B-complex vitamins, vitamin A, vitamin C, niacin, thiamine, and pyridoxine, as well as carvacrol, terpinene, cymene, and caryophyllene, which make it an excellent choice for medicinal purposes [19,38]. The aerial parts of the *Satureja* species, such as *S. hortensis*, contain essential oil that is widely used in the medicine, food, and health industries for therapeutic purposes. Scientific studies have highlighted several pharmacological properties of *Satureja*, such as antispasmodic, antioxidant, antimicrobial, antidiarrheal, and sedative properties [37,39–41]. Its beneficial effects on hypertension have also been discussed [42].

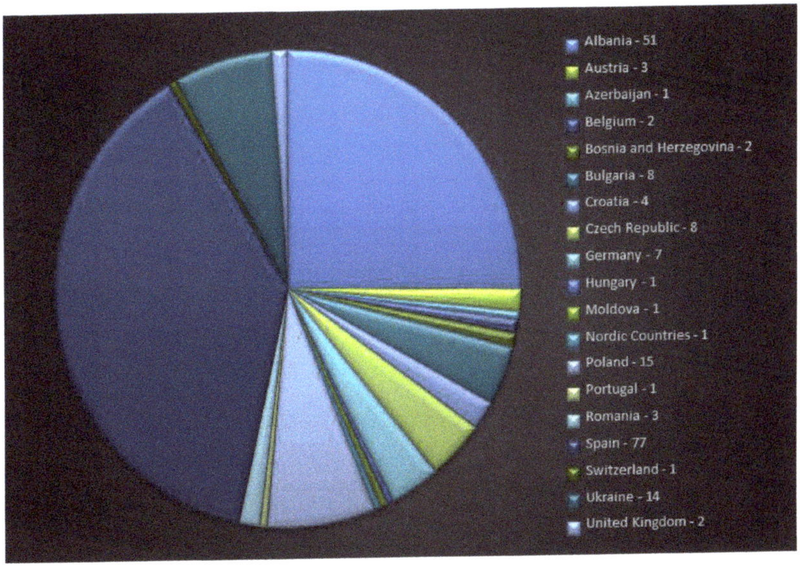

Figure 2. *T. vulgaris* L. accession availability at the European level (https://eurisco.ipk-gatersleben.de/apex/eurisco_ws/r/eurisco/taxon-search-results?p26_genus=THYMUS&p26_species=VULGARIS, accessed on 14 February 2023).

The savory essential oil contains two major compounds, thyme and carvacrol, that have antiseptic, antifungal, and antibacterial properties [3]. The concentration and composition of secondary metabolites in savory oils, such as y-terpinene, p-cymene, carvacrol methyl ether, and caryophyllene, play significant ecological roles as they possess insecticidal, antifungal, and antibacterial properties [43–45].

At the European level, the total number of *S. hortensis* L. accessions is 269, with Bulgaria (56), Romania (54), Germany (38), and Hungary (32) being the countries with the highest number of accessions uploaded to the EURISCO database (Figure 3).

3.5. Aromatic Plant Biological Activities and Stress Resistance

Various elements, such as genetic and ecological factors, have a notable impact on the chemical components of medicinal plants and their physiological and morphological characteristics [46]. Diverse environmental circumstances can be the primary cause of variability in morphological characteristics, which may induce alterations in the phenotype in the short term and in its genotype in the long term [47]. The presence of abiotic environmental stressors, such as salinity and drought, can inhibit the plant growth and development [48,49].

Essential oils perform a crucial function in plant defense by serving as antiviral, antibacterial, antimycotic, and insecticidal agents and deterring herbivores [50]. Due to these properties, diverse plant essential oils may serve as remedial or auxiliary agents in the pharmaceutical sector [51] and function as fragrances, seasonings, and natural preservatives in the food industry [52]. Finally, these oils may also serve as eco-friendly and biodegradable substances for protecting plants in agriculture [53].

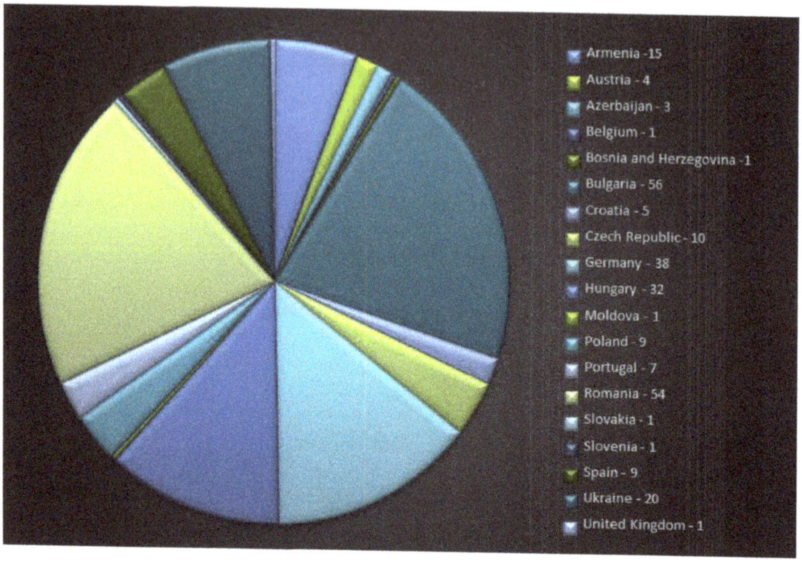

Figure 3. *S. hortensis* L. accession availability at the European level (https://eurisco.ipk-gatersleben.de/apex/eurisco_ws/r/eurisco/taxon-search-results?p26_genus=SATUREJA&p26_species=HORTENSIS accessed on 14 February 2023).

3.6. *O. basilicum L.—Biological Activities and Stress Resistance*

Basil is regarded as a functional food plant due to its abundant secondary metabolites and antioxidant traits that are believed to enable oxidative stress disease prevention. The constituents of basil's essential oil (EO) are known to differ significantly based on the genetic factors (cultivar, origin, season, chemotype, and phonological stage), environmental factors (climatic conditions, agricultural practices, and postharvest processes), and the combined effects of these factors [4,54]. Oxygenated monoterpenes and phenylpropanoids are the primary chemical compounds present in the *Ocimum* genus, while linalool, eugenol, methyl chavicol, methyl cinnamate, methyl eugenol, and geraniol are some of the significant constituents identified in various *O. basilicum* cultivars and chemotypes [55]. These constituents act as potent antioxidants by scavenging free radicals and functioning as electron donors, making them effective in safeguarding the plants against pathogens and predators. At the cellular level, they can also protect cells from the adverse effects of ROS arising from different abiotic stressors [56].

Plants that have a high concentration of antioxidants can resist damage caused by ROS. These antioxidants can also function as protective substances [57,58]. The existence of phenolic compounds relies on various factors, such as the type of soil, plant species, genetics, growth stage, and location [59]. The primary phenolic compounds responsible for the antioxidant effects in basil are caffeic acid (CA) and rosmarinic acid (RA), which is an ester of CA. These compounds are mostly produced in roots and leaves [60]. CA is an active participant in plant physiology and stress tolerance mechanisms [56]. RA has been utilized as an anti-inflammatory, anti-proliferative, and chemoprotective agent [61]. The synthesis of RA in plants occurs through the phenylpropanoid and tyrosine-derived pathways. Phenylalanine ammonia-lyase (PAL) is the principal enzyme in the phenylpropanoid pathway, catalyzing the transformation of l-phenylalanine to trans-cinnamic acid and ammonia. Research has confirmed that PAL plays a role in RA biosynthesis [62]. An increase in PAL activity induced by stress could be the starting point for the cells to adapt to drought conditions [63]. Phosphorus (P) is an essential element to produce secondary metabolites in plants, and its availability can affect the amount and composition of phenolic compounds. A competition exists between the production of phenolic compounds and the required proteins for growth.

Regarding the response of *O. basilicum* L. to different stressors, Table 2 presents the main anatomical, physiological, and molecular changes, as well as some contributing factors that could occur.

Table 2. *O. basilicum* L. response to different stressors.

Type of Stress		Anatomical, Physiological, and Molecular Changes	Contributing Factors	References
Abiotic	Salinity	Higher Na^+ concentrations	Increased MDA accumulation	[64]
			Enhanced proline content	[4]
		Photosynthetic pigments decrease	Chlorophyllase enzyme activity enhancement	[65]
		Induces essential oil production	Higher oil gland density	[4,65,66]
	Drought	Plant growth process is inhibited	Constrained cell elongation and differentiation	[67,68]
		Disruption of main metabolic processes	Chlorophyll reduction	[69]
			Photosynthesis inhibation	[69]
			Cell division suppresion	[69]
		Protein complexes imbalance	Chlorophyll a and b depletion	[70,71]
		Chlorophyllase activity enhancement		
		Photosynthesis inhibition	Stomatal blockage	[72]
			RubisCO enzyme activity cut	[73]
		Cell osmotic adjustment	Proline accumulation	[69]
			CO_2 assimilation	[74]
Biotic	Twospotted spider mite (*Tetranychus urticae* Koch)	Small chlorotic spots	Lower concentrations of nitrogen, phosphorous, and protein	[75]
		Cell physiology disruption	Photosynthesis reduction and phytotoxic compounds injection	[75,76]
		Plasma membrane potential change	Cytosolic free Ca^{2+} changes	[75]
		Oxidative damage	Increase in cellular concentration of reactive oxygen species and, subsequently, of H_2O_2	[75]

Salinity stress has various negative impacts on plants, including reduced growth and water content [77]. Additionally, salinity stress results in reduced soil moisture content and limited water absorption from the soil, leading to osmotic stress in plants [66]. However, the osmotic potential significantly improves. Moreover, in many plants, salt stress increases the levels of cell free radicals to a point where it can damage the membrane, further intensifying the effects of the stress. The sensitivity of plants to salt stress can be measured by malondialdehyde (MDA), which is a widely used parameter for estimating lipid peroxidation in plant tissue that increases under oxidative stress. At the cellular level, osmotic stress modifies the properties and composition of the membrane lipids. MDA accumulation increased in the severity of NaCl stress in the leaves of summer savory (*S. hortensis* L.) and *O. basilicum* L. The MDA content serves as an indicator of oxidative stress resulting from membrane lipid peroxidation, and it can be reduced by lowering lipid peroxidation and increasing the activity of antioxidant enzymes in salt-affected plants [64].

The composition and amount of essential oils can be influenced by environmental factors, including salinity stress, which can have a negative impact on plant growth and osmotic balance [78]. Under salinity stress, proline plays a crucial role in osmotic adjustment, helping to maintain the osmotic balance of the plant and enhance its tolerance to salt stress [79]. Proline also acts as an ROS scavenger, protein stabilizer, and osmoprotectant [80]. The accumulation of proline during drought stress has been shown to be related to improved plant performance, probably due to its antioxidant properties and its ability to stabilize macromolecules [81,82].

Regarding the drought stress, the accumulation of secondary metabolites is a defensive mechanism employed by plants to cope with stress by altering their cellular metabolism to overcome various challenges [83].

The concentration of leaf chlorophyll (Ch) is a crucial physiological characteristic that directly impacts a plant's photosynthetic ability. Both chlorophyll-a and chlorophyll-b levels in sweet basil plants decreased when they were deprived of water. The amount of chlorophyll in leaves, which is an indicator of plant vigor, is influenced by a variety of environmental factors. The reduction in chlorophyll content due to the water scarcity could be linked to the generation of ROS in cells [84], which have negative effects on plants in stressful situations. The plant water status is an indicator of their response to water scarcity, with higher relative water contents indicating a healthy plant condition [85]. In experiments involving various water supplies, the accurate evaluation of the plant water status is crucial. The water content of plants can be expressed per unit of fresh or dry weight (or, less commonly, per unit of leaf area) [86]. Particularly, the leaf relative water content (RWC) is utilized as a dependable measure of a plant's susceptibility to dehydration [87]. To reduce water loss through the leaves, the transpiration rate can be adjusted, and the leaf area can be limited. Typically, plants restrict water loss through their foliage by closing stomata, which reduces the rate of transpiration from the leaves. Nevertheless, under conditions of water scarcity, the root water uptake may be more critical in mitigating the damage caused by drought stress than controlling water loss through the leaves [88].

Employing plant growth regulators (PGRs) represents a viable means of enhancing plant resilience to stress, alongside other techniques such as selective breeding and genetic modification. There are numerous compounds that can alleviate drought-related stress in plants [89]. Salicylic acid, or 2-hydroxybenzoic acid, is a phenolic compound with hormone-like properties that can disrupt plant growth regulation, particularly when confronted with diverse stresses [90]. Additionally, it may trigger various physiological and biochemical functions in plants.

Moreover, salicylic acid (SA) has the ability to hinder catalase (CAT) function, which could result in the accumulation of hydrogen peroxide (H_2O_2). This, in turn, may stimulate the operation of ROS-detoxifying enzymes and the production of antioxidant metabolites [91].

The ability of plants to adapt to environmental pressures, such as drought or salinity, requires temporary and long-lasting reductions in transpiration water flow. This physiological reaction is influenced by both inherent and induced genetic factors [92]. Stomata are of paramount importance in transpiration management, regulating plant water loss through density on the leaf surface and the mechanism of closure in response to environmental stimuli.

On the biotic nature stressors, mites prevent harm to the epidermis, thus reducing the chances of the leaf surface detecting the attack and postponing the plant's reaction. This pressure results in a modification of the plasma membrane potential and consequent alterations in the concentration of free cytosolic Ca^{2+}, which set off a signal that initiates a series of reactions [93]. One of the primary effects of biotic stress encountered by the plant is a rise in the cellular levels of ROS, which are then transformed into H_2O_2.

3.7. *T. vulgaris* L.—Biological Activities and Stress Resistance

T. vulgaris L. displays antimicrobial, anti-inflammatory, antioxidant, and immunomodulatory properties, being effective against a variety of ailments related to the respiratory, cardiovascular, and nervous systems, among others [94,95]. These effects are ascribed to phenolic acids, other phenols, and particularly the plant's essential oil. The essential oil is composed mainly of thymol, carvacrol, geraniol, α-terpineol, 4-thujanol, linalool, 1,8-cineole, myrcene, γ-terpinene, and p-cymene. The key components' abundance varies greatly depending on the plant chemotype, with the thymol chemotype being the most widespread [96,97]. This versatile plant finds widespread use in the food and pharmaceutical industries [51,52]. It is also a promising agent for crop protection and storage preservation [52]. From an economic perspective, the plant's biomass yield and essential oil quality are crucial, and various factors can affect them, including crop nutrition, manure application, water stress, seasonal variations, or processing [98,99].

When using the DPPH free radical scavenging method, *T. vulgaris* demonstrated a robust antioxidant activity of approximately 85% [100–102]. A similar outcome was observed regarding the antioxidant activity of the thyme methanol extract [103].

The disk diffusion method revealed that the extract exhibited very strong inhibition (20 mm inhibition zone) against various bacteria, including *Bacillus subtilis*, *Staphylococcus aureus*, *Pseudomonas aeruginosa*, and *Candida albicans* [104]. However, its inhibitory activity against *Candida tropicalis* was found to be moderate [13].

The antimicrobial activity of essential oils (EOs) is dependent on their chemical constituents. The antimicrobial activity of the analyzed EO is likely associated with the presence of phenolic compounds (thymol) and terpene hydrocarbons (γ-terpinene) [105,106]. It is believed that p-Cymene, the third major component, has synergistic effects with thymol and γ-terpinene [107], which may contribute to the observed antimicrobial activity. Furthermore, several studies have indicated that EOs exhibit stronger antimicrobial activity than their major constituents or mixtures [108,109], suggesting that minor components may have synergistic effects and emphasizing the importance of all components in relation to the biological activity of EOs [13].

The *T. vulgaris* essential oil demonstrated potent antimicrobial and antibiofilm effects, with MIC values ranging from 0.0625% to 2% v/v [110]. It also exhibited lower minimum inhibitory concentration values compared with the antibiotics tested on eradicating *Candida* genus biofilm [111]. When utilized in the vapor phase, it could serve as a viable alternative to antimicrobials in the food industry due to the lower concentration of EO required compared with the liquid phase contact effect [112].

The thyme essential oil contains elevated levels of TPC along with strong radical scavenging ability against DPPH, ABTS, and linoleic acid radicals, in addition to iron chelating capabilities. This positive correlation highlights the antioxidant potential of TEO to combat various oxidation systems and prevent oxidative damage [113].

At a concentration of 0.1 mg/mL, thyme oil and a CNC-based formulation of thyme white oil demonstrated complete larvicidal activity against *Aedes albopictus* [114].

The composition of bioactive secondary metabolites synthesized by medicinal plants varies widely depending on the species, having a profound impact on their relationship with endophytic microorganisms [115]. *Bacillus* spp. connected with *T. vulgaris*, such as *Bacillus sonorensis* (EGY05), *Bacillus tequilensis* (EGY21), and *Bacillus mojavensis* (EGY25), generated plant-growth-stimulating substances such as auxin, fixed nitrogen, soluble phosphate and iron, and lytic enzymes such as chitinase, cellulase, protease, and lipase. These bacteria may provide novel tactics to alleviate salt stress [116].

Phenolic substances and their corresponding enzymes (PAL and PPO) could potentially serve as protective factors against drought stress [6] that impacts the water status of leaves, pigments, and stomatal conductance, leading to the inhibition of photosynthesis [6]. The PPOs, in particular, play a role in various processes such as the Mehler reaction, the photoreduction of molecular oxygen by PSI, the regulation of oxygen levels in plastids, and the generation of the phenylpropanoid pathway [117]. The decrease in chlorophyll-a, chlorophyll-b, and total chlorophyll concentrations on thyme plants experiencing water stress may be viewed as an important regulatory measure to prevent excessive light absorption and to restrict the over-reduction of the photosynthetic electron transport chain, thereby limiting the generation of ROS [118]. ROS have been shown to be produced by both biotic and abiotic stresses [119], and these molecules are responsible for most of the oxidative damage to biological structures, including DNA, RNA, amino acids, proteins, and lipids [120].

Hydrogen peroxide (H_2O_2) is one of the most stable ROS which is produced in plant cells during different physiological processes, including photosynthesis, photorespiration, and, to a relatively lesser extent, respiration; it plays an important role as a signaling molecule under stressful conditions [121].

Malondialdehyde (MDA) is also widely known as a biochemical marker to increase the activity of ROS and the oxidative stress in plant tissues under adverse conditions. It is considered the most final product of lipid peroxidation and an important indicator of the oxidative damage that could occur in the cellular membrane under different stress conditions [7].

In plants, mutations on the epigenetic regulator histone deacetylase-6 (HDA-6) appear to improve survival in drought conditions [122]. This response is associated with the expression of genes involved in acetic acid biosynthesis. Therefore, in conditions of water stress, there would be a relationship between HAD-6 and the regulation of genes involved in acetic acid synthesis [123].

Preserving the integrity of cellular membranes during stressful circumstances is deemed a crucial aspect of any salinity adaptation mechanisms. The percentage of electrolyte leakage (ELP) indicates the level of injury to cell membranes. The application of salicylic acid (SA) amplified the ion leakage in thyme seedlings exposed to salt stress at greater concentrations, indicating that SA concentrations play a crucial role in saline environments.

The potential of ascorbic acid in mitigating and modifying the effects of salt stress on plants is well known. As a rule, its concentration is higher in leaves compared with other plant parts and is 5–10 times higher than that of glutathione [124]. In addition, the ascorbic acid's antioxidant role has been confirmed [125]. Therefore, plants require high endogenous levels of ascorbic acid to regulate various processes of plant metabolism in addition to countering oxidative stress. Endogenous levels of ascorbic acid can be elevated by exogenously administering ascorbic acid via the rooting medium, as a foliar spray or as seed priming. It plays a significant role in photosynthesis, specifically by regulating the redox state of photosynthetic electron carriers through the Mehler peroxidase reaction with ascorbate peroxidase and acting as a co-factor for violaxanthin deep oxidase, which is involved in xanthophyll cycle-mediated photoprotection [126]. As a result, in plants treated with ascorbic acid, high levels of pigments can work synergistically with the ascorbic acid to provide an efficient barrier against oxidation under salinity stress. Ascorbic acid can mitigate the detrimental effects of salinity by increasing the auxin and gibberellin content while reducing abscisic acid levels [127], which may help protect the photosynthetic apparatus and subsequently increase photosynthetic pigments.

During stressful periods, the accumulation of compatible osmolytes, such as proline, can serve as an appropriate marker for heavy metal contamination. In addition, proline may exhibit antioxidative properties that safeguard the cells from the detrimental effects of ROSs induced by Cd contamination due to a conducive environment for Cd sequestration and phytochelatin synthesis [128].

The main anatomical, physiological, and molecular changes of *T. vulgaris* related to different stressors are presented in Table 3.

Table 3. *T. vulgaris* L. response to different stressors.

Type of Stress		Anatomical, Physiological, and Molecular Changes	Contributing Factors	References
Abiotic	Salinity	Nutritional imbalance in plant tissues	Electrolyte leakage increase	[103]
		Reduction of the photosynthetic capacity	CO_2 assimilation reduction	[129]
	Drought	Mitigate cell division, elongation, and differentiation	Decreases cell turgor	[7,130]
			Minimizes enzyme activities	[7,130]
			Decreases energy supply	[7,130]
		Photosynthetic processes reduction	Lower level of relative water content (RWC)	[131,132]
		Affects the level of endogenous phytohormones	Alters relations between ABA, ethylene, GA3, cytokinins, and auxins	[7,133–135]
		Reduces concentration of chlorophyll a and b and total chlorophyll	Increases ROS production	[7,135–137]
		Carotenoids concentration reduction	Enhancement of ABA hormone	[7,133]
		H_2O_2 and lipid peroxidation enhancement	MDA concentration increase	[7]
		Adjustment of osmotic potential	Increased soluble sugars, proline, and free amino acid concentrations	[7]
		Considerable synthesis of total soluble phenols and phenylalanine ammonia-lyase (PAL)	Enhanced specific activity of PPO (polyphenol oxidase)	[7]
		Plant gene expression adjustment	Increased HDA-6 levels	[138–140]
		Secondary metabolites boost	Phe, Trp, and Asn amino acids	[141]
	Cd contamination (seeds)	Phytochelatin synthesis	Osmolytes (proline) accumulation	[142]
		ROS production enhancement	Increase in MDA content	[142]
Biotic	*Aphis serpylli* Koch	Cell physiology disruption	Carvacrol, Geraniol, and Thymol monoterpenes mitigate attack	[143,144]
			Linalol enhances attack	[143,144]

3.8. *S. hortensis L.—Biological Activities and Stress Resistance*

The EO extracted from summer savory contains a significant amount of carvacrol, which plays a crucial role in various biological activities, such as antimicrobial, antioxidant, antidiabetic, antihyperlipidemic, antispasmodic, antinociceptive, anti-inflammatory, antiproliferative, sedative, and reproduction stimulatory effects [145]. The EO content in different species of this genus is more than 5%, the major oil constituents being carvacrol, thymol, γ-terpinene, and borneol [146]. The chemical composition of the plant extracts is influenced by several factors, including the plant part used, harvest time, extraction method, plant cultivar or genotype, geographical location, storage, and climatic conditions [39].

Regarding its antimicrobial properties, summer savory volatile oils exhibit actions on cell membranes, causing interference, destabilization, and consequent effects on the phospholipid bilayer and enzyme activity [39]. The inhibitory effect of volatile oil against bacteria and fungi can be attributed to the higher content of biologically active compounds from the monoterpenes group, particularly terpinene, thymol, and carvacrol, where thymol has significant inhibitory activity against *S. aureus*, carvacrol and p-cymene against *Escherichia coli*, and γ-terpinene against *C. albicans* and *S. aureus* [17]. Thymol and carvacrol have increased activity against bacterial strains, while γ-terpinene and p-cymene are active against fungal strains [21,147]. The volatile oil extracted from *S. hortensis* L. has a broad antimicrobial spectrum, exhibiting inhibitory effects against 25 bacterial, eight fungal, and one yeast species [20]. Its activity against *E. coli*, *Salmonella typhimurium*, *S. aureus*, *Listeria monocytogenes*, and *Pseudomonas putida* isolated strains was also demonstrated [19]. The volatile oil has a higher concentration of antimicrobial compounds compared with the extracts [20,148].

The antioxidant activity of summer savory essential oil (SHEO) could be ascribed to the abundant content of carvacrol, γ-terpinene, p-cymene, and thymol compounds, which are known for their antioxidant properties [149,150]. Meanwhile, the components of extracts from *S. hortensis* (rosmarinic acid, caffeic acid, naringenin, quercetin, apigenin, kaempferol, luteolin, chlorogenic acid, rutin, and apigenin-glycoside) are also recognized for their antioxidant potential [151,152]. Due to their antioxidant activity, natural extracts derived from *S. hortensis* are being considered for use in the meat industry, with water leaf extract found to increase the shelf life of ground beef [153]. They can also be utilized as an antioxidant in mayonnaise formulations [154]. The presence of monoterpenes, such as carvacrol, cymene, and thymol, in the essential oil of *S. hortensis* suggests its potential for antimicrobial activities against food, plant, and human pathogens [155]. The antimicrobial mechanism involves damage to the integrity of the cell membrane, leading to the leakage of ions and other cell components and eventual death. At the same time, the antimicrobial properties of individual components of the essential oil are being evaluated [15].

The potential use of *S. hortensis* essential oil (SHEO) as a natural herbicide against two widely spread weeds, *Amaranthus retroflexus* and *Chenopodium album*, was also assessed [156]. The aerial parts of the plant were used during the fruit stage by hydrodistillation, and it was found to be rich in carvacrol and γ-terpinene (determined by GC-MS to be 55.66% and 31.98%, respectively). The essential oil was formulated as a nanoemulsion with a concentration of 5 mL/L, with an observed herbicidal activity both in laboratory conditions (at a nanoemulsion concentration of 1 mL/L) and in greenhouse conditions (at a nanoemulsion concentration of 4 mL/L).

There is a lack of data on the potential applications of *S. hortensis* essential oil in cancer treatment, although several other *Satureja* species have exhibited anticancer properties. For example, *S. intermedia* essential oil has shown potential against oesophageal squamous cell carcinoma and human bladder carcinoma cell lines, while *S. spicigera* has shown promise against Rectosigmoid adenocarcinoma cells, human epithelial colorectal adenocarcinoma cells, mouse embryo fibroblast cells, and ductal carcinoma cells. *S. sahendica* essential oil has demonstrated anticancer properties against breast cancer cells, fibroblast-like kidney cells, colon adenocarcinoma cells, and choriocarcinoma cells, while *S. montana* essential oil has shown potential against colon adenocarcinoma cells. These findings have been reported in various studies [17,155].

Some of the most important anatomical, physiological, and molecular changes of *S. hortensis* L. related to different stressors are presented in Table 4.

Table 4. *S. hortensis* L. response to different stressors.

Type of Stress		Anatomical, Physiological and Molecular Changes	Contributing Factors	References
Abiotic	Salinity	Growth decrese	High osmotic proficiency	[49,66]
			Salt ions toxicity	[66]
			Cytokinin cutoff	[66]
			Enhanced inhibitor production	[66]
			Decreasing water and nutrient uptake	[49]
		Higher Na^+ concentrations	Lipid peroxidation increase	[157]
			Enhanced membrane damage	[157]
			Electrolyte leakage	[158]
			Increased MDA accumulation	[158]
			TPC, TSC, proline, and essential oil enhancement	[49]
		Chlorophyll content	Free oxygen radicals exposure/peroxidation	[49,159]
		Decreased transpiration rate	Gas exchange mittigation	[159]
		Imbalance in plant tissues	Reduced Ca and K	[49]
			Increased Cl and Na concentration	[49]
	Drought	Mitigate cell division, elongation, and differentiation	Decreases cell turgor	[160]
			Decreased relative water content (RWC)	
			Minimizes enzyme activities	[160]
			Decreases energy supply	[160]
			Reduced the plant height and the number of subsidiary branches	[43]
			Intensified malondialdehyde (MDA), H2O2, and proline contents	[43]
			Improved total chlorophyll, chlorophyll a and b, and carotenoid contents	[43]
Biotic	*Botrytis cinerea* Pers.	Necrosis and narrowing tissues	Endopolygalacturonase content enhancement	[161–163]
			Pectin degradation	

During stress periods, terpene emissions and related attracting mechanisms can indirectly contribute to plant defense mechanisms [164]. Furthermore, certain volatile compounds may act as airborne signals that can either directly or indirectly trigger systemic resistance and defense responses in neighboring plants [165,166].

Methyl jasmonate (MJ) treatment has been found to up-regulate genes involved in Jasmonate biosynthesis, secondary metabolism, and cell wall formation, as well as genes that encode stress-protective and defense proteins. Conversely, genes that are involved in photosynthesis, such as ribulose bisphosphate carboxylase/oxygenase, chlorophyll a/b-binding protein, and light-harvesting complex II, are down regulated [167].

Gibberellin may have a potential role in aiding plants to adapt to stress [168,169]. Thus, the external application of it may reduce the negative impacts of salinity, while also enhancing growth under saline conditions, as evidenced by increased nutrient uptake, dry weights, plant height, leaf area, and yield, mitigating NaCl-induced growth inhibition [170,171]. Specifically, gibberellin application increased transpiration rate, relative water content, chlorophyll b, total chlorophyll, and xanthophyll content under salinity stress conditions for savory plants. The external application of gibberellin may enhance plant growth by elevating endogenous gibberellin levels [168]. Overall, gibberellin plays a crucial role in boosting plant growth, pigment synthesis, and photosynthesis rate under salt stress conditions [169].

In plants, the contents of hydrogen peroxide (H_2O_2) and malondialdehyde (MDA) reduced when inoculated with bacteria under well-watered and water stress conditions due to a significant increase in the expression of antioxidant enzymes, leading to a decline in MDA levels and electrolyte leakage [172]. Within plant cells, certain compounds such as lipid-soluble antioxidants (e.g., α-tocopherol and carotenoids), water-soluble reductants (e.g., glutathione and ascorbate), and antioxidant enzymes can provide protection against the harmful effects of ROS [173]. Moreover, the accumulation of certain osmolytes (e.g., proline) in plant cells can aid in scavenging free radicals and safeguarding enzymes [174].

3.9. Basil, Thyme, and Summer Savory Morphological Response to Different Biotic and Abiotic Stressors

All the three *Lamiaceae* herbs have developed several morphological adaptations that enable them to cope with biotic and abiotic stresses, such as leathery leaves, pubescence, and deep root systems.

Leathery leaves, one of the main adaptations, are characterized by a thick cuticle and more sclerenchyma tissue than other types of leaves, with a pronounced retaining water role, reducing water loss through transpiration, while conferring resistance to drought stress [175]. This adaptation allows these herbs to survive in arid conditions by maintaining hydration. Leathery leaves also help the plants to cope with high temperatures, which can cause dehydration and damage to the plant tissues. Moreover, the amount of epicuticular wax, a waxy substance found on the surface of leaves, was positively correlated with drought tolerance in these herbs [175]. The ability to tolerate water stress is linked to minor modifications in cellular biochemistry due to the buildup of compatible solutes and particular proteins that can be swiftly triggered by osmotic stress [176]. Water scarcity impacts plant development at diverse degrees, from the cell to the tissue level [177].

The presence of pubescence on the surface of these herbs' leaves represents another significant mechanism of resistance (antixenosis) [175]. These tiny hairs act as a physical barrier against herbivores, such as insects and grazing animals, and reduce water loss through transpiration. They also help the plants to reflect some of the sunlight and reduce the amount of energy absorbed by the leaves, which protects them from photoinhibition or damage caused by excessive sunlight [175]. The phenylpropenes, along with terpenoids, are the primary components of essential oils that are released from the glandular trichomes of various *Lamiaceae* species [178]. In the particular case of basil, it predominantly synthesizes and stores eugenol and methyl chavicol in its glandular trichomes [179].

In addition, the three herbs have developed deep root systems, which enable them to access water and nutrients from deeper layers of soil and provide stability to the plant by anchoring it firmly in the ground. The deep roots allow the plants to survive in nutrient-poor soil and to withstand periods of drought stress by accessing water that is not available to shallower rooted plants [175]. Deep roots also provide stability to the plant by anchoring it firmly in the ground, protecting it from wind and water erosion. The root length and surface area of basil, thyme, and summer savory are positively correlated with their ability to cope with drought stress. In addition, a high root-to-shoot ratio, which is an adaptation that allows them to absorb water more efficiently and store it in their roots for future use, was observed [175].

3.10. Breeding Perspectives Regarding the Adaptability to the Main Abiotic Stressors

Genetic predisposition represents a fundamental requirement for enhancing the quality and yield of essential oils, with variety selection and plant breeding as additional factors. Induced polyploidization is one of the plant breeding techniques that can affect a plant's genome, phenotype, physiology, and metabolome, enabling us to develop novel genotypes with better morphological, physiological, and biochemical properties.

Abiotic stressors such as drought, salinity, and extreme temperatures pose major challenges for herb cultivation. To address these challenges, research needs to focus on developing new cultivars of basil, thyme, and summer savory that are more resilient to abiotic stressors. One approach has been to identify genetic markers and traits associated with stress tolerance and breed for improved adaptability to stressful conditions. Studies have identified candidate genes and quantitative trait loci (QTLs) associated with drought tolerance in basil, thyme, and summer savory and used these genetic markers to develop new cultivars with improved drought resistance.

Another promising approach has been to explore the use of plant growth-promoting rhizobacteria (PGPR) and other beneficial microorganisms to enhance the stress tolerance of these herbs. Recent studies have shown that the application of PGPR can improve the growth, yield, and quality of basil, thyme, and summer savory under stressful conditions and enhance their resistance to pests and diseases.

In addition to phenotypic screening, molecular and biochemical approaches have been used to elucidate the mechanisms of stress tolerance in basil, thyme, and summer savory, as well as the identification of the key compounds involved. Numerous investigations have highlighted the significance of aquaporins in the plant stress response. TaTIP2;2 functions as an inhibitor of drought and salinity stress, its reaction not being reliant on ABA [180]. The transcriptomic and metabolomic responses of thyme plants to heat stress were investigated, and several candidate genes and metabolites associated with thermotolerance, including heat shock proteins, proline, and flavonoids, were identified [181]. Similarly, the volatile compounds and antioxidant activity of summer savory leaves under drought stress were analyzed, and the conclusions were that some volatile terpenes, such as γ-terpinene and carvacrol, were positively correlated with drought tolerance and antioxidant capacity [43].

To identify the compounds important for stress resistance in basil, thyme, and summer savory, several approaches have been used, such as metabolomics, transcriptomics, and proteomics. Analyses of basil leaves' metabolites exposed to a water deficit found that the accumulation of phenolic acids and flavonoids was associated with drought tolerance [182]. Additionally, when the proteome of summer savory leaves under drought stress was investigated, it was found that the up-regulated proteins were related to photosynthesis, antioxidant defense, and stress response [183].

Regarding the germplasm availability, summer savory genotypes were evaluated for their tolerance to cold stress, and a promising candidate ("Mutika") was identified, exhibiting better root volume, aerial part and total fresh weights, stem height, and flower number under low-temperature conditions [184].

3.11. Breeding Perspectives Regarding the Adaptability to the Main Biotic Stressors

Pests (aphids, thrips, spider mites, and whiteflies) are major biotic stressors that can affect basil, thyme, and summer savory crops. They feed on plant sap, causing stunted growth, leaf curling, and discoloration, together with a reduction in herb yield quantity and quality. Moreover, these pests can transmit viral diseases, such as tomato spotted wilt virus (TSWV), which can cause severe damage to the crops [185]. Breeding for pest-resistant cultivars using genetic markers could be a promising solution to reduce the use of insecticides and mitigate the effects of insect pests on the crops.

Diseases, such as fungal and bacterial infections, are also significant biotic stressors that can affect basil, thyme, and savory crops. The most common fungal diseases in these herbs include powdery mildew, downy mildew, and gray mold, which can cause leaf wilting, yellowing, and necrosis. Triggering defense genes against specific pathogens is influenced by distinct environmental factors. This indicates the involvement of intricate signaling pathways that empower plants to identify and safeguard against various stressors, including pathogenic threats [186,187]. Therefore, there is a need for breeding programs that focus on developing disease-resistant varieties of these crops. However, due to the limitations of traditional breeding, such as time-consuming and limited genetic diversity, new breeding technologies, such as genome editing, could provide an efficient and more precise method to introduce disease resistance into these crops [188].

Herbivores, such as deer, rabbits, and rodents, can also cause significant damage to basil, thyme, and savory crops. These herbivores can reduce crop yield by feeding on the plants and transmit plant diseases through their saliva [183]. Developing herbivore-resistant varieties of these crops can help reduce the impact of herbivores on crop production.

Weed competition is another important biotic stress that can affect the growth and yield of basil, thyme, and savory crops. Weeds compete with the herbs for nutrients, water, and sunlight, leading to reduced yield and quality of the crops. Furthermore, weeds can also act as hosts for pests and diseases, increasing their population and spread in the crop field. However, the excessive use of herbicides can have negative impacts on the environment and human health. Therefore, breeding for herbicide-resistant cultivars could be an important strategy in future breeding perspectives.

In terms of germplasm availability, when *Ocimum* accessions were screened for resistance to downy mildew caused by *Peronospora belbahrii*, the conclusion was that some genotypes, such as "Spice", exhibited significantly lower disease severity and higher yield than others [189].

By overexpressing a gene encoding a chalcone synthase in basil plants, an increased resistance to Fusarium wilt caused by *F. oxysporum*, as well as higher levels of glyphosate-resistant basil and thyme flavonoids and phenolic acids were observed [190].

MYB and MYC proteins play a critical role in plants' ability to cope with unfavorable environmental conditions. AtMYB30 functions as an activator of the hypersensitive cell death program upon pathogenic attack [191], while AtMYB33 and AtMYB101 are associated with ABA-mediated reactions to environmental cues [192]. AtMYB96 regulates water scarcity and disease resistance by acting through the ABA signaling pathway [193], and AtMYB15 is involved in enhancing cold stress tolerance [194].

Breeding for biotic stress resistance in basil, thyme, and savory crops is essential to ensure their sustainable cultivation and production. In recent years, there have been significant advancements in molecular breeding techniques, such as marker-assisted selection (MAS) and genomic selection (GS), which can accelerate the breeding process and improve the efficiency of selecting stress-tolerant traits [195]. Moreover, the identification of stress-responsive genes and pathways in these herbs can provide valuable targets for genetic engineering and biotechnological approaches to enhance their resistance to biotic stressors [192]. The breeding of basil with resistance to *F. oxysporum* has shown promising results [196].

Molecular markers and genetic engineering techniques can be used to accelerate the breeding process and identify genes responsible for biotic stress resistance in basil, thyme, and savory. Several resistance genes from basil (Pb1A and Pb1A′), which are responsible for resistance to downy mildew, a common fungal disease, were successfully transferred by researchers [197]. Similarly, genetic engineering can be used to develop herbicide-resistant cultivars.

4. Conclusions

Selecting and breeding basil, thyme, and summer savory genotypes with enhanced tolerance to biotic and abiotic stresses is crucial for their sustainable and profitable cultivation. Phenotypic screening, molecular and biochemical approaches, and genetic engineering or exogenous application of bioactive compounds are effective strategies for identifying and enhancing stress tolerance in these herbs. Identification of key compounds and genes involved in stress resistance can also provide valuable insights for further improvement of these important crops.

The research suggests that there is a significant potential for breeding and genetic improvement, as well as the use of microbial-based strategies, to enhance the adaptability of basil, thyme, and summer savory to abiotic stressors. These efforts could have important implications for the sustainability and productivity of herb cultivation in a changing climate and could help to ensure a reliable supply of these valuable herbs for food, medicinal, and other applications.

Author Contributions: Conceptualization, D.I.A., M.C., P.M.B., C.B. and O.C.M.; methodology, D.I.A., M.C., P.M.B., C.B. and O.C.M.; validation, C.B.; resources, P.M.B. and C.B.; data curation, D.I.A.; writing—original draft preparation, D.I.A. and M.C.; writing—review and editing, D.I.A., M.C., P.M.B., C.B. and O.C.M.; visualization D.I.A., M.C., P.M.B. and C.B.; supervision, D.I.A. and C.B.; project administration, P.M.B.; funding acquisition, P.M.B. All authors have read and agreed to the published version of the manuscript.

Funding: The authors acknowledge the in-kind and cash support by the Vegetable Research and Development Station (VRDS), Bacau, Romania, Project no. 529/2018.

Institutional Review Board Statement: Not applicable.

Informed Consent Statement: Not applicable.

Data Availability Statement: Availability Statements are available in section "MDPI Research Data Policies" at https://www.mdpi.com/ethics accessed on 7 February 2023. No new data were created.

Acknowledgments: This work was carried out in frame of Project no. 529/2018 National Project funded from the State budget through "Gheorghe Ionescu-Sisesti" Academy of Agricultural and Forestry Sciences and the ADER program funded by the Ministry of Agriculture and Rural Development.

Conflicts of Interest: Not the case applicable.

References

1. Dragland, S.; Senoo, H.; Wake, K.; Holte, K.; Blomhoff, R. Several Culinary and Medicinal Herbs Are Important Sources of Dietary Antioxidants. *J. Nutr.* **2003**, *133*, 1286–1290. [CrossRef]
2. Danesi, F.; Elementi, S.; Neri, R.; Maranesi, M.; D'antuono, L.F.; Bordoni, A. Effect of Cultivar on the Protection of Cardiomyocytes from Oxidative Stress by Essential Oils and Aqueous Extracts of Basil (*Ocimum basilicum* L.). *J. Agric. Food Chem.* **2008**, *56*, 9911–9917. [CrossRef]
3. Ahmad, P.; Prasad, M.N.V. (Eds.) *Abiotic Stress Responses in Plants: Metabolism, Productivity and Sustainability*; Springer Science & Business Media: Berlin/Heidelberg, Germany, 2011.
4. Farsaraei, S.; Moghaddam, M.; Pirbalouti, A.G. Changes in growth and essential oil composition of sweet basil in response of salinity stress and superabsorbents application. *Sci. Hortic.* **2020**, *271*, 109465. [CrossRef]
5. Wang, W.-X.; Vinocur, B.; Altman, A. Plant responses to drought, salinity and extreme temperatures: Towards genetic engineering for stress tolerance. *Planta* **2003**, *218*, 1–14. [CrossRef] [PubMed]
6. Ibrahim, M.F.M.; Bondok, A.M.; Al-Senosy, N.K.; Younis, R.A.A. Stimulation Some of Defense Mechanisms in Tomato Plants under Water Deficit and Tobacco mosaic virus (TMV). *World J. Agric. Sci.* **2015**, *11*, 289–302.
7. Farag, R.E.; Abdelbar, O.H.; Shehata, S.A. Impact of Drought Stress on Some Growth, Biochemical and Anatomical Parameters of *Thymus vulgaris* L. *Arab. Univ. J. Agric. Sci.* **2019**, *27*, 37–50. [CrossRef]
8. Kiczorowska, B.; Klebaniuk, R.; Bakowski, M.; Al-Yasiry, M.H. Culinary Herbs-the Nutritive Value and Content of Minerals. *J. Elem.* **2015**, *20*, 599–608. [CrossRef]
9. Gebrehiwot, H.; Bachetti, R.; Dekebo, A. Chemical composition and antimicrobial activities of leaves of sweet basil (*Ocimum basilicum* L.). *Herb. Int. J. Basic Clin. Pharmacol.* **2015**, *4*, 869–875. [CrossRef]
10. Politeo, O.; Jukic, M.; Milos, M. Chemical composition and antioxidant capacity of free volatile aglycones from basil (*Ocimum basilicum* L.) compared with its essential oil. *Food Chem.* **2007**, *101*, 379–385. [CrossRef]
11. Hussain, A.I.; Anwar, F.; Sherazi, S.T.H.; Przybylski, R. Chemical composition, antioxidant and antimicrobial activities of basil (*Ocimum basilicum*) essential oils depends on seasonal variations. *Food Chem.* **2008**, *108*, 986–995. [CrossRef]
12. Guillén, M.; Manzanos, M. Study of the composition of the different parts of a Spanish *Thymus vulgaris* L. plant. *Food Chem.* **1998**, *63*, 373–383. [CrossRef]
13. Borugă, O.; Jianu, C.; Mişcă, C.; Goleţ, I.; Gruia, A.T.; Horhat, F.G. *Thymus vulgaris* essential oil: Chemical composition and antimicrobial activity. *J. Med. Life* **2014**, *7*, 56–60. [PubMed]

14. Ravindran, P.N.; Pillai, G.S.; Divakaran, M. Other Herbs and Spices: Mango Ginger to Wasabi. In *Handbook of Herbs and Spices*; Peter, K.V., Ed.; Woodhead Publishing: Cambridge, UK, 2012; Volume 2, pp. 557–582.
15. Fierascu, I.; Dinu-Pirvu, C.E.; Fierascu, R.C.; Velescu, B.S.; Anuta, V.; Ortan, A.; Jinga, V. Phytochemical Profile and Biological Activities of *Satureja hortensis* L.: A Review of the Last Decade. *Molecules* **2018**, *23*, 2458. [CrossRef] [PubMed]
16. Skubij, N.; Dzida, K.; Jarosz, Z.; Pitura, K.; Jaroszuk-Sierocińska, M. Nutritional value of savory herb (*Satureja hortensis* L.) and plant response to variable mineral nutrition conditions in various phases of development. *Plants* **2020**, *9*, 706. [CrossRef] [PubMed]
17. Hamidpour, R.; Hamidpour, S.; Hamidpour, M.; Shahlari, M.; Sohraby, M. Summer Savory: From the Selection of Traditional Applications to the Novel Effect in Relief, Prevention, and Treatment of a Number of Serious Illnesses Such as Diabetes, Cardiovascular Disease, Alzheimer's Disease, and Cancer. *J. Tradit. Complement. Med.* **2014**, *4*, 140–144. [CrossRef]
18. Tepe, B.; Cilkiz, M. A Pharmacological and Phytochemical Overview onSatureja. *Pharm. Biol.* **2015**, *54*, 375–412. [CrossRef]
19. Mihajilov-Krstev, T.; Radnović, D.; Kitić, D.; Zlatković, B.; Ristić, M.; Branković, S. Chemical Composition and Antimicrobial Activity of *Satureja hortensis* L. *Essential Oil. Open Life Sci.* **2009**, *4*, 411–416. [CrossRef]
20. Mahboubi, M.; Kazempour, N. Chemical composition and antimicrobial activity of *Satureja hortensis* and Trachyspermum copticum essential oil. *Iran. J. Microbiol.* **2011**, *3*, 194–200.
21. Farzaneh, M.; Kiani, H.; Sharifi, R.; Reisi, M.; Hadian, J. Chemical Composition and Antifungal Effects of Three Species of Satureja (*S. Hortensis, S. Spicigera,* and *S. Khuzistanica*) Essential Oils on the Main Pathogens of Strawberry Fruit. *Postharvest Biol. Technol.* **2015**, *109*, 145–151. [CrossRef]
22. Saeidnia, S.; Gohari, A.R.; Manayi, A.; Kourepaz-Mahmoodabadi, M. *Satureja: Ethnomedicine, Phytochemical Diversity and Pharmacological Activities*; Springer: Berlin/Heidelberg, Germany, 2016; pp. 31–40.
23. Makri, O.; Kintzios, S. *Ocimum* sp. (Basil): Botany, Cultivation, Pharmaceutical Properties, and Biotechnology. *J. Herbs Spices Med. Plants* **2008**, *13*, 123–150. [CrossRef]
24. Adham, A.N. Comparative extraction methods, phytochemical constituents, fluorescence analysis and HPLC validation of rosmarinic acid content in Mentha piperita, Mentha longifolia and Ocimum basilicum. *J. Pharmacogn. Phytochem.* **2015**, *3*, 130–139.
25. Shirazi, M.T.; Gholami, H.; Kavoosi, G.; Rowshan, V.; Tafsiry, A. Chemical composition, antioxidant, antimicrobial and cytotoxic activities of *T agetes minuta* and *O cimum basilicum* essential oils. *Food Sci. Nutr.* **2014**, *2*, 146–155. [CrossRef]
26. Koseki, P.M.; Villavicencio, A.L.C.H.; Brito, M.S.; Nahme, L.C.; Sebastião, K.I.; Rela, P.R.; de Almeida-Muradian, L.B.; Mancini-Filho, J.; Freitas, P.C. Effects of irradiation in medicinal and eatable herbs. *Radiat. Phys. Chem.* **2002**, *63*, 681–684. [CrossRef]
27. Hakkim, F.L.; Shankar, C.G.; Girija, S. Chemical Composition and Antioxidant Property of Holy Basil (*Ocimum sanctum* L.) Leaves, Stems, and Inflorescence and Their in Vitro Callus Cultures. *J. Agric. Food Chem.* **2007**, *55*, 9109–9117. [CrossRef]
28. McClatchey, W. The ethnopharmacopoeia of Rotuma. *J. Ethnopharmacol.* **1996**, *50*, 147–156. [CrossRef] [PubMed]
29. Stahl-Biskup, E.; Venskutonis, R.P. Thyme. In *Handbook of Herbs and Spices*; Woodhead Publishing: Sawston, UK, 2012; pp. 499–525.
30. Thompson, J.; Charpentier, A.; Bouguet, G.; Charmasson, F.; Roset, S.; Buatois, B.; Vernet, P.; Gouyon, P.-H. Evolution of a Genetic Polymorphism with Climate Change in a Mediterranean Landscape. *Proc. Natl. Acad. Sci. USA* **2013**, *110*, 2893–2897. [CrossRef]
31. Stahl-Biskup, E.; Sáez, F. *Thyme: The Genus Thymus*; Taylor & Francis: London, UK, 2002; ISBN 0-415-28488-0.
32. Rowland, L.S.; Smith, H.K.; Taylor, G. The potential to improve culinary herb crop quality with deficit irrigation. *Sci. Hortic.* **2018**, *242*, 44–50. [CrossRef]
33. Komaki, A.; Hoseini, F.; Shahidi, S.; Baharlouei, N. Study of the effect of extract of *Thymus Vulgaris* on anxiety in male rats. *J. Tradit. Complement. Med.* **2016**, *6*, 257–261. [CrossRef]
34. Dauqan, E.M.A.; Abdullah, A. Medicinal and Functional Values of Thyme (*Thymus vulgaris* L.) Herb. *J. Appl. Biol. Biotechnol.* **2017**, *5*, 17–22. [CrossRef]
35. Fachini-Queiroz, F.C.; Kummer, R.; Estevão-Silva, C.F.; Carvalho, M.D.D.B.; Cunha, J.M.; Grespan, R.; Bersani-Amado, C.A.; Cuman, R.K.N. Effects of Thymol and Carvacrol, Constituents of *Thymus vulgaris* L. Essential Oil, on the Inflammatory Response. *Evid. Based Complement. Altern. Med.* **2012**, *2012*, 657026. [CrossRef] [PubMed]
36. Nikolić, M.; Glamočlija, J.; Ferreira, I.C.F.R.; Calhelha, R.C.; Fernandes, Â.; Marković, T.; Marković, D.; Giweli, A.; Soković, M. Chemical composition, antimicrobial, antioxidant and antitumor activity of *Thymus serpyllum* L., *Thymus algeriensis* Boiss. and Reut and *Thymus vulgaris* L. Essential Oils. *Ind. Crops Prod.* **2014**, *52*, 183–190. [CrossRef]
37. Gontaru, L.; Plander, S.; Simándi, B. Investigation of *Satureja hortensis* L. as a possible source of natural antioxidants. *Hung. J. Ind. Chem.* **2008**, *1-2*, 36. [CrossRef]
38. Jadczak, D. Effect of sowing date on the quantity and quality of the yield of summer savory (*Satureja hortensis* L.) grown for a bunch harvest. *Herba Pol.* **2007**, *53*, 22–27.
39. Hadian, J.; Ebrahimi, S.N.; Salehi, P. Variability of morphological and phytochemical characteristics among *Satureja hortensis* L. *Accessions of Iran. Ind. Crops Prod.* **2010**, *32*, 62–69. [CrossRef]
40. Hajhashemi, V.; Zolfaghari, B.; Yousefi, A. Antinociceptive and Anti-Inflammatory Activities of *Satureja hortensis* Seed Essential Oil, Hydroalcoholic and Polyphenolic Extracts in Animal Models. *Med. Princ. Pract.* **2011**, *21*, 178–182. [CrossRef]
41. Güllüce, M.; Sökmen, M.; Daferera, D.; Ağar, G.; Özkan, H.; Kartal, N.; Polissiou, M.; Sökmen, A.; Şahin, F. In Vitro Antibacterial, Antifungal, and Antioxidant Activities of the Essential Oil and Methanol Extracts of Herbal Parts and Callus Cultures of *Satureja hortensis* L. *J. Agric. Food Chem.* **2003**, *51*, 3958–3965. [CrossRef]

42. Svoboda, K. Investigation of volatile oil glands of *Satureja hortensis* L. (summer savory) and phytochemical comparison of different varieties. *Int. J. Aromather.* **2003**, *13*, 196–202. [CrossRef]
43. Mohammadi, H.; Dashi, R.; Farzaneh, M.; Parviz, L.; Hashempour, H. Effects of beneficial root pseudomonas on morphological, physiological, and phytochemical characteristics of *Satureja hortensis* (Lamiaceae) under water stress. *Braz. J. Bot.* **2016**, *40*, 41–48. [CrossRef]
44. Cappellari, L.D.R.; Santoro, M.V.; Nievas, F.; Giordano, W.; Banchio, E. Increase of secondary metabolite content in marigold by inoculation with plant growth-promoting rhizobacteria. *Appl. Soil Ecol.* **2013**, *70*, 16–22. [CrossRef]
45. Farzaneh, V.; Carvalho, I.S. A review of the health benefit potentials of herbal plant infusions and their mechanism of actions. *Ind. Crop. Prod.* **2015**, *65*, 247–258. [CrossRef]
46. Heywood, V.H. The Conservation of Genetic and Chemical Diversity in Medicinal and Aromatic Plants. *Biodiversity* **2002**, 13–22. [CrossRef]
47. Saito, K.; Matsuda, F. Metabolomics for Functional Genomics, Systems Biology, and Biotechnology. *Annu. Rev. Plant Biol.* **2010**, *61*, 463–489. [CrossRef]
48. Flowers, T.J.; Muscolo, A. Introduction to the Special Issue: Halophytes in a changing world. *AoB Plants* **2015**, *7*, plv020. [CrossRef]
49. Estaji, A.; Roosta, H.R.; Rezaei, S.A.; Hosseini, S.S.; Niknam, F. Morphological, physiological and phytochemical response of different *Satureja hortensis* L. accessions to salinity in a greenhouse experiment. *J. Appl. Res. Med. Aromat. Plants* **2018**, *10*, 25–33. [CrossRef]
50. Bakkali, F.; Averbeck, S.; Averbeck, D.; Idaomar, M. Biological effects of essential oils—A review. *Food Chem. Toxicol.* **2008**, *46*, 446–475. [CrossRef]
51. Kokoska, L.; Kloucek, P.; Leuner, O.; Novy, P. Plant-Derived Products as Antibacterial and Antifungal Agents in Human Health Care. *Curr. Med. Chem.* **2019**, *26*, 5501–5541. [CrossRef]
52. Pandey, A.K.; Kumar, P.; Singh, P.; Tripathi, N.N.; Bajpai, V.K. Essential Oils: Sources of Antimicrobials and Food Preservatives. *Front. Microbiol.* **2017**, *7*, 2161. [CrossRef]
53. Pavela, R.; Benelli, G. Essential Oils as Ecofriendly Biopesticides? Challenges and Constraints. *Trends Plant Sci.* **2016**, *21*, 1000–1007. [CrossRef]
54. Moghaddam, M.; Pirbalouti, A.G.; Mehdizadeh, L.; Pirmoradi, M.R. Changes in composition and essential oil yield of Ocimum ciliatum at different phenological stages. *Eur. Food Res. Technol.* **2014**, *240*, 199–204. [CrossRef]
55. Pirbalouti, A.G.; Malekpoor, F.; Salimi, A.; Golparvar, A. Exogenous application of chitosan on biochemical and physiological characteristics, phenolic content and antioxidant activity of two species of basil (Ocimum ciliatum and Ocimum basilicum) under reduced irrigation. *Sci. Hortic.* **2017**, *217*, 114–122. [CrossRef]
56. Riaz, U.; Kharal, M.A.; Murtaza, G.; Zaman, Q.U.; Javaid, S.; Malik, H.A.; Aziz, H.; Abbas, Z. Prospective Roles and Mechanisms of Caffeic Acid in Counter Plant Stress: A Mini Review. *Pak. J. Agric. Res.* **2018**, *32*, 8–19. [CrossRef]
57. Zare, M.; Ganjeali, A.; Lahouti, M. Rosmarinic and caffeic acids contents in Basil (*Ocimum basilicum* L.) are altered by different levels of phosphorus and mycorrhiza inoculation under drought stress. *Acta Physiol. Plant.* **2021**, *43*, 26. [CrossRef]
58. Meot-Duros, L.; Magné, C. Antioxidant activity and phenol content of *Crithmum maritimum* L. Leaves. *Plant Physiol. Biochem.* **2009**, *47*, 37–41. [CrossRef]
59. Scagel, C.F.; Lee, J. Phenolic Composition of Basil Plants Is Differentially Altered by Plant Nutrient Status and Inoculation with Mycorrhizal Fungi. *Hortscience* **2012**, *47*, 660–671. [CrossRef]
60. Kwee, E.M.; Niemeyer, E.D. Variations in phenolic composition and antioxidant properties among 15 basil (*Ocimum basilicum* L.) cultivars. *Food Chem.* **2011**, *128*, 1044–1050. [CrossRef]
61. Srivastava, S.; Conlan, X.A.; Adholeya, A.; Cahill, D.M. Elite hairy roots of *Ocimum basilicum* as a new source of rosmarinic acid and antioxidants. *Plant Cell Tissue Organ Cult. (PCTOC)* **2016**, *126*, 19–32. [CrossRef]
62. Kim, Y.B.; Kim, J.K.; Uddin, M.R.; Xu, H.; Park, W.T.; Tuan, P.A.; Li, X.; Chung, E.; Lee, J.-H.; Park, S.U. Metabolomics Analysis and Biosynthesis of Rosmarinic Acid in Agastache rugosa Kuntze Treated with Methyl Jasmonate. *PLoS ONE* **2013**, *8*, e64199. [CrossRef]
63. Hazzoumi, Z.; Moustakime, Y.; Elharchli, E.H.; Joutei, K.A. Effect of arbuscular mycorrhizal fungi (AMF) and water stress on growth, phenolic compounds, glandular hairs, and yield of essential oil in basil (*Ocimum gratissimum* L.). *Chem. Biol. Technol. Agric.* **2015**, *2*, 10. [CrossRef]
64. Delavari, P.M.; Baghizadeh, A.; Enteshari, S.H.; Kalantari, K.M.; Yazdanpanah, A.; Mousavi, E.A. The effects of salicylic acid on some of biochemical and morphological characteristic of *Ocimum basilicucm* under salinity stress. *Aust. J. Basic Appl. Sci.* **2010**, *4*, 4832–4845.
65. Heidari, M. Effects of salinity stress on growth, chlorophyll content and osmotic components of two basil (*Ocimum basilicum* L.) genotypes. *Afr. J. Biotechnol.* **2011**, *11*, 379–384. [CrossRef]
66. Gupta, B.; Huang, B. Mechanism of Salinity Tolerance in Plants: Physiological, Biochemical, and Molecular Characterization. *Int. J. Genom.* **2014**, *2014*, 701596. [CrossRef]
67. Tardieu, F.; Reymond, M.; Hamard, P.; Granier, C.; Muller, B. Spatial distributions of expansion rate, cell division rate and cell size in maize leaves: A synthesis of the effects of soil water status, evaporative demand and temperature. *J. Exp. Bot.* **2000**, *51*, 1505–1514. [CrossRef]

68. Dos Santos, M.S.; Costa, C.A.S.; Gomes, F.P.; do Bomfim Costa, L.C.; de Oliveira, R.A.; da Costa Silva, D. Effects of water deficit on morphophysiology, productivity and chemical composition of Ocimum africanum Lour (Lamiaceae). *Afr. J. Agric. Res.* **2016**, *11*, 1924–1934. [CrossRef]
69. Damalas, C.A. Improving drought tolerance in sweet basil (*Ocimum basilicum*) with salicylic acid. *Sci. Hortic.* **2018**, *246*, 360–365. [CrossRef]
70. Santos, C.V. Regulation of chlorophyll biosynthesis and degradation by salt stress in sunflower leaves. *Sci. Hortic.* **2004**, *103*, 93–99. [CrossRef]
71. Yang, C.W.; Wang, P.; Li, C.Y.; Shi, D.C.; Wang, D.L. Comparison of effects of salt and alkali stresses on the growth and photosynthesis of wheat. *Photosynthetica* **2008**, *46*, 107–114. [CrossRef]
72. Cornic, G. Drought stress inhibits photosynthesis by decreasing stomatal aperture–not by affecting ATP synthesis. *Trends Plant Sci.* **2000**, *5*, 187–188. [CrossRef]
73. Maroco, J.P.; Rodrigues, M.L.; Lopes, C.; Chaves, M.M. Limitations to leaf photosynthesis in field-grown grapevine under drought—metabolic and modelling approaches. *Funct. Plant Biol.* **2002**, *29*, 451–459. [CrossRef] [PubMed]
74. Alves, A.A.; Setter, T.L. Abscisic acid accumulation and osmotic adjustment in cassava under water deficit. *Environ. Exp. Bot.* **2004**, *51*, 259–271. [CrossRef]
75. Golan, K.; Kot, I.; Górska-Drabik, E.; Jurado, I.G.; Kmieć, K.; Łagowska, B. Physiological Response of Basil Plants to Twospotted Spider Mite (Acari: Tetranychidae) Infestation. *J. Econ. Èntomol.* **2019**, *112*, 948–956. [CrossRef]
76. Gomez, K.S.; Oosterhuis, D.M.; Rajguru, S.N. and Johnson, D.R. Molecular biology and physiology. Foliar antioxidant enzyme responses in cotton after aphid herbivory. *J. Cotton Sci.* **2004**, *8*, 99–104.
77. Attia, H.; Arnaud, N.; Karray, N.; Lachaâl, M. Long-term effects of mild salt stress on growth, ion accumulation and superoxide dismutase expression of Arabidopsis rosette leaves. *Physiol. Plant.* **2008**, *132*, 293–305. [CrossRef] [PubMed]
78. Ben Taarit, M.; Msaada, K.; Hosni, K.; Hammami, M.; Kchouk, M.E.; Marzouk, B. Plant growth, essential oil yield and composition of sage (*Salvia officinalis* L.) fruits cultivated under salt stress conditions. *Ind. Crop. Prod.* **2009**, *30*, 333–337. [CrossRef]
79. Estrada, B.; Aroca, R.; Maathuis, F.J.M.; Barea, J.M.; Ruiz-Lozano, J.M. Arbuscular mycorrhizal fungi native from a Mediterranean saline area enhance maize tolerance to salinity through improved ion homeostasis. *Plant Cell Environ.* **2013**, *36*, 1771–1782. [CrossRef]
80. Trovato, M.; Mattioli, R.; Costantino, P. Multiple roles of proline in plant stress tolerance and development. *Rend. Lincei* **2008**, *19*, 325–346. [CrossRef]
81. Seki, M.; Umezawa, T.; Urano, K.; Shinozaki, K. Regulatory metabolic networks in drought stress responses. *Curr. Opin. Plant Biol.* **2007**, *10*, 296–302. [CrossRef]
82. Zhang, X.; Ervin, E.; Evanylo, G.; Haering, K. Impact of Biosolids on Hormone Metabolism in Drought-Stressed Tall Fescue. *Crop. Sci.* **2009**, *49*, 1893–1901. [CrossRef]
83. Ekren, S.; Sönmez, Ç.; Özçakal, E.; Kurttaş, Y.S.K.; Bayram, E.; Gürgülü, H. The effect of different irrigation water levels on yield and quality characteristics of purple basil (*Ocimum basilicum* L.). *Agric. Water Manag.* **2012**, *109*, 155–161. [CrossRef]
84. Schütz, M.; Fangmeier, A. Growth and yield responses of spring wheat (*Triticum aestivum* L. cv. Minaret) to elevated CO_2 and water limitation. *Environ. Pollut.* **2001**, *114*, 187–194. [CrossRef]
85. Verslues, P.E.; Agarwal, M.; Katiyar-Agarwal, S.; Zhu, J.; Zhu, J.-K. Methods and concepts in quantifying resistance to drought, salt and freezing, abiotic stresses that affect plant water status. *Plant J.* **2006**, *45*, 523–539. [CrossRef] [PubMed]
86. Jones, H.G. Monitoring plant and soil water status: Established and novel methods revisited and their relevance to studies of drought tolerance. *J. Exp. Bot.* **2006**, *58*, 119–130. [CrossRef]
87. Sánchez-Rodríguez, E.; Rubio-Wilhelmi, M.; Cervilla, L.M.; Blasco, B.; Rios, J.J.; Rosales, M.A.; Romero, L.; Ruiz, J.M. Genotypic differences in some physiological parameters symptomatic for oxidative stress under moderate drought in tomato plants. *Plant Sci.* **2010**, *178*, 30–40. [CrossRef]
88. Aroca, R.; Porcel, R.; Lozano, J.M.R. Regulation of root water uptake under abiotic stress conditions. *J. Exp. Bot.* **2011**, *63*, 43–57. [CrossRef]
89. Liu, H.; Wang, X.; Wang, D.; Zou, Z.; Liang, Z. Effect of drought stress on growth and accumulation of active constituents in Salvia miltiorrhiza Bunge. *Ind. Crop. Prod.* **2011**, *33*, 84–88. [CrossRef]
90. Joseph, B.; Jini, D.; Sujatha, S. Insight into the Role of Exogenous Salicylic Acid on Plants Grown under Salt Environment. *Asian J. Crop. Sci.* **2010**, *2*, 226–235. [CrossRef]
91. Chen, Z.; Silva, H.; Klessig, D.F. Active oxygen species in the induction of plant systemic acquired resistance by salicylic acid. *Science* **1993**, *262*, 1883–1886. [CrossRef]
92. Maggio, A.; Zhu, J.-K.; Hasegawa, P.M.; Bressan, R.A. Osmogenetics: Aristotle to *Arabidopsis*. *Plant Cell* **2006**, *18*, 1542–1557. [CrossRef]
93. Agut, B.; Pastor, V.; Jaques, J.A.; Flors, V. Can Plant Defence Mechanisms Provide New Approaches for the Sustainable Control of the Two-Spotted Spider Mite *Tetranychus urticae*? *Int. J. Mol. Sci.* **2018**, *19*, 614. [CrossRef]
94. Marchese, A.; Orhan, I.E.; Daglia, M.; Barbieri, R.; Di Lorenzo, A.; Nabavi, S.F.; Gortzi, O.; Izadi, M.; Nabavi, S.M. Antibacterial and antifungal activities of thymol: A brief review of the literature. *Food Chem.* **2016**, *210*, 402–414. [CrossRef]

95. Salehi, B.; Mishra, A.P.; Shukla, I.; Sharifi-Rad, M.; del Mar Contreras, M.; Segura-Carretero, A.; Fathi, H.; Nasrabadi, N.N.; Kobarfard, F.; Sharifi-Rad, J. Thymol, thyme, and other plant sources: Health and potential uses. *Phytother. Res.* **2018**, *32*, 1688–1706. [CrossRef]
96. Mandal, S.; DebMandal, M. Thyme (*Thymus vulgaris* L.) Oils. In *Essential Oils in Food Preservation, Flavor and Safety*; Chapter 94; Elsevier Inc.: Amsterdam, The Netherlands, 2016; pp. 825–834. [CrossRef]
97. Trindade, H.; Pedro, L.G.; Figueiredo, A.C.; Barroso, J.G. Chemotypes and terpene synthase genes in Thymus genus: State of the art. *Ind. Crop. Prod.* **2018**, *124*, 530–547. [CrossRef]
98. Askary, M.; Behdani, M.A.; Parsa, S.; Mahmoodi, S.; Jamialahmadi, M. Water stress and manure application affect the quantity and quality of essential oil of *Thymus daenensis* and *Thymus vulgaris*. *Ind. Crop. Prod.* **2018**, *111*, 336–344. [CrossRef]
99. Pavela, R.; Žabka, M.; Vrchotová, N.; Tříska, J. Effect of foliar nutrition on the essential oil yield of Thyme (*Thymus vulgaris* L.). *Ind. Crop. Prod.* **2018**, *112*, 762–765. [CrossRef]
100. Galovičová, L.; Borotová, P.; Valková, V.; Vukovic, N.L.; Vukic, M.; Štefániková, J.; Ďúranová, H.; Kowalczewski, P.; Čmiková, N.; Kačániová, M. *Thymus vulgaris* Essential Oil and Its Biological Activity. *Plants* **2021**, *10*, 1959. [CrossRef]
101. Punya, H.N.; Mehta, N.; Chatli, M.K.; Wagh, R.; Panwar, H. In-vitro Evaluation of Antimicrobial and Antioxidant Efficacy of Thyme (*Thymus vulgaris* L.) Essential Oil. *J. Anim. Res.* **2019**, *9*, 443–449. [CrossRef]
102. Kulisic, T.; Radonić, A.; Milos, M. Antioxidant Properties of Thyme (*Thymus vulgaris* L.) and Wild Thyme (*Thymus serpyllum* L.) Essential Oils. *Ital. J. Food Saf.* **2005**, *17*, 315–324.
103. Bistgani, Z.E.; Hashemi, M.; DaCosta, M.; Craker, L.; Maggi, F.; Morshedloo, M.R. Effect of salinity stress on the physiological characteristics, phenolic compounds and antioxidant activity of *Thymus vulgaris* L. and *Thymus daenensis* Celak. *Ind. Crop. Prod.* **2019**, *135*, 311–320. [CrossRef]
104. Al Maqtari, M. Chemical Composition and Antimicrobial Activity of Essential Oil of *Thymus Vulgaris* from Yemen. *Turk. J. Biochem.* **2011**, *36*, 342–349.
105. Rota, M.C.; Herrera, A.; Martínez, R.M.; Sotomayor, J.A.; Jordán, M.J. Antimicrobial activity and chemical composition of *Thymus vulgaris*, *Thymus zygis* and *Thymus hyemalis* essential oils. *Food Control* **2008**, *19*, 681–687. [CrossRef]
106. Imelouane, B.; Amhamdi, H.; Wathelet, J.P.; Ankit, M.; Khedid, K.; El Bachiri, A. Chemical Composition and Antimicrobial Activity of Essential Oil of Thyme (*Thymus vulgaris*) from Eastern Morocco. *Int. J. Agric. Biol.* **2009**, *11*, 205–208.
107. Gallucci, M.N.; Oliva, M.; Casero, C.; Dambolena, J.; Luna, A.; Zygadlo, J.; Demo, M. Antimicrobial Combined Action of Terpenes against the Food-Borne Microorganisms *Escherichia Coli*, *Staphylococcus Aureus* and *Bacillus Cereus*. *Flavour Fragr. J.* **2009**, *24*, 348–354. [CrossRef]
108. Gill, A.O.; Delaquis, P.; Russo, P.; Holley, R.A. Evaluation of Antilisterial Action of Cilantro Oil on Vacuum Packed Ham. *Int. J. Food Microbiol.* **2002**, *73*, 83–92. [CrossRef]
109. Mourey, A.; Canillac, N. Anti-Listeria Monocytogenes Activity of Essential Oils Components of Conifers. *Food Control* **2002**, *13*, 289–292. [CrossRef]
110. Al-Shuneigat, J.; Al-Sarayreh, S.; Al-Saraireh, Y.; Al-Qudah, M.; Al-Tarawneh, I.; Albataineh, E. Effects of Wild *Thymus Vulgaris* Essential Oil on Clinical Isolates Biofilm-Forming Bacteria. *IOSR J. Dent. Med. Sci.* **2014**, *13*, 62–66. [CrossRef]
111. Jafri, H.; Ahmad, I. Thymus Vulgaris Essential Oil and Thymol Inhibit Biofilms and Interact Synergistically with Antifungal Drugs against Drug Resistant Strains of Candida Albicans and Candida Tropicalis. *J. De Mycol. Médicale* **2020**, *30*, 100911. [CrossRef]
112. Reyes-Jurado, F.; Cervantes-Rincón, T.; Bach, H.; López-Malo, A.; Palou, E. Antimicrobial Activity of Mexican Oregano (Lippia Berlandieri), Thyme (Thymus Vulgaris), and Mustard (Brassica Nigra) Essential Oils in Gaseous Phase. *Ind. Crops Prod.* **2019**, *131*, 90–95. [CrossRef]
113. Aljabeili, H.S.; Barakat, H.; Abdel-Rahman, H.A. Chemical Composition, Antibacterial and Antioxidant Activities of Thyme Essential Oil (Thymus Vulgaris). *Food Nutr. Sci.* **2018**, *9*, 433–446. [CrossRef]
114. Shin, J.; Na, K.; Shin, S.; Seo, S.-M.; Youn, H.J.; Park, I.-K.; Hyun, J. Biological Activity of Thyme White Essential Oil Stabilized by Cellulose Nanocrystals. *Biomolecules* **2019**, *9*, 799. [CrossRef]
115. Chaparro, J.M.; Badri, D.V.; Vivanco, J.M. Rhizosphere Microbiome Assemblage Is Affected by Plant Development. *ISME J.* **2013**, *8*, 790–803. [CrossRef]
116. Abdelshafy Mohamad, O.A.; Ma, J.-B.; Liu, Y.-H.; Zhang, D.; Hua, S.; Bhute, S.; Hedlund, B.P.; Li, W.-J.; Li, L. Beneficial Endophytic Bacterial Populations Associated with Medicinal Plant *Thymus Vulgaris* Alleviate Salt Stress and Confer Resistance to Fusarium Oxysporum. *Front. Plant Sci.* **2020**, *11*, 44. [CrossRef] [PubMed]
117. Saraswathi, S.G.; Paliwal, K. Drought induced changes in growth, leaf gas exchange and biomass production in Albizia lebbeck and Cassia siamea seedlings. *J. Environ. Biol.* **2011**, *32*, 173–178. [PubMed]
118. Mazars, C.; Thuleau, P.; Lamotte, O.; Bourque, S. Cross-Talk between Ros and Calcium in Regulation of Nuclear Activities. *Mol. Plant* **2010**, *3*, 706–718. [CrossRef] [PubMed]
119. Reddy, A.R.; Chaitanya, K.V.; Vivekanandan, M. Drought-Induced Responses of Photosynthesis and Antioxidant Metabolism in Higher Plants. *J. Plant Physiol.* **2004**, *161*, 1189–1202. [CrossRef]
120. Johnson, S.M.; Doherty, S.J.; Croy, R.R.D. Biphasic Superoxide Generation in Potato Tubers. A Self-Amplifying Response to Stress. *Plant Physiol.* **2003**, *131*, 1440–1449. [CrossRef] [PubMed]
121. Slesak, I.; Libik, M.; Karpinska, B.; Karpinski, S.; Miszalski, Z. The Role of Hydrogen Peroxide in Regulation of Plant Metabolism and Cellular Signalling in Response to Environmental Stresses. *Acta Biochim. Pol.* **2007**, *54*, 39–50. [CrossRef]

122. Catala, A. Lipid Peroxidation. Available online: https://www.intechopen.com/books/2553 (accessed on 3 April 2023).
123. Ashapkin, V.V.; Kutueva, L.I.; Aleksandrushkina, N.I.; Vanyushin, B.F. Epigenetic Mechanisms of Plant Adaptation to Biotic and Abiotic Stresses. *Int. J. Mol. Sci.* **2020**, *21*, 7457. [CrossRef]
124. Smirnoff, N. Ascorbic Acid: Metabolism and Functions of a Multi-Facetted Molecule. *Curr. Opin. Plant Biol.* **2000**, *3*, 229–235. [CrossRef]
125. Müller-Moulé, P.; Golan, T.; Niyogi, K.K. Ascorbate-Deficient Mutants of Arabidopsis Grow in High Light despite Chronic Photooxidative Stress. *Plant Physiol.* **2004**, *134*, 1163–1172. [CrossRef]
126. Smirnoff, N.; Wheeler, G.L. Ascorbic Acid in Plants: Biosynthesis and Function. *Crit. Rev. Plant Sci.* **2000**, *19*, 267–290. [CrossRef]
127. Khan, T.; Mazid, M.; Mohammad, F. A Review of Ascorbic Acid Potentialities against Oxidative Stress Induced in Plants. *J. Agrobiol.* **2011**, *28*, 97–111. [CrossRef]
128. Siripornadulsil, S.; Traina, S.; Verma, D.P.; Sayre, R.T. Molecular Mechanisms of Proline-Mediated Tolerance to Toxic Heavy Metals in Transgenic Microalgae. *Plant Cell* **2002**, *14*, 2837–2847. [CrossRef]
129. Agar, H.; Galatali, S.; Ozkaya, D.E.; Kaya, E. A Primary Study: Investigation of the in Vitro Salt Stress Effects on Development in Thymus Cilicicus Boiss. & Bal. *Glob. J. Bot. Sci.* **2022**, *10*, 23–27. [CrossRef]
130. Aziz, E.E.; Hendawy, S.F. Effect of soil type and irrigation intervals on plant growth, essential oil yield and constituents of *Thymus Vulgaris* plant. *Thymus Plant* **2008**, *4*, 443–450.
131. Kaur, H.; Kohli, S.K.; Khanna, K.; Bhardwaj, R. Scrutinizing the Impact of Water Deficit in Plants: Transcriptional Regulation, Signaling, Photosynthetic Efficacy, and Management. *Physiol. Plant.* **2021**, *172*, 935–962. [CrossRef]
132. Haworth, M.; Killi, D.; Materassi, A.; Raschi, A.; Centritto, M. Impaired Stomatal Control Is Associated with Reduced Photosynthetic Physiology in Crop Species Grown at Elevated [CO_2]. *Front. Plant Sci.* **2016**, *7*, 1568. [CrossRef] [PubMed]
133. Frary, A. *Plant Physiology and Developmentplant*; Lincoln, T., Eduardo, Z., Ian Max, M., Angus, M., Eds.; Sinauer Associates Inc.: Sunderland, MA, USA, 2015; Volume 117, pp. 397–399. ISBN 978-1-60535-255-8. [CrossRef]
134. Bielach, A.; Hrtyan, M.; Tognetti, V. Plants under Stress: Involvement of Auxin and Cytokinin. *Int. J. Mol. Sci.* **2017**, *18*, 1427. [CrossRef]
135. Smirnoff, N. The Role of Active Oxygen in the Response of Plants to Water Deficit and Desiccation. *New Phytol.* **1993**, *125*, 27–58. [CrossRef]
136. Naservafaei, S.; Sohrabi, Y.; Moradi, P.; Mac Sweeney, E.; Mastinu, A. Biological Response of Lallemantia Iberica to Brassinolide Treatment under Different Watering Conditions. *Plants* **2021**, *10*, 496. [CrossRef] [PubMed]
137. Karimmojeni, H.; Rezaei, M.; Tseng, T.-M.; Mastinu, A. Effects of Metribuzin Herbicide on Some Morpho-Physiological Characteristics of Two Echinacea Species. *Horticulturae* **2022**, *8*, 169. [CrossRef]
138. Rasheed, S.; Bashir, K.; Kim, J.-M.; Ando, M.; Tanaka, M.; Seki, M. The Modulation of Acetic Acid Pathway Genes in Arabidopsis Improves Survival under Drought Stress. *Sci. Rep.* **2018**, *8*, 7831. [CrossRef]
139. Luo, M.; Cheng, K.; Xu, Y.; Yang, S.; Wu, K. Plant Responses to Abiotic Stress Regulated by Histone Deacetylases. *Front. Plant Sci.* **2017**, *8*, 2147. [CrossRef] [PubMed]
140. Ashrafi, M.; Azimi-Moqadam, M.-R.; MohseniFard, E.; Shekari, F.; Jafary, H.; Moradi, P.; Pucci, M.; Abate, G.; Mastinu, A. Physiological and Molecular Aspects of Two Thymus Species Differently Sensitive to Drought Stress. *BioTech* **2022**, *11*, 8. [CrossRef] [PubMed]
141. Moradi, P.; Ford-Lloyd, B.; Pritchard, J. Metabolic Responses of *Thymus Vulgaris* to Water Deficit Stress. *Curr. Metab.* **2018**, *6*, 64–74. [CrossRef]
142. Moori, S.; Ahmadi-Lahijani, M.J. Hormopriming Instigates Defense Mechanisms in Thyme (*Thymus Vulgaris* L.) Seeds under Cadmium Stress. *J. Appl. Res. Med. Aromat. Plants* **2020**, *19*, 100268. [CrossRef]
143. Linhart, Y.B.; Keefover-Ring, K.; Mooney, K.A.; Breland, B.; Thompson, J.D. A Chemical Polymorphism in a Multitrophic Setting: Thyme Monoterpene Composition and Food Web Structure. *Am. Nat.* **2005**, *166*, 517–529. [CrossRef]
144. Loxdale, H.D.; Balog, A. Aphid Specialism as an Example of Ecological-Evolutionary Divergence. *Biol. Rev.* **2017**, *93*, 642–657. [CrossRef]
145. Adorjan, B.; Buchbauer, G. Biological Properties of Essential Oils: An Updated Review. *Flavour Fragr. J.* **2010**, *25*, 407–426. [CrossRef]
146. Kamkar, A.; Tooryan, F.; AkhondzadehBasti, A.; Misaghi, A.; Shariatifar, N. Chemical composition of summer savory (*Satureja hortensis* L.) essential oil and comparison of antioxidant activity with aqueous and alcoholic extracts. *J. Vet. Res.* **2013**, *68*, 183–190.
147. Adiguzel, A.; Ozer, H.; Kilic, H.; Cetin, B. Screening of Antimicrobial Activity of Essential Oil and Methanol Extract of *Satureja hortensis* on Foodborne Bacteria and Fungi. *Czech J. Food Sci.* **2007**, *25*, 81–89. [CrossRef]
148. Popovici, R.A.; Vaduva, D.; Pinzaru, I.; Dehelean, C.A.; Farcas, C.G.; Coricovac, D.; Danciu, C.; Popescu, I.; Alexa, E.; Lazureanu, V.; et al. A comparative study on the biological activity of essential oil and total hydro-alcoholic extract of *Satureja hortensis* L. *Exp. Ther. Med.* **2019**, *18*, 932–942. [CrossRef]
149. Ramos, M.; Beltrán, A.; Peltzer, M.; Valente, A.J.M.; Garrigós, M.D.C. Release and antioxidant activity of carvacrol and thymol from polypropylene active packaging films. *LWT Food Sci. Technol.* **2014**, *58*, 470–477. [CrossRef]
150. Chen, Q.; Gan, Z.; Zhao, J.; Wang, Y.; Zhang, S.; Li, J.; Ni, Y. In Vitro Comparison of Antioxidant Capacity of Cumin (*Cuminum cyminum* L.) Oils and Their Main Components. *LWT Food Sci. Technol.* **2014**, *55*, 632–637. [CrossRef]

151. Chua, L.S.; Lau, C.H.; Chew, C.Y.; Ismail, N.I.; Soontorngun, N. Phytochemical Profile of Orthosiphon Aristatus Extracts after Storage: Rosmarinic Acid and Other Caffeic Acid Derivatives. *Phytomedicine* **2018**, *39*, 49–55. [CrossRef] [PubMed]
152. Lesjak, M.; Beara, I.; Simin, N.; Pintać, D.; Majkić, T.; Bekvalac, K.; Orčić, D.; Mimica-Dukić, N. Antioxidant and Anti-Inflammatory Activities of Quercetin and Its Derivatives. *J. Funct. Foods* **2018**, *40*, 68–75. [CrossRef]
153. Aksu, M.I.; Özer, H. Effects of Lyophilized Water Extract ofSatureja Hortensison the Shelf Life and Quality Properties of Ground Beef. *J. Food Process. Preserv.* **2012**, *37*, 777–783. [CrossRef]
154. Vahidyan, H.; Sahari, M.A.; Barzegar, M.; Naghdi Badi, H. Application of Zataria multiflora Boiss. and *Satureja hortensis* L. essential oils as two natural antioxidants in Mayonnaise formulated with linseed oil. *J. Med. Plants* **2012**, *11*, 69–79.
155. Jafari, F.; Ghavidel, F.; Zarshenas, M.M. A Critical Overview on the Pharmacological and Clinical Aspects of Popular Satureja Species. *J. Acupunct. Meridian Stud.* **2016**, *9*, 118–127. [CrossRef]
156. Hazrati, H.; Saharkhiz, M.J.; Niakousari, M.; Moein, M. Natural Herbicide Activity of *Satureja hortensis* L. Essential Oil Nanoemulsion on the Seed Germination and Morphophysiological Features of Two Important Weed Species. *Ecotoxicol. Environ. Saf.* **2017**, *142*, 423–430. [CrossRef] [PubMed]
157. Banu, M.N.; Hoque, M.A.; Watanabe-Sugimoto, M.; Matsuoka, K.; Nakamura, Y.; Shimoishi, Y.; Murata, Y. Proline and Glycinebetaine Induce Antioxidant Defense Gene Expression and Suppress Cell Death in Cultured Tobacco Cells under Salt Stress. *J. Plant Physiol.* **2009**, *166*, 146–156. [CrossRef]
158. Mehdizadeh, L.; Moghaddam, M.; Lakzian, A. Alleviating Negative Effects of Salinity Stress in Summer Savory (*Satureja hortensis* L.) by Biochar Application. *Acta Physiol. Plant.* **2019**, *41*, 98. [CrossRef]
159. Nikee, E.; Pazoki, A.; Zahedi, H. Influences of Ascorbic Acid and Gibberellin on Alleviation of Salt Stress in Summer Savory (*Satureja hortensis* L.). *Int. J. Biosci. (IJB)* **2014**, *5*, 245–255. [CrossRef]
160. Miranshahi, B.; Sayyari, M. Methyl jasmonate mitigates drought stress injuries and affects essential oil of summer savory. *J. Agric. Sci. Tech.* **2016**, *18*, 1635–1645.
161. Zimowska, B. Diversity of fungi occurring on savory (*Satureja hortensis* L.). *Herba Pol.* **2010**, *2*, 56.
162. Wubben, J.P.; Mulder, W.; ten Have, A.; van Kan, J.A.; Visser, J. Cloning and Partial Characterization of Endopolygalacturonase Genes from Botrytis Cinerea. *Appl. Environ. Microbiol.* **1999**, *65*, 1596–1602. [CrossRef]
163. Shah, P.; Gutierrez-Sanchez, G.; Orlando, R.; Bergmann, C. A Proteomic Study of Pectin-Degrading Enzymes Secreted by Botrytis Cinerea Grown in Liquid Culture. *PROTEOMICS* **2009**, *9*, 3126–3135. [CrossRef]
164. Zwenger, S.; Basu, C. Plant terpenoids: Applications and future potentials. *Biotechnol. Mol. Biol. Rev.* **2008**, *3*, 1.
165. Heil, M.; Silva Bueno, J.C. Within-Plant Signaling by Volatiles Leads to Induction and Priming of an Indirect Plant Defense in Nature. *Proc. Natl. Acad. Sci. USA* **2007**, *104*, 5467–5472. [CrossRef]
166. Ton, J.; D'Alessandro, M.; Jourdie, V.; Jakab, G.; Karlen, D.; Held, M.; Mauch-Mani, B.; Turlings, T.C.J. Priming by Airborne Signals Boosts Direct and Indirect Resistance in Maize. *Plant J.* **2006**, *49*, 16–26. [CrossRef]
167. Cheong, J.-J.; Do Choi, Y. Methyl jasmonate as a vital substance in plants. *Trends Genet.* **2003**, *19*, 409–413. [CrossRef]
168. Rodríguez, A.A.; Stella, A.M.; Storni, M.M.; Zulpa, G.; Zaccaro, M.C. Effects of Cyanobacterial Extracellular Products and Gibberellic Acid on Salinity Tolerance in Oryza Satival. *Saline Syst.* **2006**, *2*, 7. [CrossRef] [PubMed]
169. Maggio, A.; Barbieri, G.; Raimondi, G.; De Pascale, S. Contrasting Effects of GA3 Treatments on Tomato Plants Exposed to Increasing Salinity. *J. Plant Growth Regul.* **2009**, *29*, 63–72. [CrossRef]
170. Ashraf, M.; Karim, F.; Rasul, E. Interactive effects of gibberellic acid (GA3) and salt stress on growth, ion accumulation and photosynthetic capacity of two spring wheat (*Triticum aestivum* L.) cultivars differing in salt tolerance. *Plant Growth Regul.* **2002**, *36*, 49–59. [CrossRef]
171. Wen, F.-p.; Zhang, Z.-h.; Bai, T.; Xu, Q.; Pan, Y.-h. Proteomics Reveals the Effects of Gibberellic Acid (GA3) on Salt-Stressed Rice (*Oryza sativa* L.) Shoots. *Plant Sci.* **2010**, *178*, 170–175. [CrossRef]
172. Lu, Y.Y.; Deng, X.P.; Kwak, S.S. Over expression of CuZn superoxide dismutase (CuZn SOD) and ascorbate peroxidase (APX) in transgenic sweet potato enhances tolerance and recovery from drought stress. *Afr. J. Biotechnol.* **2015**, *9*, 8378–8391.
173. Desikan, R. Aba, Hydrogen Peroxide and Nitric Oxide Signalling in Stomatal Guard Cells. *J. Exp. Bot.* **2003**, *55*, 205–212. [CrossRef]
174. Krishnan, N.; Dickman, M.B.; Becker, D.F. Proline Modulates the Intracellular Redox Environment and Protects Mammalian Cells against Oxidative Stress. *Free. Radic. Biol. Med.* **2008**, *44*, 671–681. [CrossRef]
175. Lamalakshmi Devi, E.; Kumar, S.; Basanta Singh, T.; Sharma, S.K.; Beemrote, A.; Devi, C.P.; Chongtham, S.K.; Singh, C.H.; Yumlembam, R.A.; Haribhushan, A.; et al. Adaptation Strategies and Defence Mechanisms of Plants during Environmental Stress. In *Medicinal Plants and Environmental Challenges*; Ghorbanpour, M., Varma, A., Eds.; Springer: Cham, Switzerland, 2017; pp. 359–413. [CrossRef]
176. HongBo, S.; ZongSuo, L.; MingAn, S.; ShiMeng, S.; ZanMin, H. Investigation on Dynamic Changes of Photosynthetic Characteristics of 10 Wheat (Triticum Aestivum L.) Genotypes during Two Vegetative-Growth Stages at Water Deficits. *Colloids Surf. B Biointerfaces* **2005**, *43*, 221–227. [CrossRef]
177. Blumwald, E.; Anil, G.; Allen, G. New directions for a diverse planet. In Proceedings of the 4th international crop science congress, Brisbane, Australia, 26 September–1 October 2004. Available online: http://www.cropscience.org.au (accessed on 7 February 2023).

178. Croteau, R.B.; Davis, E.M.; Ringer, K.L.; Wildung, M.R. (-)-Menthol biosynthesis and molecular genetics. *Naturwissenschaften* **2005**, *92*, 562–577. [CrossRef]
179. Gang, D.R.; Wang, J.; Dudareva, N.; Nam, K.H.; Simon, J.E.; Lewinsohn, E.; Pichersky, E. An investigation of the storage and biosynthesis of phenylpropenes in sweet basil. *Plant Physiol.* **2001**, *125*, 539–555. [CrossRef]
180. Xu, C.; Wang, M.; Zhou, L.; Quan, T.; Xia, G. Heterologous Expression of the Wheat Aquaporin Gene TATIP2;2 Compromises the Abiotic Stress Tolerance of Arabidopsis Thaliana. *PLoS ONE* **2013**, *8*, e79618. [CrossRef] [PubMed]
181. Moradi, P.; Ford-Lloyd, B.; Pritchard, J. Metabolomic Approach Reveals the Biochemical Mechanisms Underlying Drought Stress Tolerance in Thyme. *Anal. Biochem.* **2017**, *527*, 49–62. [CrossRef]
182. Kulak, M.; Jorrín-Novo, J.V.; Romero-Rodriguez, M.C.; Yildirim, E.D.; Gul, F.; Karaman, S. Seed Priming with Salicylic Acid on Plant Growth and Essential Oil Composition in Basil (*Ocimum basilicum* L.) Plants Grown under Water Stress Conditions. *Ind. Crops Prod.* **2021**, *161*, 113235. [CrossRef]
183. Alizadeh, A.; Moghaddam, M.; Asgharzade, A.; Sourestani, M.M. Phytochemical and Physiological Response of *Satureja hortensis* L. to Different Irrigation Regimes and Chitosan Application. *Ind. Crops Prod.* **2020**, *158*, 112990. [CrossRef]
184. Bakhshian, M.; Naderi, M.R.; Javanmard, H.R.; Bahreininejad, B. The Growth of Summer Savory (Satureja Hortensis) Affected by Fertilization and Plant Growth Regulators in Temperature Stress. *Biocatal. Agric. Biotechnol.* **2022**, *43*, 102371. [CrossRef]
185. Dikova, B. Tomato spotted wilt virus on some medicinal and essential oil-bearing plants in Bulgaria. *Bulg. J. Agric. Sci.* **2011**, *17*, 306–313.
186. Bansal, K.C.; Singh, A.K.; Wani, S.H. Plastid Transformation for Abiotic Stress Tolerance in Plants. *Methods Mol. Biol.* **2012**, *913*, 351–358. [CrossRef]
187. Khan, H.; Wani, S.H. Molecular approaches to enhance abiotic stresses tolerance. In *Innovations in Plant Science and Biotechnology*; Wani, S.H., Malik, C.P., Hora, A., Kaur, R., Eds.; Agrobios (India): Jodhpur, India, 2014; pp. 111–152. ISBN 978-81-7754-553-1.
188. Gonda, I.; Faigenboim, A.; Adler, C.; Milavski, R.; Karp, M.-J.; Shachter, A.; Ronen, G.; Baruch, K.; Chaimovitsh, D.; Dudai, N. The Genome Sequence of Tetraploid Sweet Basil, Ocimum Basilicum L., Provides Tools for Advanced Genome Editing and Molecular Breeding. *DNA Res.* **2020**, *27*, dsaa027. [CrossRef] [PubMed]
189. Pyne, R.M.; Koroch, A.R.; Wyenandt, C.A.; Simon, J.E. A Rapid Screening Approach to Identify Resistance to Basil Downy Mildew (*Peronospora Belbahrii*). *HortScience* **2014**, *49*, 1041–1045. [CrossRef]
190. Bi, H.H.; Song, Y.Y.; Zeng, R.S. Biochemical and molecular responses of host plants to mycorrhizal infection and their roles in plant defence. *Allelopath. J.* **2007**, *20*, 15–28.
191. Raffaele, S.; Vailleau, F.; Leger, A.; Joubès, J.; Miersch, O.; Huard, C.; Blée, E.; Mongrand, S.; Domergue, F.; Roby, D. A MYB Transcription Factor Regulates Very-Long-Chain Fatty Acid Biosynthesis for Activation of the Hypersensitive Cell Death Response in Arabidopsis. *Plant Cell* **2008**, *20*, 752–767. [CrossRef]
192. Pérez-Clemente, R.M.; Vives, V.; Zandalinas, S.I.; López-Climent, M.F.; Muñoz, V.; Gómez-Cadenas, A. Biotechnological Approaches to Study Plant Responses to Stress. *BioMed Res. Int.* **2013**, *2013*, 654120. [CrossRef] [PubMed]
193. Seo, P.J.; Park, C.M. MYB96-mediated abscisic acid signals induce pathogen resistance response by promoting salicylic acid biosynthesis in Arabidopsis. *New Phytol.* **2010**, *186*, 471–483. [CrossRef] [PubMed]
194. Agarwal, M.; Hao, Y.; Kapoor, A.; Dong, C.-H.; Fujii, H.; Zheng, X.; Zhu, J.-K. A R2R3 Type MYB Transcription Factor Is Involved in the Cold Regulation of CBF Genes and in Acquired Freezing Tolerance. *J. Biol. Chem.* **2006**, *281*, 37636–37645. [CrossRef] [PubMed]
195. Jiang, G.-L. Molecular Markers and Marker-Assisted Breeding in Plants. *Plant Breed. Lab. Fields* **2013**, *3*, 45–83. [CrossRef]
196. Gonda, I.; Milavski, R.; Adler, C.; Abu-Abied, M.; Tal, O.; Faigenboim, A.; Chaimovitsh, D.; Dudai, N. Genome-Based High-Resolution Mapping of Fusarium Wilt Resistance in Sweet Basil. *Plant Sci.* **2022**, *321*, 111316. [CrossRef]
197. Ben-Naim, Y.; Falach, L.; Cohen, Y. Transfer of Downy Mildew Resistance from Wild Basil (Ocimum Americanum) to Sweet Basil (*O. Basilicum*). *Phytopathology* **2018**, *108*, 114–123. [CrossRef]

genes

Article

Transcriptome and Metabolome Profiling Provide Insights into Flavonoid Synthesis in *Acanthus ilicifolius* Linn

Zhihua Wu [1], Zhen Wang [2], Yaojian Xie [1], Guo Liu [1], Xiuhua Shang [1] and Ni Zhan [1,2,*]

1 Research Institute of Fast-Growing Trees, Chinese Academy of Forestry, Zhanjiang 524022, China
2 School of Life Sciences, Langfang Normal University, Langfang 065000, China
* Correspondence: 1232134@lfnu.edu.cn; Tel.: +86-0316-188-387; Fax: +86-0316-2188-387

Abstract: *Acanthus ilicifolius* is an important medicinal plant in mangrove forests, which is rich in secondary metabolites with various biological activities. In this study, we used transcriptomic analysis to obtain differentially expressed genes in the flavonoid metabolic pathway and metabolomic methods to detect changes in the types and content in the flavonoid metabolic synthesis pathway. The results showed that DEGs were identified in the mature roots vs. leaves comparison (9001 up-regulated and 8910 down-regulated), mature roots vs. stems comparison (5861 up-regulated and 7374 down-regulated), and mature stems vs. leaves comparison (10,837 up-regulated and 11,903 down-regulated). Furthermore, two *AiCHS* genes and four *AiCHI* genes were up-regulated in the mature roots vs. stems of mature *A. ilicifolius*, and were down-regulated in mature stems vs. leaves, which were highly expressed in the *A. ilicifolius* stems. A total of 215 differential metabolites were found in the roots vs. leaves of mature *A. ilicifolius*, 173 differential metabolites in the roots vs. stems, and 228 differential metabolites in the stems vs. leaves. The metabolomic results showed that some flavonoids in *A. ilicifolius* stems were higher than in the roots. A total of 18 flavonoid differential metabolites were detected in the roots, stems, and leaves of mature *A. ilicifolius*. In mature leaves, quercetin-3-O-glucoside-7-O-rhamnoside, gossypitrin, isoquercitrin, quercetin 3,7-bis-O-β-D-glucoside, and isorhamnetin 3-O-β-(2″-O-acetyl-β-D-glucuronide) were found in a high content, while in mature roots, di-O-methylquercetin and isorhamnetin were the major compounds. The combined analysis of the metabolome and transcriptome revealed that DEGs and differential metabolites were related to flavonoid biosynthesis. This study provides a theoretical basis for analyzing the molecular mechanism of flavonoid synthesis in *A. ilicifolius* and provides a reference for further research and exploitation of its medicinal value.

Keywords: transcriptomics; metabolome profiling; flavonoid metabolism; *Acanthus ilicifolius*; medicinal value

Citation: Wu, Z.; Wang, Z.; Xie, Y.; Liu, G.; Shang, X.; Zhan, N. Transcriptome and Metabolome Profiling Provide Insights into Flavonoid Synthesis in *Acanthus ilicifolius* Linn. *Genes* **2023**, *14*, 752. https://doi.org/10.3390/genes14030752

Academic Editors: Wajid Zaman and Hakim Manghwar

Received: 20 February 2023
Revised: 15 March 2023
Accepted: 17 March 2023
Published: 20 March 2023

1. Introduction

A. ilicifolius Linn, a member of the family Acanthaceae, is a shrub or small tree that can reach up to two meters in height, with sturdy stalks. It grows in mangroves and tidal areas in tropical and subtropical regions [1,2]. Its unique growth environment results in a diverse array of structurally specific secondary metabolites, which exhibit a variety of biological activities and have potential medicinal value. The leaves, roots, and whole plant of *A. ilicifolius* are used in traditional medicine in India and China. It is considered an important medicinal plant in mangroves.

The properties of *A. ilicifolius* in traditional Chinese medicine are described as cold and mild in nature [3]. It is considered effective for clearing heat and detoxifying, eliminating swelling and dispersing knots, and relieving cough and asthma [4]. In China, India, and Southeast Asian countries such as Thailand, *A. ilicifolius* is used to treat conditions including neuralgia, rheumatism, snakebite, and paralysis [5]. In Thai folk medicine, the Acanthus genus is commonly used as a laxative, anti-inflammatory treatment for rheumatoid arthritis, antipyretic, anti-abscess, and anti-ulcer agent. Its leaves are used to treat rheumatism, snakebite, paralysis, and asthma. It is often used in combination with pepper as a tonic [6]. Pharmacological studies have shown that *A. ilicifolius* has hepatoprotective, antioxidant, antitumor, antibacterial, and antiviral effects [5,7,8]. Its chemical composition is mainly benzoxazin and benzoxazolinone alkaloids [9]. Further studies have revealed that it contains various types of compounds, such as flavonoids, triterpenoids, lignans, phenylethanoid glycosides, and sterols [5]. *A. ilicifolius* contains a large number of biologically active flavonoids, and the total flavonoid content can reach 3.82% [10]. In our study, the total flavonoid content of *A. ilicifolius* roots and stems reached 4.07 mass % (Figure S1).

Flavonoids are a diverse class of polyphenolic compounds [11] exhibiting a wide range of molecular structures and biological activities. They are derived from the metabolic pathway of phenylpropanoid and can be classified into seven major groups: flavonoids, flavonols, flavanones, anthocyanins, flavanols, isoflavones, and proanthocyanidins [12]. The basic backbone of flavonoids is a C6-C3-C6 diphenylpropane structure consisting of two benzene rings (A and B rings) interconnected with three carbon atoms in the center [11]. Flavonoids have been shown to possess various pharmacological effects and biological activities, such as anti-bacterial, anti-inflammatory, anti-oxidant, anti-aging, detoxifying, anti-cancer, and cardio- and cerebrovascular protective effects [13–16].

The biosynthesis of flavonoids is a crucial area of research in the secondary metabolism of plants [17]. The flavonoid metabolic pathway is a complex and multifactor-regulated network [18,19]. The phenylalanine metabolic pathway represents an important metabolic pathway [18,20], and a large number of secondary metabolites are produced from phenylalanine or tyrosine, including flavonoids and isoflavonoids, monolignans, various phenolic acids and stilbenes [20]. The biosynthesis of plant flavonoids is predominantly governed by the concerted action of structural and regulatory genes. Structural genes encode key enzymes that catalyze the flavonoid biosynthesis pathway, such as phenylalanine deaminase, cinnamic acid-4-hydroxylase, and 4-coumarate coenzyme-A-ligase. The activity of these structural enzymes and their gene expression levels directly impact the flavonoid content in the biosynthetic pathway. However, flavonoids are evolutionarily conserved in plants, and studying their metabolic pathways and regulatory mechanisms in different plant species is of significant biological importance [19].

The current study of the biosynthesis and accumulation of secondary metabolites has made substantial progress. Still, little is known about the effects of developmental and environmental factors on the synthesis and accumulation of secondary metabolites in medicinal plants [21]. In this study, we utilized transcriptomics and metabolomics approaches to analyze differential gene and metabolite expression in the roots, stems, and leaves of *A. ilicifolius* at various developmental stages. We aimed to identify key genes closely associated with flavonoid biosynthesis, elucidate the molecular mechanisms underlying flavonoid synthesis in *A. ilicifolius*, and analyze the metabolite types and accumulation patterns. The ultimate goal was to provide a reference for further research and exploitation of the medicinal value of *A. ilicifolius*' s flavonoid products.

2. Materials and Methods

2.1. Plant Material

A. ilicifolius plants were collected from the shore of the Dangjiang River in Beihai, Guangxi (21.595° N, 109.081° E), and identified by the experts from the Beihai Mangrove Breeding Base. Their young roots, stems, and leaves (mRY, mSY, and mLY) and old/mature roots, stems, and leaves (mRO, mSO, and mLO) were sampled in April 2020, treated with liquid nitrogen, and then stored at −80 °C for total RNA extraction. To ensure the samples' consistency, *A. ilicifolius* plants were collected from the same locations, dried, and used for chemical composition analysis. Three biological replicates were set up for each sample. In this paper, mRY, mSY, and mLY represent the young roots, young stems, and young leaves, respectively, and mOR, mOS, and mOL, represent the mature roots, mature stems, and mature leaves, respectively.

2.2. Transcription Experimental Methods

The total RNA was isolated from the young and mature tissues of roots, stems, and leaves with three biological replicates using FastPure Cell/Tissue Total RNA Isolation Kit (Vazyme, Nanjing, China). The RNA concentration was detected by Nanodrop. Agilent 2100 Bioanalyzer (Agilent Technologies, Santa Clara, CA, USA) was employed for detecting RIN (RNA Integrity Number) and 28S/18S values. Then, RNA integrity and quality were measured using 1% agarose electrophoresis.

Eukaryotic mRNA was enriched with magnetic beads with Oligo (dT), and the mRNA was randomly interrupted by adding Fragmentation Buffer; the first cDNA strand was synthesized with six-base random hexamers using mRNA as a template, and the second cDNA strand was synthesized by adding buffer, dNTPs, RNase H, and DNA polymerase I. The purified double-stranded cDNA was then end-repaired, A-tailed, and connected to sequencing junctions, followed by fragment size selection with AMPure XP beads; finally, the cDNA library was enriched by PCR. Sequencing was performed with the Illumina novaseq 6000 platform (Illumina, San Diego, CA, USA).

Before data analysis, the quality of raw reads was checked using FastQC (http://www.bioinformatics.babraham.ac.uk/projects/fastqc/, accessed on 10 November 2022). The impure reads were filtered, whereas the high-quality ones were acquired using Trinity by adopting default parameters, which were later utilized for constructing the distinct consensus sequences [22].

The filtered reads were first localized to the genome. The localized reads on the genome matched transcripts and the specified *A. ilicifolius* genome (unpublished data, Genome submission: SUB12869169) were used as a reference for sequence alignment and subsequent analysis in this project.

FPKM (fragments per kilobase of transcript per million fragments mapped) values [23] were estimated to measure gene expression by Cufflinks [24]. DESeq2 [25] was used to identify differentially expressed genes (DEGs). During differentially expressed gene detection, Fold Change ≥ 1.5 and p value < 0. 01 were used as screening criteria during the expressed gene detection [26].

Subsequently, the obtained genes were mapped against NR (Non-redundant protein sequence database in GenBank) [27], Swiss-Prot (Protein Sequence Database) [28], and COG (Cluster of Orthologous Groups of proteins) [29], KOG (Clusters of orthologous groups for eukaryotic complete genomes) [30] and KEGG (Kyoto Encyclopedia of Genes and Genomes) databases were used for sequence comparison [31].

2.3. Metabolome Analysis Methods

The samples were freeze-dried in a freeze-dryer (Scientz-100F, Ningbo Xinzhi Freeze Drying Equipment Co., Ningbo, China), and each sample was accurately weighed to 0.1 g after grinding and was dissolved in 0.6 mL of 70% methanol extract. The dissolved samples were refrigerated at 4 °C overnight, during which the samples were vortexed six times to improve the extraction rate. After centrifugation of the liquid (10,000× g, 10 min), the supernatant was aspirated, and the samples were filtered through a microporous membrane (0.22 μm pore size) and stored in a feed bottle for UPLC-MS/MS analysis [32,33]. Three biological replicate samples and four mixed samples for quality control were set up for widely targeted metabolome analysis.

2.4. UPLC Conditions

The data acquisition instrumentation system consisted mainly of Ultra Performance Liquid Chromatography (UPLC) (Shim-pack UFLC SHIMADZU CBM30A, https://www.shimadzu.com.cn/, accessed on 13 November 2022) and tandem mass spectrometry (Tandem mass spectrometry, MS/MS) (Applied Biosystems 4500 QTRAP, https://www.

shimadzu.com.cn/, accessed on 25 November 2022). Tandem mass spectrometry (MS/MS) (Applied Biosystems 4500 QTRAP, http://www.appliedbiosystems.com.cn/, accessed on 14 December 2022).

Liquid chromatography was performed using a Waters ACQUITY UPLC HSS T3 C18 (Waters, Milford, MA, USA) 1.8 µm, 2.1 mm × 100 mm column. The mobile phase consisted of ultra-pure water for phase A (with 0.04% acetic acid added) and acetonitrile for phase B (with 0.04% acetic acid added). The elution gradient involved a B-phase ratio of 5% at 0.00 min, which increased linearly to 95% within 10.00 min and was maintained at 95% for 1 min. From 11.00 to 11.10 min, the B-phase ratio decreased to 5% and equilibrated with 5% until 14 min. The flow rate was 0.35 mL/min, and the column temperature was set to 40 °C. The injection volume was 4 µL.

The mass spectrometry conditions used in this study involved electrospray ionization (ESI) at 550 °C, mass spectrometry at 5500 V, curtain gas (CUR) at 30 psi, collision-activated dissociation (CAD), and a mass spectrometer at 5000 V. A high parameter was set for each analysis. In the triple quadrupole (QQQ) mass spectrometer, each ion pair was scanned and detected using optimized declustering potential (DP) and collision energy (CE) values, as previously described by Chen et al. [34].

Specific fragment ions were compared to the reference for identifying secondary metabolites and additional amino acids [35]. Metabolites that shared close fragment ions were deemed to be identical compounds. Statistical analysis of secondary metabolite data was performed using Analyst 1.6.1 software (AB SCIEX, Framingham, MA, USA). Variable importance in projection (VIP) values were determined through partial least squares discriminant analysis. The differentially changed metabolites (DCMs) were selected based on the thresholds of VIP $\geq$ 1 and absolute Log2FC (fold change) $\geq$ 1. Identified metabolites were annotated using the KEGG Compound database (http://www.kegg.jp/kegg/compound/, accessed on 15 January 2023), and annotated metabolites were mapped to the KEGG Pathway database (http://www.kegg.jp/kegg/pathway.html, accessed on 21 January 2023). Pathways with significantly regulated metabolites were mapped into MSEA (metabolite sets enrichment analysis), and their significance was determined by the hypergeometric test's *p*-values.

2.5. qRT-PCR

Real-time quantitative reverse transcription PCR (qRT-PCR) was performed using the TUREscript first Stand cDNA SYNTHESIS Kit instructions (Edelweiss). The nine genes were validated by qRT-PCR using a thermal cycler (Analytikjena-qTOWER2.2, Jena, Germany). The 20 µL reverse transcription reaction system contained RNA (2 µL), 5 × RT reaction mixture (4 µL), Rondam primer/oligodT (50 pM, 0.8 µL), TUREscript H- RTase/RI mixture (200 U/Ml, 0.8 µL), and RNase-free ddH$_2$O (12.4 µL). Reverse transcription reaction conditions: 42 °C for 40 min, 65 °C for 10 min, after the reaction, cDNA was obtained and stored at −80 °C. The 10 µL qPCR reaction system contained 2 × SYBR Green Supermix (1 × 0.5 µL), forward primer (10 µM, 0.5 µL), and reverse primer (10 µM, 0.5 µL), cDNA (N/A, 1 µL), and ddH$_2$O (3 µL). The thermal profile was comprised of two segments: 30 s at 95 °C; 5 s denaturation at 95 °C, and 30 s annealing at 60 °C for a total of 40 cycles. Every assay was repeated thrice. Primer Express 2.0 Software (PE Applied Biosystems, Foster City, CA, USA) was applied in the primer design with default parameters. Supplementary Table S1 displayed the sequences of all the primers. Excel software (Microsoft Office, Redmond, WA, USA) and $2^{-\Delta\Delta Ct}$ [36] were used for data analysis, with the Actin gene as the reference.

3. Results

3.1. Transcriptomic Analysis and Differentially Expressed Genes

To explore genes related to flavonoid metabolism in *A. ilicifolius*, transcriptomic analysis was conducted on young and mature roots, stems, and leaves. Before transcriptome sequencing data were ready for subsequent analysis, quality control of raw data was required to obtain high quality data, also called clean reads. The overall clean read counts in each sample ranged from 20,006,083 to 29,156,661, producing a total of 123.5 Gb of clean data. The sequence reads were aligned to the reference genome of *A. ilicifolius*, with more than 84.65% of reads mapped successfully. The high-quality transcriptomic results were supported by a GC concentration of 47% and a Q30 score of over 93.39% (Table S2).

The high sensitivity of transcriptomic data enabled the detection of the gene expression levels. In this study, protein-coding genes with expression levels represented as FPKM values were sequenced and found to span six orders of magnitude, ranging from 0.01 to 10,000 (Figure S2). Pearson's Correlation Coefficient R (PCC) was used to assess biological repeat correlation and to screen for reliable differentially expressed genes. The r^2 in this study was close to 1, indicating a strong correlation of duplicate samples, which facilitated the follow-up analysis (Figure 1).

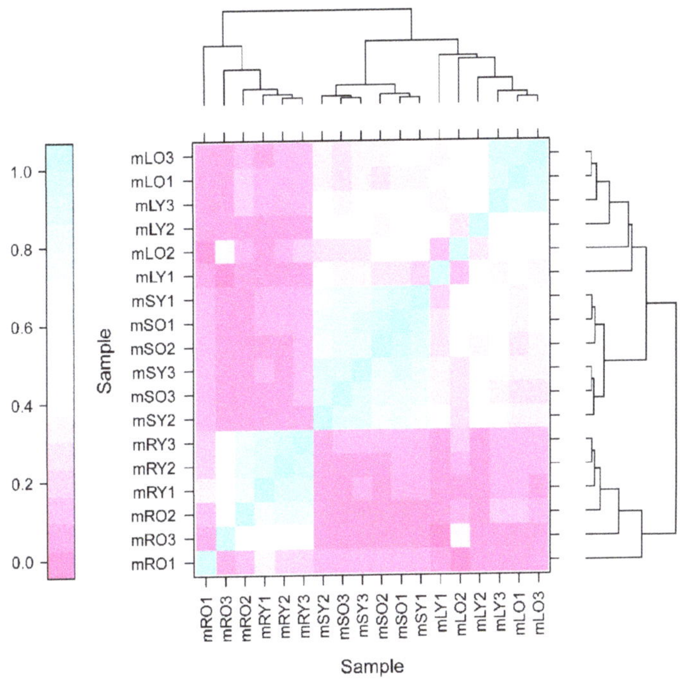

Figure 1. Heatmap of the expression correlation between two samples. Young roots, young stems, and young leaves are represented by mRY, mSY, and mLY, respectively, and mature roots, mature stems, and mature leaves are represented by mOR, mOS, and mOL, respectively. The numbers after the sample names indicate the plant numbers from different biological replicates.

This study focused on flavonoids, a class of secondary metabolites. Flavonoid accumulation was higher in the roots, stems, and leaves of mature *A. ilicifolius*, prompting an analysis of the DEGs in these organs. In mature *A. ilicifolius*, a total of 9001 DEGs were up-regulated and 8910 DEGs were down-regulated in the mature roots vs. leaves comparison (Figure 2a). Similarly, 5861 DEGs were up-regulated and 7374 DEGs were down-regulated in the mature roots vs. stems comparison (Figure 2b), and 10,837 DEGs were up-regulated and 11,903 DEGs were down-regulated in the mature stems vs. leaves comparison (Figure 2c).

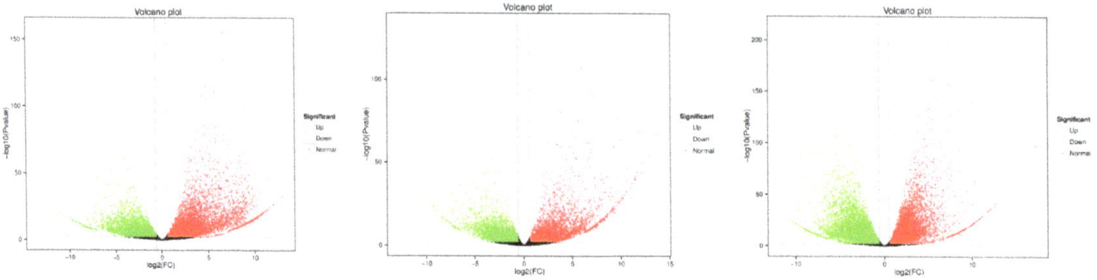

(**a**) (**b**) (**c**)

Figure 2. Volcano plots of the differential gene expression: (**a**) mRO vs. mLO; (**b**) mRO vs. mSO; (**c**) mSO vs. mLO. mOR, mOS, and mOL represent mature roots, mature stems, and mature leaves, respectively.

DEGs in mature roots vs. leaves, roots vs. stems, and stems vs. leaves of *A. ilicifolius* were mainly enriched in metabolic processes, single-organism process, cell, cell part membrane, organelle, binding, and catalytic activity (Figure 3). KEGG enrichment further indicated that the DEGs of mature roots vs. leaves and roots vs. stems of *A. ilicifolius* were mainly enriched in the plant–pathogen interaction and plant hormone signal transduction (Figure 4a,b). KEGG enrichment further indicated that DEGs of mature stems vs. leaves were mainly enriched in carbon metabolism and the biosynthesis of amino acids (Figure 4c).

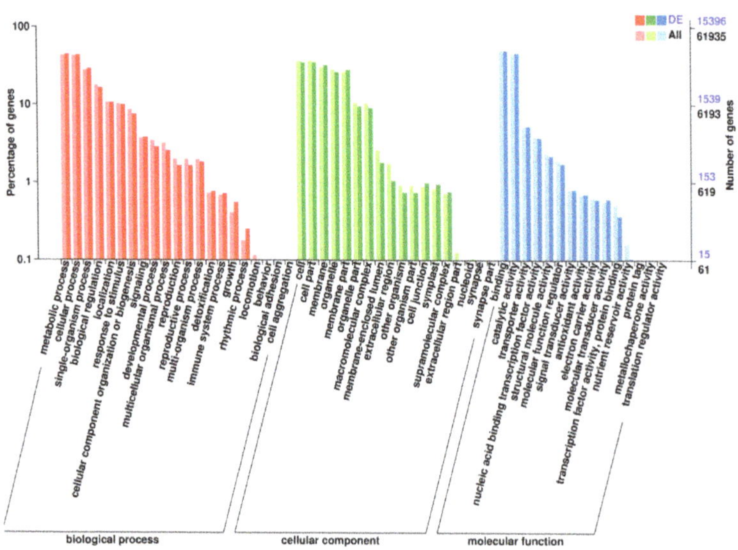

(**a**)

Figure 3. *Cont.*

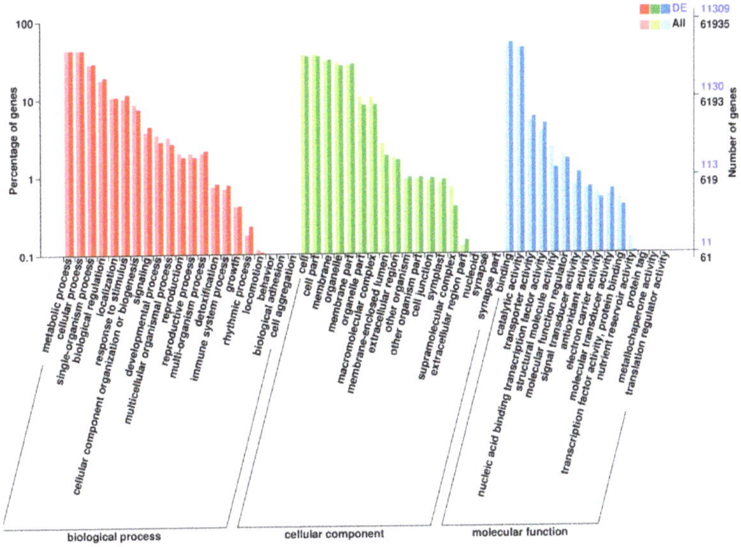

(b)

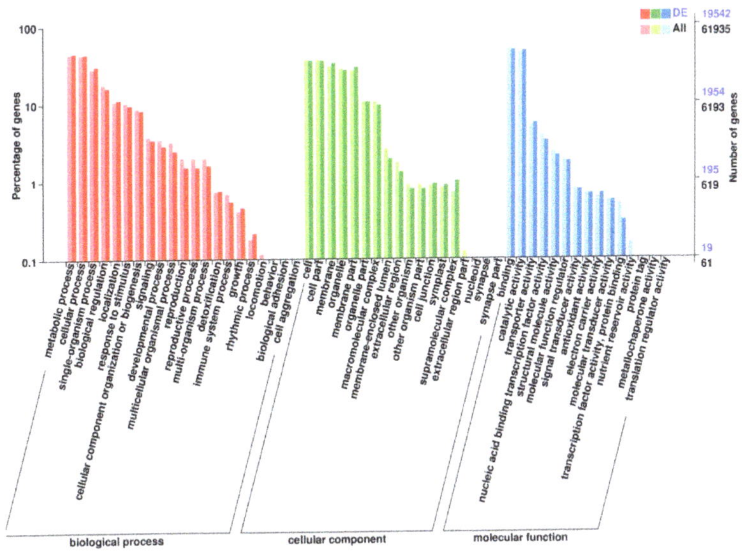

(c)

Figure 3. Statistical graph of GO annotation classification of differentially expressed genes. (**a**) mRO vs. mLO; (**b**) mRO vs. mSO; (**c**) mSO vs. mLO. mOR, mOS, and mOL represent mature roots, mature stems, and mature leaves, respectively.

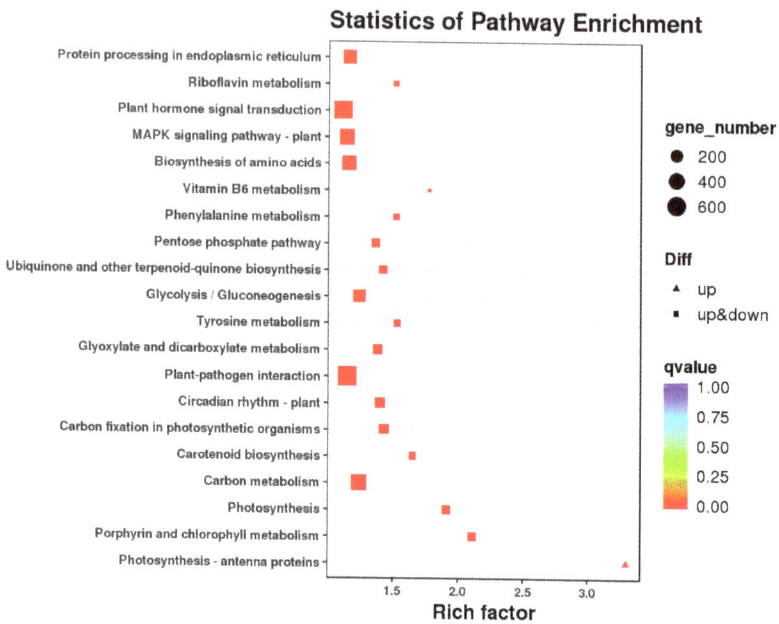

(a)

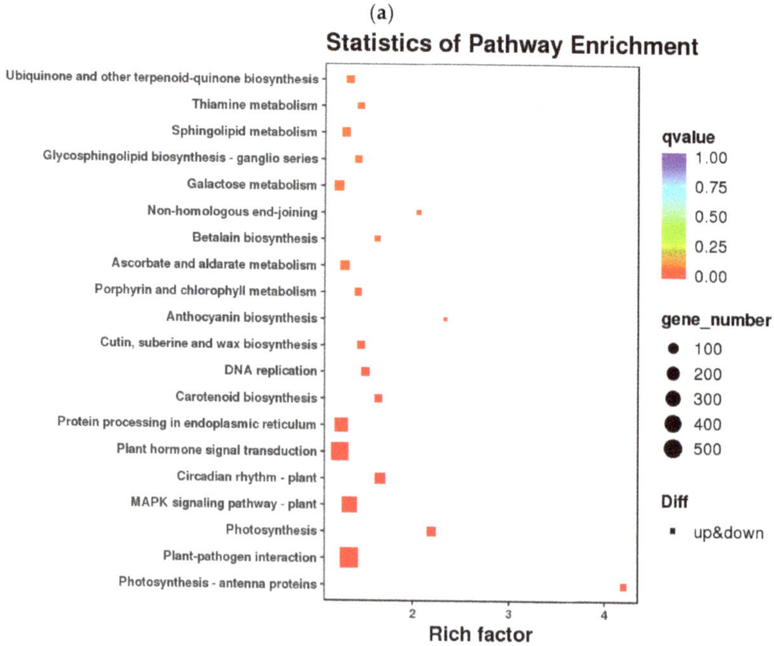

(b)

Figure 4. *Cont.*

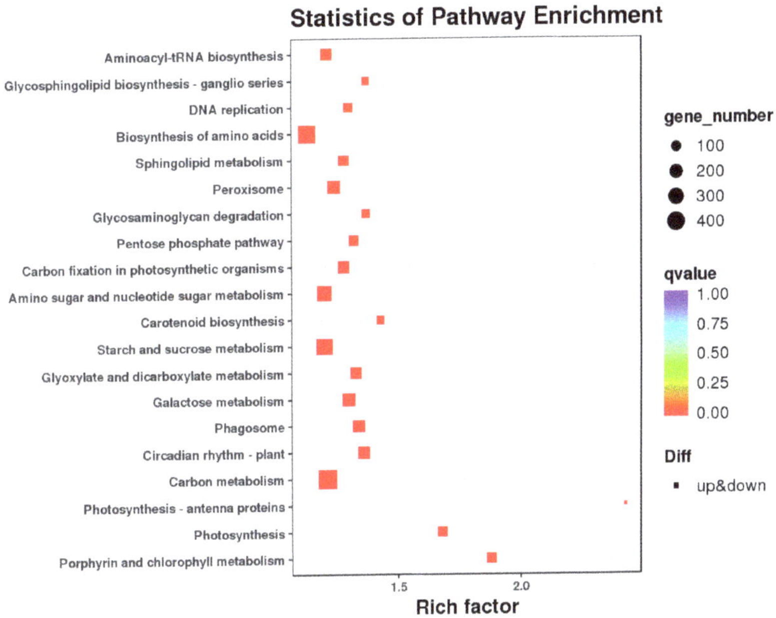

(**c**)

Figure 4. Enrichment scatters plot of differentially expressed genes in the KEGG pathway. (**a**) mRO vs. mLO; (**b**) mRO vs. mSO, (**c**) mSO vs. mLO. mOR, mOS, and mOL represent mature roots, mature stems, and mature leaves, respectively.

3.2. Flavonoid Biosynthesis Differential Genes

In mature *A. ilicifolius*, four DEGs related to flavonoid biosynthesis were identified in the roots vs. leaves comparison, with two up-regulated and two down-regulated DEGs (Figure 5a). The up-regulated DEGs were two *AiC4H* genes (Ail10G007150 and Ail34G006900), which encode cinnamate 4-hydroxylase. The down-regulated DEGs were *AiLAR* (Ail29G011710), which encodes leucoanthocyantin reducase, and *AiCHI1* (Ail07G00902), which encodes chalcone isomerase. In the roots vs. stems comparison, seven flavonoid biosynthesis-related DEGs were identified, with six being up-regulated and one down-regulated (Figure 5b). The up-regulated DEGs were two *AiCHS* genes (Ail32G006860 and Ail08G007320), which encode chalcone synthase, and four *AiCHI* genes (Ail36G002000, Ail10G010430, Ail34G010000, and Ail12G001980), which encode chalcone isomerase. The down-regulated DEG was *AiLAR* (Ail29G011710). In the stems vs. leaves comparison, eight flavonoid biosynthesis-related DEGs were identified, with two being up-regulated and six down-regulated (Figure 5c). The up-regulated DEGs were *AiCFI2* (Ail36G002000), which encodes chalcone flavanone isomerase, and *AiANR* (Ail41G004950), which encodes anthocyanidin reductase. The down-regulated DEGs were four *AiCHI* genes (Ail07G009020, Ail10G010430, Ail34G010000, and Ail31G008660), which encode chalcone isomerase, and two *AiCHS* genes (Ail0G005110 and Ail06G020390), which also encode chalcone synthase.

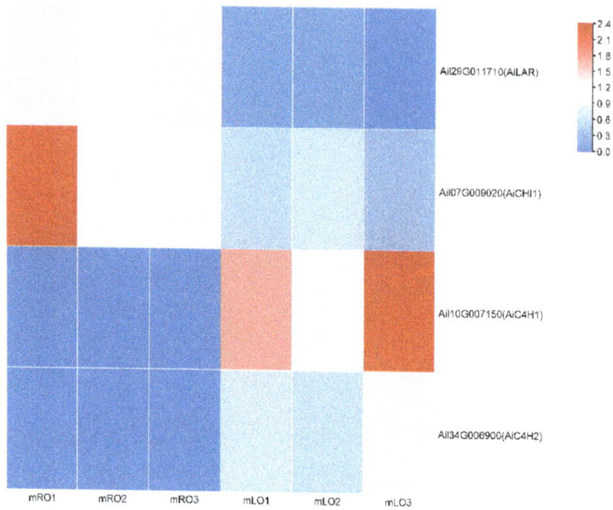

(**a**)

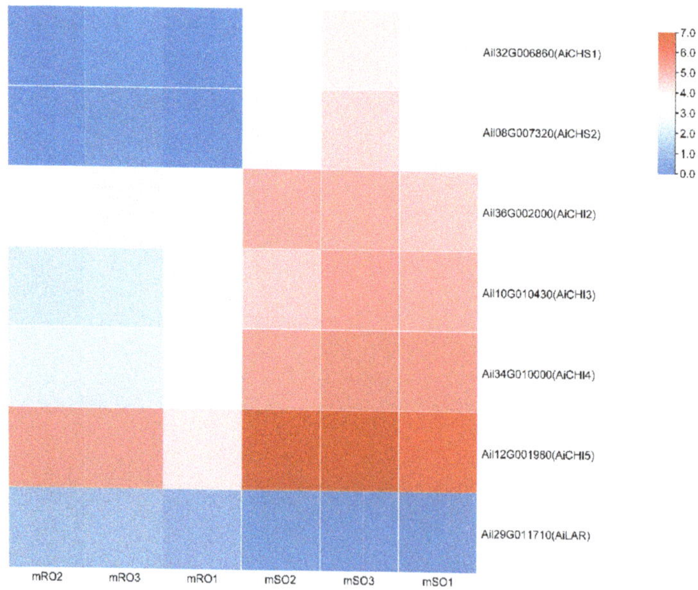

(**b**)

Figure 5. *Cont.*

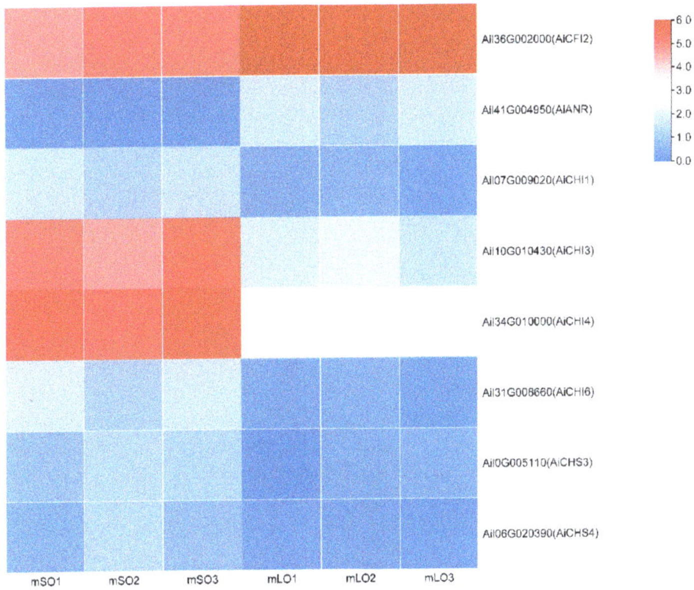

(**c**)

Figure 5. Differentially expressed genes related to flavonoid synthesis in *A. ilicifolius* parts. (**a**) mRO vs. mLO; (**b**) mRO vs. mSO, (**c**) mSO vs. mOR, mOS, and mOL represent mature roots, mature stems, and mature leaves, respectively. Each part has three biological replicates with the number following the sample name. The differential expression values of genes in the heatmap are treated with log2. Different *A. ilicifolius* genes and their related genes within the brackets are showed on the vertical axis.

The *AiLAR* gene encoding leucoanthocyanidin reductase was highly expressed in the mature roots of *A. ilicifolius* (Figure 6). *AiLAR* has been shown to reduce colorless anthocyanins and anthocyanins to flavanols, mainly catechin and epicatechin, and ultimately produce proanthocyanins through the processes of translocation, oxidation, and polymerization [37]. In contrast, *AiCHI* genes encoding chalcone isomerase were found to be highly expressed in the mature leaves and stems of *A. ilicifolius*. *CHI* catalyzes the specific cyclization of naringenin chalcone into naringenin, which is a common intermediate of several flavonoid subclasses, including flavonoids, flavanols, flavonols, anthocyanins, proanthocyanidins, terephthalic acid, and isoflavones [38].

3.3. *qRT-PCR*

The nine DEGs were randomly selected, and transcriptional data showed that the gene expression was greater in the mature roots compared with the mature stems and lowest in mature leaves. These were further screened for qRT-PCR analysis, and the results were consistent with the transcriptome results (Figure 7).

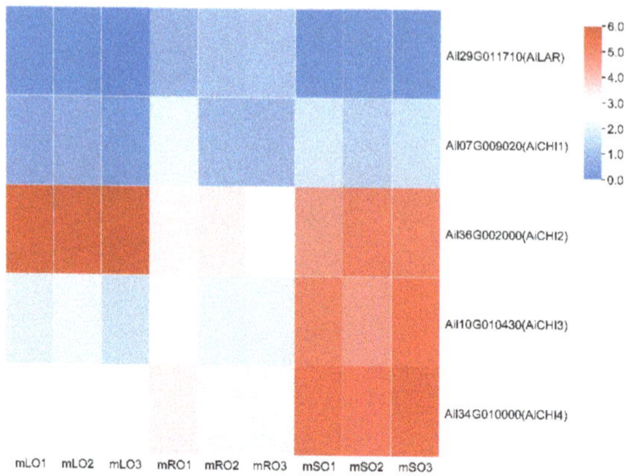

Figure 6. Expression heatmap of flavonoid synthesis-related differentially expressed genes in *A. ilicifolius* mature tissues. mOR, mOS, and mOL represent mature roots, mature stems, and mature leaves, respectively. The differential expression of genes in the heatmap are treated with log2.

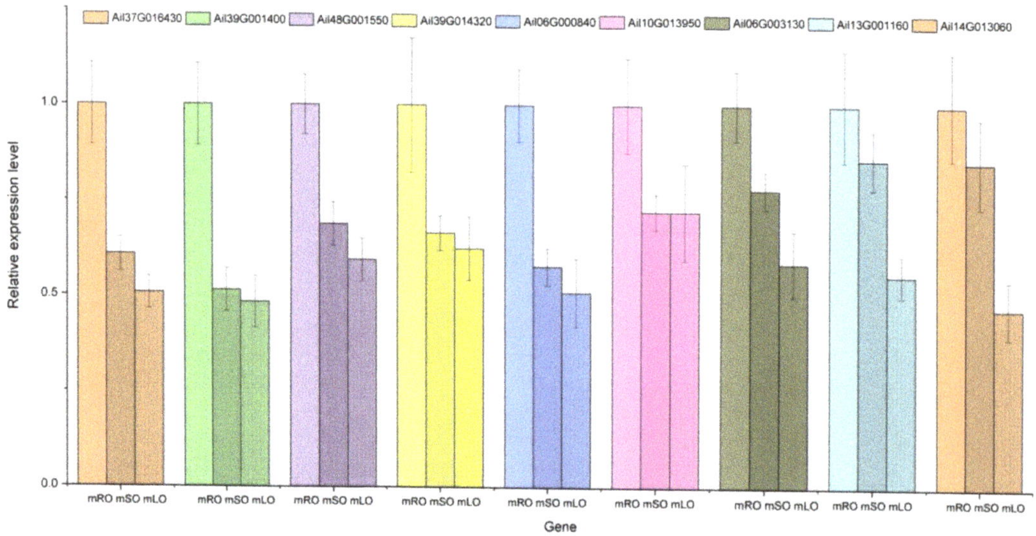

Figure 7. Expression levels of the candidate genes. The diagram distinguishes the nine genes with different colors. mOR, mOS, and mOL represent mature roots, mature stems, and mature leaves, respectively.

3.4. Metabolome Analysis and Differential Metabolites

PCA analysis classified overall variation as PC1 and PC2, contributing 35.43% and 22.11%, respectively (Figure 8a). The correlations between samples with the same organ were high, indicating a good repeatability of samples, as well as stability and reliability of the experimental data. The results of the hierarchical clustering of metabolite profiles of different samples showed that the same parts from the same stage were basically clustered into one subclass, while the root, stem, and leaf samples were clustered into three major classes, respectively (Figure 8b). The results of the metabolome analysis revealed that the metabolites in the young and mature parts were different.

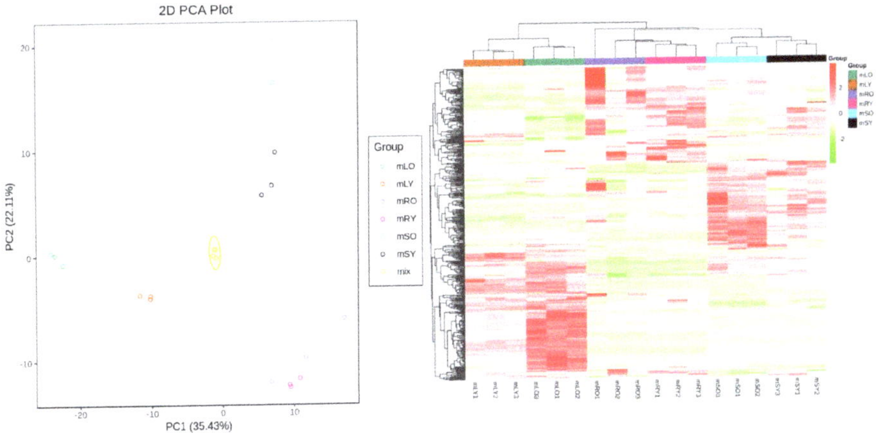

(a) (b)

Figure 8. Analysis of *A. ilicifolius* metabolome samples. (**a**) Principal component analysis (PCA) of the young and mature root, stem, and leaf samples. The ordinate represents the clustering of samples, and the abscissa represents the clustering of metabolites. (**b**) The overall clustering map of samples is based on metabolite profiles. The horizontal coordinate indicates the sample name (hierarchical clustering result), and the vertical coordinate indicates all metabolites (hierarchical clustering result). mRY, mSY, and mLY represent young roots, young stems, and young leaves, respectively, and mOR, mOS, and mOL represent mature roots, mature stems, and mature leaves, respectively.

Figure 9a shows four mixed samples represent the quality, which were prepared from a mixture of sample extracts and used to analyze the reproducibility of the samples under the same processing method. In Figure 9b, the number after the sample name represents the different biological replicate number of samples.

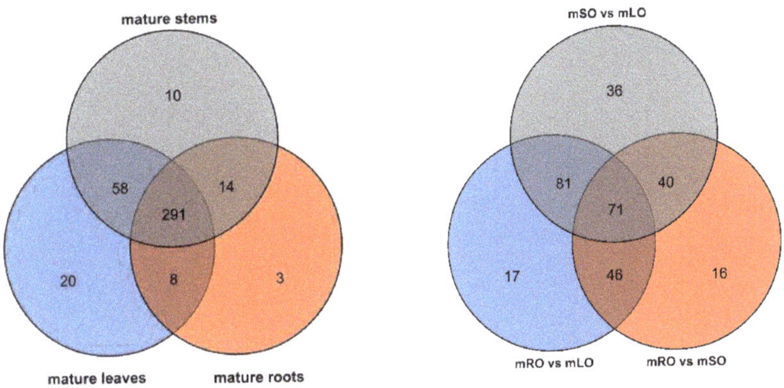

(a) (b)

Figure 9. Venn diagram of the number of compounds. (**a**) Metabolites detected in the three mature parts. (**b**) Number of differential metabolites among the three mature parts. mOR, mOS, and mOL represent mature roots, mature stem, and mature leaves, respectively.

A total of 407 metabolites were detected based on the UPLC-MS/MS detection platform and the local metabolic database (Biomarker Technologies, Beijing, China) (Table S3). Our results indicate there were 377, 316, and 373 metabolites in total that were detected for mature leaf, root, and stem (Figure 9a) Of the 215 differential metabolites in the roots vs. leaves of mature *A. ilicifolius*, (Figure 9b) 68 were flavonoids (Table S4). There were 173 differential metabolites in the roots vs. stems of mature *A. ilicifolius*. Among them, there were 59 flavonoids (Table S5). There were 228 differential metabolites in the stems vs. leaves of mature *A. ilicifolius*. Among them were 59 flavonoids (Table S6). Furthermore, the KEGG pathway enrichment analysis showed that the significantly enriched pathways were a biosynthesis of the secondary metabolism, flavonoid biosynthesis, and flavone and flavonol biosynthesis in the roots and leaves (Figure 10a); flavonoid biosynthesis in the roots and stems (Figure 10b); and flavone and flavonol biosynthesis in the stems and leaves of mature *A. ilicifolius* (Figure 10c).

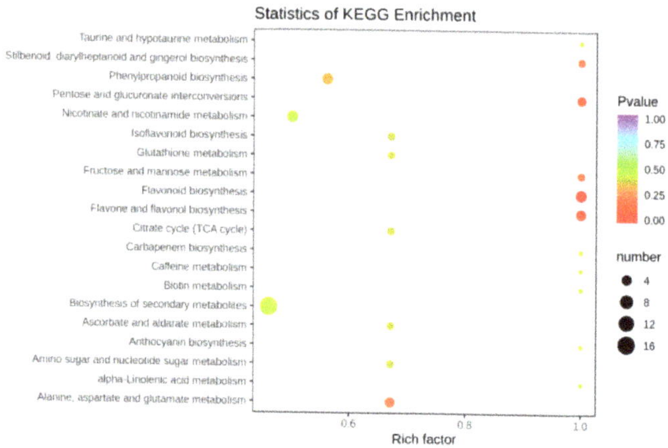

(a)

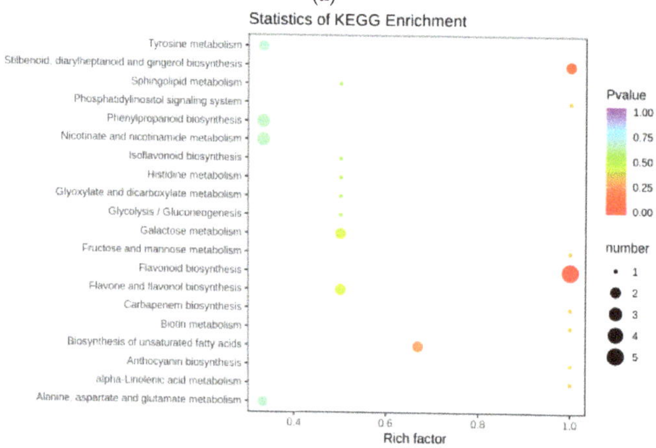

(b)

Figure 10. *Cont.*

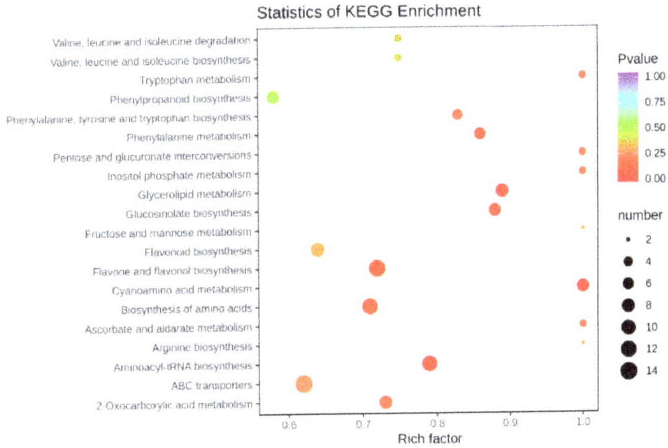

(c)

Figure 10. KEGG enrichment map of differential metabolites. (**a**) mRO vs. mLO; (**b**) mRO vs. mSO; (**c**) mSO vs. mLO. The horizontal coordinate indicates the rich factor of each pathway, the vertical coordinate is the pathway name, and the dot color is the p value; the redder it is, the more significant the enrichment. The size of the dots represents the number of differential metabolites enriched.

A total of 18 differential flavonoid metabolites were detected in the mature leaves vs. roots of *A. ilicifolius* (Figure 11). In mature leaves, quercetin-3-O-glucoside-7-O-rhamnoside, gossypitrin, isoquercitrin (quercetin 3-O-β-D-glucoside), quercetin 3,7-bis-O-β-D-glucoside, and isorhamnetin 3-O-β-(2″-O-acetyl-β-D-glucuronide) were found in high amounts, while in mature roots, di-O-methylquercetin and isorhamnetin were the major compounds (Table 1).

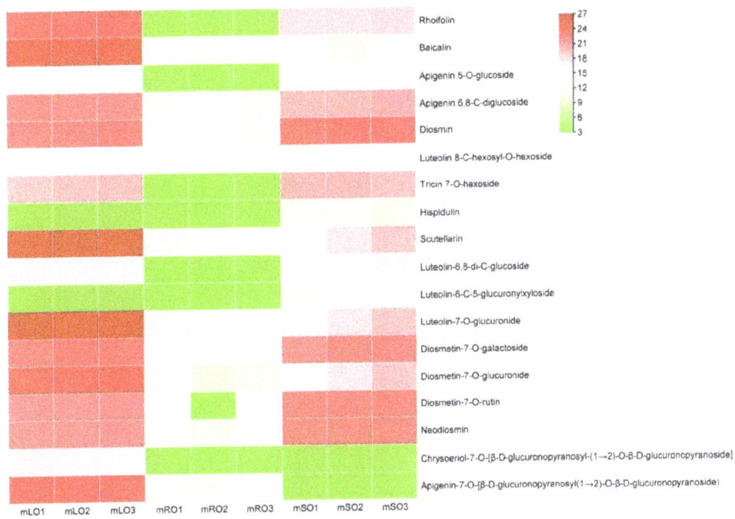

Figure 11. Flavonoid types and content heat map. Mature root, stem, and leaf samples are represented by mRO, mSO, and mLO, respectively. The flavonoid content in the heatmap is log2 processed.

Table 1. Differential metabolites associated with flavonoid synthesis within the roots and leaves of mature *A. ilicifolius*.

Index	Formula	Compounds	VIP	Log2FC	Type
GQ512006	$C_{27}H_{30}O_{16}$	Quercetin-3-O-glucoside-7-O-rhamnoside	1.24	16.60	up
mws1329	$C_{21}H_{20}O_{12}$	Gossypitrin	1.17	1.24	up
pmb3894	$C_{17}H_{14}O_{7}$	Di-O-methylquercetin	1.15	−1.59	down
pmp000580	$C_{21}H_{20}O_{12}$	Isoquercitrin(Quercetin 3-O-β-D-glucoside)	1.24	19.40	up
pmp000596	$C_{27}H_{30}O_{17}$	Quercetin 3,7-bis-O-β-D-glucoside	1.15	3.12	up
mws0066	$C_{16}H_{12}O_{7}$	Isorhamnetin	1.21	−1.87	down
pmn001645	$C_{24}H_{22}O_{14}$	Isorhamnetin 3-O-β-(2″-O-acetyl-β-D-glucuronide)	1.23	10.50	up

In the mature stems and roots of *A. ilicifolius*, quercetin-3-O-glucoside-7-O-rhamnoside, isoquercitrin (quercetin 3-O-β-D-glucoside), and quercetin 3,7-bis-O-β-D-glucoside were present in high amounts in mature stems, while quercetin 3-O-rhanosylgalactoside and bioquercetin were found in high amounts in mature roots (Table 2).

Table 2. Differential metabolites associated with flavonoid synthesis within the roots and stems of mature *A. ilicifolius*.

Index	Formula	Compounds	VIP	Log2FC	Type
GQ512006	$C_{27}H_{30}O_{16}$	Quercetin-3-O-glucoside-7-O-rhamnoside	1.32	12.80	up
Li512117	$C_{27}H_{30}O_{16}$	Quercetin 3-O-rhanosylgalactoside	1.19	−1.01	down
pmn001583	$C_{27}H_{30}O_{16}$	Bioquercetin	1.20	−1.20	down
pmp000580	$C_{21}H_{20}O_{12}$	Isoquercitrin(Quercetin 3-O-β-D-glucoside)	1.32	14.20	up
pmp000596	$C_{27}H_{30}O_{17}$	Quercetin 3,7-bis-O-β-D-glucoside	1.28	4.81	up

In the mature stems and leaves of *A. ilicifolius*, quercetin-3-O-glucoside-7-O-rhamnoside, isoquercitrin (quercetin 3-O-β-D-glucoside), and isorhamnetin 3-O-β-(2″-O-acetyl-β-D-glucuronide) were present in high amounts in mature leaves, while quercitrin, di-O-methylquercetin, quercetin 3,7-bis-O-β-D-glucoside, and isorhamnetin were the major compounds in mature stems (Table 3).

Table 3. Differential metabolites associated with flavonoid synthesis within stems and leaves of mature *A. ilicifolius*.

Index	Formula	Compounds	VIP	Log2FC	Type
GQ512006	$C_{27}H_{30}O_{16}$	Quercetin-3-O-glucoside-7-O-rhamnoside	1.16	3.84	up
mws0045	$C_{21}H_{20}O_{11}$	Quercitrin	1.07	−1.85	down
pmb3894	$C_{17}H_{14}O_{7}$	Di-O-methylquercetin	1.11	−2.21	down
pmp000580	$C_{21}H_{20}O_{12}$	Isoquercitrin(Quercetin 3-O-β-D-glucoside)	1.16	5.22	up
pmp000596	$C_{27}H_{30}O_{17}$	Quercetin 3,7-bis-O-β-D-glucoside	1.08	−1.69	down
mws0066	$C_{16}H_{12}O_{7}$	Isorhamnetin	1.14	−1.09	down
pmn001645	$C_{24}H_{22}O_{14}$	Isorhamnetin 3-O-β-(2″-O-acetyl-β-D-glucuronide)	1.17	10.5	up

3.5. Combined Analysis of Transcriptome and Metabolome Analysis

The combined analysis of the metabolome and transcriptome revealed that DEGs and differential metabolites occurred in the roots and leaves of mature *A. ilicifolius*, and were found to be enriched in 48 metabolic pathways (Figure 12a) in the roots and stems of mature *A. ilicifolius*. In the stems and leaves of mature *A. ilicifolius*, 38 metabolic pathways (Figure 12b) and 61 metabolic pathways were enriched (Figure 12c). The differential metabolites and DEGs were associated with the flavonoid, flavone, and flavonol biosynthesis.

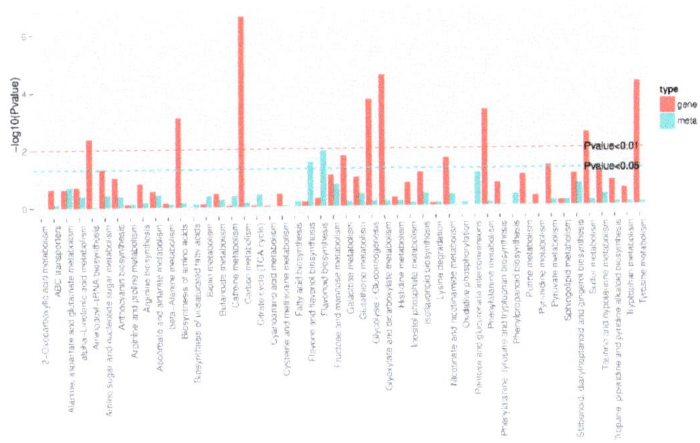

(a)

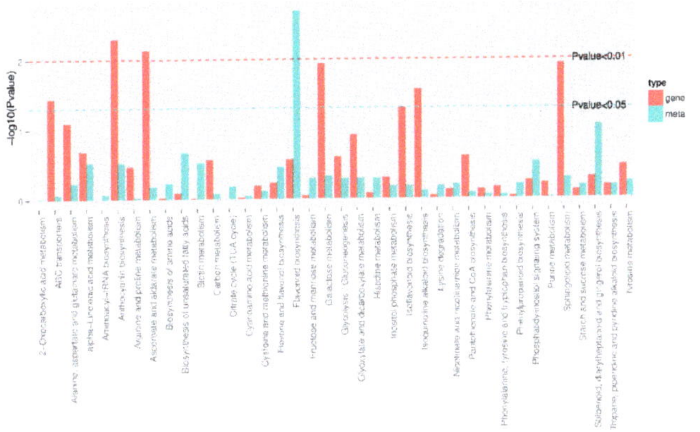

(b)

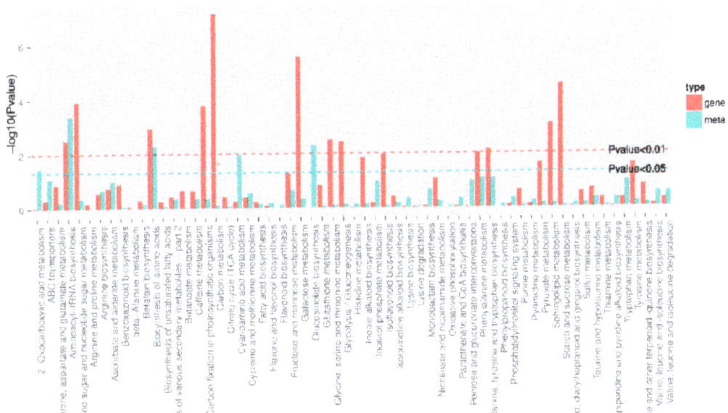

(c)

Figure 12. Column diagrams of differential metabolite and differential gene co-enrichment. (**a**) mRO vs. mLO; (**b**) mRO vs. mSO; (**c**) mSO vs. mLO.

4. Discussion

4.1. Transcriptomic Analysis of Key Genes in the Flavonoid Synthesis Pathway in A. ilicifolius

Higher plants share a common core flavonoid pathway, while distinct species frequently create specialized branches to adapt to varying environmental situations [39]. Flavonoid anabolism begins with the metabolic pathway of phenylpropanoids. Phenylalanine deaminase catalyzes the conversion of phenylalanine to cinnamic acid, which is then hydroxylated to coumaric acid by cinnamic acid-4-hydroxylase [11]. Coumaric acid is converted to coumaroyl-CoA by 4-coumarate coenzyme A ligase. Chalcone synthase then catalyzes the condensation of coumaroyl-CoA and malonyl-CoA to produce chalcone, a precursor of various flavonoids. Subsequently, chalcone enters various branching pathways to produce different flavonoid classes via a series of enzymatic reactions [40]. In the present study, in *A. ilicifolius* mature roots vs. leaves, two AiC4H up-regulated genes, which played an important role in flavonoid synthesis, belong to the upstream genes.

The flavonoid synthesis pathway is broadly divided into two phases [41,42]. The pre-synthesis stage includes chalcone synthase (CHS), chalcone isomerase (CHI), and flavanone 3-hydroxylase (F3H), which are common genes involved in all downstream flavonoid biosynthesis pathways [43]. In this study, we found that two *AiCHS* genes and four *AiCHI* genes were up-regulated in mature roots vs. stems of mature *A. ilicifolius*, and down-regulated in stems vs. leaves, and these genes were highly expressed in *A. ilicifolius* stems. While CHS plays an important role in the first stage of flavonoid biosynthesis, where it catalyzes the stepwise condensation of 4-coumaryl-CoA and malonyl-CoA into naringenin chalcone [43].

Plant flavonoids perform various biological functions, such as protecting against UV radiation, protecting plants from pathogens and herbivores, regulating auxin transport, and signaling between microorganisms and plants, and are important pigments for flowers, fruits, seeds, and leaves [43–45]. The metabolomic results of this study showed that some flavonoids in *A. ilicifolius* stems were higher than those in roots and leaves, indicating that they played an important role in the synthesis of *A. ilicifolius* flavonoids.

Naringenin is a widely distributed flavonoid intermediate that serves as a precursor for various flavonoid subclasses, including anthocyanins, proanthocyanidins, terephthalic acid, and isoflavones [38]. Naringin flavanones are hydroxylated by flavanone 3-hydroxylase (F3H) to form dihydrosanninol [46], which is further hydroxylated by flavanone 3′-hydroxylase (F3′H) and flavanone 3′5′-hydroxylase (F3′5′H), to produce dihydroquercetin and dihydromyricetin. Dihydromyricetin, dihydroquercetin, and dihydromyricetin are dihydroflavonol compounds, representing new branch points in the flavonoid synthesis pathway. Dihydroflavonols are able to catalyze the formation of flavonols under the action of flavonol synthase (FLS) [47]. Alternatively, dihydroflavonols can be reduced to colorless anthocyanins by dihydroflavonol reductase (DFR) and NADPH [48]. Further oxidation by anthocyanin synthase (ANS) and modifications by various glycosylation modifying enzymes results in the formation of different types of anthocyanins [49]. Transcriptomic data from this study revealed that the *AiLAR* gene encoded leucoanthocyantin reductase was down-regulated in mature roots vs. leaves, and in mature roots vs. stems of mature *A. ilicifolius*, and the *AiANR* gene was up-regulated in mature roots vs. stems, indicating its potential role in flavonoid synthesis in *A. ilicifolius*.

Colorless anthocyanin reductase (LAR) and anthocyanin reductase (ANR) are capable of reducing colorless anthocyanins and anthocyanins to flavanolic substances, such as catechins and epicatechins, which can then be translocated, oxidized, and polymerized to form proanthocyanidins [37].

4.2. Metabolomic Analysis of Flavonoid Content in A. ilicifolius

A. ilicifolius contains many biologically active flavonoids, and the total flavonoid extraction can reach 3.82% [10]. Studies have shown that apigenin [5], apigenin-7-O-β-D-glucuronide [50], apigenin 7-O-β-D-glucuronide methyl ester, quercetin, quercetin-3-O-β-D-glucopyranoside [51,52], mucoxanthin, acanthin, lignan-7-O-β-D-glucuronide, 7-O-α-Lglucopyranoside [51,52], mucoxanthin-rhamnopyranoside-(1→6)-O-β-D-glucopyranoside [50,53], and other flavonoids.

In our study, 68 differential flavonoids were detected in the roots vs. leaves of mature *A. ilicifolius*, 59 differential flavonoids in the roots vs. stems of mature *A. ilicifolius*, and 59 differential flavonoids in the stems vs. leaves of mature *A. ilicifolius*. Flavonoids vary greatly in content and type in different species [54]. The content of flavonoid substances also varies in different tissues or developmental stages [55]. The study results show that 18 differential flavonoid metabolites were detected in the roots, stems, and leaves of mature *A. ilicifolius*, and the active ingredients of many traditional Chinese medicines were flavonoid substances [56].

We also found that quercetin-3-O-glucoside-7-O-rhamnoside, gossypitrin, isoquercitrin (Quercetin 3-O-β-D glucoside), quercetin 3,7-bis-O-β-D-glucoside, and isorhamnetin 3-O-β-(2″-O-acetyl-β-D-glucuronide) were high in mature *A. ilicifolius* leaves, and di-O-methylquercetin and isorhamnetin were high in mature *A. ilicifolius* roots. The natural substance quercetin could inhibit the proliferation of human nasopharyngeal carcinoma cells (CNE1), causing them to undergo natural apoptosis [57]. The isorhamnetin in the extract had antioxidant, antiplatelet, and anticoagulant effects and could be used for the prevention and treatment of cardiovascular diseases [58]. In mature *A. ilicifolius* leaves, quercetin-3-O-glucoside-7-O-rhamnoside, isoquercitrin (Quercetin 3-O-β-D-glucoside), and isorhamnetin 3-O-β-(2″-O-acetyl-β-D-glucuronide) were high, and quercitrin, di-O-methylquercetin, quercetin 3,7-bis-O-β-D-glucoside, and isorhamnetin were high in the stems. Sorhamnetin-3-O-rutinoside can induce apoptosis in chronic granulomatous leukemia cells K562 by activating the activity of apoptotic factors such as caspase-8 and caspase-3 [59].

These results suggest that different parts of the plant may have unique flavonoid profiles, which could have implications for their potential medicinal or commercial applications. In addition, the metabolomic analysis revealed that some flavonoids were found in specific tissues/parts.

Currently, limited research has been conducted on the mechanism underlying flavonoid biosynthesis in *A. ilicifolius*. In our study, gene upregulation and the accumulation of metabolites (as shown in Figure 12) were not associated with several biological processes. These indicate that the biosynthesis of flavonoids is a complex process that may involve feedback loops or other regulatory mechanisms. Additionally, it is possible that many factors can influence gene expression and metabolic pathways. For example, some metabolites may be produced in response to environmental stimuli rather than as a result of the gene expression [60]. Therefore, it is important to use multiple approaches to study biological processes, including for analyzing both the gene expression and metabolite levels [61].

In the present study, differentially expressed genes were identified through transcriptome analysis, and their impact on various substances in the metabolic pathway was assessed. In addition, metabolite types and content changes within the synthetic pathway were detected using metabolomic methods. These findings are crucial for comprehensively analyzing the mechanism underlying flavonoid biosynthesis in the medicinal *A. ilicifolius*. The findings of this study provide valuable insights into the biosynthesis and diversity of flavonoids in *A. ilicifolius*, which could have implications for the development of new medicinal or commercial products.

5. Conclusions

Our study focused on *A. ilicifolius*, a significant medicinal plant in mangrove forests with valuable biological activities. We employed transcriptomic and metabolomic methods to explore changes in the flavonoid synthesis pathway, which revealed significant findings. The transcriptomic analysis identified differentially expressed genes (DEGs) with more DEGs down-regulated than up-regulated. Two *AiCHS* genes and four *AiCHI* genes were up-regulated in the mature roots vs. stems of *A. ilicifolius*, which were highly expressed in the stems. Metabolomic analysis also revealed that differential metabolites were related to flavonoid biosynthesis. Our study identified the key genes closely associated with flavonoid biosynthesis, elucidated the molecular mechanisms underlying flavonoid synthesis in *A. ilicifolius*, and analyzed the metabolite types and accumulation patterns, and will provide a reference for further research and for exploitation of the medicinal value of *A. ilicifolius*'s flavonoid products.

Supplementary Materials: The following supporting information can be downloaded at: https://www.mdpi.com/article/10.3390/genes14030752/s1. Figure S1: Total flavonoid content of roots, stems, and leaves of *A. ilicifolius*. Figure S2: Comparison of FPKM density distribution of each sample. Table S1: Sequences of primers. Table S2: RNA sequencing data and quality control. Table S3: All metabolites detected in roots, stems, and leaves of *A. ilicifolius*. Table S4: Differential metabolites in the roots vs. leaves of mature *A. ilicifolius*. Table S5: Differential metabolites in the roots vs. stems of mature *A. ilicifolius*. Table S6: Differential metabolites in the stems vs. leaves of mature *A. ilicifolius*.

Author Contributions: Z.W. (Zhihua Wu): conceived and designed the experiments, performed the experiments, analyzed the data, and wrote the paper; Z.W. (Zhen Wang): analyzed the data and wrote the paper; Y.X.: conceived and designed the experiments; G.L. and X.S.: performed the experiments and analyzed the data; N.Z.: analyzed the data, performed the experiments, wrote the paper, and authored or reviewed drafts of the paper. All authors have read and agreed to the published version of the manuscript.

Funding: This work was supported by the Key-Area Research and Development Program of Guangdong Province (no. 2020B020214001-ZKT03) and the Fundamental Research Funds for the Central Non-profit Research Institution of Chinese Academy of Forestry (CAFYBB2022MA005).

Institutional Review Board Statement: Not applicable.

Informed Consent Statement: Not applicable.

Data Availability Statement: The transcriptome raw read data that support the findings of this study are available in the NCBI BioProject database (transcriptome raw read data, SRA submission: SUB12510474; and *A. ilicifolius* genome sequence, genome submission: SUB12869169).

Acknowledgments: The authors would like to thank the editors and anonymous reviewers for their constructive comments and suggestions.

Conflicts of Interest: The authors declare no conflict of interest.

References

1. Barker, R.M. A Taxonomic Revision of Australian *Acanthaceae*. *J. Adel. Bot. Gard.* **1986**, *9*, 64–75.
2. Wu, J.; Zhang, S.; Xiao, Q.; Li, Q.; Huang, J.; Long, L.; Huang, L. Phenylethanoid and Aliphatic Alcohol Glycosides from *Acanthus ilicifolius*. *Phytochemistry* **2003**, *63*, 491–495. [CrossRef] [PubMed]
3. He, K.J.; Li, J.; Zhang, J.H. *Compilation of Medicinal Properties of Raw Herbs*; Guangdong Science and Technology Publishing House: Guangzhou, China, 2018.
4. Huang, L.Q. *National Compilation of Chinese Herbal Medicines*, 3rd ed.; People's Medical Publishing House: Beijing, China, 2014.
5. Babu, B.H.; Shylesh, B.S.; Padikkala, J. Antioxidant and Hepatoprotective Effect of *Acanthus ilicifolius*. *Fitoterapia* **2001**, *72*, 272–277. [CrossRef]
6. Babu, B.H.; Shylesh, B.S.; Padikkala, J. Tumour Reducing and Anticarcinogenic Activity of *Acanthus ilicifolius* in Mice. *J. Ethnopharmacol.* **2002**, *79*, 27–33. [CrossRef] [PubMed]
7. Huo, C.; Liang, H.; Tu, G.; Zhao, Y.; Lin, W. A New 5,11-Epoxymegastigmane Glucoside from *Acanthus ilicifolius*. *Nat. Prod. Res.* **2008**, *22*, 896–900. [CrossRef] [PubMed]
8. Minocha, P.K.; Tiwari, K.P. A Triterpenoidal Saponin from Roots of *Acanthus illicifolius*. *Phytochemistry* **1981**, *20*, 135–137. [CrossRef]
9. Kumar, K.M.S.; Puia, Z.; Barik, R.; Gorain, B.; Roy, D.; Adhikari, D.; Sen, T. The Gastroprotective Role of *Acanthus ilicifolius*—A Study to Unravel the Underlying Mechanism of Anti-Ulcer Activity. *Sci. Pharm.* **2012**, *80*, 701–717. [CrossRef]
10. Hu, H.Y.; Zhu, C.; Zhang, C.C.; Lu, C.Y. Study on Optimum Technology for Extracting Total Flavonoids from *Acanthus illicifolius* by Orthogonal Design. *Chin. J. Mar. Drugs* **2009**, *28*, 30–34.
11. Nabavi, S.M.; Šamec, D.; Tomczyk, M.; Milella, L.; Russo, D.; Habtemariam, S.; Suntar, I.; Rastrelli, L.; Daglia, M.; Xiao, J.; et al. Flavonoid Biosynthetic Pathways in Plants: Versatile Targets for Metabolic Engineering. *Biotechnol. Adv.* **2020**, *38*, 107316. [CrossRef]
12. Yonekura-Sakakibara, K.; Higashi, Y.; Nakabayashi, R. The Origin and Evolution of Plant Flavonoid Metabolism. *Front. Plant Sci.* **2019**, *10*, 943. [CrossRef]

13. García-Lafuente, A.; Guillamón, E.; Villares, A.; Rostagno, M.A.; Martínez, J.A. Flavonoids as Anti-Inflammatory Agents: Implications in Cancer and Cardiovascular Disease. *Inflamm. Res.* **2009**, *58*, 537–552. [CrossRef] [PubMed]
14. Uddin, S.J.; Grice, D.; Tiralongo, E. Evaluation of Cytotoxic Activity of Patriscabratine, Tetracosane and Various Flavonoids Isolated from the Bangladeshi Medicinal Plant *Acrostichum Aureum*. *Pharm. Biol.* **2012**, *50*, 1276–1280. [CrossRef]
15. Hudson, E.A.; Dinh, P.A.; Kokubun, T.; Simmonds, M.S.; Gescher, A. Characterization of Potentially Chemopreventive Phenols in Extracts of Brown Rice That Inhibit the Growth of Human Breast and Colon Cancer Cells. *Cancer Epidemiol. Biomark. Prev.* **2000**, *9*, 1163–1170.
16. Abbas, A.; Shah, A.; Shah, A.; Nadeem, M.; Alsaleh, A.; Javed, T.; Alotaibi, S.; Abdelsalam, N. Genome-Wide Analysis of Invertase Gene Family, and Expression Profiling under Abiotic Stress Conditions in Potato. *Biology* **2022**, *11*, 539. [CrossRef] [PubMed]
17. Liu, W.; Feng, Y.; Yu, S.; Fan, Z.; Li, X.; Li, J.; Yin, H. The Flavonoid Biosynthesis Network in Plants. *Int. J. Mol. Sci.* **2021**, *22*, 12824. [CrossRef]
18. Yang, H.; Li, H.; Li, Q. Biosynthetic Regulatory Network of Flavonoid Metabolites in Stems and Leaves of *Salvia miltiorrhiza*. *Sci. Rep.* **2022**, *12*, 18212. [CrossRef]
19. Huang, X.; Sun, G.; Li, Q.; Yan, H. Transcriptome Analysis Reveals Regulatory Networks and Hub Genes in the Flavonoid Metabolism of *Rosa roxburghii*. *Horticulturae* **2023**, *9*, 233. [CrossRef]
20. Betti, M.; Garcia-Calderon, M.; Perez-Delgado, C.M.; Credali, A.; Pal'ove-Balang, P.; Estivill, G.; Repak, M.; Vega, J.M.; Galvan, F.; Marquez, A.J. Reassimilation of Ammonium in *Lotus japonicus*. *J. Exp. Bot.* **2014**, *65*, 5557–5566. [CrossRef]
21. Li, Y.; Kong, D.; Fu, Y.; Sussman, M.R.; Wu, H. The Effect of Developmental and Environmental Factors on Secondary Metabolites in Medicinal Plants. *Plant Physiol. Biochem.* **2020**, *148*, 80–89. [CrossRef]
22. Grabherr, M.G.; Haas, B.J.; Yassour, M.; Levin, J.Z.; Thompson, D.A.; Amit, I.; Adiconis, X.; Fan, L.; Raychowdhury, R.; Zeng, Q.; et al. Full-Length Transcriptome Assembly from RNA-Seq Data without a Reference Genome. *Nat. Biotechnol.* **2011**, *29*, 644–652. [CrossRef]
23. Trapnell, C.; Roberts, A.; Goff, L.; Pertea, G.; Kim, D.; Kelley, D.R. Differential Gene and Transcript Expression Analysis of RNA-Seq Experiments with TopHat and Cufflinks. *Nat. Protoc.* **2012**, *7*, 562–578. [CrossRef]
24. Florea, L.; Song, L.; Salzberg, S.L. Thousands of Exon Skipping Events Differentiate among Splicing Patterns in Sixteen Human Tissues. *F1000Research* **2013**, *2*, 188. [CrossRef]
25. Love, M.I.; Huber, W.; Anders, S. Moderated Estimation of Fold Change and Dispersion for RNA-Seq Data with DESeq2. *Genome Biol.* **2014**, *15*, 550. [CrossRef] [PubMed]
26. Zhu, A.; Greaves, I.K.; Liu, P.-C.; Wu, L.; Dennis, E.S.; Peacock, W.J. Early Changes of Gene Activity in Developing Seedlings of Arabidopsis Hybrids Relative to Parents May Contribute to Hybrid Vigour. *Plant J.* **2016**, *88*, 597–607. [CrossRef] [PubMed]
27. Deng, Y.; Li, J.; Wu, S.; Zhu, Y.; Chen, Y.; He, F. Integrated Nr Database in Protein Annotation System and Its Localization. *Comput. Eng.* **2006**, *32*, 71–72.
28. Apweiler, R. UniProt: The Universal Protein Knowledgebase. *Nucleic Acids Res.* **2004**, *32*, 115–119. [CrossRef] [PubMed]
29. Tatusov, R.L. The COG Database: A Tool for Genome-Scale Analysis of Protein Functions and Evolution. *Nucleic Acids Res.* **2000**, *28*, 33–36. [CrossRef]
30. Koonin, E.V.; Fedorova, N.D.; Jackson, J.D.; Jacobs, A.R.; Krylov, D.M.; Makarova, K.S.; Mazumder, R.; Mekhedov, S.L.; Nikolskaya, A.N.; Rao, B.; et al. A Comprehensive Evolutionary Classification of Proteins Encoded in Complete Eukaryotic Genomes. *Genome Biol.* **2004**, *5*, R7. [CrossRef]
31. Kanehisa, M. The KEGG Resource for Deciphering the Genome. *Nucleic Acids Res.* **2004**, *32*, 277–280. [CrossRef]
32. Want, E.J.; Wilson, I.D.; Gika, H.; Theodoridis, G.; Plumb, R.S.; Shockcor, J.; Holmes, E.; Nicholson, J.K. Global Metabolic Profiling Procedures for Urine Using UPLC–MS. *Nat. Protoc.* **2010**, *5*, 1005–1018. [CrossRef]
33. Dunn, W.B.; Broadhurst, D.; Begley, P.; Zelena, E.; Francis-McIntyre, S.; Anderson, N.; Brown, M.; Knowles, J.D.; Halsall, A.; Haselden, J.N.; et al. Procedures for Large-Scale Metabolic Profiling of Serum and Plasma Using Gas Chromatography and Liquid Chromatography Coupled to Mass Spectrometry. *Nat. Protoc.* **2011**, *6*, 1060–1083. [CrossRef]
34. Chen, W.; Gong, L.; Guo, Z.; Wang, W.; Zhang, H.; Liu, X.; Yu, S.; Xiong, L.; Luo, J. A Novel Integrated Method for Large-Scale Detection, Identification, and Quantification of Widely Targeted Metabolites: Application in the Study of Rice Metabolomics. *Mol. Plant* **2013**, *6*, 1769–1780. [CrossRef] [PubMed]
35. Wang, S.; Tu, H.; Wan, J.; Chen, W.; Liu, X.; Luo, J.; Xu, J.; Zhang, H. Spatio-Temporal Distribution and Natural Variation of Metabolites in Citrus Fruits. *Food Chem.* **2016**, *199*, 8–17. [CrossRef]
36. Livak, K.; Schmittgen, T.D. Analysis of Relative Gene Expression Data Using Real Time Quantitative PCR and the $2-\Delta$Ct Method. *Methods* **2001**, *25*, 402–408. [CrossRef] [PubMed]
37. Zhou, M.; Wei, L.; Sun, Z.; Gao, L.; Meng, Y.; Tang, Y.; Wu, Y. Production and Transcriptional Regulation of Proanthocyanidin Biosynthesis in Forage Legumes. *Appl. Microbiol. Biotechnol.* **2015**, *99*, 3797–3806. [CrossRef] [PubMed]
38. Ngaki, M.N.; Louie, G.V.; Philippe, R.N.; Manning, G.; Pojer, F.; Bowman, M.E.; Li, L.; Larsen, E.; Wurtele, E.S.; Noel, J.P. Evolution of the Chalcone-Isomerase Fold from Fatty-Acid Binding to Stereospecific Catalysis. *Nature* **2012**, *485*, 530–533. [CrossRef] [PubMed]
39. García-Calderón, M.; Pérez-Delgado, C.M.; Palove-Balang, P.; Betti, M.; Márquez, A.J. Flavonoids and Isoflavonoids Biosynthesis in the Model Legume *Lotus japonicus*; Connections to Nitrogen Metabolism and Photorespiration. *Plants* **2020**, *9*, 774. [CrossRef]

40. Winkel-Shirley, B. It Takes a Garden. How Work on Diverse Plant Species Has Contributed to an Understanding of Flavonoid Metabolism. *Plant Physiol.* **2001**, *127*, 1399–1404. [CrossRef]
41. Gonzalez, A.; Zhao, M.; Leavitt, J.M.; Lloyd, A.M. Regulation of the Anthocyanin Biosynthetic Pathway by the TTG1/BHLH/Myb Transcriptional Complex in Arabidopsis Seedlings. *Plant J.* **2008**, *53*, 814–827. [CrossRef]
42. Matsui, K.; Walker, A.R. Biosynthesis and Regulation of Flavonoids in Buckwheat. *Breed. Sci.* **2020**, *70*, 74–84. [CrossRef]
43. Deng, X.; Bashandy, H.; Ainasoja, M.; Kontturi, J.; Pietiäinen, M.; Laitinen, R.A.E.; Albert, V.A.; Valkonen, J.P.T.; Elomaa, P.; Teeri, T.H. Functional Diversification of Duplicated Chalcone Synthase Genes in Anthocyanin Biosynthesis of *Gerbera Hybrida*. *New Phytol.* **2014**, *201*, 1469–1483. [CrossRef] [PubMed]
44. Jaakola, L. New Insights into the Regulation of Anthocyanin Biosynthesis in Fruits. *Trends Plant Sci.* **2013**, *18*, 477–483. [CrossRef]
45. Koskela, S.; Söderholm, P.P.; Ainasoja, M.; Wennberg, T.; Klika, K.D.; Ovcharenko, V.V.; Kylänlahti, I.; Auerma, T.; Yli-Kauhaluoma, J.; Pihlaja, K.; et al. Polyketide Derivatives Active against Botrytis Cinerea in Gerbera Hybrida. *Planta* **2011**, *233*, 37–48. [CrossRef] [PubMed]
46. Turnbull, J.J.; Nakajima, J.; Welford, R.W.D.; Yamazaki, M.; Saito, K.; Schofield, C.J. Mechanistic Studies on Three 2-Oxoglutarate-Dependent Oxygenases of Flavonoid Biosynthesis. *J. Biol. Chem.* **2004**, *279*, 1206–1216. [CrossRef]
47. Petrussa, E.; Braidot, E.; Zancani, M.; Peresson, C.; Bertolini, A.; Patui, S.; Vianello, A. Plant Flavonoids—Biosynthesis, Transport and Involvement in Stress Responses. *Int. J. Mol. Sci.* **2013**, *14*, 14950–14973. [CrossRef] [PubMed]
48. Shi, M.-Z.; Xie, D.-Y. Biosynthesis and Metabolic Engineering of Anthocyanins in Arabidopsis Thaliana. *Recent Pat. Biotechnol.* **2014**, *8*, 47–60. [CrossRef]
49. Cheng, A.-X.; Han, X.-J.; Wu, Y.-F.; Lou, H.-X. The Function and Catalysis of 2-Oxoglutarate-Dependent Oxygenases Involved in Plant Flavonoid Biosynthesis. *Int. J. Mol. Sci.* **2014**, *15*, 1080–1095. [CrossRef]
50. Huo, C.H.; Zhao, Y.Y.; Liang, H.; Lin, W.H. Studies on chemical constituents in herbs of *Acanthus ilicifolius*. *China J. Chin. Mater. Med.* **2005**, *30*, 763–765.
51. Saranya, A.; Ramanathan, T.; Kesavanarayanan, K.S.; Adam, A. Traditional Medicinal Uses, Chemical Constituents and Biological Activities of a Mangrove Plant, *Acanthus ilicifolius* Linn.: A Brief Review. *Am Eurasian J. Agric. Environ. Sci.* **2015**, *15*, 243–250.
52. Nair, A.G.R.; Pouchaname, V.J. Methylapigenin 7-O-β-D-Glucuronate-A New Flavone Glycoside from *Acanthus ilicifolius*. *Indian Chem. Soc.* **1987**, *64*, 228–229.
53. Wu, J.; Zhang, S.; Xiao, Q.; Li, Q.; Huang, J.; Long, L.; Huang, L. Megastigmane and Flavone Glycosides from *Acanthus ilicifolius*. *Pharmazie* **2003**, *58*, 363–364.
54. Zeng, G.; Li, D.; Han, Y.; Teng, W.; Wang, J.; Qiu, L.; Li, W. Identification of QTL Underlying Isoflavone Contents in Soybean Seeds among Multiple Environments. *Theor. Appl. Genet.* **2009**, *118*, 1455–1463. [CrossRef] [PubMed]
55. Saito, K.; Yonekura-Sakakibara, K.; Nakabayashi, R.; Higashi, Y.; Yamazaki, M.; Tohge, T.; Fernie, A.R. The Flavonoid Biosynthetic Pathway in Arabidopsis: Structural and Genetic Diversity. *Plant Physiol. Biochem.* **2013**, *72*, 21–34. [CrossRef] [PubMed]
56. Cui, J.; Li, J.; Dai, C.; Li, L. Transcriptome and Metabolome Analyses Revealed the Response Mechanism of Sugar Beet to Salt Stress of Different Durations. *Int. J. Mol. Sci.* **2022**, *23*, 9599. [CrossRef]
57. Chen, Z.; Hong, X.; Zeng, Z.; Zeng, Y.; Lin, Q.; Tan, Q. Study on Natural Product of Quercetin from Flos Sophorae Immaturus against Activity of Human Nasopharyngeal Carcinoma CNE1 Cells. *Genom. Appl. Biol.* **2019**, *38*, 3305–3311.
58. Skalski, B.; Lis, B.; Pecio, Ł.; Kontek, B.; Olas, B.; Żuchowski, J.; Stochmal, A. Isorhamnetin and Its New Derivatives Isolated from Sea Buckthorn Berries Prevent H2O2/Fe—Induced Oxidative Stress and Changes in Hemostasis. *Food Chem. Toxicol.* **2019**, *125*, 614–620. [CrossRef]
59. Boubaker, J.; Bhouri, W.; Sghaier, M.B.; Ghedira, K.; Franca, M.G.D.; Chekir-Ghedira, L. Ethyl Acetate Extract and Its Major Constituent, Isorhamnetin 3-O-Rutinoside, from Nitraria Retusa Leaves, Promote Apoptosis of Human Myelogenous Erythroleukaemia Cells: N. Retusa Leaf Extracts Promote Apoptosis. *Cell Prolif.* **2011**, *44*, 453–461. [CrossRef]
60. Canarini, A.; Kaiser, C.; Merchant, A.; Richter, A.; Wanek, W. Root Exudation of Primary Metabolites: Mechanisms and Their Roles in Plant Responses to Environmental Stimuli. *Front. Plant Sci.* **2019**, *10*, 157. [CrossRef]
61. Delli-Ponti, R.; Shivhare, D.; Mutwil, M. Using Gene Expression to Study Specialized Metabolism—A Practical Guide. *Front. Plant Sci.* **2021**, *11*, 625035. [CrossRef]

genes

MDPI

Article

DNA Barcoding, Phylogenetic Analysis and Secondary Structure Predictions of *Nepenthes ampullaria*, *Nepenthes gracilis* and *Nepenthes rafflesiana*

Nur Azreen Saidon [1], Alina Wagiran [1,*], Abdul Fatah A. Samad [1], Faezah Mohd Salleh [1], Farhan Mohamed [2], Jaeyres Jani [3] and Alona C Linatoc [4,5]

1 Department of Biosciences, Faculty of Science, Universiti Teknologi Malaysia, Johor Bahru 81310, Johor, Malaysia
2 Media and Games Innovation Centre, Universiti Teknologi Malaysia, Johor Bahru 81310, Johor, Malaysia
3 Borneo Medical and Health Research Center, Faculty of Medicine and Health Sciences, Universiti Malaysia Sabah, Kota Kinabalu 88400, Sabah, Malaysia
4 Faculty of Applied Sciences & Technology, Universiti Tun Hussein Onn Malaysia (UTHM), Hab Pendidikan Tinggi Pagoh, KM1, Jalan Panchor, Muar 84600, Johor, Malaysia
5 College of Forestry and Natural Resources, University of the Philippines Los Banos College, Los Baños 4031, Laguna, Philippines
* Correspondence: alina@utm.my

Abstract: Nepentheceae, the most prominent carnivorous family in the Caryophyllales order, comprises the *Nepenthes* genus, which has modified leaf trap characteristics. Although most *Nepenthes* species have unique morphologies, their vegetative stages are identical, making identification based on morphology difficult. DNA barcoding is seen as a potential tool for plant identification, with small DNA segments amplified for species identification. In this study, three barcode loci; ribulose-bisphosphate carboxylase (*rbc*L), intergenic spacer 1 (ITS1) and intergenic spacer 2 (ITS2) and the usefulness of the ITS1 and ITS2 secondary structure for the molecular identification of *Nepenthes* species were investigated. An analysis of barcodes was conducted using BLASTn, pairwise genetic distance and diversity, followed by secondary structure prediction. The findings reveal that PCR and sequencing were both 100% successful. The present study showed the successful amplification of all targeted DNA barcodes at different sizes. Among the three barcodes, *rbc*L was the least efficient as a DNA barcode compared to ITS1 and ITS2. The ITS1 nucleotide analysis revealed that the ITS1 barcode had more variations compared to ITS2. The mean genetic distance (K2P) between them was higher for interspecies compared to intraspecies. The results showed that the DNA barcoding gap existed among *Nepenthes* species, and differences in the secondary structure distinguish the *Nepenthes*. The secondary structure generated in this study was found to successfully discriminate between the *Nepenthes* species, leading to enhanced resolutions.

Keywords: DNA barcode; *Nepenthes*; *rbc*L; ITS1; ITS2; phylogenetic; barcoding gap; secondary structure predictions

Citation: Saidon, N.A.; Wagiran, A.; Samad, A.F.A.; Mohd Salleh, F.; Mohamed, F.; Jani, J.; Linatoc, A.C. DNA Barcoding, Phylogenetic Analysis and Secondary Structure Predictions of *Nepenthes ampullaria*, *Nepenthes gracilis* and *Nepenthes rafflesiana*. *Genes* **2023**, *14*, 697. https://doi.org/10.3390/genes14030697

Academic Editors: Wajid Zaman and Hakim Manghwar

Received: 7 February 2023
Revised: 2 March 2023
Accepted: 9 March 2023
Published: 11 March 2023

1. Introduction

Among the pitcher plants, the Nepentheceae family is the largest, with 120 species that produce specialized cup-shaped pitchers that attract small insects and kill them through digestive enzymes [1]. In Malaysia, there are eleven species of *Nepenthes* known to exist, including *Nepenthes gracillis*, *Nepenthes ampullaria* and *Nepenthes rafflesiana* [2,3]. According to previous reports, taxonomic classification of *Nepenthes* has been based on morphological characteristics such as shape, color, size and ornamentation [1,2,4–7]. Although this is common for *Nepenthes*, it frequently causes confusion because of characters that can be hard to find [8]. For example, taxonomic confusion in *Nepenthes* has been reported when *N. pilosa* was confused with *N. chaniana*, *N. talangensis* with *N. bongso*, and *N. lamii* with

N. vieillardii [3,7]. Human interest in *Nepenthes* extends beyond these plants' decorative nature to their therapeutic benefits, in addition to their uniqueness and beauty. The *Nepenthes* plant has traditionally been employed in traditional medicine to control the menstrual cycle, facilitate childbirth, reduce asthma, cure eye inflammation, and treat stomach ulcers, jaundice, high blood pressure, indigestion, and dysentery, and has been used as an astringent [4,9–11]. The "Jakun" community believes that the decoction of *N. ampullaria* can ease asthma attacks, while the stem can treat malaria [12,13]. Other pitcher plants have also been used in traditional delicacies [14]. Due to their purported health benefits, various studies have revealed the therapeutic value of plants employed in traditional herbal therapy.

Since their development, DNA barcodes have been viewed as the most promising approach to resolving this taxonomic issue [15]. Previous simulation studies for *Nepenthes* species using NCBI GenBank collection data showed that a single locus ITS or one coupled with plastid regions (matK) exhibited the best species discrimination with distinct barcoding gaps [16]. According to the previous literature, *trn*L and ITS DNA barcodes can be used to distinguish between the *Nepenthes* species based on their geographical origin area [17]. Meanwhile, other studies concluded that RNA secondary structure prediction is an advanced tool for species discrimination [18], and the integration of secondary structure information in species identification can significantly improve its accuracy for other plant species [18–20]. However, there are not many reports on how ITS1 and ITS2 secondary structure predictions can be used together to differentiate between species. In our study, we aimed to investigate the efficacy of *rbc*L, ITS2 and ITS1 DNA barcodes in combination with ITS1 and ITS2 secondary structure predictions in distinguishing between three *Nepenthes* species.

2. Materials and Methods

2.1. Sample Collection

The three *Nepenthes* species used in this study (*N. ampullaria, N. gracilis* and *N. rafflesiana*) were collected from a sampling trip in Gunung Janing, Kampung Peta, Endau-Rompin National Park, Johor, Malaysia (GPS location; 2.529908870220549, 103.41185691378324), and identified by Assoc Prof. Dr Alona Cuevas Linatoc. The geographical location from which the three *Nepenthes* samples were collected is presented in Figure 1. The collected *Nepenthes* species were wrapped with aluminum foil and kept in an ice box prior to extraction at the laboratory. The plant samples were cut and kept at −80 °C prior to extraction with designated labels of NA, NR and NG for *N. ampullaria, N. rafflesiana* and *N. gracilis*, respectively.

2.2. Genomic DNA Extraction and PCR Amplification

For genomic DNA extraction, frozen plant samples were first thawed before proceeding with the extraction using the commercial kit NucleoSpin Plant II (Macherey-Nagel, Duren, Germany), following the manufacturer's instructions. Approximately 100 mg of the plant sample was ground with liquid nitrogen to produce a fine powder using mortar and pestle. The plant samples were subjected to isolation steps as detailed in the protocol. Then the genomic DNA (gDNA) obtained was quantified using a Nanodrop machine (Eppendorf, Hamburg, Germany) and checked using 1% (*w*/*v*) agarose gel electrophoresis. The integrity and quality of the gDNA obtained was verified with a Gel Documentation System (BioRad). Then, the good quality gDNAwas used as a DNA template for PCR amplification. The PCRs for *rbc*L, ITS1 and ITS2 were conducted separately in 25 µL of the total reaction volume containing 12.5 µL of 2X PrimeSTAR® Max DNA Polymerase, 10 ng of gDNA for *N. gracilis, N. ampullaria* and *N. rafflesiana*, 0.625 µM of each primer set and topped up with sterile nano-pure water. The PCR was conducted using the Mastercycler® nexus gradient (Eppendorf AG, Hamburg, Germany) through three different PCR profiles according to the DNA barcodes (Table 1). The PCR products were visualized on 2% (*w*/*v*) agarose gel electrophoresis in 1X TAE buffer and later sent for sequencing using Sanger sequencing at Apical Scientific Sdn. Bhd. Sequencing was performed for both the forward and reverse directions using the same primers as those used in PCR. The details of the primers are shown in Table 2.

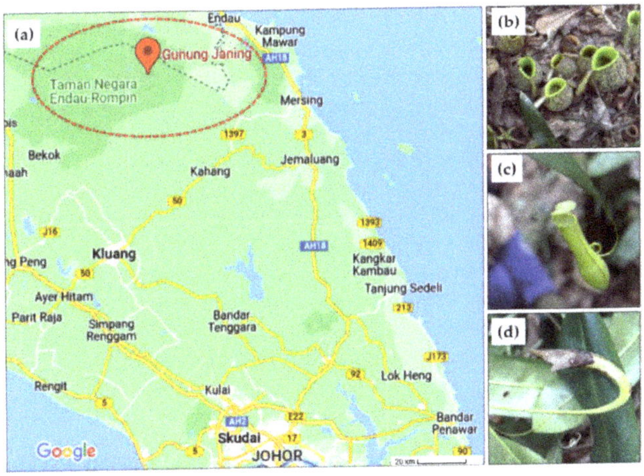

Figure 1. The geographical location (**a**) of the *Nepenthes* collected from Bukit Janing, Kampung Peta, Endau-Rompin National Park, Johor, Malaysia (GPS location; 2.529908870220549, 103.41185691378324). *Nepenthes* species collected: (**b**) *Nepenthes ampullaria*, (**c**) *Nepenthes gracilis* and (**d**) *Nepenthes rafflesiana*.

Table 1. Polymerase Chain Reaction (PCR) profiles used in this study.

Steps	No. of Cycles	*rbc*L [1]		ITS1 [2]		ITS2 [3]	
		Temperature (°C)	Duration (s)	Temperature (°C)	Duration (s)	Temperature (°C)	Duration (s)
Denaturation	35	98	10	98	10	98	10
Annealing		53	15	55	15	55	10
Elongation		72	20	72	40	72	30
Hold	1	4	∞	4	∞	4	∞

[1] ribulose-bisphosphate carboxylase (*rbc*L). [2] intergenic spacer 1 (ITS1).[3] intergenic spacer 2 (ITS2)

Table 2. List of primers used in this study.

Region	Primer Name	Primer Sequence (5′ to 3′)	Primer Length (bp)	Reference
*rbc*L	Rbcla_fwd	ATGTCACCACAAACAGAGACTAAAGC	26	[21]
	Rbclb_rvs	GTAAAATCAAGTCCACCRCG	20	
ITS1	ITS1_fwd	GGAAGGAGAAGTCGTAACAAGG	22	[21]
	ITS1_rvs	AGATATCCGTTGCCGAGAGT	20	
ITS2	ITS2_fwd	GGGGCGGATATTGGCCTCCCCTTG	24	[22]
	ITS2_rvs	GACGCTTCTCCAGACTACAAT	21	

2.3. Bioinformatics Analysis and Phylogenetic Tree

The forward and reverse sequences of the amplicon obtained from *rbc*L, ITS1 and ITS2 primers were edited using BioEdit software. Each generated consensus sequence of the forward and reverse sequences was submitted to the Basic Local Alignment Search Tool (BLAST) of the National Centre for Biotechnology Information (NCBI) for a homology search (https://blast.ncbi.nlm.nih.gov/Blast.cgi; accessed on 24 December 2022). The GenBank accession number for the generated barcode sequences were obtained after the sequences were submitted to GenBank via BankIt for *rbc*L, (https://submit.ncbi.nlm.nih.gov/about/bankit/; accessed on 24 December 2022) and the submission portal of NCBI for ITS1 and ITS2 (https://submit.ncbi.nlm.nih.gov/about/genbank/; accessed on 24 December 2022).

The BLASTn results were selected by determining the sequences with the maximum similarity score and lowest E value. The generated sequence of each barcode in the present study and its similarity was recorded as a percentage. Multiple sequence alignment was performed using Jalview v2.11.1.0 with all the obtained sequences. The phylogenetic analysis was performed following the neighbor-joining (NJ) tree and minimum evolution method with the 1000 *"Boostrap phylogeny"* test method using MEGAX software. The DNA best-fit substitution model for each dataset (Table 3) was determined prior to the NJ tree construction using MEGAX [23]. An outlier, *Dionaea muscipula* (accession number: AB072558.1), was selected in the NJ analysis to verify the identification of the three *Nepenthes* under study. Sequence divergences were calculated using the Kimura-2-parameter (K2P) distance model [24]. The calculation of the sequence divergences was implemented in MEGAX [23]. From the sequence divergence data, the extent of DNA barcoding gap/overlap was then explored as is typical for barcoding studies [25].

Table 3. DNA best-fit substitution model depicted from MEGAX software for the three DNA barcodes.

Barcode Genes	DNA Best-Fit Substitution Model
*rbc*L	Kimura-2-parameter
ITS1	Tamura-3-parameter
ITS2	Kimura-2-parameter + γ distribution

2.4. Secondary Structure Predictions

The DNA barcodes of the three *Nepenthes* were generated using a Bio-Rad DNA barcode generator (http://biorad-ads.com/DNABarcodeWeb/; accessed on 28 November 2022). To complement the tree-based methods, RNA secondary structure predictions were performed using the nucleotide sequences from ITS1 and ITS2 primers for the identification of the best potential barcodes using the rRNA database *RNAfold* WebServer v2.4.18 (http://rna.tbi.univie.ac.at//cgi-bin/RNAWebSuite/RNAfold.cgi; accessed on 28 November 2022). The results of RNA secondary prediction enhanced the resolution of the DNA barcodes.

3. Results

3.1. Amplification, Sequencing, Multiple Sequence Alignment and Species Identification

The DNA barcode primers, *rbc*L, ITS1 and ITS2, produced amplicons of 599 bp, 300–400 bp and 300–500 bp, respectively. In analyzing the sequences, *rbc*L was found to exhibit the most extensive sequence length (502–509 bp), followed by ITS1 (219–327 bp) and ITS2 (241–250 bp). All the sequences were submitted to GenBank, and the accession numbers were obtained. The top BLASTn score for the species identification of all three *Nepenthes* species is presented in Table 4. Using the BLASTn tool, the three *Nepenthes* species were identified as different *Nepenthes* species at different barcode regions of *rbcL, ITS1* and *ITS2*. In the BLAST search, *rbc*L and ITS2 genes identified all three *Nepenthes* species as *N. mirabilis* and *N. gracilis,* respectively. However, the barcode did not discriminate between the three *Nepenthes* species. Meanwhile, the ITS1 barcode identified the specimen NGits1 as *N. ventricosa*, while the specimens NAits1 and NRits1 were identified as *Nepenthes x intermedia*.

Table 4. Molecular identification of *Nepenthes* species using *rbc*L, ITS1 and ITS2 barcode genes.

Barcode Genes	Sample ID	Scientific Name	Accession Number	E Value	Query Coverage (%)	Percent Identity (%)
*rbc*L	NA*rbc*L	*N. mirabilis*	NC_041271.1	0.0	100	100
	NG*rbc*L	*N. mirabilis*	NC_041271.1	0.0	100	100
	NR*rbc*L	*N. mirabilis*	NC_041271.1	0.0	100	100
ITS1	NAits1	*Nepenthes x intermedia*	HM204899.1	1e-103	100	98.64
	NGits1	*N. ventricosa*	AB675910.1	2e-168	100	100
	NRits1	*Nepenthes x intermedia*	HM204899.1	2e-116	100	100
ITS2	NAits2	*N. gracilis*	AB675882.1	4e-104	100	95.85
	NGits2	*N. gracilis*	AB675882.1	2e-122	100	99.20
	NRits2	*N. gracilis*	AB675882.1	5e-113	100	99.14

Multiple sequence alignment (MSA) of the DNA barcode sequences (Figure 2) was used to determine that *rbc*L had the largest alignment, followed by ITS1 and ITS2 (Figure 2a). The alignment obtained for *rbc*L showed the highest similarities (100%) among the three *Nepenthes* species, whereas ITS2 showed a slight difference in *N. ampullaria, (NAits2)* with a ten-nucleotide difference between *N. gracillis* (NGits2) and *N. rafflesiana* (NRits2) (Figure 2c) at nucleotides 15, 31, 77, 80, 85, 98, 110, 115, 149 and 154. In contrast, ITS1 showed the highest difference between the three *Nepenthes* species (Figure 2b) based on the length and nucleotide composition of each sequence. Indisputably, among the three ITS1 sequences, NGits1 showed the highest variation followed by NAits1 and NRits1. Variation between NAits1 and NRits1 were based on the difference in sequence length, as NRits1 is fourteen nucleotides longer than NAits1 with a two-base difference at nucleotides 44 and 73. This corresponds with the ATGC percentage of the sequences showing that *rbc*L had the lowest GC content (44.6–45%), followed by ITS2 (63.1–66.5%) and ITS1 (66.1 to 68.5%) (Table 5). Nucleotide composition analysis also allowed us to conclude that sequences amplified using the *rbc*L barcode have the least variability since the AT and GC percentages were almost the same. Meanwhile, both ITS1 and ITS2 sequences show variations in AT and GC content, indicating more variability.

Table 5. The average AT and GC percentage nucleotide composition of *N. gracilis, N. ampullaria* and *N. rafflesiana* based on *rbc*L, ITS2 and ITS1 DNA barcodes.

DNA Barcode	Species	AT (bp)	GC (bp)	Total (bp)	AT (%)	GC (%)
*rbc*L	*N. gracilis*	280	229	509	55	45
	N. ampullaria	278	224	502	55.4	44.6
	N. rafflesiana	278	224	502	55.4	44.6
ITS2	*N. gracilis*	85	114	250	34.0	66.0
	N. ampullaria	89	152	241	36.9	63.1
	N. rafflesiana	78	155	233	33.5	66.5
ITS1	*N. gracilis*	111	216	327	33.9	66.1
	N. ampullaria	69	150	219	31.5	68.5
	N. rafflesiana	76	157	233	32.6	67.4

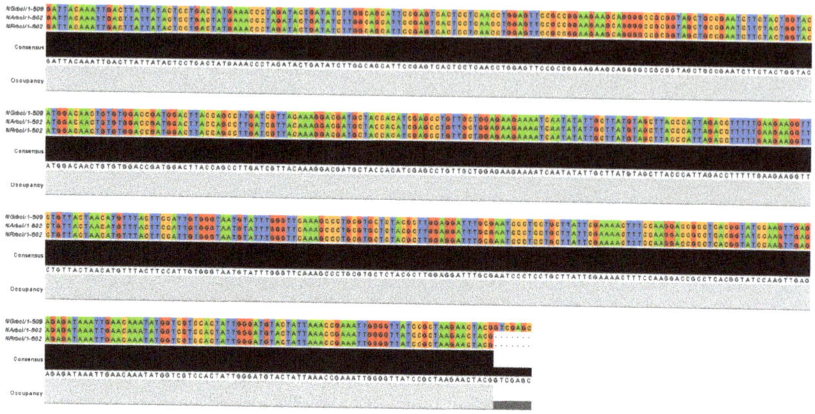

(**a**)

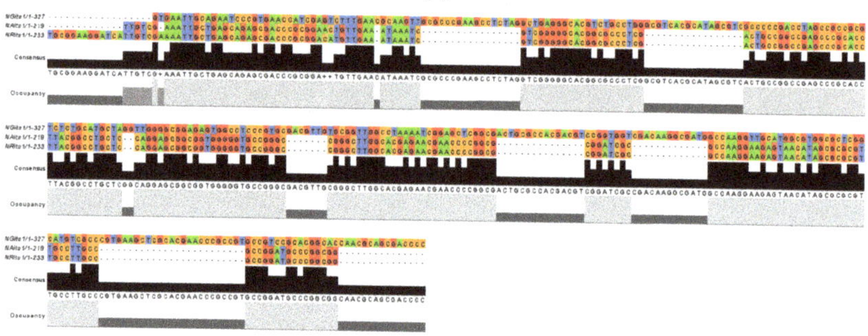

(**b**)

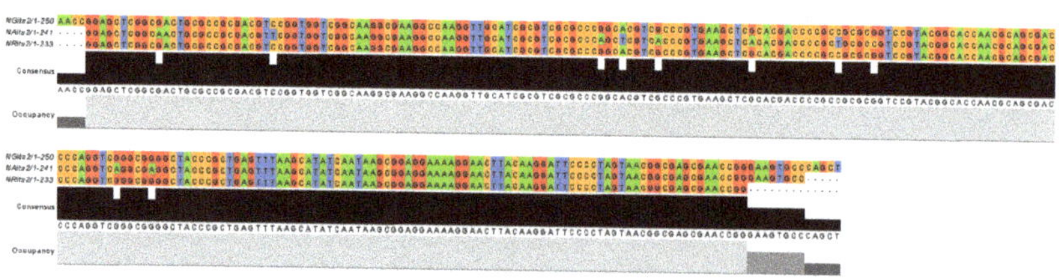

(**c**)

Figure 2. Multiple sequence alignment of *rbc*L (**a**), ITS1 (**b**) and ITS2 (**c**) barcodes.

All the sequences generated from the amplification of *rbc*L, ITS1 and ITS2 barcodes were successfully deposited into the GenBank database. The accession number for each barcode is as follows: *rbc*L: OP534746 (*N. ampullaria*), OP534747 (*N. rafflesiana*), OP534748 (*N. gracilis*); ITS1: OQ123732 (*N. ampullaria*), OQ123724 (*N. rafflesiana*), OQ123725 (*N. gracilis*); and ITS2: OQ123720 (*N. ampullaria*), OQ123721 (*N. rafflesiana*), OQ123722 (*N. gracilis*).

3.2. Phylogenetic Studies, Intraspecific Variation, Interspecific Divergence and DNA Barcoding Gap

Phylogenetic analysis using an NJ tree in the respective DNA best-fit substitution model, as shown in Table 3 with bootstrap-1000 of the three *Nepenthes* species, showed a high similarity in the BLAST search to the available sequences in the NCBI database (Figure 3). Although there was no significant variation between the generated barcode sequences from the three *Nepenthes* species when observed through multiple sequence alignment, phylogenetic analysis revealed further differences between the three *Nepenthes* under study (NA, NR and NG), especially when ITS1 and ITS2 were considered. The NJ tree of *rbc*L showed that the three species studied were in the same clade as the other species, *N. mirabilis* and *N. alata* (Figure 3a), with the same node scores of 26% for *N. gracilis*, *N. ampullaria* and *N. rafflesiana*, respectively. Meanwhile, NJ analysis of ITS1 sequences showed that the three species studied could be classified into different clades; NAits1 is in a clade with NRits1 with a node score of 78%, while NG was classified into an individual clade (Figure 3b). Further phylogenetic characterization using ITS2 revealed that NRits2 and NGits2 are related to *N. rafflesiana* (accession number: HM204904.1) since they were grouped into the same clade with node scores of 73% and 77%, respectively (Figure 3c). The NAits2 appeared individually at a clade near NRits2, NGits2 and a clade of another *N. ampullaria* species.

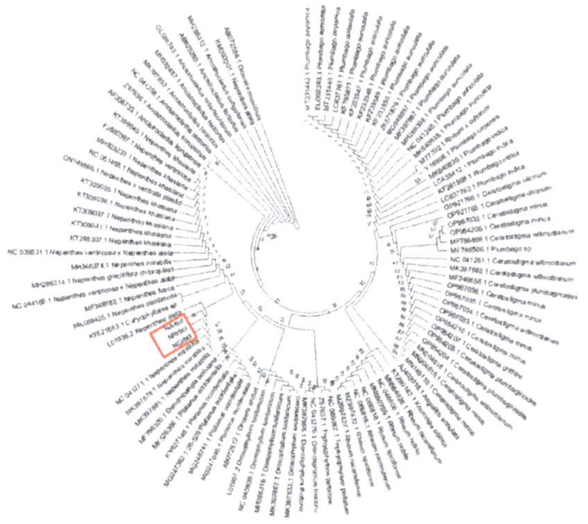

(**a**)

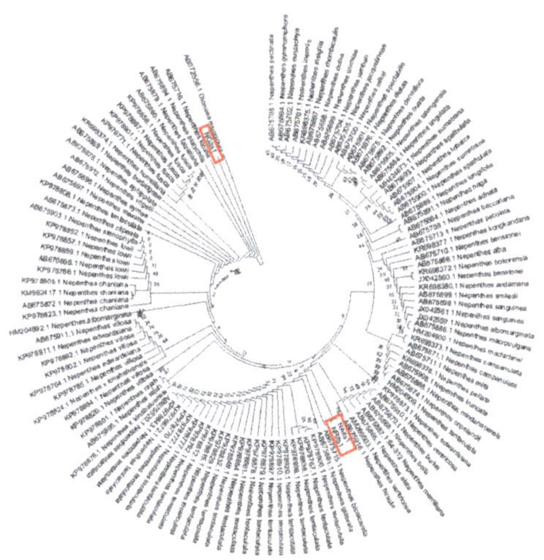

(**b**)

Figure 3. *Cont.*

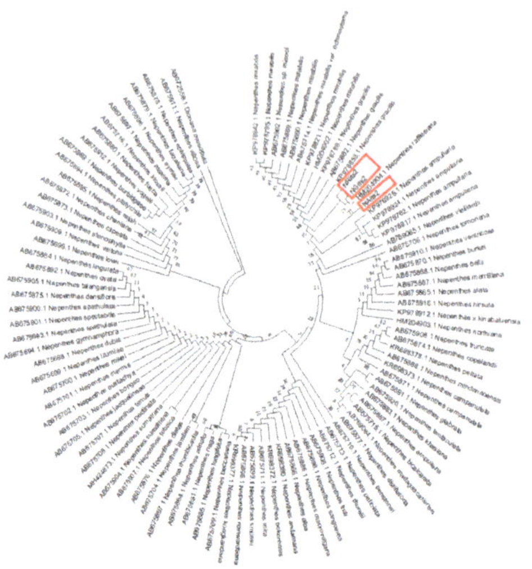

(c)

Figure 3. Neighbor joining tree of *rbc*L (**a**), ITS1 (**b**) and ITS2 (**c**) barcode primers depicting the phylogenetic analysis among *N. ampullaria, N. gracilis,* and *N. rafflesiana* (indicated in the red boxes).

Previous reports on the intraspecific and interspecific divergence among species are useful for assessing the potential of DNA barcodes [25–27]. Based on the neighbor-joining (NJ) tree of K2P distances, taxa or groups were organized to calculate the intraspecific variations and interspecific divergences among them. The *Nepenthes* species under study showed unique clades and within-species sequence divergence between 0 and 4% in *rbc*L, 0 and 5.9% in ITS2, and 0 and 26.9% in ITS1, whereas divergence between species ranged from 0 to 1% in *rbc*L, 1 to 8% in ITS2, and 1.5 to 44% in ITS1 (Figure 4a–c). The results indicate that the interspecific divergence was distinctly higher for the interspecific divergence distance for ITS2 and ITS1, and that, consequently, a clear DNA barcode gap was present (Figure 4b,c). In addition, the results obtained with ABGD analysis are consistent with the results obtained with MEGAX, evidencing DNA barcode gaps.

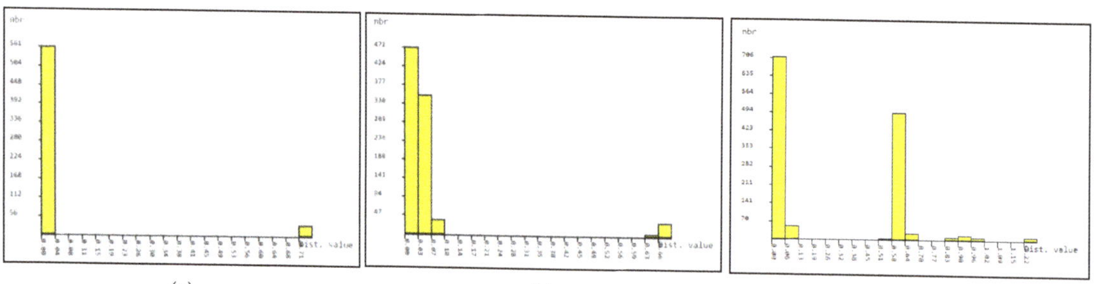

(a) (b) (c)

Figure 4. Barcode gap analysis of *N. gracilis*, *N. ampullaria*, and *N. rafflesiana* generated by Automatic Barcode Discovery Gap Discovery (https://bioinfo.mnhn.fr/abi/public/abgd/abgdweb.html, accessed on 23 January 2023) [25] for (**a**) rbcL, (**b**) ITS2 and (**c**) ITS1. Distributions of K2P distances and between each pair of specimens for the histogram of distance.

3.3. DNA Barcodes, ITS1 and ITS2 Secondary Structure Predictions

Figures 5 and 6 show the DNA barcodes and ITS1 and ITS2 secondary structure predictions based on minimum free energy (MFE). The highest MFE for NAits1, NGits1 and NRits1 was observed at 100–105 bp (Figure 5a), 120–125 bp (Figure 5b) and 115–120 bp (Figure 5c), respectively. Meanwhile, the highest MFE for both NAits2 and NRits2 was observed at 45–50 bp (Figure 6a,c), while NGits2 recorded the highest MFE at 120–125 bp (Figure 6b). DNA barcode sequences derived from ITS1 showed variations among the three *Nepenthes* species (Figure 5d–f). NGits1 exhibited the highest barcode length (328 bp), followed by NRits1 (233 bp) and NAits1 (220 bp).

Similarly, ITS1 secondary structure predictions show further variation between the three Nepenthes species. The predicted ITS1 secondary structures of the three Nepenthes showed that each structure had a central ring with different helical orientations (Figure 4g–i). Generally, the secondary structures of NAits1 and NRits1 showed a tighter configuration and pattern similarity despite having a different number of loops on the central ring and along the helices. NGits1, on the other hand, exhibited a more complex structure compared to NAits1 and NRits1, with a bigger central ring, different helix orientations and lengths, varied loop numbers and variation in angles from the spiral. The loop number, position, size and angle from the centroid were distinguishable in all three Nepenthes species.

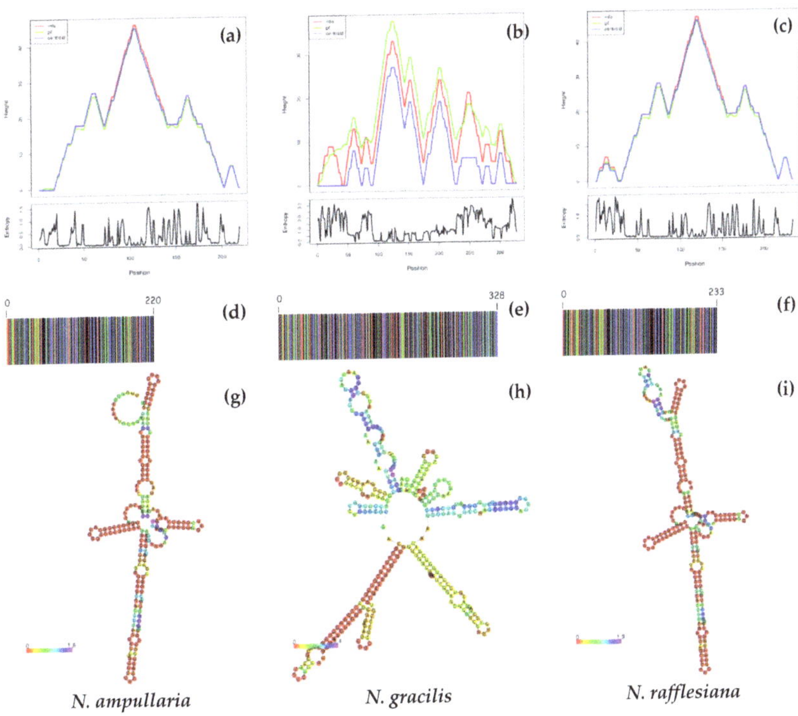

Figure 5. The variations observed in the predicted minimum free energy (MFE) (**a–c**), DNA barcodes (**d–f**), and secondary structures of the ITS1 region for the three *Nepenthes* species (**g–i**).

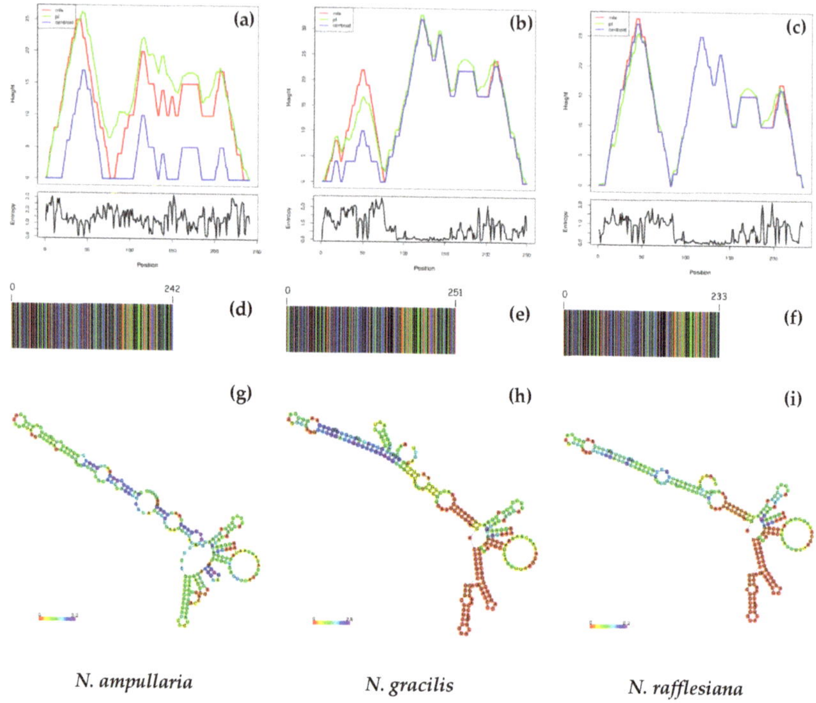

Figure 6. The variations observed in the predicted minimum free energy (MFE) (**a–c**), DNA barcodes (**d–f**), and secondary structures of the ITS2 region for the three *Nepenthes* species (**g–i**).

The predicted ITS2 secondary structure provides additional information on the differences between the three Nepenthes species by representing three slightly different structures with a central ring with different helical orientations (Figure 5g–i). In general, the predicted ITS2 secondary structures had a more uniform pattern compared to the predicted ITS1 secondary structures. All three predicted ITS2 secondary structures shared the same central and backbone pattern, but the loop number, position size and angle from the centroid were still distinguishable in all three Nepenthes species. Consequently, in ITS1′s secondary structure, NAits1 showed the most obvious pattern difference, with a larger central ring compared to NRits1 and NGits1. Based on the results, both ITS1 and ITS2 secondary structure predictions provide deeper insights into the differences between the three Nepenthes species, allowing each species to be identified as unique species and guiding comparative sequence analysis. In addition, the prediction of secondary structures can further assist in the design of species-specific RNA molecules.

4. Discussion

DNA barcoding is used in many plant biodiversity studies to identify species [28,29], discover new taxa, conserve species and enable studies of plant ecology by constructing phylogenetic trees [30]. In this study, ITS1 and ITS2 showed a greater ability for species discrimination than *rbc*L, especially with regard to secondary structure predictions. Given the universality of the *rbc*L gene, it has been proposed as a universal barcode fragment due to its ease of amplification and comparison [31]. It is widely used for phylogenetic analysis of families and subclasses in various seed plant groups [32]. However, the limitation of the *rbc*L barcode is its inability to discriminate between organisms up to the species level. Although Nepenthes does not belong to the seed plant group, the present study showed that *rbc*L could not classify the three *Nepenthes* up to the species level, indicating a poor ability to distinguish between species. This was also supported by nucleotide analysis and NJ, which only reached the genus level. This finding is consistent with earlier research showing that diversity in the *rbc*L sequence is scarce at the species level [26,33–35] and reduces its ability to be used to discriminate between species. The high universality and low resolution of *rbc*L indicates its inefficiency as a DNA barcode, which is also suggested for other plant species, such as *Acacia* [36].

The ITS region is one of the most commonly used barcodes in genus- and species-level phylogenetic analyses in eukaryotes [37,38]. A previous study had shown the application of DNA barcodes for three *Nepenthes* species (*N. ampullaria, N. rafflesiana* and *N. gracilis*), which enabled them to differentiate between them based on their geographical area of origin. The phylogenetic analysis was based on *trn*L and ITS barcodes but lacked information on the effectiveness of the two DNA barcodes in distinguishing the *Nepenthes* spp. from each other [17]. Additionally, a previous study reported that the *trn*L intron does not represent the best choice for characterizing plant species and for phylogenetic studies among closely related species [39]. The present study involved plant samples from very closely related species; thus, we made use of *rbc*L, ITS1 and ITS2 to evaluate the species variation. Our findings indicate that the *Nepenthes* species were grouped at different clades when using ITS1 (Figure 3b), with *N. gracilis* appearing individually in one clade while *N. ampullaria* and *N. rafflesiana* were grouped into another clade. These results show that ITS1 can be used to discriminate between the three *Nepenthes* species but did not effectively identify the plant samples up to the species level. In contrast, the ITS2 phylogenetic tree revealed that these three species were closely clustered together with the sequences of *N. ampullaria, N. rafflesiana* and *N. gracilis* chosen from the BLASTn search, although they did not appear in the same clade (Figure 3c).

Overall, NJ analysis showed that the phylogenetic tree method poorly distinguished between the Nepenthes species compared to distance-based analysis. A previous study showed that barcode gaps act as typical barcode data characterized by having differences between intraspecific diversity and interspecific diversity [25]. However, this was not a general feature in all groups. In this study, the histogram and ranked pairwise (K2P) distance analyzed using ABGD programs showed that the "barcoding gap" between levels of intraspecific variation and interspecific divergence did exist for the analysis of *rbc*L, ITS2 and ITS1 (Figure 4a–c). Compared to the ITS1 barcoding gap, ITS2 has a wider barcoding gap number. In the present study, phylogenetic analysis did not ambiguously identify the species, but the predicted ITS1 and ITS2 secondary structures revealed the uniqueness of the generated DNA barcode sequences. According to a previous study, the features of the ITS2 region are conserved, which in turn provides molecular and morphological features that could be used to improve the resolution of species identification [40]. The usefulness of ITS2 sequences and the ITS2 secondary structure as genetic markers has been reported for several medicinal plant species [20,41,42]. Regarding the ITS2 secondary structure, the conservative standard structural model contained the feature of a "loop" with four "helices." Among the four "helices," helix I and helix IV had the most variations, helix II and helix III were more conserved, helix II was shorter with a pyrimidine–pyrimidine mismatch "bridge," while helix III was the longest with many branches [30,43,44].

In our study, prediction of secondary structures in all the *Nepenthes* species under study revealed diverse secondary structures with distinguishable loop numbers, positions and elevations from the centroid. This information can be applied to design species-specific primers for identifying genotypes. Prediction of secondary structure differentiation indicated variation among RNA molecules in all species when using either ITS1 or ITS2. The alignment of the primary nucleic acid sequences could be optimized and modified using the secondary ITS2 information, contributing to the accuracy and robustness of the phylogeny [45]. Previous research had shown that genetic structure uniqueness at the conserved nuclear region could be useful for developing species-specific primers, as reported in a previous study [18]. Meanwhile, little information was available on the function and secondary structure of ITS1, leaving a gap in the knowledge on whether ITS1 secondary structure prediction could enhance species identification. Consequently, the region of ITS1 may play an important role in 18SrRNA maturation [43]. A few reports suggest DNA barcodes can be used to efficiently discriminate between *Nepenthes* species [16,17,46]. However, until now, no study has reported ITS1 and ITS2 secondary structure predictions in relation to species discrimination in *Nepenthes*. Our study revealed significant variations in both ITS1 and ITS2 secondary structure predictions that enhanced species discrimination between the three *Nepenthes* species under study.

5. Conclusions

This study describes the efficiency of DNA barcode genes from *rbc*L, ITS1 and ITS2 in differentiating between three *Nepenthes* along with secondary structure predictions. Although the *rbc*L gene in the chloroplast–plastid region might be easily amplified, it has a poor species identification and discrimination ability. On the other hand, the incorporation of secondary structures in the nuclear ribosomal genes of ITS1 and ITS2 may serve as a trustworthy tool in species identification and designing species-specific primers. The findings of this investigation provide clarity on the scientific foundations for species identification, genetic preservation and the secure use of this significant species of medicinal plant. The employment of DNA barcode technologies for species delimitation in commercially and medicinally important plant species may be possible.

Author Contributions: Conceptualization, A.W., F.M.S. and F.M.; methodology, N.A.S.; software, J.J.; validation, J.J. and A.C.L.; formal analysis, N.A.S.; investigation, N.A.S.; resources, N.A.S., J.J. and A.C.L.; data curation, J.J.; writing—original draft preparation, N.A.S.; writing—review and editing, N.A.S., A.W. and F.M.S.; visualization, N.A.S.; supervision, A.W. and A.F.A.S.; project administration, A.W. and A.F.A.S.; and funding acquisition, A.W. and F.M.S. All authors have read and agreed to the published version of the manuscript.

Funding: This research was fully funded by the Ministry of Higher Education, Malaysia under UTM Transdisciplinary Research Grant (QJ130000.3554.06G59).

Institutional Review Board Statement: Not applicable.

Informed Consent Statement: Not applicable.

Data Availability Statement: Not applicable.

Acknowledgments: The staff of the lab facility, namely the Plant Biotechnology Laboratory, Faculty of Science, UTM, are gratefully acknowledged, the plant sampling assistance and approval of plant sample collection by the Johor National Johor Park Corporation, specifically at Endau-Rompin National Johor Park, Johor, Malaysia, as well as Kamarul Rahim Bin Kamarudin from UTHM for his assistance in plant sample collection and herbarium preparation.

Conflicts of Interest: The authors declare no conflict of interest.

References

1. McPherson, S.; Robinson, A.S. *Field Guide to the Pitcher Plants of Peninsular Malaysia and Indochina*; Redfern Natural History: Poole, UK, 2012.
2. Clarke, C. *Nepenthes of Borneo*; Natural History Publications (Borneo): Kota Kinabalu, Malaysia, 1997.
3. Clarke, C. Nepenthes of Sumatra, and Peninsular Malaysia. In *Nepenthes of Sumatra and Peninsular Malaysia*; Natural History Publications (Borneo): Kota Kinabalu, Malaysia, 2001; p. 336.
4. Cheek, M.; Jebb, M. Flora Malesiana, Series 1: Volume 15: Nepenthaceae. In *Flora Malesiana*; National Herbarium of the Netherlands: Wageningen, The Netherlands, 2001.
5. Jebb, M.; Cheek, M. A skeletal revision of nepenthes (Nepenthaceae). *Blumea J. Plant Taxon. Plant Geogr.* **1997**, *42*, 1–106.
6. Lestari, W.; Jumari, J.; Ferniah, R.S. Identification and Cluster Analysis of Pitcher Plant (*Nepenthes* spp.) from South Sumatera Indonesia. *Biosaintifika J. Biol. Biol. Educ.* **2018**, *10*, 245–251. [CrossRef]
7. Phillipps, A.; Lamb, A. *Pitcher-Plants of Borneo*, 1st ed.; Royal Botanic Gardens: Sydney, Australia, 1996.
8. Beveridge, N.G.P.; Rauch, C.; Ke, P.J.A.; van Vugt, R.R. A new way to identify living species of Nepenthes (Nepenthaceae): More data needed ! *Carniv. Plant Newsl.* **2013**, *42*, 122–128. [CrossRef]
9. Aung, H.H.; Chia, L.S.; Goh, N.K.; Chia, T.F.; Ahmed, A.A.; Pare, P.W.; Mabry, T.J. Phenolic constituents from the leaves of the carnivorous plant Nepenthes gracilis. *Fitoterapia* **2002**, *73*, 445–447. [CrossRef]

10. Mainasara, M.M.; Fadzelly, M.; Bakar, A.; Mohamed, M.; Linatoc, A.C.; Sanusi, S.B. Ethnomedical Knowledge of Plants Used for the Treatment of Breast Cancer by Jakun community in Kampung Peta Endau Rompin Johor, Malaysia. *J. Sci. Technol.* **2017**, *9*, 42–49.
11. Thao, N.P.; Luyen, B.T.T.; Koo, J.E.; Kim, S.; Koh, Y.S.; van Thanh, N.; Cuong, N.X.; van Kiem, P.; van Minh, C.; Kim, Y.H. In vitro anti-inflammatory components isolated from the carnivorous plant *Nepenthes mirabilis* (Lour.) Rafarin. *Pharm. Biol.* **2016**, *54*, 588–594. [CrossRef]
12. Sabran, S.F.; Mohamed, M.; Abu Bakar, M.F. Ethnomedical Knowledge of Plants Used for the Treatment of Tuberculosis in Johor, Malaysia. *Evid. -Based Complement. Altern. Med.* **2016**, *2016*, 2850845. [CrossRef] [PubMed]
13. Sanusi, S.B.; Abu Bakar, M.F.; Mohamed, M.; Sabran, S.F.; Mainasara, M.M. Ethnobotanical, phytochemical, and pharmacological properties of nepenthes species: A review. *Asian J. Pharm. Clin. Res.* **2017**, *10*, 16–19. [CrossRef]
14. Schwallier, R.; de Boer, H.J.; Visser, N.; van Vugt, R.R.; Gravendeel, B. Traps as treats: A traditional sticky rice snack persisting in rapidly changing Asian kitchens. *J. Ethnobiol. Ethnomed.* **2015**, *11*, 24. [CrossRef]
15. Liu, Y.; Zhang, M.; Chen, X.; Chen, X.; Hu, Y.; Gao, J.; Pan, W.; Xin, Y.; Wu, J.; Du, Y.; et al. Developing an efficient DNA barcoding system to differentiate between *Lilium* species. *BMC Plant Biol.* **2021**, *21*, 465. [CrossRef] [PubMed]
16. Gogoi, B.; Bhau, B.S. DNA barcoding of the genus Nepenthes (Pitcher plant): A preliminary assessment towards its identification. *BMC Plant Biol.* **2018**, *18*, 153. [CrossRef]
17. Bunawan, H.; Yen, C.C.; Yaakop, S.; Noor, N.M. Phylogenetic inferences of Nepenthes species in Peninsular Malaysia revealed by chloroplast (trnL intron) and nuclear (ITS) DNA sequences. *BMC Res. Notes* **2017**, *10*, 67. [CrossRef]
18. Acharya, G.C.; Mohanty, S.; Dasgupta, M.; Sahu, S.; Singh, S.; Koundinya, A.V.V.; Kumari, M.; Naresh, P.; Sahoo, M.R. Molecular Phylogeny, DNA Barcoding, and ITS2 Secondary Structure Predictions in the Medicinally Important Eryngium Genotypes of East Coast Region of India. *Genes* **2022**, *13*, 1678. [CrossRef]
19. Liu, Z.W.; Gao, Y.Z.; Zhou, J. Molecular authentication of the medicinal species of Ligusticum (ligustici rhizoma et radix, "gao-ben") by integrating non-coding internal transcribed spacer 2 (ITS2) and its secondary structure. *Front. Plant Sci.* **2019**, *10*, 429. [CrossRef] [PubMed]
20. Singh, D.P.; Kumar, A.; Rodrigues, V.; Prabhu, K.N.; Kaushik, A.; Mani, D.N.; Shukla, A.K.; Sundaresan, V. DNA barcoding coupled with secondary structure information enhances Achyranthes species resolution. *J. Appl. Res. Med. Aromat. Plants* **2020**, *19*, 100269. [CrossRef]
21. Fadzil, N.F.; Wagiran, A.; Mohd Salleh, F.; Abdullah, S.; Izham, M.N.H. Authenticity and Detection of Eurycoma longifolia herbal products using Bar-High Resolution Melting Analysis. *Genes* **2018**, *9*, 408. [CrossRef] [PubMed]
22. Omelchenko, D.O.; Speranskaya, A.S.; Ayginin, A.A.; Khafizov, K.; Krinitsina, A.A.; Fedotova, A.V.; Pozdyshev, D.V.; Shtratnikova, V.Y.; Kupriyanova, E.V.; Shipulin, G.A.; et al. Improved Protocols of ITS1-Based Metabarcoding and Their Application in the Analysis of Plant-Containing Products. *Genes* **2019**, *10*, 122. [CrossRef] [PubMed]
23. Tamura, K.; Peterson, D.; Peterson, N.; Stecher, G.; Nei, M.; Kumar, S. MEGA5: Molecular evolutionary genetics analysis using maximum likelihood, evolutionary distance, and maximum parsimony methods. *Mol. Biol. Evol.* **2011**, *28*, 2731–2739. [CrossRef]
24. Nei, M.; Kumar, S. *Molecular Evolution and Phylogenetics*; Oxford University Press: Oxford, UK, 2001.
25. Puillandre, N.; Lambert, A.; Brouillet, S.; Achaz, G. ABGD, Automatic Barcode Gap Discovery for primary species delimitation. *Mol. Ecol.* **2012**, *21*, 1864–1877. [CrossRef] [PubMed]
26. Newmaster, S.G.; Fazekas, A.J.; Steeves, R.A.D.; Janovec, J. Testing candidate plant barcode regions in the Myristicaceae. *Mol. Ecol. Resour.* **2008**, *8*, 480–490. [CrossRef]
27. Yu, H.; Wu, K.; Song, J.; Zhu, Y.; Yao, H.; Luo, K.; Dai, Y.; Xu, S.; Lin, Y. Expedient identification of Magnoliaceae species by DNA barcoding. *Plant OMICS* **2014**, *7*, 47–53.
28. Hollingsworth, P.M. DNA barcoding plants in biodiversity hot spots: Progress and outstanding questions. *Heredity* **2008**, *101*, 1–2. [CrossRef] [PubMed]
29. Hollingsworth, P.M.; Graham, S.W.; Little, D.P. Choosing and Using a Plant DNA Barcode. *PLoS ONE* **2011**, *6*, e19254. [CrossRef] [PubMed]
30. Schultz, J.; Müller, T.; Achtziger, M.; Seibel, P.N.; Dandekar, T.; Wolf, M. The internal transcribed spacer 2 database—A web server for (not only) low level phylogenetic analyses. *Nucleic Acids Res.* **2006**, *34*, 704–707. [CrossRef]
31. Hollingsworth, P.M.; Li, D.Z.; van der Bank, M.; Twyford, A.D. Telling plant species apart with DNA: From barcodes to genomes. *Philos. Trans. R. Soc. B Biol. Sci.* **2016**, *371*, 20150338. [CrossRef]
32. Hebert, P.D.N.; Ratnasingham, S.; DeWaard, J.R. Barcoding animal life: Cytochrome c oxidase subunit 1 divergences among closely related species. *Proc. R. Soc. B Biol. Sci.* **2003**, *270* (suppl. 1), 96–99. [CrossRef]
33. Fazekas, A.J.; Burgess, K.S.; Kesanakurti, P.R.; Graham, S.W.; Newmaster, S.G.; Husband, B.C.; Percy, D.M.; Hajibabaei, M.; Barrett, S.C.H. Multiple Multilocus DNA Barcodes from the Plastid Genome Discriminate Plant Species Equally Well. *PLoS ONE* **2008**, *3*, e2802. [CrossRef]
34. Kress, W.J.; Erickson, D.L. A Two-Locus Global DNA Barcode for Land Plants: The Coding rbcL Gene Complements the Non-Coding trnH-psbA Spacer Region. *PLoS ONE* **2007**, *2*, e508. [CrossRef] [PubMed]
35. Sass, C.; Little, D.P.; Stevenson, D.W.; Specht, C.D. DNA barcoding in the Cycadales: Testing the potential of proposed barcoding markers for species identification of Cycads. *PLoS ONE* **2007**, *2*, e1154. [CrossRef]

36. Ismail, M.; Ahmad, A.; Nadeem, M.; Javed, M.A.; Khan, S.H.; Khawaish, I.; Sthanadar, A.A.; Qari, S.H.; Alghanem, S.M.; Khan, K.A.; et al. Development of DNA barcodes for selected Acacia species by using rbcL and matK DNA markers. *Saudi J. Biol. Sci.* **2020**, *27*, 3735–3742. [CrossRef] [PubMed]
37. Coleman, A.W. ITS2 is a double-edged tool for eukaryote evolutionary comparisons. *Trends Genet.* **2003**, *19*, 370–375. [CrossRef] [PubMed]
38. Edinburgh, R.B.G.; Group, C.P.W.; Hollingsworth, P.M.; Forrest, L.L.; Spouge, J.L.; Hajibabaei, M.; Ratnasingham, S.; van der Bank, M.; Chase, M.W.; Cowan, R.S.; et al. A DNA barcode for land plants. *Proc. Natl. Acad. Sci. USA* **2009**, *106*, 12794–12797.
39. Taberlet, P.; Coissac, E.; Pompanon, F.; Gielly, L.; Miquel, C.; Valentini, A.; Vermat, T.; Corthier, G.; Brochmann, C.; Willerslev, E. Power and limitations of the chloroplast trnL (UAA) intron for plant DNA barcoding. *Nucleic Acids Res.* **2007**, *35*, e14. [CrossRef] [PubMed]
40. Feng, S.; Jiang, M.; Shi, Y.; Jiao, K.; Shen, C.; Lu, J.; Ying, Q.; Wang, H. Application of the ribosomal DNA ITS2 region of physalis (Solanaceae): DNA barcoding and phylogenetic study. *Front. Plant Sci.* **2016**, *7*, 1047. [CrossRef]
41. Anaz, M.; Sasidharan, N.; Remakanthan, A.; Dilsha, M.V. ITS2 and RNA secondary structure-based analysis reveals a clear picture on phylogeny of South Indian *Salacia* spp. *Comput. Biol. Chem.* **2021**, *91*, 107438. [CrossRef]
42. Selvaraj, D.; Ramalingam, S. Identification of morphologically similar species of Tribulus (Zygophyllaceae) by employing DNA barcodes and rRNA secondary structures. *Ecol. Genet. Genom.* **2021**, *18*, 100072. [CrossRef]
43. Coleman, A.W. Pan-eukaryote ITS2 homologies revealed by RNA secondary structure. *Nucleic Acids Res.* **2007**, *35*, 3322–3329. [CrossRef]
44. Wolf, M.; Achtziger, M.; Schultz, J.; Dandekar, T.; Müller, T. Homology modeling revealed more than 20,000 rRNA internal transcribed spacer 2 (ITS2) secondary structures. *RNA* **2005**, *11*, 1616–1623. [CrossRef] [PubMed]
45. Zheng, M.; Liu, D.; Zhang, H.; Zhang, Y. Molecular authentication of medicinal and edible plant Gnaphalium affine (cudweed herb, "Shu-qu-cao") based on DNA barcode marker ITS2. *Acta Physiol. Plant.* **2021**, *43*, 119. [CrossRef]
46. Biswal, D.K.; Debnath, M.; Konhar, R.; Yanthan, S.; Tandon, P. Phylogeny and Biogeography of Carnivorous Plant Family Nepenthaceae with Reference to the Indian Pitcher Plant Kepenthes Khasiana Reveals an Indian Subcontinent Origin of Nepenthes Colonization in Southeast Asia During the Miocene Epoch. *Front. Ecol. Evol.* **2018**, *6*, 108. [CrossRef]

genes

MDPI

Article

A Comparative Analysis of the Chloroplast Genomes of Three *Lonicera* Medicinal Plants

Chenju Yang [1,†], Ni Zhang [2,3,†], Shaoxiong Wu [1], Chunyan Jiang [1], Lian Xie [1], Feng Yang [1] and Zhengwen Yu [1,*]

1 School of Life Sciences, Guizhou Normal University, Guiyang 550025, China
2 State Key Laboratory of Functions and Applications of Medicinal Plants, Guizhou Medical University, Guiyang 550014, China
3 The Key Laboratory of Chemistry for Natural Products of Guizhou Province and Chinese Academy of Sciences/Guizhou Provincal Engineering Research Center for Natural Drugs, Guiyang 550014, China
* Correspondence: yuzhengwen2001@126.com
† These authors contributed equally to this work.

Abstract: Both Lonicerae japonicae flos and Lonicerae similis flos are important components in traditional Chinese medicine (TCM) with precious medicinal value. However, the absence of studies on their chloroplast genomes and chromatography has considerably hindered the study of their evolutionary and phylogenetic relationships. In this study, the complete chloroplast (cp) genomes of *Lonicera acuminata* Wall. and *Lonicera similis* Hemsl. were sequenced using the Illumina sequencing platform and compared with that of *Lonicera japonica* Thunb., which has been previously reported. Furthermore, the chromatographic fingerprints of the three plants were constructed using HPLC and the content of quality marker (Q-Marker) was calculated. The annotation results showed that the two chloroplast genomes were typical quadripartite structures with lengths of 155,330 bp (*L. acuminata*) and 155,207 bp (*L. similis*). A total of 126 different genes were annotated, containing 82 protein-coding genes, 36 tRNA genes, and 8 rRNA genes. The expansion and contraction of the inverted repeat (IR) regions suggested that the boundary regions of IR/SC were comparatively conserved in the three species, and six regions (trnH-GUG-psbA, rps2-rpoC2, rbcL-psaI, trnN-GUU-ndhF, rps15-ycf1, and infA) with nucleotide diversity values (Pi) of variable sites higher than 1% were identified. Phylogenetic relation indicated that *L. similis* had a closer genetic relationship with *L. japonica* than *L. acuminata*. Additionally, the chromatographic fingerprints showed that the characteristic peaks of the three medicinal plants were similar, including Neochlorogenic acid, Chlorogenic acid, 4-Dicaffeoylquinic acid, Sweroside, Secoxyloganin, Luteoloside, Isochlorogenic acid A, Isochlorogenic acid B, and Isochlorogenic acid C. The content of chlorogenic acid and total phenolic acid in *L. acuminata* (7.4633 ± 0.4461%, 14.8953 ± 0.0728%) and *L. similis* (14.1055 ± 0.2566%, 21.9782 ± 0.1331%) was much higher than that of *L. japonica* (3.9729 ± 0.0928%, 6.0964 ± 0.1228%), respectively. This study provides appropriate information for species identification, phylogeny, quality assessment, and rational use of three medicinal plants of the genus *Lonicera*.

Keywords: traditional Chinese medicine (TCM); complete chloroplast genome; phylogenetic; chromatographic fingerprint; quality marker (Q-Marker)

Citation: Yang, C.; Zhang, N.; Wu, S.; Jiang, C.; Xie, L.; Yang, F.; Yu, Z. A Comparative Analysis of the Chloroplast Genomes of Three *Lonicera* Medicinal Plants. *Genes* **2023**, *14*, 548. https://doi.org/10.3390/genes14030548

Academic Editors: Wajid Zaman and Hakim Manghwar

Received: 8 February 2023
Revised: 18 February 2023
Accepted: 20 February 2023
Published: 22 February 2023

1. Introduction

Lonicerae japonicae flos, a traditional Chinese medicine (TCM), has been extensively used for several thousand years in China [1–4]. In the 2020 edition of the *Pharmacopoeia of the People's Republic of China* (Chinese Pharmacopeia), Lonicerae japonicae flos is defined as the dried flower buds or initial flowers of *L. japonica* Thunb., which belongs to Caprifoliaceae [5]. As a famous TCM, Lonicerae japonicae flos has antiviral, anti-inflammatory, and antibacterial activity, and pharmacological studies have shown it is effective in the treatment of inflammation and infectious diseases [3,6–9]. Significantly, Lianhua Qingwen Capsule/Granule, Jinhua Qinggan Granule, and Shuanghuanglian Oral Liquid, three patented Chinese medicines containing Lonicerae japonicae flos, are recommended as the effective "Chinese Solution" to combat COVID-19 [10–16]. Recently, Lonicerae japonicae flos has been widely cultivated in Shandong and Henan provinces in China [17], though it is still far from meeting the market demand [7].

Lonicerae similis flos and Lonicerae japonicae flos belong to the same family. In the 2010 edition of the *Standard of Traditional Chinese Medicine in Sichuan Province,* Lonicerae similis flos consists of the dried flower buds or initial flowers of *L. acuminata* Wall. or *L. similis* Hemsl. [18–20]. Lonicerae similis flos is mainly distributed in Sichuan Province and widely used on account of the abundance of resources [20]. Similarly, Lonicerae similis flos is effective in clearing heat, removing toxins, and dispelling wind and heat [17].

Although Lonicerae japonicae flos and Lonicerae similis flos were divided into two herbs according to their origins, they still have close proximity in plant species, appearances, and functions, together with traditional applications and great homogeneity in light of their medicinal uses [2,3,17,20]. Meanwhile, they also contain an array of identical chemical components, including Neochlorogenic acid, Chlorogenic acid, Isochlorogenic acid A, Isochlorogenic acid B, Isochlorogenic acid C, etc. [17,20,21]. Moreover, it is commonly adulterated with Lonicerae similis flos, motivated by economic interests on the basis of higher prices of Lonicerae japonicae flos [22]. Therefore, the close relationship of Lonicerae japonicae flos and Lonicerae similis flos in terms of morphology, chemical composition, potency, and therapeutic properties is controversial, especially concerning their quality standards and interchangeability.

It is difficult to distinguish Lonicerae japonicae flos and Lonicerae similis flos based on the similarity of their morphology, thus causing a mixture of the two species, which to some extent affects the safety and efficacy of the medicines base on Lonicerae japonicae flos. However, previous studies have mainly focused on morphology, chemical composition, and pharmacological activity; morphological studies are still the main means of species identification, thus causing inaccurate interspecific identification of the species within the genus [2,3,17,20,23,24]. The structure of the plant cp genome is a typical quadripartite structure, including two copies of IR regions which are segregated by a large single-copy (LSC) region and a small single-copy (SSC) region [25–27]. The sizes of the cp genomes of different species vary depending on IR contraction or expansion [28,29]. As next-generation sequencing technologies are progressing rapidly, the cost of obtaining genomes is now lower and faster than traditional Sanger sequencing techniques [30]. As a result, cp genome-scale data are increasingly being used to infer phylogenetic relationships at elevated levels of taxonomy, and great progress has been made even in lower taxonomic groups [31]. The cp genome has haploid genetics, conservative structuring, a comparatively small genome, and a sluggish burst rate of mutations, which makes it an ideally suited model for the molecular identification of species, for studies of genetic diversity, and for revealing phylogenetic relationships [32–34]. Furthermore, chromatographic fingerprint and Q-Marker content have become important indicators for the identification and quality evaluation of TCM [5,35,36], which will greatly increase the accuracy of interspecific identification and provide a reliable basis for the identification and quality evaluation of the herbs in question.

In the present study, we aimed to obtain the cp DNA sequences of *L. acuminata* and *L. similis* by using high-throughput Illumina pair-end sequencing data, and compared the two cp genomes with previously reported studies, gene number, type, and repeat sequences of Lonicerae japonicae flos. Moreover, phylogenetic analysis was performed by using whole cp genomes of 25 plants. Finally, the chromatographic fingerprints of the three plants were constructed and the content of quality marker (Q-Marker) was calculated using HPLC. This study provides a valuable resource for species identification, quality evaluation, and genetic improvement of the genus *Lonicera*.

2. Results

2.1. Cp Genome Organizations

We used high-throughput Illumina paired-end sequencing data to acquire cp DNA sequences of *L. acuminata* and *L. similis*, and the annotated, complete chloroplast genomes of the two species were then deposited in GenBank under accession numbers MZ901373 and MZ241297, respectively. Three complete cp genome sequences, including the two *Lonicera* species sequenced in this study and the *L. japonica* accession from GenBank, were combined for integrated analysis. The genomes ranged in size from 155,207 (*L. similis*) to 155,330 bp (*L. acuminata*) (Figure 1). The genomes of these species have a typical quadripartite structure, with two IRs (23,760–23,777) separated by the LSC (88,859–89,149) and SSC (18,659–18,672) regions. The GC content of the *L. japonica*, *L. similis*, and *L. acuminata* cp genomes is 38.58%, 38.59%, and 38.55%, respectively (Table 1); these three species showed similar GC levels (~39%). *Lonicera* cp genomes displayed analogous gene content and order, containing 126 genes, 82 protein-coding genes, 36 tRNA genes, and 8 rRNA genes.

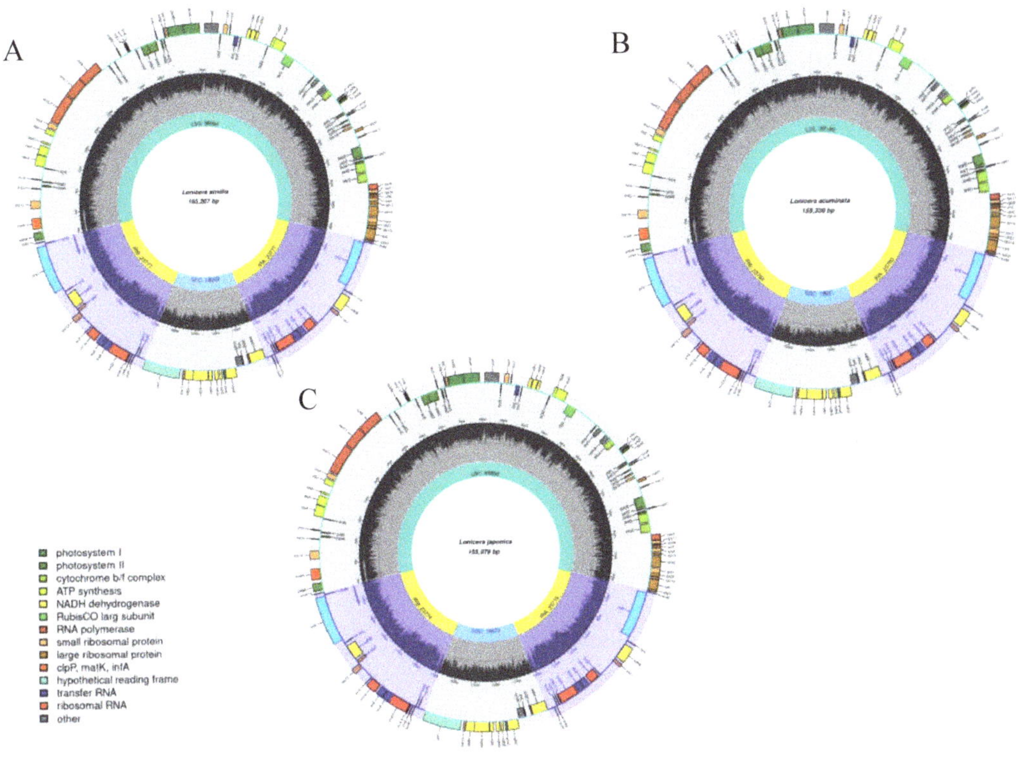

Figure 1. Gene maps of the three *Lonicera* chloroplast genomes. (**A**) *L. similis*; (**B**) *L. acuminata*; (**C**) *L. japonica*. Small single copy (SSC), large single copy (LSC), and inverted repeats (IRA, IRB) are indicated. The black of the inner circle indicates GC content, and dark gray indicates AT content.

Table 1. Summary of complete cp genomes of three *Lonicera* species.

Species	Genome Size(bp)	LSC Region (bp)	IR Region (bp)	SSC Region (bp)	Number of Genes	PCGs	tRNAs	rRNAs	GC Content (%)	Accession Number in Genbank
L. japonica	155,079	88,859	23,774	18,672	126	82	36	8	38.58	NC_026839
L. similis	155,207	88,994	23,777	18,659	126	82	36	8	38.59	MZ241297
L. acuminata	155,330	89,149	23,760	18,661	126	82	36	8	38.55	MZ901373

2.2. Repeat and Simple Sequence Repeat (SSR) Analyses

A comparative analysis of the repeats in the three *Lonicera* cp genomes suggested that the three types of repeats (tandem, forward, and palindromic repeats) were similar in number and distribution (Figure 2a). *L. japonica* presented the highest frequency of tandem repeats (72) and forward repeats (37), while *L. acuminata* had the lowest with (53) and (33), respectively. Palindromic repeats did not differ much between *L. japonica* (12), *L. similis* (14), and *L. acuminata* (13). The tandem repeats in the three *Lonicera* cp genomes ranged from 10 to 30 bp in length (Figure 2b), while the forward and palindromic repeats were mostly between 36 and 65 bp in length (Figure 2c,d).

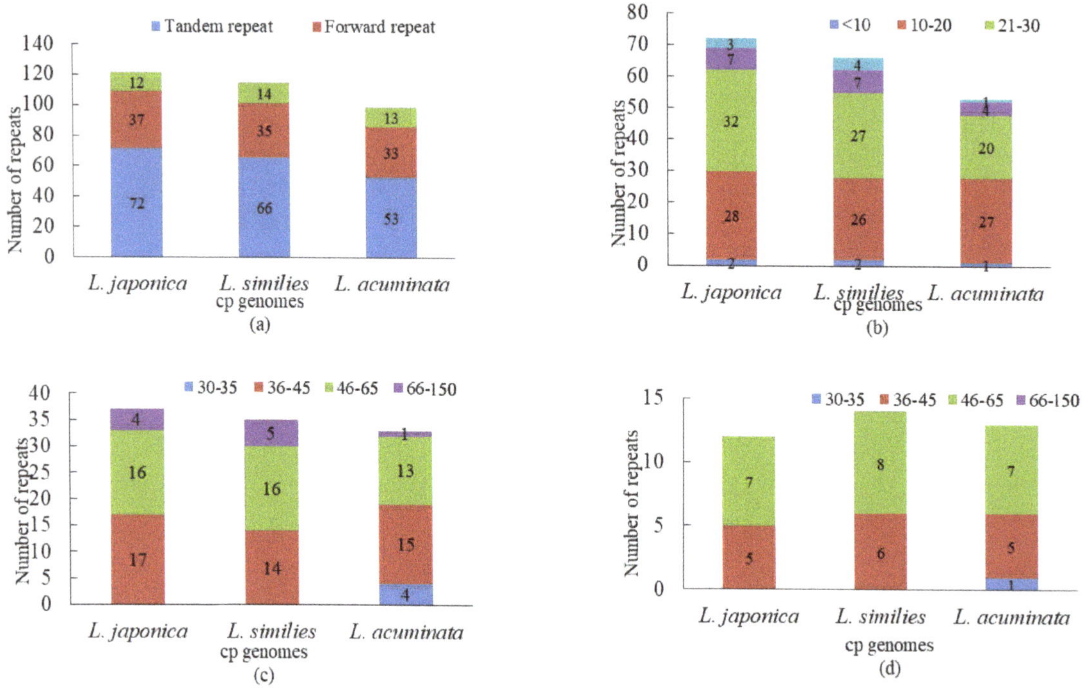

Figure 2. Repeat sequences analysis of three cp genomes. (**a**) Repeat types in three cp genomes; (**b**) tandem repeats in three cp genomes; (**c**) forward repeats in three cp genomes; (**d**) palindromic repeats in three cp genomes. In (**a**), different colors show different repeat types; in (**b**–**d**), different colors show different lengths. The ordinate represents the number of repeats; the abscissa represents three cp genomes.

The cp genome is uniparentally inherited and SSRs share a high level of variation within the same species [27]. As such, they are used as molecular markers in developmental research and also help to identify species [37]. MISA detected 155 SSRs in *L. japonica*, comprising 109 mono-, 32 di-, 2 tri-, 9 tetra-, and 3 hexanucleotide repeats. Here, a grand total of 155 SSRs were detected in the *L. similis* cp genome; among these SSRs, there were 109, 32, 2, 8, and 4 for mono-, di-, tri-, tetra-, and hexa-nucleotide repeats, respectively. No pentanucleotide repeats were discovered. A total of 155 SSRs, comprising 112 mono-, 31 di-, 4 tri-, and 8 tetra-nucleotide repeats were observed in the *L. acuminata* cp genome, while no penta- and hexa-nucleotide repeats were found (Figure 3).

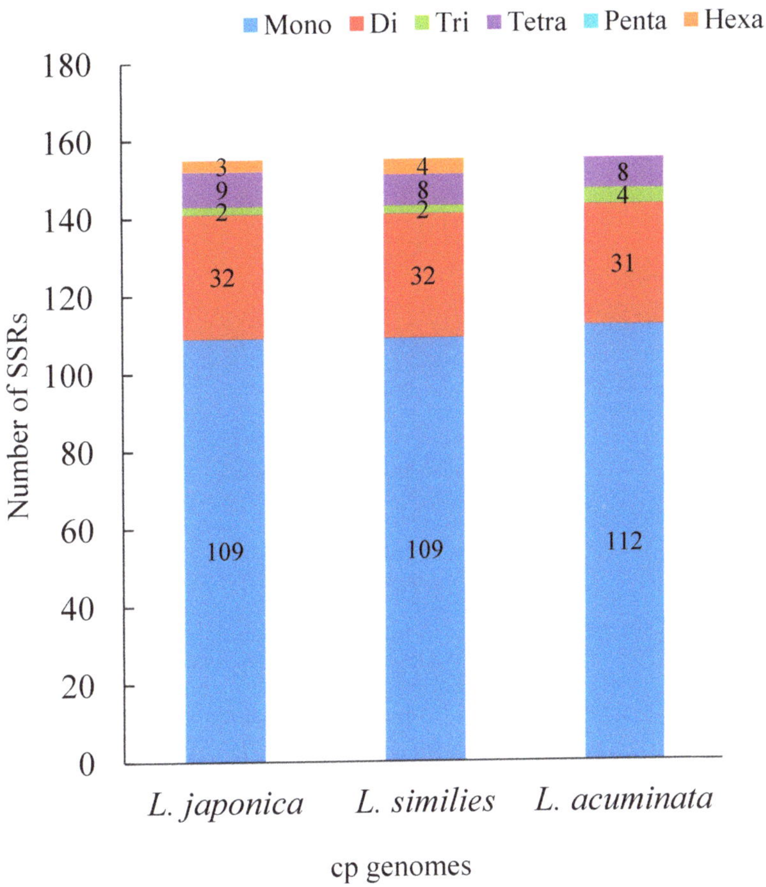

Figure 3. SSR analysis of three cp genomes. The ordinate represents the number of SSRs; the abscissa represents three cp genomes.

SSRs of the three cp genomes were composed of mono- and di- repeat motifs. In the *L. japonica* and *L. similis* cp genomes, there are the same number of mono- repeats and di- repeats: 109 and 32, respectively. In the *L. acuminata* cp genome, the mono- repeats and di- repeats number 112 and 31, respectively. However, the number of three or more oligonucleotide repeats is comparatively low.

2.3. *IR Expansion and Contraction*

It is reported that the chloroplast genome of angiosperms is conserved in structure and size [37]; the differences in IRs may reflect phylogenetic history [38]. Here, we selected three species of the family Caprifoliaceae and compared the sizes and junctions of their LSC, SSC, and IR regions. Although the lengths of the IR regions, ranging from 23,760 bp to 23,777 bp, varied little among the three species, some differences in the IR expansions and contractions were observed. As shown in Figure 4, the rpl2 gene of all three species was located in the LSC region, the rpl23 gene was located at the LSC/IRb border expanded 121 bp into the IRb region, and the ndhF gene of the three species extended 8 bp into the IRb region. For the SSC/IRa boundaries in *L. japonica*, *L. similis*, and *L. acuminata*, the distances were 220 bp, 208 bp, and 208 bp from the ycf1 gene to the boundary, respectively (Figure 4).

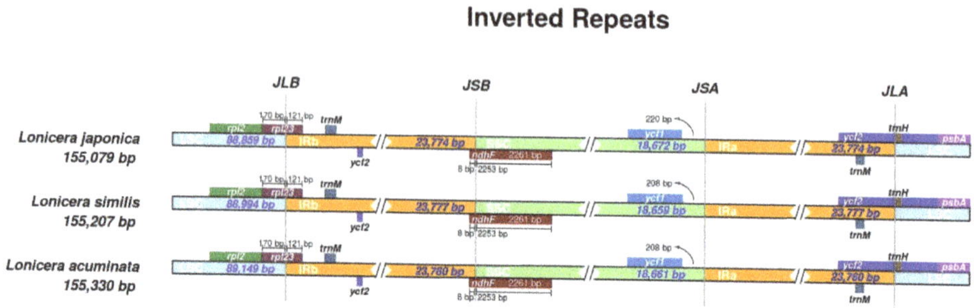

Figure 4. Comparative analyses of the boundary regions (LSC, SSC, and IR) among three chloroplast genomes of *Lonicera*. Gene names are indicated in boxes, and their lengths in junction sites are displayed above the boxes.

2.4. Sequence Divergence Analysis

The three complete cp genomes were compared with mVISTA using *L. japonica* as a reference. As shown in Figure 5, high sequence conservation across the three *Lonicera* cp genomes was observed, especially in gene regions. Variant loci in intergenic regions were distinguished higher than those in the gene regions. Most of the highly variable regions were in the conserved non-coding sequences (CNS) region, included rps16-trnQ, ycf1-trnN, psaA-ycf3, rbcL-accD, psaJ-clpP, and petB-petD, etc. In order to clarify the variation in the higher regions, the nucleotide diversity values (Pi) were calculated using the DnaSP v.6.10 software (Figure 6). The variation in intergenic regions ranged from 0 to 2.23%, with an average of 0.36%, which was twofold higher than that in the CDS regions (0.16% on average). Five divergent loci in intergenic regions (trnH-GUG-psbA, rps2-rpoC2, rbcL-psaI, trnN-GUU-ndhF, and rps15-ycf1) and one in the CDS regions (infA) had a Pi exceeding 1%. These six divergence hotspot regions should be applied to the development of molecular markers for phylogenetic and phylogeographic analyses, as well as plant identification of *Lonicera* species.

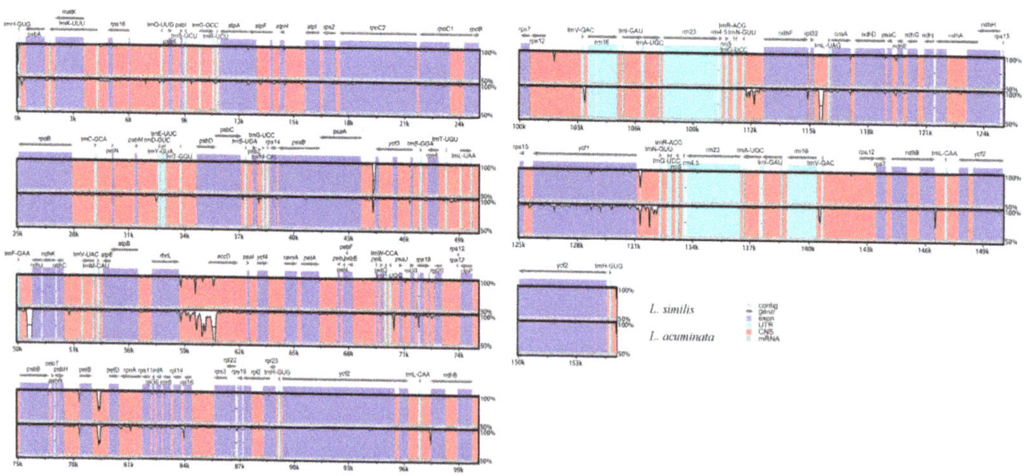

Figure 5. Sequence alignment of three chloroplast genomes in the Caprifoliaceae family performed with mVISTA using annotation of *L. japonica* as reference. The top arrow shows transcription direction; genome regions are color coded as exon, Untranslated Region (UTR), mRNA, and conserved non-coding sequences (CNS). The x-axis represents the coordinates in the cp genome while the y-axis represents percentage identity within 50–100%.

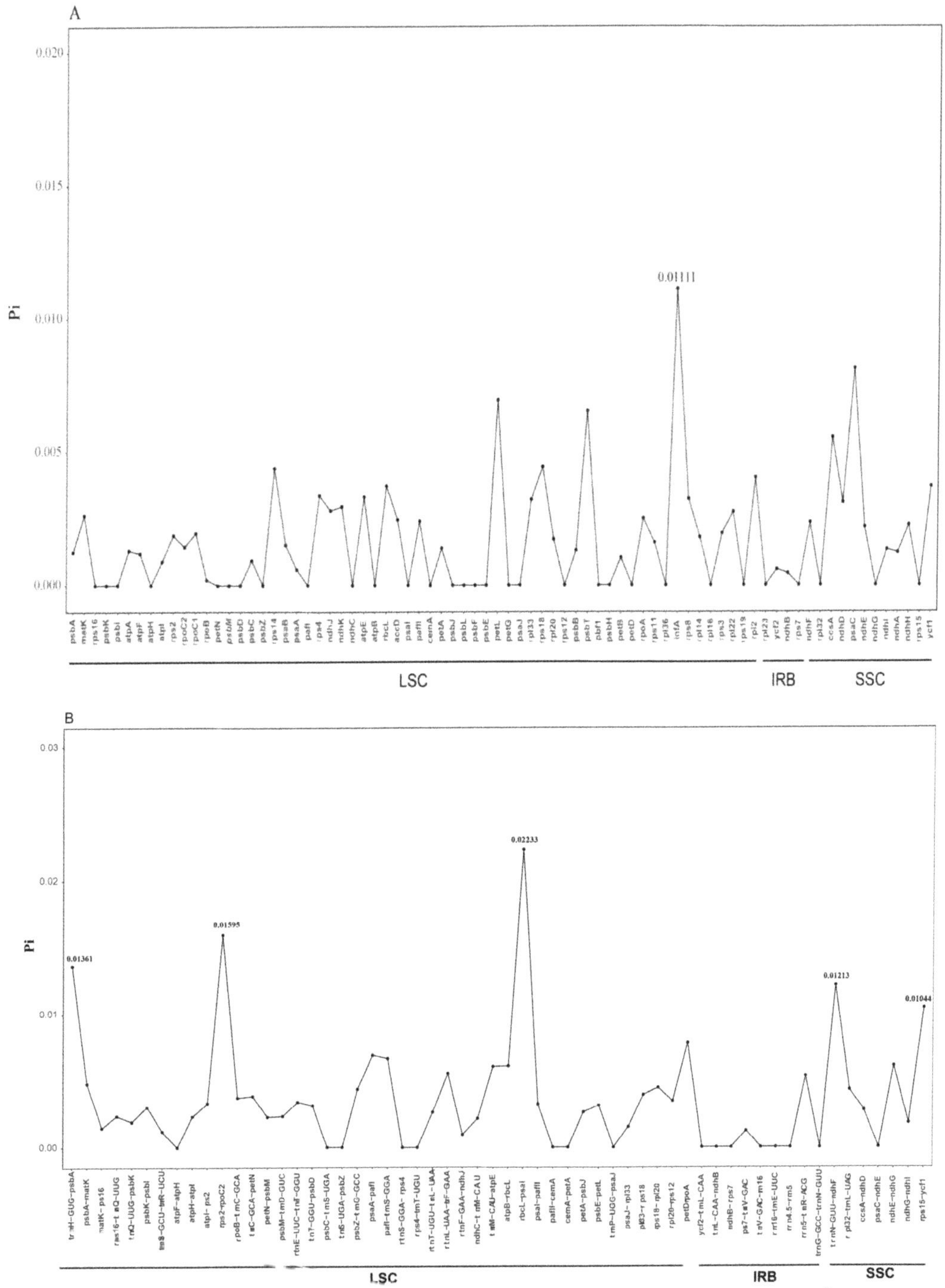

Figure 6. The nucleotide variability (Pi) values were compared among *L. acuminata*, *L. similis*, and *L. japonica*. (**A**) CDS region; (**B**) Intergenic spacer region.

2.5. Phylogenetic Analysis

To further analyze the relationships between *L. acuminata*, *L. similis*, and *L. japonica*, 22 chloroplast genome sequences of the family Caprifoliaceae and one outgroup (*Antirhea chinensis*) were selected and downloaded from GenBank to construct phylogenetic trees with *L. acuminata* and *L. similis*, adopting the maximum-likelihood (ML) method. The ML tree based on the GTR + G + I optimal model was constructed through IQ-tree (Version 2.04) with 1000 bootstrap replicates [39]. In our study, the selected species only came from six genera of the family Caprifoliaceae. Figure 7 clearly displays that all species generated six major clades, which correspond to the six selected genera. All *Lonicera* were clustered together in one monophyletic group, between *L. acuminata*, *L. similis*, and *L. japonica* of the genus *Lonicera* and family Caprifoliaceae. *L. japonica* was clustered together with *L. confusa* and *L. similis*, which implicates that *L. similis* may have a closer genetic relationship with *L. japonica* than *L. acuminata*.

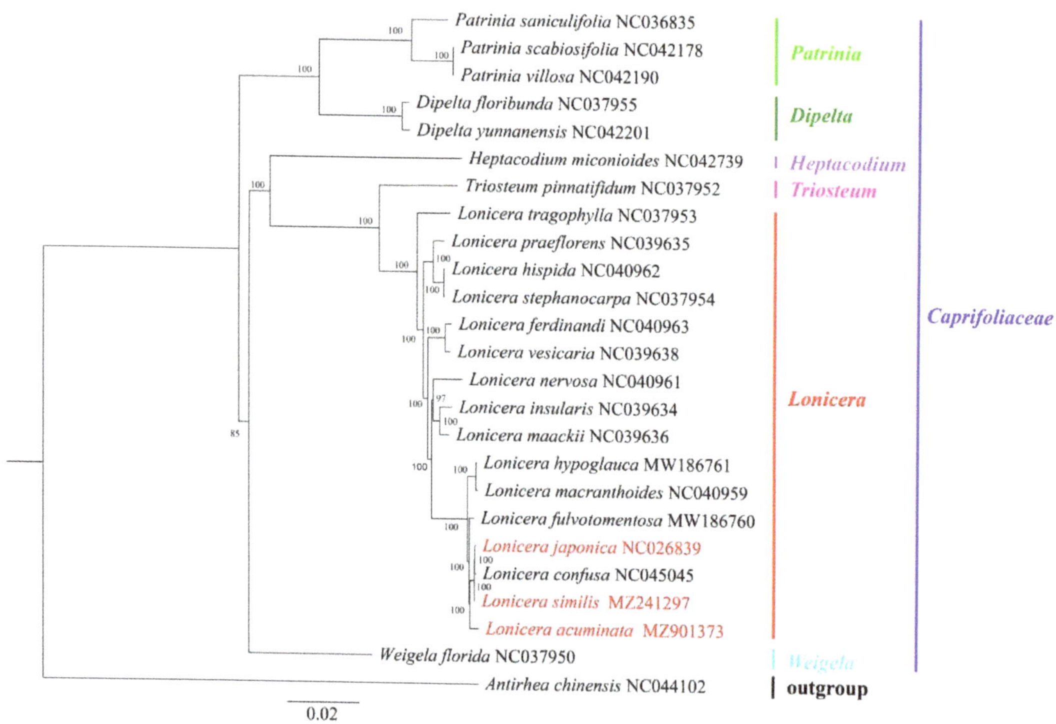

Figure 7. Maximum-likelihood tree based on the complete chloroplast genome sequences of 25 species. GenBank accession numbers are described in the figure. Shown next to the nodes are bootstrap support values based on 1000 replicates.

2.6. Plant Morphology

It can be seen in Table 2 that the main medicinal parts (flower buds and flowers) of Lonicerae japonicae flos and Lonicerae similis flos have a fresh scent and a light and slightly bitter taste, but they differ more markedly in color, length, and upper swollen parts. In our study, *L. japonica* was yellow-white in color, *L. similis* was yellow-brown, and *L. acuminata* was yellow-green. For the medicinal parts of the three plants (flower buds and flowers), *L. similis* was the longest but shortest in upper diameter, *L. acuminata* was the longest in upper diameter, and *L. japonica* was in the middle.

Table 2. The comparison of Botanical traits of Lonicerae japonicae flos and Lonicerae similis flos.

Botanical Traits	*L. japonica*	*L. similis*	*L. acuminata*
Color and glandular hairs	Surface yellow-white or green-white, densely pubescent	Surface yellow-green, green-brown or yellow-brown, covered with long, short coarse or glandular hairs, some glabrous	Surface yellow-green, brown-yellow, pale-purple to purplish-brown, glabrous or sparse
Shape	Rod-shaped, slightly curved, 2–3 cm long, upper diameter about 3 mm, lower 1.5 mm	Slender rod-shaped, slightly curved, 3–6 cm long, slightly inflated above, upper diameter 1.8–2 mm, lower 1.2–1.5 mm	Short rod-shaped, 1–2 cm long, distally expanded, upper diameter 1.5–3.5 mm, lower 0.6–1.5 mm
Smell and taste	Fresh scent, light and slightly bitter taste	Fresh scent, light and slightly bitter taste	Fresh scent, light and slightly bitter taste
Flower buds and flowers	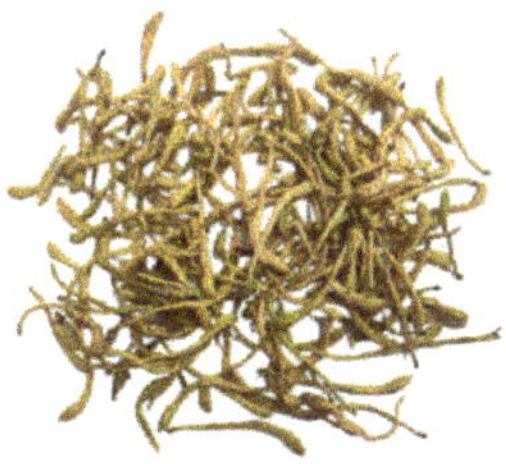		
Reference	National Pharmacopoeia Committee, 2020	Si Chuan Food and Drug Administration, 2011	Si Chuan Food and Drug Administration, 2011

2.7. HPLC Fingerprints and the Content of Main Q-Marker

In terms of chemical composition, HPLC fingerprints of the three species were obtained by high performance liquid chromatography (HPLC) analysis at 240 nm with reference to the *Chinese Pharmacopoeia* [5] and superimposed on the control (Figure 8). From the fingerprints, we can clearly see that the main active ingredients of Lonicerae similis flos and Lonicerae japonicae flos are basically the same, including Neochlorogenic acid, Chlorogenic acid, 4-Dicaffeoylquinic acid, Sweroside, Secoxyloganin, Luteoloside, Isochlorogenic acid A, Isochlorogenic acid B, Isochlorogenic acid C, etc., but the content of main Q-Marker differs (Table 3). The evaluation standards for *L. japonica* in the *Chinese Pharmacopoeia* [5] require it to contain not less than 1.5% chlorogenic acid, not less than 3.8% total phenolic acid, and not less than 0.050% Luteoloside. Here, the content of chlorogenic acid and total phenolic acid in *L. acuminata* (7.4633 $\pm$ 0.4461%, 14.8953 $\pm$ 0.0728%) and *L. similis* (14.1055 $\pm$ 0.2566%, 21.9782 $\pm$ 0.1331%) was much higher than that of *L. japonica* (3.9729 $\pm$ 0.0928%, 6.0964 $\pm$ 0.1228%), respectively. The content of Luteoloside in *L. similis* (0.0291 $\pm$ 0.0044%) was lower than that specified in the pharmacopoeia, but it is noteworthy that the content of chlorogenic acid, its main purgative and detoxifying component, was higher than that of all other species at 14.8953 $\pm$ 0.0728%.

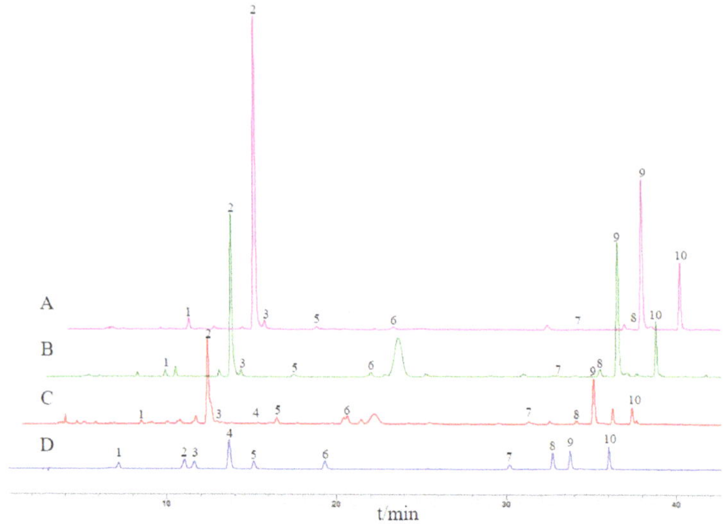

Figure 8. HPLC fingerprints of mixed control and sample. *L. similis* (**A**), *L. acuminata* (**B**), *L. japonica* (**C**), mixed control (**D**). Neochlorogenic acid (1), Chlorogenic acid (2), 4-Dicaffeoylquinic Acid (3), Caffeic acid (4), Sweroside (5), Secoxyloganin (6), Luteoloside (7), Isochlorogenic acid B (8), Isochlorogenic acid A (9), Isochlorogenic acid C (10).

Table 3. The content of main Q-Marker in Lonicerae japonicae flos and Lonicerae similis flos (%).

Species	NA	CA	4-DA	SW	LS	IA-B	IA-A	IA-C	TPC
L. japonica	0.0773 ± 0.007	3.9729 ± 0.0928	0.0584 ± 0.0041	0.1835 ± 0.0086	0.0549 ± 0.0001	0.0227 ± 0.0004	1.7260 ± 0.0161	0.3975 ± 0.0461	6.0964 ± 0.1228
L. similis	0.3098 ± 0.0013	**14.8953 ± 0.0728**	0.2749 ± 0.0515	0.0729 ± 0.0062	0.0291 ± 0.0044	0.0603 ± 0.0014	5.4567 ± 0.0278	1.6262 ± 0.0882	21.9782 ± 0.1331
L. acuminata	0.2462 ± 0.0400	7.4631 ± 0.4461	0.1519 ± 0.0507	0.0852 ± 0.0011	0.2248 ± 0.0256	0.0613 ± 0.0024	5.4267 ± 0.3922	1.2156 ± 0.2027	14.1055 ± 0.2566

Neochlorogenic acid (NA), Chlorogenic acid (CA), 4-Dicaffeoylquinic Acid (4-DA), Sweroside (SW), Luteoloside (LS), Isochlorogenic acid B (IA-B), Isochlorogenic acid A (IA-A), Isochlorogenic acid C (IA-C), Total phenolic acid (TPC).

3. Materials and Methods

3.1. Plant Material and DNA Extraction

L. similis was collected in the town of Dangwu, Huaxi District, Guiyang City (26°23′34.49″ N, 106°35′56.89″ E, 1158 m above sea level), Guizhou Province, China, and *L. acuminata* was harvested from the village of Hekou, the town of Muxi, Muchuan County, Sichuan Province, China (28°51′7.40″ N, 103°50′55.46″ E, 1114 m altitude) by Chenju Yang. According to the macroscopic morphological characteristics of the specimen, the species was determined [40]. Total genomic DNA of two species was extracted using an E.Z.N.A® plant DNA kit (FEIYANG, Guangzhou, China). For each sample a combined total of 1000 ng of DNA served as input to the DNA sample preparation. The DNA library was established utilizing the TruseqTM RNA Sample Prep Kit. Whole DNA was utilized to generate libraries of an average insert size of 400 bp. An Illumina platform was used to sequence the library preparation and obtained approximately 3 GB of 150 bp paired-end reads, which were saved in fastq format.

3.2. Assembly and Annotation

FastQC was used to examine the acquired raw data for analysis of quality, including single base content quality, base content mapping, GC content distributions, and quality of sequence bases. After removal of the adapter sequences, program GetOrganelle was used to assemble the filtered reads [41] using *L. japonica* (GenBank accession number: NC_026839) for the original genome reference, and the genome of the assembled chloroplast was annotated by the GeSeq online package [42]. Finally, the accurate, annotated, complete chloroplast genomes of two *Lonicera* species were deposited to GenBank with accession numbers MZ241297 and MZ901373. Here, the document we use was further annotated based on the accession number.

3.3. Characterization of Repeat Sequences and SSRs

Repeat sequences, comprising the forward and palindromic, were identified by REPuter [43] with a Hamming distance set at 3 bp and a minimum repeat size set at 30 bp. Tandem repeats were analyzed using the Tandem Repeats Finder with default parameters [44]. SSRs were detected using MISA with the following thresholds [45]: eight repeat units for mono SSRs, four repeat units for di- and trinucleotide repeat SSRs, and three repeat units for tetra-, penta-, and hexanucleotide repeat SSRs.

3.4. Comparative Genome Analysis and Sequence Variation

The program mVISTA in the Shuffle-LAGAN mode was applied to compare the cp genomes of Lonicerae japonicae flos (*L. japonica*) and Lonicerae similis flos (*L. acuminata* and *L. similis*) [46]. The variable character for coding and non-coding region with an aligned length of more than 200 bp was obtained based on the method of Zhang et al. [47]. The nucleotide variability was calculated with DnaSP 6.10 [48]. The online software IRSCOPE (https://irscope.shinyapps.io/irapp/, accessed on 31 October 2022) was adopted to carry out boundary analysis on the IR regions of the three species [49].

3.5. Phylogenetic Analysis

The chloroplast genome sequences of 22 Caprifoliaceae family and one outgroup (*A. chinensis*) were downloaded from GenBank to build phylogenetic trees by maximum-likelihood (ML) with *L. acuminata* and *L. similis*. The ML tree was constructed using IQ-tree (Version 2.04, Trifinopoulos J., Vienna , Austria) based on the GTR + G + I optimal model with 1000 bootstrap replicates [39,50].

3.6. HPLC Analyses

The sample solutions were obtained according to the method described in *Chinese Pharmacopoeia Commission* (2020) [5]. HPLC analyses were carried out through an Agilent 1260 Infinity II series system (Agilent Technologies, Santa Clara, CA, USA) consisting of a G1311B pump, a G4212B DAD detector, and a G1329B autosampler. J&KCHEMICA C18 column (250 × 4.6 mm, 5 μm) was adopted for the analyses. The mobile phase consisted of A (acetonitrile), B (0.1% phosphoric acid solution), and C (methanol). The gradient mode was as follows: 0~10 min, 6~10% A; 10~25 min, 10~14% A; 25~35 min, 14~24% A; 35~40 min, 24~27% A; 40~45 min, 27~35% A; 45~50 min, 35~46% A; 50~55 min, 46~51% A; 55~57 min, 51~6% A; 57~70 min, 90% B; C 4% for all times. The flow rate was 1.0 mL/min. The detection wavelength was set at 240, 326, and 350 nm. The column temperature was set at 30 °C and sample volume was set at 5 μL.

4. Discussion

In this study, we conducted a comprehensive comparative analysis of Lonicerae similis flos and Lonicerae japonicae flos. The cp genomes of the three species ranged in length from 155,079 (*L. japonica*) to 155,330 bp (*L. acuminata*); they had a typical quadripartite structure, usually consisting of an LSC, an SSC, and two IR regions, which is consistent with highly conserved features of the cp genome of most angiosperms [25]. Compared with previous studies [33,51,52], chloroplast genomes of *Lonicera* plants differ little in terms of genome organization, number of genes, and types. In a comparative analysis of repeat sequences in the cp genomes, *L. japonica* (a total of 121 repeats, 72 tandem, 37 forward, 12 palindromic repeats) and *L. similis* (a total of 115 repeats, 66 tandem, 35 forward, 14 palindromic repeats) repeats were similar, while *L. acuminata* (a total of 99 repeats, 53 tandem, 33 forward, 13 palindromic repeats) had shorter fragment repeats. The three cp genomes had similar repeat unit lengths, with tandem repeats ranging from 10 to 30 bp and forward and palindromic repeats ranging from 36 to 65 bp. The cp genome is uniparentally inherited and SSRs have a high level of variation within the same species [27]. As such, they are used as the molecular indicators in developmental studies and also help to identify species [37]. SSRs are commonly used as genetic markers in community genetics and evolutionary studies [31]. The most enriched were mononucleotide repeats, next to dinucleotide repeats. As a whole, tetranucleotide repeats were slightly more abundant than trinucleotide and hexanucleotide repeats, and in all three cp genomes pentanucleotide repeats were rare. The three species shared similar IR expansion and contraction, and in sequence alignments had higher variance genes in the CNS region, including rps16-trnQ, ycf1-trnN, psaA-ycf3, rbcL-accD, psaJ-clpP, and petB-petD, etc. The nucleotide diversity values (Pi) showed that six regions (trnH-GUG-psbA, rps2-rpoC2, rbcL-psaI, trnN-GUU-ndhF, rps15-ycf1, and infA) had a Pi exceeding 1%, and these six divergence hotspot regions should be useful for developing molecular markers. In the phylogenetic tree, *L. japonica* may be more closely related to *L. similis* than *L. acuminata*. Our evolutionary tree formed three main branches, which was consistent with previous research [33]. At the same time, plants in the same growing environment may be more closely related.

Despite their different origins, they have close proximity of plant species, appearances, and functions, together with traditional applications [5,19]. Studies have shown that chromatographic fingerprint analysis provides a more rational method for the identification and quality assessment of traditional Chinese herbal medicines [35,36], which can effectively identify the authenticity, merit, and demerit of traditional Chinese medicines and can increase the accuracy of species identification. In this study, we used high performance liquid chromatography (HPLC) to obtain HPLC fingerprints of three species and calculated their contents with reference to the *Chinese Pharmacopoeia* [5]. HPLC fingerprints showed that Lonicerae similis flos and Lonicerae japonicae flos shared the same characteristic peaks, except for Lonicerae similis flos where Caffeic acid was not detected (peak 4), but the contents of quality marker (Q-Marker) were different. The content of Luteoloside (0.0291 $\pm$ 0.0044%) in *L. similis* was lower than that specified in the pharmacopoeia for Lonicerae japonicae flos, but it is noteworthy that the content of chlorogenic acid and total phenolic acid in *L. acuminata* (7.4633 $\pm$ 0.4461%, 14.8953 $\pm$ 0.0728%) and *L. similis* (14.1055 $\pm$ 0.2566%, 21.9782 $\pm$ 0.1331%) was much higher than that of *L. japonica* (3.9729 $\pm$ 0.0928%, 6.0964 $\pm$ 0.1228%), respectively. Chlorogenic acid is known as "plant gold", which has hypotensive effects, antioxidant, and antibacterial activities [53–57]. It is may be one of the potential active ingredients of Lonicerae japonicae flos in the treatment of COVID-19 [58]. With such high chlorogenic acid content, *L. similis* may be selected as a better genetic line for superior breeding and a better source species for chlorogenic acid. These results also provide a basis for future in-depth studies on the synthesis mechanisms of high chlorogenic acid in *L. similis*.

5. Conclusions

In this study, the cp genomes of Lonicerae similis flos (*L. acuminata* and *L. similis*) were sequenced and annotated through high-throughput sequencing technology. We compared the cp genomes of these three species, and bioinformatics analysis revealed that the structure and gene content of the cp genomes of these three *Lonicerae* were highly similar and conserved, suggesting a close relationship between them. We identified 155 SSR loci that could potentially be used as molecular markers to study the diversity of the genus *Lonicera*. Six different mutation hot spots (trnH-GUG-psbA, rps2-rpoC2, rbcL-psaI, trnN-

GUU-ndhF, rps15-ycf1, and infA) with high nucleotide variability (Pi) could be considered as potential DNA barcodes of *Lonicera* species. In our ML tree, all *Lonicera* were clustered together in one monophyletic group and were closely related to the *Triosteum* plants. The HPLC fingerprints of Lonicerae japonicae flos and Lonicerae similis flos increased the accuracy of species identification. The content of Q-Marker showed great variation; the content of chlorogenic acid and phenolic acids (chlorogenic acid, isochlorogenic acid A, isochlorogenic acid C) in Lonicerae similis flos is much higher than that in Lonicerae japonicae flos. It is noteworthy that the content of chlorogenic acid was as high as $14.8953 \pm 0.0728\%$ in *L. similis*, which can be considered as a raw material for this compound. Therefore, it is important to explore the molecular mechanism of high chlorogenic acid in *L. similis*.

This study of the cp genomes of three *Lonicera* provides valuable information for the species, enriches cp genomic data of *Lonicera*, and gives genetic resources for further species identification and phylogenetic studies of the genus. Moreover, chromatographic fingerprint analysis is an effective supplementary method to identify three TCMs, and the differences in content of Q-marker will provide a reference for the quality evaluation and rational use of Chinese herbal medicines of the genus *Lonicera*. In future research, we will perfect the data related to the quality standard of Lonicerae similis flos to provide reference for bringing it into the pharmacology of the People's Republic of China.

Author Contributions: C.Y.: Conception and design, Investigation, Data interpretation, Writing—original draft, Writing—review and editing; N.Z.: Data analysis, Chart production, Writing—original draft, Writing—review and editing; S.W.: Investigation, Data analysis; C.J.: Investigation, Data interpretation; L.X.: Investigation, Data interpretation; F.Y.: Investigation, Data interpretation; Z.Y.: Conception and design, Writing—review and editing, Funding acquisition, Supervision. All authors have read and agreed to the published version of the manuscript.

Funding: This research was funded by the National Natural Science Foundation of China [Grants No. 32060068 and No. U1812401] and Natural science research project of Guizhou Provincial Department of Education [Qianjiaoji [2022] 009]. And The APC was funded by 32060068.

Institutional Review Board Statement: Not applicable.

Informed Consent Statement: Not applicable.

Data Availability Statement: The mentioned chloroplast genome data that support the findings of this study are openly available in GenBank of NCBI at https://www.ncbi.nlm.nih.gov, accessed on 18 May 2021.

Conflicts of Interest: The authors declare no conflict of interest.

References

1. Li, H.; Lu, C. *Lonicera japonica Thunb (Jinyinhua, Honey Suckle)*; Springer: Vienna, Austria, 2015.
2. Li, Y.; Cai, W.; Weng, X.; Li, Q.; Wang, Y.; Chen, Y.; Zhang, W.; Yang, Q.; Guo, Y.; Zhu, X.; et al. Lonicerae Japonicae Flos and Lonicerae Flos: A Systematic Pharmacology Review. *Evid.-Based Complement. Altern. Med.* **2015**, *2015*, 905063. [CrossRef]
3. Li, Y.; Li, W.; Fu, C.; Song, Y.; Fu, Q. Lonicerae japonicae flos and Lonicerae flos: A systematic review of ethnopharmacology, phytochemistry and pharmacology. *Phytochem. Rev.* **2020**, *19*, 1–61. [CrossRef]
4. Tang, Y.-R.; Zeng, T.; Yuan, H.-W.; Li, B.; Peng, C.-Y.; Wang, S.-C.; Jian, Y.-Q.; Qin, Y.; Choudhary, M.I.; Wang, W. Lonicerae Flos: A review of chemical constituents and biological activities. *Digit. Chin. Med.* **2018**, *1*, 173–188. [CrossRef]
5. National Pharmacopoeia Committee. *The Pharmacopoeia of the People's Republic of China: One*; The 2020 Edition; China Medical Science and Technology Press: Beijing, China, 2020; pp. 230–232.
6. Kang, O.-H.; Choi, Y.-A.; Park, H.-J.; Lee, J.-Y.; Kim, D.-K.; Choi, S.-C.; Kim, T.-H.; Nah, Y.-H.; Yun, K.-J.; Kim, Y.-H.; et al. Inhibition of trypsin-induced mast cell activation by water fraction of Lonicera japonica. *Arch. Pharm. Res.* **2004**, *27*, 1141–1146. [CrossRef]
7. Hou, Y.; Jiang, J.-G. Origin and concept of medicine food homology and its application in modern functional foods. *Food Funct.* **2013**, *4*, 1727–1741. [CrossRef]
8. Kao, S.-T.; Liu, C.-J.; Yeh, C.-C. Protective and immunomodulatory effect of flos Lonicerae japonicae by augmenting IL-10 expression in a murine model of acute lung inflammation. *J. Ethnopharmacol.* **2015**, *168*, 108–115. [CrossRef]
9. Shi, Z.; Liu, Z.; Liu, C.; Wu, M.; Su, H.; Ma, X.; Zang, Y.; Wang, J.; Zhao, Y.; Xiao, X. Spectrum-Effect Relationships Between Chemical Fingerprints and Antibacterial Effects of Lonicerae Japonicae Flos and Lonicerae Flos Base on UPLC and Microcalorimetry. *Front. Pharmacol.* **2016**, *7*, 12. [CrossRef]

10. Li, L.-C.; Zhang, Z.-H.; Zhou, W.-C.; Chen, J.; Jin, H.-Q.; Fang, H.-M.; Chen, Q.; Jin, Y.-C.; Qu, J.; Kan, L.-D. Lianhua Qingwen prescription for Coronavirus disease 2019 (COVID-19) treatment: Advances and prospects. *Biomed. Pharmacother.* **2020**, *130*, 110641. [CrossRef]
11. Zhang, Q.; Rong, G.; Meng, Q.; Yu, M.; Xie, Q.; Fang, J. Outlining the keyword co-occurrence trends in Shuanghuanglian injection research: A bibliometric study using CiteSpace III. *J. Tradit. Chin. Med. Sci.* **2020**, *7*, 189–198. [CrossRef]
12. López-Alcalde, J.; Yan, Y.; Witt, C.M.; Barth, J. Current State of Research About Chinese Herbal Medicines (CHM) for the Treatment of Coronavirus Disease 2019 (COVID-19): A Scoping Review. *J. Altern. Complement. Med.* **2020**, *26*, 557–570. [CrossRef]
13. Shi, M.; Peng, B.; Li, A.; Li, Z.; Song, P.; Li, J.; Xu, R.; Li, N. Broad Anti-Viral Capacities of Lian-Hua-Qing-Wen Capsule and Jin-Hua-Qing-Gan Granule and Rational use Against COVID-19 Based on Literature Mining. *Front. Pharmacol.* **2021**, *12*, 640782. [CrossRef]
14. Chu, L.; Huang, F.; Zhang, M.; Huang, B.; Wang, Y. Current status of traditional Chinese medicine for the treatment of COVID-19 in China. *Chin. Med.* **2021**, *16*, 63. [CrossRef]
15. Huang, K.; Zhang, P.; Zhang, Z.; Youn, J.Y.; Wang, C.; Zhang, H.; Cai, H. Traditional Chinese Medicine (TCM) in the treatment of COVID-19 and other viral infections: Efficacies and mechanisms. *Pharmacol. Ther.* **2021**, *225*, 107843. [CrossRef]
16. Tsang, H.F.; Chan, L.W.C.; Cho, W.C.S.; Yu, A.C.S.; Yim, A.K.Y.; Chan, A.K.C.; Ng, L.P.W.; Wong, Y.K.E.; Pei, X.M.; Li, M.J.W.; et al. An update on COVID-19 pandemic: The epidemiology, pathogenesis, prevention and treatment strategies. *Expert Rev. Anti-infective* **2021**, *19*, 877–888. [CrossRef]
17. Qiu, L.J.; Suo, C.X.; Pan, L.Y.; Li, G.W.; Hu, Q.P.; Sun, D.M. The quality evaluation of Lonicerae japonicae Flos, Lonicerae Flos and Lonicerae similis Flos based on multivariate statistical analysis. *J. Guangdong Pharm. Univ.* **2020**, *36*, 620–626.
18. Yang, A.W.; Ma, Y.Y.; Deng, W.; Mi, X.Q.; Zheng, G.Y. Observation of pollen grains of Chuanyinhua with scanning electron microscope. *Pharm. Clin. Chin. Materia Medica* **2010**, *1*, 10–76.
19. Si Chuan Food and Drug Administration. *Sichuansheng Zhongyaocai Biaozhun (2010 Edition)*; Sichuan Science and Technology Press: Chengdu, China, 2011; pp. 84–90.
20. Suo, C.X.; Cheng, X.R.; Pan, L.Y.; Qiu, Y.J.; Wu, W.P.; Li, G.W. Comparative Study of Lonicera L.-derived Medicinal Materials: Lonicerae Similis Flos, Lonicerae Japonicae Flos, and Lonicerae Flos. *Mod. Chin. Med.* **2021**, *23*, 1747–1754.
21. Rao, W.W.; Huang, H.Y. Comparative analyses of the main active ingredients of Lonicerae japonicae flos and Lonicerae flos. *Int. J. Trad. Chin. Med.* **2016**, *38*, 926–931.
22. Ye, H.X.; Li, L. Advance in Germplasm Identification Technology of Lonicera Speices. *Hortic. Seed* **2013**, *2013*, 59–62.
23. Zhang, S.H.; Yuan, Q.Z.; Cai, G.X.; Feng, X.Y.; Shao, Y. Fingerprint of different polarity extracts from superfine powder of Lonicera japonica and its correlation of antibacterial activity. *Chin. Tradit. Herb Drugs* **2011**, *42*, 2226–2230.
24. Pferschy-Wenzig, E.-M.; Ortmann, S.; Atanasov, A.G.; Hellauer, K.; Hartler, J.; Kunert, O.; Gold-Binder, M.; Ladurner, A.; Heiß, E.H.; Latkolik, S.; et al. Characterization of Constituents with Potential Anti-Inflammatory Activity in Chinese Lonicera Species by UHPLC-HRMS Based Metabolite Profiling. *Metabolites* **2022**, *12*, 288. [CrossRef]
25. Wicke, S.; Schneeweiss, G.M.; Depamphilis, C.W.; Müller, K.F.; Quandt, D. The evolution of the plastid chromosome in land plants: Gene content, gene order, gene function. *Plant Mol. Biol.* **2011**, *76*, 273–297. [CrossRef]
26. Wang, M.; Cui, L.; Feng, K.; Deng, P.; Du, X.; Wan, F.; Weining, S.; Nie, X. Comparative Analysis of Asteraceae Chloroplast Genomes: Structural Organization, RNA Editing and Evolution. *Plant Mol. Biol. Rep.* **2015**, *33*, 1526–1538. [CrossRef]
27. Liang, C.; Wang, L.; Lei, J.; Duan, B.; Ma, W.; Xiao, S.; Qi, H.; Wang, Z.; Liu, Y.; Shen, X.; et al. A Comparative Analysis of the Chloroplast Genomes of Four Salvia Medicinal Plants. *Engineering* **2019**, *5*, 907–915. [CrossRef]
28. Wang, R.-J.; Cheng, C.-L.; Chang, C.-C.; Wu, C.-L.; Su, T.-M.; Chaw, S.-M. Dynamics and evolution of the inverted repeat-large single copy junctions in the chloroplast genomes of monocots. *BMC Evol. Biol.* **2008**, *8*, 36. [CrossRef]
29. Raubeson, L.A.; Peery, R.; Chumley, T.W.; Dziubek, C.; Fourcade, H.M.; Boore, J.L.; Jansen, R.K. Comparative chloroplast genomics: Analyses including new sequences from the angiosperms Nuphar advena and Ranunculus macranthus. *BMC Genom.* **2007**, *8*, 174. [CrossRef]
30. Lu, R.-S.; Li, P.; Qiu, Y.-X. The Complete Chloroplast Genomes of Three Cardiocrinum (Liliaceae) Species: Comparative Genomic and Phylogenetic Analyses. *Front. Plant Sci.* **2017**, *7*, 2054. [CrossRef]
31. Yang, Y.; Zhou, T.; Duan, D.; Yang, J.; Feng, L.; Zhao, G. Comparative Analysis of the Complete Chloroplast Genomes of Five Quercus Species. *Front. Plant Sci.* **2016**, *7*, 959. [CrossRef]
32. Shi, X.; Xu, W.; Wan, M.; Sun, Q.; Chen, Q.; Zhao, C.; Sun, K.; Shu, Y. Comparative analysis of chloroplast genomes of three medicinal Carpesium species: Genome structures and phylogenetic relationships. *PLoS ONE* **2022**, *17*, e0272563. [CrossRef]
33. Gu, L.; Hou, Y.; Wang, G.; Liu, Q.; Ding, W.; Weng, Q. Characterization of the chloroplast genome of Lonicera ruprechtiana Regel and comparison with other selected species of Caprifoliaceae. *PLoS ONE* **2022**, *17*, e0262813. [CrossRef]
34. Zhu, A.; Guo, W.; Gupta, S.; Fan, W.; Mower, J.P. Evolutionary dynamics of the plastid inverted repeat: The effects of expansion, contraction, and loss on substitution rates. *New Phytol.* **2016**, *209*, 1747–1756. [CrossRef]
35. Xie, P.; Chen, S.; Liang, Y.-Z.; Wang, X.; Tian, R.; Upton, R. Chromatographic fingerprint analysis—A rational approach for quality assessment of traditional Chinese herbal medicine. *J. Chromatogr. A* **2006**, *1112*, 171–180. [CrossRef]
36. Wagner, H.; Bauer, R.; Melchart, D.; Xiao, P.G.; Staudinger, A. *Chromatographic Fingerprint Analysis of Herbal Medicines*; Springer: Berlin/Heidelberg, Germany, 2011.

37. Yaradua, S.S.; Alzahrani, D.A.; Albokhary, E.J.; Abba, A.; Bello, A. Complete chloroplast genome sequence of Justicia flava: Genome comparative analysis and phylogenetic relationships among Acanthaceae. *Biomed Res. Int.* **2019**, *2019*, 17. [CrossRef]
38. Wang, S.; Yang, C.; Zhao, X.; Chen, S.; Qu, G.-Z. Complete chloroplast genome sequence of Betula platyphylla: Gene organization, RNA editing, and comparative and phylogenetic analyses. *BMC Genom.* **2018**, *19*, 950. [CrossRef]
39. Nguyen, L.-T.; Schmidt, H.A.; Von Haeseler, A.; Minh, B.Q. IQ-TREE: A Fast and Effective Stochastic Algorithm for Estimating Maximum-Likelihood Phylogenies. *Mol. Biol. Evol.* **2015**, *32*, 268–274. [CrossRef]
40. Ullah, F.; Gao, Y.; Sari, I.; Jiao, R.-F.; Saqib, S.; Gao, X.-F. Macro-Morphological and Ecological Variation in *Rosa sericea* Complex. *Agronomy* **2022**, *12*, 1078. [CrossRef]
41. Jin, J.-J.; Yu, W.-B.; Yang, J.-B.; Song, Y.; Depamphilis, C.W.; Yi, T.-S.; Li, D.-Z. GetOrganelle: A fast and versatile toolkit for accurate de novo assembly of organelle genomes. *Genome Biol.* **2020**, *21*, 241. [CrossRef]
42. Tillich, M.; Lehwark, P.; Pellizzer, T.; Ulbricht-Jones, E.S.; Fischer, A.; Bock, R.; Greiner, S. GeSeq—Versatile and accurate annotation of organelle genomes. *Nucleic Acids Res.* **2017**, *45*, W6–W11. [CrossRef]
43. Kurtz, S.; Choudhuri, J.V.; Ohlebusch, E.; Schleiermacher, C.; Stoye, J.; Giegerich, R. REPuter: The manifold applications of repeat analysis on a genomic scale. *Nucleic Acids Res.* **2001**, *29*, 4633–4642. [CrossRef]
44. Benson, G. Tandem repeats finder: A program to analyze DNA sequences. *Nucleic Acids Res.* **1999**, *27*, 573–580. [CrossRef]
45. Thiel, T.; Michalek, W.; Varshney, R.; Graner, A. Exploiting EST databases for the development and characterization of gene-derived SSR-markers in barley (*Hordeum vulgare* L.). *Theor. Appl. Genet.* **2003**, *106*, 411–422. [CrossRef]
46. Frazer, K.A.; Pachter, L.; Poliakov, A.; Rubin, E.M.; Dubchak, I. VISTA: Computational tools for comparative genomics. *Nucleic Acids Res.* **2004**, *32* (Suppl. 2), W273–W279. [CrossRef]
47. Zhang, Y.-J.; Ma, P.-F.; Li, D.-Z. High-Throughput Sequencing of Six Bamboo Chloroplast Genomes: Phylogenetic Implications for Temperate Woody Bamboos (Poaceae: Bambusoideae). *PLoS ONE* **2011**, *6*, e20596. [CrossRef]
48. Rozas, J.; Ferrer-Mata, A.; Sánchez-DelBarrio, J.C.; Guirao-Rico, S.; Librado, P.; Ramos-Onsins, S.E.; Sánchez-Gracia, A. DnaSP 6: DNA Sequence Polymorphism Analysis of Large Data Sets. *Mol. Biol. Evol.* **2017**, *34*, 3299–3302. [CrossRef]
49. Amiryousefi, A.; Hyvönen, J.; Poczai, P. IRscope: An online program to visualize the junction sites of chloroplast genomes. *Bioinformatics* **2018**, *34*, 3030–3031. [CrossRef]
50. Hu, H.; Liu, B.; Liang, Y.; Ye, J.; Saqib, S.; Meng, Z.; Lu, L.; Chen, Z. An updated Chinese vascular plant tree of life: Phylogenetic diversity hotspots revisited. *J. Syst. Evol.* **2020**, *58*, 663–672. [CrossRef]
51. Wu, S.; Jiang, C.; Feng, X.; Yang, C.; Yu, Z. The complete chloroplast genome of Lonicera similis Hemsl. and its phylogenetic analysis. *Mitochondrial DNA Part B* **2021**, *6*, 3151–3153. [CrossRef]
52. Jiang, C.; Wu, S.; Feng, X.; Yang, C.; Yu, Z. The complete chloroplast genome of Lonicera pampaninii Levl. and its phylogenetic analysis. *Mitochondrial DNA Part B* **2021**, *6*, 3025–3027. [CrossRef]
53. Del Moral, R. On the variability of chlorogenic acid concentration. *Oecologia* **1972**, *9*, 289–300. [CrossRef]
54. Lou, Z.; Wang, H.; Zhu, S.; Ma, C.; Wang, Z. Antibacterial Activity and Mechanism of Action of Chlorogenic Acid. *J. Food Sci.* **2011**, *76*, M398–M403. [CrossRef]
55. Naveed, M.; Hejazi, V.; Abbas, M.; Kamboh, A.A.; Khan, G.J.; Shumzaid, M.; Ahmad, F.; Babazadeh, D.; FangFang, X.; Modarresi-Ghazani, F.; et al. Chlorogenic acid (CGA): A pharmacological review and call for further research. *Biomed. Pharmacother.* **2018**, *97*, 67–74. [CrossRef]
56. Sato, Y.; Itagaki, S.; Kurokawa, T.; Ogura, J.; Kobayashi, M.; Hirano, T.; Sugawara, M.; Iseki, K. In vitro and in vivo antioxidant properties of chlorogenic acid and caffeic acid. *Int. J. Pharm.* **2011**, *403*, 136–138. [CrossRef]
57. Zhao, Y.; Wang, J.; Ballevre, O.; Luo, H.; Zhang, W. Antihypertensive effects and mechanisms of chlorogenic acids. *Hypertens. Res.* **2012**, *35*, 370–374. [CrossRef]
58. Han, Y.Q. Mechanism of Tanreqing Capsule on treatment of coronavirus disease 2019 based on network pharmacology. *Chin. Tradit. Herb Drugs* **2020**, *51*, 2967–2976.

genes

MDPI

Article

Sequence Characterization of ITS Regions of Immortelle *Helichrysum italicum* (Roth) G. Don from the East Adriatic Coast

Matjaž Hladnik, Alenka Baruca Arbeiter and Dunja Bandelj *

Faculty of Mathematics, Natural Sciences and Information Technologies, University of Primorska, Glagoljaška 8, SI-6000 Koper, Slovenia
* Correspondence: dunja.bandelj@upr.si

Abstract: The immortelle (*Helichrysum italicum* (Roth) G. Don) is a typical perennial plant of natural vegetation in the Mediterranean region, and due to secondary metabolites with several biological properties (anti-inflammatory, antioxidant, antimicrobial, and anti-proliferative), it has become an important species for essential oil production, especially in the cosmetic industry. To increase the production of highly priced essential oils, it has been moved to cultivated fields. However, due to the lack of highly characterized planting material, there is a great need for genotype identification, and to provide a link with chemical profiles and geographic origin as a basis for the identification of local superior genotypes. The aims of the study were to characterize the ITS (ribosomal internal transcribed spacer) regions, ITS1 and ITS2, in samples from the East Adriatic region to determine the possibility of using these regions for plant genetic resources identification. Genetic variation was observed when comparing the ITS sequence variants of samples from the North-East Adriatic and the South-East Adriatic. Some rare and unique ITS sequence variants can be helpful for identifying specific populations from different geographical regions.

Keywords: *Helichrysum*; barcodes; internal transcribed spacers; traceability; planting material

Citation: Hladnik, M.; Baruca Arbeiter, A.; Bandelj, D. Sequence Characterization of ITS Regions of Immortelle *Helichrysum italicum* (Roth) G. Don from the East Adriatic Coast. *Genes* **2023**, *14*, 480. https://doi.org/10.3390/genes14020480

Academic Editors: Wajid Zaman and Hakim Manghwar

Received: 29 December 2022
Revised: 11 February 2023
Accepted: 11 February 2023
Published: 14 February 2023

1. Introduction

The immortelle (*Helichrysum italicum* (Roth) G. Don (basionym = *Gnaphalium italicum* Roth)) (abbreviated as HI), known also by the synonyms 'everlasting' and 'curry plant', is a typical perennial plant of natural vegetation in the Mediterranean region, where it is traditionally used in folk medicine for healing respiratory, digestive, and skin inflammatory conditions [1–4]. Due to the strong anti-inflammatory, antioxidant, antimicrobial, and antiproliferative activities of the plant extracts and essential oil, the latter has especially become a valuable and highly prized ingredient in the perfume and cosmetic industries [4,5]. The importance of the phytochemical composition and the bioactive properties of essential oil is reflected in numerous recently performed analyses [6–12]. The recognized importance of the immortelle plant has led to the establishment of commercial plantations in Croatia, Slovenia, Bosnia and Herzegovina, Montenegro, and Serbia. The planting materials for these plantations usually originate from wild plants collected in natural growing sites, which are further vegetatively or generatively (via seeds) propagated in local nurseries. The overexploitation of wild populations in continental Dalmatia, and in the islands Pag and Krk (HRV) have already caused negative consequences for the natural growing sites [13]. Reduced genetic diversity, as was observed with *Tanacetum cinerariifolium* (Trevir.) Sch. Bip., another commercially interesting aromatic species, will be even more evident in the future [14].

In addition, the complex taxonomy of *Helichrysum* species, HI subspecies (four subspecies were proposed by Herrando-Moraira et al. [3]: *H. italicum* subsp. *italicum* (abbreviated as HII), *H. italicum* subsp. *microphyllum* (Willd.) Nyman (abbreviated as HIM), *H. italicum* subsp. *siculum* (Jord. and Fourr.) Galbany, L. Sáez and Benedí (abbreviated as HIS), and *H. italicum* subsp. *tyrrhenicum* (Bacch., Brullo, and Giusso) Herrando, J. M. Blanco, L. Sáez, and Galbany (abbreviated as HIT)) and the lack of taxonomic classification knowledge can lead to the propagation of incorrect plant material. The planting material of HI available in the market usually lacks the exact taxonomic classification, or is mislabeled and is sometimes offered as other species rather than HI, which is the most appreciated for essential oil production (personal observations). In such cases, the products from HI do not meet the industry requirements. The problems related to authenticity and traceability can cause economic losses in agricultural production, since the quality and biological properties of the plants are strongly influenced by genotype [15]. The latter supports the need to develop an effective traceability system that will allow for differentiation among planting material, the identification of local genotypes (for the protection of local products' typicality and geographical origin), and the development of new cultivars with supreme quality properties, as well as the prevention of possible adulterations or the contamination of original plants with other herbs [16].

Analyses on a DNA level have become essential in identifying plant genetic resources, reference planting material, and studying genetic diversity. However, the lack of genomic data aggravates the development of reliable DNA markers. Recently, microsatellite markers were developed for HII, which showed the potential for discriminating samples from distinct geographical locations [17]. However, due to the complex taxonomic classification of *Helichrysum* species and HI subspecies, there is still a need for the development of new additional markers for species confirmation, as well as for controlling the origin and traceability of the plant material.

DNA barcodes, derived either from chloroplast or nuclear genomes, have been frequently used for species identification and phylogenetic analysis in the last decade. Among nuclear regions, internal transcribed spacers (ITS1 and ITS2) of the ribosomal RNA gene cluster (18S-5.8S-26S) were the most commonly analyzed regions across the plants [18]. The comparison of seven candidate DNA markers (*psbA-trnH, matK, rbcL, rpoC1, ycf5*, ITS1, and ITS2) from medicinal plant species revealed that the second internal transcribed spacer (ITS2) of nuclear ribosomal DNA represents the most suitable region for DNA barcoding applications. Its high discrimination ability (92.7%) was demonstrated in more than 6600 plant samples belonging to 4800 species from 753 distinct genera [19].

To increase the power of discrimination of closely related samples, ETS (external transcribed spacer) was used in combination with ITS regions in some studies with plant material belonging to the tribe Gnaphalieae (Asteraceae) [20], and species from the HAP (*Helichrysum-Anaphalis-Pseudognaphalium*) clade [21,22] and *Helichrysum* species [21,23–25].

It was observed that the ITS and ETS trees reflected a greater congruence with phylogenetic relationships as inferred based on morphological characters, compared to chloroplast markers [21], despite drawbacks such as intraindividual/interspecific variation and the possibility of novel recombinant types generation after species crossing (reviewed by Smissen et al. [26] and Nie et al. [20]).

Dominant markers are promising tools for studying genetic diversity, because of the simultaneous amplification of a larger number of DNA loci without prior sequence knowledge for primer development, in contrast to codominant markers. Recently, AFLP markers were used to study the genetic diversity of HI samples from 18 populations sampled along the eastern Adriatic region [14]. Based on low, though significant, genetic differentiation among the populations, extensive gene flow between populations was hypothesized. AFLP markers were also used to study differentiation between HII and HIM samples from different locations in the Mediterranean basin. The geographic structure was observed in analyzed samples, but AFLP markers did not separate the HII and HIM samples [27]. The main drawback of the dominant markers is the difficult and time-consuming scoring of genetic profiles, which limits their use for identification purposes.

The sequences of different DNA barcode regions are publicly available and could serve as a reference for the taxonomic classification and the origin identification of a defined genotype. Therefore, the aim of our study was to analyze the ITS regions of HI from the East Adriatic coast, to obtain information about the genetic variability of these ITS regions, and to determine their capability for the differentiation of samples from distinct geographical regions.

Previously mentioned studies of Galbany-Casals et al. [27] and Herrando-Moraira et al. [25], who analyzed species from the *Helichrysum pendulum* complex, and other species from *Helichrysum* sect. *Stoechadina* with ETS and *rpl32-trnL* cpDNA, also including samples of HII from Bosnia and Herzegovina, and Croatia, but these studies were not focused on the East Adriatic region.

The last aim of the study was to test the possibility of discriminating HII from the East Adriatic region from other closely related species and subspecies. To test this hypothesis, a phylogenetic analysis with additional ITS sequences from the NCBI Nucleotide database, i.e., two additional HI subspecies (HIS, HIT), *Helichrysum serotinum* subsp. *serotinum* (abbreviated as HSS) and *Helichrysum serotinum* subsp. *picardii* (Boiss. and Reut.) Galbany, L. Sáez and Benedí (abbreviated as HSP), and *Helichrysum litoreum* Guss. (abbreviated as HL) as other representatives of *H.* sect. *Stoechadina*, and selected *Helichrysum* species from *Helichrysum* sect. *Helichrysum* and *Helichrysum* sect. *Virginea*, was performed.

2. Materials and Methods

In total, 51 samples of HII were included in the analysis. Plants from natural habitats were sampled in Croatia and Montenegro. Two plants grown in city flower beds were sampled in Izola, Slovenia (hereby indicated as ornamental plants). In addition, four plants grown from the purchased seeds (labeled as certified seed) with a certificate that they belong to HI, were also included in the analysis (Table 1).

Table 1. List of sampling locations and number of HII samples from each collection, including samples from certified seeds (HL).

Sampling Location	Country	Origin	Number of Samples
Bibići	Croatia	Natural habitat	2
Cavtat	Croatia	Natural habitat	10
Cres island	Croatia	Natural habitat	1
Črišnjeva	Croatia	Natural habitat	1
Kamenjak peninsula (north, middle, and south parts)	Croatia	Natural habitat	11
Kraljevica	Croatia	Natural habitat	1
Plomin	Croatia	Natural habitat	5
Pelješac peninsula	Croatia	Natural habitat	6
Border crossing point Sitnica	Montenegro	Natural habitat	2
Kruševica	Montenegro	Natural habitat	2
Luštica peninsula	Montenegro	Natural habitat	1
Sutorman	Montenegro	Natural habitat	7
Izola	Slovenia	Ornamental	2
Certified seeds	Slovenia	Certified seeds	4

Morphological evaluation and distinct genetic profiles of these plants confirmed that they belong to HL (avowed substitute for *Helichrysum angustifolium* (Lam.) DC [28]) [17]. HL, compared to HI, is a very robust plant with longer leaves and many more capitula, and it is usually less glandular than HI on the abaxial sides of the leaves and on the phyllaries [1,28]. The amplified regions of plants from certified seeds of HL were compared with the obtained sequences from HII samples to identify potential polymorphisms among two different species.

Shoot tips with leaves were collected in the field and immediately preserved in a saturated NaCl-CTAB solution [29], or dried in a silica gel until DNA extraction. DNA was extracted with the protocol described by Japelaghi et al. [30], with the only difference being that phenol:chloroform:isoamyl alcohol was used instead of chloroform:isoamyl alcohol.

The DNA was stored in the laboratory of the Faculty of Mathematics, Natural Sciences and Information Technologies, University of Primorska.

ITS1, 5.8S, and ITS2 regions were simultaneously amplified using the primers ITS1 and ITS4 [31]. The PCR reaction in a final volume of 50 μL contained 1× PCR buffer, 50 ng of DNA, 1.5 mM concentration of MgCl2, 0.8 mM of each dNTP, 0.5 μM of each primer, and 1.25 U of *Taq* DNA polymerase (Promega, Mannheim, Germany). The PCR protocol was as follows: a 5 min denaturation step at 94 °C, followed by 13 cycles of 35 s at 93 °C, 55 s at 53 °C, and 45 s at 72 °C. In the next 13 cycles, the denaturation and annealing steps remained the same, while the elongation step was prolonged to 59 s per cycle; in the last 9 cycles, the annealing step was 118 s. The amplification ended with the last elongation step at 72 °C for 10 min.

In order to confirm different variants of the ITS sequences in the same samples, observed as double peaks in the electropherograms, the PCR products were cloned into the pGEM®-T Vector (Promega, Madison, WI, USA) with the following procedure: the PCR products were excised from 1.7% agarose gel and cleaned with a Silica Bead DNA Gel Extraction Kit (Thermo Fisher Scientific, Waltham, MA, USA). pGEM®-T Vector ligation reactions were performed in a final volume of 5.5 μL containing 2 μL of purified PCR product, 2.5 μL of 2× Rapid Ligation Buffer, 0.5 μL of pGEM®-T, and 0.5 μL of ligase, and were incubated at 4 °C overnight.

The ligation mixture (3 μL) was introduced into competent *Escherichia coli* XL-10 Gold cells (Agilent Technologies, La Jolla, CA, USA) (100 μL), incubated for 20 min on ice, and transformed using the heat shock method at 42 °C for 1 min in a water bath and for 5 min on ice. A total of 900 μL of liquid SOC media (without antibiotic) was then added to the cells, followed by incubation for 45 min at 37 °C to generate the antibiotic resistance. Two different aliquots of transformation (20 μL and 70 μL) were plated onto LB agar plates containing the antibiotic, and incubated at 37 °C for one day. An LB agar medium (1 L) was made using a standard procedure, using 35 g of LB Agar High Salt2 L1706 (Duchefa, Haarlem, Neatherlands), 10 g of Daishin agar D1004 (Duchefa), 0.2 mM of IPTG (Duchefa), 40 μg ml-1 of X-GAL (Duchefa), and 150 mg l-1 of antibiotic carbenicillin (Duchefa).

For colony PCR, 10 white colonies from each LB agar plate (with the exception of sample 6) were picked with fine pipette tips, dissolved in 80 μL of TdE buffer, and heated at 96 °C for 8 min. The DNA was amplified via PCR reaction in a final volume of 20 μL containing 5 μL of DNA (dissolved in TdE buffer in the previous step), 2 μL of 10× PCR Buffer with $(NH_4)_2SO_4$, 2 mM concentration of $MgCl_2$, 0.8 mM of each dNTP, 0.5 μM of primer SP6, 0.5 μM of primer T7, and 0.5 U of *Taq* DNA polymerase (EP0401) (Thermo Fisher Scientific, Vilnius, Lithuania). After an initial denaturation of 3 min at 94 °C, the amplification protocol was 34 cycles 30 s at 94 °C, 30 s at 55 °C, and 1 min at 72 °C. The amplification was concluded with an elongation step at 72 °C for 8 min.

Amplified DNA (5 μL) was analyzed on a 1.5% agarose gel containing Midori Green (NIPPON Genetics, Düren, Germany). The rest of the reaction mixture of all successfully amplified samples was purified and used for sequencing analysis.

PCR products were cleaned with Exonuclease I (Thermo Fisher Scientific, Vilnius, Lithuania) and FastAP Thermosensitive Alkaline Phosphatase (Thermo Fisher Scientific). A DNA sequencing reaction was performed in a final volume of 7 μL, and it contained 5 μL of PCR product, 2 U of Exonuclease I, 0.5 U of FastAP Thermosensitive Alkaline Phosphatase, and 1.4 μL of 1× PCR buffer (Promega). The reaction was incubated for 45 min at 37 °C and ended at 80 °C for 15 min. The fragments were sequenced from both directions using BigDye Terminator v3.1 (Thermo Fisher Scientific) sequencing chemistry. The sequencing reaction was cleaned using the EDTA and ethanol precipitation methods, and analyzed on a 3130 Genetic Analyser (Thermo Fisher Scientific, Applied Biosystems, Tokyo, Japan).

Bases were called using KB Basecaller, implemented in Sequencing analysis software v5.1 (Applied Biosystems). The program CodonCode Aligner 6.02 (CodonCode Corporation, Dedham, MA, USA) was used to align the forward and reverse sequences. The resulting consensus sequences for each individual were aligned using ClustalW, implemented in the MEGA package 7.0 [32,33].

Dnasp v.6.12.03 [34] was used to determine the number of sequence variants and the number of variable sites (singleton variable sites and parsimony informative sites).

A barcoding gap analysis with all our sequences (HII and HL) was performed with Kimura 2-parameter (K2P) genetic distances using the barcodingR [35] package.

A network with ITS1-5.8S-ITS2 sequence variants was created with the TCS method [36], implemented in the PopArt software [37]. An input Nexus file was prepared with MEGA package 7.0. HII sequences (accession numbers from KJ159118.1 to KJ159126.1) were included in the construction of the network as well. For the network analysis, sampling locations were grouped into the North-East Adriatic and the South-East Adriatic. Sequence variants of plants from certified seeds and from ornamental plants were considered separately. The distribution of sequence variants across geographical locations was presented with a map, generated with R version 4.2.1, and R packages maps (3.4.0) and mapplots (1.5.1). Newly identified sequence variants of HII have been deposited in the NCBI Nucleotide database (accession numbers from OP874600 to OP874610), whereas sequence variants identified only at HL were assigned to GenBank accession numbers from OQ330863 to OQ330867.

Phylogenetic analysis was performed with ITS sequence variants (concatenated ITS1 and ITS2 regions; 5.8S rDNA sequence was excluded from the analysis) of HII obtained in this study, and with additional sequences of selected species/subspecies belonging to *H.* sect. *Stoechadina* (*Helichrysum crassifolium* D. Don (abbreviated as HC), *Helichrysum heldreichi* Boiss. (abbreviated as HH), HIS, HIT, *Helichrysum massanellanum* Herrando, J. M. Blanco, L. Sáez and Galbany (abbreviated as HM), HSS and HSP, *Helichrysum stoechas* (L.) Moench (abbreviated as HSt)), *H.* sect. *Helichrysum* (*Helichrysum thianschanicum* Regel (abbreviated as HT) and *Helichrysum orientale* (L.) Vaill. (abbreviated as HO)), and *H.* sect. *Virginea* (*Helichrysum sibthorpii* Rouy (abbreviated as HSi)), which were downloaded from the NCBI Nucleotide database. Species from *H.* sect. *Helichrysum* and *H.* sect. *Virginea* were included as a control for our study, since they formed different phylogenetic clades in previous studies [21,24]. Among the existing HI subspecies with ITS sequences available in the NCBI Nucleotide database, HII, HIT, and HIS were included in the analysis. Most of the sequences belonged to HII from Corsica (FR). However, in view of the revised classification of the HI subspecies [3] and after additional verification of the origin of the sequenced samples deposited in the NCBI database, some taxonomic annotations were updated, considering the latest taxonomic classification. Sample HM244710.1 originated from Corsica (source: GenBank); thus, its previous classification, HIM, was corrected to HIT, since the subsp. *microphyllum* is restricted to Crete (GR) only [3]. In addition, the sequence AY445196.1 originally annotated to HI and collected in Sicily (IT) [23] represents HIS. The new classification assumes that HIS is endemic to Sicily [3]. Another sample, AY445195.1, collected on Massanella massif (Majorcan mountain) [23], previously classified as HIM, was assigned to a newly described species, HM [3].

The best available model of DNA substitution for the data was identified with the Akaike Information Criterion (AIC), as implemented in MrModeltest2 v2.4 [38]. A HKY model with the gamma distribution (G) shape parameter 0.1770 was identified as the most appropriate model. Phylogenetic analysis using the Bayesian inference (BI) and Maximum Likelihood (ML) was performed with MrBayes v3.2.7a [39] and PhyML v3.3 [40], respectively. Four Monte Carlo Markov chains were run simultaneously for 5×10^6 generations, with the resulting trees sampled every 500 generations. Bayesian posterior probabilities (BPP) were used to assess the branch support of the BI tree. Trace files were imported in TRACER v1.7.1 to examine the trace plots and effective sample size (ESS) estimates [41]. The following flags were used for the phyml analysis: -d 'nt' -b 1000 -m HKY85 -t 1.7025 -a 0.1770 –leave_duplicates. Branch support in the ML tree was estimated using 1000 bootstrap replicates. The resulting trees were compared using a Shimodaira–Hasegawa (SH) test [42] implemented in PAUP v4.0a168 [43], based on likelihood scores. The following command was used: Lscore all/nst = 6 rmatrix = estimate basefreq = estimate rates = gamma shape = estimate pinvar = estimate Shtest = yes bootreps = 1000. Trees were rooted with species *Helichrysum litorale* Bolus (KT865507).

3. Results and Discussion

Immortelle is a relatively new plant in agricultural cultivation, enabling the production of highly prized essential oil and the development of products with high added value in the cosmetic industry. According to our own experience, there is confusion regarding the correct naming and labeling of seedlings in the market. The main reasons for this are the abundance of different *Helichrysum* species and HI subspecies present in the Mediterranean region, the complex taxonomical classification, the presence of hybrids, and the transfer of planting material between countries.

In this study, 51 HII samples, including 49 samples from their natural growing sites in the North-East Adriatic and the South-East Adriatic regions, and two ornamental samples from city flower beds, were used for the amplification of the ITS1-5.8S-ITS2 regions to determine whether the geographical origins of the samples can be identified based on polymorphisms of the ITS sequences. The geographical region may have an influence on the development of unique populations with special genetic backgrounds, which can be reflected in the quality of immortelle products. From this point of view, it is important to obtain information about genetic resources for the successful introduction of plants in the agriculture ecosystem. In addition, four samples obtained from purchased seeds with a certificate that they belong to HI, but later determined as HL [17], were also included to test the ability of ITS loci to detect not true-to-type plant material. HL is present in the East Adriatic region, and it grows together with HI in certain areas, leading to the development of intermediate specimens [1].

An examination of the electropherograms revealed the presence of up to two double peaks at 27 samples. After the reamplification of PCR products with a DNA cloning technique, up to four different sequence variants were identified in one sample. In total, 93 sequences (considering all sequence variants per sample) were obtained in the analyzed samples.

An intra-individual variation of ITS sequences was commonly observed [44]. In addition, intraindividual variation was observed for the ETS marker [25], which is a part of the same transcriptional unit.

After the multiple sequence alignment, 25 bp were removed from the beginning of the ITS1 region, since the first few bases were not called with adequate quality scores in a few samples. However, the entire ITS2 region was sequenced in all samples. In total, 18 ITS sequence variants with 12 polymorphic sites were identified (Supplementary Table S1). In ITS1, three singleton variable sites and five parsimony informative sites were observed, whereas in the ITS2 region, there were two singleton variable sites and one parsimony informative site. One sample from certified seed showed a mutation in the 5.8S gene, and an insertion of 2 bp was observed in one sample from Kamenjak in the ITS2 region. The distribution of the ITS1-5.8S-ITS2 sequence variants across sampling geographical regions is shown in Figure 1.

A TCS network analysis was performed to identify relationships among the ITS1-5.8S-ITS2 sequence variants (Figure 2). Sequence variant S10 with the insertion was considered as S2, since PopArt masked gaps.

The most frequently represented sequence variant S1 was observed in all geographical regions, as well as in plants from certified seeds (previously classified as HL) and in ornamental plants. Fifteen sequences were derived from S1 (including sequences from Corsica's samples) with one mutational step. Sequence variant S2 was observed in samples from the North East and South-East Adriatic regions only. Sequence S3 was observed in samples from the East Adriatic region and Corsica, whereas sequences S4 and S5 were identified only in the South-East Adriatic region. Twelve sequence variants were observed only once in different samples (seven sequence variants were identified only once in the samples from the South-East Adriatic, and two sequence variants were observed only once in the samples from the North-East Adriatic, respectively). Five unique sequence variants (S14–S18) were observed in samples from certified seeds (HL), whereas sequences S19–S25 were identified in Corsica only.

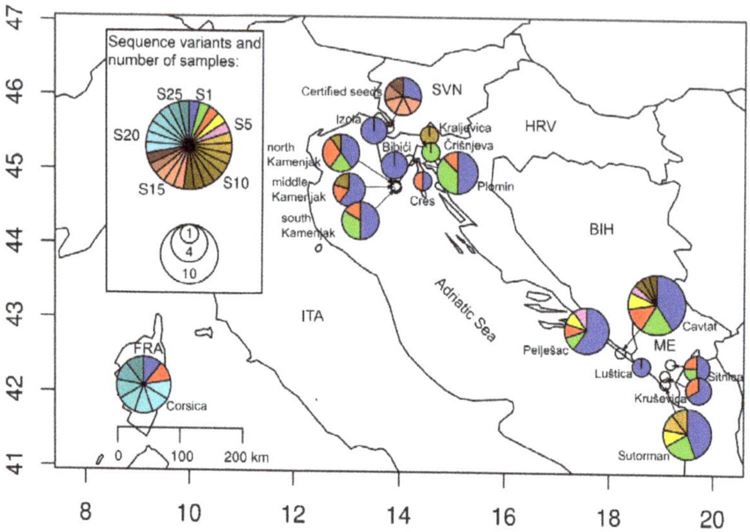

Figure 1. Distribution of identified ITS1-5.8S-ITS2 sequence variants across the North-East and South-East Adriatic regions. Sequences from Corsica were obtained from the NCBI Nucleotide database. The radii of pie charts indicate the number of samples per location (from 1 to 10), whereas the sizes of the slices present the relative abundances of sequence variants. The most frequently observed sequence variants, labeled 1 to 5, were marked with distinct colors, whereas sequences identified at single samples were marked with similar color shades, based on their origins (S6 to S13 North-East and South-East Adriatic region, S14 to S18 certified seeds (identified as HL) and S19 to S25 GenBank sequences from Corsica plants).

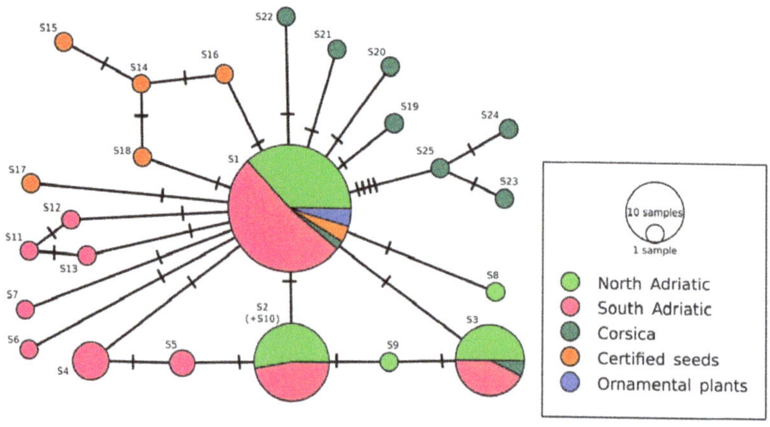

Figure 2. HII TCS network with ITS1-5.8S-ITS2 sequence variants. Sequences with labels from S1 to S5 indicate the most frequently observed sequence variants (S1, S2, S3, S4, and S5 were identified in 46, 17, 14, 4, and 2 samples, respectively). Hatch marks present the number of mutations. Sequences from Corsica were downloaded from the NCBI Nucleotide database (accession numbers from KJ159118 to KJ159126).

Bayesian phylogenetic (BP) and maximum likelihood (ML) analyses were performed in order to infer the relationship of HII ITS sequence variants from distinct geographical locations, and the relationship to other selected closely related species from the *Helichrysum* genus. The SH test did not reveal a significant difference between the BP and ML tree (the ML tree is presented in Figure 3).

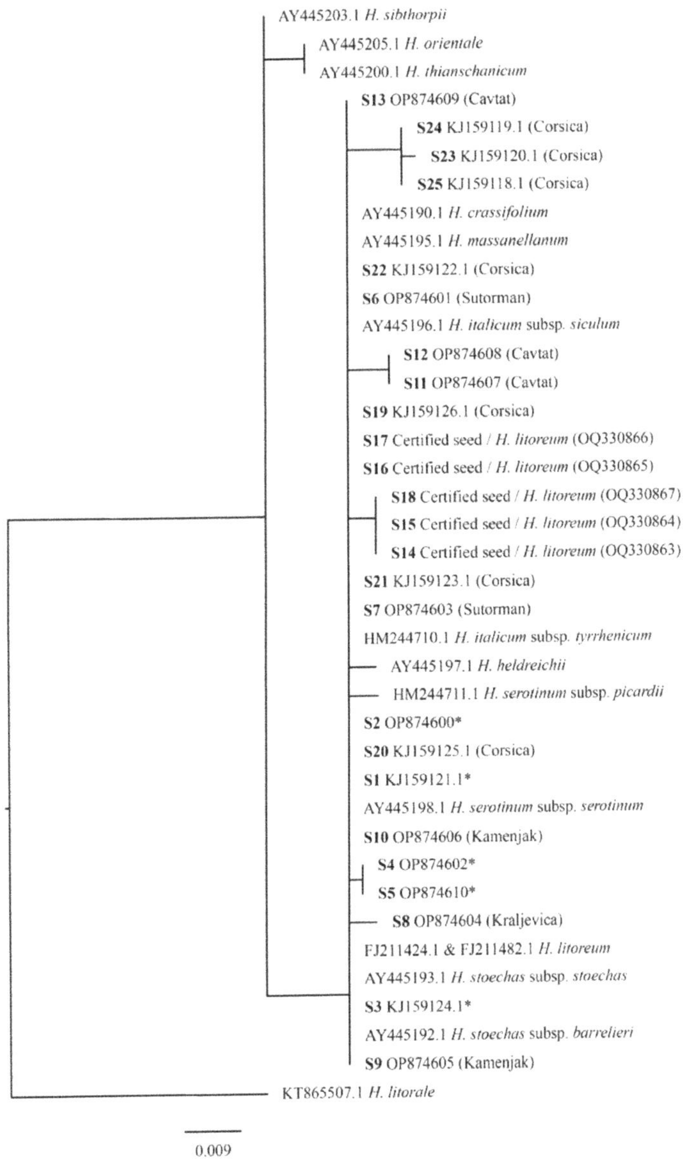

Figure 3. Maximum likelihood tree with the ITS sequences. *H. litorale* species was used as a root. Sequence variants S1 to S5 of HII (marked with an asterix) were found in two or more samples from different locations (sequences with GenBank accession numbers represent a reference sequence), whereas the other ITS sequence variants were found in separate samples.

Based on ML phylogenetic analysis, HT and HO (belonging to *H.* sect. *Helichrysum*) were separated from HSi (belonging to *H.* sect. *Virginea*), whereas all HI subspecies, as well as *H. serotinum* subspecies, *H. stoechas* subspecies, HH, HL, HM, and HC were placed in the same clade, with subclades and leaves being presented as multifurcations. All previously mentioned taxa (except HT, HO, and HSi) belong to a monophyletic *H.* sect. *Stoechadina*, which generally has a western-central Mediterranean distribution [21,24].

The discrimination of species from the *H.* sect. *Stoechadina* was not observed since HII ITS sequences were in groups with different *Helichrysum* species, and the same ITS sequences were identified in different taxonomic units (HC (AY445190.1), HIS (AY445196.1), HSt subsp. *stoechas* (AY445193.1), HM (AY445195.1), HL (FJ11424.1 and FJ211482.1), and sequence S1 (including HL from certified seeds (Table S1)), which was found in samples from all locations). Sequences of HSt subsp. *barrelieri* (AY445192.1) and HSS (AY445198.1) were duplicates of the sequence S2, characteristic for the North and South Adriatic.

A low resolution of the nrDNA phylogenies, based on ITS or ETS alone, was reported for the Mediterranean-Asiatic *Helichrysum* group [24]. In addition, morphologically distinct taxa from close geographic origins were grouped together. This could be a consequence of historical and contemporary hybridizations [1,21,25]. Galbany-Casals et al. [27] indicated that specimens from some localities appear to be intermediate between HII and either HIM or HIS. The same authors performed a comparison of the chloroplast *psbA-trnH* sequence barcode of 81 samples of different subspecies of HI (subsp. *italicum*, subsp. *microphyllum*, and subsp. *siculum*) and other members of sect. *Stoechadina* (species HL, HSP, and HSS), and showed that some chloroplast haplotypes were shared across species (e.g., haplotype G was identified at HII, HIM, HL, and HSS). The same was observed in the group of the New Zealand endemic species of Gnaphalieae, where the chloroplast *psbA-trnH* marker was used, and the obtained haplotypes did not correspond with the taxonomic units [26].

In addition, AFLP markers and ITS sequences, together with the morphological data, supported the existence of hybrids between HO and HSt [45].

However, since the region-specific ITS sequence variants were observed in our study, the results were in agreement with Galbany-Casals et al. [24], as they also observed a correlation of nrDNA markers with geographic distribution. Additionally, Herrando-Moraira et al. [25] observed that in the case of ETS, most of the variation (83.9%) could be attributed to the differences between populations, which supported the presence of the phylogeographical structure for this marker. Region-specific ITS sequence variants could be supported with the recently published study of Ninčević et al. [14], which showed a clustering of HI populations with AFLP markers into two distinct clusters, i.e., (1) populations from the north (Kvarner Bay, HRV) and (2) populations from the central and southern parts of Dalmatia (HRV).

With the ITS sequence polymorphisms, we were not able to discriminate the HI subspecies. The latter could be explained with the hypothesis of Galbany-Casals et al. [27], where due to the widespread distribution of HI, it is probably not possible to define discrete populations because of gradual clinal variation.

Despite the low discrimination resolution observed among the analyzed sequences, distinct ITS1-5.8S-ITS2 sequence variants were observed in two plants obtained from certified seeds (HL). However, two plants shared a sequence variant that was identical to the most common sequence variant (S1), and the same was noticed for the ITS reference sequences of HL from the NCBI Nucleotide database. A barcoding gap analysis with our sequences (HII and HL) with Kimura 2-parameter (K2P) genetic distances revealed no gap between intraspecies and interspecies genetic distances. One of the reasons for this could be the different number of sequence variants compared (HII: 86 vs. HL: 7) in this study. Therefore, an additional analysis of the plant material is needed to confirm that the obtained distinct sequence variants are characteristic of HL. It can be mentioned that HL and HI can be better differentiated using microsatellites, as a comparison of their DNA profiles revealed 8 unique alleles and 14 unique allelic combinations at multiple loci in the HL sample [17].

4. Conclusions

The study demonstrated that the same ITS sequences are shared by different HI individuals from different geographic regions, and even by different species of *H.* sect. *Stoechadina*, making this marker not useful as a barcode for the differentiation of *Helichrysum* species. However, some rare, localized, and uncommon sequence variants can be helpful to identify specific populations or geographic variants. In addition, new ITS1-5.8S-ITS2 regions of 11 HII and 5 sequence variants of HL from the East Adriatic region have been deposited in the NCBI Nucleotide database. In the future, it is expected that an increased number of accessible sequences in the NCBI Nucleotide database and the new markers developed for HI will contribute to a better resolution of the species and subspecies of this important aromatic plant.

Supplementary Materials: The following supporting information can be downloaded at: https://www.mdpi.com/article/10.3390/genes14020480/s1, Table S1. List of identified mutations in the observed sequence variants.

Author Contributions: Conceptualization, D.B.; methodology, D.B. and M.H.; formal analysis, M.H.; investigation, A.B.A. and M.H.; resources, A.B.A., D.B. and M.H.; data curation, M.H.; writing—original draft preparation, A.B.A., D.B. and M.H.; writing—review and editing, A.B.A., D.B. and M.H.; visualization, M.H.; supervision, D.B.; project administration, D.B.; funding acquisition, D.B. All authors have read and agreed to the published version of the manuscript.

Funding: This research was financially supported by the Slovenian Research Agency, the research program "Conservation biology, from molecules to ecosystem" (P1-0386), and the applied project, "Production, isolation and formulation of health beneficial substances from *Helichrysum Italicum* for application in cosmetic industry" (L2-4430), founded by the Slovenian Research Agency and Afrodita Ltd.

Institutional Review Board Statement: Not applicable.

Informed Consent Statement: Not applicable.

Data Availability Statement: The representative sequence variants have been deposited in the NCBI Nucleotide database under accession numbers OP874600 to OP874610 (*H. italicum* subsp. *italicum*), and from OQ330863 to OQ330867 (*H. litoreum*).

Acknowledgments: The authors are grateful to their research colleagues Z. Šatović (University of Zagreb, Faculty of Agriculture, Croatia) and M. Čizmović (University of Montenegro, Biotechnical Faculty, Montenegro), who provided invaluable support for collecting plant material in natural habitats on the Eastern Adriatic coast.

Conflicts of Interest: The authors declare no conflict of interest. The funders had no role in the design of the study; in the collection, analyses, or interpretation of data; in the writing of the manuscript; or in the decision to publish the results.

References

1. Galbany-Casals, M.; Sáez, L.; Benedí, C. A Taxonomic Revision of *Helichrysum* Sect. *Stoechadina (Asteraceae, Gnaphalieae). Can. J. Bot.* **2006**, *84*, 1203–1232. [CrossRef]
2. Antunes Viegas, D.; Palmeira-De-Oliveira, A.; Salgueiro, L.; Martinez-De-Oliveira, J.; Palmeira-De-Oliveira, R. *Helichrysum Italicum*: From Traditional Use to Scientific Data. *J. Ethnopharmacol.* **2014**, *151*, 54–65. [CrossRef] [PubMed]
3. Herrando-Moraira, S.; Blanco-Moreno, J.M.; Sáez, L.; Galbany-Casals, M. Re-Evaluation of the *Helichrysum Italicum* Complex (Compositae: Gnaphalieae): A New Species from Majorca (Balearic Islands). *Collect. Bot.* **2016**, *35*, e009. [CrossRef]
4. Ninčević, T.; Grdiša, M.; Šatović, Z.; Jug-Dujaković, M. *Helichrysum Italicum* (Roth) G. Don: Taxonomy, Biological Activity, Biochemical and Genetic Diversity. *Ind. Crops Prod.* **2019**, *138*, 111487. [CrossRef]
5. Andreani, S.; Uehara, A.; Blagojević, P.; Radulović, N.; Muselli, A.; Baldovini, N. Key Odorants of Industrially-Produced *Helichrysum Italicum* Subsp. *Italicum Essential Oil. Ind. Crops Prod.* **2019**, *132*, 275–282. [CrossRef]
6. Leonardi, M.; Ambryszewska, K.E., Melai, B.; Flamini, G.; Cioni, P.L.; Parri, F.; Pistelli, L. Essential-Oil Composition of *Helichrysum Italicum* (Roth) G.Don Ssp. *Italicum from Elba Island (Tuscany, Italy). Chem. Biodivers.* **2013**, *10*, 343–355. [CrossRef] [PubMed]
7. Cávar Zeljković, S.; Šolić, M.E.; Maksimović, M. Volatiles of *Helichrysum Italicum* (Roth) G. Don from Croatia. *Nat. Prod. Res.* **2015**, *29*, 1874–1877. [CrossRef]

8. Cui, H.; Zhao, C.; Lin, L. Antibacterial Activity of *Helichrysum Italicum* Oil on Vegetables and Its Mechanism of Action. *J. Food Process. Preserv.* **2015**, *39*, 2663–2672. [CrossRef]
9. Djihane, B.; Wafa, N.; Elkhamssa, S.; Pedro, H.J.; Maria, A.E.; Mohamed Mihoub, Z. Chemical Constituents of *Helichrysum Italicum* (Roth) G. Don Essential Oil and Their Antimicrobial Activity against Gram-Positive and Gram-Negative Bacteria, Filamentous Fungi and Candida Albicans. *Saudi Pharm. J.* **2017**, *25*, 780–787. [CrossRef]
10. Gismondi, A.; Di Marco, G.; Canini, A. *Helichrysum Italicum* (Roth) G. Don Essential Oil: Composition and Potential Antineoplastic Effect. *S. Afr. J. Bot.* **2020**, *133*, 222–226. [CrossRef]
11. Jerković, I.; Rajić, M.; Marijanović, Z.; Bilić, M.; Jokić, S. Optimization of Supercritical CO_2 Extraction of Dried *Helichrysum Italicum* Flowers by Response Surface Methodology: GC-MS Profiles of the Extracts and Essential Oil. *Sep. Sci. Technol.* **2016**, *51*, 2925–2931. [CrossRef]
12. Fraternale, D.; Flamini, G.; Ascrizzi, R. In Vitro Anticollagenase and Antielastase Activities of Essential Oil of *Helichrysum Italicum* Subsp. *Italicum (Roth) G. Don. J. Med. Food* **2019**, *22*, 1041–1046. [CrossRef]
13. Pohajda, I.; Dragun, G.; Visković, L.P. *Smilje*; Horvat, H., Ed.; Savjetodavna služba: Zagreb, Croatia, 2015; ISBN 978-953-6763-51-1.
14. Ninčević, T.; Jug-Dujaković, M.; Grdiša, M.; Liber, Z.; Varga, F.; Pljevljakušić, D.; Šatović, Z. Population Structure and Adaptive Variation of *Helichrysum Italicum* (Roth) G. Don along Eastern Adriatic Temperature and Precipitation Gradient. *Sci. Rep.* **2021**, *11*, 24333. [CrossRef] [PubMed]
15. Melito, S.; Sias, A.; Petretto, G.L.; Chessa, M.; Pintore, G.; Porceddu, A. Genetic and Metabolite Diversity of Sardinian Populations of *Helichrysum Italicum*. *PLoS ONE* **2013**, *8*, e79043. [CrossRef] [PubMed]
16. Mishra, M. Adulteration in Commercially Important Raw Medicinal Plants of Vidarbha Region of Maharshtra: Awareness among Consumers and Changing Behavior of End Users. *Indian J. Soc. Res.* **2016**, *57*, 159–168.
17. Baruca Arbeiter, A.; Hladnik, M.; Jakše, J.; Bandelj, D. First Set of Microsatellite Markers for Immortelle (*Helichrysum Italicum* (Roth) G. Don): A Step towards the Selection of the Most Promising Genotypes for Cultivation. *Ind. Crops Prod.* **2021**, *162*, 113298. [CrossRef]
18. Edger, P.P.; Tang, M.; Bird, K.A.; Mayfield, D.R.; Conant, G.; Mummenhoff, K.; Koch, M.A.; Pires, J.C. Secondary Structure Analyses of the Nuclear RRNA Internal Transcribed Spacers and Assessment of Its Phylogenetic Utility across the Brassicaceae (Mustards). *PLoS ONE* **2014**, *9*, e101341. [CrossRef]
19. Chen, S.; Yao, H.; Han, J.; Liu, C.; Song, J.; Shi, L.; Zhu, Y.; Ma, X.; Gao, T.; Pang, X.; et al. Validation of the ITS2 Region as a Novel DNA Barcode for Identifying Medicinal Plant Species. *PLoS ONE* **2010**, *5*, e8613. [CrossRef]
20. Nie, Z.L.; Funk, V.A.; Meng, Y.; Deng, T.; Sun, H.; Wen, J. Recent Assembly of the Global Herbaceous Flora: Evidence from the Paper Daisies (Asteraceae: Gnaphalieae). *New Phytol.* **2016**, *209*, 1795–1806. [CrossRef] [PubMed]
21. Galbany-Casals, M.; Unwin, M.; Garcia-Jacas, N.; Smissen, R.D.; Susanna, A.; Bayer, R.J. Phylogenetic Relationships in *Helichrysum* (Compositae: Gnaphalieae) and Related Genera: Incongruence between Nuclear and Plastid Phylogenies, Biogeographic and Morphological Patterns, and Implications for Generic Delimitation. *Taxon* **2014**, *63*, 608–624. [CrossRef]
22. Andrés-Sánchez, S.; Verboom, G.A.; Galbany-Casals, M.; Bergh, N.G. Evolutionary History of the Arid Climate-Adapted *Helichrysum* (Asteraceae: Gnaphalieae): Cape Origin and Association between Annual Life-History and Low Chromosome Numbers. *J. Syst. Evol.* **2019**, *57*, 468–487. [CrossRef]
23. Galbany-Casals, M.; Garcia-Jacas, N.; Susanna, A.; Sáez, L.; Benedí, C. Phylogenetic Relationships in the Mediterranean *Helichrysum* (Asteraceae, Gnaphalieae) Based on Nuclear RDNA ITS Sequence Data. *Aust. Syst. Bot.* **2004**, *17*, 241–253. [CrossRef]
24. Galbany-Casals, M.; Garcia-Jacas, N.; Sáez, L.; Benedí, C.; Susanna, A. Phylogeny, Biogeography, and Character Evolution in Mediterranean, Asiatic, and Macaronesian *Helichrysum* (Asteraceae, Gnaphalieae) Inferred from Nuclear Phylogenetic Analyses. *Int. J. Plant Sci.* **2009**, *170*, 365–380. [CrossRef]
25. Herrando-Moraira, S.; Carnicero, P.; Blanco-Moreno, J.M.; Sáez, L.; Véla, E.; Vilatersana, R.; Galbany-Casals, M. Systematics and Phylogeography of the Mediterranean *Helichrysum* Pendulum Complex (Compositae) Inferred from Nuclear and Chloroplast DNA and Morphometric Analyses. *Taxon* **2017**, *66*, 909–933. [CrossRef]
26. Smissen, R.D.; Breitwieser, I.; Ward, J.M. Phylogenetic Implications of Trans-Specific Chloroplast DNA Sequence Polymorphism in New Zealand Gnaphalieae (Asteraceae). *Plant Syst. Evol.* **2004**, *249*, 37–53. [CrossRef]
27. Galbany-Casals, M.; Blanco-Moreno, J.M.; Garcia-Jacas, N.; Breitwieser, I.; Smissen, R.D. Genetic Variation in Mediterranean *Helichrysum Italicum* (Asteraceae; Gnaphalieae): Do Disjunct Populations of Subsp. Microphyllum Have a Common Origin? *Plant Biol.* **2011**, *13*, 678–687. [CrossRef]
28. Aghababyan, M.; Greuter, W.; Mazzola, P.; Raimondo, F.M. On the Taxonomy and Nomenclature of *Gnaphalium Angustifolium* Lam. and *Helichrysum Litoreum* Guss. (Compositae). *Bocconea* **2009**, *23*, 157–163.
29. Rogstad, S.H. Saturated NaCl-CTAB Solution as a Means of Field Preservation of Leaves for DNA Analyses. *Taxon* **1992**, *41*, 701–708. [CrossRef]
30. Japelaghi, R.H.; Haddad, R.; Garoosi, G.-A. Rapid and Efficient Isolation of High Quality Nucleic Acids from Plant Tissues Rich in Polyphenols and Polysaccharides. *Mol. Biotechnol.* **2011**, *49*, 129–137. [CrossRef]
31. White, T.J.; Bruns, T.; Lee, S.; Taylor, J. Amplification and Direct Sequencing of Fungal Ribosomal RNA Genes for Phylogenetics. In *PCR Protocols, a Guide to Methods and Applications*; Academic Press: London, UK, 1990; pp. 315–322.
32. Kumar, S.; Stecher, G.; Tamura, K. MEGA7: Molecular Evolutionary Genetics Analysis Version 7.0 for Bigger Datasets. *Mol. Biol. Evol.* **2016**, *33*, 1870–1874. [CrossRef]

33. Thompson, J.D.; Higgins, D.G.; Gibson, T.J. CLUSTAL W: Improving the Sensitivity of Progressive Multiple Sequence Alignment through Sequence Weighting, Position-Specific Gap Penalties and Weight Matrix Choice. *Nucleic Acids Res.* **1994**, *22*, 4673–4680. [CrossRef] [PubMed]
34. Rozas, J.; Ferrer-Mata, A.; Sanchez-DelBarrio, J.C.; Guirao-Rico, S.; Librado, P.; Ramos-Onsins, S.E.; Sanchez-Gracia, A. DnaSP 6: DNA Sequence Polymorphism Analysis of Large Data Sets. *Mol. Biol. Evol.* **2017**, *34*, 3299–3302. [CrossRef] [PubMed]
35. Zhang, A.; Hao, M.; Yang, C.; Shi, Z. BarcodingR: An Integrated r Package for Species Identification Using DNA Barcodes. *Methods Ecol. Evol.* **2017**, *8*, 627–634. [CrossRef]
36. Clement, M.; Posada, D.; Crandall, K.A. TCS: A Computer Program to Estimate Gene Genealogies. *Mol. Ecol.* **2000**, *9*, 1657–1659. [CrossRef] [PubMed]
37. Leigh, J.W.; Bryant, D. Popart: Full-Feature Software for Haplotype Network Construction. *Methods Ecol. Evol.* **2015**, *6*, 1110–1116. [CrossRef]
38. Nylander, J.A.A. *MrModeltest v2*; Program Distributed by the Author; Evolutionary Biology Centre, Uppsala University: Uppsala, Sweden, 2004.
39. Ronquist, F.; Teslenko, M.; Van Der Mark, P.; Ayres, D.L.; Darling, A.; Höhna, S.; Larget, B.; Liu, L.; Suchard, M.A.; Huelsenbeck, J.P. MrBayes 3.2: Efficient Bayesian Phylogenetic Inference and Model Choice across a Large Model Space. *Syst. Biol.* **2012**, *61*, 539–542. [CrossRef]
40. Guindon, S.; Dufayard, J.F.; Lefort, V.; Anisimova, M.; Hordijk, W.; Gascuel, O. New Algorithms and Methods to Estimate Maximum-Likelihood Phylogenies: Assessing the Performance of PhyML 3.0. *Syst. Biol.* **2010**, *59*, 307–321. [CrossRef]
41. Rambaut, A.; Drummond, A.J.; Xie, D.; Baele, G.; Suchard, M.A. Posterior Summarization in Bayesian Phylogenetics Using Tracer 1.7. *Syst. Biol.* **2018**, *67*, 901–904. [CrossRef]
42. Shimodaira, H.; Hasegawa, M. Multiple Comparisons of Log-Likelihoods with Applications to Phylogenetic Inference. *Mol. Biol. Evol.* **1999**, *16*, 1114. [CrossRef]
43. Swofford, D.L. *PAUP*. Phylogenetic Analysis Using Parsimony (*and Other Methods)*; Version 4; Sinauer Associates: Sunderland, MA, USA, 2003.
44. Song, J.; Shi, L.; Li, D.; Sun, Y.; Niu, Y.; Chen, Z.; Luo, H.; Pang, X.; Sun, Z.; Liu, C.; et al. Extensive Pyrosequencing Reveals Frequent Intra-Genomic Variations of Internal Transcribed Spacer Regions of Nuclear Ribosomal DNA. *PLoS ONE* **2012**, *7*, e43971. [CrossRef]
45. Galbany-Casals, M.; Carnicero-Campmany, P.; Blanco-Moreno, J.M.; Smissen, R.D. Morphological and Genetic Evidence of Contemporary Intersectional Hybridisation in Mediterranean *Helichrysum* (Asteraceae, Gnaphalieae). *Plant Biol.* **2012**, *14*, 789–800. [CrossRef]

genes

Article

Identification of CmbHLH Transcription Factor Family and Excavation of *CmbHLHs* Resistant to Necrotrophic Fungus *Alternaria* in *Chrysanthemum*

Yifeng Ding [1,†], Xiaomeng Wang [1,†], Dandan Wang [1], Liwei Jiang [1], Jing Xie [1], Tianle Wang [1], Lingyu Song [1] and Xiting Zhao [1,2,3,*]

1 Department of Biological Sciences, College of Life Sciences, Henan Normal University, Xinxiang 453007, China
2 Engineering Technology Research Center of Nursing and Utilization of Genuine Chinese Crude Drugs in Henan Province, Xinxiang 453007, China
3 Engineering Laboratory of Biotechnology for Green Medicinal Plant of Henan Province, Xinxiang 453007, China
* Correspondence: 041121@htu.edu.cn or zhaoxt0411@126.com; Tel.: +86-182-3739-1085 or +86-135-6988-6182
† These authors contributed equally to this work.

Abstract: *Chrysanthemum morifolium* Ramat. 'Huaihuang' is a traditional Chinese medicinal plant. However, a black spot disease caused by *Alternaria* sp., a typical necrotrophic fungus, has a serious damaging influence on the field growth, yield, and quality of the plant. 'Huaiju 2#' being bred from 'Huaihuang', shows resistance to *Alternaria* sp. bHLH transcription factor has been widely studied because of their functions in growth development, signal transduction, and abiotic stress. However, the function of bHLH in biotic stress has rarely been studied. To characterize the resistance genes, the CmbHLH family was surveyed in 'Huaiju 2#'. On the basis of the transcriptome database of 'Huaiju 2#' after *Alternaria* sp. inoculation, with the aid of the *Chrysanthemum* genome database, 71 *CmbHLH* genes were identified and divided into 17 subfamilies. Most (64.8%) of the CmbHLH proteins were rich in negatively charged amino acids. CmbHLH proteins are generally hydrophilic proteins with a high aliphatic amino acid content. Among the 71 CmbHLH proteins, five *CmbHLHs* were significantly upregulated by *Alternaria* sp. infection, and the expression of *CmbHLH18* was the most significant. Furthermore, heterologous overexpression of *CmbHLH18* could improve the resistance of *Arabidopsis thaliana* to necrotrophic fungus *Alternaria brassicicola* by enhancing callose deposition, preventing spores from entering leaves, reducing ROS accumulation, increasing the activities of antioxidant enzymes and defense enzymes, and promoting their gene expression levels. These results indicate that the five *CmbHLHs*, especially *CmbHLH18*, may be considered candidate genes for resistance to necrotrophic fungus. These findings not only increase our understanding of the role *CmbHLHs* play in biotic stress but also provide a basis by using *CmbHLHs* to breed a new variety of *Chrysanthemum* with high resistance to necrotrophic fungus.

Keywords: *Chrysanthemum morifolium* Ramat.; bHLH transcription factor; *CmbHLHs*; resistance; necrotrophic fungus

Citation: Ding, Y.; Wang, X.; Wang, D.; Jiang, L.; Xie, J.; Wang, T.; Song, L.; Zhao, X. Identification of CmbHLH Transcription Factor Family and Excavation of *CmbHLHs* Resistant to Necrotrophic Fungus *Alternaria* in *Chrysanthemum*. *Genes* **2023**, *14*, 275. https://doi.org/10.3390/genes14020275

Academic Editors: Wajid Zaman and Hakim Manghwar

Received: 22 December 2022
Revised: 15 January 2023
Accepted: 16 January 2023
Published: 20 January 2023

1. Introduction

Chrysanthemum morifolium Ramat., a perennial herb with a wide variety of flower types, has been widely used in cut flowers, potting, and garden greening [1]. *C. morifolium* 'Huaijuhua' produced mainly in Jiaozuo city of Henan province, China, is one of the Four Famous Huai Herbals; it has been cultivated for more than three thousand years in China. Its annual planting area is approximately 800 hm^2 with an output value of one hundred and ninety million Yuan [2]. 'Huaijuhua' is rich in three kinds of medicinal active ingredients, namely luteoloside, chlorogenic acids, and 3, 5-dicaffeoyl-quinic acid [3]. As a traditional medicinal Chinese *Chrysanthemum*, besides its medicinal value, it can also be used as ornamental and drinking tea [3]. However, 'Huaijuhua' is subjected to year-round propagation through cuttings and division propagation, so it is susceptible to a variety of pathogens during the growth stage. Our previous investigation showed that a black spot disease caused by *Alternaria* sp., a typical necrotrophic fungus, has a serious negative impact on the field growth, yield, and quality of 'Huaijuhua' [4]. The incidence of black spot was 30% to 80%, even up to 100% in the field planting of 'Huaijuhua' [5]. After six years of selection and cultivation, a new resistant cultivar 'Huaiju $2^{\#}$' has been bred. This makes it possible to characterize potential resistance genes from 'Huaiju $2^{\#}$'.

Recently, bHLH (basic helix-loop-helix) transcription factors (TFs) have been widely studied because of their functions in signal transduction, biosynthesis, response to various stresses, and growth regulation of plants [6]. bHLH TFs are named for having a highly conserved bHLH domain. The bHLH domain of this family contains about 50–60 amino acid residues with the N-terminal basic region (b) and the C-terminal helix-loop-helix (HLH) [6]. The HLH region can contribute to the formation of homodimers or heterodimers between proteins [7,8]. Leu-23 and Leu-52 are necessary for the formation of dimer complexes of these identified members of the bHLH family [9,10].

Since bHLH proteins were first found in maize [11], they have been successively identified in other plants, including *Arabidopsis*, rice, poplar, potato, cabbage, and so on [12]. However, the bHLH family in *Chrysanthemum* is still unknown.

Studies have shown that bHLH regulates multiple physiological progresses of plants. It is widely involved in all kinds of signal transduction mediated by plant hormones, such as gibberellic acid (GA), abscisic acid (ABA), and brassinosteroids [10,13,14]. bHLH has also been involved in response to a variety of abiotic stresses, including drought, cold, and salt stress, as well as regulation of iron homeostasis [15–18]. bHLHs are also related to biological processes, including the biosynthesis of anthocyanins, flavonoids, and ethylene [19,20]. They also take part in the growth regulation of plants, including cell elongation, the growth of fruit and flowers, the development of root hairs and stomata, and senescence of the leaf [21–25]. However, the functions of bHLH in biotic stress are still rarely studied, with a few exceptions in some model plants and crops. For example, heterologous overexpression of *TabHLH060* significantly enhanced *Arabidopsis* sensitivity to the biotrophic pathogen *Pst.* DC3000 [26]. The defense of *Phytophthora nicotiana* of tobacco was associated with 28 NbbHLHs [27], and overexpression of *GhbHLH171* activated the synthesis and signaling pathway of JA and improved the resistance of cotton to the necrotrophic pathogen *V. dahlia* [28].

According to the transcriptome database (Accession number: PRJNA448499) of 'Huaiju $2^{\#}$' after *Alternaria* sp. inoculation, using the *Chrysanthemum* genome database (http://www.amwayabrc.com/zh-cn/index.html, accessed on 22 January 2022), CmbHLHs were identified in this study. Their phylogenetic relationships were analyzed. Candidate *CmbHLHs* of resistance to *Alternaria* sp. were confirmed. Furthermore, based on the expression patterns of the *CmbHLHs* under inoculation with *Alternaria* sp., this study also focused on the role of *CmbHLH18* in enhancing the resistance of *Arabidopsis* to the typical necrotrophic fungus, *Alternaria brassicicola*, through heterologous overexpression. This will provide a basis for a deeper understanding of the functions of bHLHs in *Chrysanthemum*.

2. Materials and Methods

2.1. Plant Materials and Pathogenic Strains

Following the culture method of Zhao et al. [3], both *Chrysanthemum morifolium* 'Huaiju $2^{\#}$' (Appreciation of Traditional Chinese Medicine in Henan Province: 2016002) and *Arabidopsis thaliana* (Columbia, Col-0) used in this study were grown in 13 cm diameter pots containing a peat-vermiculite (v/v = 1:1) mixture in the controlled greenhouse at 20–25 °C, 40–60% relative humidity and under a LD cycle (14 h light/10 h dark), 60 mE$\cdot$s^{-1} m^{-2} light intensity.

Alternaria sp. (Strain: HQJH10092301; GenBank accession number: KF688111) and *A. brassicicola* (Strain: BCHB13101606; GenBank accession number: OP295488) were used as pathogenic strains to inoculate the plants. The plants and fungi are maintained for research purposes at the Engineering Technology Research Center of Nursing and Utilization of Genuine Chinese Crude Drugs in Henan Province, Henan Normal University, Xinxiang, China.

2.2. Screening and Physicochemical Properties Analysis of the CmbHLH TF Family

Sequences annotated as the bHLH in the 'Huaiju 2#' transcriptome database (Accession number: PRJNA448499) were aligned to the *Chrysanthemum* genome database using BLAST on the http://www.amwayabrc.com/zh-cn/index.html (accessed on 22 January 2022). The retrieved candidate genes were checked through the CD-HIT website server (http://weizhongli-lab.org/cd-hit/, accessed on 9 November 2021) to delete redundant sequences [29]. NCBI CDD (https://www.ncbi.nlm.nih.gov/Structure/bwrpsb/bwrpsb.cgi, accessed on 10 November 2021) and SMART (http://smart.embl-heidelberg, accessed on 11 November 2021) were used to remove sequences without conserved bHLH domains, incomplete CDS sequences were also removed, and sequences with typical bHLH domains were selected as putative bHLH TF family members [30,31].

Based on the identified amino acid sequence of the CmbHLH protein, the ExPASY online program (https://www.expasy.org/, accessed on 15 November 2021) was used to predict the physicochemical properties of the CmbHLH TFs [32].

2.3. Multiple Sequence Alignment, Phylogenetic, and Conserved Motifs Analysis of CmbHLH Family Proteins

To analyze the phylogenetic relationship of CmbHLH proteins, the conserved motifs of the CmbHLH proteins were searched and analyzed using the MEME online tool (http://meme.nbcr.net/meme/tools/meme, accessed on 15 March 2022) (search criteria: motif length was set to 10–100, the maximum number of retrieved motifs was set to 40, and the default parameters were used for the rest) [33].

Using the NJ method, a neighbor-joining phylogenetic tree was constructed based on the alignment of the identified CmbHLHs with AtbHLHs from *Arabidopsis* using MEGA 7.0 with 1000 bootstraps (the others with default parameters) [34]. Protein sequences of 156 AtbHLH family members were downloaded from TAIR (https://www.Arabidopsis.org/index.jsp, accessed on 12 November 2021) [35]. A phylogenetic tree was drawn using iTOL (http://itol.embl.de/, accessed on 9 November 2021) to group the CmbHLH proteins [36].

2.4. Excavation of Candidate CmbHLHs Resistant to Necrotrophic Fungi and Their Cis-Element Analysis

Based on the selection criteria for differential genes [37], the upregulated expression of *CmbHLHs* in response to *Alternaria* sp. inoculation were taken as potential candidate *CmbHLHs* resistant to necrotrophic fungi. Putative promoter sequences (2000 bp upstream of ATG) of the candidate genes were retrieved from the *Chrysanthemum* genome database (http://www.amwayabrc.com/zh-cn/index.html, accessed on 9 May 2021) (Table S1). *cis*-acting elements in the promoter sequences were predicted using the online software PlantCARE (http://bioinformatics.psb.ugent.be/webtools/plantcare/html/, accessed on 10 May 2021) [38]. The results were visualized and mapped using TBtools software v0.6673 [39].

2.5. Isolation and Arabidopsis Transformation of CmbHLH18

The total RNAs of 'Huaiju 2#' leaves were extracted using an RNAiso Plus kit (TaKaRa, Beijing, China) and purified using an ND-ONE-W spectrophotometer (Thermo, Waltham, MA, USA) according to the manufacturer's protocol. Reverse transcription into cDNA was performed using a HiScript II 1st Strand cDNA Synthesis Kit (Vazyme, Nanjing, China) in a 20 μL reaction. cDNA was synthesized using specific primers (Table S6) with 500 ng of total RNA as the template.

The cDNA was used as a template and primer pair (CmbHLH18-F and CmbHLH18-R) (Table S6) were designed to amplify the coding area of *CmbHLH18*. The PCR program was conducted as follows: 5 min at 94 °C; 35 cycles of 94 °C for 30 s, 56 °C for 30 s and 72 °C for 60 s; 10 min at 72 °C.

The coding DNA sequence of *CmbHLH18* (GenBank accession number: OP 313866) was constructed into the Super1300-35S vector [3]. The recombinant vector was introduced into *Agrobacterium* strains and transformed into *Arabidopsis* plants to produce plants overexpressing *CmbHLH18* using the floral dip method [40]. After three generations of Hygromycin B resistance screening, a small amount of *Arabidopsis* leaf genomic DNA was extracted for PCR amplification using the cetyltrimethylammonium bromide (CTAB) method, and a PCR reaction system (50 µL) was prepared consisting of 100 ng of template DNA, 2 µL of primer super1300-F, 2 µL of primer CmbHLH18-V-R, 25 µL of 2× Pfu MasterMix (Dye) (Vazyme Biotech, Nanjing, China), and 19 µL of ddH_2O. The PCR reaction program consisted of the following steps: 94 °C for 5 min, 35 cycles of 94 °C for 30 s, 56 °C for 30 s, 72 °C for 60 s, and finally, 72 °C for 10 min. Electrophoresis using 1% agarose was performed to separate the PCR products, and a gel imaging system was used to detect and identify transgenic *Arabidopsis* plants. This was also identified by RT-qPCR with the primer pair of CmbHLH18-At-F and CmbHLH18-At-R (Table S6), following Section 2.9 in this part.

2.6. Inoculation of Necrotrophic Fungi and Sampling

According to the method of Zhao et al. [3], *Alternaria* sp. and *A. brassicicola* were activated. When plants of 'Huaiju 2#' grew to 8–10 leaves, the middle and upper leaves were punctured with a needle (approximately 0.41 mm diameter) and inoculated with spores of *Alternaria* sp. [3]. When *Arabidopsis* plants grew to 4 weeks, the spore solution of *A. brassicicola* was sprayed on leaves until droplet runoff occurred [41]. The two spores were suspended and diluted to a concentration of 10^7 spores per milliliter in sterile distilled water (SDW) using the hemocytometer technique. After inoculation, plants were kept in a dark incubation chamber for 48 h at 25 °C with 100 % relative humidity (RH). Then, the plants were grown in the greenhouse with 40–60% RH, 20–25 °C, 60 $mE{\cdot}s^{-1}$ m^{-2} light intensity, and 14 h light/10 h dark light cycle for 'Huaiju 2#' [3], 10 h light/14 h dark light cycle for *Arabidopsis*.

Leaves of 'Huaiju 2#' were collected as samples at 0 d (before inoculation) and 1, 3, and 5 days post inoculation (dpi) with *Alternaria* sp. for expression profile analysis of candidate *CmbHLHs*. Leaves of the transgenic and WT *Arabidopsis* lines were collected at 0 d (before inoculation) and 1 dpi, 3 dpi, and 5 dpi with *A. brassicicola* for gene expression analysis and activity determination of antioxidant enzymes and defense enzymes.

2.7. Morphology and Disease Severity Index (DSI) after Inoculation with A. brassicicola in CmbHLH18 Transgenic Arabidopsis Lines

We divided the infected leaves into five grades following the method of Xu [42]: grade 0 (lesion size ≤ 5%), grade 1 (5% < lesion size ≤ 25%), grade 2 (25% < lesion size ≤ 50%), grade 3 (50% < lesion size ≤ 75%) and grade 4 (lesion size > 75%). At 5 dpi, 15 plants from each line were randomly selected, and the number of leaves of each grade was counted. The DSI was calculated with the following formula.

$$DSI = \frac{\sum(\text{Grade of disease} \times \text{Number of diseased leaves of this grade})}{150} \times 100\%$$

2.8. Histochemical Staining and Microscopic Analysis after Inoculation with A. brassicicola in Arabidopsis and CmbHLH18 Transgenic Arabidopsis Lines

Evans Blue [43] was used to observe necrotic cells at 0, 5 dpi and fungal growth at 3, 5 dpi. Aniline blue staining [44] was used to observe callose deposition at 0 and 0.5 dpi. At 0 and 0.5 dpi, nitro blue tetrazolium (NBT) and diaminobenzidine (DAB) staining were used to detect the superoxide anion (O_2^-) and hydrogen peroxide (H_2O_2) of the leaves, respectively [45].

2.9. Determination of Enzyme Activity and Gene Expression Analysis of Arabidopsis Lines

Following the literature, we quantified the activities of antioxidant enzymes, including superoxide dismutase (SOD) and catalase (CAT), and defense enzymes, including phenylalanine ammonia lyase (PAL), peroxidase (POD), chitinase (CHT) and β-1,3-glucanase (GLU) in leaves of transgenic and WT *Arabidopsis* plants were measured following Liu et al. [46].

The method of obtaining cDNA is the same as 2.5; cDNA was used as a sample for the RT-qPCR of genes. RT-qPCR was done using an AceQ RT-qPCR SYBR Green Master Mix (Vazyme, Nanjing, China). *CiUBI* (Cse2.0_LG8) was used as an internal reference for the expression level of *CmbHLH18* in various tissues of *Chrysanthemum*. *AtActin* (823805) was selected as an internal reference for the expression level of genes post-inoculation in *Arabidopsis*. Specific primers were designed using primer 5.0 software [3] (Table S6). The expression levels in RT-qPCR were calculated using the $2^{-\triangle\triangle CT}$ method [47].

2.10. Statistical Analysis

A one-way ANOVA and Student's *t*-test were performed using SPSS 18.0 software (IBM, Armonk, NY, USA). One-way ANOVA tests were performed to test the DSI of the *Arabidopsis* lines after inoculation of *A. brassicicola*. Student's *t*-tests were used to test the gene expression level analysis and activities of antioxidant and defense enzymes after *Alternaria* sp. inoculation.

3. Results

3.1. Identification of the CmbHLH Transcription Factor Family

3.1.1. Hydrophilicity and High Aliphatic Amino Acid Content in 71 CmbHLH TFs Identified in *Chrysanthemum*

Our sequence analysis identified 71 proteins with a typical bHLH domain (Figure S1) and designated as CmbHLH1, 2, 3 ... 71, and their physical and chemical properties are summarized in Table 1. For these 71 CmbHLH proteins, the predicted polypeptide sequence lengths ranged from 158 to 937; the predicted molecular weights ranged from 17.67–103.49 kDa, and their theoretical isoelectric points (pIs) ranged between 4.75–9.2. Among them, the pIs of 15 proteins were greater than 7.5, while the pIs of 46 proteins were lower than 6.5. Most (64.8%) of these CmbHLH proteins were rich in negatively charged amino acids. The results of the protein hydropathicity analysis showed that the grand average of hydropathicity (GRAVY) of all 71 sequences was lower than 0, which indicated that the bHLH proteins in *Chrysanthemum* are generally hydrophilic proteins. The greater II (the instability index) value, the more unstable the protein. Therefore, most of the CmbHLH proteins were apparently unstable; the least stable of them was CmbHLH3 (CHR00034167-RA). Furthermore, these CmbHLHs had relatively high aliphatic amino acid (AI) contents, with CmbHLH22 (CHR00078227-RA) containing the highest number of AIs. Taken together, these analyses showed that *Chrysanthemum* harbors a wide array of bHLH TFs with hydrophilicity and high aliphatic amino acid content.

Table 1. Physiochemical properties of the CmbHLH TF family.

Name	ID	aa	MW ($\times 10^4$)	pI	II	AI	GRAVY
CmbHLH1	CHR00023546-RA	272	3.062	7.79	49.89	81.36	−0.436
CmbHLH2	CHR00043772-RA	307	3.483	6.41	52.99	73.65	−0.578
CmbHLH3	CHR00034167-RA	328	3.700	4.8	62.02	74.85	−0.588
CmbHLH4	CHR00082406-RA	244	2.729	5.99	28.9	80.7	−0.441
CmbHLH5	CHR00073020-RA	417	4.630	5.59	59.66	54.68	−0.792
CmbHLH6	CHR00087928-RA	411	4.581	6.82	57.63	61.39	−0.763
CmbHLH7	CHR00079719-RA	468	5.246	6.35	45.31	75.79	−0.551
CmbHLH8	CHR00010518-RA	480	5.293	6.09	47.99	58.06	-0.739
CmbHLH9	CHR00002838-RA	398	4.480	8.64	60.66	76.68	−0.589

Table 1. *Cont.*

Name	ID	aa	MW (×10^4)	pI	II	AI	GRAVY
CmbHLH10	CHR00029628-RA	609	6.830	4.92	47.81	87.87	−0.329
CmbHLH11	CHR00053551-RA	361	4.081	6.08	61.37	54.88	−0.923
CmbHLH12	CHR00027604-RA	278	3.093	8.41	29.44	80.22	−0.53
CmbHLH13	CHR00020071-RA	410	4.558	9.29	47.53	76.56	−0.437
CmbHLH14	CHR00001268-RA	306	3.452	5.73	61.53	78.73	−0.423
CmbHLH15	CHR00066657-RA	453	5.049	6.33	32.95	63.07	−0.728
CmbHLH16	CHR00085490-RA	937	10.349	5.36	51.72	70.52	−0.551
CmbHLH17	CHR00014650-RA	350	3.838	6.19	50.86	61.57	−0.693
CmbHLH18	CHR00061827-RA	280	3.096	5.07	52.56	74.18	−0.44
CmbHLH19	CHR00004349-RA	427	4.836	5.51	51.06	77.07	−0.594
CmbHLH20	CHR00030629-RA	291	3.261	9.08	60.77	56.98	−0.832
CmbHLH21	CHR00050993-RA	310	3.481	6.76	46.73	84.00	−0.414
CmbHLH22	CHR00078227-RA	201	2.288	6.3	48.15	109.6	−0.098
CmbHLH23	CHR00016706-RA	534	6.031	5.9	38.2	80.24	−0.398
CmbHLH24	CHR00089135-RA	376	4.183	4.87	52.25	66.68	−0.716
CmbHLH25	CHR00058509-RA	262	2.931	6.62	46.86	59.12	−0.844
CmbHLH26	CHR00075848-RA	169	1.871	8.3	54.21	76.21	−0.576
CmbHLH27	CHR00021504-RA	294	3.272	7.01	48.12	66.33	−0.805
CmbHLH28	CHR00085488-RA	567	6.347	5.99	59.9	85.31	−0.386
CmbHLH29	CHR00066393-RA	592	6.578	6.42	41.98	79.34	−0.503
CmbHLH30	CHR00049557-RA	191	2.134	6.2	45.33	90.79	−0.458
CmbHLH31	CHR00020620-RA	501	5.364	5.56	58.67	57.84	−0.734
CmbHLH32	CHR00040132-RA	399	4.398	5.3	55.48	59.65	−0.624
CmbHLH33	CHR00025933-RA	251	2.872	8.31	43.43	67.17	−0.74
CmbHLH34	CHR00067596-RA	319	3.529	6.32	39.39	87.12	−0.37
CmbHLH35	CHR00085567-RA	249	2.785	7.65	37.57	90.44	−0.363
CmbHLH36	CHR00032751-RA	326	3.705	6.47	51.81	80.77	−0.557
CmbHLH37	CHR00051320-RA	200	2.273	6	58.68	83.3	−0.385
CmbHLH38	CHR00052362-RA	277	3.206	6.51	48.97	86.86	−0.588
CmbHLH39	CHR00032387-RA	309	3.414	6.26	62	54.56	−0.984
CmbHLH40	CHR00068388-RA	211	2.379	4.75	55.92	42.99	−1.068
CmbHLH41	CHR00062368-RA	280	3.096	5.07	52.56	74.18	−0.44
CmbHLH42	CHR00005466-RA	319	3.574	6.77	46.83	78.84	−0.399
CmbHLH43	CHR00028154-RA	215	2.422	6.46	49.22	86.56	−0.66
CmbHLH44	CHR00041708-RA	322	3.517	5.84	44.78	58.73	−0.793
CmbHLH45	CHR00076371-RA	310	3.501	7.29	43.91	70.81	−0.701
CmbHLH46	CHR00027603-RA	368	4.127	6.07	29.5	78.59	−0.551
CmbHLH47	CHR00053214-RA	318	3.554	6.98	55.24	84.37	−0.614
CmbHLH48	CHR00031690-RA	572	6.428	5.22	43.38	83.50	−0.306
CmbHLH49	CHR00022486-RA	354	3.957	5.81	61.58	79.35	−0.638
CmbHLH50	CHR00077133-RA	398	4.381	6.1	58.84	71.28	−0.572
CmbHLH51	CHR00002812-RA	323	3.562	6.33	49.77	67.28	−0.541
CmbHLH52	CHR00091912-RA	360	3.774	5.98	48.34	64.28	−0.498
CmbHLH53	CHR00040106-RA	445	4.818	5.44	61.72	56.36	−0.612
CmbHLH54	CHR00053312-RA	240	2.746	7.01	35.36	71.88	−0.659
CmbHLH55	CHR00073646-RA	293	3.244	6.26	35.57	65.87	−0.744
CmbHLH56	CHR00078290-RA	391	4.097	5.98	50.74	69.44	−0.423
CmbHLH57	CHR00063973-RA	238	2.673	8.49	51.21	92.69	−0.329
CmbHLH58	CHR00023344-RA	255	2.752	5.28	56.8	69.65	−0.636
CmbHLH59	CHR00015822-RA	172	1.906	4.89	44.96	80.41	−0.412
CmbHLH60	CHR00050362-RA	517	5.822	9.16	49.15	56.44	−0.992
CmbHLH61	CHR00025279-RA	268	2.892	5.77	60.24	54.55	−0.83
CmbHLH62	CHR00061024-RA	418	4.680	5.69	51.93	85.50	−0.364
CmbHLH63	CHR00086566-RA	158	1.767	5.86	55.74	78.92	−0.717
CmbHLH64	CHR00076271-RA	240	2.699	9.2	33.72	85.21	−0.556
CmbHLH65	CHR00065282-RA	185	2.105	8.52	37.98	90.11	−0.437
CmbHLH66	CHR00005538-RA	317	3.586	8.59	59.45	85.52	−0.466

Table 1. *Cont.*

Name	ID	aa	MW (×10⁴)	pI	II	AI	GRAVY
CmbHLH67	CHR00079802-RA	392	4.363	7.15	57.19	62.45	−0.627
CmbHLH68	CHR00060897-RA	244	2.806	5.07	61.31	85.08	−0.498
CmbHLH69	CHR00064856-RA	229	2.593	5.31	57.55	70.83	−0.717
CmbHLH70	CHR00031761-RA	327	3.679	8.15	51.81	64.71	−0.779
CmbHLH71	CHR00029899-RA	311	3.510	8.28	61.74	77.78	−0.446

Note: ID were obtained from *Chrysanthemum* genome database (http://www.amwayabrc.com/index.html, accessed on 9 May 2021). aa, protein length, MW (molecular weight; kDa), PI (theoretical isoelectric point), II (the instability index), AI (aliphatic index), and GRAVY (grand average of hydropathicity) were obtained from ExPASY (http://web.expasy.org/protparam/, accessed on 15 November 2021).

3.1.2. Phylogenetic Differences between Groups of bHLHs from *Chrysanthemum* and *Arabidopsis*

Phylogenetic analysis of the DNA-binding domain of the 71 CmbHLHs as well as 156 bHLHs from *Arabidopsis* suggested that there were 17 groups, none of which belonged to clade Ib, Va, VIIa (1), VII(a+b), VIIIc, XIV, and XV in *Chrysanthemum* (Figure 1, Table S2). Five bHLHs were unique to *Chrysanthemum*, resulting in their classification as "Orphans". These results showed that the CmbHLH TF family probably expanded and contracted during the evolutionary process in *Chrysanthemum*. Motif analysis of the CmbHLH proteins showed that 40 putative motifs among the 71 CmbHLH proteins were identified (Figure S2). Members of the same subfamily of CmbHLH had similar conservative motifs and numbers, and their location distribution was relatively conservative. This may help to analyze the phylogenetic relationship between bHLH TFs.

3.1.3. Domains of CmbHLHs Identified by Variation-Conserved Residues

To better understand the function of the 71 CmbHLH TFs identified in *Chrysanthemum*, their domains and DNA-binding capacities were analyzed. There were 24 amino acid residues that were highly conserved in >50% of the candidate proteins (Figure 2), and the R13, L23, L42, and L52 residues were conserved in >80% of the proteins (Table S3). The results of the domain analysis suggested that these conserved sites were likely involved in bHLH function. Further analysis showed that the basic regions of CmbHLHs were composed of 10–13 amino acids and contained a typical H5-E9-R13 sequence motif (His5-Glu9-Arg13) for binding DNA, which is necessary for bHLH to function as a transcription factor [48]. In particular, E9 and R13 residues have a function in the recognition and binding of E-box motifs in the promoter regions of target genes.

3.2. *Excavation of Candidate CmbHLHs' Resistance to Necrotrophic Fungus Alternaria*

Five CmbHLHs were highly upregulated in the transcriptome (Accession number: PRJNA448499) when plants were inoculated with *Alternaria* sp. and were potentially involved in the resistance to necrotrophic fungi. The five candidate genes were named *CmbHLH16*, *CmbHLH18*, *CmbHLH28*, *CmbHLH30*, and *CmbHLH60* (Table S4). CmbHLH16, CmbHLH28, and CmbHLH30 belong to the IVd subfamily, CmbHLH18 belongs to the IVa subfamily, and CmbHLH60 belongs to the IX subfamily. They contain a total of 15 motifs (Figure S2).

3.2.1. Analysis of Promoters of the Five Candidate *CmbHLHs*

To comprehend the transcriptional regulating function of the 71 CmbHLH TFs identified in *Chrysanthemum*, taken the five UP *CmbHLHs* as representative, sequence analysis identified 2000 bp fragments upstream of these five genes with their respective sequences in the *Chrysanthemum* genome database (http://www.amwayabrc.com/index.html, accessed on 9 May 2021). Then, a promoter analysis was conducted to identify the potential function of *cis*-acting elements within these regions (Figure S3; Table 2) and found that three major types of *cis*-elements were present in the promoters of the UP *CmbHLHs*. The first type is responsive to plant hormones and includes ABRE, AuxRR, CGTCA/TGACG, P-box, and

TCA elements, which are sensitive to ABA, IAA, JA, GA, and SA, respectively. The second type is responsive to defense and stress, including TC-rich repeats (defense and stress responsiveness, which are only in the promoter of *CmbHLH18*), MBS (drought stress), LTR (low-temperature stress), and ARE (anaerobic stress) elements. The third type participates in plant growth and development, such as that of the GCN4_Motif in endosperm expression. Our finding of hormone-responsive and defense-responsive promoter elements suggested that the *CmbHLHs* may have been upregulated specifically through these elements during *Alternaria* sp. infection and prompted us to investigate whether these elements could be related to an enhanced disease resistance phenotype.

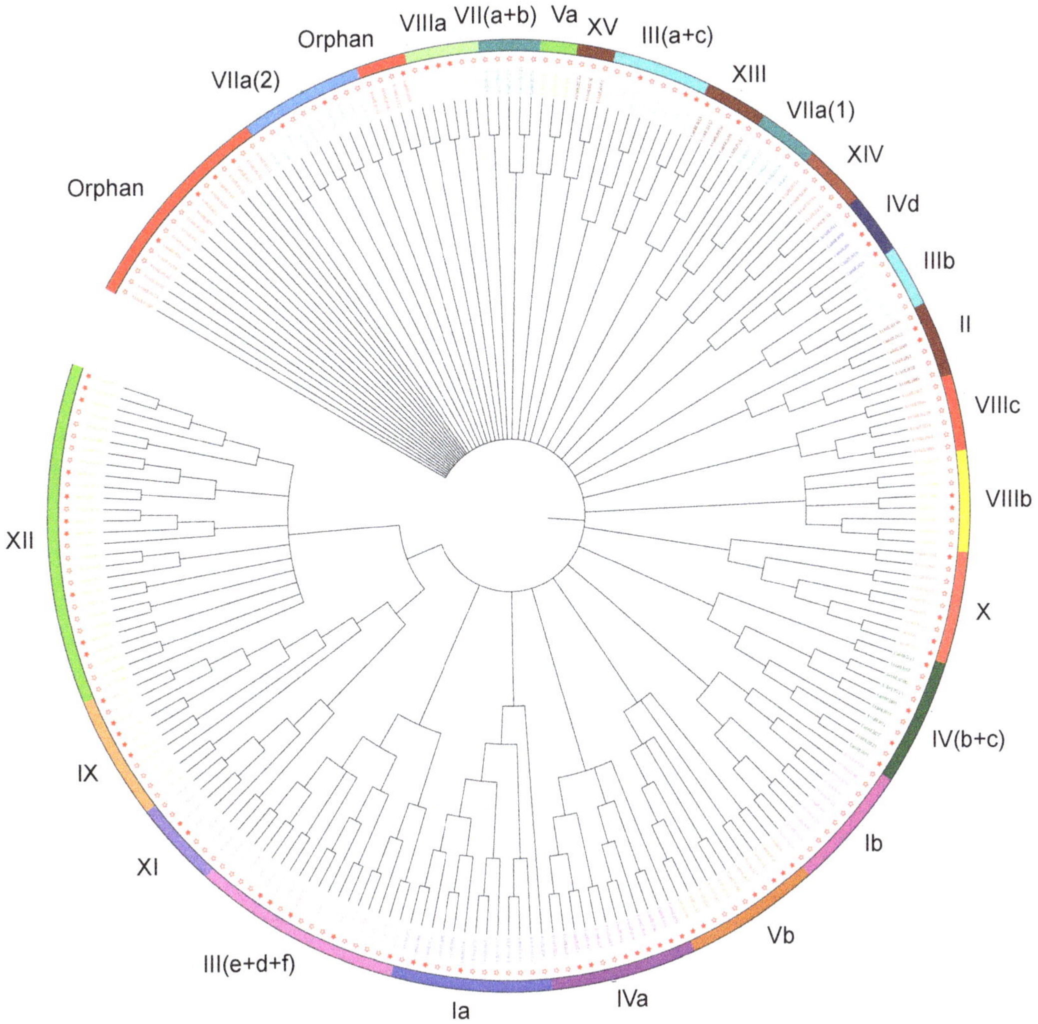

Figure 1. Phylogenetic relationship of the bHLH TF family among *Arabidopsis* and *Chrysanthemum*. Different colored arcs are used to distinguish different groups or subgroups. The hollow star and red solid star represent *Arabidopsis* and *Chrysanthemum*, respectively.

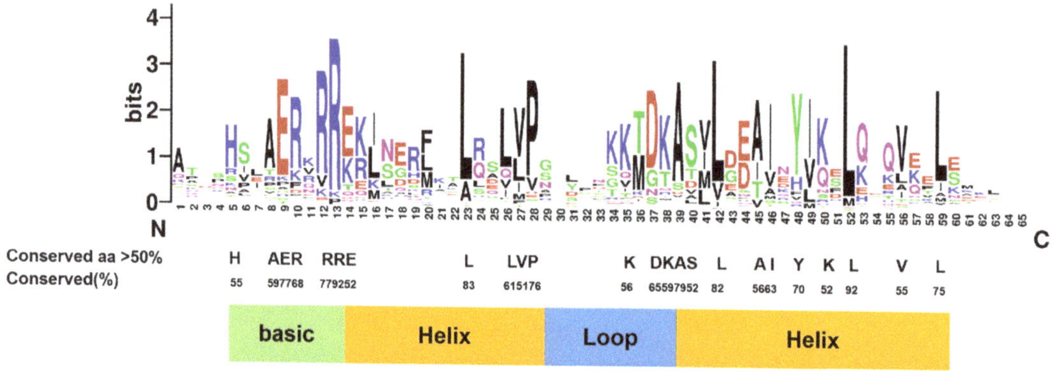

Figure 2. Sequence logos of the CmbHLH domains. The height of the amino acids at each site indicates the conservatism of the corresponding amino acids. Capitalized letters below the sequence logos indicate that more than 50% of amino acids are consistent in the conserved domain of all CmbHLH proteins.

Table 2. Classification of *cis*-elements in the promoter sequences of the five UP *CmbHLHs*.

Items	Name of *cis*-Element	Sequence	Position from Translation Start Site	Function
Hormone response	ABRE	ACGTG/CACGTG/GACACGTGGC	+477	*Cis*-acting element involved in the ABA responsiveness
	CGTCA-motif	CGTCA	−1760	*Cis*-acting element involved in the MeJA-responsiveness
	P-box	CCTTTTG	+1172	GA-responsive element
	TCA-element	CCATCTTTTT	−905	*Cis*-acting element involved in SA responsiveness
	TGACG-motif	TGACG	+1760	*Cis*-acting element involved in the MeJA-responsiveness
	AuxRR-core	TGTCTCAATAAG	−287	*Cis*-acting element involved in auxin responsiveness
Defense and stress response	TC-rich repeats	GTTTTCTTAC	−515	*Cis*-acting element involved in defense and stress responsiveness
	ARE	AAACCA	−1728 − 1711 − 1162 +338	*Cis*-acting element essential for the anaerobic induction
	LTR	CCGAAA	+1616 + 1963	*Cis*-acting element involved in low-temperature responsiveness
	MBS	CAACTG	−1234	MYB binding site involved in drought-inducibility
Growth and development	GCN4-motif	TGAGTCA	+1694	*Cis*-regulatory element involved in endosperm expression

3.2.2. Expression Analysis of the Five up *CmbHLHs* in Response to Necrotrophic Fungus *Alternaria*

In order to better understand whether and how these promoter elements may contribute to transcriptional regulation of these five UP *CmbHLHs* to mediate a disease-resistant phenotype, we next inoculated *Chrysanthemum* with *Alternaria* sp. Then, their expression levels were examined (Figure 3). The results showed that these *CmbHLHs* were significantly induced by inoculation with *Alternaria* sp. compared with Mock. More specifically, *CmbHLH18* expression peaked at 1 dpi and was 9.12 times higher than that of the Mock, whereas the expression of *CmbHLH16*, *CmbHLH30*, *CmbHLH28*, and *CmbHLH60* reached their highest levels at 3 dpi (reaching 6.72-, 6.58-, 5.61-, and 3.93-fold, respectively, that of the Mock).

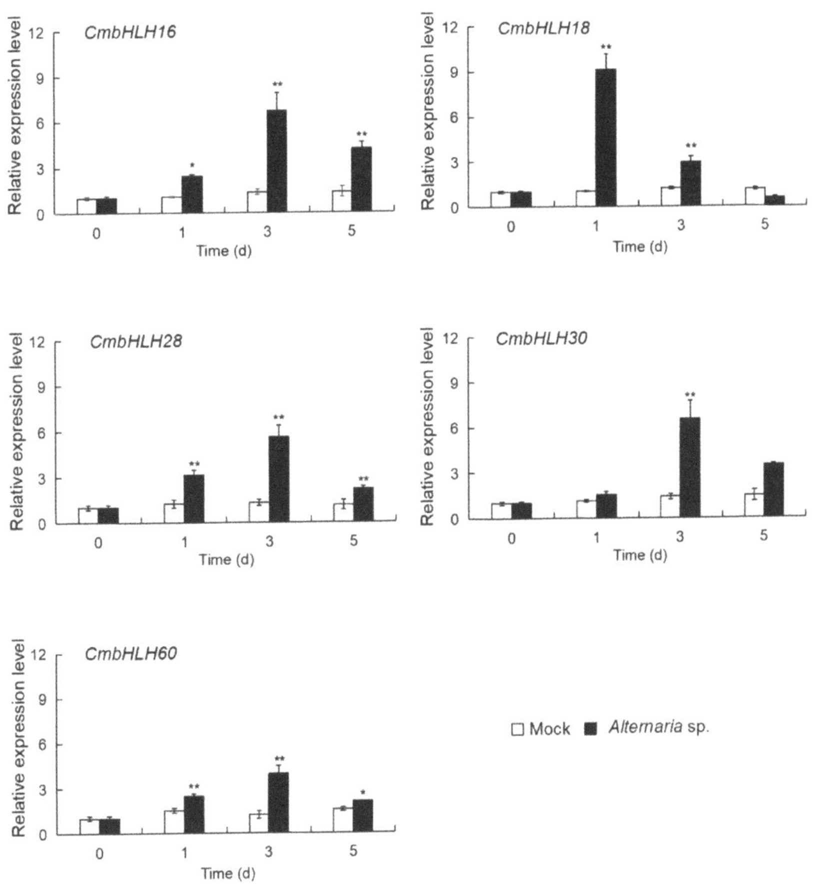

Figure 3. Expression level analysis of the five UP *CmbHLHs* after *Alternaria* sp. inoculation. Asterisks represent statistically significant differences for each time interval between *Alternaria* sp. versus Mock using the *t*-test ($n = 3$; *, $p < 0.05$; **, $p < 0.01$).

These results indicated that these five UP *CmbHLHs* might play a positive regulatory role in response to *Alternaria* sp. In particular, the expression of *CmbHLH18* was strongly upregulated by *Alternaria* sp. (the necrotrophic fungus) infection. Therefore, *CmbHLH18* is mostly expected to become a candidate gene resistance against the necrotrophic fungus *Alternaria* in plants.

3.2.3. *CmbHLH18* Enhances Resistance against Necrotrophic Fungus *Alternaria* in *Arabidopsis*

In order to further verify the function of *CmbHLH18* resistance against the necrotrophic fungus in plants, we isolated *CmbHLH18* (GenBank accession number: OP 313866) and transformed it into *Arabidopsis* (Col). Transgenic plants were identified using PCR and RT-qPCR (Figure 4a,b). Transgenic plants show PCR products of 516 bp, while normal plants contain the amplified region with 0 bp in size (Figure 4a). There were no significant differences in morphology between the WT and transgenic plants before *A. brassicicola* inoculation. However, at 5 dpi, DSI was significantly higher in the WT than in transgenic plants (Figure 4c,d). This is reflected in cell activities being higher in transgenic plants than the WT (Figure 5a, Table S5).

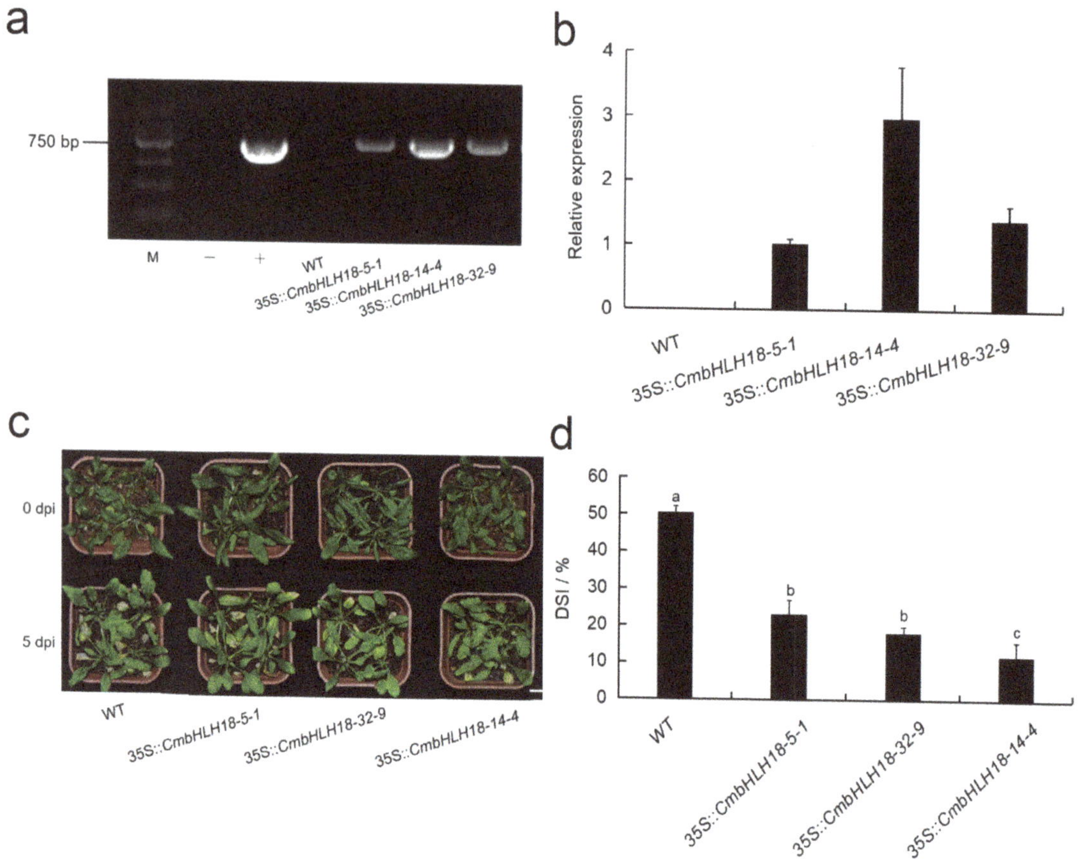

Figure 4. Identification of *CmbHLH18*-transformed *Arabidopsis* plants and morphology and DSI statistics. (**a**,**b**) PCR and RT-qPCR detection of *CmbHLH18* in transformed *Arabidopsis*. M: DNA DL2000 Marker; (**c**) Morphologic change of *CmbHLH18* transformed *Arabidopsis* lines after *A. brassicicola* inoculation (Bar = 1 cm); (**d**) DSI of different *Arabidopsis* lines at 5 dpi. The different letters signified significant differences at $p < 0.05$ by one-way ANOVA.

The speed of spore growth of inoculated leaves at 3 dpi and 5 dpi was significantly slower in transgenic plants than in WT plants (Figure 5b). The leaves of transgenic plants showed more callose deposition compared with WT plants at 0.5 dpi with *A. brassicicola* (Figure 5c).

To further analyze the effect of *CmbHLH18* on ROS accumulation in transformed *Arabidopsis* leaves after *A. brassicicola* inoculation, NBT staining and DAB staining were conducted. Transgenic plants of *Arabidopsis* after *A. brassicicola* inoculation exhibited a lower accumulation of ROS than the wild type plants at 0.5 dpi (Figure 6), as indicated by the NBT staining and DAB staining.

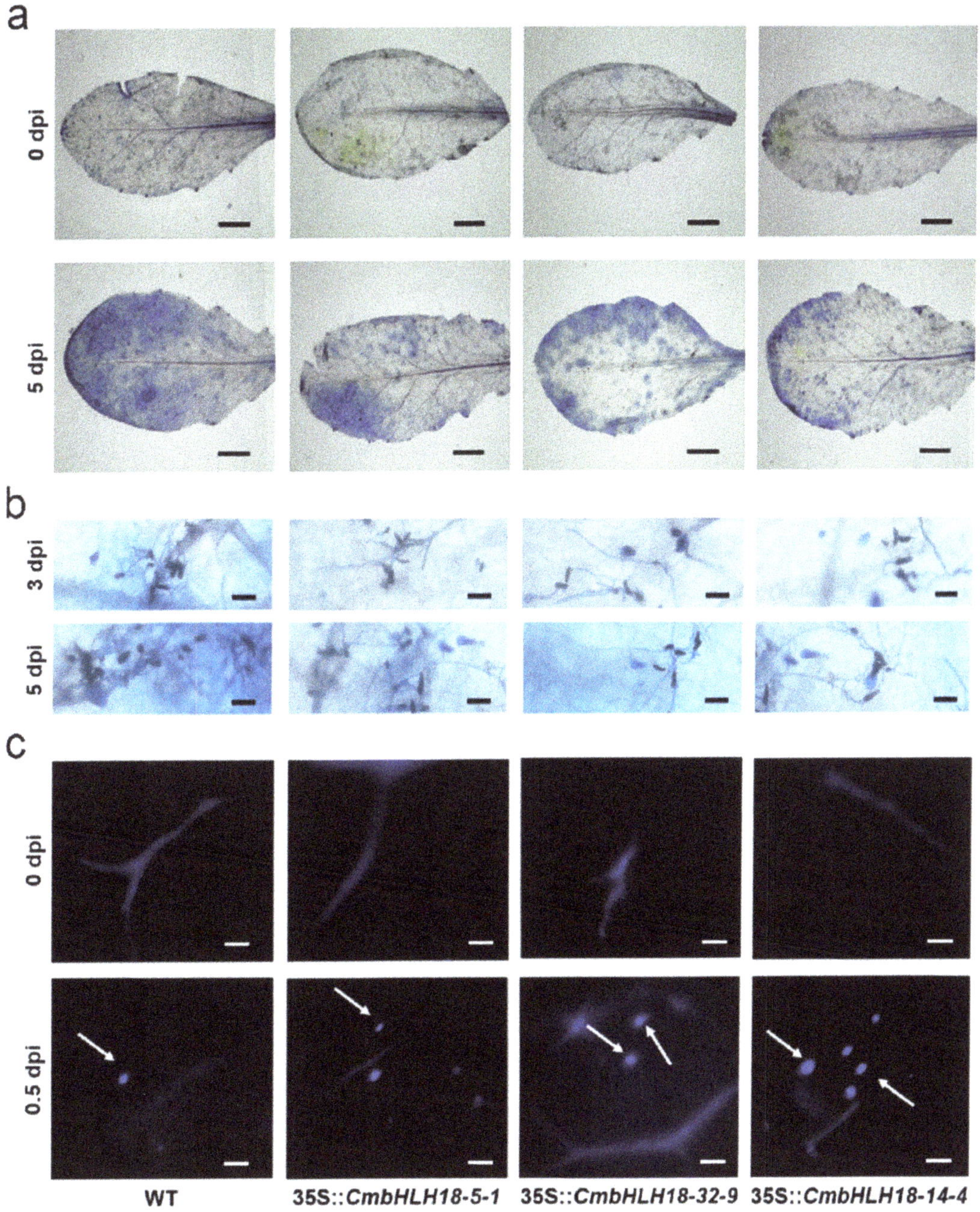

Figure 5. Effect of *CmbHLH18* on the cell activity of leaves, growth of fungi, and callose deposition of leaves in *Arabidopsis* after inoculation with *A. brassicicola*. (**a**) Microscopic visualization of necrotic cells in leaves (Bar = 1 cm); (**b**) observation of fungi growth (Bar = 50 μm); (**c**) Detection of callose deposition in leaves (Bar = 100 μm).

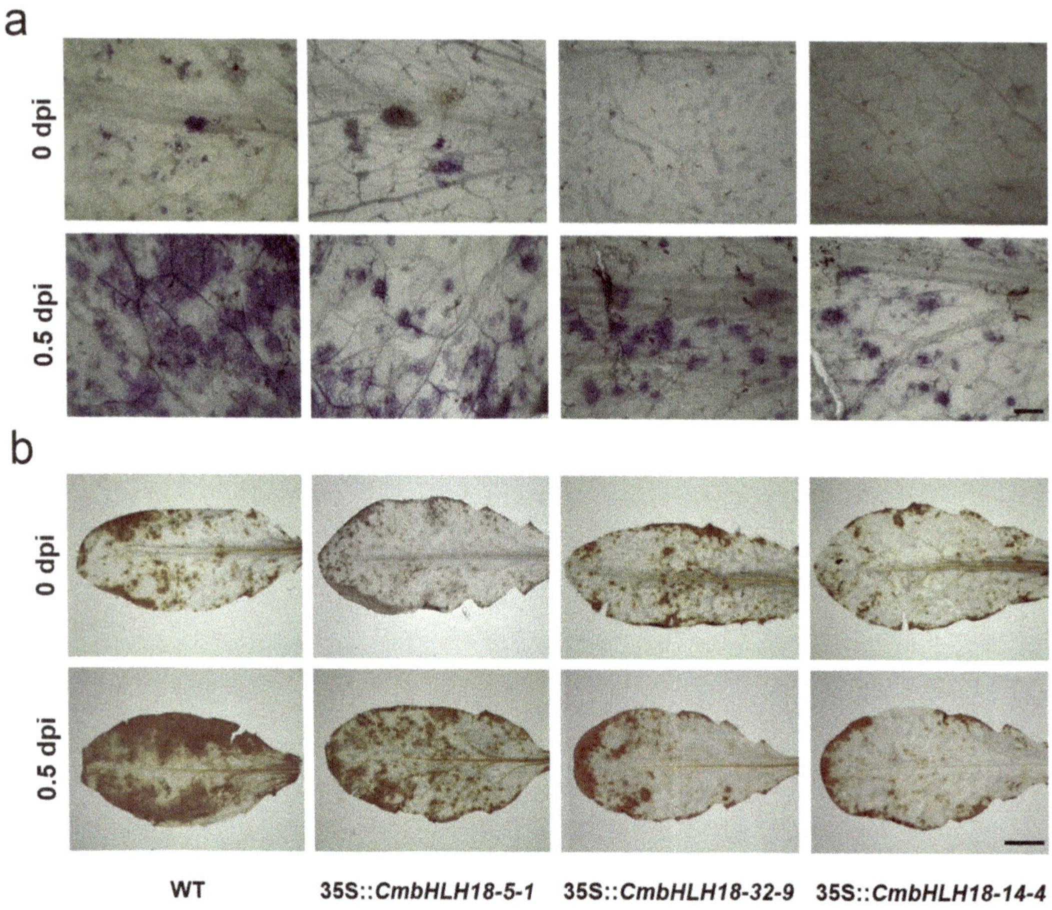

Figure 6. Effect of *CmbHLH18* on ROS accumulation in leaves in *Arabidopsis* after inoculation with *A. brassicicola*. (**a**) Accumulation of O_2^- (NBT staining) (Bar = 1cm); (**b**) Accumulation of H_2O_2 (DAB staining) (Bar = 100 μm).

The profiling of gene expression and activities of enzymes involved in ROS scavenging and plant defense suggested that after inoculation, there was a significant increase in ROS scavenging and defense abilities in transgenic plants (Figures 7 and 8). This indicated that *CmbHLH18* enhanced resistance to *A. brassicicola* by increasing the ability of ROS scavenging and the activity of defense enzymes in *Arabidopsis*.

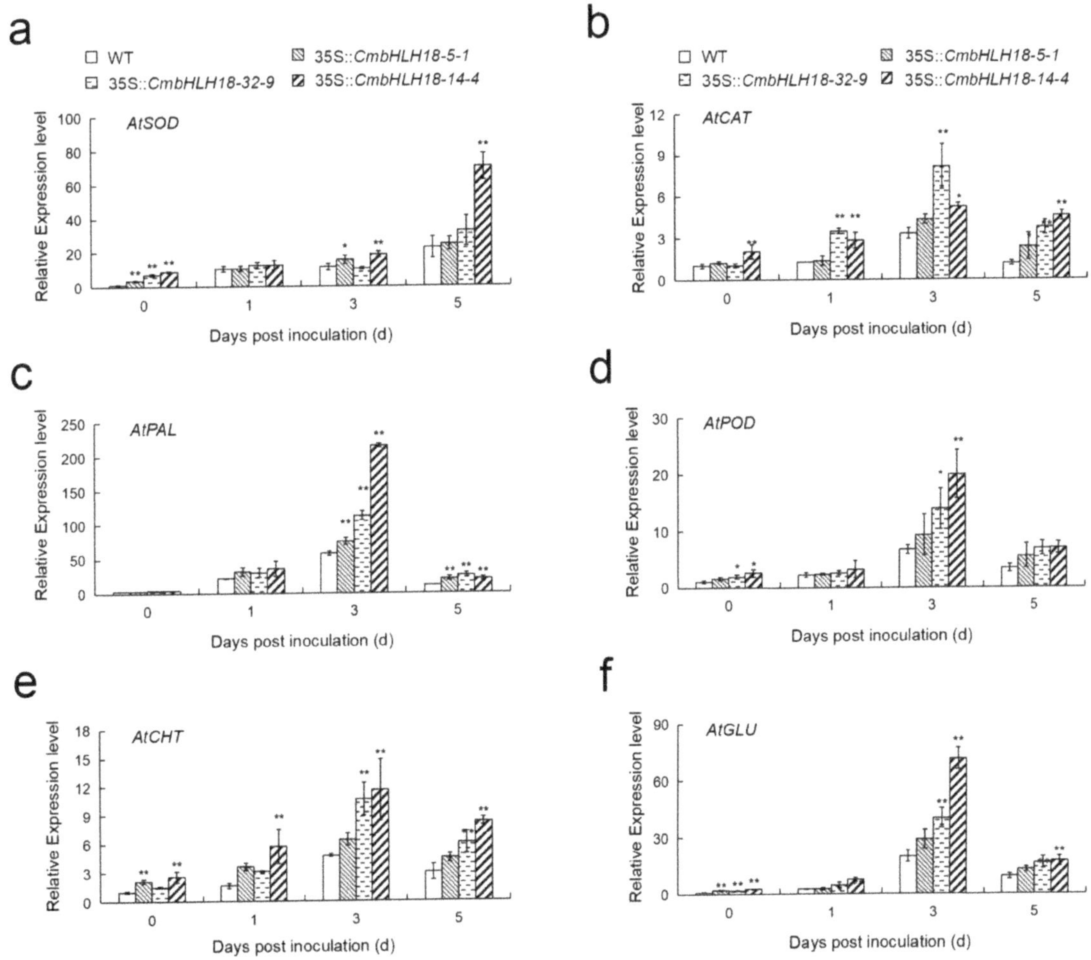

Figure 7. Effect of *CmbHLH18* on the expression level of antioxidant and defense enzyme genes in *Arabidopsis* after inoculation with *A. brassicicola*. (**a**,**b**) Expression levels of antioxidant enzyme genes; (**c**–**f**) Expression levels of defense enzyme genes. Asterisks represent statistically significant differences between different lines by *t*-test ($n = 3$; *, $p < 0.05$; **, $p < 0.01$).

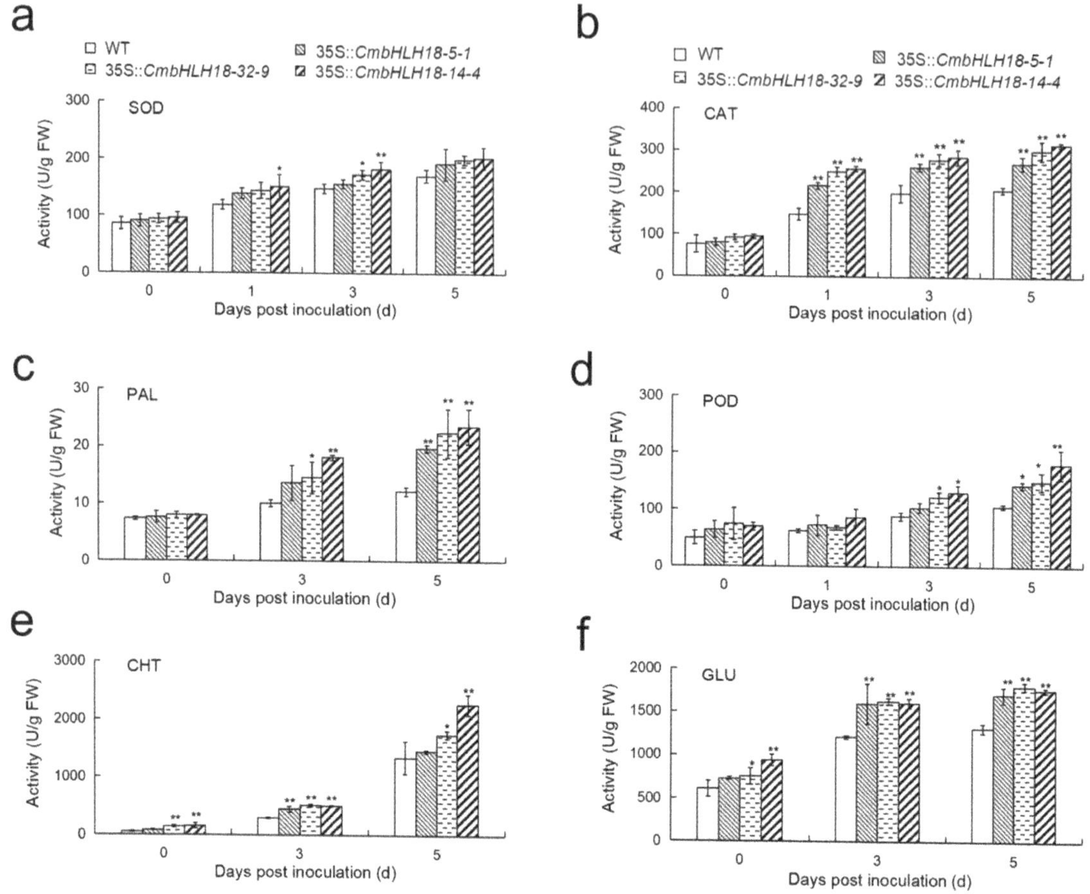

Figure 8. Effect of *CmbHLH18* on the activities of antioxidant and defense enzymes in *Arabidopsis* after inoculation with *A. brassicicola*. (**a**,**b**) Activities of antioxidant enzymes; (**c**–**f**) Activities of defense enzymes. Asterisks represent statistically significant differences between different lines by *t*-test ($n = 3$; *, $p < 0.05$; **, $p < 0.01$).

4. Discussion

The bHLH TF family exists widely in plants and animals and is one of the largest transcription factor families in plants [26,27]. This study reports the gene family for the first findings in *Chrysanthemum*. Our sequence analysis of genes identified 71 bHLHs of 17 subfamilies in *Chrysanthemum*, which is significantly smaller than that reported in *Arabidopsis* (162 bHLH TFs into 21 subfamilies, [7]), white poplar (202 bHLH TFs into 25 subfamilies, [17]), and rice (167 bHLH TFs into 22 subfamilies, [49]). Therefore, gene duplications occur more in some species than in others, such as *Chrysanthemum*, sacred lotus [50], and watermelon [51].

In general, genes with the same or similar structures are clustered in the same subfamily and may have a similar function [52]. For example, TT8 of the bHLH IIIf subfamily in *Arabidopsis* and TaPpb1 of the bHLH IIIf subfamilies in *Triticum aestivum L.* can both regulate anthocyanin synthesis [19,53]. AtbHLH3 of the bHLH IIId subfamily in *Arabidopsis* and GmbHLH3 of the bHLH IIId subfamily in *Glycine max (Linn.) Merr.* can both regulate jasmonate-induced leaf senescence [54,55]. Therefore, cluster and comparative analysis of the phylogenetic tree classification results between CmbHLH proteins and the bHLH of other species with known functions can be helpful to predict the functions of some genes in *Chrysanthemum* (Figure 1). In other cases, there are some differences in function, even if these genes are clustered into the same subfamily. For example, NFL of the bHLH IIIb subfamily in *Arabidopsis* can promote flowering, but OsbHLH61 of the bHLH IIIb subfamily in *Oryza sativa L.* can increase resistance to the brown planthopper [56,57]. The GL3 of the bHLH IIIf subfamily in *Arabidopsis* can promote root epidermal development, which differs from the function of TaPpb1 in the same subfamily [19,58]. AtbHLH18 (belonging to IVa) achieved iron homeostasis by promoting JA-induced degradation of the Ferritin (FIT) protein and inhibiting iron absorption in *Arabidopsis* [59]. There may be a great difference in function between CmbHLH18 and AtbHLH18 in function.

The Weblogo diagram obtained by multi-sequence alignment of the CmbHLH domain shows that the bHLH domain is conserved in all CmbHLHs, which is similar to the *Arabidopsis* bHLH domain (Figure 2). In the CmbHLH domain, two amino acid residues, Leu-23 and Leu-52, are relatively conservative, accounting for 83% and 92%, respectively. The two residues are necessary for the formation of a homologous or heterodimer [9]. This indicates that, on the one hand, CmbHLH proteins probably form homodimers by itself, and on the other hand, CmbHLH proteins probably also interact with other TFs, such as R2R3-MYB, to form heterodimers [9]. Understanding its domains will deepen our understanding of the function.

bHLH TF family has a variety of biological functions [26–28]. However, there are still many unknowns about the function of bHLH TFs in biotic stress. The comparison of the *cis*-acting elements on the promoters of the five *CmbHLHs* showed that there are *cis*-elements involving plant hormone responsiveness, stress responsiveness, and plant growth and development. SA plays a crucial role in plant defense and is generally involved in the activation of defense responses against biotrophic and hemi-biotrophic pathogens, as well as the establishment of systemic acquired resistance [60]. By contrast, JA is usually associated with a defense against necrotrophic pathogens and herbivorous insects [61]. Interestingly, the TC-rich repeats were unique to the promoter of *CmbHLH18*, which was located at 515 bp upstream of ATG and was involved in defense and stress responsiveness (Figure S3; Table 2). This may be related to the fact that *CmbHLH18* can significantly respond to the necrotrophic fungus *Alternaria* sp. infection in *Chrysanthemum*. Surprisingly, *CmbHLH18* may also enhance the resistance of *Arabidopsis* to *A. brassicicola* (Figure 4c), which is another typical necrotrophic fungus of *Alternaria*.

Callose is generally believed to positively regulate the stress response of plants [62–65]. Under abiotic and biotic stress, plants often trigger the accumulation of callose [64]. Callose deposition in the cell wall, plasmodesmata, and sieve pores controls the cell wall permeability, prevents further penetration of the pathogen into the tissue, and reduces the loss of cellular water and solute to maintain the internal stabilities of cells [66]. In this study, leaves of *CmbHLH18*-transformed *Arabidopsis* had more callose than WT after *A. brassicicola* inoculation, indicating that *CmbHLH18* could enhance the defense of *Arabidopsis* against *A. brassicicola* by inducing callose deposition.

When plants are subjected to stress (including biotic and abiotic stress), a large amount of ROS (O_2^-, H_2O_2, etc.) is generated in the cells. If ROS is not cleared in time, it damages plant tissues. *CmbHLH18*-transformed *Arabidopsis* plants reduce the content of O_2^- and H_2O_2 in response to fungal attacks. This is consistent with Yao et al. [67], who showed that overexpression of *FtbHLH2* reduced the amount of ROS and increased the endurance of cold stress in *Arabidopsis*. In this study, the heterologous overexpression of *CmbHLH18* enhanced resistance to *A. brassicicola* in *Arabidopsis* by decreasing the content of O_2^- and H_2O_2 (Figure 6).

Plants can scavenge ROS and be protected from oxidative stress damage through an antioxidant protecting system involving SOD and CAT [68]. Heterologous overexpression of *VvbHLH1* in *Arabidopsis* [69] and *NtbHLH123* in tobacco [16] both improved the antioxidant enzyme system and increased the antioxidant capacity of transgenic plants. Our results showed a similar pattern, suggesting that *CmbHLH18* could decrease the amount of ROS, reduce plant damage, and enhance resistance to *A. brassicicola* by activating the expression of antioxidant genes and increasing the activities of antioxidant enzymes.

Plants can resist pathogen invasion through lignification. PAL and POD are key enzymes in the lignin synthesis pathway [70]. CHT and GLU are pathogenesis-related (PR) enzymes that can destroy fungal cell walls and prevent the invasion of pathogenic fungus [71]. Studies have shown that the expression levels of PAL, POD, CHT, and GLU are significantly increased by *Magnaporthe grisea* in rice [72]. In this study, after inoculation with *A. brassicicola*, the activities of PAL, POD, CHT, and GLU in transgenic leaves were significantly higher than those in WT leaves (Figure 8), and the corresponding gene expression levels in transgenic leaves were also higher than those in WT leaves (Figure 7).

5. Conclusions

We first identified 71 CmbHLHs genes of 17 subfamilies, and the genes in 'Huaiju $2^{\#}$' are generally hydrophilic proteins and high aliphatic amino acid content. Additionally, combining the results of RT-qPCR analysis after *Alternaria* sp. inoculation, we found that *CmbHLH16, 18, 28, 30,* and *60* may be involved in resistance to biotic necrotrophic fungus. They all have hormone-responsive elements that are associated with enhanced disease resistance in their *cis*-acting elements, and *CmbHLH18* may be a gene resistant to necrotrophic fungus in *Chrysanthemum*.

Heterologous overexpression of *CmbHLH18* can enhance the resistance to necrotrophic fungus in *Arabidopsis* by a series of physiological and genetic activities, inducing callose deposition and increasing the activities and gene expression levels of antioxidant and defense enzymes.

The identification of CmbHLHs in this study will help to further identify candidate genes for resistance to necrotrophic fungus, elucidate the molecular mechanism of resistance to necrotrophic fungus in *Chrysanthemum*, and use *CmbHLHs* to breed a new variety of *Chrysanthemum* with high resistance to necrotrophic fungus.

Supplementary Materials: The following supporting information can be downloaded at: https://www.mdpi.com/article/10.3390/genes14020275/s1, Figure S1: Multiple sequence alignment of CmbHLH family proteins. The four-color modules at the top of the figure are represented, respectively: red represents the N-terminal basic region of bHLH, yellow, and green represent two amphiphilic regions, and blue represents a variable length loop. Process the data into pictures using Clustal W. Figure S2: Motif composition of the CmbHLH TF family. (a) Phylogenetic tree of CmbHLHs; (b) conserved motifs of the CmbHLHs; (c) sequence composition of each motif in CmbHLHs. Figure S3: Visualization of *cis*-elements in the promoter sequences of *CmbHLHs*. Different colors represent different *cis*-acting elements in the promoter sequences of *CmbHLHs*. Table S1: Promoter sequences of candidate resistant genes to necrotrophic fungus. Table S2: Number of bHLH TF in each subfamily of *Chrysanthemum* and *Arabidopsis*. Table S3: Proportion of conserved amino acids at each point of CmbHLH protein. Table S4: Selection criteria for differential genes. Table S5: Relative activity of leaf cells of WT and transgenic *Chrysanthemum*. Table S6: Primers used in this study.

Author Contributions: Conceptualization, X.Z.; methodology, Y.D.; investigation, L.J.; formal analysis, T.W.; software, J.X. and L.S.; data curation, writing-original draft preparation, Y.D., D.W. and X.W.; writing-review and editing, X.Z.; funding acquisition, X.Z. All authors have read and agreed to the published version of the manuscript.

Funding: This research was funded by the Joint Funds of the National Natural Science Foundation of China under Grant Agreement, grant number No. U1704120, and the National Natural Science Foundation of China, grant number No. 31372105, and the Doctoral foundation of Henan Normal University, grant number No. QD18076.

Institutional Review Board Statement: Not applicable.

Informed Consent Statement: Not applicable.

Data Availability Statement: The datasets supporting the results of this article are included within the article.

Acknowledgments: We are very grateful to Jianhua Li (Hope College in USA) and John Hugh Snyder (National Institute of Biological Sciences of China, Beijing) for reading the earlier version of the paper and providing helpful comments.

Conflicts of Interest: The authors declare no conflict of interest.

References

1. Yasemin, S.; Köksal, N.; Özkaya, A.; Yener, M. Growth and physiological responses of "*Chrysanthemum paludosum*" under salinity stress. *J. Biol. Environ. Sci.* **2017**, *11*, 59–66.
2. Li, J.M.; Li, H.Q.; Hu, S.X. Investigation on the present situation of commodity types of Chrysanthemum. *China Mod. Med.* **2016**, *23*, 93–96.
3. Zhao, X.T.; Song, L.Y.; Jiang, L.W.; Zhu, Y.T.; Gao, Q.H.; Wang, D.D.; Xie, J.; Lv, M.; Liu, P.; Li, M.J. The integration of transcriptomic and transgenic analyses reveals the involvement of the SA response pathway in the defense of Chrysanthemum against the necrotrophic fungus *Alternaria* sp. *Hortic. Res.* **2020**, *7*, 80. [CrossRef] [PubMed]
4. Zhao, X.T.; Wang, M.; Wang, T.L.; Li, M.J. Isolation and identification of the pathogen of black spot disease of Chrysanthemum huai. *Acta Hortic. Sin.* **2015**, *42*, 174–182. [CrossRef]
5. Ding, S.M.; Yang, Y.C.; Zhang, H.H. A preliminary report on the occurrence and prevention of Chrysanthemum black spot in some areas of Shandong province. *Guangdong Landsc. Archit.* **2005**, *6*, 31–35.
6. Sun, X.; Wang, Y.; Sui, N. Transcriptional regulation of bHLH during plant response to stress. *Biochem. Biophys. Res. Commun.* **2018**, *503*, 397–401. [CrossRef]
7. Toledo-Ortiz, G.; Huq, E.; Quail, P.H. The *Arabidopsis* basic/helix-loop-helix transcription factor family. *Plant Cell* **2003**, *15*, 1749–1770. [CrossRef]
8. Ferre-D'amare, A.R.; Pognonec, P.; Roeder, R.G.; Burley, S.K. Structure and function of the b/HLH/Z domain of USF. *EMBO J.* **1994**, *13*, 180–189. [CrossRef]
9. Wang, D.D. Identification of bHLH Transcription Factor Family and Heterologous Disease Resistance Function of *CmbHLH18* Gene in *Chrysanthemum morifolium* Ramat. Master's Thesis, Henan Normal University, Xinxiang, China, May 2021.
10. Tian, H.N.; Guo, H.Y.; Dai, X.M.; Cheng, Y.X.; Zheng, K.J.; Wang, X.P.; Wang, S.C. An ABA down-regulated bHLH transcription repressor gene, *bHLH129* regulates root elongation and ABA response when overexpressed in *Arabidopsis*. *Sci. Rep.* **2015**, *5*, 17587. [CrossRef]
11. Zhang, T.T.; Lv, W.; Zhang, H.S.; Ma, L.; Li, P.H.; Ge, L.; Li, G. Genome-wide analysis of the basic Helix-Loop-Helix (bHLH) transcription factor family in maize. *BMC Plant Biol.* **2018**, *18*, 1–14. [CrossRef]
12. Salih, H.; Tan, L.; Htet, N.N.W. Genome-wide identification, characterization of bHLH transcription factors in Mango. *Trop. Plant Biol.* **2021**, *14*, 72–81. [CrossRef]
13. Friedrichsen, D.M.; Nemhauser, J.; Muramitsu, T.; Maloof, J.N.; Alonso, J.; Ecker, J.R.; Furuya, M.; Chory, J. Three redundant brassinosteroid early response genes encode putative bHLH transcription factors required for normal growth. *Genetics* **2002**, *162*, 1445–1456. [CrossRef]
14. Zhang, T.; Qu, Y.X.; Wang, H.B.; Wang, J.J.; Song, A.P.; Hu, Y.H.; Chen, S.M.; Jiang, J.F.; Chen, F.D. The heterologous expression of a Chrysanthemum TCP-P transcription factor *CmTCP14* suppresses organ size and delays senescence in *Arabidopsis thaliana*. *Plant Physiol. Biochem.* **2017**, *115*, 239–248. [CrossRef]
15. Seo, J.S.; Joo, J.; Kim, M.J.; Kim, Y.K.; Nahm, B.H.; Song, S.L.; Cheong, J.J.; Lee, J.S.; Kim, J.K.; Choi, Y.D. OsbHLH148, a basic helix-loop-helix protein, interacts with OsJAZ proteins in a jasmonate signaling pathway leading to drought tolerance in rice. *Plant J.* **2011**, *65*, 907–921. [CrossRef] [PubMed]
16. Zhao, Q.; Xiang, X.H.; Liu, D.; Yang, A.G.; Wang, Y.Y. Tobacco transcription factor NtbHLH123 confers tolerance to cold stress by regulating the NtCBF pathway and reactive oxygen species homeostasis. *Front. Plant Sci.* **2018**, *9*, 381. [CrossRef] [PubMed]
17. Zhao, K.; Li, S.X.; Yao, W.J.; Zhou, B.R.; Li, R.H.; Jiang, T.B. Characterization of the basic helix-loop-helix gene family and its tissue-differential expression in response to salt stress in poplar. *PeerJ* **2018**, *6*, e4502. [CrossRef] [PubMed]
18. Zhao, M.; Song, A.P.; Li, P.L.; Chen, S.M.; Jiang, J.F.; Chen, F.D. A bHLH transcription factor regulates iron intake under Fe deficiency in Chrysanthemum. *Sci. Rep.* **2014**, *4*, 6694. [CrossRef]
19. Jiang, W.H.; Liu, T.X.; Nan, W.Z.; Jeewani, D.C.; Niu, Y.L.; Li, C.L.; Wang, Y.; Shi, X.; Wang, C.; Wang, J.H.; et al. Two transcription factors TaPpm1 and TaPpb1 co-regulate anthocyanin biosynthesis in purple pericarps of wheat. *J. Exp. Bot.* **2018**, *69*, 2555–2567. [CrossRef]
20. Alessio, V.M.; Cavaçana, N.; Dantas, L.L.D.B.; Lee, N.; Hotta, C.T.; Imaizumi, T.; Menossi, M. The FBH family of bHLH transcription factors controls ACC synthase expression in sugarcane. *J. Exp. Bot.* **2018**, *69*, 2511–2525. [CrossRef]
21. Ikeda, M.; Fujiwara, S.; Mitsuda, N.; Ohme-Takagi, M. A triantagonistic basic helix-loop-helix system regulates cell elongation in *Arabidopsis*. *Plant Cell* **2012**, *24*, 4483–4497. [CrossRef]
22. Liljegren, S.J.; Roeder, A.H.K.; Kempin, S.A.; Gremski, K.; Østergaard, L.; Guimil, S.; Reyes, D.K., Yanofsky, A.M.F. Control of fruit patterning in *Arabidopsis* by INDEHISCENT. *Cell* **2004**, *116*, 843–853. [CrossRef] [PubMed]

23. Szecsi, J.; Joly, C.; Bordji, K.; Varaud, E.; Cock, J.M.; Dumas, C.; Bendahmane, M. BIGPETALp, a bHLH transcription factor is involved in the Control of *Arabidopsis* petal size. *EMBO J.* **2006**, *25*, 3912–3920. [CrossRef]
24. Karas, B.; Amyot, L.; Johansen, C.; Sato, S.; Tabata, S.; Kawaguchi, M.; Szczyglowski, K. Conservation of *Lotus* and *Arabidopsis* basic helix-loop-helix proteins reveals new players in root hair development. *Plant Physiol.* **2009**, *151*, 1175–1185. [CrossRef] [PubMed]
25. Kanaoka, M.M.; Pillitteri, L.J.; Fujii, H.; Yoshida, Y.; Bogenschutz, N.L.; Takabayashi, J.; Zhu, J.K.; Torii, K.U. *SCREAM/ICE1* and *SCREAM2* specify three cell-state transitional steps leading to *Arabidopsis* stomatal differentiation. *Plant Cell* **2008**, *20*, 1775–1785. [CrossRef] [PubMed]
26. Wang, F.T.; Lin, R.M.; Feng, J.; Qiu, D.W.; Chen, W.Q.; Xu, S.C. Wheat bHLH transcription factor gene, *TabHLH060*, enhances susceptibility of transgenic *Arabidopsis thaliana* to *Pseudomonas syringae*. *Physiol. Mol. Plant Pathol.* **2015**, *90*, 123–130. [CrossRef]
27. Yu, J.; Ai, G.; Shen, D.; Chai, C.; Jia, Y.; Liu, W.; Dou, D. Bioinformatical analysis and prediction of *Nicotiana benthamiana* bHLH transcription factors in *Phytophthora parasitica* resistance. *Genomics* **2018**, *111*, 473–482. [CrossRef]
28. He, X.; Zhu, L.F.; Wassan, G.M.; Wang, Y.J.; Miao, Y.H.; Shaban, M.; Hu, H.Y.; Sun, H.; Zhang, X.L. *GhJAZ2* attenuates cotton resistance to biotic stresses via the inhibition of the transcriptional activity of *GhbHLH171*. *Mol. Plant Pathol.* **2018**, *19*, 896–908. [CrossRef]
29. Huang, Y.; Niu, B.F.; Gao, Y.; Fu, L.M.; Li, W.Z. CD-HIT Suite: A web server for clustering and comparing biological sequences. *Bioinformatics* **2010**, *26*, 680–682. [CrossRef]
30. Yang, M.Z.; Derbyshire, M.K.; Yamashita, R.A.; Marchler-Bauer, A. NCBI's conserved domain database and tools for protein domain analysis. *Curr. Protoc. Bioinform.* **2020**, *69*, e90. [CrossRef]
31. Letunic, I.; Bork, P. 20 years of the SMART protein domain annotation resource. *Nucleic Acid. Res.* **2018**, *46*, 493–496. [CrossRef]
32. Artimo, P.; Jonnalagedda, M.; Arnold, K.; Baratin, D.; Csardi, G.; Castro, E.D.; Duvaud, S.; Flegel, V.; Fortier, A.; Gasteiger, E.; et al. ExPASY: SIB bioinformatics resource portal. *Nucleic Acid. Res.* **2012**, *40*, 597–603. [CrossRef]
33. Li, H.T.; Gao, W.L.; Xue, C.L.; Zhang, Y.; Liu, Z.G.; Zhang, Y.; Meng, X.W.; Liu, M.J.; Zhao, J. Genome-wide analysis of the bHLH gene family in Chinese jujube (*Ziziphus jujuba* Mill.) and wild jujube. *BMC Genom.* **2019**, *20*, 568. [CrossRef] [PubMed]
34. Kumar, S.; Stecher, G.; Tamura, K. MEGA7: Molecular evolutionary genetics analysis version 7.0 for bigger datasets. *Mol. Biol. Evol.* **2016**, *33*, 1870–1874. [CrossRef] [PubMed]
35. Lamesch, P.; Berardini, T.Z.; Li, D.H.; Swarbreck, D.; Wilks, C.; Sasidharan, R.; Muller, R.; Dreher, K.; Alexander, D.L.; Garcia-Hernandez, M.; et al. The *Arabidopsis* Information Resource (TAIR): Improved gene annotation and new tools. *Nucleic Acid. Res.* **2012**, *40*, 1202–1210. [CrossRef] [PubMed]
36. Letunic, I.; Bork, P. Interactive Tree of Life (iTOL): An online tool for phylogenetic tree display and annotation. *Bioinformatics* **2007**, *23*, 127–128. [CrossRef]
37. Meyer, M.; Fehling, H.; Matthiesen, J.; Lorenzen, S.; Schuldt, K.; Bernin, H.; Mareen, Z.; Lender, C.; Ernst, T.; Ittrich, H.; et al. Overexpression of differentially expressed genes identified in non-pathogenic and pathogenic entamoeba histolytica clones allow identification of new pathogenicity factors involved in amoebic liver abscess formation. *PLoS Pathog.* **2016**, *12*, e1005853. [CrossRef]
38. Lescot, M.; Dehais, P.; Thijs, G.; Marchal, K.; Moreau, Y.; Peer, Y.V.; Rouzéet, P.; Rombauts, S. PlantCARE, a database of plant *cis*-acting regulatory elements and a portal to tools for in silico analysis of promoter sequences. *Nucleic Acid. Res.* **2002**, *30*, 325–327. [CrossRef]
39. Chen, C.J.; Chen, H.; Zhang, Y.; Thomas, H.R.; Frank, M.H.; He, Y.H.; Xia, R. TBtools: An integrative toolkit developed for interactive analyses of big biological data. *Mol. Plant* **2020**, *13*, 1194–1202. [CrossRef]
40. Ghedira, R.; Buck, S.D.; Ex, F.V.; Angenon, G.; Depicker, A. T-DNA transfer and T-DNA integration efficiencies upon *Arabidopsis thaliana* root explant cocultivation and floral dip transformation. *Planta* **2013**, *238*, 1025–1037. [CrossRef]
41. Miles, A.K.; Newman, T.K.; Gultzow, D.L.; Parfitt, S.C.; Drenth, A.; Smith, M.W. Commercial-scale *Alternaria* brown spot resistance screening as the first step in breeding new mandarins for Australia. *Acta Hortic.* **2015**, *1065*, 971–978. [CrossRef]
42. Xu, G.J. Studies on Seedling's Resistance of Related Species of Chrysanthemum to *Alternaria* Leaf Spot and Genetic Transformation of Chrysanthemum with *hrfA* gene. Ph.D. Thesis, Nanjing Agricultural University, Nanjing, China, June 2009.
43. Liu, N.; Lin, Z.F. Use of evans blue for testing cell viability of intact leaves of plant. *J. Plant Physiol.* **2011**, *47*, 570–574. [CrossRef]
44. Kuritatashiro, A.; Hayashi, N.; Oyanagi, T.; Sasamoto, H. New factors for protoplast-callose-fiber formation in salt-tolerant mangrove plants, *Avicennia alba* and *Bruguiera sexangula* and analysis of fiber substructures. *J. Plant Stud.* **2021**, *9*, 1. [CrossRef]
45. Shi, J.; Fu, X.Z.; Peng, T.; Huang, X.S.; Fan, Q.J.; Liu, J.H. Spermine pretreatment confers dehydration tolerance of citrus in vitro plants via modulation of antioxidative capacity and stomatal response. *Tree Physiol.* **2010**, *30*, 914–922. [CrossRef]
46. Liu, H.X.; Jiang, W.B.; Bi, Y.; Luo, Y.B. Postharvest BTH treatment induces resistance of peach (*Prunus persica* L. cv. Jiubao) fruit to infection by *Penicillium expansum* and enhances activity of fruit defense mechanisms. *Postharvest Biol. Technol.* **2005**, *35*, 263–269. [CrossRef]
47. Livak, K.J.; Schmittgen, T.D. Analysis of relative gene expression data using real-time quantitative PCR and the $2^{-\triangle\triangle CT}$ method. *Methods* **2001**, *25*, 402–408. [CrossRef] [PubMed]
48. Niu, X.; Guan, Y.X.; Chen, S.K.; Li, H.F. Genome-wide analysis of basic helix-loop-helix (bHLH) transcription factors in *Brachypodium distachyon*. *BMC Genom.* **2017**, *18*, 619. [CrossRef]

49. Li, X.X.; Duan, X.P.; Jiang, H.X.; Sun, Y.J.; Tang, Y.P.; Yuan, Z.; Guo, J.K.; Liang, W.Q.; Chen, L.; Yin, J.Y.; et al. Genome-wide analysis of basic/helix-loop-helix transcription factor family in rice and *Arabidopsis*. *Plant Physiol.* **2006**, *141*, 1167–1184. [CrossRef]
50. Hudson, K.A.; Hudson, M.E. The basic helix-loop-helix transcription factor family in the sacred lotus, *Nelumbo Nucifera*. *Trop. Plant Biol.* **2014**, *7*, 65–70. [CrossRef]
51. He, J.; Gu, X.R.; Wei, C.H.; Yang, X.Z.; Zhang, X. Identification and expression analysis under abiotic stresses of the bHLH transcription factor gene family in Watermelon. *Acta Hortic. Sin.* **2016**, *43*, 281. [CrossRef]
52. Li, Z.; Gao, Z.Q.; Li, R.H.; Xu, Y.; Kong, Y.Z.; Zhou, G.K.; Meng, C.X.; Hu, R.B. Genome-wide identification and expression profiling of HD-ZIP gene family in *Medicago truncatula*. *Genomics* **2020**, *112*, 3624–3635. [CrossRef]
53. Nesi, N.; Debeaujon, I.; Jond, C.; Pelletier, G.; Caboche, M.; Lepiniec, L. The *TT8* gene encodes a basic helix-loop-helix domain protein required for expression of *DFR* and *BAN* genes in *Arabidopsis Siliques*. *Plant Cell* **2000**, *12*, 1863–1878. [CrossRef] [PubMed]
54. Qi, T.C.; Wang, J.J.; Huang, H.; Liu, B.; Gao, H.; Liu, Y.L.; Song, S.S.; Xie, D.X. Regulation of jasmonate-induced leaf senescence by antagonism between bHLH subgroup IIIe and IIId factors in *Arabidopsis*. *Plant Cell* **2015**, *27*, 1634–1649. [CrossRef] [PubMed]
55. Wang, T.L.; Wang, S.; Wang, Y.; Li, J.W.; Yan, F.; Liu, Y.J.; Zhao, L.; Wang, Q.Y. Jasmonic acid-induced inhibition of root growth and leaf senescence is reduced by *GmbHLH3*, a soybean bHLH transcription factor. *Can. J. Plant Sci.* **2020**, *100*, 477–487. [CrossRef]
56. Sharma, N.; Xin, R.J.; Kim, D.H.; Sung, S.; Lange, T.; Hu, E. NO FLOWERING IN SHORT DAY (NFL) is a bHLH transcription factor that promotes flowering specifically under short-day conditions in *Arabidopsis*. *Development* **2016**, *143*, 682–690. [CrossRef]
57. Wang, M.L.; Yang, D.Y.; Ma, F.L.; Zhu, M.L.; Shi, Z.Y.; Miao, X.X. OsHLH61-OsbHLH96 influences rice defense to brown planthopper through regulating the pathogen-related genes. *Rice* **2019**, *12*, 9. [CrossRef]
58. Bernhardt, C.; Lee, M.M.; Gonzalez, A.; Zhang, F.; Lloyd, A.; Schiefelbein, J. The bHLH genes *GLABRA3* (*GL3*) and *ENHANCER OF GLABRA3* (*EGL3*) specify epidermal cell fate in the *Arabidopsis* root. *Development* **2003**, *130*, 6431–6439. [CrossRef]
59. Cui, Y.; Chen, C.L.; Cui, M.; Zhou, W.J.; Wu, H.L.; Ling, H.Q. Four IVa bHLH transcription factors are novel interactors of FIT and mediate JA inhibition of iron uptake in *Arabidopsis*. *Mol. Plant* **2018**, *11*, 1166–1183. [CrossRef]
60. Grant, M.; Lamb, C. Systemic immunity. *Curr. Opin. Plant Biol.* **2006**, *9*, 414–420. [CrossRef]
61. Bari, R.; Jones, J.D.G. Role of plant hormones in plant defence responses. *Plant Mol. Biol.* **2009**, *69*, 473–488. [CrossRef]
62. Flors, V.; Ton, J.; Jakab, G.; Mauch-Mani, B. Abscisic acid and callose: Team players in defence against pathogens? *J. Phytopathol.* **2005**, *153*, 377–383. [CrossRef]
63. Krzeslowska, M. The cell wall in plant cell response to trace metals: Polysaccharide remodeling and its role in defense strategy. *Acta Physiol. Plant* **2011**, *33*, 35–51. [CrossRef]
64. Luna, E.; Pastor, V.; Robert, J.; Flors, V.; Mauch-Mani, B.; Ton, J. Callose deposition: A multifaceted plant defense response. *Mol. Plant Microbe Interact.* **2011**, *24*, 183–193. [CrossRef]
65. Hunter, K.; Kimura, S.; Rokka, A.; Tran, H.C.; Toyota, M.; Kukkonen, J.P.; Wrzaczek, M. CRK2 enhances salt tolerance by regulating callose deposition in connection with PLDα1. *Plant Physiol.* **2019**, *180*, 2004–2021. [CrossRef] [PubMed]
66. Wang, B.; Andargie, M.; Fang, R. The function and biosynthesis of callose in high plants. *Heliyon* **2022**, *8*, e09248. [CrossRef] [PubMed]
67. Yao, P.F.; Sun, Z.X.; Li, C.L.; Zhao, X.R.; Li, M.F.; Deng, R.Y.; Huang, Y.J.; Zhao, H.X.; Chen, H.; Wu, Q. Overexpression of *Fagopyrum tataricum FtbHLH2* enhances tolerance to cold stress in transgenic *Arabidopsis*. *Plant Physiol. Biochem.* **2018**, *125*, 85–94. [CrossRef] [PubMed]
68. Zeng, H.Q.; Xu, H.R.; Wang, H.; Chen, H.; Wang, G.Q.; Bai, Y.J.; Wei, Y.X.; Shi, H.T. LSD3 mediates the oxidative stress response through fine-tuning APX2 activity and the NF-YC15-GSTs module in cassava. *Plant J.* **2022**, *110*, 1447–1461. [CrossRef]
69. Wang, F.B.; Zhu, H.; Chen, D.H.; Li, Z.J.; Peng, R.H.; Yao, Q.H. A grape bHLH transcription factor gene, *VvbHLH1*, increases the accumulation of flavonoids and enhances salt and drought tolerance in transgenic *Arabidopsis thaliana*. *Plant Cell Tissue Organ Cult.* **2016**, *125*, 387–398. [CrossRef]
70. Gayoso, C.; Pomar, F.; Novo-Uzal, E.; Merino, F.; de Ilárduya, Ó.M. The *Ve*-mediated resistance response of the tomato to *Verticillium dahliae* involves H_2O_2, peroxidase and lignins and drives *PAL* gene expression. *BMC Plant Biol.* **2010**, *10*, 232. [CrossRef]
71. Guo, Y.J.; Zhou, J.X.; Zhang, J.R.; Zhang, S.Y. Chitosan combined with sodium silicate treatment induces resistance against rot caused by *Alternaria alternata* in postharvest jujube fruit. *J. Phytopathol.* **2019**, *167*, 451–460. [CrossRef]
72. Sasaki, K.; Iwai, T.; Hiraga, S.; Kuroda, K.; Seo, S.; Mitsuhara, I.; Miyasaka, A.; Iwano, M.; Ito, H.; Matsui, H.; et al. Ten rice peroxidases redundantly respond to multiple stresses including infection with rice blast fungus. *Plant Cell Physiol.* **2004**, *45*, 1442–1452. [CrossRef]

genes

MDPI

Article

The Complete Chloroplast Genome of An *Ophiorrhiza baviensis* Drake Species Reveals Its Molecular Structure, Comparative, and Phylogenetic Relationships

Mai Huong Pham [1], Thu Hoai Tran [1], Thi Dung Le [1], Tung Lam Le [1], Ha Hoang [1] and Hoang Ha Chu [1,2,*]

[1] Institute of Biotechnology (IBT), Vietnam Academy of Science & Technology (VAST), Hanoi 100000, Vietnam
[2] Faculty of Biotechnology, Graduate University of Science and Technology, VAST, Hanoi 100000, Vietnam
* Correspondence: chuhoangha@ibt.ac.vn

Abstract: *Ophiorrhiza baviensis* Drake, a flowering medical plant in the Rubiaceae, exists uncertainly within the *Ophiorrhiza* genus' evolutionary relationships. For the first time, the whole chloroplast (cp) genome of an *O. baviensis* Drake species was sequenced and annotated. Our findings demonstrate that the complete cp genome of *O. baviensis* is 154,770 bp in size, encoding a total of 128 genes, including 87 protein-coding genes, 8 rRNAs, and 33 tRNAs. A total of 59 SSRs were screened in the studied cp genome, along with six highly variable loci, which can be applied to generate significant molecular markers for the *Ophiorrhiza* genus. The comparative analysis of the *O. baviensis* cp genome with two published others of the *Ophiorrhiza* genus revealed a high similarity; however, there were some notable gene rearrangements in the *O. densa* plastome. The maximum likelihood phylogenetic trees were constructed based on the concatenation of the *rps16* gene and the *trnL-trnF* intergenic spacer sequence, indicating a close relationship between the studied *O. baviensis* and other *Ophiorrhiza*. This study will provide a theoretical molecular basis for identifying *O. baviensis* Drake, as well as species of the *Ophiorrhiza* genus, and contribute to shedding light on the chloroplast genome evolution of Rubiaceae.

Keywords: *Ophiorrhiza baviensis*; chloroplast genome; comparative analysis; phylogeny

Citation: Pham, M.H.; Tran, T.H.; Le, T.D.; Le, T.L.; Hoang, H.; Chu, H.H. The Complete Chloroplast Genome of An *Ophiorrhiza baviensis* Drake Species Reveals Its Molecular Structure, Comparative, and Phylogenetic Relationships. *Genes* **2023**, *14*, 227. https://doi.org/10.3390/genes14010227

Academic Editors: Wajid Zaman and Hakim Manghwar

Received: 30 November 2022
Revised: 19 December 2022
Accepted: 7 January 2023
Published: 15 January 2023

1. Introduction

The chloroplast (cp) is an essential organelle in photosynthetic plant and microbial cells that produces energy to feed the cell through photosynthesis [1]. Each chloroplast contains its own ribosomes and a separate genome from the cell's nuclear genome, ranging in size from 20 to 160 kilobase pairs (kp). The cp genome is uniparentally inherited with a quadripartite structure consisting of one large single-copy (LSC) region, one small single-copy (SSC) region, and two inverted repeat regions (IRs) of the same length [2]. As a result of the small size of the cp genome, which contains only around 100 to 120 protein coding genes, chloroplasts are often the first target for sequencing in evolutionary analysis, barcoding, and meta-barcoding [2]. In the NCBI Genbank database at present, there are more than 1000 cp genomes of plant species. However, this number is very small compared to the existing plant diversity on the planet, which raises the need to collect and store sequences of uncharacterized species.

For medicinal plants such as *Ophiorrhiza baviensis*, the potential for exploitation and the need for systematic classification are even more essential. *O. baviensis* is a species of flowering plant in the Rubiaceae family, first described scientifically by Drake in 1895, and re-identified by Wu et al. [3]. Information on the ecology and genomic characteristics of this species is extremely limited, with only four sequences of *O. baviensis*—the gene junctions *trnL-trnF* (#MH626989.1), *rps16* (#MH626923.1), the external transcribed spacer (ETS) (#MH626743.1), and *ITS* (#MH626804.1)—available on the Genbank database of the National Center for Biotechnology Information (USA) (NCBI). Each sequence is less than

1000 base pairs (bp) in size, only two of which belong to the chloroplast genome. Thus, there is a need to study the entire chloroplast genome of *O. baviensis* species for taxonomy and diversity assessment, as well as chloroplast genome characterization, conservation, and future research. With an estimated chloroplast genome size of 154 kb, the potential for exploiting genomic information on the *O. baviensis* chloroplast genome is very large.

Recently, PacBio sequencing technology has been applied to sequence cp genomes, and there have been studies demonstrating the superior ability of PacBio in de novo assembly with 99% accuracy; moreover, as the repeatability increases, this can exceed 99.9% [4]. PacBio sequencing is also a great technology in resolving gaps in rRNA, i.e., internal transcribed spacer (ITS) regions and the surrounding regions to obtain accurate molecular biology information for species identification. For the first time, we report a new complete chloroplast genome of *O. Baviensis* Drake from Vietnam and compare it with previously published *Ophiorrhiza* complete chloroplast genome data to evaluate the genome organization, phylogenetic relationships, and conserved genetic resources.

2. Materials and Methods

2.1. Sample Collection and Chloroplast Genome Sequencing

O. baviensis samples were collected in Ba Vi National Park, Hanoi, Vietnam in August 2022 (code number: Xacan 01), 1217.6 m, 21°3′32″ N; 105°4′58″ E (Figure 1). The voucher specimens were placed in the herbarium of the Institute of Ecology and Biological Resources (HN), Hanoi, Vietnam. Fresh leaves with the same code number were used to extract genomic DNA.

Figure 1. *O. baviensis* Drake. (**A**) Habitat; (**B**) Morphological characteristic of infructescence in side view; Photos by Thu Hoai Tran.

2.2. DNA Extraction and Chloroplast Genome Sequencing

We treated samples prior to extraction with the Chloroplast Isolation Kit (ab234623-Abcam, Cambridge, UK) for cp enrichment to increase the cpDNA concentration. The total DNA was extracted by the GeneAll®Exgene™ Plant SV mini kit using the enriched samples (including both genomic DNA and cp DNA). The extracted DNA integrity was evaluated by electrophoresis on a 0.8% agarose gel for 45 minutes at 120 V, and the DNA concentration was measured by Nanodrop 2000 (Thermo, Waltham, MA, USA) and Qubit 2.0 devices to ensure quality for library preparation and sequencing on the Pacbio system according to the manufacturer's instructions.

The total DNA was fragmented and the DNA damage from fragmentation, as well as the 5′/3′ ends, underwent repair using the SMRTbell Damage Repair Kit SPv3 (#100-992-200, Pacific Biosciences, Menlo Park, CA, USA) before being attached to PacBio adapters. Products without adapters are rejected by the Exo III and Exo VII enzymes. The SMRTbell library was cleaned with Ampure PB beads (Beckman Coulter, Brea, CA, USA) and checked for length and concentration using the Bioanalyzer 2100. Subsequently, it was cleaned and sized using BluePippin (SageScience, Beverly, MA, USA) with a gel concentration of 0.75% to filter out library DNA fragments above 20 kb in length. The library was lastly checked for size and fragmentation with the Bioanalyzer 2100 before loading to the SMRT Cell (#101-008-000, PacBio).

The prepared library was loaded on one chip and sequenced on a PacBio SEQUEL system at the National Key Laboratory for Gene Technology, Institution of Biotechnology (Hanoi, Vietnam). SMRTbell library was attached with polymerase and purified using the Sequel Binding Internal Ctrl Kit 2.0 (#101-400-900, PacBio) and the SMRTbell Clean Up Column v2 Kit-Dif (101-184-100, PacBio) according to the procedure generated by the Sample Setup software included in the SMRTLink portal version 5.1.

2.3. Genome Assembly and Annotation

Total DNA was sequenced using the PacBio platform. Sequences derived from the cp genome were identified through the pbmm2 program using the cp genome of the reference *Ophiorrhiza* species (accession number: NC_057496.1) obtained from the Genbank database [5]. Then, the Hierarchical Genome Assembly Process version 4 (HGAP4) software was used to assemble the cp genome [6]. Protein-coding genes and RNA were annotated by the GeSeq webtool [7], while tRNAscan-SE software version 2.0 was applied to verify the tRNA genes [8]. The OrganellarGenomeDRAW (OGDRAW) web-tool was selected to generate the circular gene map [9]. Repeat elements were identified using two approaches. The web-based MISA finder was used for detecting microsatellites in nucleotide sequences, with the following parameters: 10 repeats for mono-, 5 for di-, 4 for tri-, and 3 for tetra-, penta-, and hexa-nucleotide SSRs [10]. Size comparison of the SSRs among the SSRs of each type was used to count polymorphic SSRs. The size and pattern of repeats in the cp genome were identified using the REPuter with the following set of parameters: minimum repeat size 20 bp, hamming distance 3 kb, and 90% or more sequence similarity [11].

2.4. Genome Comparison and Phylogenetic Identification

For cp genome comparison, we collected available cp genomes of *Ophiorrhiza* species (*O. pumila*—NC_057496.1 and *O. densa*—NC_058252.1) from the GenBank database (https://www.ncbi.nlm.nih.gov/genbank/, accessed on 15 November 2022). The overall genome structure, gene content, genome size, and number of repeats across the genomes were compared. The entire cp genome sequences of the *Ophiorrhiza* species were aligned through MAFFT software with default parameters and visualized in the mVISTA webtool with the LAGAN mode [12]. We used the annotated cp genome of the project as the reference genome in the mVISTA diagram. Subsequently, Irscope was used to visualize and compare the contiguous region between the large and small single-copy, along with the inverted repeat regions of the genomes. We also examined codon usage bias and sequence divergence via computational nucleotide diversity (Pi) analysis among cp genomes in DnaSP software version 6.12.03 [13]. For the sequence divergence analysis, we applied a window size of 600 bp with a step size of 200 bp.

A concatenation of the *rps16* gene and *trnL-trnF* intergenic spacer sequences from the *Ophiorrhiza* species and two *Xanthophytum* species of the Rubiaceae family from the Genbank database was used to identify the phylogenetic relationships of the studied *O. Baviensis* Drake. The nucleotide sequences were aligned with MAFFT software with default parameters [14] before the maximum likelihood (GTR+CAT model) phylogenetic tree was constructed using FastTree [15] with a 1000 bootstrap and visualized by FigTree software version 1.4.4 (http://tree.bio.ed.ac.uk/software/figtree/, accessed on 1 July 2021).

3. Results

3.1. Chloroplast Genome Assembly and Annotation

Using the PacBio SEQUEL I system, 28,402,467,862 bp of raw sequence data were generated with a mean read length of 1938 bp, an N50 contig size of 2412 bp, and approximately 9% of the raw reads belonging to the *O. baviensis* cp genome with 158 × coverage. The resequencing assembly resulted in a circular cp genome size of 154,770 bp (Figure 2), and the percentage of GC content was 37.6%. As reported in most angiosperm cp genomes, the assembled *O. baviensis* Drake plastome demonstrated the typical quadripartite structure consisting of four regions, LSC (84,626 bp), SSC (18,574 bp), and a pair of inverted repeats (IRs 25,685 bp).

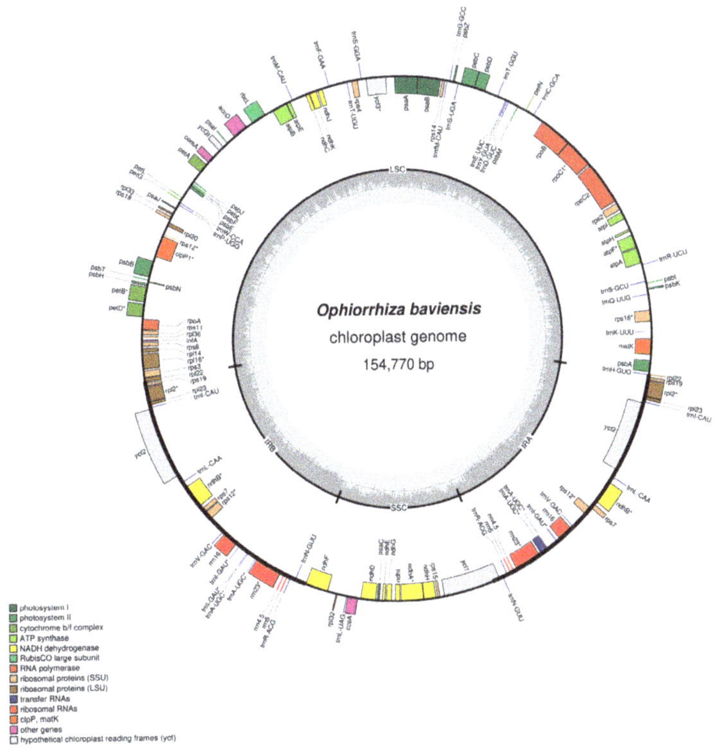

Figure 2. Chloroplast genome map of *O. baviensis* Drake in Vietnam. Genes shown inside the circle are transcribed clockwise, whereas genes outside are transcribed counterclockwise. The light gray inner circle shows the AT content, the dark gray corresponds to the GC content.

In addition, the annotation results from GeSeq and tRNAscan-SE revealed that the *O. baviensis* Drake cp genome possessed a total of 128 genes, of which there were 87 protein-coding genes, 33 tRNA genes, and 8 rRNA genes (16S, 23S, 5S, and 4.5S) (Table 1). The annotated gene models were assigned into three major groups based on their functions. Regarding the photosynthesis-related gene category, there were 44 genes encoding the subunits of ATP synthase, cytochrome complex, photosystem I and II, and putative NADPH dehydrogenase, along with the large subunit of Rubisco related to the photosynthetic electron transport chain. The other 76 genes were functionally characterized in the transcription and translation processes. The majority were tRNA genes, and the others were rRNA genes and genes encoding DNA-dependent RNA polymerase, the subunits of the ribosome, and ribosome proteins. The remaining nine genes were classified in the category of other genes, consisting of five genes with reported functions in RNA processing (*matK*), c-type cytochrome synthesis (*ccsA*), fatty acid synthesis (*accD*), carbon metabolism (*cemA*), and proteolysis (*clpP*). In addition, four genes encoding the conserved reading frames (*ycf1, ycf2, and ycf3*) were also annotated in the cp genome.

Table 1. Summary of the chloroplast genome of *O. baviensis* Drake species.

Genome Features	*O. baviensis* Drake
Genome size (bp)	154,770 bp
LSC size (bp)	84,826
SSC size (bp)	18,574
IR size (bp)	25,685
GC content (%)	37.6
No. of genes	128
No. of PCGs	87
No. of tRNA	33
No. of rRNA	8

Otherwise, each IR region of the *O. baviensis* cp genome was annotated to comprise 18 genes (all 4 rRNA genes, 7 tRNA genes, 1 NADH-dehydrogenase protein-coding gene, 4 ribosomal protein-coding genes, and 2 other genes). There were 17 cp genes that harbored introns, among which 15 genes (*atpF*, *rpl2* (×2), *rpl16*, *ndhA*, *ndhB* (×2), *rpoC1*, *rps12*, *rps16*, *trnA-UGC* (×2), *trnG-GCC*, and *trnI-GAU* (×2)) contained a single intron, while two genes (*ycf3*, *clpP*) had double introns (Table 2).

Table 2. Gene composition of *O. baviensis* Drake chloroplast genome.

Category of Genes	Group of Genes	Name of Genes
Photosynthesis	Subunits of ATP synthase	*atpA, atpB, atpE, atpFa, atpH, atpI*
	Subunits of NADH-dehydrogenase	*ndhAa, ndhB (×2)a, ndhC, ndhD, ndhE, ndhF, ndhG, ndhH, ndhI, ndhJ, ndhK*
	Subunits of cytochrome b/f complex	*petL, petB, petG, petA, petD, petN*
	Subunits of photosystem I	*psaJ, psaC, psaA, psaI, psaB*
	Subunits of photosystem II	*psbA, psbB, psbC, psbD, psbE, psbF, psbH, psbJ, psbK, psbL, psbM, psbN, psbT, psbZ*
	Subunit of rubisco	*rbcL*
Transcription and translation	Large subunit of ribosome	*rpl14, rpl16a, rpl2 (×2)a, rpl20, rpl22, rpl23 (×2), rpl32, rpl33, rpl36*
	DNA-dependent RNA polymerase	*rpoB, rpoA, rpoC1a, rpoC2*
	Small subunit of ribosomal proteins	*rps11, rps12(×2)a, rps14, rps15, rps16a, rps18, rps19 (×2), rps2, rps3, rps4, rps7 (×2), rps8*
	rRNA genes	*rrn23S (×2), rrn16S (×2), rrn5S (×2), rrn4.5S (×2)*
	tRNA genes	*trnA-UGC (×2)a, trnC-GCA, trnD-GUC, trnE-UUC, trnF-GAA, trnG-GCCa, trnH-GUG, trnI-GAU (×2)a, trnL-CAA (×2), trnL-UAG, trnN-GUU (×2), trnP-UGG, trnQ-UUG, trnR-ACG (×2), trnR-UCU, trnS-GCU, trnS-GGA, trnS-UGA, trnT-GGU, trnT-UGU, trnV-GAC (×2), trnW-CCA, trnY-GUA*
Other genes	Translational initiation factor	*infA*
	Subunit of acetyl-CoA-carboxylase (fatty acid synthesis)	*accD*
	c-type cytochrome synthesis gene	*ccsA*
	Envelope membrane protein (carbon metabolism)	*cemA*
	Protease	*clpPb*
	Maturase (RNA processing)	*matK*
	Conserved open reading frames	*ycf1, ycf2 (×2), ycf3b*

Genes marked with the sign are the gene with a single (a) or double (b) introns and duplicated genes (×2).

3.2. Repeat Sequences and Codon Analysis

A total of 59 simple sequence repeats (SSRs) were investigated in the *O. baviensis* Drake chloroplast genome via the MISA web-tool. Almost all of the screened repeats were mono repeats (composed of A/T and C) with the size ranging from 10 to 16 bp (Figure 3A). Two di-, five tri-, seven tetra-, and three penta-nucleotide SSRs were found in the *O. baviensis* Drake plastid. A total of 53 SSRs were classified as simple based SSRs and the six remaining SSRs presented in a compound formation. The majority of SSR types were discovered in the LSC, while the IR regions included the smallest number of SSRs (Figure 3B).

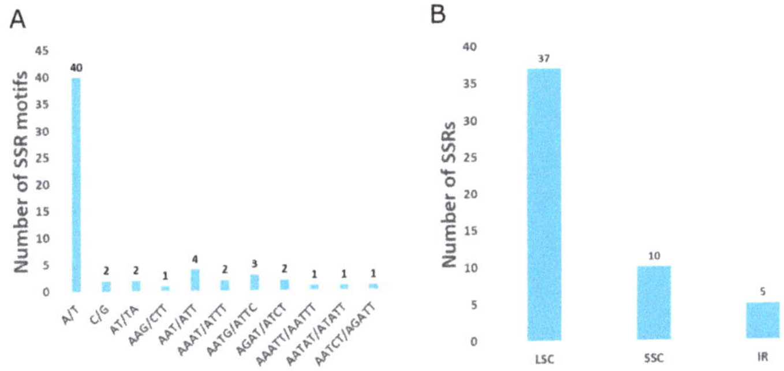

Figure 3. Analysis of single sequence repeats (SSRs) of the *O. baviensis* Drake chloroplast genome. (**A**) Number of identified SSR sequence motifs; (**B**) Frequency of repeat types in LSC, SSC, and IR regions.

The cp genome of *O. baviensis* Drake was annotated to possess 49 long repeats including 9 palindromic repeats, along with 12 forward and 22 reverse repeats. There was only one complement repeat (Figure 4). The unit size of the detected repeats ranged from 20 to 58 bp, while a majority of the repeat size (67%) was shorter than 30 bp.

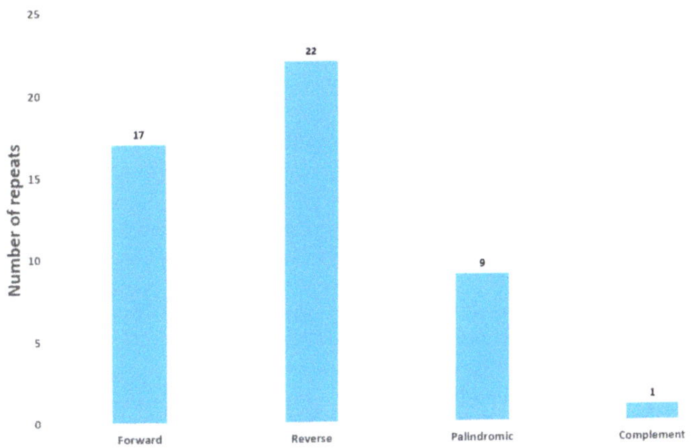

Figure 4. Repeat analysis of *O. baviensis* Drake chloroplast genome.

The codon usage frequency of 64 protein-coding genes was evaluated for three cp genomes: *O. baviensis* Drake and two other available *Ophiorrhiza* species. The total number of codons found in the coding regions was 51,517, while the A- and U-ending were found more frequently than the G/C-ending (Table 3). Leucine was the most prevalent among the 20 amino acids with a percentage of 10.46% (5068 codons), followed by serine with 9.95% (4817 codons). Meanwhile, the rarest was tryptophan with a total of 681 codons accounting for approximately 1.4%. A total of 30 codons exhibited the codon usage bias (RSCU < 1), while 32 codons were observed to be more frequent than the expected usage at equilibrium (RSCU > 1) (Table 3). The usage frequency for the start codons AUG and UGG (methionine and tryptophan) exhibited no bias (RSCU = 1).

Table 3. Relative synonymous codon usage (RSCU) for protein-coding genes in *O. baviensis*.

Codon	AA	Frequency	RCSU	Codon	AA	Frequency	RCSU	Codon	AA	Frequency	RCSU
UAA	*	1259	1.22	AUC	I	1205	0.80	CGG	R	420	0.75
UAG	*	825	0.80	AUA	I	1471	0.98	AGA	R	1093	1.94
UGA	*	1004	0.98	AAA	K	2050	1.35	AGG	R	627	1.12
GCU	A	446	1.23	AAG	K	982	0.65	UCU	S	1113	1.39
GCC	A	351	0.97	UUA	L	1040	1.23	UCC	S	982	1.22
GCA	A	401	1.11	UUG	L	1095	1.30	UCA	S	824	1.03
GCG	A	250	0.69	CUU	L	1063	1.26	UCG	S	622	0.77
UGU	C	679	1.20	CUC	L	653	0.77	AGU	S	747	0.93
UGC	C	449	0.80	CUA	L	737	0.87	AGC	S	529	0.66
GAU	D	1012	1.42	CUG	L	480	0.57	ACU	T	668	1.13
GAC	D	413	0.58	AUG	M	856	1.00	ACC	T	651	1.10
GAA	E	1337	1.43	AAU	N	1779	1.38	ACA	T	647	1.09
GAG	E	537	0.57	AAC	N	800	0.62	ACG	T	406	0.68
UUU	F	2212	1.20	CCU	P	611	1.03	GUU	V	784	1.38
UUC	F	1481	0.80	CCC	P	618	1.04	GUC	V	411	0.72
GGU	G	540	0.96	CCA	P	726	1.23	GUA	V	682	1.20
GGC	G	383	0.68	CCG	P	414	0.70	GUG	V	402	0.71
GGA	G	747	1.33	CAA	Q	987	1.40	UGG	W	681	1.00
GGG	G	577	1.03	CAG	Q	420	0.60	UAU	Y	1345	1.36
CAU	H	880	1.36	CGU	R	376	0.67	UAC	Y	637	0.64
CAC	H	414	0.64	CGC	R	280	0.50				
AUU	I	1830	1.22	CGA	R	576	1.02				

* Stop codon.

3.3. Chloroplast Genome Comparison

To characterize genomic divergence, the percentage of sequence identity was evaluated for three *Ophiorrhiza* species with the functional annotation of *O. baviensis* Drake as a reference. The comparison using the mVISTA program revealed that the gene organization among the three species was highly similar and there were several regions of sequence variation (Figure 5). The results exhibited a higher frequency of divergence in the LSC and SSC regions than in the IR regions. Moreover, the coding regions of the three cp genomes were observed to be more conserved, whereas a majority of the detected variations were screened in the conserved non-coding sequences (CNS). Among the protein-coding gene sequences, the highly disparate genes consisted of *matK, rpoC2, rpoB, clpP, rpl16, ndhF, ndhA,* and *ycf1*.

The sliding window analysis indicated that the average polymorphism information (Pi) values of the LSC (Pi = 0.005635) and SSC (Pi = 0.007472) regions were greater than that of the IR (Pi = 0.001285) regions, which showed that most of the variations were located in the LSC and SSC regions (Figure 6). Of the three *Ophiorrhiza* species, the average value of nucleotide diversity (Pi) was 0.00441.

3.4. IR Contraction and Expansion in the Chloroplast Genome

The IR/LSC and IR/SSC boundaries of three *Ophiorrhiza* cp genomes were compared using the IRscope program. Overall, the results indicated that the region size, gene organization, and gene content showed a high similarity among the cp genome of *O. baviensis* and *O. pumila* (Figure 7). On the other hand, the *O. densa* cp genome showed several variants with the two abovementioned *Ophiorrhiza* species. The size of IR regions ranged from 25,684 bp (*O. baviensis* Drake) to 26,066 bp (*O. pumila*), and the size of IR of *O. densa* was 25,701 bp. The *rpl22* gene was located within the LSC region with a 102 bp overlap with the IRb for *O. baviensis* and *O. pumila*, while *O. densa* showed a 347 bp overlap of the *rps3* gene in this boundary. Apart from *O. densa*, the *ndhF* gene was detected on the boundary of the SSC and IRb region. The border across IRa and SSC was found in the *ycf1* gene with 1438, 1316, and 730 bp tail sections of the gene placed in the IRa of *O. densa*, *O. pumila*, and *O. baviensis*, respectively (Figure 7). The IRa and LSC boundary showed the presence of the *trnH* gene in the forward strand of all three species and the *rpl22* gene in the reverse strand of *O. baviensis* and *O. densa*. The results of the IR analysis indicated extensive contraction and expansion of the IR regions in the three species.

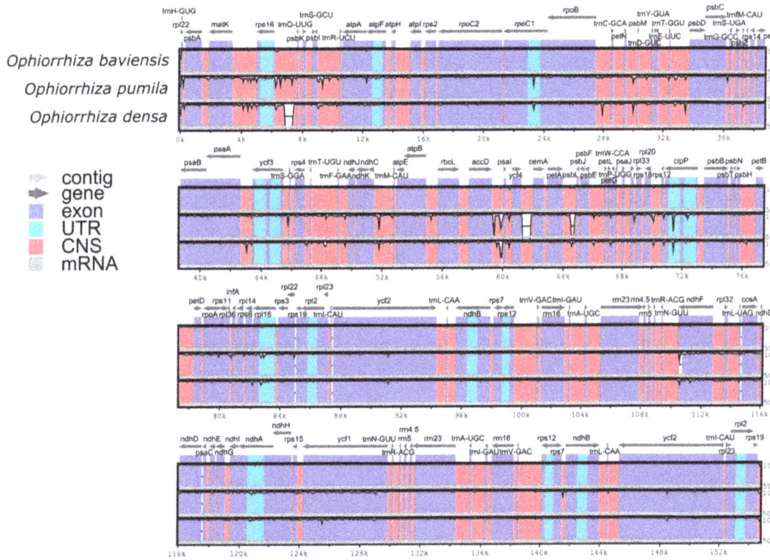

Figure 5. Complete chloroplast genome alignments of the three *Ophiorrhiza* species. The horizontal axis indicates the coordinates within the chloroplast genome. The vertical scale indicates the percent identity within 50–100%. Annotated genes are displayed along the top.

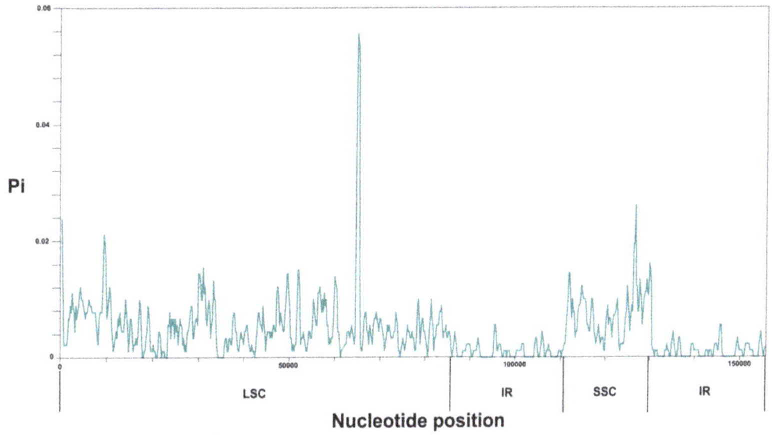

Figure 6. Nucleotide diversity (Pi) values among the three *Ophiorrhiza* species. X-axis: the position in the genome; Y-axis: Pi value. Pi, polymorphism information.

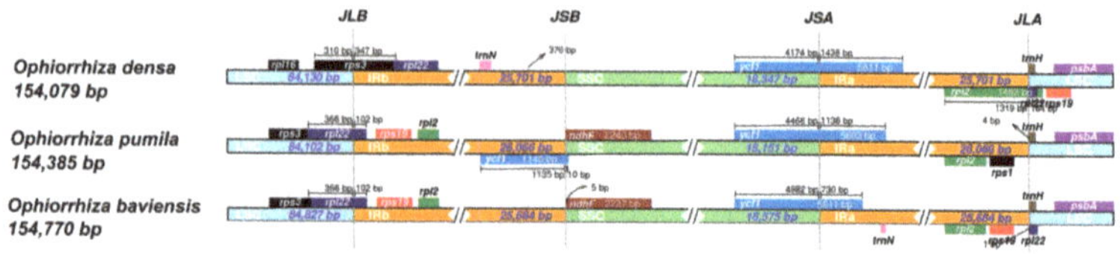

Figure 7. Comparison of LSC, IR, and SSC junction positions among the three *Ophiorrhiza* chloroplast genomes. JLB (junction IRb/LSC), JSB (junction IRb/SSC), JSA (junction IRa/SSC), JLA (junction IRa/LSC).

3.5. Phylogenetic Inference

The number of available sequences of *O. baviensis* on the Genbank databases, especially belonging to the cp genome, is limited (only the *rps16* gene and the *trnL-trnF* intergenic spacer). Therefore, we extracted these sequences from the assembled cp genome and used them to access the phylogenetic relationship of the studied *O. baviensis* at the species level. Figure 8 shows the phylogenetic resolution based on the concatenated sequence between the *rps6* gene and the *trnL-trnF* intergenic spacer with a high bootstrap value of 92% between the studied *O. baviensis* Drake and the reference *O. baviensis* voucher Averyanov & al. VH940 (AAU) (Accession number: MH626923.1). With a bootstrap value of 100%, all eight *Ophiorrhiza* species were grouped separately from the two *Xanthophytum* species as an outgroup. In the case of barcoding among the Rubiaceae family, the combined *rps16-trnL-F* intergenic spacer sequences provided a high capacity for phylogenetic resolution.

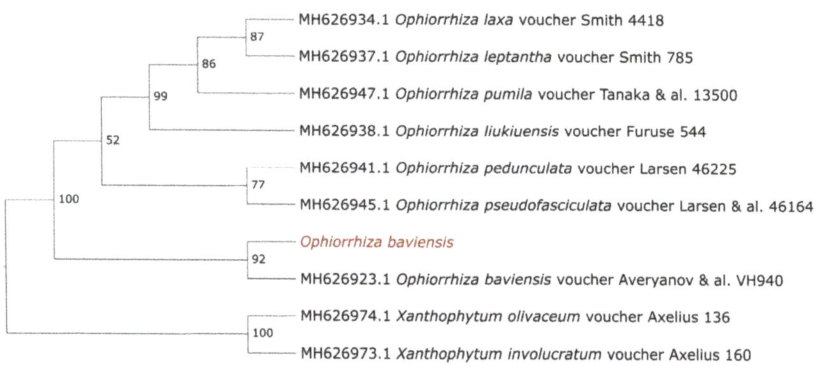

Figure 8. The maximum likelihood phylogenetic trees based on the concatenated sequences between the *rps16* genes and the *trnL-trnF* intergenic spacer. Numbers on the branches indicate bootstrap percentage after 1000 replications in constructing the tree. The species investigated in this study are colored in red.

4. Discussion

Rubiaceae is a family of flowering plants containing 620 genera with approximately 13,500 species over the world, which makes it the fourth-largest angiosperm family. Over 300 cp genomes in the Rubiaceae family have been published in the Genbank database until now, only three of which belong to *Ophiorrhiza*. The genus *Ophiorrhiza* consists of about 200–300 species mainly distributed in humid tropical forests from East India to the Western Pacific, and from South China to Northern Australia [16,17]. Bioactive compounds from this family, such as quinine, emetine, caffeine, and camptothecin are of major pharmaceutical importance; thus, many species in the genus *Ophiorrhiza* are of interest [18]. In the present study, we sequenced and annotated the entire cp genome of a Vietnamese medicinal plant.

Angiosperm cp genomes have a highly conserved gene order and gene content with 127–134 genes found across the chloroplast genomes. The analyzed *O. baviensis* cp genomes demonstrated the typical quadripartite structure and showed the expected size range (~154 kb) for *Ophiorrhiza* species and the conserved gene contents. Our gene annotation results were similar to the genetic properties of angiosperm chloroplast genomes. The number of genes present in the cp genome from *O. baviensis* was 128, of which, 17 genes included one or two introns. In addition, the deletions of the *petB* and *petD* introns were observed in the studied *O. baviensis* cp genome, which also occurred in *O. pumila* species. Introns play an important role in gene expression regulation. Recent research has revealed gene or intron loss in chloroplast genomes [19–21], among which *petB* and *petD* intron loss was reported in many angiosperms [22].

In addition to two copies of IR regions, 49 small repeats were found to be located within coding and non-coding regions of the *O. baviensis* plastome. The cp genome includes numerous dispersed repeats, which are supposed to be biomarkers of mutational hotspots [23,24]. The repeat number is similar to the data of other species belonging to the Rubiaceae family [25,26]. Repeats are closely related to angiosperm plastome reconstruction and can be assumed as recognition signals of recombination because of their potential to generate secondary structures. In this study, the similar number of repeats in comparison with previous estimates might not demonstrate inter- and intra-specific plastome recombination. In higher plants, SSRs are identified as crucial molecular markers for the investigations of population variation due to their distinct uniparental inheritance, and they are commonly used to evaluate genetic diversity and population structure in evolutionary studies [27–29]. In total, 59 SSRs were screened in the *O. baviensis* cp genomes with strong A/T bias. These repeats play a significant role for generating genetic markers in *O. baviensis* species, which may be applied to assess the variation at the intraspecific level in phylogenetic and ecological studies.

Comparative analyses on *O. baviensis* and two available *Ophiorrhiza* cp genomes were implemented to explore the plastome structure in the taxa. The cp genome size of the three *Ophiorrhiza* ranged from 154,079 bp (*O. densa*) to 154,770 bp (*O. baviensis*), the figure for *O. pumila* was 154,385 bp. Gene organization and codon usage patterns exhibited high conservation, which could be applicable for further population genetics and phylogenetic studies. Moreover, the three *Ophiorrhiza* cp genomes were less variable in their coding regions than in their noncoding regions, which is consistent with the common pattern in most angiosperms [30] (Figure 5). Codon usage preference is closely related to gene expression and can affect the level of mRNA and proteins in the genome [31–33]. The most prevalent amino acid in the *Ophiorrhiza* was leucine (Leu), which has also been commonly detected in the other angiosperms. The high similarity in codon usage may indicate that these *Ophiorrhiza* species underwent similar environmental pressure through their evolutionary processes. The *Ophiorrhiza* cp genomes indicated that the RSCU values of most codons ending in A/U were greater than 1, which may be caused by a bias toward a high A/T ratio in composition. Additionally, we investigated that the partial sequences of the *ycf1* gene along with five intergenic spacers (IGSs), including *petA-psbJ, trnH-GUG-psbA, trnS-GCU-trnR-UCU, psbM-trnD-GUC,* and *ndhC-trnM-CAU*, had relatively high nucleotide diversity values (Pi > 0.015). These divergence regions could be studied to provide molecular markers for DNA barcoding and phylogenetic research in *Ophiorrhiza*.

While the three plastomes showed an approximate similarity in genome size, the size of the structural regions exhibited significant differences in a detailed comparison of junction sites (Figure 7). The regions of the cp genome frequently undergo length variations during the evolution of terrestrial plants, which leads to the emergence of many boundary features [34]. The expansion and contraction of the boundaries between IRs and the single-copy (SC) regions are the primary causes of the size change in cp genomes and influence the evolution rate of cp genomes [35,36]. Our finding revealed that the boundary-gene set of the *Ophiorrhiza* species included *rpl22, rps19, ndhF, ycf1,* and *trnH*. Several notable gene rearrangements were observed in the *O. densa* plastome; these were the presence of the *rps3* gene at the JLB instead of the *rpl22* gene, the expansion of the *rpl2* gene to the JLA, and the absence of the *rps19* gene in the IR regions. Expansion and contraction, as well as variation, at the junction of the SC–IR regions were characterized, suggesting that gene organization in the IR regions can report the distance between species to some extent.

The majority of taxonomic levels of plant phylogenetic connections have been demonstrated using complete chloroplast genomes and protein-coding genes [37,38]. The current study provides the phylogeny of the *Ophiorrhiza* genus based on the combined *rps16-trnL-F* intergenic spacer sequences. The previous study of Razafimandimbison and Rydin demonstrated that *O. baviensis* had been resolved as a sister relationship with *O. japonica* and *O. hayatana* [39]. In terms of species classification, the phylogenetic tree based on the concatenation of the *rps16* gene and the *trnL-F* intergenic spacer indicated the close relationship between the studied plant and the *O. baviensis* voucher Averyanov & al. VH940 (AAU) with a high bootstrap value of 92%. This approach showed effectiveness in the classification of the lower taxonomic levels among the Rubiaceae family. Further, the combination of these barcodes can lead to better species classification compared to the results from a single gene [39]. This study will help to clarify the evolutionary position of *O. baviensis* in the *Ophiorrhiza* genus, as well as offering applicable cp genome data for further research into the genesis and diversification of the Rubiaceae family. Overall, our phylogenetic investigation of the *O. baviensis* cp genome was successful in discovering the intrageneric connections within the *Ophiorrhiza* genus.

5. Conclusions

In this study, the first complete chloroplast genome of an *O. baviensis* Drake species from Vietnam was characterized and compared with two other published *Ophiorrhiza* plastomes. The assembly resulted in a whole cp genome of 154,770 bp in size. According to the comparative result, the structure and gene content of three *Ophiorrhiza* cp genomes exhibited a high similarity, and the SC-IR junction analysis revealed the expansion and contraction of IR regions. Additionally, the phylogenetic tree indicated close relationships between our novel cp genome sequence and other *Ophiorrhiza* species. This study provides the potential to employ cp genomes for enhancing species classification and genetic source conservation during further study of the Rubiaceae family.

Author Contributions: Conceptualization, H.H.C.; sampling, T.H.T.; methodology, T.H.T., H.H. and T.D.L.; software, M.H.P.; validation, T.L.L.; formal analysis, M.H.P.; data curation, T.L.L.; writing—original draft preparation, M.H.P.; writing—review and editing, H.H.C.; visualization, M.H.P.; supervision, H.H.C. and T.H.T.; project administration, H.H.C. All authors have read and agreed to the published version of the manuscript.

Funding: This work was supported by the project of the Vietnam Academy of Science and Technology (VAST): "Sequencing and characterizing the chloroplast genome of an *Ophiorrhiza baviensis* species by PacBio SMRT next-generation sequencing technology for genetic classification and conservation" (project no. CSCL08.02/22-22).

Institutional Review Board Statement: Not applicable.

Informed Consent Statement: Not applicable.

Data Availability Statement: This complete chloroplast genome of *O. Baviensis* Drake has been deposited at DDBJ/ENA/GenBank under the accession number OP902221.

Acknowledgments: We thank Khang Sinh Nguyen, researcher at the Institute of Ecology and Biological Resources, Vietnam Academy of Science and Technology for authenticating the taxonomic identification of the plant samples.

Conflicts of Interest: The authors declare no conflict of interest.

References

1. Neuhaus, H.E.; Emes, M.J. Nonphotosynthetic Metabolism in Plastids. *Annu. Rev. Plant Biol.* **2000**, *51*, 111. [CrossRef]
2. Bendich, A.J. Circular Chloroplast Chromosomes: The Grand Illusion. *Plant Cell* **2004**, *16*, 1661–1666. [CrossRef]
3. Lei, W.; Liu, W.-J.; Nguyen, K.S. Revision of Three Taxa of Ophiorrhiza (Rubiaceae) from China. *Phytotaxa* **2019**, *387*, 129–139. [CrossRef]

4. Rhoads, A.; Au, K.F. PacBio Sequencing and Its Applications. *Genom. Proteom. Bioinform.* **2015**, *13*, 278–289. [CrossRef] [PubMed]
5. PacificBiosciences. Pbmm2: A Minimap2 Frontend for PacBio Native Data Formats. Available online: https://github.com/PacificBiosciences/pbmm2 (accessed on 10 January 2021).
6. Chin, C.-S.; Alexander, D.H.; Marks, P.; Klammer, A.A.; Drake, J.; Heiner, C.; Clum, A.; Copeland, A.; Huddleston, J.; Eichler, E.E.; et al. Nonhybrid, Finished Microbial Genome Assemblies from Long-Read SMRT Sequencing Data. *Nat. Methods* **2013**, *10*, 563–569. [CrossRef]
7. Tillich, M.; Lehwark, P.; Pellizzer, T.; Ulbricht-Jones, E.S.; Fischer, A.; Bock, R.; Greiner, S. GeSeq—Versatile and Accurate Annotation of Organelle Genomes. *Nucleic Acids Res.* **2017**, *45*, W6–W11. [CrossRef] [PubMed]
8. Chan, P.P.; Lin, B.Y.; Mak, A.J.; Lowe, T.M. TRNAscan-SE 2.0: Improved Detection and Functional Classification of Transfer RNA Genes. *Nucleic Acids Res.* **2021**, *49*, 9077–9096. [CrossRef] [PubMed]
9. Lohse, M.; Drechsel, O.; Bock, R. OrganellarGenomeDRAW (OGDRAW): A Tool for the Easy Generation of High-Quality Custom Graphical Maps of Plastid and Mitochondrial Genomes. *Curr. Genet.* **2007**, *52*, 267–274. [CrossRef] [PubMed]
10. Beier, S.; Thiel, T.; Münch, T.; Scholz, U.; Mascher, M. MISA-Web: A Web Server for Microsatellite Prediction. *Bioinformatics* **2017**, *33*, 2583–2585. [CrossRef] [PubMed]
11. Kurtz, S.; Schleiermacher, C. REPuter: Fast Computation of Maximal Repeats in Complete Genomes. *Bioinformatics* **1999**, *15*, 426–427. [CrossRef]
12. Frazer, K.A.; Pachter, L.; Poliakov, A.; Rubin, E.M.; Dubchak, I. VISTA: Computational Tools for Comparative Genomics. *Nucleic Acids Res.* **2004**, *32*, W273–W279. [CrossRef]
13. Rozas, J.; Ferrer-Mata, A.; Sánchez-DelBarrio, J.C.; Guirao-Rico, S.; Librado, P.; Ramos-Onsins, S.E.; Sánchez-Gracia, A. DnaSP 6: DNA Sequence Polymorphism Analysis of Large Data Sets. *Mol. Biol. Evol.* **2017**, *34*, 3299–3302. [CrossRef] [PubMed]
14. Katoh, K.; Rozewicki, J.; Yamada, K.D. MAFFT Online Service: Multiple Sequence Alignment, Interactive Sequence Choice and Visualization. *Brief. Bioinform.* **2019**, *20*, 1160–1166. [CrossRef] [PubMed]
15. Price, M.; Dehal, P.; Arkin, A. FastTree 2—Approximately Maximum-Likelihood Trees for Large Alignments. *PLoS ONE* **2010**, *5*, e9490. [CrossRef]
16. Chen, T.; Taylor, C. Ophiorrhiza. In *Flora of China*; Press, S., Ed.; Beijing & Missouri Botanical Garden Press: St. Louis, MO, USA, 2011; Volume 19, pp. 258–282.
17. Lei, W.; Tan, Y.; Hareesh, V.S.; Liu, Q. Ophiorrhiza Macrocarpa (Rubiaceae), a New Viviparous Species from Yunnan, Southwestern China. *Nord. J. Bot.* **2018**, *36*, njb-01637. [CrossRef]
18. Hamzah, A.S. Isolation, Characterization and Biological Activities of Chemical Constituents of Ophiorrhiza and Hedyotis Species. Ph.D. Dissertation, Universiti Pertanian Malaysia, Serdang, Malaysia, 1994.
19. GAO, L.; SU, Y.-J.; WANG, T. Plastid Genome Sequencing, Comparative Genomics, and Phylogenomics: Current Status and Prospects. *J. Syst. Evol.* **2010**, *48*, 77–93. [CrossRef]
20. Frailey, D.C.; Chaluvadi, S.R.; Vaughn, J.N.; Coatney, C.G.; Bennetzen, J.L. Gene Loss and Genome Rearrangement in the Plastids of Five Hemiparasites in the Family Orobanchaceae. *BMC Plant Biol.* **2018**, *18*, 30. [CrossRef]
21. Oyebanji, O.; Zhang, R.; Chen, S.-Y.; Yi, T.-S. New Insights Into the Plastome Evolution of the Millettioid/Phaseoloid Clade (Papilionoideae, Leguminosae). *Front. Plant Sci.* **2020**, *11*, 151. [CrossRef]
22. Li, X.; Li, Y.; Sylvester, S.P.; Zang, M.; El-Kassaby, Y.A.; Fang, Y. Evolutionary Patterns of Nucleotide Substitution Rates in Plastid Genomes of Quercus. *Ecol. Evol.* **2021**, *11*, 13401–13414. [CrossRef]
23. Abdullah; Mehmood, F.; Shahzadi, I.; Ali, Z.; Islam, M.; Naeem, M.; Mirza, B.; Lockhart, P.J.; Ahmed, I.; Waheed, M.T. Correlations among Oligonucleotide Repeats, Nucleotide Substitutions, and Insertion–Deletion Mutations in Chloroplast Genomes of Plant Family Malvaceae. *J. Syst. Evol.* **2021**, *59*, 388–402. [CrossRef]
24. Liu, Q.; Li, X.; Li, M.; Xu, W.; Schwarzacher, T.; Heslop-Harrison, J.S. Comparative Chloroplast Genome Analyses of Avena: Insights into Evolutionary Dynamics and Phylogeny. *BMC Plant Biol.* **2020**, *20*, 406. [CrossRef] [PubMed]
25. Ly, S.N.; Garavito, A.; De Block, P.; Asselman, P.; Guyeux, C.; Charr, J.-C.; Janssens, S.; Mouly, A.; Hamon, P.; Guyot, R. Chloroplast Genomes of Rubiaceae: Comparative Genomics and Molecular Phylogeny in Subfamily Ixoroideae. *PLoS ONE* **2020**, *15*, e0232295. [CrossRef] [PubMed]
26. Amenu, S.G.; Wei, N.; Wu, L.; Oyebanji, O.; Hu, G.; Zhou, Y.; Wang, Q. Phylogenomic and Comparative Analyses of Coffeeae Alliance (Rubiaceae): Deep Insights into Phylogenetic Relationships and Plastome Evolution. *BMC Plant Biol.* **2022**, *22*, 88. [CrossRef] [PubMed]
27. Varshney, R.K.; Sigmund, R.; Börner, A.; Korzun, V.; Stein, N.; Sorrells, M.E.; Langridge, P.; Graner, A. Interspecific Transferability and Comparative Mapping of Barley EST-SSR Markers in Wheat, Rye and Rice. *Plant Sci.* **2005**, *168*, 195–202. [CrossRef]
28. Dong, W.; Liu, H.; Xu, C.; Zuo, Y.; Chen, Z.; Zhou, S. A Chloroplast Genomic Strategy for Designing Taxon Specific DNA Mini-Barcodes: A Case Study on Ginsengs. *BMC Genet.* **2014**, *15*, 138. [CrossRef]
29. Provan, J.; Powell, W.; Hollingsworth, P.M. Chloroplast Microsatellites: New Tools for Studies in Plant Ecology and Evolution. *Trends Ecol. Evol.* **2001**, *16*, 142–147. [CrossRef] [PubMed]
30. Yang, C.-H.; Liu, X.; Cui, Y.-X.; Nie, L.-P.; Lin, Y.-L.; Wei, X.-P.; Wang, Y.; Yao, H. Molecular Structure and Phylogenetic Analyses of the Complete Chloroplast Genomes of Three Original Species of Pyrrosiae Folium. *Chin. J. Nat. Med.* **2020**, *18*, 573–581. [CrossRef]
31. Zhou, M.; Guo, J.; Cha, J.; Chae, M.; Chen, S.; Barral, J.M.; Sachs, M.S.; Liu, Y. Non-Optimal Codon Usage Affects Expression, Structure and Function of Clock Protein FRQ. *Nature* **2013**, *495*, 111–115. [CrossRef]

32. Somaratne, Y.; Guan, D.-L.; Wang, W.-Q.; Zhao, L.; Xu, S.-Q. The Complete Chloroplast Genomes of Two Lespedeza Species: Insights into Codon Usage Bias, RNA Editing Sites, and Phylogenetic Relationships in Desmodieae (Fabaceae: Papilionoideae). *Plants* **2020**, *9*, 51. [CrossRef]
33. Lyu, X.; Liu, Y. Nonoptimal Codon Usage Is Critical for Protein Structure and Function of the Master General Amino Acid Control Regulator CPC-1. *Mbio* **2020**, *11*, e02605-20. [CrossRef]
34. Ding, S.; Dong, X.; Yang, J.; Guo, C.; Cao, B.; Guo, Y.; Hu, G. Complete Chloroplast Genome of Clethra Fargesii Franch., an Original Sympetalous Plant from Central China: Comparative Analysis, Adaptive Evolution, and Phylogenetic Relationships. *Forests* **2021**, *12*, 441. [CrossRef]
35. Kim, K.-J.; Lee, H.-L. Complete Chloroplast Genome Sequences from Korean Ginseng (Panax Schinseng Nees) and Comparative Analysis of Sequence Evolution among 17 Vascular Plants. *DNA Res. Int. J. Rapid Publ. Rep. Genes Genomes* **2004**, *11*, 247–261. [CrossRef] [PubMed]
36. Zhang, H.; Li, C.; Miao, H.; Xiong, S. Insights from the Complete Chloroplast Genome into the Evolution of *Sesamum indicum* L. *PLoS ONE* **2013**, *8*, e80508. [CrossRef]
37. De Las Rivas, J.; Lozano, J.J.; Ortiz, A.R. Comparative Analysis of Chloroplast Genomes: Functional Annotation, Genome-Based Phylogeny, and Deduced Evolutionary Patterns. *Genome Res.* **2002**, *12*, 567–583. [CrossRef]
38. Moore, M.J.; Bell, C.D.; Soltis, P.S.; Soltis, D.E. Using Plastid Genome-Scale Data to Resolve Enigmatic Relationships among Basal Angiosperms. *Proc. Natl. Acad. Sci. USA* **2007**, *104*, 19363–19368. [CrossRef] [PubMed]
39. Razafimandimbison, S.G.; Rydin, C. Molecular-Based Assessments of Tribal and Generic Limits and Relationships in Rubiaceae (Gentianales): Polyphyly of Pomazoteae and Paraphyly of Ophiorrhizeae and Ophiorrhiza. *Taxon* **2019**, *68*, 72–91. [CrossRef]

genes

MDPI

Article

Transcriptome Analysis of CYP450 Family Members in *Fritillaria cirrhosa* D. Don and Profiling of Key *CYP450s* Related to Isosteroidal Alkaloid Biosynthesis

Rui Li [1,2], Maotao Xiao [1,2], Jian Li [2,3], Qi Zhao [1,2], Mingcheng Wang [2,4,*] and Ziwei Zhu [2,4,*]

1 College of Food and Biological Engineering, Chengdu University, Chengdu 610106, China
2 Engineering Research Center of Sichuan-Tibet Traditional Medicinal Plant, Chengdu 610106, China
3 School of Basic Medical Sciences, Chengdu University, Chengdu 610106, China
4 Institute for Advanced Study, Chengdu University, Chengdu 610106, China
* Correspondence: wangmingcheng@cdu.edu.cn (M.W.); zhuziwei@cdu.edu.cn (Z.Z.)

Abstract: *Fritillaria cirrhosa* D. Don (known as Chuan-Bei-Mu in Chinese) can synthesize isosteroidal alkaloids (ISA) with excellent medicinal value, and its bulb has become an indispensable ingredient in many patented drugs. Members of the cytochrome P450 (CYP450) gene superfamily have been shown to play essential roles in regulating steroidal alkaloids biosynthesis. However, little information is available on the P450s in *F. cirrhosa*. Here, we performed full-length transcriptome analysis and discovered 48 *CYP450* genes belonging to 10 clans, 25 families, and 46 subfamilies. By combining phylogenetic trees, gene expression, and key *F. cirrhosa* ISA content analysis, we presumably identify seven *FcCYP* candidate genes, which may be hydroxylases active at the C-22, C-23, or C-26 positions in the late stages of ISA biosynthesis. The transcript expression levels of seven *FcCYP* candidate genes were positively correlated with the accumulation of three major alkaloids in bulbs of different ages. These data suggest that the candidate genes are most likely to be associated with ISA biosynthesis. Finally, the subcellular localization prediction of FcCYPs and transient expression analysis within *Nicotiana benthamiana* showed that the FcCYPs were mainly localized in the chloroplast. This study presents a systematic analysis of the CYP450 gene family in *F. cirrhosa* and provides a foundation for further functional characterization of the *CYPs* involved in ISA biosynthesis.

Keywords: *Fritillaria cirrhosa* D. Don; cytochrome P450; isosteroidal alkaloid; full-length transcriptome

Citation: Li, R.; Xiao, M.; Li, J.; Zhao, Q.; Wang, M.; Zhu, Z. Transcriptome Analysis of CYP450 Family Members in *Fritillaria cirrhosa* D. Don and Profiling of Key *CYP450s* Related to Isosteroidal Alkaloid Biosynthesis. *Genes* **2023**, *14*, 219. https://doi.org/10.3390/genes14010219

Academic Editors: Wajid Zaman and Hakim Manghwar

Received: 27 December 2022
Revised: 10 January 2023
Accepted: 13 January 2023
Published: 14 January 2023

1. Introduction

F. cirrhosa D. Don, a member of the Liliaceae family, is a well-known medicinal plant in China. The bulb of this plant, named "Chuan-Bei-Mu" in Chinese, was first introduced in Divine Farmer's Materia Medica (Shennong Bencao Jing) [1]. It is used as a medicine and healthcare material in many traditional Chinese medicine (TCM) prescriptions [2]. It has been used to treat chronic respiratory disorders such as asthma, cough, lung cancer, and tuberculosis [3]. Modern pharmacological studies have indicated that the bulb of *F. cirrhosa* contained a variety of bioactive isosteroidal alkaloids (ISA) such as imperialine, verticinone, verticine, delavine, peimisine, etc., which are mainly responsible for the anti-asthmatic, antioxidant, antitumor or anti-inflammatory action ascribed to them [4–7]. Imperialine is a vital ingredient of these *F. cirrhosa* ISA and is frequently utilized as a pharmacopoeia reference standard for the quality assessment of "Chuan-Bei-Mu" due to its significant pharmacological properties [8].

Currently, only extremely low quantities of *F. cirrhosa* isosteroidal alkaloids (FISA), between 0.02% and 0.03% of dry weight bio-mass, can be recovered from "Chuan-Bei-Mu." [9]. The Fritillaria plants belong to a critically endangered species and are recognized as a Class II protected species in China's 'Wild Official Species under Protection of the State according to Wild Medicine Material Protection Rules' (http://www.cites.org.cn/, accessed on 7 August 2021). Accordingly, further investigation of its pharmaceutical utility is limited by supply shortage [10]. A more promising approach for obtaining structurally complex natural products is the combination of genetic and metabolic engineering [11–14]. Because of the impact of the environment, Fritillaria is represented in China by various cultivars and variations, each of which has distinct qualities [15]. A multi-species metabolomics investigation revealed that *F. cirrhosa* contains abundant steroidal alkaloids, such as imperialine, suggesting that *F. cirrhosa* D. Don might be considered a suitable plant material for research on FISA biosynthesis [14].

FISAs exhibit a distinctive steroidal architecture with five or six carbocyclic or heterocyclic rings on the C27 steroidal carbon skeleton [1]. FISAs are nitrogen-containing specialized metabolites used as chemophenetic markers in Fritillaria (liliaceae) [16]. In the first step, isosteroidal backbones are produced by the cytosolic mevalonic acid (MVA) and methylerythritol phosphate (MEP) pathways. ISA is then produced via initial cyclization of the central intermediate cycloartenol to cholesterol, followed by hydroxylation, oxidation, and transamination [14]. The majority of the research has been devoted to the cloning and control of functional genes upstream of the terpenoid biosynthesis pathway, including the 3-hydroxy-3-methylglutaryl coenzyme A reductase gene (HMGR), farnesyl diphosphate synthase (FPS), squalene synthase (SS), and cycloartenol synthase (CAS) [15]. Secondary metabolic pathways following cycloartenol generation are rarely studied, particularly the downstream CYP450 genes responsible for C-22, C-23, and C-26 hydroxylation in the steroidal alkaloid biosynthesis pathway.

Here, we used various bioinformatic tools to identify and analyze CYP450 superfamily genes from the full-length transcriptome of *F. cirrhosa*, including gene nomenclature, evolution, structure, and expression patterns. More importantly, we identified candidate CYPs potentially involved in ISA biosynthesis using co-expression analyses and previous reports that were validated by quantitative polymerase chain reaction (qPCR) analyses. Last, the subcellular localization of candidate FcCYP450s were predicted using online tool, two of which were analyzed in *N. benthamiana* leaves. Our study on *CYPs* genes in *F. cirrhosa* provides essential information for further exploration into their physiological and biochemical functions, as well as their application in improving ISA content in Fritillaria plants.

2. Materials and Methods

2.1. Plant Materials, RNA Extraction, and Sequencing

The plants of *F. cirrhosa* D. Don were harvested inside the wild in July 2020 at Kangding city, northwestern China (placed at 30°3′44.9″ N, 101°58′3.81″ E, altitude 4300 m). The specimen (No. F-200725–1) was deposited at the Engineering Research Center of the Sichuan-Tibet Traditional Medicinal Plant, Chengdu, China. We certify that the research program complies with all applicable institutional, national, and international guidelines and legislation, and that we have authorization to collect *F. cirrhosa*. After collection, the bulbs were rinsed with cold distilled water, sterilized for in vivo culture, frozen in liquid nitrogen, and kept at −80 °C until they were needed again.

An RNA Kit (Vazyme, Nanjing, China) was used to extract the total RNA from the bulbs, and it was then processed in accordance with the manufacturer's instructions. The purity and concentration of the RNA were determined using a Nanodrop microspectrophotometer (Thermo Scientific, Waltham, DE, USA) and an Agilent 2100 bioanalyzer (Agilent Technologies, Santa Clara, USA). PacBio Iso-Seq library construction followed the official protocol described by Pacific Biosciences (Menlo Park, CA, USA). The PCR products were size selected using BluePippin™ Size Selection System (Sage Science, Beverly, MA, USA), and fragments with 0.5–6 kb were retained. A large-scale PCR was performed to construct the next SMRTbell library. Finally, the PacBio RS II platform was used by Berry Biotechnology Co. (Beijing, China) to complete the sequencing of the four SMRT cells.

2.2. Bioinformatics Analysis

For comprehensive functional annotation, the full-length transcripts were compared against public protein databases, including the NCBI nonredundant protein (Nr) database (http://www.ncbi.nlm.nih.gov, accessed on 5 July 2022), Swiss-Prot protein database (https://web.expasy.org/docs/swiss-prot_guideline.html, accessed on 5 July 2022), Kyoto Encyclopedia of Genes and Genomes (KEGG, http://www.genome.jp/kegg/kegg2.html, accessed on 10 July 2022), the COG/KOG database (http://www.ncbi.nlm.nih.gov/COG, accessed on 10 July 2022), and TrEMBL using BLASTx with a cut-off E-value of ≤10–5 (https://blast.ncbi.nlm.nih.gov/; version 2.2.23, accessed on 18 July 2022). Gene Ontology (GO) annotation was analyzed by Blast2GO software (https://www.blast2go.com/; version 4.4, accessed on 18 July 2022) and GO function categories were performed using WEGO software (http://wego.genomics.org.cn/; version 1.0, accessed on 22 July 2022).

2.3. Identification, Sequence Analyses and Phylogenetic of FcCYPs

The predicted CYPs were first searched for using keywords against the aforementioned annotations. The National Center for Biotechnology Information (NCBI)'s BLASTx tool was used to manually examine each potential CYPs sequence. After filtering out repeated results, the coding sequences of the resultant subjects were retrieved. ORF Finder (https://www.ncbi.nlm.nih.gov/orffinder/, accessed on 27 July 2022) was used to identify the open reading frame (ORF) sequences of CYPs. The nucleotide sequences of the ORF were translated into amino acid sequences using the ExPASY translation tool (https://web.expasy.org/translate/, accessed on 2 August 2022). The theoretical isoelectric point and molecular weight of each identified CYPs protein were predicted using the "Compute pI/MW" tool in the ExPASy server (https://web.expasy.org/protparam/, accessed on 2 August 2022). The protein subcellular localization of CYPs was predicted using WoLF PSORT (https://www.genscript.com/wolf-psort.html, accessed on 8 August 2022). The names of the CYPs proteins were assigned by Prof. David Nelson [17]. The sequences were divided into A-type, which included only the CYP71 clan, and non-A-type, which included all other clans based on CYP family membership. The CYPs proteins identified here were selected as query sequences to do the BLASTp (https://blast.ncbi.nlm.nih.gov/; version 2.2.23, accessed on 8 August 2022) searches against all available sequences in the Nr database. The full-length sequences of FcCYPs were used to create a phylogenetic tree with other known plant CYPs sequences in public databases using the neighbor-joining (NJ) method with MEGA 11 software at 1000 bootstrap replicates and default settings (Substitution type: Amino acid; Model/Method: Jones-Taylor-Thornton (JTT) model; Rates among Sites: Uniform Rates; Gaps/Missing Data Treatment: Complete deletion).

2.4. RNA Extraction and qRT-PCR Analysis

Total RNA was isolated from the bulbs as described above. We examined transcript expression patterns across different tissues (bulb, stem, and leaf). Gene expression levels were determined by qRT-PCR using ChamQ SYBR qPCR Master Mix (Vazyme) on a Bio-Rad CFX96 system (Bio-Rad, Hercules, CA, USA).18S ribosomal RNA was selected as the reference gene. All the primers used are listed in Table S1. The relative expression was calculated using the $2^{-\Delta\Delta Ct}$ method. The experiment was performed with three biological and technical replicates.

2.5. Determination of Alkaloids by HPLC-ELSD

The working standard solutions for the calibration curves were created by diluting the stock solutions of imperialine (420 g/mL), peimine (413 g/mL), and peiminine (406 g/mL) with methanol. The samples were prepared for alkaloid analysis, as described by Ma [18]. The HPLC-ELSD conditions were used to determine the alkaloids, and a liquid chromatograph (Agilent 1260 Infinity, Waldbronn, Germany) equipped with an Alltech 3300 evaporative light-scattering detector was used. The chromatographic separations were carried out a WondaSil C18-WR column (250 mm × 4.6 mm, 5 μm; Agilent) at a column temperature of 30 °C. The column was eluted with a mixture of methanol (mobile phase A) and 0.02% diethylamine (mobile phase B) at a flow rate of 1.0 mL/min. The following elution conditions were used: 0–10 min, 60% A to 75% A; 10–25 min, 75% A to 82% A; and 25–45 min, 82% A to 90% A. The ELSD's drift tube temperature was set to 60 °C, and nitrogen was used as the carrier gas at a flow rate of 1.2 L/min, gain value of 2, and injection volume of 20 μL. The calibration curves for each compound were used to determine the concentration based on an external standard method.

2.6. Subcellular Localization Analysis

We chose the *CYP90A1* and *CYP90B27* genes and used PCR-amplifying the ORFs for these genes without a stop codon while using specific primers with corresponding enzyme sites (Table S2) to assess representative CYP450 subcellular localization. The isolated PCR products were then subjected to sequence validation by Sangon Biotech (Shanghai, China), after which they were inserted into the pCAMBIA1300-35S-YFP vector upstream of the enhanced yellow fluorescent protein (YFP) in order to produce p35S::CYP450-YFP vectors. These recombinant plasmids were transformed into *N. benthamiana* along with the OsRac3-mCherry plasmid using an *Agrobacterium tumefaciens*-mediated transient transformation system. The plasma membrane localization marker was the fusion protein OsRac3-mCherry as previously described [19]. The cultured *A. tumefaciens* cells (OD600 = 0.4) were infiltrated into the leaves of *N. benthamiana*, which were grown in a greenhouse for 4 weeks at 23 °C under a 16 h light/8 h dark cycle. The same process using the empty vector pCAMBIA1300-35S-YFP was used as the control. Leaves expressing the resultant YFP fusion proteins were visualized 48 h after infiltration using a Nikon Eclipse Ni-U microscope (Nikon, Tokyo, Japan).

3. Results

3.1. SMRT Sequencing, Similarity Analysis, and Functional Annotation

The sequencing platform PACBIO RS II was used to carry out single-molecule real-time (SMRT) sequencing. A total of 71.35 Gb raw reads were generated and 47,579,643 subreads were obtained after filtering, among which 907,484 reads of inserts (ROIs) were successfully extracted with mean lengths of 1772 bp and 48 passes. In total, 55,751 full-length consensus isoforms, including 55,101 polished high-quality (HQ) and 650 low-quality (LQ) transcripts, were produced.

To obtain a comprehensive annotation of the *F. cirrhosa* transcriptome, a set of 55,101 polished high-quality isoforms was annotated by searching five databases (GO, KOG, Nr, KEGG, and Swiss-Prot). In total, 23,388 isoforms were annotated. Moreover, the specific details of the overall functional annotation are described in Table S3. In addition, the 70 unannotated isoforms may represent novel *F. cirrhosa* genes.

The gene functions of the identified isoforms were classified using GO enrichment analysis according to their molecular function, cellular component, and biological process terms (Table S4). Among them, biological processes (30,678 genes) comprised the majority of the GO terms. A significant portion of the isoforms were assigned to GO categories such as cellular processes, metabolic processes, and catalytic activities, which are crucial processes in plants and are involved in metabolite biosynthesis. According to the KOG classification, 14,083 isoforms were identified (Figure S1). The largest group was the cluster for general function prediction (2420).

3.2. Identification and Analyses of FcCYPs Sequences

A total of 189 sequences encoding putative CYPs were selected from the *F. cirrhosa* full-length transcriptome by keyword searching and bioinformatics methods, of which 44 lacked the complete signature motifs of P450, and only the longest translation form was retained to present each gene. Eventually, a total of 48 *FcCYP* genes were chosen for the following analyses (Table 1). According to P450 nomenclature, all 48 FcCYP genes were classified into 25 families and 46 subfamilies. They can be divided into two types: A-type

(9 families) with 23 genes and non-A type (16 families) with 25 genes. The largest family among them is the CYP71 family, which has 23 members. In contrast, the CYP51, CYP711, CYP95, CYP63, CYP40, and CYP710 families each have just one gene. The deduced FcCYP protein sequences ranged in length from 195 amino acids (CYP728B29) to 689 amino acids (CYP97A10), corresponding to a molecular mass of 21.69 to 76.98 kDa for the proteins.

Table 1. List of 48 CYP450s of F. cirrhosa identified in this study.

No	Gene ID	Clan	Family	Subfamily	Amino Acid Residues	MW (kDa)	pI	Subcellular Localization
1	27614	71	701	CYP701A6	429	49,317.84	5.73	cytoplasmic
2	25847	71	71	CYP71A1	495	56,263.57	6.78	chloroplast
3	25128	71	71	CYP71A1	496	56,013.99	8.22	chloroplast
4	25333	71	71	CYP71A2	511	56,277.12	6.89	chloroplast
5	24267	71	71	CYP71A25	503	57,395.89	7.69	chloroplast
6	25022	71	71	CYP71A4	498	56,964.15	7.72	vacular membrane
7	19214	71	71	CYP71A9	490	55,311.46	8.42	chloroplast
8	23625	71	71	CYP71D381	495	55,669.95	9.04	chloroplast
9	23436	71	71	CYP71D55	510	58,297.56	9.08	chloroplast
10	27190	71	71	CYP71D8	498	56,365.75	6.48	chloroplast
11	24560	71	73	CYP73A1	505	58,138.45	8.52	plasma membrane
12	22595	71	73	CYP73A100	518	58,650.4	8.6	plasma membrane
13	23153	71	73	CYP73A16	504	57,796.28	9.24	plasma membrane
14	22888	71	75	CYP75B137	503	55,336.92	6.88	chloroplast
15	23199	71	81	CYP81E8	507	56,385.25	7.15	chloroplast
16	24881	71	81	CYP81E8	509	57,543.15	9.06	chloroplast
17	25104	71	81	CYP81F3	494	55,327.71	6.19	chloroplast
18	23883	71	81	CYP81Q32	513	57,720.7	7.26	chloroplast
19	21801	71	84	CYP84A1	515	57,470.57	6.44	chloroplast
20	22019	71	93	CYP93A3	502	56,432.31	5.96	chloroplast
21	25162	71	89	CYP89A2	507	57,250.83	8.82	chloroplast
22	23387	71	89	CYP89A9	498	56,047.58	8.41	chloroplast
23	23217	71	98	CYP98A2	510	57,981.92	8.61	chloroplast
24	14155	72	72	CYP72A14	522	59,146.97	9.19	chloroplast
25	18891	72	72	CYP72A15	516	59,380.97	8.72	cytoplasmic
26	23630	72	72	CYP72A219	518	59,496.39	9.47	chloroplast
27	24173	72	72	CYP72A616	522	59,004.81	9.37	chloroplast
28	25879	72	734	CYP734A6	515	58,157.25	9.44	chloroplast
29	20540	85	707	CYP707A1	468	52,762.44	9.15	chloroplast
30	24431	85	707	CYP707A2	470	53,332.54	8.96	chloroplast
31	23919	85	85	CYP85A1	468	53,353.24	8.65	chloroplast
32	23991	85	90	CYP90A1	510	57,539.32	8.7	cytoplasmic
33	25638	85	90	CYP90B27	483	54,036.79	8.93	chloroplast
34	22207	85	90	CYP90D2	482	54,659.17	8.11	chloroplast
35	24036	86	704	CYP704C1	507	58,078.96	8.42	chloroplast
36	24455	71	77	CYP77A4	509	57,610.25	8.77	chloroplast
37	22541	86	94	CYP94A1	508	57,163.09	9.09	chloroplast
38	27487	86	94	CYP94N1	499	55,985.22	8.47	chloroplast
39	26422	86	94	CYP94B1	489	54,906.75	8.32	chloroplast
40	26493	86	94	CYP94C1	495	55,621.93	6.43	chloroplast
41	23522	86	96	CYP96A15	499	56,957.64	9.08	chloroplast
42	15644	86	97	CYP97A3	614	68,383.46	5.86	chloroplast
43	4877	95	95	CYP95	853	94,883.01	11.48	nucleus
44	26361	710	710	CYP710A11	495	56,010.29	7.19	plasma membrane
45	20822	711	711	CYP711A1	526	58,970.43	9.31	plasma membrane
46	32983	40	40	CYP40	369	40,848.45	5.49	cytoskeleton
47	18441	51	51	CYP51	488	55,580.08	8.13	chloroplast
48	10385	63	63	CYP63	689	76,552.99	10.67	nucleus

To explore the evolutionary relationship between FcCYPs, an unrooted NJ tree with 48 full-length FcCYP sequences was built (Figure 1). According to phylogenetic analysis, FcCYPs were classified into 10 clans, including six single-family clans (CYP95, CYP710, CYP711, CYP40, CYP51, and CYP63) and four multi-family clans (CYP71, CYP86, CYP85, and CYP72, in descending order, by number). Genes from the same clan were grouped together to form a single clade. For example, 86 clans, comprising eight CYP450s belonging to five families, were clustered into one clade. The CYP71 clan is regarded a special cluster with the most members.

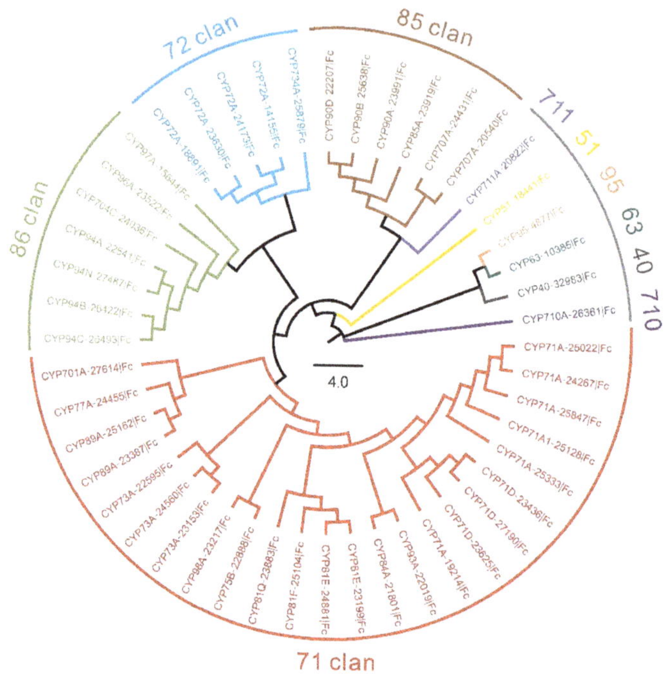

Figure 1. Phylogenetic analysis of predicted CYP450s in *F. cirrhosa*.

3.3. *Identification of Candidates CYPs Involved in ISA Biosynthesis*

Four pathways, namely the mevalonate, phytosterol, cholesterol, and core ISA pathways, are involved in ISA biosynthesis [14,15]. We specifically evaluated CYP450s involved in the synthesis of FISA. To date, 47 *CYP450s* have been found to play a role in sterol biosynthesis [20–22]. The CYP51G1, CYP94N, and CYP94A subfamilies are the primary CYP450 gene families involved in steroid diversification, with the CYP90 family being the largest multifunctional C-22/23 hydroxylation family involved in steroid biosynthesis [23–25]. To identify the most relevant CYP450 unigenes involved in FISA biosynthesis, we performed BLASTp searches that compared FcCYP450s to 51 CYP450s from other plant species known to be involved in steroid metabolism. This analysis showed that CYP90B (contig 25638) exhibited 73.72% sequence identity with *Veratrum californicum* CYP90B27v1, which is a steroid C-22 hydroxylase involved in verazine production [26]. BLASTp analysis also indicated that contig22207 exhibited 63.03% identity with *Oryza sativa* CYP90D2, which catalyzes the C-23 hydroxylation of brassinosteroid [27].

To further investigate the functional roles and evolutionary relationships of these FcCYPs, NJ phylogenetic trees were constructed using 48 CYPs and plants known to be involved in steroid metabolism (Figure 2). Phylogenetic analyses revealed that FcCYP94A-22541 and FcCYP94N-27487 belong to the CYP94 family and are most closely related to *V. californicum* CYP94N1v2, which encodes the C-26 hydroxylation/oxidation enzyme involved in cyclopamine production [26]. Four contigs (CYP72A-18891, CYP72A-23630, CYP72A-24173, and CYP72A-14155) clustered in the CYP72A group and were most closely related to *Solanum lycopersicum* GAMEs [28], *Trigonella foenumgraecum* CYP72A613, and *Paris polyphylla* CYP72A616 [21], which are involved in steroidal glycoalkaloid and diosgenin biosynthesis.

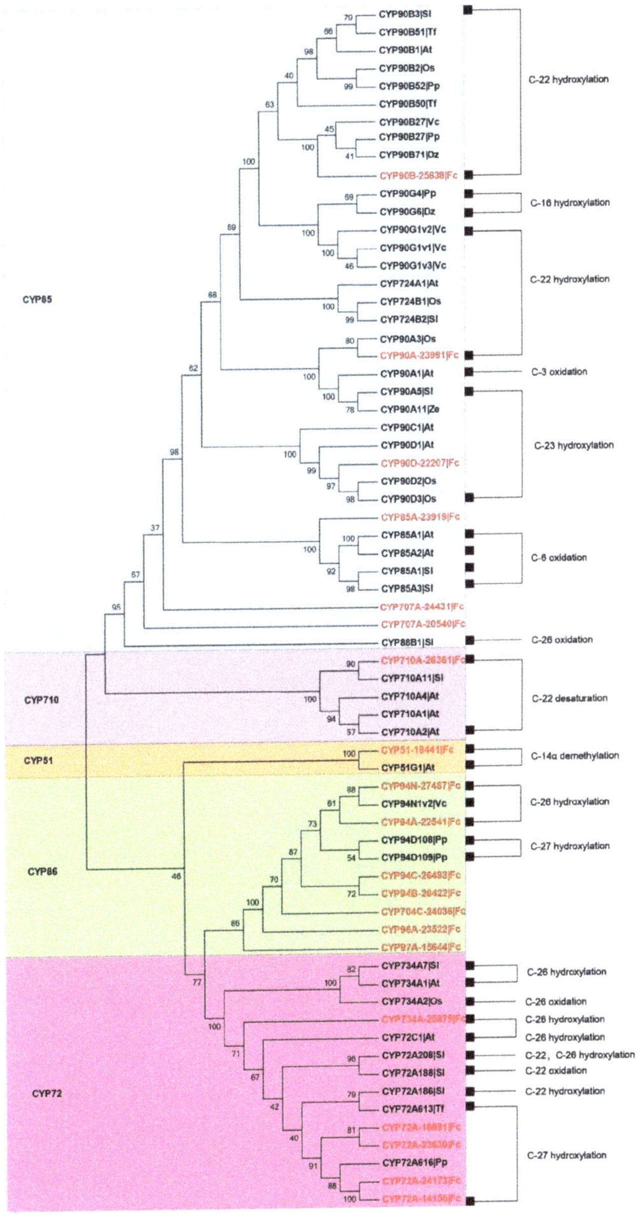

Figure 2. Phylogenetic tree of previously characterized sterol biosynthesis *CYP450s* and those *F. cirrhosa CYP450s* isolated in this study (in red). The known biochemical activities of *P450s* are indicated on the right.

3.4. Analysis of Candidate CYPs Gene Expression and Key FISA Accumulation in Bulbs of Different Ages

As the CYP90, CYP72, and CYP94 family genes are involved in steroid metabolism, we examined their expression in relation to alkaloid accumulation in bulbs of various ages. We found that all the selected genes were transcribed more in bulbs from 3 year-old *F. cirrhosa* plants than in bulbs from 1 year-old plants (Figure 3). To verify whether the accumulation of FISA at various ages in *F. cirrhosa* was consistent with the expression levels of these genes, we used HPLC-ELSD to examine the contents of three major FISA (imperialine, peimine, and peininine) in the bulbs of *F. cirrhosa* from the two different age groups (Figure 4 and Table S5). We found that the accumulation of the three FISA in bulbs cultivated for 3 years was significantly higher than that in bulbs cultivated for 1 year, similar to the trend of the relative expression levels of the candidate CYPs genes. This result suggests that FcCYP proteins are correlated with the FISA biosynthesis.

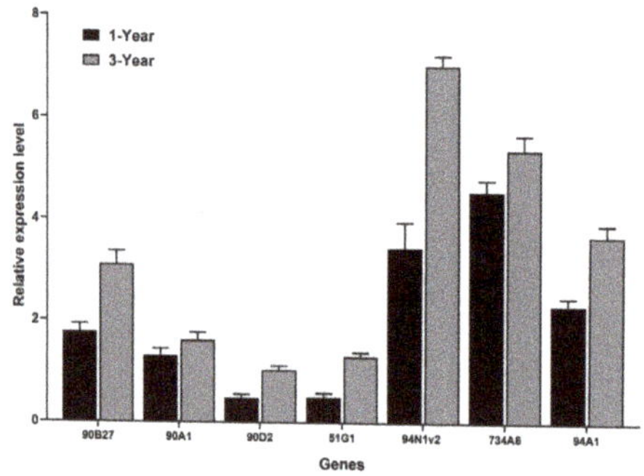

Figure 3. The relative expression levels of candidate *CYPs* in *F. cirrhosa* bulbs of 1 year of age and 3 years of age. The error bars indicate ± SDs (n = 3).

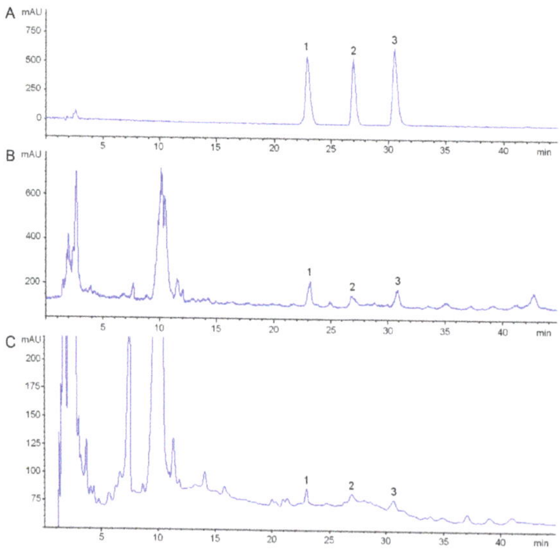

Figure 4. Representative HPLC-ELSD chromatographic profiles of FISA in the bulbs of *F. cirrhosa*. (**A**) HPLC-ELSD chromatographic profiles of ISA standards. (**B**) HPLC-ELSD chromatographic profiles of FISA in the bulbs from 3-year-old *F. cirrhosa*. (**C**) HPLC-ELSD chromatographic profiles of FISA in the bulbs from 1-year-old *F. cirrhosa* bulbs. 1, imperialine; 2, peininine; 3, peimine.

3.5. Assessment of the Subcellular Localization of Two CYP450 Fusion Proteins

Most plant CYP450s target the endoplasmic reticulum (ER) and are occasionally associated with plastids, plasma membranes, and other organelles [29,30]. We used an online tool WoLF PSORT to predict the subcellular localization of 48 FcCYP proteins. Most FcCYP proteins were located in membrane-bound chloroplasts (36), some in the plasma membrane (5), and a few in the cytoplasm (3), nucleus (2), cytoskeleton (1), and vacuolar membrane (1) (Table 1).

To determine the subcellular localization of *F. cirrhosa* CYP450 proteins, we transiently expressed CYP90A1-YFP and CYP90B27-YFP fusion proteins in *N. benthamiana* epidermal cells. OsRac3, a plasma membrane localization marker [19], was also co-transformed with FcCYP fusion proteins. As shown in Figure 5, the fluorescence of CYP90A1-YFP and CYP90B27-YFP partially overlapped with that of OsRac3-mCherry. Meanwhile, the fluorescence of both seemed to be in the chloroplast, some other plastids, and cytoplasm, which was consistent with the Wolfpsort predictions. These results suggest that FcCYP proteins function by targeting different organelles in the cytoplasm.

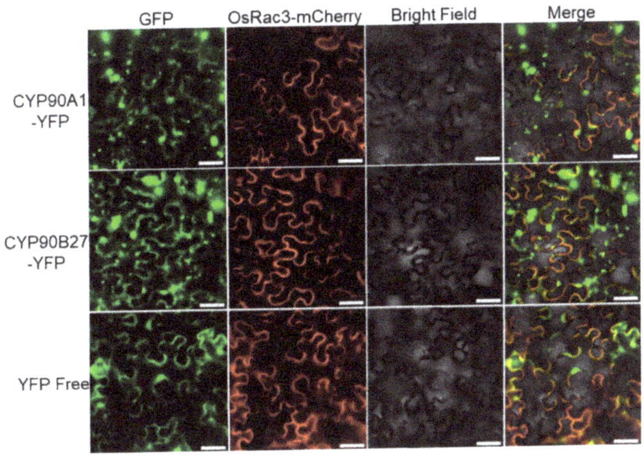

Figure 5. Subcellular localization of CYP90A1 and CYP90B27 in *N. benthamiana* leaves. Transiently expressing p35S::CYP90A1-YFP or p35S::CYP90B27-YFP fusion protein in the leaves of *N. benthamiana* by *A. tumefaciens*. Transient expression of p35S::YFP was used as the control. Subcellular localizations of the fused proteins were analyzed by fluorescence microscopy at 48 h after infiltration. The OsRac3-mCherry fusion protein is a plasma membrane marker. Scale bars = 50 μm.

4. Discussion

F. cirrhosa is a well-known medicinal herb containing imperialine (a steroidal alkaloid), health products, and landscape, and its bulb is a pharmaceutical and breeding organ. Previous studies on *F. cirrhosa* have mainly focused on cultivation, breeding, phytochemistry, and pharmacology [31]. Because of its highly heterogeneous, low-abundance repeat-derived DNA and large genomes, assembling the genome of *F. cirrhosa* is difficult, time-consuming, and expensive [32]. As a result, little information on the genetic characteristics has been obtained. Although RNA-Seq can be used to explore and analyze differentially expressed genes, the sequences are frequently incomplete. As a result, PacBio Iso-Seq is a more effective and cost-effective method for directly producing a comprehensive transcriptome with precise genetic structural characteristics for *F. cirrhosa*.

In comparison to previous gene discovery studies in *F. cirrhosa* using cDNA cloning or Illumina sequencing [11,13,14], we generated a comprehensive transcriptome dataset. Our Iso-Seq data generated a total of 55,101 deredundant high-quality isoforms, and the full-length transcripts could be directly used for future gene discovery research without additional PCR amplification. Using a combination of SMRT analyses, phylogenetic trees, and gene expression level analyses, we recharacterized the crucial genes encoding CYPs involved in FISA synthesis.

Steroidal alkaloids (SA) are the most common active ingredients in the Liliaceae family, and their biosynthetic pathways of SA have been investigated in many plants, such as *Paris* [24], *Veratrum* [26], and *Fritillaria* species [15]. The CYP51 clan member genes, which descended from a sterol-metabolizing CYP51 ancestor, are thought to be the most ancient CYP450s in the SA biosynthetic pathway. In the present analysis, we identified only one CYP51 clan member (CYP51G1). In *V. californicum* Jervine-type SA synthesis, three specific cytochromes P450 (CYP90B27, CYP94N1, and CYP90G1) exhibit hydroxylation/oxidation activity at the C-22 and C-26 positions. In *F. cirrhosa*, peimisine is a Jervine-type SA, and BLAST and phylogenetic tree analysis showed that FcCYP90B-25638 obtained in this study shares high identity with VcCYP90B27 [26]. CYP90B subfamily members can encode cytochrome P450s that participate in the C-22 hydroxylation of cholesterol [24,26]. In the CYP86 clan, FcCYP94A-22541 and FcCYP94N-27487 are highly similar to VcCYP94N1v2, which encodes steroid C-26 hydroxylase/oxidase. Imperialine is classified as a cevanine-type alkaloid with a ring scaffold typical of cholesterol, and includes a series of hydroxylation/oxidation reactions at C-22 and C-26. Therefore, FcCYP90B-25638, FcCYP94A-22541, and FcCYP94N-27487 were identified as candidate CYPs involved in FISAs biosynthesis.

We then performed a co-expression analysis of key FISAs with gene expression levels in bulbs of *F. cirrhosa* of different ages. We measured the contents of key FISAs in *F. cirrhosa* of various ages, including imperialine, peimine, and peininine. We predicted candidate CYPs that are highly relevant to FISAs. We conducted qRT-PCR to analyze the gene expression levels and found that all the CYPs genes were expressed at higher levels in bulbs in the late growth stage than in the early stage, in line with steroid alkaloid content, which is also in accordance with the levels of alkaloids in the growth stages of *F. taipaiensis* and *F. cirrhosa* [33,34]. This finding suggests that the selected CYPs may be involved in the formation of hydroxylation/oxidase of FISAs. The prediction and analysis of the subcellular localization of FcCYP proteins showed that FcCYP proteins can function in the chloroplast, plasma membrane, endoplasmic reticulum, nucleus, and some plastids. They play key roles in the biosynthetic pathways of SA in different organelles. However, the functions and mechanisms of these conversions remain unclear. We plan to verify these conjectures in *Saccharomyces cerevisiae* heterologous expression in a follow-up study.

5. Conclusions

In this study, we identified 48 putative CYP450s with complete cytochrome P450 domains in the full-length transcriptome of *F. cirrhosa*. According to the classification criteria, 48 CYP450s were classified into 10 clans consisting of 25 families and 46 subfamilies. Following that, we conducted phylogenetic analysis to characterize 20 CYP450s identified as being involved in FISAs biosynthesis. The qRT-PCR and HPLC-ELSD results showed that all of the predicted candidate *CYP450s* genes expressed similarly with key FISAs concentrations in bulbs of different growth ages. Finally, the subcellular localizations of the two CYP450 candidates were investigated. Together, this study provides thorough understanding of the CYP450 gene families in *F. cirrhosa* and will help in figuring out how the CYP450 family functions in this and related species. Further research is required to confirm the role of the selected candidate *CYP450s* genes in FISA biosynthesis.

Supplementary Materials: The following supporting information can be downloaded at: https://www.mdpi.com/article/10.3390/genes14010219/s1, Figure S1: KOG function classification of consensus sequence; Table S1: Sequences of the primers used in this study; Table S2: Details of overall functional annotation; Table S3: GO annotation of the full-length deredundant high-quality transcripts; Table S4: A list of 47 previously reported plant CYP450s involved in steroid biosynthesis; Table S5: Method validation results and content of 3 components in *F. cirrhosa*.

Author Contributions: M.X. and Z.Z. performed the experiments; R.L., Z.Z. and M.W. analyzed the data and wrote the manuscript; R.L., Q.Z., J.L. and Z.Z. conceived and designed this study, wrote and revised the manuscript, conducted the project administration, and acquired the funding grants. All authors have read and agreed to the published version of the manuscript.

Funding: This study was funded by the National Natural Science Foundation of China (grant no. 32100312 and 32270407), Natural Science Foundation of Sichuan (grant no. 2022NSFSC0583), Sichuan Provincial Science Fund for Distinguished Young Scholars (grant no. 2022JDJQ0053), and Aba Autonomous Prefecture application technology (grant no. R22YYJSYJ0015).

Institutional Review Board Statement: Not applicable.

Informed Consent Statement: Not applicable.

Data Availability Statement: All datasets used in this study are available on FigShare at the link: https://doi.org/10.6084/m9.figshare.21900003.v1.

Conflicts of Interest: The authors declare no conflict of interest. The funders had no role in the design of the study; in the collection, analyses, or interpretation of data; in the writing of the manuscript; or in the decision to publish the results.

References

1. Da-Cheng, H.; Xiao-Jie, G.; Pei-Gen, X.; Yong, P. Phytochemical and biological research of Fritillaria medicine resources. *Chin. J. Nat. Med.* **2013**, *11*, 330–344.
2. Cunningham, A.; Brinckmann, J.; Pei, S.-J.; Luo, P.; Schippmann, U.; Long, X.; Bi, Y.-F. High altitude species, high profits: Can the trade in wild harvested *Fritillaria cirrhosa* (Liliaceae) be sustained? *Chin. J. Nat. Med.* **2018**, *223*, 142–151. [CrossRef] [PubMed]
3. Quan, Y.; Li, L.; Yin, Z.; Chen, S.; Yi, J.; Lang, J.; Zhang, L.; Yue, Q.; Zhao, J. Bulbus *Fritillariae cirrhosae* as a Respiratory Medicine: Is There a Potential Drug in the Treatment of COVID-19? *Front Pharmacol.* **2021**, *12*, 784335. [CrossRef] [PubMed]
4. Wang, D.; Zhu, J.; Wang, S.; Wang, X.; Ou, Y.; Wei, D.; Li, X. Antitussive, expectorant and anti-inflammatory alkaloids from Bulbus *Fritillariae cirrhosae*. *Chin. J. Nat. Med.* **2011**, *82*, 1290–1294. [CrossRef] [PubMed]
5. Wang, D.; Yang, J.; Du, Q.; Li, H.; Wang, S. The total alkaloid fraction of bulbs of *Fritillaria cirrhosa* displays anti-inflammatory activity and attenuates acute lung injury. *Front. Pharmacol.* **2016**, *193*, 150–158. [CrossRef]
6. Wang, D.; Wang, S.; Feng, Y.; Zhang, L.; Li, Z.; Ma, J.; Luo, Y.; Xiao, W. Antitumor effects of Bulbus *Fritillariae cirrhosae* on Lewis lung carcinoma cells in vitro and in vivo. *Chin. J. Nat. Med.* **2014**, *54*, 92–101. [CrossRef]
7. Liu, S.; Yang, T.; Ming, T.W.; Gaun, T.K.W.; Zhou, T.; Wang, S.; Ye, B. Isosteroid alkaloids with different chemical structures from *Fritillariae cirrhosae* bulbus alleviate LPS-induced inflammatory response in RAW 264.7 cells by MAPK signaling pathway. *Chin. J. Nat. Med.* **2020**, *78*, 106047. [CrossRef]
8. Wang, Y.; Hou, H.; Ren, Q.; Hu, H.; Yang, T.; Li, X. Natural drug sources for respiratory diseases from Fritillaria: Chemical and biological analyses. *Chin. Med.* **2021**, *16*, 40. [CrossRef]
9. Wang, D.; Wang, S.; Du, Q.; Wang, N.; Liu, S.; Wang, X.; Jiang, J. Optimization of extraction and enrichment of steroidal alkaloids from bulbs of cultivated *Fritillaria cirrhosa*. *Biomed Res. Int.* **2014**, *2014*, 258402.
10. Mathela, M.; Kumar, A.; Sharma, M.; Goraya, G.S. Hue and cry for *Fritillaria cirrhosa* D. Don, a threatened medicinal plant in the Western Himalaya. *Discovery* **2021**, *2*, 38.
11. Chao, S.; Yongqiao, S.; Jingyuan, S.; Chenji, L.; Xiwen, L.; Xiaowei, Z.; Ying, L.; Songnian, H.; Hongmei, L.; Yingjie, Z. Discovery of genes related to steroidal alkaloid biosynthesis in *Fritillaria cirrhosa* by generating and mining a dataset of expressed sequence tags (ESTs). *Plant Physiol.* **2011**, *5*, 5307–5314.
12. Zhao, Q.; Li, R.; Zhang, Y.; Huang, K.; Wang, W.; Li, J. Transcriptome analysis reveals in vitro-cultured regeneration bulbs as a promising source for targeted *Fritillaria cirrhosa* steroidal alkaloid biosynthesis. *3 Biotech* **2018**, *8*, 191. [CrossRef]
13. Kumar, P.; Acharya, V.; Warghat, A.R.J.P. Comparative transcriptome analysis infers bulb derived in vitro cultures as a promising source for sipeimine biosynthesis in *Fritillaria cirrhosa* D. Don (Liliaceae, syn. Fritillaria roylei Hook.)-High value Himalayan medicinal herb. *Chin. J. Nat. Med.* **2021**, *183*, 112631. [CrossRef]
14. Lu, Q.; Li, R.; Liao, J.; Hu, Y.; Gao, Y.; Wang, M.; Li, J.; Zhao, Q. Integrative analysis of the steroidal alkaloids distribution and biosynthesis of bulbs *Fritillariae cirrhosae* through metabolome and transcriptome analyses. *BMC Genomics* **2022**, *23*, 511. [CrossRef]
15. Qu, A.; Wu, Q.; Su, J.; Li, C.; Yang, L.; Wang, Z.; Wang, Z.; Li, Z.; Ruan, X.; Zhao, Y. A Review on the Composition and Biosynthesis of Alkaloids and on the Taxonomy, Domestication, and Cultivation of Medicinal Fritillaria Species. *Agron. J.* **2022**, *12*, 1844. [CrossRef]
16. Li, H.-J.; Jiang, Y.; Li, P. Chemistry, bioactivity and geographical diversity of steroidal alkaloids from the Liliaceae family. *Nat. Prod. Rep.* **2006**, *23*, 735–752. [CrossRef]
17. Nelson, D.R. The cytochrome p450 homepage. *Hum. Genomics* **2009**, *4*, 59. [CrossRef]

18. Ma, B.; Ma, J.; Li, B.; Tao, Q.; Gan, J.; Yan, Z. Nutrition. Effects of different harvesting times and processing methods on the quality of cultivated *Fritillaria cirrhosa* D. Don. *Food Sci. Nutr.* **2021**, *9*, 2853–2861. [CrossRef]
19. Tao, Y.; Zou, T.; Zhang, X.; Liu, R.; Chen, H.; Yuan, G.; Zhou, D.; Xiong, P.; He, Z.; Li, G. Secretory lipid transfer protein OsLTPL94 acts as a target of EAT1 and is required for rice pollen wall development. *Plant J.* **2021**, *108*, 358–377. [CrossRef]
20. Malhotra, K.; Franke, J. Cytochrome P450 monooxygenase-mediated tailoring of triterpenoids and steroids in plants. *Beilstein J. Org. Chem.* **2022**, *18*, 1289–1310. [CrossRef]
21. Christ, B.; Xu, C.; Xu, M.; Li, F.-S.; Wada, N.; Mitchell, A.J.; Han, X.-L.; Wen, M.-L.; Fujita, M.; Weng, J.-K. Repeated evolution of cytochrome P450-mediated spiroketal steroid biosynthesis in plants. *Nat. Commun.* **2019**, *10*, 3206. [CrossRef] [PubMed]
22. Hansen, C.C.; Nelson, D.R.; Møller, B.L.; Werck-Reichhart, D. Plant cytochrome P450 plasticity and evolution. *Mol. Plant* **2021**, *14*, 1244–1265. [CrossRef] [PubMed]
23. Szeliga, M.; Ciura, J.; Tyrka, M. Representational difference analysis of transcripts involved in jervine biosynthesis. *Life* **2020**, *10*, 88. [CrossRef] [PubMed]
24. Hua, X.; Song, W.; Wang, K.; Yin, X.; Hao, C.; Duan, B.; Xu, Z.; Su, T.; Xue, Z. Effective prediction of biosynthetic pathway genes involved in bioactive polyphyllins in Paris polyphylla. *Commun. Biol.* **2022**, *5*, 50. [CrossRef]
25. Sonawane, P.D.; Pollier, J.; Panda, S.; Szymanski, J.; Massalha, H.; Yona, M.; Unger, T.; Malitsky, S.; Arendt, P.; Pauwels, L. Plant cholesterol biosynthetic pathway overlaps with phytosterol metabolism. *Nat. Plants* **2016**, *3*, 16205. [CrossRef]
26. Augustin, M.M.; Ruzicka, D.R.; Shukla, A.K.; Augustin, J.M.; Starks, C.M.; O'Neil-Johnson, M.; McKain, M.R.; Evans, B.S.; Barrett, M.D.; Smithson, A. Elucidating steroid alkaloid biosynthesis in *Veratrum californicum*: Production of verazine in Sf9 cells. *Plant J.* **2015**, *82*, 991–1003. [CrossRef]
27. Sakamoto, T.; Ohnishi, T.; Fujioka, S.; Watanabe, B.; Mizutani, M. Rice CYP90D2 and CYP90D3 catalyze C-23 hydroxylation of brassinosteroids in vitro. *Plant Physiol. Biochem.* **2012**, *58*, 220–226. [CrossRef]
28. Itkin, M.; Heinig, U.; Tzfadia, O.; Bhide, A.; Shinde, B.; Cardenas, P.; Bocobza, S.; Unger, T.; Malitsky, S.; Finkers, R. Biosynthesis of antinutritional alkaloids in solanaceous crops is mediated by clustered genes. *Science* **2013**, *341*, 175–179. [CrossRef]
29. Schuler, M.A.; Duan, H.; Bilgin, M.; Ali, S. Arabidopsis cytochrome P450s through the looking glass: A window on plant biochemistry. *Phytochem. Rev.* **2006**, *5*, 205–237. [CrossRef]
30. Magwanga, R.O.; Lu, P.; Kirungu, J.N.; Dong, Q.; Cai, X.; Zhou, Z.; Wang, X.; Hou, Y.; Xu, Y.; Peng, R. Knockdown of cytochrome P450 genes Gh_D07G1197 and Gh_A13G2057 on chromosomes D07 and A13 reveals their putative role in enhancing drought and salt stress tolerance in *Gossypium hirsutum*. *Genes* **2019**, *10*, 226. [CrossRef]
31. Chen, T.; Zhong, F.; Yao, C.; Chen, J.; Xiang, Y.; Dong, J.; Yan, Z.; Ma, Y.J.; Medicine, A. A systematic review on traditional uses, sources, phytochemistry, pharmacology, pharmacokinetics, and toxicity of *Fritillariae cirrhosae* bulbus. *Evid. Based Complement Alternat. Med.* **2020**, *2020*, 1536534. [CrossRef]
32. Kelly, L.J.; Renny-Byfield, S.; Pellicer, J.; Macas, J.; Novák, P.; Neumann, P.; Lysak, M.A.; Day, P.D.; Berger, M.; Fay, M.F. Analysis of the giant genomes of *Fritillaria* (*Liliaceae*) indicates that a lack of DNA removal characterizes extreme expansions in genome size. *New Phytol.* **2015**, *208*, 596–607. [CrossRef]
33. Peng, R.; Ma, P.; Mo, R.; Sun, N.J. Analysis of the bioactive components from different growth stages of *Fritillaria taipaiensis* PY Li. *Acta Pharm. Sin. B* **2013**, *3*, 167–173. [CrossRef]
34. Geng, Z.; Liu, Y.; Gou, Y.; Zhou, Q.; He, C.; Guo, L.; Zhou, J.; Xiong, L.J. Metabolomics study of cultivated Bulbus *Fritillariae cirrhosae* at different growth stages using UHPLC-QTOF-MS coupled with multivariate data analysis. *Phytochem. Anal.* **2018**, *29*, 290–299. [CrossRef]

genes

MDPI

Article

Genome-Wide Identification, Characterization, and Expression Analysis of GRAS Gene Family in Ginger (*Zingiber officinale* Roscoe)

Shuming Tian [1,2], Yuepeng Wan [1], Dongzhu Jiang [1], Min Gong [1,2], Junyao Lin [1], Maoqin Xia [1], Cuiping Shi [1], Haitao Xing [3,*] and Hong-Lei Li [1,*]

1 College of Landscape Architecture and life Science/Institute of special Plants, Chongqing University of Arts and Sciences, Chongqing 402168, China
2 College of Biology and Food Engineering, Chongqing Three Gorges University, Chongqing 404020, China
3 Chongqing Key Laboratory of Economic Plant Biotechnology, Chongqing University of Arts and Sciences, Chongqing 402160, China
* Correspondence: xinght@cqwu.edu.cn (H.X.); lihonglei215@cqwu.edu.cn (H.-L.L.)

Abstract: GRAS family proteins are one of the most abundant transcription factors in plants; they play crucial roles in plant development, metabolism, and biotic- and abiotic-stress responses. The GRAS family has been identified and functionally characterized in some plant species. However, this family in ginger (*Zingiber officinale* Roscoe), a medicinal crop and non-prescription drug, remains unknown to date. In the present study, 66 GRAS genes were identified by searching the complete genome sequence of ginger. The GRAS family is divided into nine subfamilies based on the phylogenetic analyses. The GRAS genes are distributed unevenly across 11 chromosomes. By analyzing the gene structure and motif distribution of GRAS members in ginger, we found that the GRAS genes have more than one *cis*-acting element. Chromosomal location and duplication analysis indicated that whole-genome duplication, tandem duplication, and segmental duplication may be responsible for the expansion of the GRAS family in ginger. The expression levels of GRAS family genes are different in ginger roots and stems, indicating that these genes may have an impact on ginger development. In addition, the GRAS genes in ginger showed extensive expression patterns under different abiotic stresses, suggesting that they may play important roles in the stress response. Our study provides a comprehensive analysis of GRAS members in ginger for the first time, which will help to better explore the function of GRAS genes in the regulation of tissue development and response to stress in ginger.

Keywords: ginger; GRAS family; gene expression; phylogenetic analyses

Citation: Tian, S.; Wan, Y.; Jiang, D.; Gong, M.; Lin, J.; Xia, M.; Shi, C.; Xing, H.; Li, H.-L. Genome-Wide Identification, Characterization, and Expression Analysis of GRAS Gene Family in Ginger (*Zingiber officinale* Roscoe). *Genes* **2023**, *14*, 96. https://doi.org/10.3390/genes14010096

Academic Editors: Hakim Manghwar and Wajid Zaman

Received: 30 November 2022
Revised: 15 December 2022
Accepted: 16 December 2022
Published: 29 December 2022

1. Introduction

Transcription factors (TFs) are extraordinary proteins that participate in regulating plant growth, development, signal transduction, and resistance to biotic and abiotic and stresses [1]. TFs modulate the expression of target genes by combining with their *cis*-acting regulatory elements [2,3]. The identification and analysis of TFs are fundamental in functional genomic research. Since the first transcription factor was discovered in maize [4], more than 3000 TFs have been demonstrated to be involved in various physiological processes and regulatory networks in plants [2]. Among them, some important transcription factors, including AP2/EREBP, WRKY, bZIP, MADS, MBS, ARF, HB, SBP, and others, have been well studied [5–8].

The GRAS gene family is an important plant-specific transcription factor that is named after the three earliest identified members: GAI (gibberellic acid insensitive), RGA (repressor of GAI), and SCR (scarecrow) [9,10]. These proteins were first found in bacteria and were transferred into plants via the mechanism of lateral gene transfer [11]. In general,

GRAS proteins contain 400 to 700 amino acid (aa) residues, which are mainly located in the nucleus [12]. The C-terminal region of GRAS genes is very conserved and forms the GRAS domain. The GRAS domain has five motifs: VHIID motif, PFYRE motif, Leucine heptad repeat II (LHR II), Leucine heptad repeat I (LHR I), and the SAW motif [13]. According to related studies, the GRAS gene family has been identified in a variety of plants so far, including poplar, Arabidopsis, rice, plum, and pine [14–17]. There are 35 and 61 members of the GRAS gene family in the model plant Arabidopsis and rice genomes, respectively. These GRAS proteins can be divided into at least eight subfamilies: SCR, LAS, DELLA, PAT1, SHR, HAM, SCL3, and LISCL [18,19]. Based on the accumulation of data on the GRAS family, more than 13 clades were identified in plants [20–22]. The evolutionary history of the GRAS family in plants has been studied extensively so far, including the model species, Arabidopsis [19] and tobacco [23], as well as the common crops rice [23], maize [24], sweet potato [25], cotton [5], sativa [19], castor [26], poplar [27], cabbage [24], and grape [28].

The members of GRAS family play important roles in diverse processes such as plant growth and development, and tolerance to biotic or abiotic stresses [29–34]. Each subfamily of GRAS has unique functions, whereas the proteins from the same subfamily have similar functions. Genes of the DELLA subfamily have been shown to be blockers of the gibberellin (GA) signaling pathway, e.g., GAI, RGA, and RGL [32]. GAI and RGA act as negative regulators of GA signaling and control various plant developmental processes, including seed development and vegetative reproductive-phase transition in Arabidopsis [35]. It has been reported that members of the SCR and SHR subfamilies can form the SCR/SHR complex, which is involved in regulating the radial growth of root [36]. The subfamily LAS genes are always in the hub of the gene regulatory network and can downregulate GA content [37]. When *At*LAS was knocked out in Arabidopsis, lateral roots could not be formed [31]. The apical meristem organization and growth of *Capsicum annuum* were regulated by members of the subfamily HAM. The PAT1 subfamily plays important roles in activating the ROS scavenging system, which could improve tolerance to salinity, drought, and cold [38,39]. SCL proteins display unexpected tolerance to diverse environmental stresses; for example, SCL14 is involved in promoting the resistance of Arabidopsis through activating stress-inducible promoters [40]. The drought and salt resistance were enhanced by the overexpression of the *PeSCL7* gene in Arabidopsis [41]. Moreover, when the *ZmSCL7* gene was overexpressed in tobacco, plant resistance to salt was significantly improved.

Ginger (*Zingiber officinale* Roscoe) is an herbaceous perennial plant that belongs to the family Zingiberaceae. Ginger is one of the most widely cultivated crops in numerous countries, including China, India, and Nigeria. The global ginger production in 2019 was up to 4.08 million tons according to data from the FAO. The rhizome of ginger is a popular spice and traditional Eastern medicine [42,43]. The bioactive components of ginger have shown therapeutic properties as an agent for treating rheumatoid arthritis, asthma, and other ailments [44,45]. A number of TFs have been reported to be crucial in regulating the rhizome development and abiotic stress. For example, AUX/IAA and MADS box protein respond to rhizome initiation and development [46]. The ZoAP2/ERF family genes play a crucial role in tolerance to cold, drought, and salt [47]. However, the function of the GRAS family in ginger remains unknown. Based on the published whole genome of ginger [48], this study identified and analyzed the ginger GRAS family genes by using bioinformatic methods. The present systematic research provided some insights that could be used in elucidating the function of ginger GRAS.

2. Materials and Methods

2.1. Genome-Wide Identification of GRAS Genes in Ginger

The full genome sequence and annotation data of *Arabidopsis thaliana* and rice were obtained from the Phytozome 13. The ginger genome was sequenced by our group and was published in *Horticulture Research* [48]. Moreover, all the Arabidopsis and rice GRAS genes were obtained from Plant TFDB, based on previous reports. In order to identify out all the possible GRAS genes in ginger, the BLASTP search against all the protein sequences of ginger was performed using all the GRAS sequences of both Arabidopsis and rice as inquires. GRAS sequences of ginger were extended by combining gene annotation information and BLASTP search. To confirm the accuracy of predicted ginger proteins, we used the NCBI website to evaluate whether the candidates contain no GRAS structural domains or less than half the length of a typical GRAS structural domain (350 aa) [49]. Those candidates lacking a typical conserved GRAS domain were treated as non-redundant GRAS member and excluded. The ExPASy software (https://web.expasy.org/protparam/, accessed on 16 December 2022) was used to analyze the composition, chemical, and physical characterization of ginger GRAS proteins [50]. WoLF PSORT (http://wolfpsort.org, accessed on 16 December 2022) was used to predicate the subcellular localization of the identified GRAS proteins.

2.2. Phylogenetic Analysis of the GRAS Genes

The phylogenetic analysis was conducted based on all of the GRAS protein sequences of ginger, *A. thaliana* and rice. The Clustal X software was used to align the GRAS sequences through the default parameters. To investigate the relationships of GRAS members, the phylogenetic analysis was performed using the maximum likelihood (ML) method in MEGA 11 (https://www.megasoftware.net/, accessed on 21 May 2022), with 1000 bootstrap replications [51]. Subsequently, these GRAS sequences were classified into different subfamilies on the basis of the obtained phylogenetic tree. At last, the phylogenetic tree was edited and visualized.

2.3. Chromosomal Location and Gene Duplication

The chromosome positions of 66 GRAS genes were obtained from ginger transcriptome of the research group. Duplicate gene pairs were generated by BLAST program through aligning the paralogs of ginger GRAS. The multiple collinear scanning tool kits (MCScanX) was used to analyze the gene duplications and collinearity relationships of GRAS genes [52]. If both coverage and similarity of aligned genes are greater than 75%, potential gene duplication events will be hypothesized to occurred among these genes. Tandem duplications were defined as duplicated genes with less than one intervening on a single chromosome. If not, they were treated as segmental repeats in the present study. We used TBtools software to draw the physical map of GRAS genes and display the distribution of GRAS genes on different ginger chromosomes.

2.4. Gene Structure and Conserved Motif Analysis of GRAS Genes in Ginger

In order to determine the features in the ginger GRAS sequences, multiple sequence alignment of 66 GRAS proteins was obtained by using Jalview software v2.10.5 with default parameters and realigned using Clustal. The conserved motif domains of ginger GRAS proteins were examined by using the MEME suite tool (http://meme.nbcr.-net/meme/intro.html, accessed on 23 May 2022) from the website [53]. The exon–intron structure of ginger GRAS was analyzed by using the TBtools based on the gff3 file of ginger genome. The patterns of gene structure and conversed motif were visualized by TBtools.

2.5. Expression Analysis of Ginger GRAS Genes

The transcriptome data was used to reveal the expression patterns of GRAS genes in different tissues. Young leaves of ginger after salt, heat and cold treatments were collected for RNA-seq. Total RNA was extracted by using the TRIzol kit (Invitrogen, Carlsbad, CA, USA) an mRNA purification kit (Promega, Beijing, China) was used to purify mRNA from total RNA, according to the manufacturers' protocols. For each sample, an amount of 20 µg RNA was enriched by using oligo (dT) magnetic beads. Then, these RNAs were digested into short fragments. The first- and second-strand cDNA were synthesized and purified at the BGI (Shenzhen, China). The purified fragments were linked to the sequencing adaptors. The RNA-seq was carried out on an Illumina Hiseq 2000 sequencing system.

Differentially expressed genes were identified by a rigorous algorithm. The significance of gene expression difference was judged by a criteria of the absolute value of log2 ratio > 1 and $p < 0.001$ based on the *Z. officinale* genome sequence.

QRT-PCR was used to detect the expression of selected GRAS genes in response to abiotic-stress. The qRT-PCR primers were designed by using Primer Primer 5 software (http://frodo.wi.mit.edu/, accessed on 23 May 2022) (Supplementary Materials Table S1). The TUB2 gene is always used as an internal reference gene; it was found to be expressed in most of the tissues with stable expression levels. The TUB2 gene was used as an internal control in this study. The PCR program was was performed as follows: an initial denaturation of 95 °C for 30 s, a second denaturation of 95 °C for 10 s 40 cycles, and 60 °C for 30 s. Each reaction was carried out with three biological replicates. The $2^{-\triangle\triangle CT}$ method was used to calculated the relative expression level of each GRAS gene [54].

2.6. cis-Elements and Target miRNAs Analysis of Ginger GRAS

For each *ZoGRAS* gene, 2 kb (kilo-base) DNA sequences in upstream regions were extracted from the ginger genome. The *cis*-regulatory elements were searched and analyzed in this 2 kb region by using the online server of PlantCARE (http://bioinformatics.psb.ugent.be/webtools/plantcare/html/, accessed on 23 May 2022) and TBtools v1.0971 [52]. The web-based psRNA Target Server (http://zhaolab.org/psRNATarget, accessed on 27 December 2022) was used to determine the potential miRNAs that targeting the *ZoGRAS* genes with default parameters.

3. Results

3.1. Identification and Physicochemical Properties of GRAS Genes in Ginger

In total, 66 GRAS proteins in ginger were identified by a combination of methods (Supplementary Materials Table S2). The complete GRAS domain (PF03514) were found in all of the ginger GRAS proteins. The physical and chemical characteristics of ginger GRAS proteins were analyzed by using ExPasy. Among the 66 GRAS proteins, ZoGRAS#34 is the smallest, with 381 aa; the largest is ZoGRAS#13, with 3159 aa. The molecular weight of the ginger GRAS proteins ranged from 14.66 kDa (ZoGRAS#34) to 117.37 kDa (ZoGRAS#13). The pI ranged from 4.81 (ZoGRAS#34) to 9.65 (ZoGRAS#37). The numerical range of molecular weight and PI agree with that of other species, indicating the conservation of GRASs in different species. Subcellular localization analysis showed that a total of 30 GRAS proteins were located in the nuclear region, 16 in the cytoplasm, 1 in the intercellular filaments, and 19 in the chloroplasts (Supplementary Materials Table S3).

3.2. Multiple Sequence Alignment, Phylogeny, and Classification of GRAS Genes

In order to detect the relationships among GRAS family proteins, the GRAS domains of 70 rice, 48 Arabidopsis and 66 ginger GRASs were collected. The aligned GRAS data were used to construct the phylogeny using the Maximum Likelihood method in MEGA (Figure 1 and Supplementary Materials Table S2). According to the relationship with known rice and Arabidopsis homologs, and the clustering patterns here, ginger GRAS proteins were classified into nine subfamilies: SCL3, DELLA, HAM, LISCL, SHR, PAT1, LAS, SCR, and DLT. SHR and PAT1 were the most abundant subfamilies with 12 and 13 GRAS genes, respectively. The other subfamilies, namely SCL, DELLA, HAM, LISCL, LAS, SCR, and DLT, had 4, 6, 11, 7, 2, 9, and 2 GRAS genes, respectively.

3.3. Chromosome Distribution of Ginger GRAS Genes

According to the physical location of genes in the ginger genome, the chromosomal positions of the GRAS gene were depicted (Figure 2). Our results showed that the 66 GRAS genes distributed unevenly on the 11 ginger chromosomes. Most of the *ZoGRAS* located in the areas with high gene density. It is worth noting that Chr04 contained the most GRAS genes (9 genes). Both Chr03 and Chr01 have 8 GRAS gene members. Some chromosomes (e.g., Chr06, Chr07 and Chr06) have less than 5 GRAS genes.

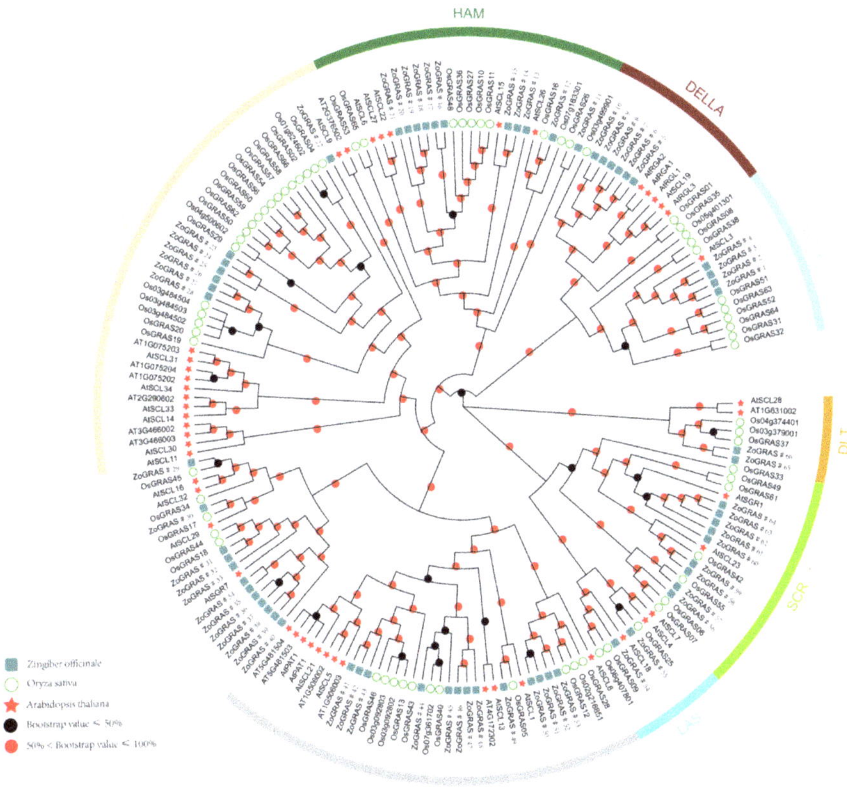

Figure 1. A rootless phylogenetic tree representing 66 GRAS proteins in ginger, rice, and Arabidopsis. The arcs in different color indicate different subfamilies of GRAS. Ginger GRAS proteins with the prefix "Zo" indicate "*Zingiber officinale*".

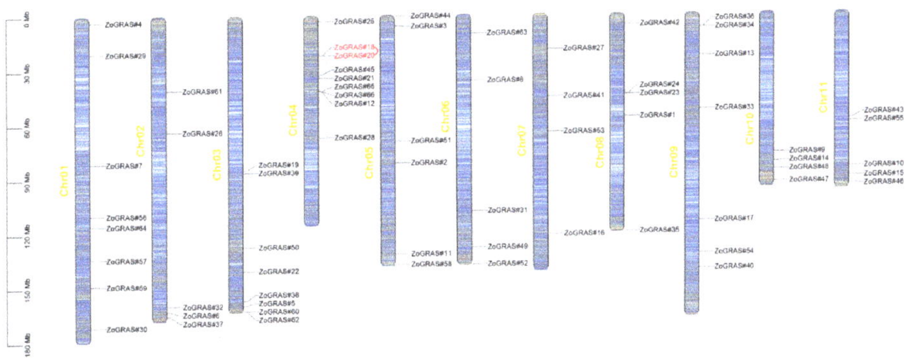

Figure 2. The chromosomal distribution patterns of ginger GRAS genes. GRAS gene pairs resulted from tandem duplication event are connected with red lines. Characters in yellow are the chromosome names. The color gradient of chromosomes from blue to red represent gene density from low to high. Blank area represents lack of gene distribution on chromosomes.

3.4. Gene Structure and Motif Analysis of Ginger GRAS

Exon/intron gene structure plays an important role in gene family evolution. GRAS intron and exon structure were obtained by comparing the genomic DNA sequences (Figure 3). The positioin and number of exons and introns of all GRAS genes were analyzed. Our analysis showed that 77.3% of the GRAS genes had no intron. Nine *Zo*GRAS members contain one or more introns. GRAS13 contains 15 introns. Three introns were found in *Zo*GRAS49, *Zo*GRAS55, and *Zo*GRAS15.

Figure 3. Gene structure of GRAS genes in ginger. Yellow boxes represent untranslated regions, green boxes represent exons, black lines represent introns.

There are 10 conserved motifs (Motif 1–10) in ginger GRAS proteins. The number of motifs vary among ginger GRAS proteins. The conserved motifs of all GRAS proteins were mainly found in the C-terminus, and the types and positions of conserved motifs contained in the same subfamily of proteins were very similar (Figure 4). For example, the DELLA subfamily proteins all contain conserved motifs, and the SCR subfamily proteins all contain conserved motifs. In ginger, Motif-1, Motif-2, and Motif-7 were identified in most of the GRAS proteins. ZoGRAS#34 contains 3 conserved motifs: Motif 3, Motif 5, and Motif 8. ZoGRAS#39 and ZoGRAS#38 contain 4 motifs: Motif 1, Motif 2, Motif 7, and Motif 9. ZoGRAS#60 contains 5 motifs: Motif 2, Motif 3, Motif 6, Motif 8, and Motif 9. ZoGRAS#36 contains 5 motifs: Motif 1, Motif 2, Motif 7, Motif 8, and Motif 9. The number of conserved motifs on the rest of GRAS proteins in ginger ranged from 6 to 10. Based on the analysis of the structure and conserved motifs of GRAS proteins in different subclades, it is clear that members within the same subclade are relatively conserved. Thus, the functions of genes in the same subclade may be similar.

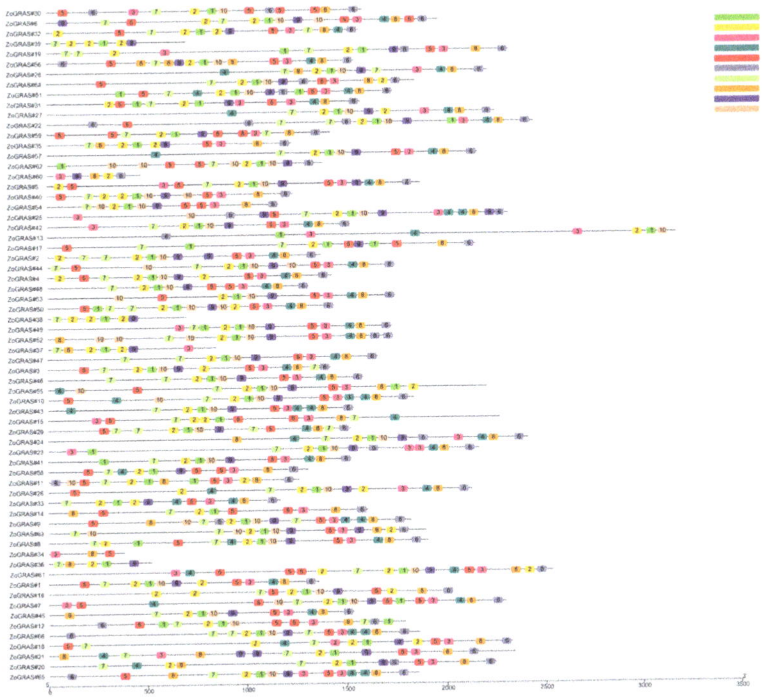

Figure 4. Motif composition of GRAS proteins in ginger. Colored boxes represent motifs detected in this study.

3.5. GRAS Promoter cis-Acting Elements and Target miRNA Analysis

In order to further study the regulation mechanism of ginger GRAS genes under abiotic stress, the upstream 2 kb sequences of 66 GRAS genes were captured from ginger genome to analyze the cis-acting elements. The results show that there are approximately 30 cis-acting elements that can function efficiently, and cis-acting elements that can be efficiently expressed, as well as elements with well-defined functions, were analyzed and explained. Hormone response elements, including ABRE (abscisic acid responsive), P-box, GARE-motif and TATC-box (gibberellin responsive), TCA-element and SARE (salicylic acid responsive), TGACG-motif and CGTCA-motif (MeJA-responsive), TGA-box (auxin-responsive), and Unnamed_1 (phytochrome) (Figure 5), were identified. Abiotic-stress-response elements include TC-rich repeats (defense and stress responses), GC-motif (hypoxia-specific induction), MBS (drought inducibility), WUN-motif (wound responsive element), DRE, and LTR (dehydration, low temperature, and salt-stress responsiveness). Response elements to light include the TCT-motif, ACE, GT1-motif, G-Box, Sp1, 3-AF1 binding site, AAAC-motif, 4cl-CMA1b, MRE, Box 4, ATCT-motif, ATC-motif, CAG-motif, TCCC-motif, LAMP-element, GATA-motif, ACE, I-box, AE-box, ACA-motif, AT1-motif, BoxII, chs-CMA1a, chs-Unit 1 m1, C-box, chs-CMA2a, Gap-motif, GA-motif, L-box, Pc-CMA2c, sbp-CMA1c, and TCT-motif. Other response elements include the CCAAT-box, Unnamed__1, AT-rich element, Box-III, HD-Zip 3 (protein binding site), NON-box (meristem expression) and CAT-box, TATA-box, motif I (root specificity), CAAT-box and A-box (promoter and enhancer), AT-rich sequence (ecitator), MBST (flavonoid biosynthesis gene regulation), MSA-like (cell cycle), GCN4_motif (endosperm expression), ARE (anaerobic induction), O2-site (zein metabolic regulation), HD-Zip 1 (differentiation of palisade mesophyll cells), circadian (circadian rhythm adjust) and RY-element (seed-specific regulation). A WUN-motif associated with wound reactivity was found in both *Zo*GRAS#19 and

*Zo*GRAS#11. The RY element, a seed-specific element, is found only in *Zo*GRAS#49 and *Zo*GRAS#3. Motif I is a root-specific regulated cis-element found in *Zo*GRAS#53. MBST is a flavonoid biosynthesis gene-regulated cis-element found in *Zo*GRAS#42, *Zo*GRAS#58, *Zo*GRAS#26, and *Zo*GRAS#20. MSA, which is associated with the cell cycle, is found in *Zo*GRAS#25, *Zo*GRAS#48, *Zo*GRAS#47, *Zo*GRAS#46, *Zo*GRAS#66, and *Zo*GRAS#65. Presumptive target gene prediction analyses showed that 19 target miRNAs for 11 *Zo*GRAS genes were predicted (Supplementary Materials: Table S4).

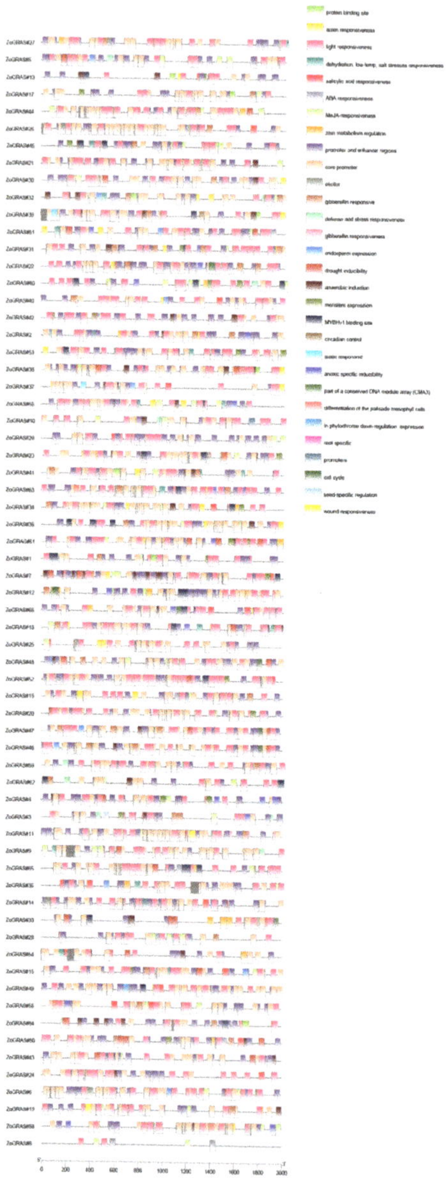

Figure 5. Cis-acting elements of ginger GRAS promoters. Different types of cis-acting elements are marked by different colors.

3.6. GO Annotation of GRAS Protein Sequences

To understand the functions of GRAS proteins in different biological processes in ginger, a GO annotation analysis of ZoGRAS genes were performed. The results suggest that GRAS proteins may be involved in many biological, cellular, and molecular processes. Most GRAS proteins have the function of protein-binding and transcriptional regulation. The major GRAS proteins are involved in the regulation of RNA biosynthesis processes, nucleic acid template transcription, and template transcription. There are some genes involved in hormonal regulation, such as gibberellin regulation. Some genes were also found to be involved in regulating plant organs development and abiotic stresses response. The results showed that 50 of the 66 GRAS proteins have organic cyclic compound binding, specific binding, and transcriptional regulatory activities. The biological process analysis of GRAS genes showed that GRAS genes function in biometabolic, abiotic, and biotic-stress response (Figure 6).

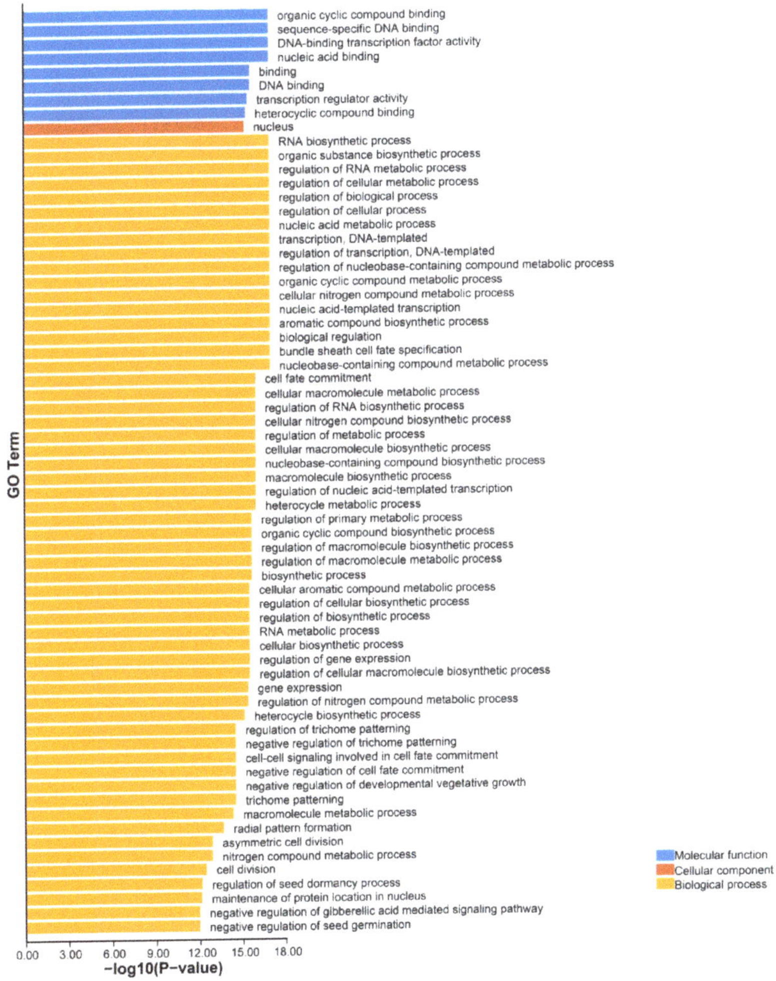

Figure 6. GO annotation of GRAS protein sequences.

3.7. Synteny and Evolutionary Analysis of GRAS Genes

In all organisms, gene replication happens frequently, causing functional changes of new genes from previous genes. In a previous study, tandem duplication event was defined as more than one gene family member arising within a 200 kb intergenic region. In contrast, segmental duplication is common in plants. Because the process of polyploidization make these plants retain many large blocks of duplicated chromosomes. In general, the copy number increase of gene family in plants was contributed by tandem and segmental duplications. Many ginger GRAS genes exist on different chromosomes in ginger, indicating that the GRAS gene family is highly conserved (Figure 7). In this study, two GRAS genes resulted from one tandem duplication event (*Zo*GRAS#18/*Zo*GRAS#20) were found in ginger chromosome (Chr04). In total, 50 ginger GRAS genes associated with 25 segmental duplication events were detected, which indicate that segmental duplication might drive the evolution of ginger GRAS family. The segmental events were assessed to have occurred 4.66 million years ago (Ma).

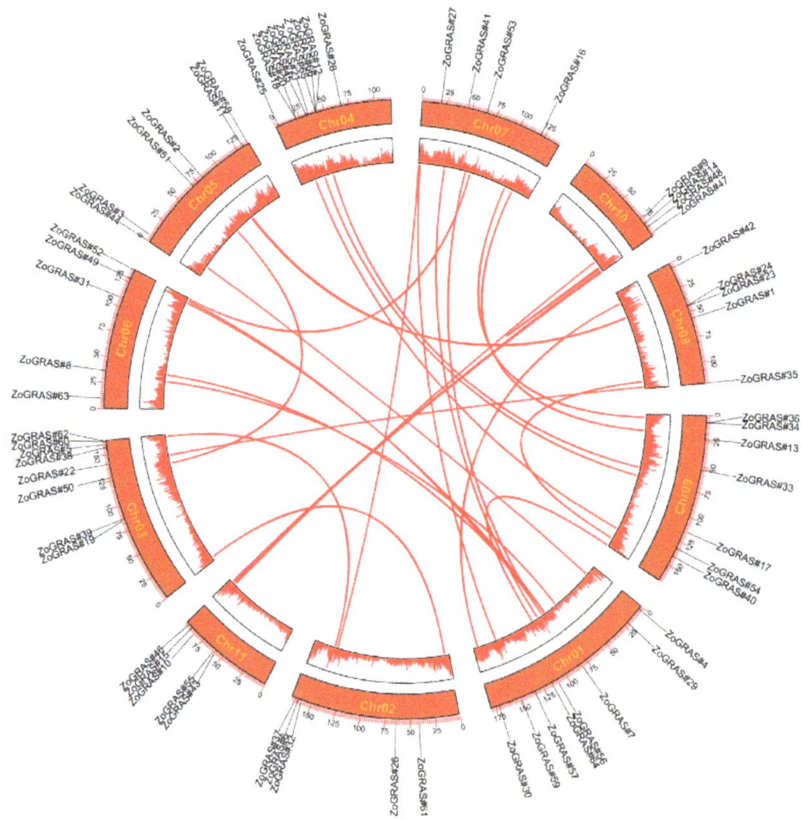

Figure 7. Diagram of chromosomal relationship of GRAS genes in ginger. The red lines represent duplicated GRAS gene pairs in ginger. The chromosome names of ginger are shown in the middle of each chromosome.

To explore the relationships of the ginger GRAS family with other plants, two comparative collinear maps of ginger related to Arabidopsis and barley were constructed (Figure 8). Homology of ginger GRAS genes were found in both Arabidopsis and barley. In total, 6 GRAS genes of ginger displayed syntenic relationships with those in Arabidopsis. However, no syntenic gene pairs between barley and ginger were found.

3.8. Expression Patterns of Ginger GRAS Genes in Response to Abiotic Stress

In order to study the potential functions of the GRAS genes under various non-biological stresses, RNA-seq data under heat, cold, and salt treatments were used to detect their expression levels. In total, 66 GRAS genes were found differentially expressed in at least one stress treatment. The expression of 12 randomly selected genes were also detected by using qRT-PCR under cold- and heat-stress treatments. All the 12 selected genes are significantly induced by stress at one or more time points (Figure 9). In general, genetic response was slowe under cold conditions. The gene expression level increased gradually under low-temperature conditions and peaked at 12 h or 24 h (Figure 10).

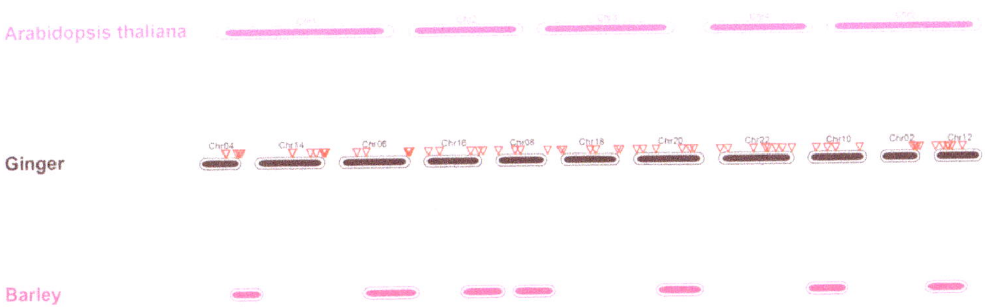

Figure 8. Collinearity analysis of ginger, Arabidopsis, and Barley. The red inverted triangles represents the *ZoGRAS* genes.

3.9. GRAS Gene Expression Profile of Ginger in Different Tissues

In order to explore the potential functions of the ginger GRAS genes in different developmental stages of ginger organs, RNA-seq data were used to examine their expression patterns (Figure 11). A total of 39 of the 66 GRAS genes were found expressed in all samples (FPKM > 0), and 27 GRAS genes exhibited constitutive expression (FPKM > 1 in all samples). Some genes showed preferential expression in the tissues detected. For example, one gene in the meristem of stems (*ZoGRAS#45*), two genes in leaves (*Zo*GRAS#9), a couple of genes in mature florescences (*Zo*GRAS#9 and *Zo*GRAS#45), and three genes in the roots (*Zo*GRAS#44, *Zo*GRAS#9, and *Zo*GRAS#45) exhibited highest expression level. The expression level of some genes showed a obvious trend in different development stages of ginger organs. Such as, the expression level of *ZoGRAS#9* gradually decreased, whereas, the expression level of *ZoGRAS#19* showed a increasing trend (Figure 11). The transcriptional abundance of GRAS genes varied in different organs, suggesting that GRAS genes exhibit diverse functions during the growth and development of ginger.

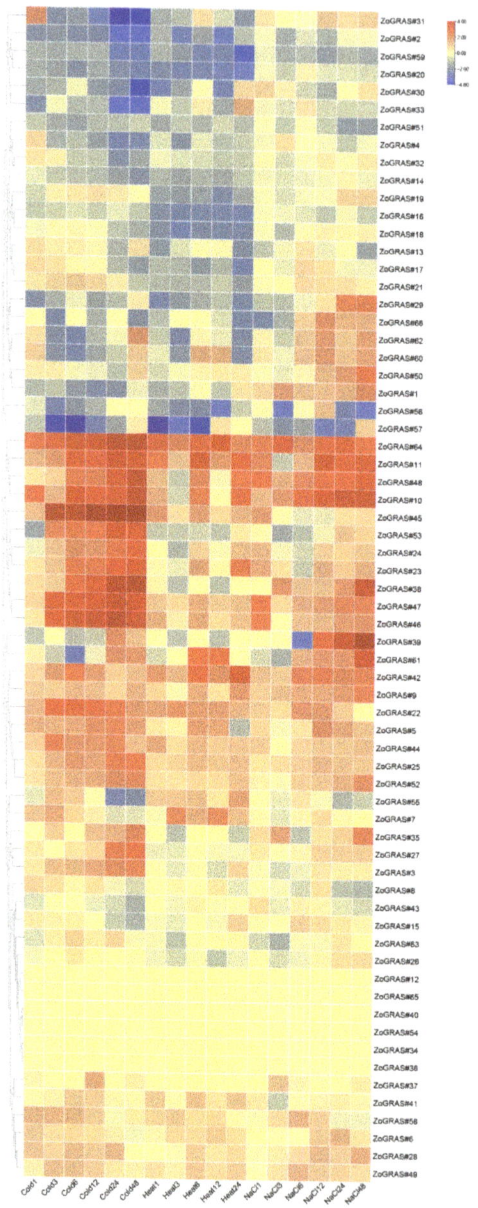

Figure 9. GRAS gene expression profile in ginger, including hierarchical clustering of GRAS gene expression profiles in ginger leaves under different stress conditions.

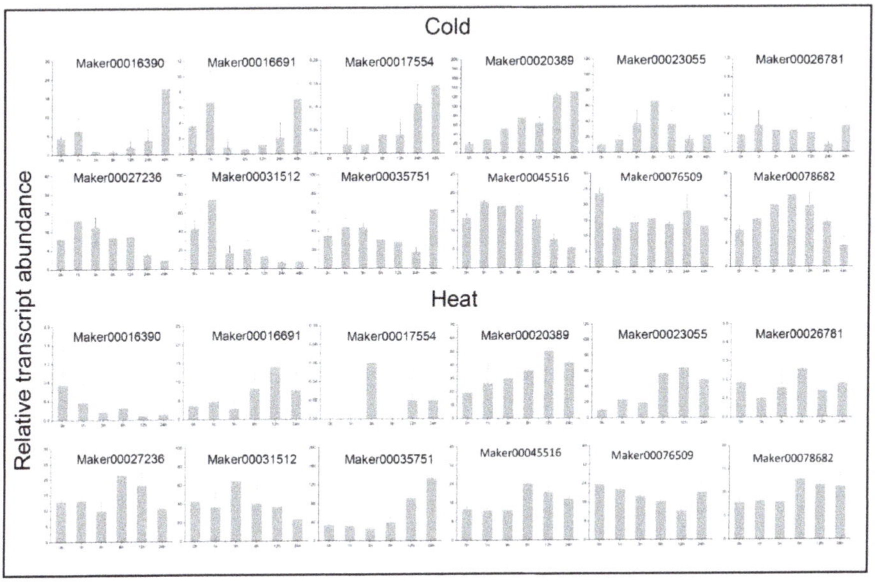

Figure 10. QRT-PCR expression analysis of GRAS genes under abiotic stresses. The data has been normalized, and the bar represents the standard deviation.

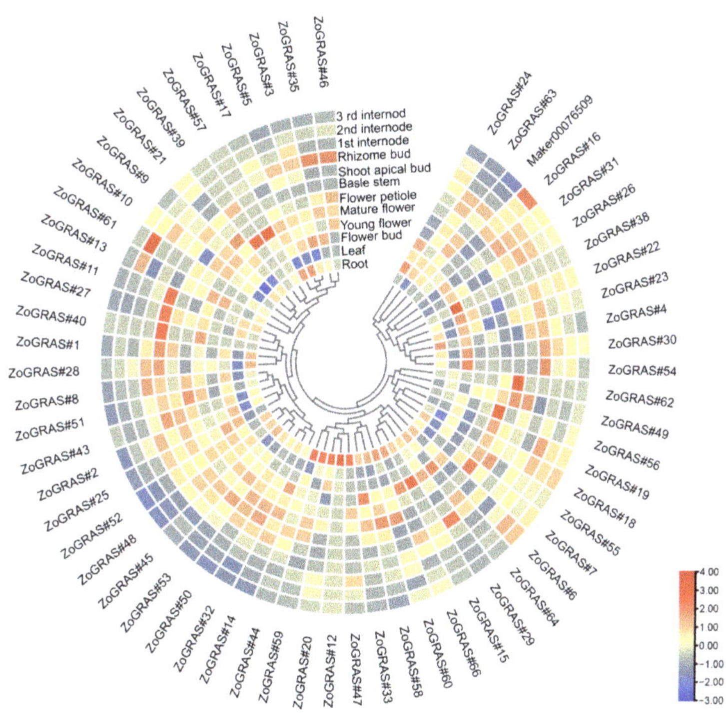

Figure 11. GRAS gene expression profile of ginger. Hierarchical clustering of GRAS gene expression profiles of ginger in 12 samples, including different tissues and developmental stages.

4. Discussion

Transcription factors play important roles in plant growth and development, response to adversity stress, and various aspects of plant life activities. Previous studies have also shown that the expression of GRAS family members in different organs e.g., roots, stems, leaves, flowers, and fruits varies during plant growth and development, thus indicating that the GRAS gene family is involved in all stages of plant growth and development [29–34]. Moreover, the GRAS gene family plays a key regulatory role in response to environmental stresses [3,20,26,28]. Nevertheless, the GRAS genes in ginger have not been reported to date. In this study, we searched GRAS genes in the ginger genome and identified 66 GRAS family genes, which is more than the number of GRAS genes in Arabidopsis (48) and tomato (53), but less than the number of GRAS genes in rice (70) [55]. Previous studies have suggested that GRAS proteins may expand after the divergence of higher and lower plants, and their numbers vary widely in different plants [3,24]. Comparative analyses have shown that amplification of the GRAS gene family is associated with whole-genome duplications, tandem duplications, and segmental duplications [12]. Tandem repeats were detected in Arabidopsis, tomato, rice, and poplar [22,27]. In ginger, a whole-genome duplication event occurred during the evolutionary history [48]. In this study, chromosome mapping showed that *Zo*GRAS genes are distributed on all the 11 chromosomes of ginger, whereas the number of *Zo*GRAS ranges from 4 to 9. A total of 1 tandem duplication event and 25 segmental duplication events in connection with 52 GRAS genes were determined. Thus, expansion of the *Zo*GRAS family in ginger might be the result of whole-genome duplication, segmental duplication, and tandem duplications.

In this study, the GRAS genes of Arabidopsis, rice, and ginger were divided into nine clades. The relationship between ginger and *Arabidopsis thaliana* suggests the origin and diversity of GRAS. The results indicate that the *Zo*GRAS genes originated and differentiated in the ancestor of monocotyledons (ginger) and dicotyledons (*A. thaliana*). The PAT1 subfamily contains the largest number of ginger GRAS genes (15). The genes of the PAT1 subfamily play an important role in plant phytochrome and defense signaling pathways [10,56]. SHR and SCR subfamilies are involved in maintaining root apical meristem and regulating the root morphology in Arabidopsis [57]. Therefore, 19 *Zo*GRAS genes (12 homologous to *At*SHR and 7 homologous to *At*SCR) may have similar functions. The HAM subfamily, consisting of four *Zo*GRAS proteins, could be involved in shoot meristem formation [32]. In ginger, six *Zo*GRAS genes were identified as DELLA subfamily members. Studies have shown that the DELLA subfamily plays a negative regulatory role in the GA pathway and could lead to the dwarfing of plant growth [58]. The phylogenetic analysis of the GRAS gene family provides a theoretical basis for further study of the functional genome of ginger.

Introns play a very important role in plants' evolution and undergo loss and gain during the evolutionary history [59]. The GRAS gene has been identified in 67% of individual exon-less genes in Arabidopsis [27], 84% in poplar [60], and 77.4% in tomato [49]. The proportion of intron-less GRAS genes in ginger was higher (~77.3%) than in Populus (54.7%), rice (55%), and Arabidopsis (67.6%), but lower than in Plum (82.2%) [24,61,62]. Intron-less genes have also been identified in other large gene families, e.g., the small auxin-up RNA (Saur) gene family and the DEAD box RNA helicase F-box TF family [63,64]. Plant GRAS genes were hypothesized to originate from prokaryotic genes through a mechanism of horizontal gene transfer and duplicate during evolution, which could explain the scenery of abundant GRAS genes without introns [27,65]. Intron-less genes may be able to respond quickly to stress and regulate growth and developmental processes in plants [66–68]. Thus, many *Zo*GRAS members tend to respond rapidly to environmental changes. We analyzed 10 different conserved motifs in the *Zo*GRAS family and found that members in the same group usually have a similar motif composition and transcriptional regulators, whereas the motif compositions and distributions in the N-terminus vary remarkably among different *Zo*GRAS subgroups. In evolutionary processes, mutations of non-conservative amino acids cause motifs' variation, which contributes to the distinct and diverse gene functions of GRAS genes.

GRAS proteins exhibit multiple functions and play important roles in plant growth and a variety of physiological processes, including GA signal transduction and hormone metabolism regulation in response to biotic and abiotic stresses [13,69]. The DELLA protein plays a negative role in the endogenous plant hormone gibberellin, leading to the reduced sensitivity of plants to gibberellin, resulting in a range of phenomena such as dwarfism during growth [70]. In addition, the study of the PATI family revealed the response of the plant PATI family to light [10]. Both PAT1 and SCL13 were found to act in the light signaling pathway of Arabidopsis and control the basic plant hairs through photoreceptor transduction signals [56]. Tissue-specific expression patterns indicated that most of the ZoGRAS genes were expressed in all tested tissues. In the GO enrichment analysis, the genotype ends are subject to functional diversity, resulting in the majority of ginger GRAS proteins playing important roles in many different biological processes. Based on the analysis of the promoted region, we found growth-related, abiotic-responsive, and hormone-responsive cis-elements in the promoted region. These results suggest that ginger GRAS transcription factors play important roles in hormone signaling pathways. The comparison of GRAS homologs in different species, including expression patterns and protein sequences, has enhanced our understanding of the role of these GRAS genes in ginger.

Transcriptional regulators belonging to the consent subcluster share a recent common evolutionary origin and have molecularly functionally related conserved motifs. Highly homologous genes between ginger and rice and Arabidopsis were used to predict gene function in ginger. No studies have been conducted to analyze the expression pattern of ginger GRAS family genes under abiotic stress. In this study, we analyzed the expression patterns of ginger GRAS family genes in different tissues, different development stages, and different abiotic stress treatment with transcriptome and real-time qPCR. In Arabidopsis, a GRAS family transcription factor called SCARECROW-LIKE28 (SCL28) plays a critical role in determining cell size [71]. We found that *ZoGRAS#66* and *ZoGRAS#65* in ginger are closely related to *AtSCL28*, thus indicating that these two genes may be involved in regulating cell size, which is associated with the expansion of the ginger rhizome. *AtSCL15*, a member of the HAM subfamily, is involved in salt- and cold-stress responses [72]. In particular, three genes of the HAM subfamily in ginger, namely *ZoGRAS#13–15*, were also found to be significantly differently expressed in cold- and salt-stress treatment. Previous studies have shown that *AtSCL14* genes of the LISCL subfamily in *A. thaliana* can enhance stress resistance through upregulating the expression of stress-response genes [66]. Similarly, four genes (*ZoGRAS#23–25,27*) of the LISCL subfamily were significantly upregulated under cold stress. A member of the PAT1 subfamily *AtSCL13* showed strong tolerance to drought-, cold-, and salt-stress treatments [66,73]. In this study, six *ZoGRAS* genes of the PAT1 subfamily were found to be differently expressed. Among them, *ZoGRAS#42* and *ZoGRAS#48* were significantly upregulated under heat treatment. *ZoGRAS#48*, *ZoGRAS#52*, and *ZoGRAS#53* increased significantly after cold and salt treatments. In general, functions of *ZoGRAS* genes overlap in tissue development and response to various stresses; this finding is in agreement with previous studies, as reviewed by Waseem et al. [3]. These results lay a foundation for further study of stress response and additional functions of ZoGRAS gene family members.

5. Conclusions

In this study, a comprehensive analysis of the GRAS family genes in ginger was performed, and a total of 66 full-length GRAS genes were identified. Based on the phylogenetic tree and the distribution of conserved motifs, the GRAS family was divided into 10 subfamilies with highly similar gene structures and motif compositions in the same subfamily or subgroup. The phylogenetic comparison and collinear analysis of different plant GRAS genes provided important clues for the evolutionary characteristics of ginger GRAS genes. The GRAS gene plays an important role in the growth and development of ginger. The phylogeny and gene expression analysis and abiotic stress treatment during the development of ginger rhizome will provide a reference for the functional analysis of the GRAS gene. Taken together, these results provide a valuable resource for a better understanding of the biological role of the GRAS gene in ginger. In the future, various advanced approaches, e.g., yeast hybrids, co-immunoprecipitation, and yeast pull-down assay, are needed to uncover the mechanism of GRAS TFs in ginger development and stress.

Supplementary Materials: The following supporting information can be downloaded at: https://www.mdpi.com/article/10.3390/genes14010096/s1, Table S1: A member of the GRAS transcription factor family in ginger, Arabidopsis, and rice; Table S2: Q-PCR and primer list; Table S3: Subcellular localization of GRAS gene in *Z. officinale* Roscoe; Table S4: miRNA targets of GRAS genes in *Z. officinale* Roscoe.

Author Contributions: H.-L.L. and H.X. conceived the study; S.T., D.J., J.L., Y.W., M.G., J.L. and C.S. performed the experiments; H.-L.L., S.T. and Y.W. analyzed the data; H.-L.L., S.T., D.J., J.L., Y.W., M.G., J.L., M.X. and C.S. contributed reagents/materials/analysis tools; H.-L.L., S.T. and H.X. wrote the paper; H.-L.L. and H.X. edited the paper. All authors have read and agreed to the published version of the manuscript.

Funding: This research was funded by the Scientific and Technological Research Program of Chongqing Municipal Education Commission (KJZD-M202101301), the National Natural Science Foundation of China (32270237), the Foundation for High-Level Talents of Chongqing University of Arts and Sciences (P2021YL11 and R2018STZ25), and Chongqing Science and Technology support projects (CSTB2022NSCQ-MSX1263).

Institutional Review Board Statement: Not applicable.

Informed Consent Statement: Not applicable.

Acknowledgments: We would like to thank Zhiduan Chen (IBCAS) for helping to read and revise the manuscript.

Conflicts of Interest: The authors declare no conflict of interest.

References

1. Khan, M.; Ali, S.; Manghwar, H.; Saqib, S.; Ullah, F.; Ayaz, A.; Zaman, W. Melatonin Function and Crosstalk with Other Phytohormones under Normal and Stressful Conditions. *Genes* **2022**, *13*, 1699. [CrossRef] [PubMed]
2. Dutta, M.; Saha, A.; Moin, M.; Kirti, P.B. Genome-Wide Identification, Transcript Profiling and Bioinformatic Analyses of GRAS Transcription Factor Genes in Rice. *Front. Plant Sci.* **2021**, *12*, 777285. [CrossRef] [PubMed]
3. Waseem, M.; Nkurikiyimfura, O.; Niyitanga, S.; Jakada, B.H.; Shaheen, I.; Aslam, M.M. GRAS transcription factors emerging regulator in plants growth, development, and multiple stresses. *Mol. Biol. Rep.* **2022**, *49*, 9673–9685. [CrossRef] [PubMed]
4. Paz-Ares, J.; Ghosal, D.; Wienand, U.; Peterson, P.A.; Saedler, H. The regulatory c1 locus of Zea mays encodes a protein with homology to myb proto-oncogene products and with structural similarities to transcriptional activators. *EMBO J.* **1987**, *6*, 3553–3558. [CrossRef]
5. Zhang, B.; Liu, J.; Yang, Z.E.; Chen, E.Y.; Zhang, C.J.; Zhang, X.Y.; Li, F.G. Genome-wide analysis of GRAS transcription factor gene family in *Gossypium hirsutum* L. *BMC Genom.* **2018**, *19*, 348. [CrossRef]
6. Shen, Z.; Lin, Y.; Zou, Q. Transcription factors–DNA interactions in rice: Identification and verification. *Briefings Bioinform.* **2019**, *21*, 946–956. [CrossRef]
7. Sidhu, N.S.; Pruthi, G.; Singh, S.; Bishnoi, R.; Singla, D. Genome-wide identification and analysis of GRAS transcription factors in the bottle gourd genome. *Sci. Rep.* **2020**, *10*, 14338. [CrossRef]
8. Peng, H.; Hu, H.; Xi, K.; Zhu, X.; Zhou, J.; Yin, J.; Guo, F.; Liu, Y.; Zhu, Y. Silicon Nanoparticles Enhance Ginger Rhizomes Tolerance to Postharvest Deterioration and Resistance to *Fusarium solani*. *Front. Plant Sci.* **2022**, *13*, 816143. [CrossRef]
9. Pysh, L.D.; Wysocka-Diller, J.W.; Camilleri, C.; Bouchez, D.; Benfey, P.N. The GRAS gene family in Arabidopsis: Sequence characterization and basic expression analysis of the SCARECROW-LIKE genes. *Plant J.* **1999**, *18*, 111–119. [CrossRef]
10. Bolle, C. The role of GRAS proteins in plant signal transduction and development. *Planta* **2004**, *218*, 683–692. [CrossRef]
11. Zhang, D.; Iyer, L.M.; Aravind, L. Bacterial GRAS domain proteins throw new light on gibberellic acid response mechanisms. *Bioinformatics* **2012**, *28*, 2407–2411. [CrossRef] [PubMed]
12. Zhu, X.; Wang, B.; Wei, X. Genome wide identification and expression pattern analysis of the GRAS family in quinoa. *Funct. Plant Biol.* **2021**, *48*, 948. [CrossRef] [PubMed]
13. Cenci, A.; Rouard, M. Evolutionary Analyses of GRAS Transcription Factors in Angiosperms. *Front. Plant Sci.* **2017**, *8*, 273. [CrossRef] [PubMed]

14. Ko, J.-H.; Yang, S.H.; Park, A.H.; Lerouxel, O.; Han, K.-H. ANAC012, a member of the plant-specific NAC transcription factor family, negatively regulates xylary fiber development in *Arabidopsis thaliana*. *Plant J.* **2007**, *50*, 1035–1048. [CrossRef] [PubMed]
15. Zhong, R.; Lee, C.; Ye, Z.-H. Global Analysis of Direct Targets of Secondary Wall NAC Master Switches in Arabidopsis. *Mol. Plant* **2010**, *3*, 1087–1103. [CrossRef]
16. He, X.-J.; Mu, R.-L.; Cao, W.-H.; Zhang, Z.-G.; Zhang, J.-S.; Chen, S.-Y. AtNAC2, a transcription factor downstream of ethylene and auxin signaling pathways, is involved in salt stress response and lateral root development. *Plant J.* **2005**, *44*, 903–916. [CrossRef]
17. Kikuchi, K.; Ueguchi-Tanaka, M.; Yoshida, K.T.; Nagato, Y.; Matsusoka, M.; Hirano, H.-Y. Molecular analysis of the NAC gene family in rice. *Mol. Genet. Genom.* **2000**, *262*, 1047–1051. [CrossRef]
18. Lee, M.-H.; Kim, B.; Song, S.-K.; Heo, J.-O.; Yu, N.-I.; Lee, S.A.; Kim, M.; Kim, D.G.; Sohn, S.O.; Lim, C.E.; et al. Large-scale analysis of the GRAS gene family in *Arabidopsis thaliana*. *Plant Mol. Biol.* **2008**, *67*, 659–670. [CrossRef]
19. Tian, C.; Wan, P.; Sun, S.; Li, J.; Chen, M. Genome-Wide Analysis of the GRAS Gene Family in Rice and Arabidopsis. *Plant Mol. Biol.* **2004**, *54*, 519–532. [CrossRef]
20. Liu, Y.; Huang, W.; Xian, Z.; Hu, N.; Lin, D.; Ren, H.; Chen, J.; Su, D.; Li, Z. Overexpression of *SlGRAS40* in Tomato Enhances Tolerance to Abiotic Stresses and Influences Auxin and Gibberellin Signaling. *Front. Plant Sci.* **2017**, *8*, 1659. [CrossRef]
21. To, V.-T.; Shi, Q.; Zhang, Y.; Shi, J.; Shen, C.; Zhang, D.; Cai, W. Genome-Wide Analysis of the GRAS Gene Family in Barley (*Hordeum vulgare* L.). *Genes* **2020**, *11*, 553. [CrossRef] [PubMed]
22. Huang, W.; Xian, Z.; Kang, X.; Tang, N.; Li, Z. Genome-wide identification, phylogeny and expression analysis of GRAS gene family in tomato. *BMC Plant Biol.* **2015**, *15*, 209. [CrossRef] [PubMed]
23. Chen, Y.; Tai, S.; Wang, D.; Ding, A.; Sun, T.; Wang, W.; Sun, Y. Homology-based analysis of the GRAS gene family in tobacco. *Genet. Mol. Res.* **2015**, *14*, 15188–15200. [CrossRef] [PubMed]
24. Song, X.-M.; Liu, T.-K.; Duan, W.-K.; Ma, Q.-H.; Ren, J.; Wang, Z.; Li, Y.; Hou, X.-L. Genome-wide analysis of the GRAS gene family in Chinese cabbage (*Brassica rapa* ssp. pekinensis). *Genomics* **2014**, *103*, 135–146. [CrossRef]
25. Chen, Y.; Zhu, P.; Wu, S.; Lu, Y.; Sun, J.; Cao, Q.; Li, Z.; Xu, T. Identification and expression analysis of *GRAS* transcription factors in the wild relative of sweet potato *Ipomoea trifida*. *BMC Genom.* **2019**, *20*, 911. [CrossRef]
26. Xu, W.; Chen, Z.; Ahmed, N.; Han, B.; Cui, Q.; Liu, A. Genome-Wide Identification, Evolutionary Analysis, and Stress Responses of the GRAS Gene Family in Castor Beans. *Int. J. Mol. Sci.* **2016**, *17*, 1004. [CrossRef]
27. Liu, X.; Widmer, A. Genome-wide Comparative Analysis of the GRAS Gene Family in Populus, Arabidopsis and Rice. *Plant Mol. Biol. Rep.* **2014**, *32*, 1129–1145. [CrossRef]
28. Grimplet, J.; Agudelo-Romero, P.; Teixeira, R.T.; Martinez-Zapater, J.M.; Fortes, A.M. Structural and Functional Analysis of the GRAS Gene Family in Grapevine Indicates a Role of GRAS Proteins in the Control of Development and Stress Responses. *Front. Plant Sci.* **2016**, *7*, 353. [CrossRef]
29. Silverstone, A.L.; Ciampaglio, C.N.; Sun, T.-P. The Arabidopsis *RGA* gene encodes a transcriptional regulator repressing the gibberellin signal transduction pathway. *Plant Cell* **1998**, *10*, 155–169. [CrossRef]
30. Ikeda, A.; Ueguchi-Tanaka, M.; Sonoda, Y.; Kitano, H.; Koshioka, M.; Futsuhara, Y.; Matsuoka, M.; Yamaguchi, J. Slender Rice, a Constitutive Gibberellin Response Mutant, Is Caused by a Null Mutation of the SLR1 Gene, an Ortholog of the Height-Regulating Gene GAI/RGA/RHT/D8. *Plant Cell* **2001**, *13*, 999–1010. [CrossRef]
31. Greb, T.; Clarenz, O.; Schäfer, E.; Müller, D.; Herrero, R.; Schmitz, G.; Theres, K. Molecular analysis of the LATERAL SUPPRESSOR gene in Arabidopsis reveals a conserved control mechanism for axillary meristem formation. *Genes Dev.* **2003**, *17*, 1175–1187. [CrossRef] [PubMed]
32. Stuurman, J.; Jäggi, F.; Kuhlemeier, C. Shoot meristem maintenance is controlled by a *GRAS*-gene mediated signal from differentiating cells. *Genes Dev.* **2002**, *16*, 2213–2218. [CrossRef] [PubMed]
33. Di Laurenzio, L.; Wysocka-Diller, J.; E Malamy, J.; Pysh, L.; Helariutta, Y.; Freshour, G.; Hahn, M.G.; A Feldmann, K.; Benfey, P.N. The SCARECROW Gene Regulates an Asymmetric Cell Division That Is Essential for Generating the Radial Organization of the Arabidopsis Root. *Cell* **1996**, *86*, 423–433. [CrossRef] [PubMed]
34. Morohashi, K.; Minami, M.; Takase, H.; Hotta, Y.; Hiratsuka, K. Isolation and Characterization of a Novel GRAS Gene That Regulates Meiosis-associated Gene Expression. *J. Biol. Chem.* **2003**, *278*, 20865–20873. [CrossRef] [PubMed]
35. Tyler, L.; Thomas, S.G.; Hu, J.; Dill, A.; Alonso, J.M.; Ecker, J.R.; Sun, T.-P. DELLA Proteins and Gibberellin-Regulated Seed Germination and Floral Development in Arabidopsis. *Plant Physiol.* **2004**, *135*, 1008–1019. [CrossRef]
36. Cui, H.; Levesque, M.P.; Vernoux, T.; Jung, J.W.; Paquette, A.J.; Gallagher, K.L.; Wang, J.Y.; Blilou, I.; Scheres, B.; Benfey, P.N. An Evolutionarily Conserved Mechanism Delimiting SHR Movement Defines a Single Layer of Endodermis in Plants. *Science* **2007**, *316*, 421–425. [CrossRef]
37. Zhang, Q.Q.; Wang, J.G.; Wang, L.Y.; Wang, J.F.; Wang, Q.; Yu, P.; Bai, M.Y.; Fan, M. Gibberellin repression of axillary bud formation in Arabidopsis by modulation of DELLA-SPL9 complex activity. *J. Integr. Plant Biol.* **2020**, *62*, 421–432. [CrossRef]
38. Yuan, Y.; Fang, L.; Karungo, S.K.; Zhang, L.; Gao, Y.; Li, S.; Xin, H. Overexpression of VaPAT1, a GRAS transcription factor from *Vitis amurensis*, confers abiotic stress tolerance in Arabidopsis. *Plant Cell Rep.* **2015**, *35*, 655–666. [CrossRef]
39. Zhang, S.; Li, X.; Fan, S.; Zhou, L.; Wang, Y. Overexpression of HcSCL13, a *Halostachys caspica* GRAS transcription factor, enhances plant growth and salt stress tolerance in transgenic Arabidopsis. *Plant Physiol. Biochem.* **2020**, *151*, 243–254. [CrossRef]
40. Fode, B.; Siemsen, T.; Thurow, C.; Weigel, R.; Gatz, C. The *Arabidopsis* GRAS Protein SCL14 Interacts with Class II TGA Transcription Factors and Is Essential for the Activation of Stress-Inducible Promoters. *Plant Cell* **2008**, *20*, 3122–3135. [CrossRef]

41. Ma, H.-S.; Liang, D.; Shuai, P.; Xia, X.-L.; Yin, W.-L. The salt- and drought-inducible poplar GRAS protein SCL7 confers salt and drought tolerance in *Arabidopsis thaliana*. *J. Exp. Bot.* **2010**, *61*, 4011–4019. [CrossRef] [PubMed]
42. Kaushal, M.; Gupta, A.; Vaidya, D.; Gupta, M. Postharvest Management and Value Addition of Ginger (*Zingiber officinale* Roscoe): A Review. *Int. J. Environ. Agric. Biotechnol.* **2017**, *2*, 397–412. [CrossRef] [PubMed]
43. Li, H.; Liu, Y.; Luo, D.; Ma, Y.; Zhang, J.; Li, M.; Yao, L.; Shi, X.; Liu, X.; Yang, K. Ginger for health care: An overview of systematic reviews. *Complement. Ther. Med.* **2019**, *45*, 114–123. [CrossRef] [PubMed]
44. Ahmad, S.B.; Rehman, M.U.; Amin, I.; Arif, A.; Rasool, S.; Bhat, S.A.; Afzal, I.; Hussain, I.; Bilal, S.; Mir, M. A Review on Pharmacological Properties of Zingerone (4-(4-Hydroxy-3-methoxyphenyl)-2-butanone). *Sci. World J.* **2015**, *2015*, 1–6. [CrossRef]
45. Mao, Q.-Q.; Xu, X.-Y.; Cao, S.-Y.; Gan, R.-Y.; Corke, H.; Beta, T.; Li, H.-B. Bioactive Compounds and Bioactivities of Ginger (*Zingiber officinale* Roscoe). *Foods* **2019**, *8*, 185. [CrossRef]
46. Koo, H.; McDowell, E.T.; Ma, X.; A Greer, K.; Kapteyn, J.; Xie, Z.; Descour, A.; Kim, H.; Yu, Y.; Kudrna, D.; et al. Ginger and turmeric expressed sequence tags identify signature genes for rhizome identity and development and the biosynthesis of curcuminoids, gingerols and terpenoids. *BMC Plant Biol.* **2013**, *13*, 27. [CrossRef]
47. Xing, H.; Jiang, Y.; Zou, Y.; Long, X.; Wu, X.; Ren, Y.; Li, Y.; Li, H.-L. Genome-wide investigation of the AP2/ERF gene family in ginger: Evolution and expression profiling during development and abiotic stresses. *BMC Plant Biol.* **2021**, *21*, 561. [CrossRef]
48. Li, H.-L.; Wu, L.; Dong, Z.; Jiang, Y.; Jiang, S.; Xing, H.; Li, Q.; Liu, G.; Tian, S.; Wu, Z. Haplotype-resolved genome of diploid ginger (*Zingiber officinale*) and its unique gingerol biosynthetic pathway. *Hortic. Res.* **2021**, *8*, 189. [CrossRef]
49. Niu, Y.; Zhao, T.; Xu, X.; Li, J. Genome-wide identification and characterization of GRAS transcription factors in tomato (*Solanum lycopersicum*). *PeerJ* **2017**, *5*, e3955. [CrossRef]
50. Artimo, P.; Jonnalagedda, M.; Arnold, K.; Baratin, D.; Csardi, G.; de Castro, E.; Duvaud, S.; Flegel, V.; Fortier, A.; Gasteiger, E.; et al. ExPASy: SIB bioinformatics resource portal. *Nucleic Acids Res.* **2012**, *40*, W597–W603. [CrossRef]
51. Kumar, S.; Stecher, G.; Li, M.; Knyaz, C.; Tamura, K. MEGA X: Molecular Evolutionary Genetics Analysis across Computing Platforms. *Mol. Biol. Evol.* **2018**, *35*, 1547–1549. [CrossRef] [PubMed]
52. Chen, C.J.; Chen, H.; Zhang, Y.; Thomas, H.R.; Frank, M.H.; He, Y.H.; Xia, R. TBtools: An Integrative Toolkit Developed for Interactive Analyses of Big Biological Data. *Mol. Plant* **2020**, *13*, 1194–1202. [CrossRef] [PubMed]
53. Bailey, T.L. STREME: Accurate and versatile sequence motif discovery. *Bioinformatics* **2021**, *37*, 2834–2840. [CrossRef] [PubMed]
54. Livak, K.J.; Schmittgen, T.D. Analysis of relative gene expression data using real-time quantitative PCR and the $2^{-\Delta\Delta CT}$ method. *Methods* **2001**, *25*, 402–408. [CrossRef]
55. Jin, J.; Tian, F.; Yang, D.-C.; Meng, Y.-Q.; Kong, L.; Luo, J.; Gao, G. PlantTFDB 4.0: Toward a central hub for transcription factors and regulatory interactions in plants. *Nucleic Acids Res.* **2016**, *45*, D1040–D1045. [CrossRef]
56. Bolle, C.; Koncz, C.; Chua, N.-H. PAT1, a new member of the GRAS family, is involved in phytochrome A signal transduction. *Genes Dev.* **2000**, *14*, 1269–1278. [CrossRef]
57. Heo, J.-O.; Chang, K.S.; A Kim, I.; Lee, M.-H.; Lee, S.A.; Song, S.-K.; Lee, M.M.; Lim, J. Funneling of gibberellin signaling by the GRAS transcription regulator SCARECROW-LIKE 3 in the *Arabidopsis* root. *Proc. Natl. Acad. Sci. USA* **2011**, *108*, 2166–2171. [CrossRef]
58. Sun, T.-P. Gibberellin-GID1-DELLA: A Pivotal Regulatory Module for Plant Growth and Development. *Plant Physiol.* **2010**, *154*, 567–570. [CrossRef]
59. Doerks, T.; Copley, R.R.; Schultz, J.; Ponting, C.P.; Bork, P. Systematic Identification of Novel Protein Domain Families Associated with Nuclear Functions. *Genome Res.* **2002**, *12*, 47–56. [CrossRef]
60. Zhang, N.; Yang, J.; Wang, Z.; Wen, Y.; Wang, J.; He, W.; Liu, B.; Si, H.; Wang, D. Identification of Novel and Conserved MicroRNAs Related to Drought Stress in Potato by Deep Sequencing. *PLoS ONE* **2014**, *9*, e95489. [CrossRef]
61. Lu, J.; Wang, T.; Xu, Z.; Sun, L.; Zhang, Q. Genome-wide analysis of the GRAS gene family in Prunus mume. *Mol. Genet. Genom.* **2014**, *290*, 303–317. [CrossRef]
62. Abarca, D.; Pizarro, A.; Hernández, I.; Sánchez, C.; Solana, S.P.; del Amo, A.; Carneros, E.; Díaz-Sala, C. The GRAS gene family in pine: Transcript expression patterns associated with the maturation-related decline of competence to form adventitious roots. *BMC Plant Biol.* **2014**, *14*, 354. [CrossRef]
63. Aubourg, S.; Kreis, M.; Lecharny, A. The DEAD box RNA helicase family in *Arabidopsis thaliana*. *Nucleic Acids Res.* **1999**, *27*, 628–636. [CrossRef]
64. Jain, M.; Tyagi, A.K.; Khurana, J.P. Genome-wide analysis, evolutionary expansion, and expression of early auxin-responsive SAUR gene family in rice (*Oryza sativa*). *Genomics* **2006**, *88*, 360–371. [CrossRef]
65. Guo, Y.; Wu, H.; Li, X.; Li, Q.; Zhao, X.; Duan, X.; An, Y.; Lv, W.; An, H. Identification and expression of GRAS family genes in maize (*Zea mays* L.). *PLoS ONE* **2017**, *12*, e0185418. [CrossRef]
66. Sang, Y.; Liu, Q.; Lee, J.; Ma, W.; McVey, D.S.; Blecha, F. Expansion of amphibian intronless interferons revises the paradigm for interferon evolution and functional diversity. *Sci. Rep.* **2016**, *6*, 29072. [CrossRef]
67. Jain, M.; Khurana, P.; Tyagi, A.K.; Khurana, J.P. Genome-wide analysis of intronless genes in rice and Arabidopsis. *Funct. Integr. Genom.* **2007**, *8*, 69–78. [CrossRef]
68. Fan, Y.; Yan, J.; Lai, D.; Yang, H.; Xue, G.; He, A.; Guo, T.; Chen, L.; Cheng, X.-B.; Xiang, D.-B.; et al. Genome-wide identification, expression analysis, and functional study of the GRAS transcription factor family and its response to abiotic stress in sorghum [*Sorghum bicolor* (L.) Moench]. *BMC Genom.* **2021**, *22*, 509. [CrossRef]

69. Zentella, R.; Zhang, Z.-L.; Park, M.; Thomas, S.G.; Endo, A.; Murase, K.; Fleet, C.M.; Jikumaru, Y.; Nambara, E.; Kamiya, Y.; et al. Global Analysis of DELLA Direct Targets in Early Gibberellin Signaling in *Arabidopsis*. *Plant Cell* **2007**, *19*, 3037–3057. [CrossRef]
70. Wang, F.; Zhu, D.; Huang, X.; Li, S.; Gong, Y.; Yao, Q.; Fu, X.; Fan, L.-M.; Deng, X.W. Biochemical Insights on Degradation of *Arabidopsis* DELLA Proteins Gained From a Cell-Free Assay System. *Plant Cell* **2009**, *21*, 2378–2390. [CrossRef]
71. Nomoto, Y.; Takatsuka, H.; Yamada, K.; Suzuki, T.; Suzuki, T.; Huang, Y.; Latrasse, D.; An, J.; Gombos, M.; Breuer, C.; et al. A hierarchical transcriptional network activates specific CDK inhibitors that regulate G2 to control cell size and number in Arabidopsis. *Nat. Commun.* **2022**, *13*, 1–16. [CrossRef]
72. Zeng, X.; Ling, H.; Chen, X.; Guo, S. Genome-wide identification, phylogeny and function analysis of GRAS gene family in *Dendrobium catenatum* (Orchidaceae). *Gene* **2019**, *705*, 5–15. [CrossRef]
73. Torres-Galea, P.; Huang, L.-F.; Chua, N.-H.; Bolle, C. The GRAS protein SCL13 is a positive regulator of phytochrome-dependent red light signaling, but can also modulate phytochrome A responses. *Mol. Genet. Genom.* **2006**, *276*, 13–30. [CrossRef]

genes

MDPI

Article

Comparative Gene Enrichment Analysis for Drought Tolerance in Contrasting Maize Genotype

Syed Faheem Anjum Gillani [1], Adnan Rasheed [2], Asim Abbasi [3], Yasir Majeed [1], Musawer Abbas [4], Muhammad Umair Hassan [5], Sameer H. Qari [6], Najat Binothman [7], Najla Amin T. Al Kashgry [8], Majid Mahmood Tahir [9] and Yunling Peng [1,]*

1 Gansu Provincial Key Lab of Arid Land Crop Science, Lanzhou 730070, China
2 College of Agronomy, Hunan Agricultural University, Changsha 410128, China
3 Department of Environmental Sciences, Kohsar University, Murree 47150, Pakistan
4 Department of Agronomy, University of Agriculture Faisalabad, Faisalabad 38000, Pakistan
5 Research Center on Ecological Sciences, Jiangxi Agricultural University, Nanchang 330029, China
6 Genetics and Molecular Biology Central Laboratory, Department of Biology, Aljumum University College, Umm Al-Qura University, Makkah 24243, Saudi Arabia
7 Department of Chemistry, College of Sciences and Arts, King Abdulaziz University, Rabigh 21911, Saudi Arabia
8 Department of Biology, College of Science, Taif University, P.O. Box 11099, Taif 21944, Saudi Arabia
9 Department of Soil and Environmental Sciences, Faculty of Agriculture, University of Poonch, Rawalakot 12350, Pakistan
* Correspondence: pengyunlingpyl@163.com

Abstract: Drought stress is a significant abiotic factor influencing maize growth and development. Understanding the molecular mechanism of drought tolerance is critical to develop the drought tolerant genotype. The identification of the stress responsive gene is the first step to developing a drought tolerant genotype. The aim of the current research was to pinpoint the genes that are essential for conserved samples in maize drought tolerance. In the current study, inbred lines of maize, 478 and H21, a drought-tolerant and susceptible line, were cultivated in the field and various treatments were applied. The circumstances during the vegetative stage (severe drought, moderate drought and well-watered environments) and RNA sequencing were used to look into their origins. In 478, 68%, 48% and 32% of drought-responsive genes (DRGs) were found, with 63% of DRGs in moderate drought and severe drought conditions in H21, respectively. Gene ontology (GO) keywords were explicitly enriched in the DRGs of H21, which were considerably over-represented in the two lines. According to the results of the GSEA, "phenylpropanoid biosynthesis" was exclusively enriched in H21, but "starch and sucrose metabolism" and "plant hormone signal transduction" were enhanced in both of the two lines. Further investigation found that the various expression patterns of genes linked to the trehalose biosynthesis pathway, reactive oxygen scavenging, and transcription factors, may have a role in maize's ability to withstand drought. Our findings illuminate the molecular ways that respond to lack and offer gene resources for maize drought resistance. Similarly, SNP and correlation analysis gave us noticeable results that urged us to do the same kind of analysis on other crops. Additionally, we isolated particular transcription factors that could control the expression of genes associated to photosynthesis and leaf senescence. According to our findings, a key factor in tolerance is the equilibrium between the induction of leaf senescence and the preservation of photosynthesis under drought.

Keywords: GO analysis; maize; GSEA; SNP; PCR; drought tolerant

Citation: Gillani, S.F.A.; Rasheed, A.; Abbasi, A.; Majeed, Y.; Abbas, M.; Hassan, M.U.; Qari, S.H.; Binothman, N.; Al Kashgry, N.A.T.; Tahir, M.M.; et al. Comparative Gene Enrichment Analysis for Drought Tolerance in Contrasting Maize Genotype. *Genes* **2023**, *14*, 31. https://doi.org/10.3390/genes14010031

Academic Editors: Wajid Zaman and Hakim Manghwar

Received: 21 November 2022
Revised: 12 December 2022
Accepted: 12 December 2022
Published: 22 December 2022

1. Introduction

Maize (*Zeal mays* L.) is a significant cereal crop. Maize production around the world is adversely impacted by drought stress. The breeding of drought-tolerant genotypes is a practical way of addresing the water shortage problem. This calls for a thorough understanding of the sub-atomic elements involved in the reactions caused by drought pressure in maize, as well as genomic means such as genomic determination, hereditary designing, and genomic alteration for drought tolerant improvement. Plants' response to stress can be categorized into four stages: alert stage, aggregation stage, support stage, and fatigue stage. The sub-atomic reactions to the drought pressure of plants include more administrative and useful assets [1].

The transcription factor (TFs) is a significant controller protein that can unequivocally direct initiation and the restraint of target qualities. Then, plants reduce water loss as a result of partially closing stomata, effectively taking up water from underground by root foundation, and regulating metabolic cycles to correspond with existing carbon reserve. With the increment of drought, osmolytes, such as proline and solvent sugars, would aggregate to keep up with the cell turgor pressure. Moreover, various antioxidants, for example, superoxide dismutase (SOD, catalyze), reduce the toxic effect activity in reactive oxygen species (ROS) [2].

The stacking bearing connections inside the receiver wall organization, for example, caused specific revision under the activity of the cell wall's changing proteins, and the guideline of the phenylpropanoid pathway prompted a high collection of lignins in the cell wall. Quality delivery network policies play an essential role in plant drought reactions. Various transcriptions are responsible in various crops. In two wheat genotypes, different responses to drought stress were found bases on transcription analysis. [3].

The results showed that in wheat the auxin and abscisic acid (ABA) receptor related to ROS and the biosynthesis of the cell wall was up regulated, while ethylene receptors were down regulated. In maize, the moderate response to drought were linked to DNA and cell cycle. It was shown that qualities engaged with metabolic pathways and the biosynthesis of optional digestion showed a reaction to lack, and the statement of a few qualities connected with ABA signal transduction were prompted under drought [4]. The plant roots detect changes in water and soil oxygen content. It was shown that exceptional drought-responsive components were associated with various parts of the root zone in maize. The qualities connected with ROS and carbon digestion, and qualities linked to layer carrier, were essentially improved in drought-responsive qualities of 0–3 mm root peak, with 3–7 mm upon root chimp, separately. Comparable outcomes were likewise acquired in another research study utilizing two deep-rooted maize lines. Cell reinforcement chemical qualities were demonstrated to assume the essential parts in drought tolerance in the root transcriptomes analysis of essential root under drought stress, conducted by RNA sequencing, and the results demonstrated the expression of genes under low water stress. A few members from the MAPKKK family enhanced drought tolerance in leaf and maize stems, recommending the significance of the MAPKKK family under drought stress [5].

Plant drought resilience is generally assessed by variances in aggregates and physiological and bio-chemical reactions. Drought plants can keep up with constant morphological features throughout delayed times of drought pressure by maintaining water-holding capability. According to one perspective, drought-resistant lines have some development to do to increase their capacity to keep moisture content in their root system. The entire root development of maize may be made to conduct less research on the soil profile and extend into more soil by reducing the number of particular cortical cells in maize, while simultaneously improving their size. This allows maize to uptake greater water from the soil. The thickness of the leaves may be reduced, and the opening of the stomata in the plants can be regulated to increase the plant's resistance to drought [6].

Furthermore, during conditions of drought, plants are able to maintain the homeostasis of their cells via the process of osmoregulation, which is orchestrated by soluble sugars. This helps to decrease the amount of water that is likely to be present in the cells. Water stress also mandates the peroxide-rummaging catalyst structure to eradicate the quantity of responsive oxygen species incited by drought. This is necessary since ROS may cause damage to the cell film structures, which can ultimately lead to cell death. The activity of the cell reinforcement (nonenzymatic and enzymatic) framework addresses an important file for evaluating maize's resistance to drought. Biochemical and physiological alterations between deep-rooted lines and transgenic lines have allowed for the identification of drought-tolerant lines. Despite this, an exclusive explanation for maize's drought resistance could not be provided by these advances, since different types had different genetic basis [7].

It is realistic that the drought reaction of plants includes a chain of reactions. In this approach, it is planned that reading the article would help agricultural plants become more resilient to drought by providing a realistic illustration of these features. To date a couple of droughts, and tolerant features, for example, H21 and 478 in maize, were distinguished by conventional sequencing strategies and practically depicted. Many key qualities were suggested in drought-tolerant maize; however, an overall significant stretch was expected to explain the elements of these drought reaction qualities. RNA-seq is an incredible technique for the enormous scope ID of drought-responsive attributes with minimal expense, high throughput, and high responsiveness. It can work with mining crucial drought-tolerant qualities in plants, e.g., maize [8].

For instance, research findings have demonstrated that the overexpression of cell wall oxidative metabolism characteristics makes it possible for maize recombinant inculcated lines to procure water stress flexibility under water shortages; the attributes associated with cell wall reformation are involved with water stress chemical reactions in a drought resistant corn line; and the compositions of proteins and carbohydrates are associated with water stress support. Through the process of lighting seeds from an intrinsic maize line, we were able to create a deficit resistant line for this study. To investigate the components of drought opposition of H21, we have analyzed physiological, bio-chemical, and transcriptomic changes between H21 and 478 lines. Our outcomes would feature the drought-tolerant marks of H21 and add to the distinguishing resilient and useful investigations of novel drought resilience qualities in maize [9].

2. Materials and Methods

2.1. Plant Material and Experimental Design

Seeds of maize inbred lines, drought-tolerant (478) and drought-sensitive (H21) were given by College of Agronomy, Gansu Agricultural University, Lanzhou, China. Seeds were sown in pots holding standard preserving soil. The pots were kept in a climate that was sustained at a temperature of 25 $\pm$ 5 degrees Celsius, with a relative humidity of 65 $\pm$ 5 percent, and a light/dark cycle of 16/8 h. All the plants were adequately sprinkled to keep the moistness level persistent till 10 days after emergence; after that, drought treatments began. Treatments included two different degrees of drought conditions (no stress and stress), as well as three different levels of BR (0, 25, and 50 mL). To simulate the effects of drought, an application of fifty milliliters of 20 percent PEG 6000 every 3 days was made, whereas the control area received applications of pure water. For the transcriptome study, plants were chosen after they had reached one of four fully formed leaf stages. Samples were then kept at -80 degrees Celsius in liquid nitrogen for 7 days. The experiment was laid out in a completely randomized design (CRD), and there were three separate replications of the experiment.

2.2. RNA Extraction, Construction of Library, and Sequencing

Pure and simple RNA was detached using Trizol reagent (Thermo Fisher, Carlsbad, CA, USA, 15596018). Total RNA quantity and quality was determined using the Bioanalyzer 2100 and RNA 6000 Nano LabChip Kit (Agilent, Santa Clara, CA, USA, 5067-1511) and RNA assays from the first order with a RIN greater than 7, 0 was used in building sequencing libraries. After removal of the whole RNA, the mRNA residual in the purified RNA (5 µg) was cleaned by two cycles of filtration with Dynabeads Oligo (dT) (Thermo Fisher, Carlsbad, CA, USA). Once filtered, the mRNA was cleaved into shorter fractions using higher temperature divalent cations (using the NEB Magnesium RNA Fragmentation Module, cat. e6150, Ipswich, MA, USA) for 5–7 min at 94 °C. The cut RNA fragments were then reverse disassembled to generate cDNA using the Super ScriptTM II reverse transcriptase enzyme (Invitrogen, cat. 1896649, Carlsbad, CA, USA). The cDNA was then used to combine the second desert DNA named U with DNA polymerase I from *E. coli* (NEB, cat.m0209, USA) and RNase H (NEB, cat.m0297, USA) and dUTP solution (Thermo Fisher, Carlsbad, CA, USA, USA, ref. R0133). The crude ends of each thread receive an A-base coating, which prepares the fibers for tightening into the taped connectors.

Each particular connector was constructed with a T-base shade so that it could ligate to the A-followed segmented DNA and complete the connection. AM Pure XP dabs were used for the purpose of carrying out size verification after twofold document connections had been affixed to the various components. Following the power labile UD Genzyme (NEB, cat.m0280, USA) treatment of the U-named second-deserted DNAs, the ligated things were escalated with PCR using the following conditions: starting denaturation at 95 °C for 3 min; 8 examples of denaturation at 98 °C for 15 s, treating at 60 °C for 15 s, and increase at 72 °C for 30 s; and a short time later last extension at The expansion sizes of the cDNA libraries that came before this one varied, on average, from 300 to 50 base pairs in length. We finished the 2×150 bp matched end sequencing (PE150) utilizing an Illumina NovaseqTM 6000 after the shipper advised that we do so (LC-Bio Technology Co., Ltd., Hangzhou, China).

2.3. *Sequence and Filtering of Clean Reads*

A cDNA library was created by design and sequenced utilizing pooled RNA from experiments conducted using the Illumina Novaseq TM 6000 succession stage. We used the Illumina paired-end RNA-seq method to sequence the transcriptome; readings generated from the sequencing equipment comprise raw read out with devices or low-quality bases, which have an impact on the subsequent assembly and analysis (Martin, 2011). As a result, readings were further filtered using Cutadapt to provide high-quality clean reads. The following were the criteria: eliminating reads that contain adapters, poly A and poly, more than 5% of unknown nucleotides (N), low-quality reads that contain more than 20% of low-quality (Q-value 20) bases, and reads that contain adapters. Then, FastQC was used to assess the sequence's quality. It was noted that the data's Q20, Q30, and GC-content had not been tainted. Following that, readings with clean endings totaling Gigabase Pairs were produced [7]. The NCBI Gene Expression Omnibus (GEO) databases or the NCBI Short Read Archive (SRA) databases have received the raw sequencing data along with accession numbers. These two archives each have an accession number.

3. Methods

3.1. *Differently Expressed Genes (DEGs)*

Differentially expressed genes refer to the up-regulated genes and down-regulated genes between samples or after different treatments of the same sample. The genes are usually screened in terms of fold difference and significant level. The consistency fold FC $\geq$ 2 or FC $\geq$ 0.5 (that is, the absolute value of log 2 FC $\geq$ 1) and q value < 0.05 ($|\log 2\ fc| \geq 1$ & $q < 0.05$) as the standard (multi-group comparison without differential fold, the genes with q < 0.05 were screened for genes with statistical differences among multiple groups), and the genes screened out were considered to be differentially expressed genes, and the significant column in the result file was displayed as "yes". You can also change the above two conditional thresholds for screening differentially expressed genes to obtain differentially expressed genes that align with your experimental design (using the multi-column screening function of excel).

The input data for the differential expression analysis of genes (transcripts) is the reads count data of genes (transcripts). The *p*-value calculation model based on the negative binomial distribution is used to calculate the *p*-value. Repeat (comparison between samples) using Edger for analysis; multi-group comparison using edgeR for analysis; using BH to correct *p* value to obtain q value (FDR value, p.adj value); the fold difference is the average expression level of the experimental group The value was divided by the mean expression level of the control group [10]. The raw sequencing data were produced on the Illumina HiSeq 2000 (Illumina, San Diego, CA, USA).

3.2. GSEA Analysis

Enrichment analysis based on a hypergeometric test generally needs to use the information of differentially expressed genes, that is, a change threshold needs to be set to determine which genes are significantly different and which genes are not very different (general or common thresholds may not be applicable to all systems). Traditional enrichment analysis may yield few or no results when single gene changes are weak. Gene Set Enrichment Analysis (GSEA) may more thoroughly describe the protective mechanism of a system component and efficiently compensate for standard enrichment analysis's inability to properly mine the information from minor genes, which can interact with traditional enrichment analysis. Replenish. Due to the different principles of formal enrichment analysis, the results of GSEA analysis are different from those of conventional enrichment analysis [11].

3.3. GO Analysis

Gene Ontology (GO) is an internationally consistent gene function cataloging system that runs vigorously reorganized standard vocabularies to systematically define the properties of genes and gene products in organisms. GO has three ontologies, which represent the molecular function (mf), cellular component (cc), and biological process (bp) of genes, respectively. The elementary component of GO is Word (entry, node, entry), and each term relates to a characteristic. Biological process: A biological process accomplished by various molecular activities, such as cell cycle, cell cycle arrest, cell cycle checkpoint, and positive execution of the execution phase of apoptosis. Processes such as positive regulation of the execution phase of apoptosis. Cell components: the location of the cell structure where the gene product performs its function, such as cytoplasm, nucleus, lysosome, and other structures. Molecular function: the activity of gene products at the molecular level, such as protein binding, ubiquitin-protein transferase activity, G protein-coupled receptor activity, etc. [12].

3.4. SNP Analysis

The full name of SNP is Single Nucleotide Polymorphisms, which refers to the genetic markers formed by the variation of a single nucleotide on the genome, with a large number and rich polymorphisms. Variations of single nucleotides in the genome include substitutions, transversions, deletions, and insertions. The diversity of single nucleotide base morphology can be divided into substitution (transition, CT, GA on its complementary strand) and transversion (transversion, CA, GT, CG, AT). SNPs with substitutional variants account for about 2/3, and the occurrence probability of other variants is similar. SNPs seem most commonly on CG sequences, most of which are C-to-T conversions, because C in CG is often methylated and becomes thymine after spontaneous deamination. A SNP is a single nucleotide variant with a frequency greater than 1% [13].

Depending on where the SNP is placed in the gene, it may be classified as either being in the non-coding area of the gene, the spacer portion of the gene (which is the portion between genes), or the coding section of the gene. These classifications can be found in the table below. Single nucleotide polymorphisms (SNPs) not in the coding regions of genes may still affect gene splicing, transcription factor binding, messenger RNA degradation, or RNA sequences in non-coding parts. The expression of a gene affected by this single nucleotide polymorphism (SNP) is called an expression of a single nucleotide polymorphism (ESNP). It can occur either upstream or downstream of the gene. The number of SNPs (coding SNPs, cSNPs) located in the coding region of genes is relatively small as, inside exons, the modification rate is only 1/5 of that of adjacent sequences. However, it is of great significance in studying hereditary diseases, so the study of cSNP is more concerned [14].

From the perception of the impact on the genetic traits of organisms, cSNP can be divided into two types: one is synonymous cSNP (synonymous cSNP), that is, the change of the coding sequence caused by the SNP does not affect the amino acid sequence of the protein it translates. The mutated base has the same meaning as the unmutated base. The other is non-synonymous cSNP (non-synonymous cSNP), which means that the base sequence can change the line of the protein translated based on it, thus affecting the protein function. Such changes are frequently the direct cause of changes in biological traits. About half of the cSNPs are non-synonymous cSNPs. InDel (insertion-deletion) refers to the insertion and deletion of small fragments in the sample relative to the reference genome, which may contain one or more bases. Based on the transcriptome level, SNP loci in coding regions were analyzed. According to the Hisat2 alignment results of each sample and the reference genome, each sample's possible SNP and INDEL information was obtained by mpileup processing with Samtools software and then annotated with ANNOVAR [15].

3.5. Correlations Analysis

The association analysis of the gene expression information of the samples can better judge the clustering situation between the samples. The greater the correlation coefficient between the samples, the better the clustering of the samples. The Pearson correlation coefficient between samples and PCA (Principal Component Analysis) allows the repeatability between samples to be understood, which helps to exclude outlier samples. The abscissa and ordinate in the Pearson correlation coefficient graph are for each piece. The depth of color indicates the size of the correlation coefficient of the two samples. The closer to red (the coefficient is closer to 1) the higher the correlation; the closer to white the lower the correlation. Some sample types (such as clinical samples, animal models, tissue fluids, cell pools with low knockdown efficiency, etc.) may cause the correlation of samples within a group to be lower than those between groups due to heterogeneity [16].

3.6. QPCR Analysis

The Rotor-Gene 3000 detection equipment (Corbett Research, Singapore) was used to conduct the PCR experiments. One-Step PrimeScriptTM RT-PCR Kit (Perfect Real Time) was used (Takara, Shiga, Japan). The following ingredients were used in each set of reactions: 10 L of 2 One Step RT-PCR Buffer III, 0.4 L of each of the following: MCMVf (10 M), MCMVr (10 M), 0.8 L of probe (10 M), 0.4 L of EX TaqTM (Takara) HS (5 U/L), 0.4 L of PrimeScriptTM RT Enzyme Mix II, 1.0 L of total RNA or 1.0 L of The following amplification responses were carried out: forty cycles of 95 °C for 5 s, followed by 60 °C for 20 s. Six distinct reactions, including a water control, were used to assess the specificity of this TaqMan test. Under optimal reaction circumstances, the TaqMan probe was only able to identify strong fluorescent signals after reactions involving samples, whereas the indications after four more samples and the water control were superposed to the baseline. By contrasting the signals at various levels, these samples may be distinguished from the four other corn samples. Agarose gel electrophoresis was used to further evaluate the PCR results. While the four other assays did not show the unexpected weak band above the 67-bp band, the assessment with the taster did show the probable band of 67 bp.

3.7. Statistical Analysis

Statistical significances between drought treatment and control were tested using the—Newman–Keuls method at $p < 0.05$ by IBM® SPSS®, 22 and data was generated by Sigma Plot 12.5.

4. Results

The results obtained after detailed analysis are as follows.

4.1. Differently Expressed Genes (DEGs)

Using the sequence map (SAM) files that were generated by Tophat2, aligned reads of the genes were computed in order to determine the genes that have differentially expressed proteins. Then, using the R program "DESeq2, differential gene expression analysis were carried out comparing the two sets of samples. Gene with | log 2 (fold changes) | > 1 and false discovery rate (FDR) < 0.02 were recognized as differentially expressed. To detect the DRGs, four comparison groups, i.e., LMC_LMD (Lv28 under moderate drought control vs. Lv28 under moderate drought), LSC_LSD (Lv28 under severe drought control vs. Lv28 under severe drought), HMC_HMD (H21 under moderate drought control vs. 478 under moderate drought), and HSC_HSD (H21 under severe drought control vs. 478 under severe drought) were included Figure 1. To identify genes responding to drought in the two lines, differential expression analysis was performed by comparing the gene expression profiles between different drought conditions.

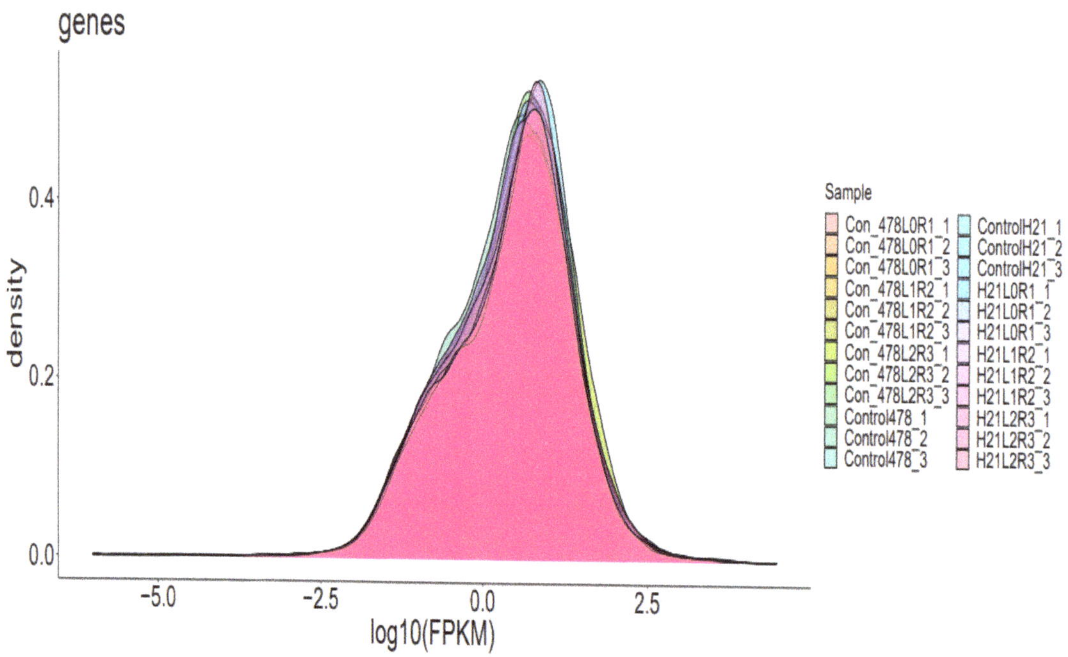

Figure 1. Visual representation of genes expression density of maize, showing a sharp peak at high density.

In total, 100% and 51% of genes showed drought response under MD and SD in 478, respectively, among which 28% and 26% were up-regulated under MD and SD, while 16% and 12% were down-regulated, respectively, Figure 2. In the drought tolerant line H21, 68% and 3 3% genes showed response to drought stress under MD and SD, respectively, among which 62% and 59% genes were up-regulated, while 16% and 15% genes were down-regulated, respectively.

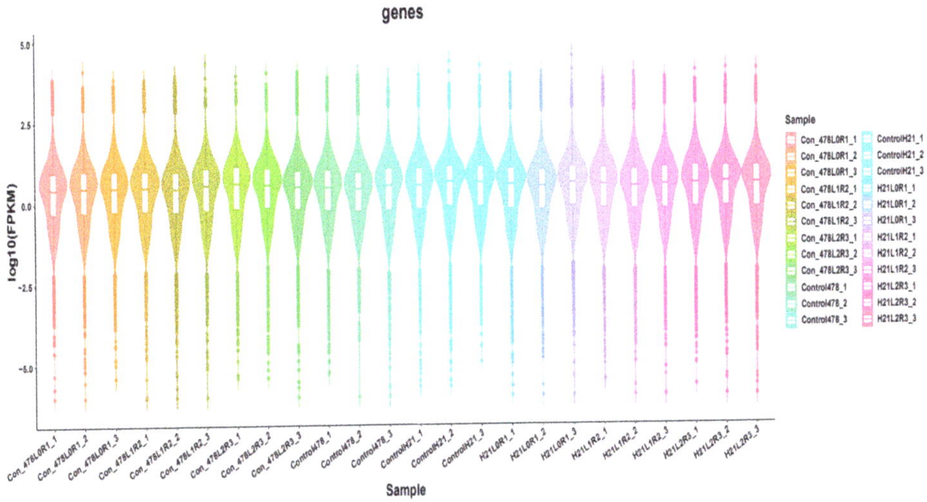

Figure 2. Different samples are analyzed simultaneously and are presented in the form of violin.

4.2. GO Analysis

The significance of GO enhancement analysis showed that GO terms were improved in the DRGs of the samples, as shown in Figure 3, with 15 in the metaphysics of "natural cycle", nine in the cosmology of "sub-atomic capability", and seven in the philosophy of "cell part". There were many GO terms enhanced in the samples, including eight organic cycle terms, which were primarily about pressure reaction (e.g., "reaction to boost", "reaction to stretch", "reaction to abiotic improvement", "reaction to biotic upgrade", "reaction to endogenous improvement", and "reaction to outside boost"). Be that as it may, these pressure reaction-related GO terms (with the exception of "reaction to boost") were missing in H21 under MD. Likewise, "metabolic sugar cycle" and "extracellular district" were over-addressed in every one of the four DRG sets (LMD, LSD, HMD, and HSD), showing that two lines had a piece of comparative drought tolerant components.

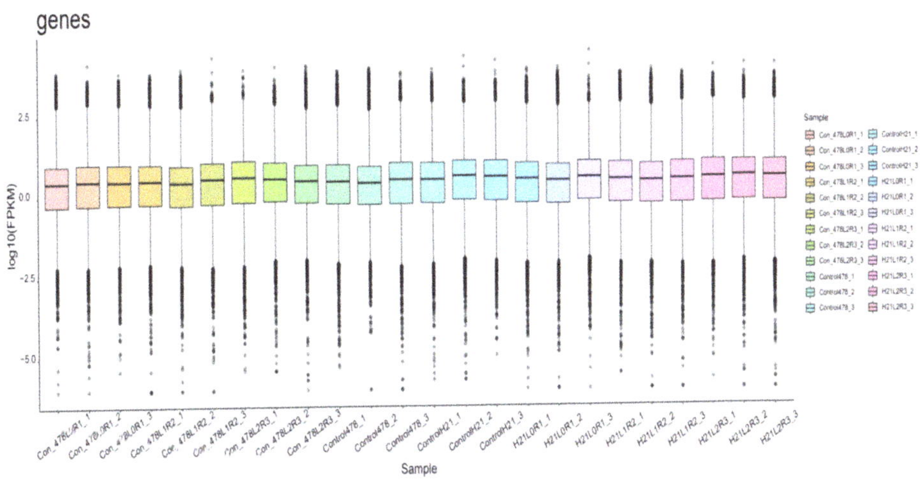

Figure 3. Different samples portraying GO analysis and each sample is mentioned as individual gene.

Oddly, many GO terms were particularly advanced in the DRGs of the drought tolerant line H21, including six natural interaction terms (e.g., "lipid metabolic interaction", "cell cycle", and "formative cycle"), six atomic capability terms (e.g., "record factor action", "record controller movement", "kinase action", and "transferase action"), and one cell part terms ("cytoskeleton").

These outcomes showed that the support of root improvement, the dependability of cytoskeleton, and the propulsion of different administrative frameworks under drought might add to the drought resilience of H21. The Kyoto reference book of qualities and genomes (KEGG) improvement examination showed a sum of 10 metabolic pathways were addressed in the two lines (Figure 4).

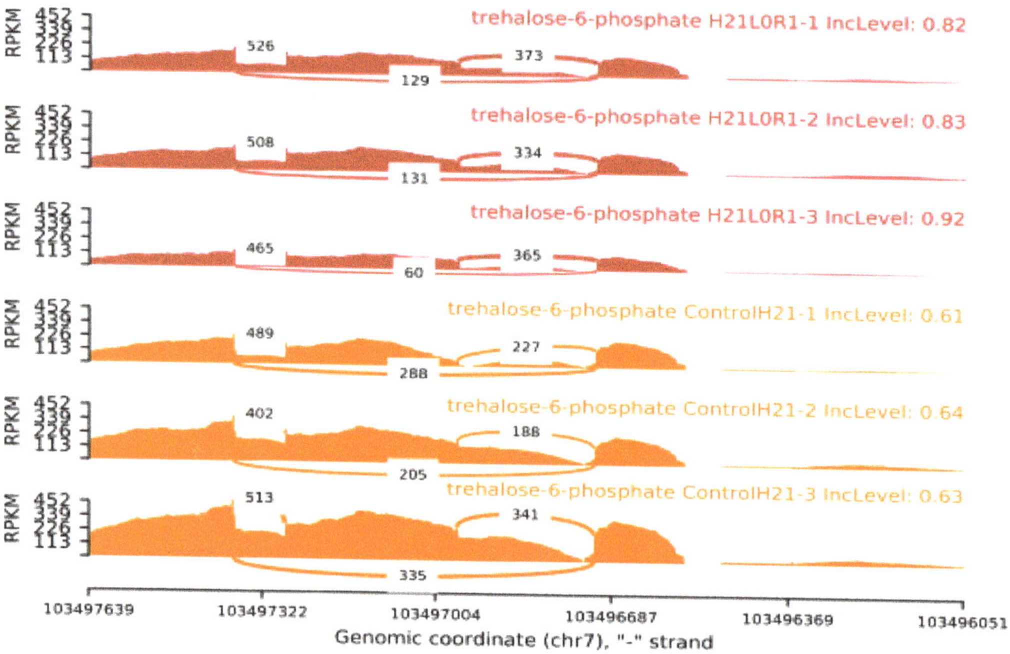

Figure 4. GO analysis different peaks are shown clearly in the figure with different sample sizes.

"Plant chemical sign transduction" and "starch and sucrose digestion" were completely advanced in both of the two lines, yet the importance levels in H21 were generally higher than those in 478. Similarly, the pathway of "phenylpropanoid biosynthesis" was just fundamentally enhanced in H21 under MD and SD. These outcomes showed that these three metabolic pathways could assume an essential part in drought tolerant of H21.

4.3. GSEA Method

GSEA analysis is based on the expression information of all genes, sorts the genes with Signal2Noise as the standard (the default is in descending order), and analyzes whether a specific gene set (such as all genes in a pathway or all genes in a GO Term) is more ranked among all genes. Anterior or posterior (whether there is a statistically significant and consistent difference between two biological states or two groups, the calculation of Signal2Noise is generally the experimental group relative to the control group, so forward and following refer to, respectively (Figure 5).

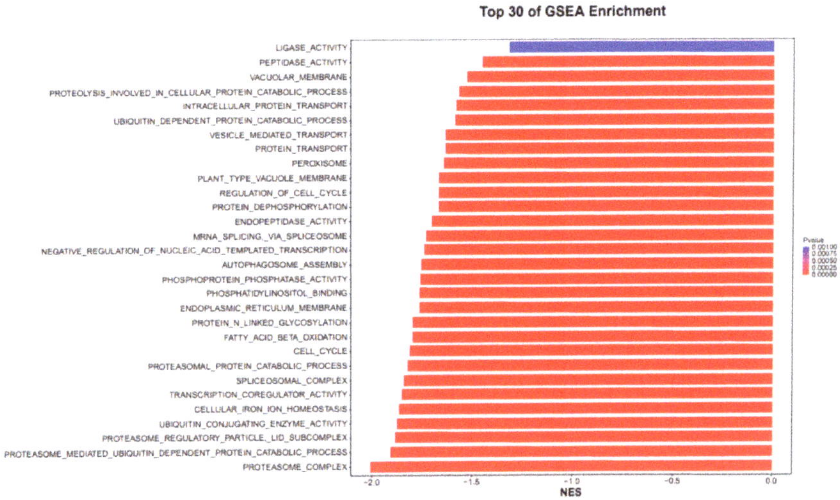

Figure 5. GSEA analysis top 30 samples of GSEA enrichment analysis are shown in the form of a horizontal bar graph.

Whether the gene set is highly expressed in the experimental group or the control group, score the pathway or term where the gene set is located, and the score is called the ES (enrichment score) value. The permutation test was performed based on the gene set, and the significant p value was calculated. Finally, the standardized ES value (NES value) was corrected by various tests to obtain the FDR value. Gene sets with |NES| > 1, Nymphal < 0.05, and FDR.qval < 0.25 are generally considered meaningful (when there are fewer samples per group, the replacement type is gene sets, and a stricter FDR cutoff can be used, such as 0.05). GSEA analysis was performed only on gene sets of size 15–500, as neither too large nor too small (the number of genes contained in the gene set) were of analytical significance (Figure 6).

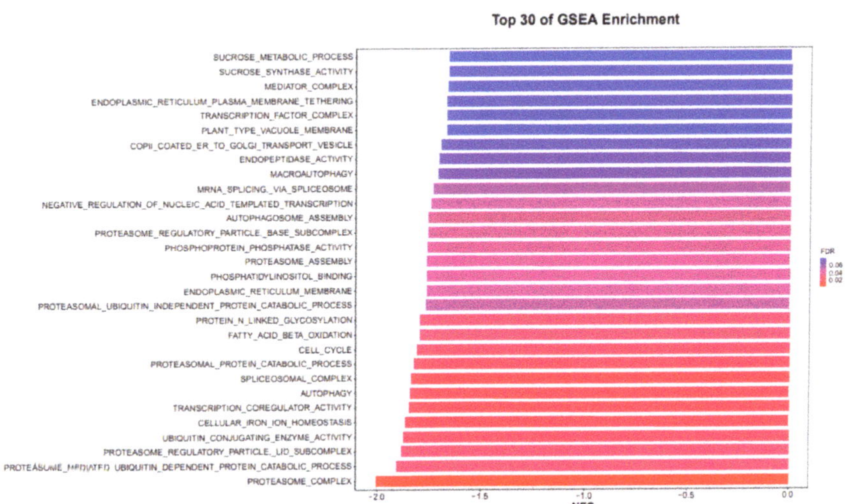

Figure 6. GSEA analysis top 30 samples of GSEA enrichment analysis are shown in the form of a horizontal bar graph.

Using the developmental time series data, finding genes that may be coregulated and/or engaged in the same biological processes is possible. Because of this, we divided the expressed genes into 30 co-expression modules, each containing genes with a similar pattern of expression (Figure 6). Based on the MapMan annotation of 16,657 expressed genes, overrepresented functional categories in each module were found. The 30 modules were split into the first, second H21, and third 478 phases based on the time of peak expression (Figure 6). The first stage, which involved significant physiological changes, occurred when seeds emerged from dormancy. The genes, which are connected to RNA processing and cell vesicle movement, were strongly expressed in dry seeds. The gene expression peaked at H21, but following 478 it rapidly decreased. The genes known as H21 are involved in RNA binding, protein synthesis and degradation, stress, the hormones ethylene and abscisic acid (ABA), and abundant storage proteins during late embryogenesis. Further, 478 has excess functions linked to stress, mitochondrial electron transport, and lipid breakdown.

4.4. SNP

The maize genome has a variety of SNPs. Each chromosome has an average of 525 SNPs, ranging from 327 on chromosome 10 to 926 on chromosome 1. (Figure 7). This SNPs distribution throughout the genome matched earlier studies in maize that used the same kind of marker. High-density marker genotyping, one of the primary molecular marker systems, enables the simultaneous analysis of markers that are extensively dispersed across the genome. SNP markers offer more comprehensive genome coverage than other markers, including RFLP, AFLP, and SSR. The degree of the marker's polymorphism, which is reflected in the genetic variety among the genotypes being studied, determines how informative the marker is (Table 1).

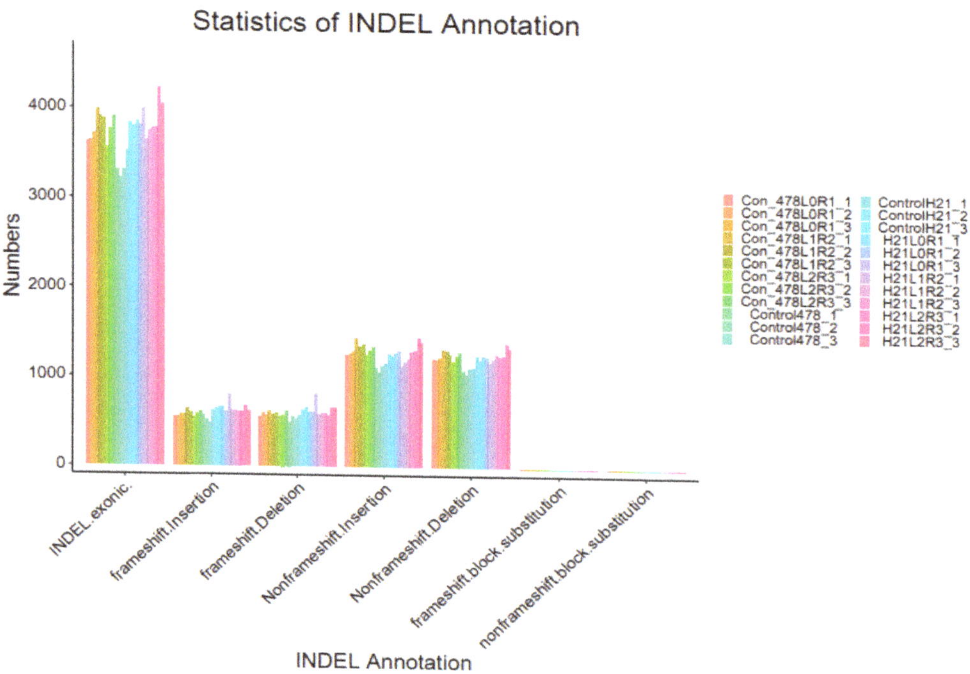

Figure 7. SNP analysis results INDEL analysis are shown in the form of a bar graph.

Table 1. Results of SNP analysis.

Sample	SNV	SNV (Gene Region)	INDEL	INDEL (Gene Region)
Con_478L0R1_1	317,447	109,631	22,372	3621
Con_478L0R1_2	335,578	112,446	23,550	3638
Con_478L0R1_3	330,907	111,822	23,177	3711
Con_478L1R2_1	351,518	117,124	25,118	3987
Con_478L1R2_2	349,311	116,003	24,412	3906
Con_478L1R2_3	346,695	116,230	24,528	3875
Con_478L2R3_1	311,727	109,236	21,552	3560
Con_478L2R3_2	308,668	111,474	22,214	3766
Con_478L2R3_3	310,894	114,079	22,806	3908
Control478_1	259,539	100,510	18,424	3318
Control478_2	249,938	95,025	17,361	3226
Control478_3	299,808	104,763	20,935	3316
ControlH21_1	284,771	105,851	20,496	3518
ControlH21_2	347,813	109,906	24,376	3841
ControlH21_3	312,458	106,749	22,326	3805
H21L0R1_1	344,214	111,197	23,577	3847
H21L0R1_2	343,231	111,465	23,266	3817
H21L0R1_3	367,762	110,892	24,537	3990
H21L1R2_1	314,609	104,783	21,679	3643
H21L1R2_2	314,381	106,501	22,467	3746
H21L1R2_3	335,042	109,192	23,055	3781
H21L2R3_1	300,528	107,825	21,661	3775
H21L2R3_2	387,144	117,458	26,906	4221
H21L2R3_3	353,192	113,905	24,772	4048

Figure 1 shows that the 5252 SNPs employed throughout the genomes of 293 inbred lines were highly informative. PIC varied from 0.092 to 0.375, with a mean of 0.297, whereas MAF ranged from 0.051 to 0.5, with a mean of 0.284. When analyzing the genetic characteristics of Chinese maize germplasm using SNP marker data, comparable PIC range values were discovered. In their examination of CIMMYT's tropical and temperate maize inbred lines, PIC scores between 0.25 and 0.5 show multiallelic (Figure 8).

Due to the bi-allelic nature of SNPs, where the supreme PIC value is 0.375, higher quartile PIC values, as those discovered in our study, can be regarded as being very informative. According to these standards, 65.6% of the markers used in this investigation were extremely informative (Figure 9). The average value of MAF is used to measure the level of genetic difference in the population. Generally speaking, higher MAF are recommended to increase the average allelic differentiation. A genetic distance matrix was created between all pairings of inbred line based on the 5252 polymorphic SNPs (Table 1).

The genetic distance matrix-based cluster analysis produced four unique clusters, as illustrated in Figure 2. Separating the dendrogram clusters allowed for the closest match between the number of lines grouped and the total number of lines of the formerly recognized heterotic groupings. The statistically substantial cophenetic correlation coefficient ($r = 0.953$; $p = 0.0001$; 10.000) permutations showed that the cluster analysis suited the underlying genetic distance matrix rather well. According to analysis, a precise dendrogram is essential for breeders to organize their genetic material. Results showed positive cophenetic values when inferences regarding genetic diversity using microsatellites (Figure 10).

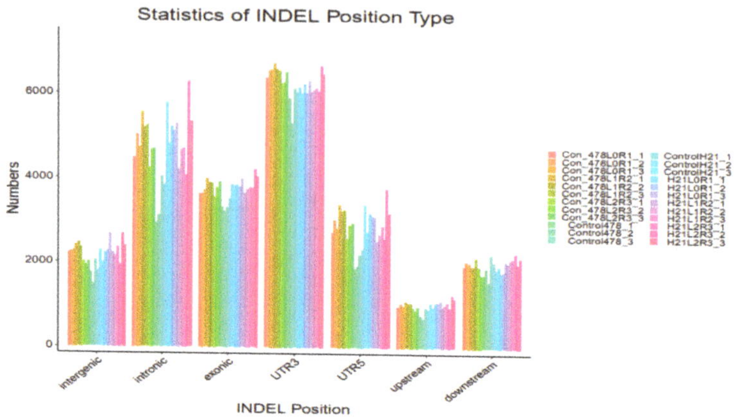

Figure 8. SNP analysis results INDEL analysis are shown in the form of a bar graph.

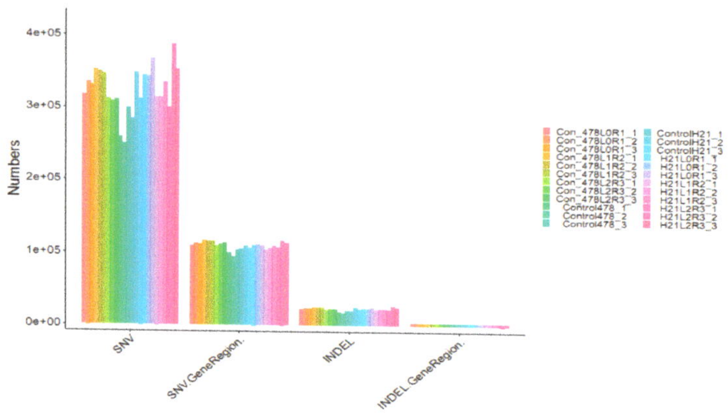

Figure 9. SNP analysis results INDEL analysis are shown in the form of a bar graph.

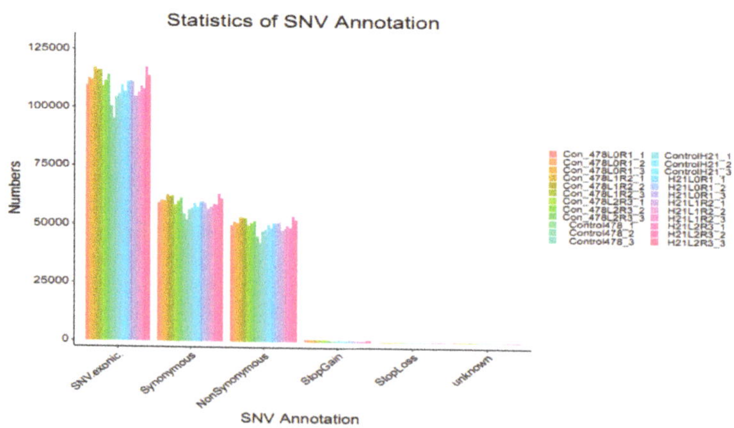

Figure 10. SNP analysis results SNV analysis are revealed in the form of a bar graph.

In assumption, 10 (3.4 percent) of the corn inbred lines examined in this study were categorized differently from the prior heterotic group categorization. This demonstrates the significance of combining traditional breeding with molecular breeding for a genetic diversity study that is more precise (Table 1). The categorizing of the contradictions in marker-based genotypes may be inaccurate owing to pedigree information mistakes or genetic drift during the inbreeding process. (Figure 11).

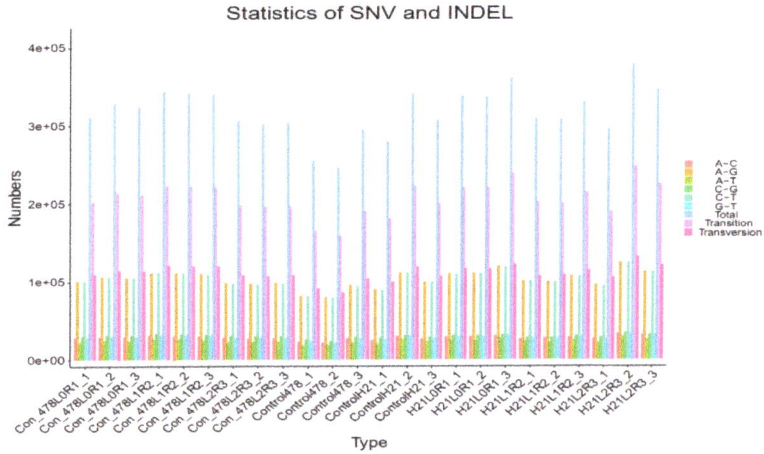

Figure 11. SNP analysis results SNV analysis are shown in the form of a bar graph.

4.5. Correlations Analysis

It was found that each plant had a respectable amount of root, with the genotype H21 having the greatest number by around 30%. The specific number of productions measured capacity roots was 1.10 with genotype H21 having the most significant number of 5% production estimated roots. The most elevated new root yield was kept in H21 while most elevated dry root yield was kept in 478. Recorded dry matter constituent went from 23.57% in 478 to 59.45% in H21. Dry root yield was intensely and primarily connected with the number of production potential roots, the number of roots per plant, the number of roots harvested, the weight of the roots, and the number of new roots produced.

New root yield was fundamentally related to roots per plant, production roots and collect list. Protein content was deeply corresponded with gather file and dry matter substance, and adversely associated with a number of productions measured roots, roots per plant, and dry matter substance. The general commitment of the different qualities to the genotype execution was made sense of by guideline part examination, Figure 12.

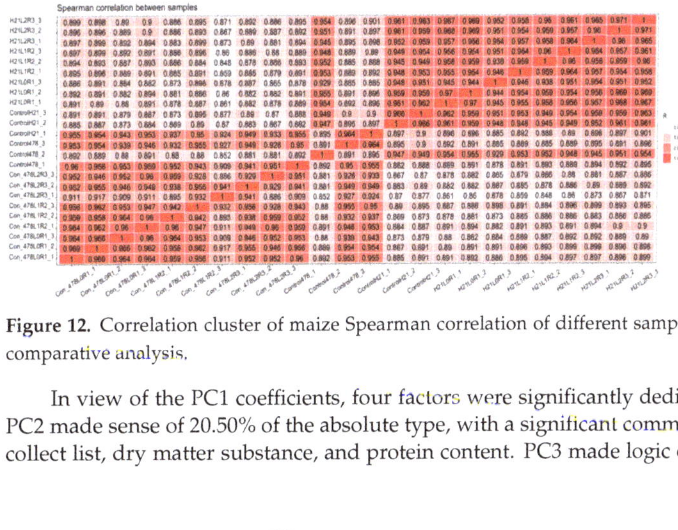

Figure 12. Correlation cluster of maize Spearman correlation of different samples that showed a comparative analysis.

In view of the PC1 coefficients, four factors were significantly dedicated to variety. PC2 made sense of 20.50% of the absolute type, with a significant commitment from the collect list, dry matter substance, and protein content. PC3 made logic of 14.03% of the absolute variety with a significant commitment from production roots, collecting wet and dry matter substance. A careful analysis of the changes that took place in the preliminary spots north of CIAT and Quilichao over the course of two years revealed that genotype was particularly crucial for the number of roots produced per plant, the weight of the roots, the number of new roots produced, the number of dry roots produced, the protein content, and the dry matter content of the properties that were evaluated (Figure 12).

The joined examination of the difference in the two areas of CIAT preliminary locales is introduced in Figure 12. There were very major variances in the direct effects of the genotype on root per plant, new root production, dry root produce, and dry matter content. Connection among genotype and area was critical for just dry matter substance. After the adjustments were made to the production and the protein content, every one of the qualities showed a decent coefficient of assurance (R2) of 0.99 across the three conditions. There was a profoundly critical area impact for every one of the characteristics.

5. Discussion

The current study has resulted in the development of a maize freak, H21 by 478, that is resistant to drought. In comparison to its wild type 478, it showed a higher level of drought resistance when it was subjected to drought conditions. Between H21 and 478, there continued to be significant differences in the analytical files (Figure 1). The relative water content of the leaf, often known as RWC, is a metric that is commonly used to identify the drought resistance of several agricultural crops, including corn. Under drought conditions, H21 showed a higher RWC in its leaves than 478 did; for example, it had a higher ability to maintain high moisture content. The amino acid proline and the sugar that is easily dissolved both play important roles in the osmotic guideline, which is connected with drought tolerance (Figure 2). When exposed to drought, the leaves of the H21 variety contain much more proline and soluble carbohydrates than those of the 478 variety. When maize is subjected to drought stress during the transformation from somatic to reproductive development, the plant shows a considerable reduction in grain output. In order to create drought-resistant varieties, it is essential to have a comprehensive understanding of the biochemical and gene regulatory networks involved in corn's tolerance to drought conditions during these various stages of development. This will allow for the development of cultivars that are more resistant to drought. Even with recent advances in molecular biology tools, the complex adaptive mechanisms that underlie water deficiency stress resistance from cellular proliferation to vegetative plant growth continue to be an enigma [17].

A better productivity of the antioxidative safeguard framework might safeguard photosynthetic colors, proteins, and genetic material from abundant ROS loss. Feline action is conversely relative to MDA content under serious drought tolerance. The examiner of CAT action and MDA content demonstrated that H21 experienced a smaller amount of ROS harm than 478 under drought. Light dispersion in plant shade can be improved in the field by diminishing H21 content (Figure 3). Then again, nitrogen and energy saved by a fading H21 combination would upgrade the particular reactions to drought. In this manner, low H21 content in the leaves of H21 may be a piece of its drought variation methodology [18]. More significant, the levels of RWC, osmolyte collection, cell reinforcement exercises, photosynthetic effectiveness, and a lower degree of MDA added to the drought tolerance of H21 than 478 (Figure 4).

Evaluating the transcriptome state and mining genes associated to drought tolerance in maize may both be accomplished successfully with the use of genome-wide gene expression profiling, performed by RNA-Seq [5]. Numerous genes that are sensitive to drought or that exhibit genotypically varied expression across the two lines with divergent resistance to dry conditions were found in our investigation [11]. The opposite transcriptional regulatory predisposition that reacted to water stress in maize and the diversified gene function enhancement between drought resistant and susceptible lines providing a picture of the regulatory network in maize's transcriptome that was linked to water stress. This was shown by the fact that there was a noteworthy variance among the two types of lines. The overrepresentation of transcriptional regulators is common in most cases, particularly sensitive genes and genotypically differently expressed genes, in the tolerant line revealed a varied mechanism for maize drought tolerance [6].

Solvent sugars such as sucrose, galactose, maltotriose, and oligosaccharides, are the types of sugars that accept dynamic fractions in osmotic adjustment when there is a drought. In drought-tolerant wheat, the expanded delivery of dissoluble sugar blend-related qualities made sense of its drought tolerance. In the current research, the solvent sugar content in H21 was greater than that in 478 under the influence and drought disorders (Figure 5). Obviously, H21 has an effective drought transformation system mostly founded on the osmoregulation of dissolvable sugar [5]. In the present research, the discharge of photosynthesis-related qualities in H21 was practically unaffected, hence H21 could compose solvent sugar all the more effectively under drought conditions. In order to identify important regulatory genes and gene co-expression pathways that are involved in the maize plant's response to environmental stress, we used RNA-seq based technique to undertake a thorough relative transcriptome study of drought-resistant lines from the somatic to propagative development stages. In order to support the RNA-seq data, we also measured photosynthetic parameters. In addition, a functional validation study using qRT-PCR provided evidence for the differential expression of the genes that were discovered. Our insights not only improve our knowledge of the processes that enable maize to survive the effects of drought stress, but they also provide a genetic resource that is both efficient and cost-effective for the genetic improvement of corn (Figure 6).

At the level of the transcriptome, H21 guided the activity of a specific component of the sugar solution. Glucose has the potential to form trehalose through a 1,1-glycosidic bond, so mitigating the damage produced by high contents of trehalose-6-phosphate under conditions of prolonged drought. In the current study, phosphatase's behavior was restrained in 478, while their expression displayed a sharp fluctuation under drought (log 2 overlay changes from 3.46 to 7.63), indicating that the increased substance of solvent sugars in H21 probably will not be expected to the improved trehalose combination (Figure 7). Additionally, both 478 and H21's -amylase outputs were activated by dryness, indicating that the leaves' starch could be broken down into maltose [19].

Using two breeding lines of corn that had varied levels of tolerance to droughts, researchers used RNA-Seq technology to investigate an amount of DNA that were susceptible to drought, as well as proteins that showed differential expression across variants [2]. Between the resistant and susceptible lines, researchers discovered significant differences in the regulatory patterns and operational abundance of these proteins. The fact that drought resistance and resilience were both enhanced by the overrepresentation of certain transcription factors in the tolerant line [18] demonstrates how important these variables are. Our findings subsidize to an improved considerate of the molecular processes involved in drought response and to the identification of candidate genes for future research in the quest to improve maize's resistance to drought [14].

In summary, H21 was able to accumulate more solvent sugars in leaves and keep up with the consistence of photosynthesis during dry conditions, which resulted in it being more drought resistant than 478. The contribution of characteristics associated to the biosynthesis of cell walls to the preservation of water in H21. The incorporation of a synchronized multienzyme complex in the process of polysaccharide production provided regions of stability for the combination of a cell wall (Figure 8). The increased supply of glycosyltransferase contributed to the drought-induced cytokinin deficiency in *Arabidopsis thaliana* freaks. The investigation of GO development revealed that in H21 a number of DEGs were associated with cell wall connection. This was notably the case with the development family, which was induced by a variety of abiotic stressors and ABA. It was observed that increasing stomatal expression levels decreased their thickness (Figure 9).

Moreover, expansions could decrease water misfortune by ceasing cell wall movement and hardening cell structures [15]. There was a strong correlation between drought resistance in maize and both the lignin concentration and the middle outcome of the phenylpropanoid mix route (Figure 11). The accumulation of caffeic abrasive and p-coumaric abrasive in the xylem sap influenced the growth of maize leaves during periods of drought by having an effect on lignin production. The abrasive effects of p-coumaric acid might rapidly lignify tissues when placed under strain and further enhance adaptability when exposed to drought. In the most recent study, KEGG and GO enhancement studies revealed that lignin-related metabolic reactions, such as the phenylpropanoid blend route, were improved in H21 under drought conditions. This was shown to be the case in the plant. The most likely scenario is that the phenylpropanoid production pathway and its components, including caffeic corrosive and p-coumaric corrosive, were accountable for reaction of H21 to drought (Figure 12).

Plants' ability to withstand drought is a complicated process with varied responses. At the molecular scale, as well as the cellular scale, a wide array of proteins are engaged in the process of adapting to and responding to drought. The genes that code for functional proteins and the genes that regulate those proteins may be used to classify these genetic traits [9]. There are many different types of structural proteins, some of which include transporters, detoxifying enzymes, osmolyte biosynthesis enzymes, late embryogenesis abundant (LEA) proteins, and others [15]. In order to regulate transcription factors, a regulatory network configuration that incorporates stress detection, cell signaling, and functional gene regulation has been designed [16].

A major objective of the study was to demonstrate that classifying maize lines using high-density molecular marker data is a precise approach that does not require extensive field-combining ability studies [8]. The current study's findings are consistent with past genetic diversity studies that have found SNPs to be a valuable tool for inbred grouping lines into genetically related groups and for guiding hybrid crossings between members of various groups to provide superior yield performance. SNPs are also helpful for genotyping and integrating new inbred lines in a new cluster analysis, which can help to breed by swiftly detecting the genetic link of new inbred lines. Trace endogenous substances come in many forms as plant hormones. The loss in plant growth under drought-stress circumstances may exist due to an altered hormonal balance. Ultra-trace levels of plant hormones might play a crucial role in plant development, growth, and fast response to abiotic and biotic stressors. The Brassinosteroid C-6 Oxidase, which catalyzes the last steps of brassinosteroid production, is encoded by the Brassinosteroid-deficient dwarf1 gene. It is also interestingly linked to documented height loci in maize. Plants that overexpress ABA, which is necessary for drought resistance, show improved drought tolerance. However, ABA 8′-hydroxylase hydroxylates ABA to 8′-hydroxy- abscisate and $NADP^{+}$, depleting ABA to lower ABA levels and resulting in phenotypes ABA-deficient.

Comparing the categorization of maize lines conducted earlier using pedigree and breeding data to the classification achieved using SNPs, revealed good accuracy. Although the majority of the genotypes in this research fell into one of the four heterotic categories that the company's breeding program had previously developed, there were be seen, demonstrating the large amount of genetic diversity present in maize germplasm [9]. Furthermore, the genotyping information derived from this work may be used to further hybrid prediction models, allowing for more precise identification of hybrids, and increasing breeding effectiveness [12].

6. Conclusions

We can state with confidence that the results of our in-depth comparative study of the gene enrichment analysis of two distinct maize hybrids to drought stress at the 478-development stage are presented here. Genetically, in comparison to the sensitive line 478, the drought-resistant line H21 had a greater number of enzymes, lower levels of oxidative stress, and improved cell water holding capacity. Overall, the genome our RNA-seq data identified many differentially expressed genes as being asserted in response to drought, with several of these genes being particularly disclosed in H21. We found that the expression of genes related to earlier identified pathways involved in the drought stress response, such as those involved in the manufacture of secondary metabolites, TF regulation, detoxification, and stress resistance, increased in response to drought stress. Our results provide a foundation for future research utilizing targeted cloning and aid in our thoughtful investigation of the mechanisms governing maize's drought resilience.

Author Contributions: Conceptualization, S.F.A.G., Y.P. and A.R.; methodology, S.F.A.G., Y.M., A.A. and M.A.; software, S.F.A.G. and A.R.; validation, Y.P. and M.U.H.; formal analysis, S.F.A.G. and S.H.Q.; investigation, Y.P. and N.B.; resources, Y.P. and N.A.T.A.K.; data curation, S.F.A.G., A.A. and A.R.; writing—original draft preparation, S.F.A.G., A.A., M.A., M.M.T. and A.R.; writing—review and editing, S.F.A.G., Y.M., A.A., N.A.T.A.K. and N.B.; visualization, Y.P. and M.U.H.; supervision, Y.P.; project administration, Y.P.; funding acquisition, Y.P. All authors have read and agreed to the published version of the manuscript.

Funding: This research was supported by the National Key research and development project (2022YFD1201804), Talent Fuxi Project of Gansu Agricultural University, (No. GAUFX-47102Y09), and the Lanzhou Science and Technology Bureau (No. 2020-RC-122).

Institutional Review Board Statement: Not applicable.

Informed Consent Statement: Not applicable.

Data Availability Statement: Not applicable.

Acknowledgments: The authors also would like to thank the deanship of Scientific Research at Umm Al-Qura University for supporting this work by Grant code (22UQU4281560DSR04).

Conflicts of Interest: The authors declare no conflict of interest.

References

1. Dossa, K.; Li, D.; Wang, L.; Zheng, X.; Liu, A.; Yu, J.; Wei, X.; Zhou, R.; Fonceka, D.; Diouf, D. Transcriptomic, biochemical and physio-anatomical investigations shed more light on responses to drought stress in two contrasting sesame genotypes. *Sci. Rep.* **2017**, *7*, 8755. [CrossRef] [PubMed]
2. Ereful, N.C.; Liu, L.-Y.; Greenland, A.; Powell, W.; Mackay, I.; Leung, H. RNA-seq reveals differentially expressed genes between two indica inbred rice genotypes associated with drought-yield QTLs. *Agronomy* **2020**, *10*, 621. [CrossRef]
3. Faghani, E.; Gharechahi, J.; Komatsu, S.; Mirzaei, M.; Khavarinejad, R.A.; Najafi, F.; Farsad, L.K.; Salekdeh, G.H. Comparative physiology and proteomic analysis of two wheat genotypes contrasting in drought tolerance. *J. Proteom.* **2015**, *114*, 1–15. [CrossRef] [PubMed]
4. Fracasso, A.; Trindade, L.M.; Amaducci, S. Drought stress tolerance strategies revealed by RNA-Seq in two sorghum genotypes with contrasting WUE. *BMC Plant Biol.* **2016**, *16*, 115. [CrossRef] [PubMed]
5. Ghodke, P.; Khandagale, K.; Thangasamy, A.; Kulkarni, A.; Narwade, N.; Shirsat, D.; Randive, P.; Roylawar, P.; Singh, I.; Gawande, S.J. Comparative transcriptome analyses in contrasting onion (*Allium cepa* L.) genotypes for drought stress. *PLoS ONE* **2020**, *15*, e0237457. [CrossRef] [PubMed]
6. Hao, L.-Y.; Liu, X.-Y.; Zhang, X.-J.; Sun, B.-C.; Cheng, L.; Zhang, D.-F.; Tang, H.-J.; Li, C.-H.; Li, Y.-X.; Shi, Y.-S. Genome-wide identification and comparative analysis of drought related genes in roots of two maize inbred lines with contrasting drought tolerance by RNA sequencing. *J. Integr. Agric.* **2020**, *19*, 449–464. [CrossRef]
7. Kang, Z.; Babar, M.A.; Khan, N.; Guo, J.; Khan, J.; Islam, S.; Shrestha, S.; Shahi, D. Comparative metabolomic profiling in the roots and leaves in contrasting genotypes reveals complex mechanisms involved in post-anthesis drought tolerance in wheat. *PLoS ONE* **2019**, *14*, e0213502. [CrossRef] [PubMed]
8. Kumar, J.; Gunapati, S.; Kianian, S.F.; Singh, S.P. Comparative analysis of transcriptome in two wheat genotypes with contrasting levels of drought tolerance. *Protoplasma* **2018**, *255*, 1487–1504. [CrossRef] [PubMed]
9. Liu, X.; Zhang, X.; Sun, B.; Hao, L.; Liu, C.; Zhang, D.; Tang, H.; Li, C.; Li, Y.; Shi, Y. Genome-wide identification and comparative analysis of drought-related microRNAs in two maize inbred lines with contrasting drought tolerance by deep sequencing. *PLoS ONE* **2019**, *14*, e0219176. [CrossRef] [PubMed]
10. Luo, M.; Zhao, Y.; Wang, Y.; Shi, Z.; Zhang, P.; Zhang, Y.; Song, W.; Zhao, J. Comparative proteomics of contrasting maize genotypes provides insights into salt-stress tolerance mechanisms. *J. Proteome Res.* **2018**, *17*, 141–153. [CrossRef] [PubMed]
11. Wang, N.; Zhao, J.; He, X.; Sun, H.; Zhang, G.; Wu, F. Comparative proteomic analysis of drought tolerance in the two contrasting Tibetan wild genotypes and cultivated genotype. *BMC Genom.* **2015**, *16*, 432. [CrossRef] [PubMed]
12. Wang, W.; Qin, Q.; Sun, F.; Wang, Y.; Xu, D.; Li, Z.; Fu, B. Genome-wide differences in DNA methylation changes in two contrasting rice genotypes in response to drought conditions. *Front. Plant Sci.* **2016**, *7*, 1675. [CrossRef] [PubMed]

13. Wang, X.; Zenda, T.; Liu, S.; Liu, G.; Jin, H.; Dai, L.; Dong, A.; Yang, Y.; Duan, H. Comparative proteomics and physiological analyses reveal important maize filling-kernel drought-responsive genes and metabolic pathways. *Int. J. Mol. Sci.* **2019**, *20*, 3743. [CrossRef] [PubMed]
14. Yang, L.; Fountain, J.C.; Wang, H.; Ni, X.; Ji, P.; Lee, R.D.; Kemerait, R.C.; Scully, B.T.; Guo, B. Stress sensitivity is associated with differential accumulation of reactive oxygen and nitrogen species in maize genotypes with contrasting levels of drought tolerance. *Int. J. Mol. Sci.* **2015**, *16*, 24791–24819. [CrossRef] [PubMed]
15. Zenda, T.; Liu, S.; Wang, X.; Jin, H.; Liu, G.; Duan, H. Comparative proteomic and physiological analyses of two divergent maize inbred lines provide more insights into drought-stress tolerance mechanisms. *Int. J. Mol. Sci.* **2018**, *19*, 3225. [CrossRef] [PubMed]
16. Zenda, T.; Liu, S.; Wang, X.; Liu, G.; Jin, H.; Dong, A.; Yang, Y.; Duan, H. Key maize drought-responsive genes and pathways revealed by comparative transcriptome and physiological analyses of contrasting inbred lines. *Int. J. Mol. Sci.* **2019**, *20*, 1268. [CrossRef] [PubMed]
17. Zeng, W.; Peng, Y.; Zhao, X.; Wu, B.; Chen, F.; Ren, B.; Zhuang, Z.; Gao, Q.; Ding, Y. Comparative proteomics analysis of the seedling root response of drought-sensitive and drought-tolerant maize varieties to drought stress. *Int. J. Mol. Sci.* **2019**, *20*, 2793. [CrossRef] [PubMed]
18. Zhang, X.; Hong, M.; Wan, H.; Luo, L.; Yu, Z.; Guo, R. Identification of key genes involved in embryo development and differential oil accumulation in two contrasting maize genotypes. *Genes* **2019**, *10*, 993. [CrossRef] [PubMed]
19. Zhang, X.; Liu, X.; Zhang, D.; Tang, H.; Sun, B.; Li, C.; Hao, L.; Liu, C.; Li, Y.; Shi, Y. Genome-wide identification of gene expression in contrasting maize inbred lines under field drought conditions reveals the significance of transcription factors in drought tolerance. *PLoS ONE* **2017**, *12*, e0179477. [CrossRef] [PubMed]

genes

MDPI

Article

Population Genetics and Anastomosis Group's Geographical Distribution of *Rhizoctonia solani* Associated with Soybean

Aqleem Abbas [1], Xiangling Fang [1,*], Shehzad Iqbal [2], Syed Atif Hasan Naqvi [3], Yasir Mehmood [3], Muhammad Junaid Rao [4,*], Zeshan Hassan [5], Roberto Miño Ortiz [6], Alaa Baazeem [7], Mahmoud Moustafa [8,9], Sulaiman Alrumman [8] and Sally Negm [10,11]

1 State Key Laboratory of Grassland Agroecosystems, Key Laboratory of Grassland Livestock Industry Innovation, Ministry of Agriculture and Rural Affairs, College of Pastoral Agriculture Science and Technology, Lanzhou University, Lanzhou 730000, China
2 College of Plant Sciences and Technology, Huazhong Agricultural University, Wuhan 430070, China
3 Department of Plant Pathology, Faculty of Agricultural Sciences and Technology, Bahauddin Zakariya University, Main Campus, Bosan Road, Multan 60800, Pakistan
4 State Key Laboratory for Conservation and Utilization of Subtropical Agro-Bioresources, Guangxi Key Laboratory of Sugarcane Biology, College of Agriculture, Guangxi University, Nanning 530021, China
5 College of Agriculture, Bahauddin Zakariya University, Multan, Bahadur Sub Campus, Layyah 31200, Pakistan
6 Instituto de Ciencias Biológicas, Universidad de Talca, Talca 3460000, Chile
7 Department of Biology, College of Science, Taif University, Taif 21944, Saudi Arabia
8 Department of Biology, Faculty of Science, King Khalid University, Abha 62529, Saudi Arabia
9 Department of Botany and Microbiology, Faculty of Science, South Valley University, Qena 83523, Egypt
10 Department of Life Sciences, College of Science and Art Mahyel Aseer, King Khalid University, Abha 62529, Saudi Arabia
11 Unit of Food Bacteriology, Central Laboratory of Food Hygiene, Ministry of Health, Branch in Zagazig, Sharkia 44511, Egypt
* Correspondence: xlf@lzu.edu.cn (X.F.); mjunaidrao@webmail.hzau.edu.cn (M.J.R.)

Citation: Abbas, A.; Fang, X.; Iqbal, S.; Naqvi, S.A.H.; Mehmood, Y.; Rao, M.J.; Hassan, Z.; Ortiz, R.M.; Baazeem, A.; Moustafa, M.; et al. Population Genetics and Anastomosis Group's Geographical Distribution of *Rhizoctonia solani* Associated with Soybean. *Genes* **2022**, *13*, 2417. https://doi.org/10.3390/genes13122417

Academic Editors: Wajid Zaman and Hakim Manghwar

Received: 25 November 2022
Accepted: 15 December 2022
Published: 19 December 2022

Abstract: *Rhizoctonia solani* is a species complex composed of many genetically diverse anastomosis groups (AG) and their subgroups. It causes economically important diseases of soybean worldwide. However, the global genetic diversity and distribution of *R. solani* AG associated with soybean are unknown to date. In this study, the global genetic diversity and distribution of AG associated with soybean were investigated based on rDNA-ITS sequences deposited in GenBank and published literature. The most prevalent AG, was AG-1 (40%), followed by AG-2 (19.13%), AG-4 (11.30%), AG-7 (10.43%), AG-11 (8.70%), AG-3 (5.22%) and AG-5 (3.48%). Most of the AG were reported from the USA and Brazil. Sequence analysis of internal transcribed spacers of ribosomal DNA separated AG associated with soybean into two distinct clades. Clade I corresponded to distinct subclades containing AG-2, AG-3, AG-5, AG-7 and AG-11. Clade II corresponded to subclades of AG-1 subgroups. Furthermore, AG and/or AG subgroups were in close proximity without corresponding to their geographical origin. Moreover, AG or AG subgroups within clade or subclades shared higher percentages of sequence similarities. The principal coordinate analysis also supported the phylogenetic and genetic diversity analyses. In conclusion, AG-1, AG-2, and AG-4 were the most prevalent AG in soybean. The clade or subclades corresponded to AG or AG subgroups and did not correspond to the AG's geographical origin. The information on global genetic diversity and distribution will be helpful if novel management measures are to be developed against soybean diseases caused by *R. solani*.

Keywords: soybean; *R. solani*; anastomosis groups; distribution; genetic diversity; phylogeny

1. Introduction

Soybean (*Glycine max* L.) is one of the world's most significant oilseed crops, accounting for 25% of all edible oil production [1]. About 176.6 million tons of soybeans are produced over 75.5 million hectares of fertile land each year [2,3]. *R solani* Kuhn [teleomorph, *Thanatephorus cucumeris* (Frank) Donk] is a serious threat to soybean production worldwide. The fungus causes blights (foliar and web), pre-and post-emerging damping-off, root and hypocotyl rot diseases of soybean [4,5]. These diseases caused massive yield losses in soybeans all over the world. For example, in Brazil and the United States, only foliar blight has resulted in 30 to 69% yield losses [6–9]. Moreover, these diseases are difficult to control because of soil born nature of *R. solani* and the broad host range [4]. Fungicides have been widely used to manage these diseases [10,11]. However, fungicides have caused severe environmental and health concerns. The most cost-effective and environmentally sustainable option to manage *R. solani* is breeding resistant cultivars [12]. However, understanding the genetic diversity of *R. solani* is critical if novel management measures, such as developing *Rhizoctonia* -resistant cultivars, are to be developed. *R. solani* exhibits tremendous genetic diversity and is classified into different anastomosis groups (AG). To date, 14 anastomosis groups (AG 1 to 13 and AG-BI) have been identified based on the fusion of hyphae, morphology, virulence (pathogenicity), physiology, and DNA homology [13,14]. Some of the AG have been further divided into subgroups based on anastomosis frequency, physiological and morphological features, pathogenic, bimolecular, biochemical, genetic, and DNA homology characteristics [15,16].For example, AG-1 has been divided into six subgroups: IA, IB, IC, ID, IE, and IF [17]. Similarly, AG-4 has been divided into three subgroups: HGI, HGII, and HGIII [18], and AG-2 has been divided into nine subgroups such as 1, 2, t, Nt, 2IIIB, 2IV, 2LP, 3 and 4 [18]. AG-2, AG-4, AG-5, AG-3, AG-7, and AG-11 are causing damping-off and root, and hypocotyls rot, whereas AG-1 is responsible for the foliar and web blight of soybean [5,19–22].

To evaluate genetic diversity and to characterize AG of *R. solani*, numerous molecular markers were used, which included inter-simple sequence repeats (ISSR) [23], simple sequence repeats (SSR) [24,25], single nucleotide polymorphisms (SNPs) [26], amplified fragment length polymorphism (AFLP) [27], restriction fragment length polymorphisms (RFLP) [28], randomly amplified polymorphic DNA (RAPD) markers [29], electrophoretic karyotype [30], DNA-DNA hybridization [31] and sequence analysis of rDNA ITS1-5.8 S-ITS2region [32–34]. Among these molecular markers, sequence analysis of rDNA ITS1-5.8 S-ITS2 region, because of their presence of multiple copies of tandem repeats within the genome of all fungi, has proven to be a more powerful tool for genetic diversity and phylogenetic studies of AG and AG subgroups of *R. solani* [15,18,34–37]. Moreover, it validates the grouping of AG on the basis of classical hyphal anastomosis reactions [38,39]. Furthermore, the rDNA ITS1-5.8S-ITS2 region sequences evolve quickly and are bordered by highly conserved nucleotide sequences [40]. They are found between the 18S and 5.8S rRNA genes (ITS1) and the 5.8S and 28S rRNA genes (ITS2) [38,41].

Information on the genetic diversity and distribution of *R. solani* AG associated with soybean of a particular country is available [1,42–44]. However, there was no attempt to investigate the global genetic diversity and distribution of *R. solani* AG associated with soybean. Considering the genetic diversity and different diseases causing abilities of *R. solani* AG on soybean mentioned above, the current study was aimed (1) To determine most frequently reported and dominant AG associated with soybean; (2) To explore the genetic diversity of AG based on rDNA ITS1-5.8S-ITS2 sequence analysis; (3) To determine the relationship between geographical origin and genetic diversity of AG.

2. Materials and Methods

2.1. Data Collection

Soybean (*G. max* L.) infected by *R. solani* was targeted in this study. Relevant literature was searched in Google Scholar (http://scholar.google.com; accessed on 11 November 2021), Web of Science (http://apps.webofknowledge.com; accessed on 12 November 2021),

China National Knowledge Infrastructure (CNKI) (https://www.cnki.net; accessed on 12 November 2021), Scholarly, Academic Information Navigator (CiNii) (http://ci.nii.ac.jp/en; accessed on 13 November 2021), PubMed (https://pubmed.ncbi.nlm.nih.gov/; accessed on 14 November 2021), and Scopus (https://www.elsevier.com/en-gb/solutions/scopus; accessed on 15 November 2021) using the main keywords "soybean-*R. solani*, " and the "soybean-anastomosis groups (AG)" solely and in combination. The search for literature was limited from January 2001 to October 2021. To be included in this study, the literature had to meet the following criteria: (i) Only articles published in peer-reviewed journals were chosen; (ii) Articles mentioning the accession numbers of AG and the sequencing data of AG was publicly available in GenBank; (iii) Articles mentioning the geographical origin, isolates and AG that could cause symptoms on soybean; (iv) Articles mentioning about pathogen isolation from the soil (e.g., rhizosphere soil, topsoil), root, and shoot of the symptomatic soybean plants. The articles from the databases mentioned above were imported into the EndNote X9 software to acquire information of AG, isolate, geographical origin, and isolation sources. The information on AG, isolates, geographical origin, isolation sources were compiled as shown in Table S1. The primary isolation sources included diseased roots, shoots of soybean crop, and soil (e.g., rhizosphere soil, topsoil) surrounding symptomatic soybean. To create a dataset of all publicly available sequences from the rDNA ITS1-5.8S-ITS2 region associated with *R. solani* AG, we queried National Center for Biotechnology Information (NCBI) GenBank(https://www.ncbi.nlm.nih.gov/genbank; accessed on 12 November 2021) and downloaded all of the sequences for the following studies.

2.2. Characterization of the Distribution and Frequency of Anastomosis Groups Assoiated with Soybean

The frequency of AG with known sequences in GenBank from the published literature was calculated using the formula with modification for this study; relative frequency $(F) = 100 \times (n/N)$, in which n = the number of each AG/AG subgroup reported in the published literature and N = the total number of all AG/AG subgroup reported in the published literature [19,45]. AG showed higher frequency was considered most frequently reported or highly distributed AG or AG subgroup.

2.3. Sequences Alignments

The nucleotide sequences of the rDNA ITS1-5.8S-ITS2 region, representing AG were edited and assembled using BioEdit v. 7.2.5 with manual adjustment [46]. The sequences were aligned using the Clustal W algorithm in the Molecular Evolutionary Genetics Analysis (MEGA v. 7.0.26) software, and obvious errors were adjusted manually [47]. The resultant alignments were imported into BioEdit v. 7.2.5 and adjusted manually by visual examination [46]. A sequence of *Atheliarolfsii* FSR-052 (GenBank Accession No. AY684917) (anamorph, *Sclerotiumrolfsii*) was used as the outgroup [38,39,41–48].

2.4. Phylogenetic Analysis

Before conducting phylogenetic analysis, best-fit substitution model selection of the aligned sequences was carried out using the jModelTest v. 2.1.6 package program [49] with model selection strictly based on the Akaike Information Criterion (AIC) estimateand Bayesian Information Criterion (BIC) [50,51]. The Tamura-Nei [52] model was suggested by jModelTest v. 2.1.6. The best-fit substitution model for the phylogenetic trees was mentioned in Table S2. Phylogenetic trees on the multiple alignments were constructed using MEGA v. 7.0.26. The phylogenetic trees were built using Maximum Likelihood (ML) [53], Neighbor-Joining (NJ) [54] and Maximum-parsimony (MP) [55]. Rates among sites were selected as G (γ distributed) for both ML and NJ. The partial deletion for ML and NJ was set as gap/missing data treatment with a 95% site coverage cut-off, and Nearest-neighbor interchange (NNI) was selected for the heuristic method. The MP analysis was obtained using the Close-Neighbor-Interchange algorithm [55]. Bootstrapping of

1000 random samples from various sequence alignments was used to test each phylogenetic tree's robustness. Gaps and missing data were removed from all positions. Only nodes with bootstrap values of 70% or higher were shown in the phylogenetic trees. Phylogenetic trees were visualized using the Interactive Tree of Life (iTOLv. 6(http://itol.embl.de/; accessed on 16 November 2021) [56] and Fig Tree v. 1.4.4 (http://tree.bio.ed.ac.uk/software/figtree/; accessed on 17 November 2021) and edited in Adobe Illustrator® CS5 v. 15.0.0 (San Jose, CA, USA).

2.5. Principal Coordinate Analysis (PCoA) and Sequence Similarities

Pairwise percentages of sequence similarities of all the isolates within AG and AG subgroups and among the AG and AG subgroups were calculated with the MatGAT v. 2.0 program [57]. Principal coordinate analyses (PCoA) were conducted on pairwise sequences similarity matrix to investigate clustering of AG and AG subgroups using paleontological statistics software package for education and data analysis (PAST v. 4.03) with Gower similarity index [58].

3. Results

3.1. Distribution of Anastomosis Groups

Nine anastomosis groups (AG) such as AG-1, AG-2, AG-3, AG-4, AG-5, AG-6, AG-7, AG-9 and AG-11, were associated with soybean. According to the geographical distribution, most AG were reported from the United States and Brazil while only a few AG were reported from Japan, India, Canada and Taiwan (Figure 1). Among the AG, AG-1 was the most prevalent and frequently reported AG with a relative frequency of 40%, followed by AG-2 (19.13%), AG-4 (11.30%), AG-7 (10.43%), AG-11 (8.70%), AG-3 (5.22%), AG-5 (3.48%) and AG-6 and AG-9 each with a frequency of 0.87% (Table 1). Similarly, among the AG subgroups associated with soybean, AG-1-IA was frequently reported AG (33.91%), followed by AG-2-2IIIB (12.17%) and each of AG-4-HGII and AG-4-HGIII with a frequency of 4.35%. On soybeans, all of these AG were pathogenic (Table 1). AG-1 caused severe foliar and web blight of soybean, while the rest of the AG were reported to cause damping-off, root, and hypocotyl rot of soybean (Table 1).

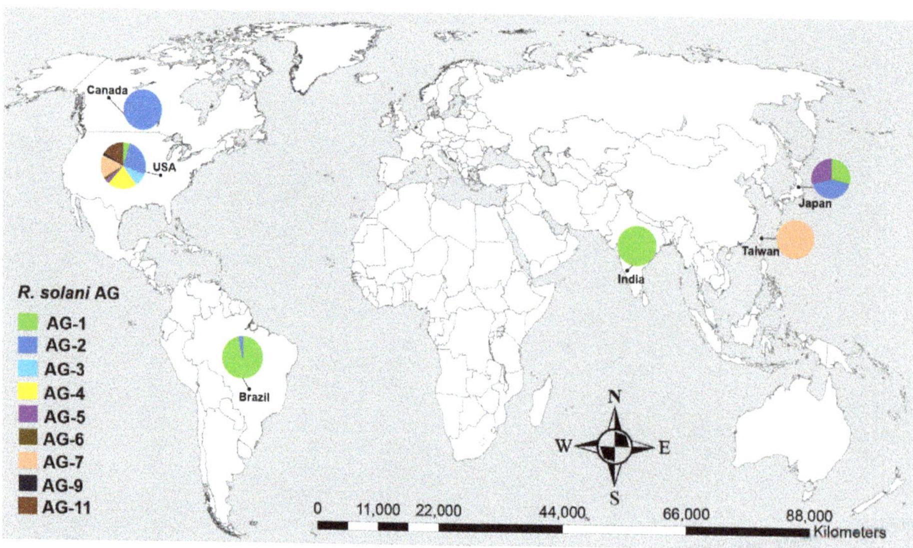

Figure 1. Geographical distribution of *R. solani* anastomosis groups (AG) associated with soybean with known sequences in GenBank. Data in pie charts indicate the number of AG reported in each geographical origin.

Table 1. Frequency of *R. solani* AG associated with soybean across the geographical origins.

AG	AG Subgroups	Geographical Origin								
		USA	Brazil	Canada	Taiwan	India	Japan	Total	[a]F	
AG-1	AG-1-IA	1	36				2	39	33.91	40.00
	AG-1-IB	1				2		3	2.61	
	AG-1-IC	1						1	0.87	
	AG-1-IF		1					1	0.87	
	1					1		1	0.87	
	AG-1-ID		1					1	0.87	
AG-2	AG-2-1	3						3	2.61	19.13
	AG-2-2	1	1					2	1.74	
	AG-2-3						3	3	2.61	
	AG-2-2IIIB	12		2				14	12.17	
AG-3		6						6	5.22	5.22
AG-4	AG-4	1						1	0.87	11.30
	AG-4-HGI	2						2	1.74	
	AG-4-HGII	5						5	4.35	
	AG-4-HGIII	5						5	4.35	
AG-5		2					2	4	3.48	3.48
AG-6		1						1	0.87	0.87
AG-7		10			2			12	10.43	10.43
AG-9		1						1	0.87	0.87
AG-11		10						10	8.70	8.70
	Total	62	39	2	2	3	7			

[a]F (Relative frequency) = 100 × (n / N), in which n = the number of each AG/AG subgroup and N = the total number of all AG/AG subgroups.

3.2. Genetic Diversity of Anastomosis Groups

Initially, 115 *R. solani* AG associated with soybean were collected from published literature in this study (Table 1). Only 102 AG with known isolate names, pathogenicity and geographic origins were used to explore genetic diversity and phylogeny (Figures 2–4). Using NJ, ML and MP methods, sequence analysis clustered AG into two major clades (Figures 2–4). The 102 AG clustered into two distinct clades, with clade I including 62 AG and clade II including 40 AG (Figures 2–4). The 62 AG in clade I further clustered into five distinct subclades (Ia, Ib, Ic, Id, and Ie), with sufficient bootstrap support for each subclade (Figures 2–4). Clade Ia included isolates of AG-11 (subclade Ia-1) and isolates of AG-5 (subclade Ia-2), clade Ib included isolates of AG-2, clade Ic included isolates of AG-3, clade Id included isolates of AG-4, and clade Ie included isolates of AG-7. Subclades Ia-1 (isolates of AG-11) and Ia-2 (AG-5) within Ia were closely spaced subclades in the tree. Even the subgroups of AG within these subclades clustered separately. For example, subclade clade Ib was further subdivided into three subclades (Ib-1, Ib-2, and Ib-3). Ib-1 included isolates of AG-2-1, Ib-2 included AG-2-2 and Ib-3 included isolates of AG-2-2IIIB. Similarly, clade Id was further subdivided into four subclades, i.e., subclade Id-1 (containing isolates of AG-4-HGI), subclade Id-2 (isolates of AG-4-HGII), subclade Id-3 (AG-4-HGIII), and subclade Id-4 (AG-4-4). The 40 AG in clade II also clustered into two distinct subclades (IIa and IIb). Subclade IIa only included isolates of AG-1-IC and AG-1-IB, while subclade IIb included 38 isolates of AG-1-IA (Figures 2–4).

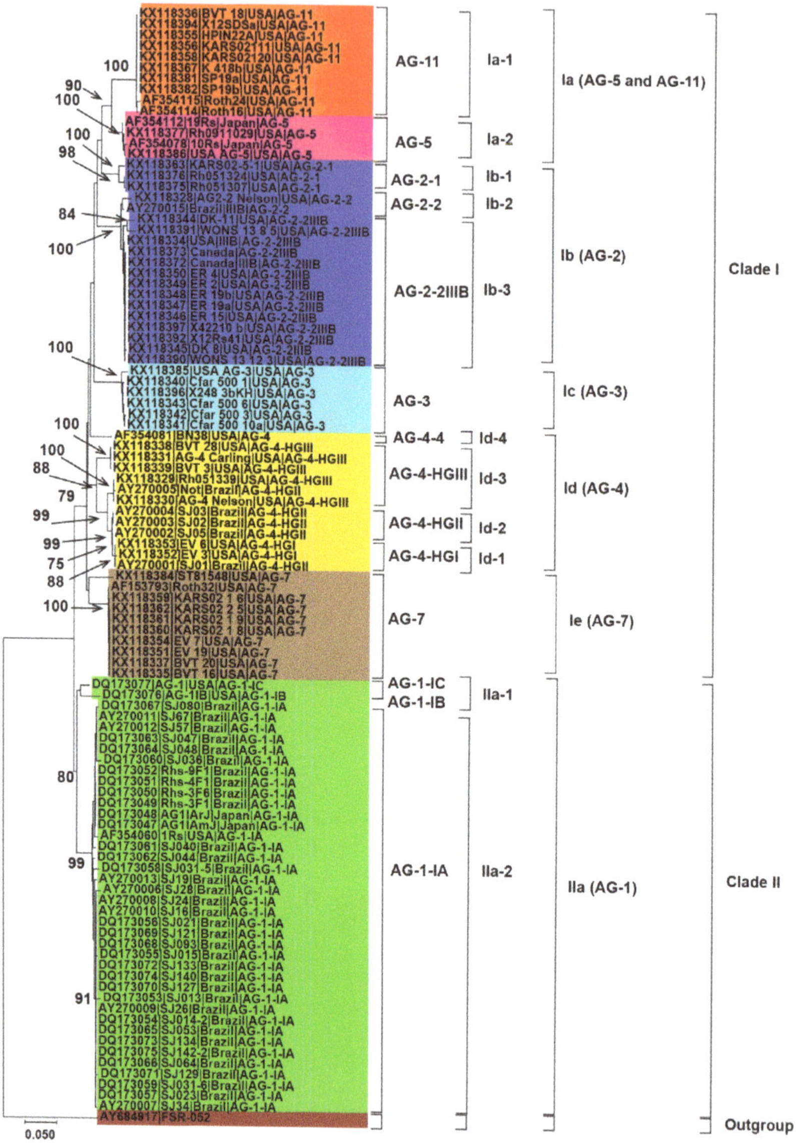

Figure 2. Genetic relatedness among the AG from soybean. Neighbor-Joining (NJ) analysis was used to build the trees. The accession numbers are followed by isolate, geographical origin and AG. Different colors show clades and/or subclades associated with AG. Thousands of replications were used to bootstrap tree branches. Numbers at nodes indicate bootstrap values, and only those ≥70 are shown. The outgroup, *Athelia rolfsii* FSR-052 (GenBank Accession No. AY684917), was used to root the tree. Scale bar indicates a genetic distance of 0.05 for horizontal branch length.

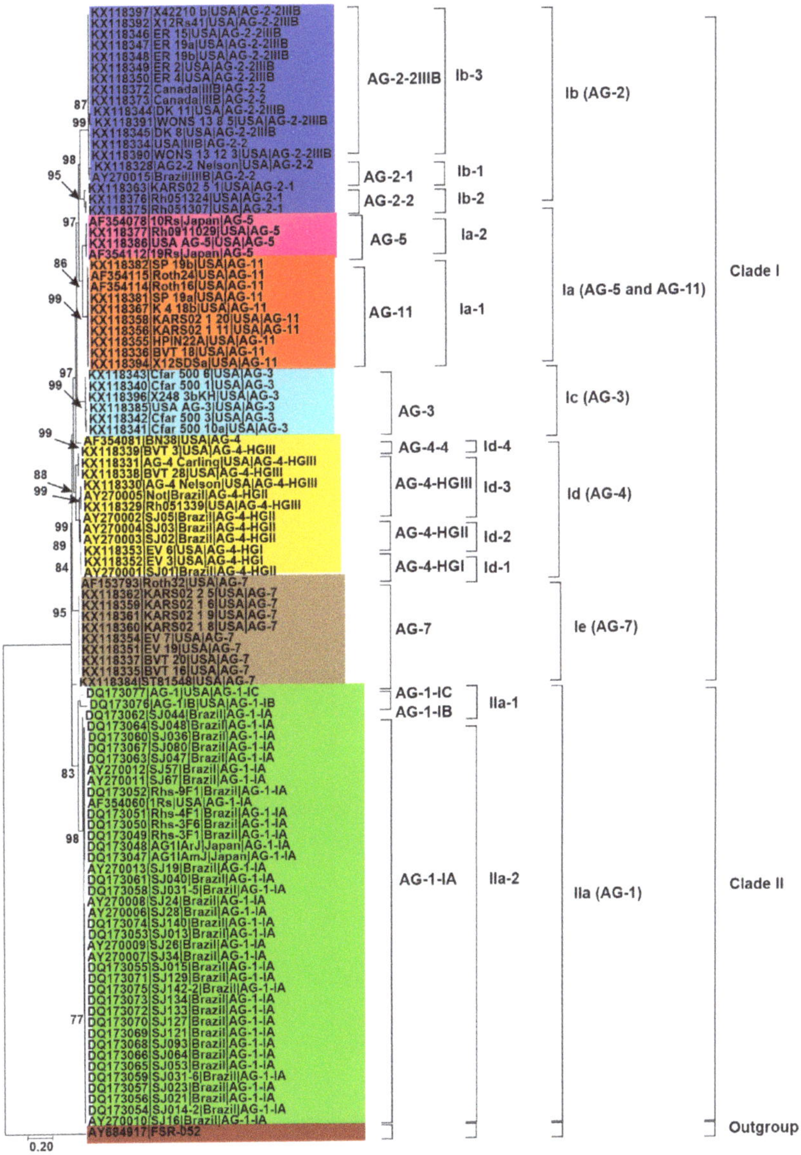

Figure 3. Genetic relatedness among the AG from soybean. Maximum likelihood (ML) analysis was used to build the trees. The accession numbers are followed by isolate, geographical origin and AG. Different colors show clades and/or subclades associated with AG. Thousands of replications were used to bootstrap tree branches. Numbers at nodes indicate bootstrap values, and only those ≥70 are shown. The outgroup, *Athelia rolfsii* FSR-052 (GenBank Accession No. AY684917), was used to root the tree. Scale bar indicates a genetic distance of 0.2 for horizontal branch length.

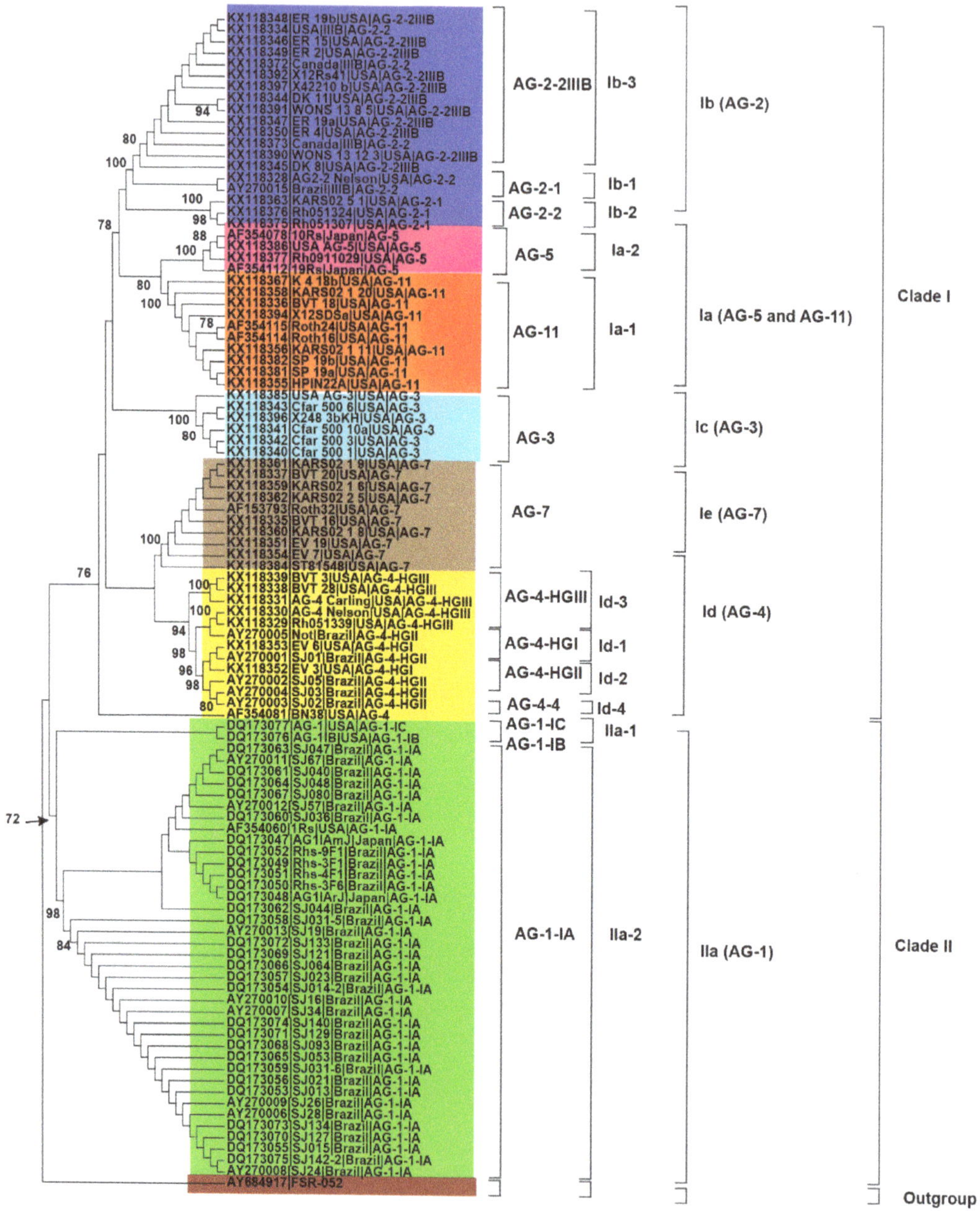

Figure 4. Genetic relatedness among the AG from soybean. Maximum parsimony (MP) analysis was used to build the trees. The accession numbers are followed by isolate, geographical origin and AG. Different colors show clades and/or subclades associated with AG. 1000 replications were used to bootstrap tree branches. Numbers at nodes indicate bootstrap values, and only those ≥70 are shown. The outgroup, *Athelia rolfsii* FSR-052 (GenBank Accession No. AY684917), was used to root the tree.

3.3. Relationship between Genetic Diversity of Anastomosis Groups and Their Geographic Origin

Besides, closely related AG or AG subgroups associated with soybean were clustered together regardless of the geographical origin from where they had been identified (Figures 2–4). For example, isolates of AG-1-IA in subclade IIb from Brazil, Japan, and the USA clustered together (Figures 2–4). Similarly, isolates of AG-5 from the USA and Japan, in subclade Ia-2, clustered together. Moreover, isolates of AG-2-2 from the USA and Brazil in the subclade Ib-2 clustered together (Figures 2–4). In conclusion, closely related AG associated with soybean were clustered together regardless of the geographical origin from where they had been identified.

3.4. Genetic Relatedness among and within Clades and Subclades Representing Anast Mosis Groups

ML, MP and NJ phylogenetic trees showed that AG or AG subgroups were clustered together regardless of the geographical origin. Therefore the percentage of sequence similarities within and among clades and subclades of 108 AG without their corresponding geographical origin, was established by direct pairwise comparisons (Table 2). Within the proposed subclades, all the AG-11 and AG-5 closely related isolates within subclade Ia had the widest range of sequence similarities of 77 to 99.4%, followed by AG-7 closely related isolates within subclade Ie which had sequence similarities of 82.1 to 100%, AG-4 closely related isolates within subclade Id which had sequence similarities of 87.4 to 100% and AG-1 closely related isolates within subclades IIa-1 and IIa-2 which had sequence similarities of 87.5 to 100% (Table 2). All the AG-5 closely related isolates within subclade Ia-2, AG-2 closely related isolates within subclade Ib, and AG-3 closely related isolates within subclade Ic had the sequence similarity greater than 90%. The sequence similarity of AG-11 (subclade Ia-1), AG-5 (subclade Ia-2), AG-2 (subclade Ib), AG-3 (subclade Ic), AG-4 (subclade Id), and AG-7 (subclade Ie) of clade I were higher than isolates of AG-1 of clade II. There were also variations in the percentage of sequence similarities of AG and/or AG subgroups between the subclades. For example, the sequence similarity between subclade Ib-1 (isolates of AG-2-1) and subclade Ib-2 (isolates of AG-2-2) and between subclade Ib-1 (isolates of AG-2-1) and subclade Ib-3 (isolates of AG-2-2IIIB) were 83.7–86% and 81.5–85.4%, respectively. In contrast, the sequence similarity between subclade Ib-2 (isolates of AG-2-2) and subclade Ib-3 (isolates of AG-2-2IIIB) was 92 to 99%. This suggests a closer genetic relatedness between the isolates of AG-2-2 and AG-2-2IIIB than between the isolates of AG-2-1. The AG-4 (Subclade Id) isolates were further sub-clustered into four subgroups (1, HG-I, -II, and -III). The sequence similarity within AG-4 (Subclade Id) as a whole rangesfrom 81.7% to 100%; it is considerably lower than within the subgroups: 99.4, 99.6–100, 87.4–99, and 87.2% for HG-I, -II, -III, and 1 respectively. Furthermore, the percentage of sequence similarity was higher than 87.2% within an AG subgroup (IA, IB, IC, HG-1, HG-II, HG-III, 2-1, 2-2, 2-IIIB), 81.7 to 100% for different subgroups within an AG, and 69.5 to 92.5% among different AG. The above-mentioned phylogenetic trees were further explained and supported by PCoA. PCoA grouped AG into two major clades (Figure 5). Clade I included the isolates of AG-3, 4, 5, 7, and 11, and clade II included isolates of AG-1. Within the clades, subgroups of AG formed a separate cluster. For example, isolates of AG-1-IB and AG-1-IC (subclade IIa) formed distinct clusters than isolates of AG-1-1A (subclade IIb). Similarly, isolates within subclade Id such as AG-4-HGI, AG-4-HGII, and AG-4-HGIII formed separate clusters (Figure 5).

Table 2. Percentages of sequence similarities of rDNA-ITS sequences of *R. solani* AG within and between clades and subclades.

Clades and Subclades		Ia (AG-11 and AG-5)			Ib (AG-2)		Ic (AG-3)		Id (AG-4)			Ie (AG-7)	IIa (AG-1)		Outgroup
		Ia-1	Ia-2	Ib-1	Ib-2	Ib-3		Id-1	Id-2	Id-3	Id-4		IIa-1	IIa-2	
Ia (AG-11 and AG-5)	Ia-1	77–99.6													
	Ia-2	85.8–93	93–99.4												
	Ic-1	84.6–89.18	81.9–89	97.2–99.4											
Ib (AG-2)	Ib-2	80.7–85.1	76.1–84.6	83.7–86	90.9–96.5										
	Ib-3	79.4–85.7	74–85.1	81.5–85.4	92–99.1	91.6–100									
Ic (AG-3)		82.3–88.3	83.8–84.9	91.4–92.5	82.7–85.2	79.3–85.7	97.7–99.4								
	Ie-1	80.7–84.6	80.1–80.2	85.1–85.9	82–82.8	77.2–83.8	81.8–84.9	99.4 *							
Id (AG-4)	Id-2	79.3–80.8	79.4–84.2	80.7–81.9	78.3–80.5	75.8–82.5	78.3–80.6	91.1–94	99.6–100						
	Id-3	79.5–84.1	76–84.5	82.5–86	74.6–83.3	76.2–84.3	81.7–83.89	87.1–96.4	81.7–94.4	87.4–99.1					
	Id-4	79.1–82.8	79.6–83.6	81.8–83.3	75.5–77.1	72.5–77.1	81.9–82.2	82.6–82.9	86.1–86.4	82.6–83.6	87.2 *				
Ie (AG-7)		77.2–83.8	76–84.6	83.3–86.3	79.9–85.9	74.7–82.3	78.4–85.9	81.5–87.9	80–83.7	84.4–87.5	83.1–84.6	82.1–99.1			
IIa (AG-1)	IIa-1	72.5–79.8	74–80.5	74.7–77.8	69.6–75.2	69.5–77.5	73.3–76.5	76.2–77.9	81.7–83.4	74.4–78.5	79.4–81.6	74.3–85.5	87.9–100		
	IIa-2	72.8–79.4	73–83.7	75.8–80.6	69.5–80.5	70.4–80.2	73–79.8	77.9–82.9	81.9–88.8	75.7–83.3	81.7–82.8	74.7–83.6	76.3–92.2	89.5–100	
Outgroup		47.4–51.3	46.2–49.1	50–51	47.3–52.7	48.4–52.5	51.5–52.5	51.5–51.9	46.6–48	51.1–53.1	48.3 *	48.8–49.3	39.1–42.3	40.9–46	100 *

* Range could not be calculated for one isolate having one sequence.

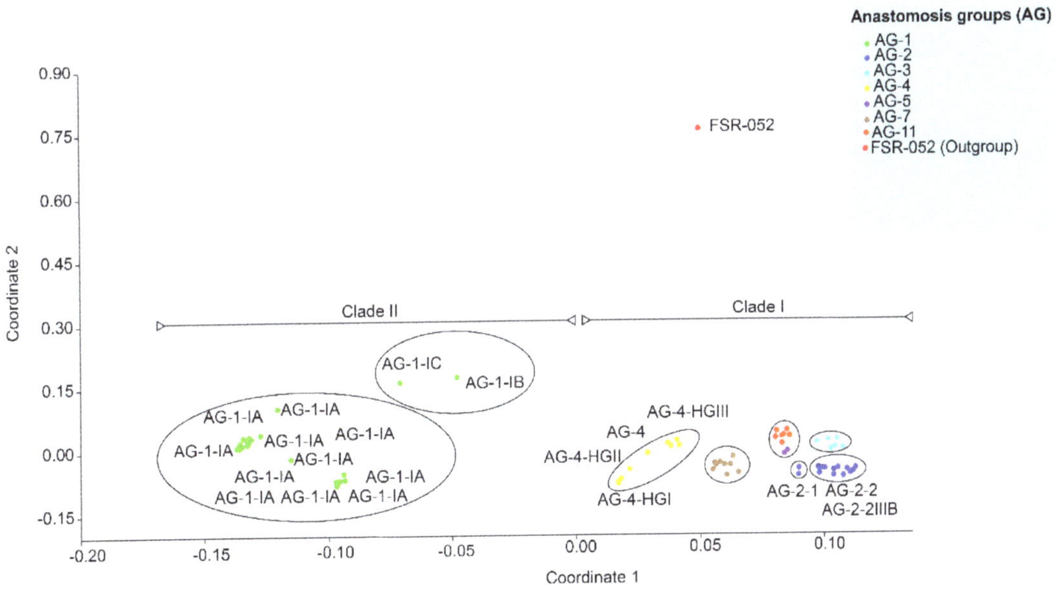

Figure 5. Principal coordinate analyses (PCoA) ordination of *R.solani* anastomosis groups. (AG) based on a Gower distance of sequence similarity matrix.

4. Discussion

R. solani is a soil borne pathogen that affects soybean worldwide and has a significant economic impact in all soybean growing countries [1,7,11,19,21,59–61]. All AG associated with soybean reported in this study belonged to *R. solani*. The frequency of AG varied substantially geographically. Most of the AG were reported from the diseased soybean plants in the USA and Brazil. More AG discoveries associated with soybean in the USA may imply an expansion of the host range and genetic diversity of *R. solani* [4]. Furthermore, AG associated with soybean might have been studied more intensively in the USA than other countries because of their greater relative importance as plant pathogen of soybean. In the USA, foliar blight caused by AG-1 and, hypocotyl and root rot caused by AG-2-2IIIB, AG-4, AG-5, AG-3, AG-7, and AG-11 caused as high as 45% soybean yield losses [4,21,61–63]. In this study, the reports of a few AG from Japan, Canada, Taiwan, Japan, and India were probably due to a lack of sampling or isolation methods. In Brazil, foliar blight, damping-off, and root rot caused by AG-1 and AG-2 resulted in an estimated 31 to 60% soybean yield loss [8,32,64]. In Canada, root rot ranked fourth among 22 diseases causing severe losses in soybean [65,66]. In India, foliar blight caused by AG-1 caused an average yield loss of 40% to 50% [1,43,67,68]. Besides, AG-2 and AG-5 have been reported to cause hypocotyl rot of soybean in Japan [44]. AG-7 is responsible for the damping-off of soybean seedlings in Taiwan [42]. In recent years, the frequency of legumes in crop rotations has increased, and also the intensive cultivation of soybean might be another reason for increasing the frequency of *R. solani* AG [4,5,21,42,61–63].

Besides, our study also revealed the most frequently reported AG from soybean. AG-1 was the most frequently reported AG from soybean, followed by AG-2, AG-4, AG-11, AG-7, AG-3, and AG-5. Frequently reported AG doesn't indicate whether it is highly pathogenic or not pathogenic on soybean. For example, AG-1-1A is highly pathogenic on soybean in Brazil; however, AG-2-2IIIB, AG-4, and AG-5 are highly pathogenic on soybean in the USA [5,19–22]. Hence, AG diversity, frequency, and distribution could be influenced by the dynamics of the host-pathogen relationship, genetic flexibility, and degree of adaptation [69]. Furthermore, crop rotation, soil types, soybean cultivars, cropping patterns, and climatic conditions of the particular region may encourage the presence of specific AG over others [70]. In addition, root-associated microbial communities also influence AG distribution [32].

The most reliable approach for phylogenetic analysis and genetic diversity of AG and AG subgroups of *R. solani* is the molecular characterization utilizing the sequences of the rDNA ITS1-5.8S-ITS2 region [5,37,38]. We were able to make conclusions about the phylogenetic relationships among AG and AG subgroups using the sequences of the rDNA ITS1-5.8S-ITS2 region from the NCBI GenBank. In this study, phylogenetic analysis based on MP, NJ and ML showed AG forms two distinct clades. Clade I included isolates of AG-2, AG-3, AG-4, AG-5, AG-7, and AG-11, whereas clade II included isolates of AG-1. Each AG forms a distinct subclade within the clades except AG-5 and AG-11, which form a distinct subclade (Ia).This suggests that isolates of AG-5 and AG-11 may be more closely related to each other. Previous studies have shown that even isolates of AG-5 of soybean clustered with the isolates of AG-11 of other legumes such as lupins [37]. Besides, our study showed that even AG subgroups form distinct subclades. For example, isolates of AG-1C and IB form a sister subclade with isolates of AG-1A within clade IIa (AG-1 isolates). Sequence analysis in previous studies revealed that AG-1-B was genetically distinct from AG-1 IA and IB [69]. Likewise, within the sub clade Ib (AG-2 isolates), isolates of AG-2-1 form a sister subclade with isolates of AG-2-2 and AG-2-IIIB. Previous studies considered AG-2 a polyphyletic with subgroups consistently forming different clades or subclades [71].

AG-2 is a highly heterogeneous AG with substantial genetic diversity and is further divided into nine subgroups such as 1, 2, t, Nt, 2IIIB, 2IV, 2LP, 3, 4 that cause rots and damping-off disease in soybean [37]. Moreover, within subclade Id (AG-4 isolates), few AG-4-HGIII isolates form a sister clade with isolates of AG-4-HGII and AG-4-HGI. Besides, most of the isolates of AG-4-HGII and AG-4-I were clustered together. This indicated that subgroups HGI and HGII were found to be more closely related than subgroup HGIII [15,17,32]. Furthermore, our study also revealed that AG did not have preferences for geographical origin. Most clades or subclades with high bootstrap support indices include AG and AG subgroups from USA, Brazil, and other countries. In a previous study, the authors of the reference [34] analyzed sequences of AG from Europe, North America, Australia and Asia associated with legumes, cereals and vegetables and found that AG did not have a preference for a geographic origin; however, some AG were found to be host-specific [72]. The authors of the reference [73–78] showed that isolates of AG from different countries are categorized under the same AG or AG subgroups. Furthermore, the pairwise distance matrix based on sequence similarities revealed that the isolates of AG within the clades and subclades shared high sequence similarities. In contrast, isolates of AG from different clades and subclades showed less similarity. Furthermore, the sequence similarity was higher than 87.2% within an AG subgroup, 81.7 to 100% for different subgroups within an AG, and 69.5 to 92.5% among different AG. These results are consistent with previous studies that assessed the sequence similarities of ITS sequences [15]. They found that sequence similarity was above 96% for the same AG subgroup, 66–100% for different subgroups within an AG, and 55–96% for AG. In addition, PCoA revealed that AG and/or AG subgroups form a separate group from each other. Previous reports showed that the sequence homology in the ITS regions was higher for isolates of the same subgroup than isolates of different subgroups within an AG and isolates of different AG [38]. Our study revealed that the rDNA-ITS sequences were clustered consistently according to their known AG and not according to geographical origin. Cluster analyses based on rDNA-ITS on sequences of *R. solani* AG and AG subgroups associated with a soybean of a specific geographical origin have already been reported [1,7,19,21,61].

5. Conclusions, Limitations and Future Directions

In conclusion, this study provides the first documentation regarding the global genetic diversity and distribution of *R. solani* AG associated with soybean. AG-1, AG-2, and AG-4 were the most prevalent and widely documented AG in soybean. AG-1 was responsible for foliar and web blight of soybean, whereas the remaining AG were causing damping-off, root and hypocotyl rot. Across geographical origin, most of the AG were reported from the USA, followed by Brazil. Phylogenetic and genetic diversity analysis revealed that AG and/or AG subgroups formed distinct clades and subclades without corresponding to geographical origin. Pairwise percentages of sequence similarities within AG and subgroups and principal coordinate analysis also support the phylogenetic and genetic diversity analysis. The rDNA ITS1-5.8S-ITS2 region has been successfully sequenced and phylogenetically analyzed to reliably separate *R. solani* isolates into several groups and subgroups that correspond to the various AG [5,15,17,18,35]. However, sequence analysis of the rDNA ITS1-5.8S-ITS2 region is not without its attendant limitations. Though the differences in the rDNA ITS1-5.8S-ITS2region are sufficiently large to differentiate the AG reliably, they could not differentiate isolates of the same AG [75]. Furthermore, researchers do not verify or validate sequences deposited in databases and repositories; depositing an incorrectly named AG is almost inevitable. Complete information about isolate name, host, or geographical origin may not be included. Besides, the rDNA ITS1-5.8S-ITS2 region would not always be ideal because of high mutation rates. Furthermore, *R. solani* is multinucleate; therefore, there is the possibility of numerous nucleotide variations in this region even in the single strain of *R. solani* [76–78]. Hence, the genetic diversity and phylogeny of AG must be augmented with additional sequences such as large-subunit rRNA (LSU) region, ß-tubulin, the largest (*RPB1*) and the second-largest (*RPB2*) subunits of RNA polymerase, translation elongation factor (*tef-1α*), the mini-chromosome maintenance protein (*mcm7*), calmodulin (*CaM*), and topoisomerase I (*top1*) gene. Furthermore, studies involving genomic, transcriptomic, proteomic and mitogenomic analysis may provide insights into the phylogeny and genetic diversity of *R. solani* AG.

Supplementary Materials: The following are available online at https://www.mdpi.com/article/10.3390/genes13122417/s1, Table S1. GenBank accession numbers of DNA sequences from the rDNA ITS1-5.8S-ITS2 region of *R. solani* AG were used to determine AG's phylogenetic relationships and genetic diversity associated with soybean, Table S2. The best models used for ML and NJ analysis in this study.

Author Contributions: Conceptualization, A.A. and X.F.; Data curation, A.A., S.I. and S.A.H.N.; Formal analysis, A.A.; Funding acquisition, M.J.R., S.A.H.N. and A.B.; Investigation, A.A.; Methodology, A.A. and X.F.; Project administration, S.A.H.N.; Resources, X.F.; Software, A.A.; Supervision, X.F. and A.B.; Validation, A.A. and X.F.; Visualization, A.A. and X.F.; Writing—original draft, A.A. and A.B.; Writing—review & editing, Y.M., Z.H., A.B., R.M.O., M.M., S.A., S.N. and S.A.H.N. All authors have read and agreed to the published version of the manuscript.

Funding: This research was funded by the National Natural Science Foundation of China (No. 31802126; 32171678) and the Basic Research Innovation-Team Program of Qinghai Provincial Science Foundation (No. 2021-ZJ-901).

Institutional Review Board Statement: Not applicable.

Informed Consent Statement: Not applicable.

Data Availability Statement: Data sets analyzed during the current study are available from the corresponding author on reasonable request.

Acknowledgments: The first author acknowledges the salary support for his Post Doc studies provided by Lanzhou University. Authors thank the Deanship of Scientific Research, King Khalid University, Saudi Arabia for funding this work under grant no. (R.G.P.2/197/43).

Conflicts of Interest: The authors declare no conflict of interest.

References

1. Surbhi, K.; Singh, K.P.; Singh, N.K.; Aravind, T. Assessment of genetic diversity among soybean genotypes differing in response to aerial blight (*Rhizoctonia solani* Kuhn) using SSR markers. *J. Phytopathol.* **2021**, *169*, 37–44. [CrossRef]
2. Lewandowska, S.; Łoziński, M.; Marczewski, K.; Kozak, M.; Schmidtke, K. Influence of priming on germination, development, and yield of soybean varieties. *Open Agric.* **2020**, *5*, 930–935. [CrossRef]

3. Michalak, I.; Lewandowska, S.; Niemczyk, K.; Detyna, J.; Bujak, H.; Arik, P.; Bartniczak, A. Germination of soybean seeds exposed to the static/alternating magnetic field and algal extract. *Eng. Life Sci.* **2019**, *19*, 986–999. [CrossRef]
4. Ajayi-Oyetunde, O.; Bradley, C. *Rhizoctonia solani:* taxonomy, population biology and management of *Rhizoctonia* seedling disease of soybean. *Plant pathol.* **2018**, *67*, 3–17. [CrossRef]
5. Ajayi-Oyetunde, O.O.; Bradley, C.A. Identification and characterization of *Rhizoctonia* species associated with soybean seedling disease. *Plant Dis.* **2017**, *101*, 520–533. [CrossRef]
6. Fenille, R.C.; De Souza, N.L.; Kuramae, E.E. Characterization of *Rhizoctonia solani* associated with soybean in Brazil. *Eur. J. Plant Pathol.* **2002**, *108*, 783–792. [CrossRef]
7. Stetina, K.C.; Stetina, S.R.; Russin, J.S. Comparison of severity assessment methods for predicting yield loss to *Rhizoctonia* foliar blight in soybean. *Plant Dis.* **2006**, *90*, 39–43. [CrossRef]
8. Ciampi, M.B.; Meyer, M.C.; Costa, M.J.; Zala, M.; McDonald, B.A.; Ceresini, P.C. Genetic structure of populations of *Rhizoctonia solani* anastomosis group-1 IA from soybean in Brazil. *Phytopathol.* **2008**, *98*, 932–941. [CrossRef]
9. Copley, T.R.; Duggavathi, R.; Jabaji, S. The transcriptional landscape of *Rhizoctonia solani* AG1-IA during infection of soybean as defined by RNA-seq. *PLoS One* **2017**, *12*, e0184095. [CrossRef]
10. Van den Heuvel, J. *Rhizoctonia Species: Taxonomy, Molecular Biology, Ecology, Pathology and Disease Control*; Sneh, B., Jabaji-Hare, S., Neate, S.M., Dijst, G.N., Eds.; Kluwer Academic: Dordrecht, The Netherlands, 1996; pp. XVI + 577.
11. Urrea, K.; Rupe, J.C.; Rothrock, C.S. Effect of fungicide seed treatments, cultivars, and soils on soybean stand establishment. *Plant Dis.* **2013**, *97*, 807–812. [CrossRef]
12. Chawla, S.; Bowen, C.R.; Slaminko, T.L.; Hobbs, H.A.; Hartman, G.L.; Nachilima, C.; Chigeza, G.; Chibanda, M.; Mushoriwa, H.; Diers, B.D.; et al. A public program to evaluate commercial soybean cultivars for pathogen and pest resistance. *Plant Dis.* **2013**, *97*, 568–578. [CrossRef] [PubMed]
13. Ogoshi, A. Ecology and pathogenicity of anastomosis and intraspecific groups of *Rhizoctoniasolani* Kuhn. *Annu. Rev. Phytopathol.* **1987**, *25*, 125–143. [CrossRef]
14. Godoy-Lutz, G.; Steadman, J.R.; Higgins, B.; Powers, K. Genetic variation among isolates of the web blight pathogen of common bean based on PCR-RFLP of the ITS-rDNA region. *Plant Dis.* **2003**, *87*, 766–771. [CrossRef] [PubMed]
15. Kuninaga, S.; Natsuaki, T.; Takeuchi, T.; Yokosawa, R. Sequence variation of the rDNA ITS regions within and between anastomosis groups in *Rhizoctonia solani. Curr. Genet.* **1997**, *32*, 237–243. [CrossRef] [PubMed]
16. Carling, D.E. Grouping in *Rhizoctonia solani* by hyphal anastomosis. In *Rhizoctonia species: Taxonomy, molecular biology, ecology, pathology, and disease control*; Sneh, B., Jabaji-Hare, S., Neate, S.M., Dijst, G.N., Eds.; Kluwer Academic: Dordrecht, the Netherlands, 1996; pp. 37–47.
17. Kuninaga, S.; Godoy-Lutz, G.; Yokosawa, R. rDNA-ITS nucleotide sequences analysis of *Thanatephoruscucumeris* AG-1 associated with web blight on common beans in Central America and Caribbean. *Ann. Phytopathol. Soc. Japan.* **2002**, *68*, 187. [CrossRef]
18. Carling, D.E.; Kuninaga, S.; Brainard, K.A. Hyphal anastomosis reactions, rDNA-internal transcribed spacer sequences, and virulence levels among subsets of *Rhizoctonia solani* anastomosis group-2 (AG-2) and AG-BI. *Phytopathol.* **2002**, *92*, 43–50. [CrossRef]
19. Zhao, G.; Ablett, G.R.; Anderson, T.R.; Rajcan, I.; Schaafsma, A.W. Anastomosis groups of *Rhizoctoniasolani* associated with soybean root and hypocotyl rot in Ontario and resistance of accession PI 442031 to different anastomosis groups. *Can. J. Plant Pathol.* **2005**, *27*, 108–117. [CrossRef]
20. Carling, D.E.; Rothrock, C.S.; MacNish, G.C.; Sweetingham, M.W.; Brainard, K.A.; Winters, S.W. Characterization of anastomosis group 11 (AG-11) of *Rhizoctonia solani. Phytopathol.* **1994**, *84*, 1387–1393. [CrossRef]
21. Nelson, B.; Helms, T.; Christianson, T.; Kural, I. Characterization and pathogenicity of *Rhizoctonia* from soybean. *Plant Dis.* **1996**, *80*, 74–80. [CrossRef]
22. Baird, R.E.; Carling, D.E.; Mullinix, B.G. Characterization and comparison of isolates of *Rhizoctonia solani* AG-7 from Arkansas, Indiana, and Japan, and select AG-4 isolates. *Plant Dis.* **1996**, *80*, 1424–1441. [CrossRef]
23. Guleria, S.; Aggarwal, R.; Thind, T.S.; Sharma, T.R. Morphological and pathological variability in rice isolates of *Rhizoctoniasolani* and molecular analysis of their genetic variability. *JJ. Plant Pathol.* **2007**, *155*, 654–661.
24. Gonzalez-Vera, A.D.; Bernardes-de-Assis, J.; Zala, M.; McDonald, B.A.; Correa-Victoria, F.; Graterol-Matute, E.J.; Ceresini, P.C. Divergence between sympatric rice-and maize-infecting populations of *Rhizoctonia solani* AG-1 IA from Latin America. *Phytopathol.* **2010**, *100*, 172–182. [CrossRef] [PubMed]
25. Padasht-Dehkaei, F.; Ceresini, P.C.; Zala, M.; Okhovvat, S.M.; Nikkhah, M.J.; McDonald, B.A. Population genetic evidence that basidiospores play an important role in the disease cycle of rice-infecting populations of *Rhizoctonia solani* AG-1 IA in Iran. *Plant Pathol.* **2013**, *62*, 49–58. [CrossRef]
26. Ciampi, M.B.; Gale, L.R.; Lemos, E.G.; Ceresini, P.C. Distinctively variable sequence-based nuclear DNA markers for multi-locusphylogeography of the soybean-and rice-infecting fungal pathogen *Rhizoctonia solani* AG-1 IA. *Genet. Mol. Biol.* **2009**, *32*, 840–846. [CrossRef]
27. Taheri, P.; Gnanamanickam, S.; Höfte, M. Characterization, genetic structure, and pathogenicity of *Rhizoctonia* spp. associated with rice sheath diseases in India. *Phytopathol.* **2007**, *97*, 373–383. [CrossRef]
28. Rosewich, U.L.; Pettway, R.E.; McDonald, B.A.; Kistler, H.C. High levels of gene flow and heterozygote excess characterize *Rhizoctonia solani* AG-1 IA (*Thanatephoruscucumeris*) from Texas. *Fungal Genet. Biol.* **1999**, *28*, 148–159. [CrossRef]

29. Wang, L.; Liu, L.M.; Wang, Z.G.; Huang, S.W. Genetic Structure and Aggressiveness of *Rhizoctoniasolani* AG 1-IA, the Cause of Sheath Blight of Rice in Southern C hina. *J. Plant Pathol.* **2013**, *161*, 753–762.
30. Keijer, J.; Houterman, P.M.; Dullemans, A.M.; Korsman, M.G. Heterogeneity in electrophoretic karyotype within and between anastomosis groups of *Rhizoctonia solani*. *Mycol. Res.* **1996**, *100*, 789–797. [CrossRef]
31. Carling, D.E.; Kuninaga, S. DNA base sequence homology in *Rhizoctoniasolani* Kuhn: Inter- and intragroup relatedness of anastomosis group-9. *Phytopathol.* **1990**, *80*, 1362–1364. [CrossRef]
32. Fenille, R.C.; Ciampi, M.B.; Kuramae, E.E.; Souza, N.L. Identification of *Rhizoctonia solani* associated with soybean in Brazil by rDNA-ITS sequences. *Fitopatol. Bras.* **2003**, *28*, 413–419. [CrossRef]
33. Ciampi, M.B.; Kuramae, E.E.; Fenille, R.C.; Meyer, M.C.; Souza, N.L.; Ceresini, P.C. Intraspecific Evolution of *Rhizoctonia solani* AG-1 IA Associated with Soybean and Rice in Brazil based on Polymorphisms at the ITS-5.8S rDNA Operon. *Eur. J. Plant Pathol.* **2005**, *113*, 183–196. [CrossRef]
34. Monazzah, M.; Esfahani, M.N.; Sattar, T.E. Genetic structure and proteomic analysis associated in potato to Rhizoctonia solani AG-3PT-stem canker and black scurf. *Physiol. Mol. Plant Pathol.* **2022**, *122*, 101905. [CrossRef]
35. Boysen, M.; Borja, M.; del Moral, C.; Salazar, O.; Rubio, V. Identification at strain level of *Rhizoctonia solani* AG4 isolates by direct sequence of asymmetric PCR products of the ITS regions. *Curr. Genet.* **1996**, *29*, 174–181. [CrossRef] [PubMed]
36. Salazar, O.; Julián, M.C.; Rubio, H.V. Phylogenetic Grouping of Cultural Types of *Rhizoctonia solani* AG 2-2 Based on Ribosomal ITS Sequences. *Mycologia* **2000**, *92*, 505–509. [CrossRef]
37. Gonzalez, D.; Carling, D.E.; Kuninaga, S.; Vilgalys, R.; Cubeta, M.A. Ribosomal DNA systematics of *Ceratobasidium* and *Thanatephorus* with *Rhizoctonia* anamorphs. *Mycologia* **2001**, *93*, 1138–1150. [CrossRef]
38. Sharon, M.; Kuninaga, S.; Hyakumachi, M.; Sneh, B. The advancing identification and classification of *Rhizoctonia* spp. using molecular and biotechnological methods compared with the classical anastomosis grouping. *Mycoscience* **2006**, *47*, 299–316. [CrossRef]
39. Ohkura, M.; Abawi, G.S.; Smart, C.D.; Hodge, K.T. Diversity and aggressiveness of *Rhizoctonia solani* and *Rhizoctonia* -like fungi on vegetables in New York. *Plant Dis.* **2009**, *93*, 615–624. [CrossRef] [PubMed]
40. Álvarez, I.; Wendel, J.F. Ribosomal ITS sequences and plant phylogenetic inference. *Mol. Phylogenet. Evol.* **2003**, *29*, 417–434. [CrossRef] [PubMed]
41. Sharon, M.; Sneh, B.; Kuninaga, S.; Hyakumachi, M.; Naito, S. Classification of *Rhizoctonia* spp. using rDNA-ITS sequence analysis supports the genetic basis of the classical anastomosis grouping. *Mycoscience* **2008**, *49*, 93–114. [CrossRef]
42. Lin, Y.C.; Tseng, M.N.; Chang, H.X. First Report of Soybean Seedling Disease caused by *Rhizoctonia solani* AG-7 in Taiwan. *Plant Dis.* **2021**, *106*, 314. [CrossRef]
43. Mamta, M.; Singh, K.P. Prevalence and severity of *Rhizoctonia* aerial blight of soybean in Uttarakhand. *Indian J. Ecol.* **2017**, *44*, 417–419.
44. Misawa, T.; Komatsu, T. Occurrence of *Rhizoctonia* root rot of soybean caused by *Rhizoctonia solani* AG-2-2IV and adzuki bean caused by *R. solani* AG-1 IB in commercial fields of Hokkaido. *Annu. Rep. Soc. Plant Prot. North Jpn.* **2011**, 50–54.
45. Arias, M.M.D.; Munkvold, G.P.; Ellis, M.L.; Leandro, L.F.S. Distribution and frequency of *Fusarium* species associated with soybean roots in Iowa. *Plant Dis.* **2013**, *97*, 1557–1562. [CrossRef] [PubMed]
46. Hall, T.A. BioEdit: A User-Friendly Biological Sequence Alignment Editor and Analysis Program for Windows 95/98/NT. *Nuclc Acids Symposium Series* **1999**, *41*, 95–98.
47. Thompson, J.D.; Higgins, D.G.; Gibson, T.J. CLUSTAL W: Improving the sensitivity of progressive multiple sequence alignment through sequence weighting, position-specific gap penalties and weight matrix choice. *Nucleic Acids Res.* **1994**, *22*, 4673–4680. [CrossRef] [PubMed]
48. Sharon, M.; Freeman, S.; Kuninaga, S.; Sneh, B. Genetic diversity, anastomosis groups and virulence of *Rhizoctonia* spp. from strawberry. *Eur. J. Plant Pathol.* **2007**, *117*, 247–265. [CrossRef]
49. Darriba, D.; Taboada, G.L.; Doallo, R.; Posada, D. jModelTest 2: More models, new heuristics and parallel computing. *Nat. Methods.* **2012**, *9*, 772. [CrossRef]
50. Akaike, H. A new look at the statistical model identification. *IEEE Trans. Autom. Control.* **1974**, *19*, 716–723. [CrossRef]
51. Neath, A.A.; Cavanaugh, J.E. The Bayesian information criterion: Background, derivation, and applications. *Wiley Interdisciplinary Reviews: Comput. Stat.* **2012**, *4*, 199–203. [CrossRef]
52. Hsieh, Y.C.; Chang, C.W.; Wang, C.J. First report of Rhizoctonia solani AG-4 HG-I causing leaf blight disease on Cattleya × hybrid in Taiwan. *J. Gen. Plant Pathol.* **2022**. [CrossRef]
53. Tamura, K.; Nei, M. Estimation of the number of nucleotide substitutions in the control region of mitochondrial DNA in humans and chimpanzees. *Mol. Biol. Evol.* **1993**, *10*, 512–526. [CrossRef]
54. Saitou, N.; Nei, M. The neighbor-joining method: A new method for reconstructing phylogenetic trees. *Mol. Biol. Evol.* **1987**, *4*, 406–425.
55. Nei, M.; Kumar, S. *Molecular Evolution and Phylogenetics*; Oxford University Press: Oxford, UK, 2000.
56. Letunic, I.; Bork, P. Interactive Tree Of Life (iTOL) v5: An online tool for phylogenetic tree display and annotation. *Nucleic Acids Res.* **2021**, *49*, W293–W296. [CrossRef]
57. Campanella, J.J.; Bitincka, L.; Smalley, J. MatGAT: An application that generates similarity/identity matrices using protein or DNA sequences. *BMC Bioinformatics* **2003**, *4*, 29. [CrossRef]
58. Gower, J.C. A general coefficient of similarity and some of its properties. *Biometrics* **1971**, *22*, 857–871. [CrossRef]

59. Rong, C.; Zheng, D.; Wei, L.; Dai, L.; Liu, S. Research Progress of Soybean Root Rot. *Chinese Agricultural Science Bulletin* **2016**, *32*, 58–62.
60. Sun, C.; Guo, R.J.; Dong, L.S. Control of soybean root rot disease with silver-carrying antimicrobial preparations. *ActaPhytophylacica Sin.* **2006**, *36*, 550–554.
61. Tachibana, H. *Rhizoctonia solani* root rot epidemic of Soybeans in Central Iowa 1967. *Plant Disease Reporter* **1968**, *52*, 613–614.
62. Liu, Z.; Sinclair, J.B. Isolates of *Rhizoctonia solani* anastomosis group 2-2 pathogenic to soybean. *Plant Dis.* **1991**, *75*, 682–687. [CrossRef]
63. Muyolo, N.G.; Lipps, P.E.; Schmitthenner, A.F. Anastomosis grouping and variation in virulence among isolates of *Rhizoctoniasolani* associated with dry bean and soybean in Ohio and Zaire. *Phytopathol* **1993**, *83*, 438–444. [CrossRef]
64. Chavarro-Mesa, E.; Ceresini, P.; Pereira, D.; Vicentini, S.; Silva, T.; Ramos-Molina, L.; Negrisoli, M.; Schurt, D.; Vieira Júnior, J.R. A broad diversity survey of *Rhizoctonia* species from the Brazilian Amazon reveals the prevalence of *R. solani* AG-1 IA on signal grass and the new record of AG-1 IF on cowpea and soybeans. *Plant Pathol* **2020**, *69*, 455–466. [CrossRef]
65. Anderson, T.R.; Tenuta, A.U. Diseases of soybean in Ontario and estimated yield losses, 1994, 1996–2000. *Can. Plant Dis. Surv.* **2001**, *81*, 133–135.
66. Chang, K.-F.; Hwang, S.-F.; Ahmed, H.U.; Strelkov, S.E.; Harding, M.W.; Conner, R.L.; McLaren, D.L.; Gossen, B.D.; Turnbull, G.D. Disease reaction to *Rhizoctonia solani* and yield losses in soybean. *Can. J. Plant Sci.* **2017**, *98*, 115–124. [CrossRef]
67. Joshi, D.; Pushpendra, K.S.; Adhikari, S.; Rani, S. Screening of soybean germplasm for important disease prevalent in North India. *Int. J. Chem. Stud.* **2018**, *6*, 2731–2733.
68. Dubey, S.C.; Tripathi, A.; Upadhyay, B.K.; Deka, U.K. Diversity of *Rhizoctoniasolani* associated with pulse crops in different agro-ecological regions of India. *World J. Microbiol. Biotechnol.* **2014**, *30*, 1699–1715. [CrossRef]
69. Godoy-Lutz, G.; Kuninaga, S.; Steadman, J.R.; Powers, K. Phylogenetic analysis of *Rhizoctoniasolani* subgroups associated with web blight symptoms on common bean based on ITS-5.8S rDNA. *J. Gen. Plant Pathol.* **2008**, *74*, 32–40. [CrossRef]
70. Muzhinji, N.; Truter, M.; Woodhall, J.W.; Van der Waals, J.E. Anastomosis groups and pathogenicity of *Rhizoctonia solani* and binucleate *Rhizoctonia* from potato in South Africa. *Plant Dis.* **2015**, *99*, 1790–1802. [CrossRef]
71. Budge, G.E.; Shaw, M.W.; Lambourne, C.; Jennings, P.; McPherson, M. Characterization and origin of infection of *Rhizoctoniasolani* associated with *Brassica oleracea* crops in the UK. *Plant Pathol.* **2010**, *58*, 1059–1070. [CrossRef]
72. González, D.; Cubeta, M.A.; Vilgalys, R. Phylogenetic utility of indels within ribosomal DNA and β-tubulin sequences from fungi in the *Rhizoctonia solani* species complex. *Mol. Phylogenet. Evol.* **2006**, *40*, 459–470. [CrossRef]
73. Toda, T.; Hyakumachi, M.; Arora, D.K. Genetic relatedness among and within different *Rhizoctoniasolani* anastomosis groups as assessed by RAPD, ERIC and REP-PCR. *Microbiol. Res.* **1999**, *154*, 247–258. [CrossRef]
74. Toda, T.; Mghalu, J.M.; Priyatomojo, A.; Hyakumachi, M. Comparison of sequences for the internal transcribed spacer region in *Rhizoctoniasolani* AG 1-ID and other subgroups of AG 1. *J. Gen. Plant Pathol.* **2004**, *70*, 270–272. [CrossRef]
75. Lehtonen, M.J.; Ahvenniemi, P.; Wilson, P.S.; German-Kinnari, M.; Valkonen, J.P.T. Biological diversity of *Rhizoctonia solani* (AG-3) in a northern potato-cultivation environment in Finland. *Plant Pathol.* **2008**, *57*, 141–151. [CrossRef]
76. Broders, K.D.; Parker, M.L.; Melzer, M.S.; Boland, G.J. Phylogenetic diversity of *Rhizoctoniasolani* associated with canola and wheat in Alberta, Manitoba, and Saskatchewan. *Plant Dis.* **2014**, *98*, 1695–1701. [CrossRef]
77. Pope, E.J.; Carter, D.A. Phylogenetic placement and host specificity of mycorrhizal isolates belonging to AG-6 and AG-12 in the *Rhizoctonia solani* species complex. *Mycologia* **2001**, *93*, 712–719. [CrossRef]
78. Gónzalez, D.; Rodriguez-Carres, M.; Boekhout, T.; Stalpers, J.; Kuramae, E.E.; Nakatani, A.K.; Vilgalys, R.; Cubeta, M.A. Phylogenetic relationships of *Rhizoctonia* fungi within the Cantharellales. *Fungal Biol.* **2016**, *120*, 603–619. [CrossRef]

genes

MDPI

Article

DNA Barcoding and *ITS2* Secondary Structure Predictions in Taro (*Colocasia esculenta* L. Schott) from the North Eastern Hill Region of India

Mayengbam Premi Devi [1,2,†], Madhumita Dasgupta [1,†], Sansuta Mohanty [3,†], Susheel Kumar Sharma [1,4], Vivek Hegde [5,6], Subhra Saikat Roy [1], Rennya Renadevan [7], Kinathi Bipin Kumar [7], Hitendra Kumar Patel [7,*] and Manas Ranjan Sahoo [1,3,*]

1 Indian Council of Agricultural Research (ICAR) Research Complex for North Eastern Hill Region, Imphal 795004, India
2 College of Agriculture, Central Agricultural University (CAU-Imphal), Kyrdemkulai 793105, India
3 Central Horticultural Experiment Station, ICAR–Indian Institute of Horticultural Research, Bhubaneswar 751019, India
4 ICAR—Indian Agricultural Research Institute, Pusa Campus, New Delhi 110012, India
5 ICAR—Central Tuber Crops Research Institute, Thiruvananthapuram 695017, India
6 ICAR—Indian Institute of Horticultural Research, Bengaluru 560089, India
7 Centre for Cellular and Molecular Biology, Hyderabad 570007, India
* Correspondence: hkpatel@ccmb.res.in (H.K.P.); manas.sahoo@icar.gov.in (M.R.S.); Tel.: +91-674-247-1867 (M.R.S.); Fax: +91-674-247-1712 (M.R.S.)
† These authors contributed equally to this work.

Citation: Devi, M.P.; Dasgupta, M.; Mohanty, S.; Sharma, S.K.; Hegde, V.; Roy, S.S.; Renadevan, R.; Kumar, K.B.; Patel, H.K.; Sahoo, M.R. DNA Barcoding and *ITS2* Secondary Structure Predictions in Taro (*Colocasia esculenta* L. Schott) from the North Eastern Hill Region of India. *Genes* **2022**, *13*, 2294. https://doi.org/10.3390/genes13122294

Academic Editors: Wajid Zaman and Hakim Manghwar

Received: 15 October 2022
Accepted: 1 November 2022
Published: 5 December 2022

Abstract: Taro (*Colocasia esculenta* L. Schott, *Araceae*), an ancient root and tuber crop, is highly polygenic, polyphyletic, and polygeographic in nature, which leads to its rapid genetic erosion. To prevent the perceived loss of taro diversity, species discrimination and genetic conservation of promising taro genotypes need special attention. Reports on genetic discrimination of taro at its center of origin are still untapped. We performed DNA barcoding of twenty promising genotypes of taro indigenous to the northeastern hill region of India, deploying two chloroplast-plastid genes, *matK* and *rbcL*, and the ribosomal nuclear gene *ITS2*. The secondary structure of *ITS2* was determined and molecular phylogeny was performed to assess genetic discrimination among the taro genotypes. The *matK* and *rbcL* genes were highly efficient (>90%) in amplification and sequencing. However, the *ITS2* barcode region achieved significant discrimination among the tested taro genotypes. All the taro genotypes displayed most similar sequences at the conserved *matK* and *rbcL* loci. However, distinct sequence lengths were observed in the *ITS2* barcode region, revealing accurate discriminations among the genotypes. Multiple barcode markers are unrelated to one another and change independently, providing different estimations of heritable traits and genetic lineages; thus, they are advantageous over a single locus in genetic discrimination studies. A dynamic programming algorithm that used base-pairing interactions within a single nucleic acid polymer or between two polymers transformed the secondary structures into the symbol code data to predict seven different minimum free energy secondary structures. Our analysis strengthens the potential of the *ITS2* gene as a potent DNA barcode candidate in the prediction of a valuable secondary structure that would help in genetic discrimination between the genotypes while augmenting future breeding strategies in taro.

Keywords: *Araceae*; DNA barcoding; genetic discrimination; *ITS2* secondary structure; molecular phylogeny; taro

1. Introduction

Taro (*Colocasia esculenta* L. Schott), which belongs to the *Araceae* family, is one of the ancient tuber crops and the most widespread food crop cultivated in the tropical to temperate world [1]. In ancient civilizations, the paddy crop was considered as weed in a flooded taro field. Today, paddy has become the staple food crop in the developed world. However, taro is still considered an orphan crop and a crop for the resource-poor [2]. Taro corms are one of the cheapest sources of carbohydrates, protein, minerals, and vitamins and are used in treatment of fungal infections, pulmonary congestion, tuberculosis, and ulcers [3]. Taro is believed to have originated in Indian subcontinent, and the north eastern hill (NEH) region of India, Indo-Myanmar in particular, is considered as the secondary place of origin of taro [4]. It was further dispersed to East and Southeast Asia, the Pacific Islands, Africa, and tropical America [5]. Taro is highly polymorphic, polygenetic, and polygeographic in nature. The phylogeographic and phylogenetic evolution of taro is still unclear [6], which makes this potential tuber crop underutilized.

The wild taro populations were widespread worldwide before its cultivation started, thus showing huge genetic diversities across geographical regions [7]. The NEH region of India is a treasure trove for taro diversities with immense genetic potential for biotic and abiotic stress tolerance. There is a scope for introgression of this untapped gene pool for the genetic improvement of taro. The rich taro diversities in the NEH region of India need special attention for the conservation of the gene pool and the prevention of perceived genetic erosion [8]. Proper identification, characterization of promising landraces, and understanding of their genetic background facilitate rapid genetic improvement in taro [9]. Several marker-assisted studies have elucidated the genetic diversity in taro; however, it is difficult to distinguish the progenitor and descendants of taro's wild relatives due to their polygenic nature [10].

Phenotypic indicators have resolved the morphological variability with baseline information on diversity. However, morphological indicators could not confer genetic variations due to spontaneous vegetative mutagenesis [11]. Molecular characterization using simple sequence repeat (SSRs) has been used to study the genetic diversity of taro worldwide [6] and understand its genetic bottleneck and secondary domestication. Genetic characterization of the taro gene pool at the chloroplast DNA (cpDNA) using restricted fragment length polymorphism (RFLP) was attempted in wild and cultivated genotypes, which formed a monophyletic group without distinct clade structures [12]. Previous attempts could not establish a phylogenetic relationship between the wild and cultivated taro gene pool [1].

Recent advancements in molecular and *omics* tools have helped overcome the limitations of morphological discriminations [13]. DNA barcoding is an advanced, robust, and efficient tool that enables species identification and delimitation [14]. DNA barcoding involves two important steps, including building the barcode library from known taxa and assigning the barcode sequence against the library for species identification following phylogenetic relationships [15]. DNA barcoding and molecular phylogeny efficiently discriminate among closely related species [16]. On the other hand, RNA secondary structures predicted from the nuclear *ITS2* sequences could validate the phylogenetic relationships derived from the short DNA sequences. RNA molecules are the basic building blocks that fold into secondary and tertiary structures with diverse functional properties. RNA structural motifs can unveil the genetic understanding of how RNA structures regulate functional elements [17]. RNA secondary structure predictions can help in understanding the RNA molecules that govern gene function and regulation. RNA secondary structure based on the minimum free energy (MFE) algorithm has higher accuracy in establishing relationships among the RNA families [18], eliminating pseudogene footprints and minimizing sequencing error in DNA barcoding [19].

A DNA barcode deploys a short gene sequence from a conserved coding and non-coding region used as a marker for species discrimination [20]. Chloroplast genome regions are highly conserved regions due to the complex uniparental inherent repetition mechanisms and are often used for phylogenetic studies [21]. The chloroplast-plastid gene *matK* encodes the maturase K enzyme [22], whereas the *rbcL* gene encodes the ribose-1,5-bisphosphate carboxylase oxidase enzyme [23]. Although there are some mixed responses in terms of amplification and sequence success rate, the *matK* gene is one of the most informative loci for establishing phylogenetic relationships [24]. The *rbcL* gene is the first single-copy gene sequenced in plants contains no intron [25]. As one of the most conserved genes in the chloroplast region, *rbcL* is widely used to establish phylogenetic relationships among the species showing too little variation at the species level [25]. However, *ITS2*, the internal transcribed spacer region of the nuclear-ribosomal region is often considered as a candidate barcode gene due to its efficient amplification ability, universal characteristics, and significant variability [26]. Combining barcode loci efficiently elucidates intra-specific variability. CBOL (2009) [20] suggests that a combination of multi-loci barcode regions (*matK+ITS*) apparently discriminates the species more efficiently than a single-locus region (*matK+rbcL*). However, the distinct variation in the *ITS* region supersedes the ambiguity that arises among the variation in the chloroplast region [27]. Characterization studies of the intra-specific variations in taro advanced breeding lines indigenous to its centre of origin, the northeastern hill region of India, using multiple barcode regions are particularly scarce [3,28,29].

The present study aimed to investigate genetic discrimination using the conserved *matK*, *rbcL*, and *ITS2* barcode genes among the twenty advanced breeding gene pools of taro developed from the landraces indigenous to the NEH region of India. In addition, we assessed the efficacy of the nuclear-plastid and nuclear region barcode markers for species delimitation among the taro genotypes. The molecular phylogeny based on the DNA barcode sequences confirms the inter-relationship among the advanced breeding lines at the conserved region. The present study also focused on MFE-based RNA secondary structure predictions concerning the *ITS2* nuclear gene for further validation of the molecular phylogeny by grouping the genotypes into various clades. The DNA barcode-anchored species delimitation would be useful for a close understanding of the genotypes, helping to augment future breeding strategies in taro.

2. Materials and Methods

2.1. Plant Materials and Experimental Site

Twenty diverse taro genotypes (Supplementary Table S1) were collected from the northeastern hill (NEH) regions of India and from the Indian Council of Agricultural Research (ICAR)-Central Tuber Crops Research Institute (CTCRI), Regional Centre, Bhubaneswar, India. The genotypes were maintained at the Langol hill research farm of the ICAR-Research Complex for the Northeastern Hill Region (ICAR RC NEHR), Manipur Centre, Imphal, India. The experimental site is located at 24° 50′ N latitude, 93° 55′ E longitude, and at an altitude of 860 m above mean sea level. The taro genotypes were grown following the package of practices recommended by the ICAR-CTCRI, Regional Centre, Bhubaneswar, India. The climatic condition of the study site is tropical with an average temperature of 22–30 °C, rainfall of 1540 mm, and a relative humidity of 85–90% during the crop growth period from April to October 2019. This study was conducted under the twinning program of the Department of Biotechnology (DBT), Ministry of Science and Technology, Govt. of India, in collaboration with the ICAR-CTCRI, Thiruvananthapuram, India, with prior approval from the competent authorities which ensure compliance with ethical regulations.

2.2. Plant Sample Preparation and Total Genomic DNA Isolation

The central newly emerged leaf samples from the twenty taro genotypes were collected in sterile Eppendorf tubes and snap-frozen in liquid nitrogen for genomic DNA (gDNA) extraction. The collected leaf tissue was homogenized by using Tissue Lyser LT (QIAGEN, Hilden, Germany) and processed for total gDNA isolation in a QIAcube automated nucleic acid extractor (QIAGEN, Hilden, Germany) using a DNeasy Plant Mini Kit (QIAGEN, New Delhi, India) and following the manufacturer's protocol. The quantification of the isolated gDNA was carried out by using a QIA expert, single-channel UV–Vis nano reader (QIAGEN, Hilden, Germany) and electrophoresed on a 0.8% agarose gel (Tarson, Mumbai, India). To enable PCR, the final concentration of the gDNA was adjusted to 50 ng μL^{-1} by dilution from a total gDNA concentration of 40-120 ng μL^{-1}.

2.3. *PCR Amplification and Purification*

PCR amplification was carried out with the *matK*, *rbcL*, and *ITS* DNA barcode primers as recommended by the Consortium for the Barcode of Life–Plant Group [20]. The *matK*, *rbcL*, and *ITS2* primers were synthesized by M/S Bioserve Biotechnologies (Hyderabad, India) Pvt. Ltd., Hyderabad, India. The primer details are given in Table 1. Each primer set was optimized by gradient PCR to obtain the annealing temperature. The PCR reaction was carried out in a volume of 25 μL consisting of 50 ng template DNA (1 μL), 12.5 μL 2 × Taq PCR master mix (QIAGEN, New Delhi, India), and 10 pM primers. The amplification was achieved under the following conditions: an initial denaturation at 95 °C for 5 min, followed by 40 cycles of denaturation at 95 °C for 1 min, annealing at 55 °C for 1 min, extension at 72 °C for 1 min, and a final extension at 72 °C for 10 min in a SimpliAmp™ thermal cycler (Applied Biosystems Life Technologies®, Waltham, MA, USA). For the detection of amplicons, PCR products were resolved on a 1.5% agarose gel, and pictures were taken on the E-box gel documentation imaging system (Vilber Lourmat, France). The PCR products were purified using a QIA quick PCR purification kit (QIAGEN, New Delhi, India) by following the manufacturer's instructions. The quality of the purified PCR products was further processed for DNA sequencing at M/S Bioserve Biotechnologies (India) Pvt. Ltd., Hyderabad, India.

Table 1. Details of the barcode primers used for PCR amplification of the taro genotypes.

Region	Primers	Sequences (5′ to 3′)	Amplicon Size	Amplification Success Rate
matK	*matK XF*	TAATTTACGATCAATTCATTC	1000 bp	90–95%
	matK 5R	GTTCTAGCACAAGAAAGTCG		
rbcL	*rbcLa F*	ATGTCACCACAAACAGAGACTAAAGC	600 bp	90%
	rbcLa R	GTAAAATCAAGTCCACCRCG		
ITS2	*ITS2 S2F*	ATGCGATA CTTGGTGTGAATTATAGAAT	300–400 bp	80–85%
	ITS2 S3R	GACGCTTCTCCAGACTACAAT		

2.4. *DNA Sequencing*

The sequencing of the purified PCR products was performed at M/S Bioserve Biotechnologies (India) Pvt. Ltd., Hyderabad, India, using Sanger sequencing (ABI Genetic Analyzer 3730, 48 capillaries, 50 cm, Thermo Fisher, Waltham, MA, USA). The obtained sequences were viewed and analyzed in FinchTV v1.4.0 (Geospiza, Denver, CO, USA).

2.5. *Bioinformatics Analysis*

The reads from forward and reverse sequences from all the PCR products amplified with *matK*, *rbcL*, and *ITS* primers were trimmed using SnapGene v5.3 (https://www.snapgene.com; accessed on 1 October 2022). The manual editing of barcode gaps was done in pairwise alignment view using BLAST, and by following nucleotide blast with the maximum similarity score and lowest E value, species identification was performed. The muscle algorithm method was carried out using ClustalW v10.1.8 (https://www.genome.jp/tools-bin/clustalw; accessed on 1 October 2022) to obtain multiple sequence alignment (MSA) using MEGA11 software (Molecular Evolutionary Genetic Analysis; https://www.megasoftware.net; accessed on 1 October 2022) [30]. By considering the transitional and transversional nucleotide substitution, the phylogenetic study was executed using the neighbor-joining tree (NJT) and minimum evolution method (MEM) with 1000 as the *"Bootstrap phylogeny"* value and the *"kimura–2–parameter"* substitution model (*d–transitions*) in MEGA11 software [31].

2.6. *RNA Secondary Structure Predictions*

The nucleotide sequences obtained from *ITS2* primers were used to predict the *RNA* secondary structure from the rRNA database of RNAfold Web Server v2.4.18 (http://rna.tbi.univie.ac.at/cgi-bin/RNAWebSuite/RNAfold.cgi; accessed on 1 October 2022) [32].

3. Results

3.1. PCR Amplification, Sequencing, and Multiple Sequence Alignment

The PCR amplification rate was higher (>90%) for the *matK* and *rbcL* genes as compared to the *ITS2* genes (>80%) [Table 1]. The *matK, rbcL,* and *ITS2* sequences of the twenty taro genotypes were queried against the NCBI database (https://www.ncbi.nlm.nih.gov; accessed on 1 October 2022) to find the regions of similarity, and the blast-edited sequences were processed further for analysis. The consensus lengths of *matK* (222–514 bp), *rbcL* (250–270 bp), and *ITS2* (313–315 bp) sequences after trimming varied among the taro genotypes. The distribution of mean nucleotide base frequencies of candidate nucleotide sequences at different coding positions (*matK, rbcL,* and *ITS2*) in taro genotypes are presented in Table 2. Sequence analysis of the *matK, rbcL,* and *ITS2* loci identified the twenty tested taro genotypes as *Colocasia esculenta* with 99.04 (*ITS2*)–100% (*matK* and *rbcL*) homology among all the taro genotypes (Table 3). The *ITS2* gene exhibited wide dispersal sequence similarity, as revealed by the sequence alignments, whereas *matK* and *rbcL* exhibited homologous sequences among the tested taro genotypes.

Table 2. The nucleotide base frequencies of candidate nucleotide sequences at different coding positions in taro plants.

Sequences	Base Contents (%)					
	A	T	G	C	AT	GC
matK	30.75	37.56	12.56	19.13	68.31	31.69
rbcL	25.80	33.41	20.28	20.51	59.20	40.80
ITS2	19.52	11.44	34.78	34.26	30.96	69.04

Table 3. BLASTn search results for *matK, rbcL,* and *ITS2* gene sequences of twenty taro (*Colocasia esculenta*) genotypes.

Voucher Name	*matK* and *rbcL*				*ITS2*			Species
	Percent Identity	E Value	Accession No. (*matK*)	Accession No. (*rbcL*)	Percent Identity	E Value	Accession No. (*ITS2*)	
RCMC-1	100%	0.0	LT995105.1	MH270468.1	99.05%	3×10^{-156}	MK961250.1	*C. esculenta*
RCMC-2	100%	0.0	LT995105.1	MH270468.1	99.05%	3×10^{-156}	MK961250.1	*C. esculenta*
RCMC-5	100%	0.0	LT995105.1	MH270468.1	99.37%	7×10^{-158}	MK961250.1	*C. esculenta*
RCMC-6	100%	0.0	LT995105.1	MH270468.1	99.37%	7×10^{-158}	MK961250.1	*C. esculenta*
RCMC-10	100%	0.0	LT995105.1	MH270468.1	99.36%	2×10^{-157}	MK961250.1	*C. esculenta*
DP-25	100%	0.0	LT995105.1	MH270468.1	100.00%	1×10^{-160}	MK961250.1	*C. esculenta*
Duradim	100%	0.0	LT995105.1	MH270468.1	99.04%	1×10^{-155}	MK961250.1	*C. esculenta*
Jhankri	100%	0.0	LT995105.1	MH270468.1	99.36%	2×10^{-157}	MK961250.1	*C. esculenta*
MK	100%	0.0	LT995105.1	MH270468.1	99.36%	2×10^{-157}	MK961250.1	*C. esculenta*
Topi	100%	0.0	LT995105.1	MH270468.1	99.68%	5×10^{-159}	MK961250.1	*C. esculenta*
Satasankha	100%	0.0	LT995105.1	MH270468.1	99.68%	5×10^{-159}	MK961250.1	*C. esculenta*
TSL	100%	0.0	LT995105.1	MH270468.1	99.05%	3×10^{-156}	MK961250.1	*C. esculenta*
P Chandel	100%	0.0	LT995105.1	MH270468.1	99.05%	3×10^{-156}	MK961250.1	*C. esculenta*
R5JH10	100%	0.0	LT995105.1	MH270468.1	99.68%	5×10^{-159}	MK961250.1	*C. esculenta*
R5MK	100%	0.0	LT995105.1	MH270468.1	99.04%	1×10^{-155}	MK961250.1	*C. esculenta*
R1B9	100%	0.0	LT995105.1	MH270468.1	99.05%	3×10^{-156}	MK961250.1	*C. esculenta*

Table 3. *Cont.*

Voucher Name	matK and rbcL				ITS2			Species
	Percent Identity	E Value	Accession No. (*matK*)	Accession No. (*rbcL*)	Percent Identity	E Value	Accession No. (*ITS2*)	
PCT1	100%	0.0	LT995105.1	MH270468.1	99.05%	3×10^{-156}	MK961250.1	*C. esculenta*
BBSL	100%	0.0	LT995105.1	MH270468.1	99.68%	5×10^{-159}	MK961250.1	*C. esculenta*
TOPI-1kR	100%	0.0	LT995105.1	MH270468.1	99.68%	5×10^{-159}	MK961250.1	*C. esculenta*
BBSR	100%	0.0	LT995105.1	MH270468.1	100.00%	1×10^{-160}	MK961250.1	*C. esculenta*

The rate of different transitional substitutions and transversional substitutions showed significant variations among the different conserved barcode regions, which indicated the differences among the taro genotypes (Table 4). *ITS2* showed higher transitional substitution (point mutations) values among the nucleotides across the regions compared to *rbcL* and *matK*. However, *matK* exhibited higher transversional substitution values followed by *rbcL* and *ITS2* (Table 4).

Table 4. Maximum likelihood estimates of the substitution matrix of *matK*, *rbcL*, and *ITS2* sequences. Rates of different transitional substitutions are in **bold**, and those of transversional substitutions are shown in *italics*.

	From/To	A	T	G	C
			matK		
matK	A	-	*12.3839*	*5.8154*	**4.9319**
	T	*10.2083*	-	**5.8096**	*4.9368*
	G	*10.2083*	**12.3715**	-	*4.9368*
	C	**10.1981**	*12.3839*	*5.8154*	-
			rbcL		
rbcL	A	-	*9.6984*	*7.2905*	**7.2164**
	T	*9.1319*	-	**7.2832**	*7.2236*
	G	*9.1319*	**9.6887**	-	*7.2236*
	C	**9.1227**	*9.6984*	*7.2905*	-
			ITS2		
ITS2	A	-	*0.5673*	*1.7239*	**18.0123**
	T	*1.0720*	-	**45.5929**	*1.7201*
	G	*1.0720*	**15.0027**	-	*1.7201*
	C	**11.2258**	*0.5673*	*1.7239*	-

3.2. Phylogenetic Studies, Replacement Rate Matrices, and DNA Barcoding

The unrooted maximum likelihood tree (MLT) [Figures 1–3] depicts the phylogenetic relationship with the branch lengths representing evolutionary distances or the *bootstrap* values in a *K2P* model. The unrooted MLT derived from *matK* sequences grouped the twenty taro genotypes into two major clades. Clade I comprises 14 genotypes, and clade II includes 6 genotypes (Figure 1). The *rbcL* sequences showed 100% homogeny among the tested genotypes, and as such could not establish significant discrimination in the present study (Figure 2). However, the ribosomal-nucleus *ITS2* gene sequences categorized the taro genotypes into five distinct clades. Clades I, II, and III each comprise five genotypes, whereas clades IV and V include two and three genotypes, respectively (Figure 3).

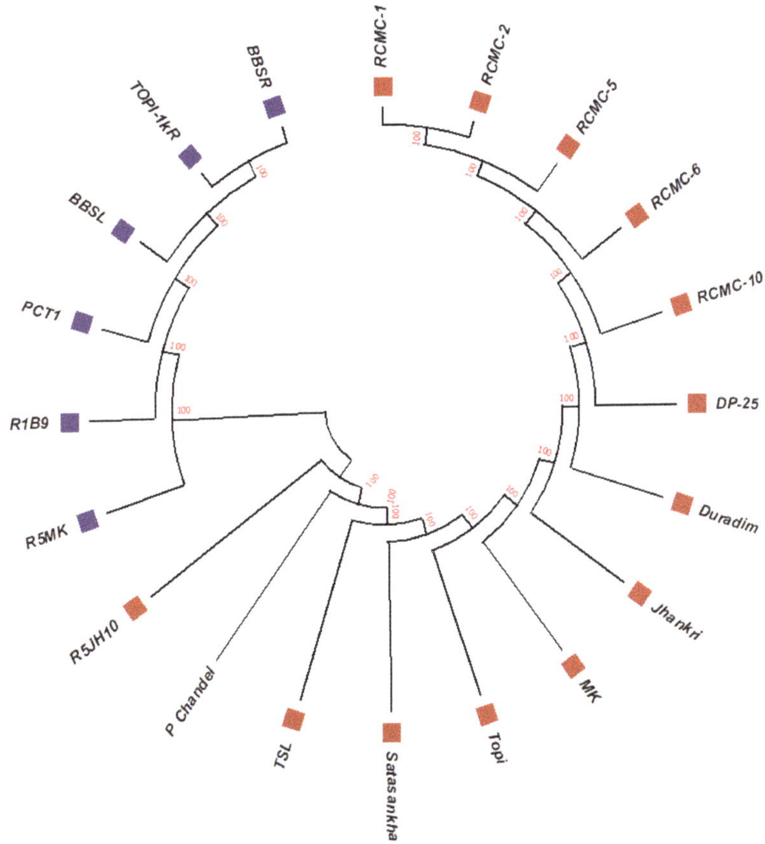

Figure 1. The unrooted maximum likelihood tree of the twenty taro genotypes based on the *matK* gene sequences (Note: Bootstrap scores of ≥50% only are shown for each branch after 1000 bootstrap replication tests).

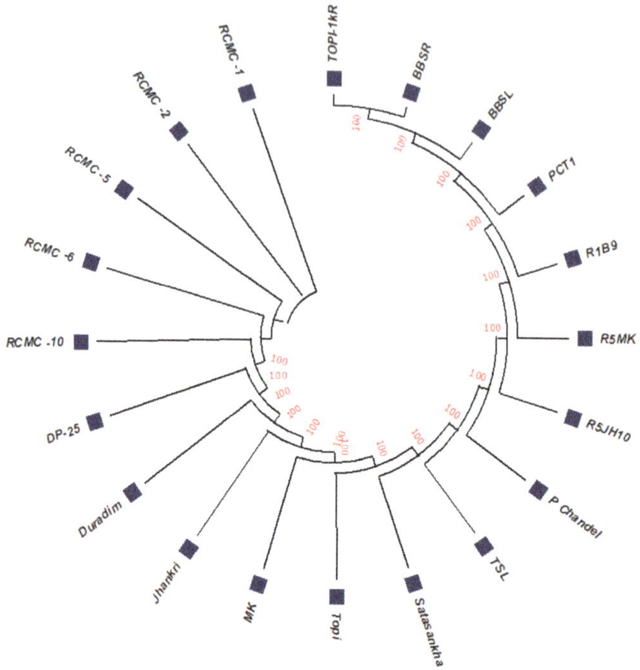

Figure 2. The unrooted maximum likelihood tree of the twenty taro genotypes based on the *rbcL* gene sequences (Note: Bootstrap scores of ≥50% only are shown for each branch after 1000 bootstrap replication tests).

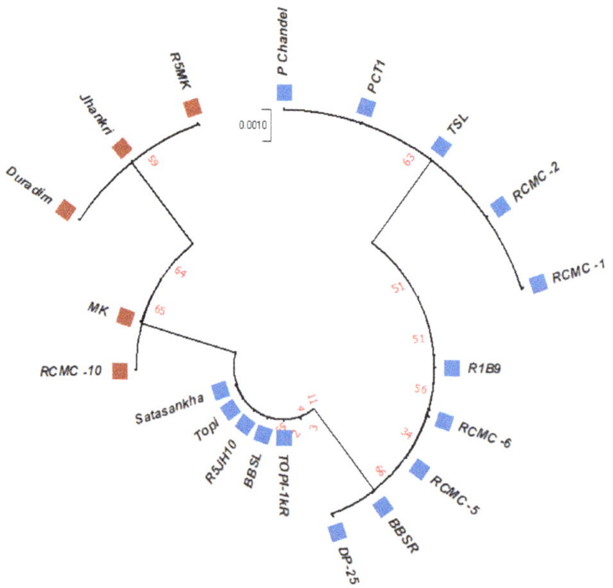

Figure 3. The unrooted maximum likelihood tree of the twenty taro genotypes based on the *ITS2* gene sequences (Note: Bootstrap scores of ≥50% only are shown for each branch after 1000 bootstrap replication tests).

3.3. *ITS2* Secondary Structure Predictions

A dynamic programming algorithm transformed the secondary structures into the symbol code data to predict the minimum free energy secondary structure. The *ITS2* loci displayed seven different e-loop motif features, which contained a ring-pin that was a typical feature of the secondary structure model of taro (Figure 4). Structures I, III, V, and VI represented a centroid ring with 3-6 hair-pin motifs and 2-6 bulge loops. However, structures II, IV, and VII displayed a centroid bulge with multi-branched loops (Figure 4). As per the *ITS2* secondary structures, the taro genotypes were grouped into seven different clades. Clade I comprises six genotypes (RCMC-1, RCMC-2, DP-25, TSL, BBSR, Satasankha), Clade II includes five genotypes (RCMC-5, Duradim, Jhankri, Topi 1kR, BBSL), and Clade III involves five genotypes (RCMC-6, P. Chandel, R5MK, R1B9, PCT1). However, the RCMC-10, Topi, MK, and R5JH10 genotypes exhibited unique secondary structures each. R5JH10 exhibited a similar structure to that of Clade I with the omission of the helix at the centroid ring (Figure 4). The *ITS2* secondary structures are the unique genetic structures at the conserved nuclear region which could be useful in discriminating the taro landraces at the genotypic level by understanding the RNA molecules.

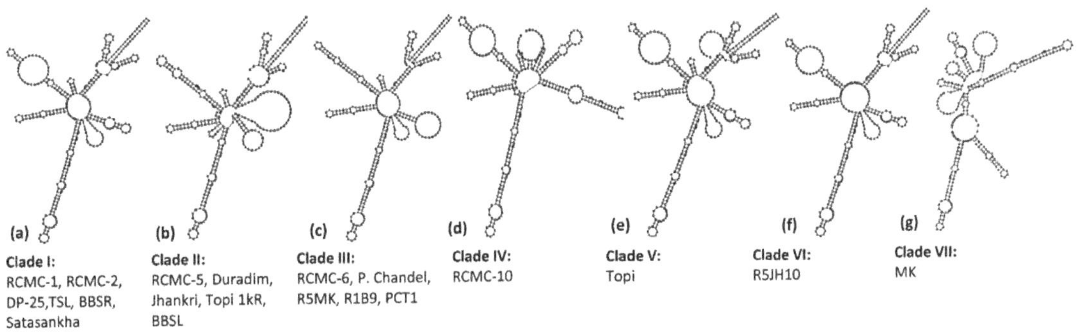

Figure 4. Seven consensus *ITS2* secondary structure predictions from the twenty taro genotypes based on the minimum free energy (MFE).

4. Discussion

Tropical root crops and the products they generate are crucial components for a sustainable agricultural system as well as key components of human and animal nutrition. The domestication of numerous perennial herbaceous species in the *Colacasia* genus calls for the use of various evolutionary biology approaches that can effectively identify and authenticate the species. Additionally, a comparison of the physiological, biochemical, and chromosomal (heritable) traits of several taro species found discrepancies in the categorization made within the genus, indicating the need for more trustworthy methods involving different genetic DNA markers with broad binomial nomenclature specificity.

4.1. DNA Barcoding Using *rbcL*, *matK*, and *ITS2* Genes

The adoption of multi-loci techniques [33] including both nuclear and chloroplast DNA barcodes has greatly improved accuracy of species identification, especially for closely related species, despite the absence of one universal barcode that can be utilized to identify relevant plant species [21]. The molecular identification of the *Colacasia* species was made possible by the nucleotide sequences of plastid genes, the inter-gene region (*rbcL*, and *matK*), and the nuclear region (*ITS2*). However, the *ITS2* sequences have a distinct way of differentiating the taro cultivars and do so more efficiently.

We used DNA barcoding analysis employing the extremely diverse *ITS2* barcoding regions to confirm the accuracy of species identification for the indicated taro specimens placed in regional NCBI banks. Even though there is extensive species coverage of all the DNA barcodes in the NCBI gene bank archive for many species of plants, such as *Colocasia esculenta*, *ITS2* enables more precise species categorization using sequence similarity (>99%). In order to better comprehend the features of the taxonomic variations of taro farmed in Brazil, a phylogenetic study was carried out, and the variability of taro was investigated by [29] using chloroplast genome sequences, such as *rbcL* and *pbsA-trnH*. In [34], the genetic diversity of taro was assessed using two chloroplast DNA barcoding markers, *rbcL* and *pbsA-trnH,* and the authors found an enhanced perception of genetic evaluation and of their characteristics. DNA barcoding analysis employing the extremely variable chloroplast *trnL* and nuclear *ITS2* DNA barcoding regions has been used to confirm the correctness of species identification for indicated Eurasian *Vicia* species [35]. However, [36] acknowledge that after a comparison of several variations in the number of species detected from phylogenetic groupings,, the genetic diversity index using the *rbcL* marker was unable to resolve the identification of yam (*Dioscorea spp.*) well, exposing the ineffectiveness of this marker in differentiating the genetic variability of yam.

Using the results of the BLAST analysis of the DNA sequences, the twenty studied genotypes were grouped into the same species with notably high correlation values, and have been identified with >99% identity (Tables 1 and 2). Furthermore, we have demonstrated that diversity studies in the taro can rely on the nuclear *ITS2* region. Although minor differences were found for a few clades that resulted from the *rbcL* and *matK* phylogenetic trees, we confirmed the topology of the genotypes and the related reference records were found to be consistent (Figures 1 and 2). Although *rbcL* and *matK* are not the best identifiers for species discrimination, the relatively conserved nuclear *ITS2* region was considered the finest marker because it allows for a more ideal interspecific alignment.

4.2. *ITS2 Marker and RNA Secondary Structure*

In order to provide insight into emerging RNA control frameworks, digitally generated RNA structures should reflect the native RNA folded state of a species. Because it limits sequencing inaccuracy and removes pseudogene imprints, RNA secondary initial design is an effective approach for species characterization [37], RNA structure is constantly found to be more conserved than the DNA sequences obtained. To the standard approach of sequence-based alignment, the homologous RNA secondary structures can operate to function equally on the binding of the base pair, and cleaving or catalyzing may be considered to yield unique nucleotides. Hence, base-pair interactions resulting in secondary or tertiary RNA structures are necessary for *ITS2* to function [38]. The accuracy of the results in identifying species under common clades or groups can be considerably increased if the common structures of these homologous sequences are measured during RNA folding and alignment. On the other hand, determining a set of RNA molecules' similar structures can be used to anticipate the essential components of their function [39].

In particular, the *ITS2* preserved structure has recently received a lot of attention and application [18]. *ITS2*-derived RNA folding and sequence-structure alignment have increasingly utilized *ITS2* sequences and their established secondary structures as templates. Our discovered RNA secondary structure was compared to all the genotypes, which revealed a "ring-pin" model structure as the general core throughout all taro genotypes, providing a superior differentiation method than other markers, such as *rbcL* and *matK*. Therefore, the *ITS2* marker is found to be an excellent alternative identifying method for integrated analysis of RNA structures and their function in evolutionary attention. Notably, the *ITS2* DNA barcoding marker will be taken forward for frequent use in phylogenetic techniques that will grow exponentially and it will become a sequence-directed genetic information contributor. The identification of these RNA secondary patterns increases the utility of the *ITS2* secondary structure in taro as a taxonomic DNA marker. Breeders of taro may select from a variety of parents for greater diversity or a hybrid for taro crop development [3]. Since the *ITS2* marker is revealed to be effective and it can be recommended as the stand-alone marker of choice for DNA barcoding of *Colocasia* accessions, especially for accurate identification, discrimination, and estimation of the genetic diversity of this important tuber crop.

5. Conclusions

This study described the efficiency of DNA barcode genes in chloroplast-plastid (*matK* and *rbcL*) and ribosomal-nuclear (*ITS2*) regions in discriminating twenty promising taro lines at the genotypic level. Although *matK* and *rbcL* markers exhibited higher amplification and sequencing rates, *ITS2* emerged as an ideal marker for DNA barcoding, phylogenetic relationship determination, and secondary structure predictions in our study. The results of this study successfully delimit the *ITS2* boundary among the taro genotypes and the associated evolutionary information. The divergent *ITS2* sequence alignments have been used for establishing the phylogenetic relationship and secondary structure predictions. The *ITS2* loci-based maximum likelihood tree grouped the twenty taro genotypes into five major groups; however, *ITS2* secondary structure predictions categorized the twenty genotypes into seven different clades. The close relationship of the taro genotypes indicated that they homogenously originated from the same geographical region. This study is the first report on genetic discrimination of taro genotypes from the center of origin of taro, the northeastern hill region of India, in particular. This study also provides insight into the scope of the development of species-specific markers for the discrimination of closely related taxa by deploying multiple barcode genes. Understanding the RNA molecules would be useful for the selection of parents in trait-specific breeding strategies for taro improvement.

Supplementary Materials: The following supporting information can be downloaded at https://www.mdpi.com/article/10.3390/genes13122294/s1: Table S1. List of twenty taro genotypes used in the study.

Author Contributions: Conceptualization, M.R.S. and M.D.; data curation, M.R.S., M.D., M.P.D., and S.M.; formal analysis, M.R.S., S.M., and M.D.; investigation, M.P.D., M.D., and S.M.; methodology, M.D., M.P.D., S.M., and M.R.S.; resources, M.R.S.; software, S.M., M.R.S., K.B.K., and H.K.P.; writing—original draft, M.R.S., S.M., M.D., and M.P.D.; writing—review and editing, M.R.S., H.K.P., S.K.S., V.H., S.S.R., and R.R. All authors have read and agreed to the published version of the manuscript.

Funding: This study was funded by the Department of Biotechnology (DBT) twinning project 'Gene expression profiling of taro (*Colocasia esculenta* L. Schott) and role of transcriptional activators of epicuticular wax in host resistance against *Phytophthora* leaf blight' (BT/PR25192/NER/95/1069/2017). The study was also supported by the Focused Basic Research (FBR) project of the Council of Scientific and Industrial Research (CSIR).

Institutional Review Board Statement: Not applicable.

Informed Consent Statement: Not applicable.

Data Availability Statement: Not applicable.

Acknowledgments: The infrastructure facility provided by the Director and Joint Director of the Indian Council of Agricultural Research (ICAR) Research Complex for NEH Region, Manipur Centre, India, is gratefully acknowledged. The taro genotypes provided by the Head of the ICAR-Central Tuber Crops Research Institute (ICAR-CTCRI), Regional Centre, Bhubaneswar, are duly acknowledged.

Conflicts of Interest: The authors declare no conflict of interest.

References

1. Ahmed, I.; Lockhart, P.J.; Agoo, E.M.G.; Naing, K.W.; Nguyen, D.V.; Medhi, D.K.; Matthews, P.J. Evolutionary origins of taro (*Colocasia esculenta*) in Southeast Asia. *Eco. Evo.* **2020**, *10*, 13530–13543. [CrossRef] [PubMed]
2. Matthews, P.J.; Ghanem, M.E. Perception gaps that may explain the status of taro (*Colocasia esculenta*) as an "orphan crop". *Plants People Planet* **2020**, *3*, 99–112. [CrossRef]
3. Mwamba, N.; Patrick, A.; Michael, P. Genetic diversity analysis in South African taro (*Colocasia esculenta*) accessions using molecular tools. *Int. J. Genet. Mol. Bio.* **2016**, *8*, 18–24. [CrossRef]
4. Matthews, P. A possible tropical wildtype taro: *Colocasia esculenta* var. *aquatilis*. *Indo. Pac. Prehis. Assn. Bull.* **1991**, *11*, 69–81.
5. Ivancic, A.; Lebot, V. *The Genetics and Breeding of Taro*; CIRAD: Montpellier, France, 2000.
6. Chair, H.; Traore, R.E.; Duval, M.F.; Rivallan, R.; Mukherjee, A.; Aboagye, L.M. Genetic diversification and dispersal of Taro (*Colocasia esculenta* (L.) Schott). *PLoS ONE* **2016**, *11*, e0157712. [CrossRef] [PubMed]
7. Mathews, P. Genetic diversity in taro and the preservation of culinary knowledge. *Ethnobot. Res. Appl.* **2004**, *2*, 55–71. [CrossRef]

8. Anbalagan, T.; Deka, B.C.; Sivakumar, S.; Rangnamei, L.; Walling, N. Colocasia (*Colocasia esculenta* L.) in Northeast India. *J. Innov. Agric.* **2022**, *9*, 1–7. [CrossRef]
9. Devi, Y.I.; Sahoo, M.R.; Mandal, J.; Dasgupta, M.; Prakash, N. Correlations between antioxidative enzyme activities and resistance to *Phytophthora* leaf blight in taro. *J. Crop. Impro.* **2020**, *35*, 250–263. [CrossRef]
10. Ahmed, I.; Matthews, P.J.; Biggs, P.J.; Naeem, M.; McLenachan, P.A.; Lockhart, P.J. Identification of chloroplast genome loci suitable for high-resolution phylogeographic studies of *Colocasia esculenta* (L.) Schott (Araceae) and closely related taxa. *Mol. Ecol. Resour.* **2013**, *13*, 929–937. [CrossRef]
11. Vinutha, K.B.; Devi, A.A.; Sreekumar, J. Morphological characterization of above ground characters of Taro (*Colocasia esculenta* (L.) Schott.) accessions from North East India. *J. Root. Crop.* **2015**, *41*, 3–11.
12. Ochiai, T.; Tahara, M.; Yoshino, H. Phylogenetic relationships of taro and allied species based on restriction fragment length polymorphisms (RFLPs) of chloroplast DNA. *Sci. Rep. Fac. Agri.* **2000**, *89*, 15–21.
13. Sahoo, M.R.; Naresh, P.; Kumari, M.; Acharya, G.C. Omics in leafy vegetables. In *Omics in Horticultural Crops*; Rout, G.R., Peter, K.V., Eds.; Academic Press: Cambridge, MA, USA, 2022; pp. 281–302. [CrossRef]
14. Hebert, P.D.N.; Cywinska, A.; Ball, S.L.; deWaard, J.R. Biological identifications through DNA barcodes. *Proc. R. Soc. B Bio. Sci.* **2003**, *270*, 313–321. [CrossRef]
15. Kress, W.J.; Erickson, D.L. DNA barcodes: Methods and protocols. *Methods Mol. Biology* **2012**, *858*, 3–8.
16. Acharya, G.C.; Mohanty, S.; Dasgupta, M.; Sahu, S.; Singh, S.; Koundinya, A.V.V.; Kumari, M.; Naresh, P.; Sahoo, M.R. Molecular phylogeny, DNA barcoding, and ITS2 secondary structure predictions in the medicinally important *Eryngium* genotypes of East Coast Region of India. *Genes* **2022**, *13*, 1678. [CrossRef]
17. Macke, T.J. RNAMotif, an RNA secondary structure definition, and search algorithm. *Nucl. Acid. Res.* **2001**, *29*, 4724–4735. [CrossRef] [PubMed]
18. Zhang, W.; Tian, W.; Gao, Z.; Wang, G.; Zhao, H. Phylogenetic utility of rRNA ITS2 sequence-structure under functional constraint. *Int. J. Mol. Sci.* **2020**, *21*, 6395. [CrossRef] [PubMed]
19. Rampersad, S.N. ITS1, 5.8S and ITS2 secondary structure modelling for intra-specific differentiation among species of the *Colletotrichum gloeosporioidessensulato* species complex. *Springer Plus* **2014**, *3*, 684. [CrossRef] [PubMed]
20. CBOL Plant Working Group. A DNA barcode for land plants. *Proc. Natl. Acad. Sci. USA* **2009**, *106*, 12794–12797. [CrossRef]
21. Yang, L.; Tian, J.; Xu, L.; Zhao, X.; Song, Y.; Wang, D. Comparative chloroplast genomes of six *Magnoliaceae* species provide new insights into intergeneric relationships and phylogeny. *Biology* **2022**, *11*, 1279. [CrossRef]
22. Kress, W.J.; Wurdack, K.J.; Zimmer, E.A.; Weigl, A.T.; Janzen, D.H. Use of DNA barcodes to identify flowering plants. *Proc. Natl. Acad. Sci. USA* **2005**, *102*, 8369–8374. [CrossRef] [PubMed]
23. Cuenoud, P.V.; Savolainen, W.; Chatrou, L.W.; Powell, M.; Grayer, R.J.; Chase, M.W. Molecular phylogenetics of Caryophyllales based on nuclear 18S rDNA and plastid *rbcL*, *atpB*and *matK* DNA sequences. *Americ. J. Bot.* **2002**, *89*, 132–144. [CrossRef]
24. Hilu, K.W.; Borsch, T.; Muller, K.; Soltis, D.E.; Soltis, P.S.; Savolainen, V.; Chase, M.W.; Powell, M.P.; Alice, L.A.; Evans, R.; et al. Angiosperm phylogeny based on *matK* sequence information. *Am. J. Bot.* **2003**, *90*, 1758–1776. [CrossRef]
25. Hollingsworth, P.M.; Graham, S.W.; Little, D.P. Choosing and using a plant DNA barcode. *PLoS ONE* **2011**, *5*, e19254. [CrossRef]
26. Cowan, R.S.; Fay, M.F. Challenges in the DNA barcoding of plant material. *Mol. Bio.* **2012**, *862*, 23–33.
27. Armenise, L.; Simeone, M.C.; Piredda, R.; Schirone, B. Validation of DNA barcoding as an efficient tool for taxon identification and detection of species diversity in Italian conifers. *Europ. J. For. Res.* **2012**, *131*, 1337–1353. [CrossRef]
28. Singh, S.D.R.; Singh, F.F.; Naresh, K.; Damodaran, V.; Srivastava, R.C. Diversity of 21 Taro (*Colocasia esculenta* (L.) Schott) accessions of Andaman Islands. *Gen. Resou. Crop. Evo.* **2012**, *59*, 821–829. [CrossRef]
29. Nunes, R.S.C.; Del Aguila, E.; Paschoalin, V.; da Silva, J. DNA barcoding assessment of the genetic diversity of varieties of taro, *Colocasia esculenta* (L.) Schott in Brazil. In *Breeding and Genetic Engineering: The Biology and Biotechnology Research*; iConcept Press Ltd.: Hong Kong, China, 2014.
30. Al-Juhani, W.S.; Khalik, K.N.A. Identification and phylogenetics study of *Arabis alpina* L. From the Kingdom of Saudi Arabia. *Pak. J. Bot.* **2021**, *53*, 1057–1064. [CrossRef] [PubMed]
31. Kumar, S.; Stecher, G.; Li, M.; Knyaz, C.; Tamura, K. MEGA X: Molecular evolutionary genetics analysis across computing platforms. *Mol. Bio. Evo.* **2018**, *35*, 1547–1549. [CrossRef] [PubMed]
32. Lorenz, R.; Bernhart, S.H.; HönerzuSiederdissen, C.; Tafer, H.; Flamm, C.; Stadler, P.F.; Hofacker, I.L. ViennaRNA Package 2.0. *Algo. Mol. Bio.* **2011**, *6*, 26. [CrossRef] [PubMed]
33. Li, H.; Xiao, W.; Tong, T.; Li, Y.; Zhang, M.; Lin, X.; Guo, X. The specific DNA barcodes based on chloroplast genes for species identification of *Orchidaceae* plants. *Sci. Rep.* **2021**, *11*, 1–15. [CrossRef]
34. Kreike, C.M.; Van-Eck, H.J.; Lebot, V. Genetic diversity of taro, *Colocasia esculenta* (L.) Schott, in Southeast Asia and the Pacific. *Theor. Appl. Genet.* **2004**, *109*, 761–768. [CrossRef] [PubMed]
35. Bosmali, I.; Lagiotis, G.; Haider, N.; Osathanunkul, M.; Biliaderis, C.; Madesis, P. DNA-based identification of Eurasian *Vicia* species using chloroplast and nuclear DNA barcodes. *Plants* **2022**, *11*, 947. [CrossRef] [PubMed]
36. Ude, G.N.; Igwe, D.O.; McCormick, J.; Ozokonkwo-Alor, O.; Harper, J.; Ballah, D.; Aninweze, C.; Chosen, O.; Okoro, M.; Ene, C.; et al. Genetic diversity and DNA barcoding of yam accessions from Southern Nigeria. *Am. J. Plant Sci.* **2019**, *10*, 179–207. [CrossRef]

37. Ankenbrand, M.J.; Keller, A.; Wolf, M.; Schultz, J.; Förster, F. ITS2 database V: Twice as much. *Mol. Biol. Evol.* **2015**, *32*, 3030–3032. [CrossRef] [PubMed]
38. Tan, Z.; Fu, Y.; Sharma, G.; Mathews, D.H. TurboFold II: RNA structural alignment and secondary structure prediction informed by multiple homologs. *Nucl. Acid. Res.* **2017**, *45*, 11570–11581. [CrossRef]
39. Seemann, S.E.; Mirza, A.H.; Hansen, C.; Bang-Berthelsen, C.H.; Garde, C.; Christensen-Dalsgaard, M.; Torarinsson1, E.; Yao, Z.Z.; Workman, C.T.; Pociot, F. The identification and functional annotation of RNA structures conserved in vertebrates. *Genome Res.* **2017**, *27*, 1371–1383. [CrossRef]

genes

Article

The Complete Chloroplast Genome of *Hypoestes forskaolii* (Vahl) R.Br: Insights into Comparative and Phylogenetic Analyses within the Tribe Justiceae

Samaila Samaila Yaradua [1,2,*] and Kowiyou Yessoufou [1]

1 Department of Geography, Environmental Management and Energy Studies, APK Campus, University of Johannesburg, Johannesburg 2006, South Africa
2 Department of Biology, Umaru Musa Yaradua University, Katsina 820102, Nigeria
* Correspondence: dryaradua@gmail.com

Abstract: *Hypoestes forskaolii* is one of the most important species of the family Acanthaceae, known for its high economic and medicinal importance. It is well distributed in the Arab region as well as on the African continent. Previous studies on ethnomedicine have reported that *H. forskaolii* has an anti-parasitic effect as well as antimalarial and anthelmintic activities. Previous studies mainly focused on the ethnomedicinal properties, hence, there is no information on the genomic architecture and phylogenetic positions of the species within the tribe Justiceae. The tribe *Justicieae* is the most taxonomically difficult taxon in Acanthoideae due to its unresolved infratribal classification. Therefore, by sequencing the complete chloroplast genome (cp genome) of *H. forskaolii*, we explored the evolutionary patterns of the cp genome and reconstructed the phylogeny of Justiceae. The cp genome is quadripartite and circular in structure and has a length of 151,142 bp. There are 130 genes (86 coding for protein, 36 coding for tRNA and 8 coding for rRNA) present in the plastome. Analyses of long repeats showed only three types of repeats: forward, palindromic and reverse were present in the genome. Microsatellites analysis revealed 134 microsatellites in the cp genome with mononucleotides having the highest frequency. Comparative analyses within Justiceae showed that genomes structure and gene contents were highly conserved but there is a slight distinction in the location of the genes in the inverted repeat and single copy junctions. Additionally, it was discovered that the cp genome includes variable hotspots that can be utilized as DNA barcodes and tools for determining evolutionary relationships in the Justiceae. These regions include: *atpH-atpI, trnK-rps16, atpB-rbcL, trnT-trnL, psbI-trnS, matK, trnH-psbA*, and *ndhD*. The Bayesian inference phylogenetic tree showed that *H. forskaolii* is a sister to the *Dicliptra* clade and belongs to Diclipterinae. The result also confirms the polyphyly of *Justicia* and inclusion of Diclipterinae within justicioid. This research has revealed the phylogenetic position of *H. forskaolii* and also reported the resources that can be used for evolutionary and phylogenetic studies of the species and the *Justicieae*.

Keywords: chloroplast genome; *Hypoestes forskaolii*; justiceae; phylogenomics

Citation: Yaradua, S.S.; Yessoufou, K. The Complete Chloroplast Genome of *Hypoestes forskaolii* (Vahl) R.Br: Insights into Comparative and Phylogenetic Analyses within the Tribe Justiceae. *Genes* **2022**, *13*, 2259. https://doi.org/10.3390/genes13122259

Academic Editors: Wajid Zaman, Hakim Manghwar and Zhiqiang Wu

Received: 10 November 2022
Accepted: 28 November 2022
Published: 30 November 2022

1. Introduction

Hypoestes forskaolii (Vahl) R.Br is a herbaceous plant and one of the most important Acanthaceaous species mainly distributed in some part of Africa and the Arab region [1]. The plant has been used in folk medicine in the treatment of various diseases like cancer, gonorrhea, heart and chest diseases [2,3]. The species is also reported to have several biological properties which includes antifungal, atrypanosomal, antileishmanial, antiplasmodial and cytotoxic properties [2,4]. A leaf decoction from the plant is used by Herdsmen to kill insect and parasite on sheep [5,6]. Fresh leaves of *H. forskaolii* are used to heal wounds and are also reported to have insecticidal activity [7]. Despite the traditional medicinal value of the plant, its complete chloroplast genome (cp genome) has not been sequenced, thus the genome evolution remains unexplored.

The tribe Justiceae is the largest tribe in Acanthaoideae (Acanthaceae) with ca. 2000 species distributed worldwide and exhibit a great diversity in morphological features. The tribe is considered the most taxonomically difficult taxon in Acanthoideae due to its unresolved tribal classification. The taxonomic problems in the tribe began after the infrafamilial classification of Acanthaceae by Lindau (in 1895). In his work, he classified *Justicia* L. and its assumed sister species in a tribe that is characterized by the presence of two stamens and two or three aperturate pollen. The problem of this classification is that he included species that don't have this character in Justiceae and placed other species with two stamen in other tribes. Since then, several studies [8–12] have been conducted to address the infratribal classification. Their work was based on morphological data, but they did not analyze their data phylogenetically. However, their work contributed immensely to providing knowledge of infrafamilial and infratribal relationships within the Acanthaceae. The major setback of the Bremekamp infrafamilial classification is that he did not account for the generic composition of his tribes [11]. McDade and her colleagues (in 2000) were the first students of Acanthaceae who employed phylogenetic analysis using the *nrITS* and the *trnL-F* region from 55 species to resolve the Justiceae tribal classification. Their results recognized four lineages and these lineages are characterized by tricolporate hexapseudocolpate pollen and also provided a base for evaluating the phylogenetic relationship of the *Justicieae* species. Since then, ref. [13] worked on *Isoglossinae*, refs. [14–16] worked on the Tetmerium lineage. The recent study that tries to evaluate the tribal classification of Justiceae, particularly the justicoid lineages is [17]. In their study, they used six DNA regions from 178 samples of Justiceae and concluded that *Justicia* is paraphyletic and called for further investigation to address the infratribal classification of the tribe.

Understanding the evolutionary links among plant species at various taxonomic hierarchies can be achieved with the help of genetic data. This is due to the presence of various protein-coding and transfer RNA genes that play a crucial role in the cells of plants. The major role of the chloroplast organelle in a plant is the conversion of carbon dioxide and water into simple sugars [18]. The chloroplast genome (cp genome) is the most informative and conserved genome in angiosperms compared with nuclear and mitochondria. The cp genome contains sufficient information for inferring phylogenetic relationships, studies of taxa diversification and comparative analyses [19]. The cp genome of land plants usually have a circular and a quandripartite structure which include a single copy region (70–80 kb), small single copy region (15–28 kb) and a pair of inverted repeat (18–30 kb) [20]. The length of the cp genome of angiosperms ranges from 140 kb to 180 kb and they contain about 132 unique genes (ca. 86 coding for protein, 4 coding for ribosomal ribonucleic acid and 32 coding for transfer ribonucleic acid) [21]. Despite the conserved nature of the cp genome, there is some degree of variation in gene content and genome structure as well as mutations [22]. This causes the sequences of different species to diverge, which may be utilized to analyse the phylogenetic relationships of various taxonomic groups [23]. Only the chloroplast genomes of six taxa of the Justiceae have been sequenced, despite the significance of the cp genomes in contemporary systematics.

Here, we have reported and characterized the cp genome of *H. forskaolii* using the Illumina Hiseq 2500 system. We have combined the newly sequenced *H. forskaolii* with the 12 cp genomes of Justiceae sampled from the public Genbank database to explore the evolutionary circumstance of Justiceae. We have also analyzed the structure of the cp genome with other Justiceae to identify regions to be used for the species authentication and genetic diversity studies of the species including population genetics studies. We have reconstructed the phylogenetic relationship of the species using the plastome sequences to infer its phylogenetic positions and attempt to resolve the infratribal classification of Justiceae.

2. Materials and Methods

A typical specimen of *H. forskaolii* was collected from Batagarawa town, Katsina, Katsina State, Nigeria. The leaves from the specimen were collected and stored in a zip lock bag containing silica gel and the specimen was taken to Herbarium Center for Biodiversity and Conservation, Umaru Musa Yaradua University for identication. The specimen was authenticated by the Curator and was given accession number UMH0145. The dried leaf material was used to extract the DNA using a Qiagen genomic DNA extraction kit, following the manufacturer's instructions.

2.1. Library Construction, Sequencing and Assembly

For the DNA sample preparations, 1.0 μg of DNA was used as input material. Following the manufacturer's instructions, the NEBNext DNA Library Prep Kit was used to create the sequencing libraries, followed by addition of indices to the sample. The genomic DNA was sheared at random into segments of 350 bp, which were then prepared for further PCR enrichment and sequencing. The resulting PCR products were subject to purification (using the AMPure XP system), then the resultant libraries were measured using real-time PCR and evaluated for size distribution using an Agilent 2100 Bio analyzer. After pooling, the qualified libraries based on projected data volume and effective concentration, were fed onto Illumina sequencers. Using PRINSEQlite v0.20.4 [24], clean reads sequences (5.2 GB) were obtained from the raw data through filtering, which were then subjected to assembly using NOVOPlasty 4.2. [25] using kmer (K-mer = 39) and the cp genome of *Justicia flava* (NC 044862.1) was used as the seed and reference.

2.2. Gene Annotation

The online tool GeSeq [26] was used to annotate all the genes present in the cp genome of *H. forskaolii* using the plastome sequence of *Justicia flava* as reference. The sequin program was used to correct the errors in the genes that were not successfully annotated in the Geseq by adjusting the position of the codons. The cp genome architecture was drawn in OGDRAW (Organellar Genome DRAW) [26].

2.3. Codon Usage Analysis

The software MEGA version 6 [27] was used to calculate the relative synonymous codon usage values (RSCU), codon usage and base composition.

2.4. Repeat Analysis

MicroSAtellite (MISA) (http://webblast.ipk-gatersleben.de/misa/index.php accessed on 24 November 2022) was used to identify Simple Sequence Repeats (SSRs) in the *H. forskaolii* cp genome [28]. For mononucleotides, dinucleotides, trinucleotides, and tetra, penta, and hexa nucleotides SSR motifs eight, five, four, and three repetitions units were used, respectively. The online software REPuter [29] was used to identify the types of long repeats present in the cp genome.

2.5. Genome Comparison

Using the annotation of *H. forskaolii* as a reference in the Shuffle-LAGAN mode [30], the cp genomes of seven species of Justiceae were compared using the tool mVISTA [31]. The online software IR scope (https://irscope.shinyapps.io/irapp/ accessed on 1 November 2022) was used to compare the border regions of the cp genome.

2.6. Characterization of Substitution Rate

To identify the genes that are under selection pressure, DNAsp version 6 [32] was used to examine synonymous (dS) and nonsynonymous (dN) substitution rate and dN/dS ratio.

2.7. Sequence Divergence

To determine sequence divergence and identify the variable regions in the cp genome, sliding window analysis was used to evaluate the diversity of nucleotide in the cp genome using DnaSP version 6 with the following parameters: 600 bp for window length and 200 bp for step size.

2.8. Phylogenetic Analysis

The cp genome of seventeen Acanthoideae species as well as four species, *Erythranthe lutea* (Phrymaceae), *Scrophularia dentata* (Scrophulariaceae), *Lysionotus pauciflorus* (Gesneriaceae) and *Tanaecium tetragonolobum* (Bignoniaceae) were downloaded from Genbank. The Acanthoideae species are the in group while the other four were used as the out group. Using MAFFT version 7 [33], the newly sequenced cp genome of *H. forskaolii* was aligned with the downloaded cp genomes. Using MrBayes version 3.2.6, a Bayesian inference strategy was used to reconstruct the phylogenetic tree [34]. The appropriate model was chosen using jModelTest version 3.7 [35].

3. Results and Discussion

3.1. Characteristics of H. forskaolii Chloroplast Genome

Previous findings revealed that the cp genome of flowering plants is extremely well-preserved in both the gene content and structural organization; however, expansion and contraction in the inverted repeat and single copy junctions are among the evolutionary events leading to variations in the cp genomes [36,37]. The cp genome of *H. forskaolii* is quadripartite and circular in structure and has a total length of 151,142 bp. The cp genome consists of a pair of inverted repeats (IRb and IRa) 25,477 bp, Small Single Copy (SSC) 17,012 and Large Single Copy 83,176 bp (Figure 1). A total of 72,852 bp make up the genome's non-coding region which is comprised of introns and intergenic spacers, and the 78,290 bp code for protein genes. The GC percentage of the LSC and SSC regions was 36.0% and 32.3%, respectively, whereas the GC content of the inverted repeats IRa and IRb was 43.4% and 43.3%, respectively (Table 1). It is discovered that the IRs have a larger proportion of GC than SSC and LSC regions.

Table 1. Nucleotide composition in the *H. forskaolii* cp genome.

Region		T(U) (%)	C (%)	A (%)	G (%)	Total (bp)
cp genome		31.2	19.4	30.8	18.7	151,142
LSC		32.4	18.5	31.16	17.5	83,176
SSC		34.0	16.8	34.0	15.2	17,012
IRA		28.4	22.5	28.3	20.9	25,477
IRB		28.2	20.8	28.4	22.5	25,477
	1st Positiom	31.2	19.4	30.1	19.0	50,381
	2nd Position	31.0	19.4	31.3	18.3	50,381
	3rd Position	31.0	19.2	30.9	18.6	50,381

The annotation of *H. forskaolii* cp genome revealed a total of 130 genes (86 protein-coding genes, 36 tRNAs genes and 8 rRNAs genes), 96 genes are present in the LSC (82 protein coding genes and 21 tRNA genes) and SSC (14 protein coding genes and 1 tRNA) while the remaining 17 genes (7 tRNAs, 4 rRNAs and 7 protein-coding genes) are repeated in the IRa and IRb regions (Table 2 and Figure 1). Almost all the protein-coding genes in the cp genome have ATG as their start codon and few of them have alternative start codon, GTG, ACG and ATC; this has been reported to occur in the cp genome of angiosperms [38–40].

Some of the coding genes present in the cp genome of *H. forskaolii* have introns. Introns are reported to occur in some of the protein-coding and tRNAs genes of flowering plants cp genomes [38,39]. Out of the 130 coding genes, 16 are characterized with one or two introns (Table 3). Among the 16 genes, six are tRNAs and 11 are protein coding genes. Ten of the intron-containing genes are located in the LSC, one gene in the SSC while the remaining five are in the inverted repeat region. ATP dependent protease subunit p gene (*clpP*) and one of the Photosysem I gene (*ycf3*) possessed two introns while the remaining 14 genes have only one intron. The tRNA, *trnK-UUU* is the gene with longest intron which is due to the inclusion of *matK* in the gene.

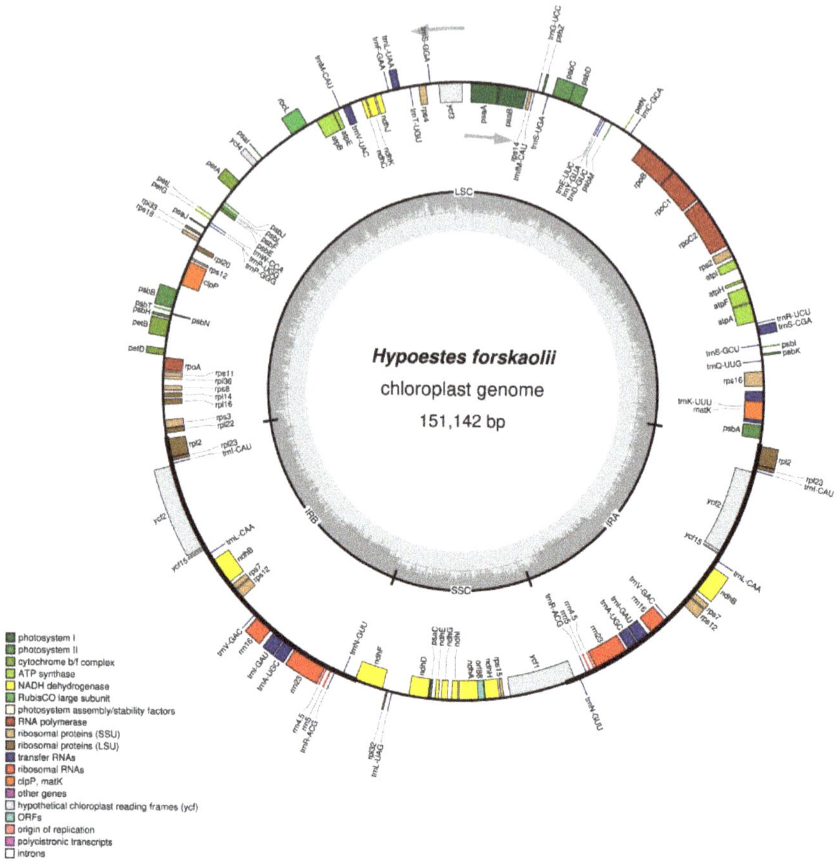

Figure 1. The structure of the *H. forskaolii* cp genome. Genes inside the circles are transcribed clockwise, while those outside the circles are transcribed counterclockwise. The colorful bar displays genes that are known to be functioning. The inner circle's dark grey and light grey colors, respectively, designate the GC and AT contents.

The frequency of a codon that encodes for a specific amino acid was compared using codon usage analysis [41]. A codon is a sequence of trinucleotides that encodes for specific amino acids that are used in protein synthesis [42]. Because of bias in mutation, codon use is a factor influencing the development of the chloroplast genome [43] and it differs between species [44]. The nucleotide sequence of the protein-coding genes (78,290 bp) was used to calculate the frequency of the codons present in the cp genome. The relatively synonymous codon usage (RSCU) of the genes in the cp genome is shown in (Table 4, Figure 2). The codon usage analysis revealed that 26,095 codons encode for the genes in the cp genome. All the 20 known amino acids are encoded by 61 codons (Figure 2). Codons coding for the amino acid leucine are more prevalent in the cp genome while codons that code for Cysteine are the less common (Table 4). The Cytosine (C) and Guanine (G) endings are more prevalent than the Thymine (T) and Adenine (A) endings; the cp genomes of other angiosperms have Thymine and Adenine endings occurring more frequently [45–47]. The result of the analysis (Table 4) revealed that codon usage bias is low in the cp genome of *H. forskaolii*. The RSCU values for 31 codons were less than 1, and are characterized with G/C endings, whereas the RSCU values for 30 codons were greater than 1, with A/T endings. Methionine and Tryptophan are the only two amino acids without codon bias with RSCU values of 1.

Table 2. Genes present in the chloroplast genome of *H. forskaolii*.

Category	Grpup of Genes	Name of Genes
RNA genes	ribosomal RNA genes (rRNA)	*rrn5, rrn4.5, rrn16, rrn23*
	Transfer RNA genes (tRNA)	*trnL-UAG., trnK-UUU* [+], *trnS-GCU, trnQ-UUG, trnV-UAC* [+], *trnR UCU, trnD-GUC, trnE-UUC, trnC-GCA, trnM-CAU, trnY-GUA, trnT-GGU, trnG-GCC, trnfM-CAU, trnS-UGA, trnT-UGU, trnL-UAA* [+], *trnF-GAA, trnW-CCA, trnP-UGG, trnP-GGG, trnL-CAA* [a], *trnV-GAC* [a], *trnI-GAU* [+,a], *trnA-UGC* [+,a], *trnR-ACG* [a], *trnN-GUU* [a], *trnH-GUG, trnG-UCC, trnS-GGA*
Ribosomal proteins	Small subunit of ribosome	*rps19, rps8, rps14, rps7* [a], *rps2, rps11, rps2, rps14, rps18, rps16* [+], *rps15, rps12* [a]
Transcription	Large subunit of ribosome	*rpl33, rpl16, rpl20, rpl32, rpl23* [a], *rpl22, rpl14, rpl36, rpl2* [+,a]
	DNA dependent RNA polymerase	*rpoC2, rpoB rpoA, rpoC1* [+],
Protein genes Other genes	Photosystem I	*psaI, psaB, psaj, psaA, psaC, ycf3* [++]
	Photosystem II	*psbZ, psbK, psbF, psbD, psbL, psbN, psbH, psbT, psbJ, psbB, psbE, psbM, psbC, psbI, psbA*
	Subunit of cytochrome	*petL, petB, petN, petG, petA, petD*
	Subunit of synthase	*atpH, atpF, atpE, atpB* [+], *atpA, atpI*
	Chloroplast envelope membrabe protien	*cemA*
	NADH dehydrogenase	*ndhK, ndhH, ndhI, ndhD, ndhB* [+,a] *ndhG, ndhF, ndhE, ndhC, ndhJ, ndhA* [+]
	Large subunit of rubisco	*rbcL*
	Subunit acetyl-coA carboxylase	*accD*
	ATP dependent protease subunit P	*clpP* [++]
	Maturase	*matK*
	C-type cytochrome systhesis	*ccsA*
	Translational initiation factor	*infA*
	Component of TIC complex	*ycf1* [a]
	Hypothetical proteins	*ycf2* [a] ,*ycf4, ycf15* [a]

[a] Duplicated genes, [+] Gene with one intron and [++] Gene with two introns.

Figure 2. The 20 amino acids and stop codons' relative synonymous codon usage (RSCU) in the chloroplast genome of *H. forskaolii's* protein-coding genes.

Table 3. Genes with introns in the *H. forskaolii* chloroplast genome and length of introns and exons.

Gene	Location	Exon I (bp)	Intron I (bp)	Exon II (bp)	Intron II (bp)	Exon III (bp)
trnK-UUU	LSC	36	2453	36		
rps16	LSC	224	912	35		
trnS-CGA	LSC	31	661	59		
atpF	LSC	470	649	140		
rpoC1	LSC	1635	784	433		
ycf3	LSC	152	710	228	683	127
trnL-UAA	LSC	36	513	49		
trnV-UAC	LSC	36	584	37		
clpP	LSC	226	631	299	749	69
petB	LSC	5	687	653		
rpl2	IR	435	658	397		
ndhB	IR	755	679	776		
rps12	IR	25	542	231		
trnI-GAU	IR	41	939	34		
trnA-UGC	IR	37	820	34		
ndhA	SSC	538	989	552		

Table 4. Codon usage of the *H. forskaolii* chloroplast genome.

Codon	Amino Acid	RSCU	tRNA	Codon	Amino Acid	RSCU	tRNA
UUC	Phe	0.67	*trnF-GAA*	UAU	Tyr	1.61	*trnY-GUA*
UUU	Phe	1.33		UAC	Tyr	0.39	
CUG	Leu	0.42		CAG	Gln	0.47	
UUA	Leu	1.84	*trnL-UAA*	UAA	Stop	1.55	
CUA	Leu	0.83		CAA	Gln	1.53	*trnQ-UUG*
CUU	Leu	1.27	*trnL-UAG*	CAU	His	1.54	*trnH-GUG*
CUC	Leu	0.4		CAC	His	0.46	
UUG	Leu	1.24	*trnL-CAA*	UAG	Stop	0.93	
AUG	Met	1	*trnM-CAU*	AAG	Lys	0.51	
AUU	Ile	1.5	*trnI-GAU*	AAU	Asn	1.54	*trnN-GUU*
GUC	Val	0.49		GAC	Asp	0.36	
AUA	Ile	0.89	*trnI-CAU*	AAA	Lys	1.49	*trnK-UUU*
AUC	Ile	0.61		AAC	Asn	0.46	
GUU	Val	1.5	*trnV-GAC*	GAU	Asp	1.64	*trnD-GUC*
UCU	Ser	1.68	*trnS-GGA*	UGU	Cys	1.46	*trnC-GCA*
GUA	Val	1.45		GAA	Glu	1.47	*trnE-UUC*
GUG	Val	0.56	*trnV-UAC*	GAG	Glu	0.53	
UCA	Ser	1.21		UGA	Stop	0.52	
UCC	Ser	0.92		UGC	Cys	0.54	
CCA	Pro	1.1		CGA	Arg	1.41	
UCG	Ser	0.67	*trnS-UGA*	UGG	Trp	1	*trnW-CCA*
CCU	Pro	1.51	*trnP-UGG*	CGU	Arg	1.2	*trnR-ACG*
CCC	Pro	0.78		CGC	Arg	0.41	*trnR-UCU*
GCA	Ala	1.14		GGA	Gly	1.58	
CCG	Pro	0.61		CGG	Arg	0.55	
ACU	Thr	1.61		AGA	Arg	1.78	
ACC	Thr	0.73		AGG	Arg	0.65	
GCC	Ala	0.67		GGC	Gly	0.37	
ACG	Thr	0.47	*trnT-UGU*	AGC	Ser	0.3	
GCU	Ala	1.75	*trnA-UGC*	GGU	Gly	1.27	
ACA	Thr	1.19	*trnT-GGU*	AGU	Ser	1.22	*trnS-GCU*
GCG	Ala	0.44		GGG	Gly	0.78	*trnG UCC*

3.2. Repeat Analyses

3.2.1. Long Repeats

Long repeats sequences present in the cp genome of *H. forskaolii* were identified using the program REPuter; from the results it was discovered that tandem repeats and three types of long repeats (forward, reverse and palindromic) were present in the plastome *H. forskaolii* (Figure 3). In total, there are 88 repeats in the cp genome of *H. forskaolii* (19 palindromic repeats, 27 forward repeats, 4 reverse repeats and 38 tandem repeats). Most of the palindromic and forward repeats sizes are between 20–29 bp, followed 10–19 bp. The length of repeated sequences in *H. forskaolii* cp genome ranges from 10 to 44 bp, are analogous to the lengths in other angiosperm cp genomes [48–50].

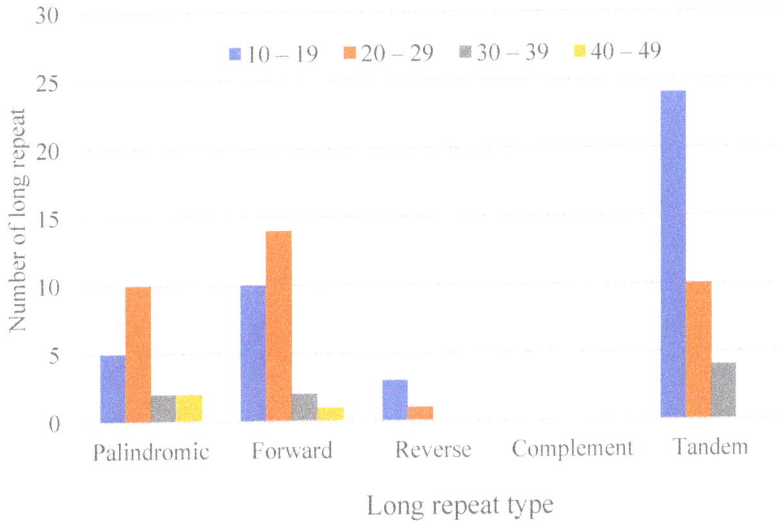

Figure 3. Types of tandem and long repeat distribution and their frequency in the chloroplast genomes of *H. forskaolii*.

We compared the frequency of repeats among seven Justiceae cp genomes, and the result indicated that only three species, namely, *D. acuminata*, *P. japonica* and *R. pectinata* contained complement, forward, reverse and palindromic repeats (Figure 4D). *C. nutans* and *P. haikangense* have the highest frequency of forward repeats (37) while *R. pectinata* has the lowest (12). *D. acuminata* and *J. procumbes* have the same number of reverse repeats, five each. *R. pectinata* has the highest number of reverse repeats (9) while *H. forskaolii* has the lowest (4). Complement repeats are found to be the less numerous type of repeat across the genome with *D. acuminata*, *P. japonica* and *R. pectinata* having one, two and five, respectively.

3.2.2. Simple Sequence Repeats (SSRs)

Simple sequence repeats (SSRs) are short repeats of sequences usually 1–6 bp that are very useful at evaluating genetic variation among species. These SSRs are present in cp genomes of angionsperms and are uniparently inherited. They are therefore employed as molecular markers in developmental research such as genetic heterogeneity investigations, and they also aid in the identification of species [51–53]. This study discovered 136 microsatellites in the cp genome of *H. forskaolii* (Table 5). Mononucleotides are the most frequent SSRs in the cp genome, constituting about 78.67%, of which majority are polythymine (41.91%) and polyadenine (30.14%); this is consistent with previous studies [53]. Among the dinucleotide only AT/AT is found in the genome. Reflecting series complementary, only two trinucleotide AAT/ATT and ATC/ATG, five tetra AAAC/GTTT, AAAG/CTTT, AAAT/ATTT, AATC/ATTG, AATT/AATT were present in the cp genome.

Penta nucleotide and hexa nucleotide were not discovered in the cp genome (Figure 4A). The LSC region harbored most of the microsatellites, followed by SSC (Figure 4B).

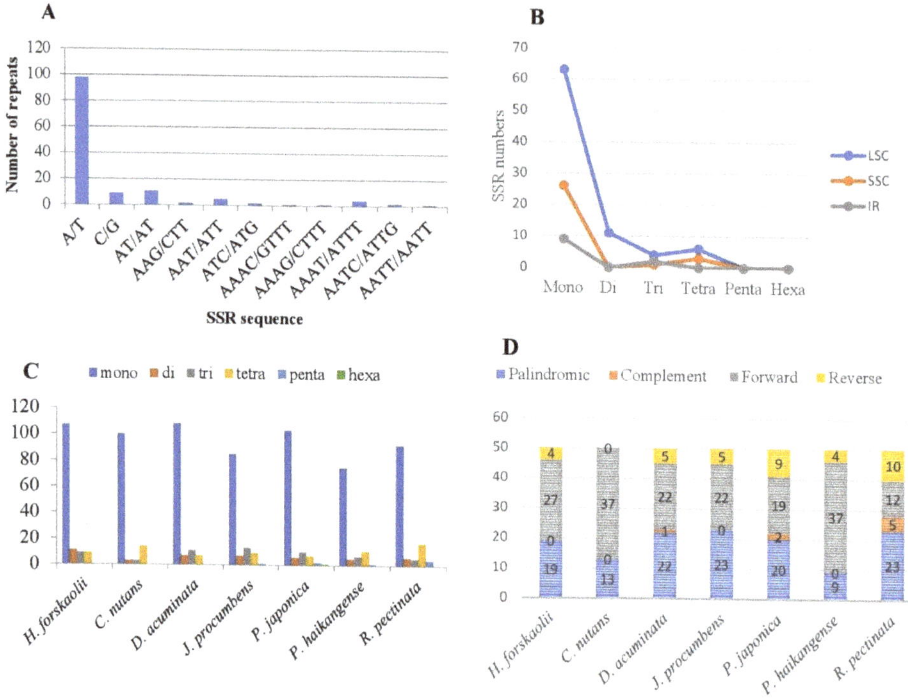

Figure 4. Repeat analyses of the sampled plastome of Justicae. (**A**): Frequency of different SSR motifs in different repeat types in *H. forskaolii* chloroplast genome. (**B**): Distribution of SSR in LSC, SSC, and IR regions. (**C**): Number of different SSR types in the chloroplast genome of seven Justiceae. (**D**): Number of different repeats in chloroplast genomes of seven species of Justiceae.

The frequency of SSRs among the cp genomes of seven species of Justiceae was analyzed (Figure 4C); the results revealed that mononucleotides repeats are the most frequent across all the cp genomes. *D. acuminata* and *H. forskaolii* are the species with the highest frequency of mononucleotide; 108 and 107, respectively. Pentanucleotides were not present in the cp genomes of *H. forskaolii*, *D. nutans* and *D. acuminata* while hexanucleotide was only present in *P. japonica*.

3.3. Comparative Analysis of Justiceae Species Cp Genome

To evaluate the level of genome divergence in Justiceae, the newly sequenced cp genome *H. forskaolii* was compared with six Justiceae species cp genomes downloaded from the GenBank. The cp genomes were aligned using mVISTA with the annotation of *H. forskaolii* as a reference. The result of the anaysis revealed that the compared cp genomes are well-preserved in terms of genome structure and gene content; however, there was some level of variation. The protein-coding regions were found to be more conserved than the introns and intergenic spacers. In terms of the four regions, the IRa and IRb were more conserved than the SSC and LSC (Figure 5). This has been reported to occur in certain taxa cp genomes in earlier studies [54,55]. The most divergent non-coding regions among the seven cp genomes are *trnH-psbA*, *trnK-rps16*, *rps16-trnQ-UUG*, *trnE-UUC-psbD*, *atpH-atpI*, *trnT-trnL*, *ndhC-trnV*, *accD-psaI*, *petA-psbJ*, *atpB-rbcL*, *rps12* and *trnL-rpl32*. A slight sequence variation was observed in the following genes *psbM*, *matK*, *ycf1*, *trnA-UGC*, *ndhH*, and *rrn16*. These regions can be used as a source of potential barcode for identification/authentication of Justiceae species as well as resources for inferring phylogenetic relationships of the Acanthoideae.

Table 5. Microsatellite in the chloroplast genome of *H. forskaolii*.

cpSSR ID	Repeat Motif	Length (bp)	No. of Repeats	SSR Start Position
1	(A) 8	8	23	4087; 8109; 9324; 14,945; 15,707; 18,082; 22,395; 41,631; 64,140; 80,149; 95,787; 110,230; 111,185; 112,039; 112,263; 114,062; 114,556; 118,479; 125,329; 136,542; 151,007; 151,074
2	(A) 9	9	9	7444; 11,725; 26,746; 43,488; 43,722; 66,819; 73,976; 88,658; 151,041
3	(A) 10	10	5	7650; 12,601; 14,734; 27,880; 132,799
4	(A) 12	12	1	15,729
5	(A) 13	13	1	7927
6	(A) 16	16	1	127,320
7	(A) 17	17	1	13,063
8	(C) 11	11	1	4477
9	(C) 12	12	1	131,580
10	(G) 8	8	2	4422; 57,979
11	(G) 9	9	1	5964
12	(G) 10	10	1	116,259
13	(G) 11	11	1	74,737
14	(G) 12	12	1	102,700
15	(G) 16	16	1	66,828
16	(T) 8	8	27	7431; 8012; 25,690; 30,519; 32,669; 35,551; 42,497; 45,300; 50,150; 59,760; 72,331; 74,672; 75,843; 81,161; 82,593; 82,893; 83,210; 83,277; 97,742; 109,264; 111,758; 112,374; 113,566; 123,714; 123,762; 123,859; 138,497
17	(T) 9	9	14	16,042; 17,941; 28,559; 31,561; 42,560; 65,592; 77,186; 83,242; 109,326; 115,581; 123,423; 123,827; 123,846; 145,625
18	(T) 10	10	9	11,804; 12,514; 30,933; 53,860; 68,645; 70,546; 101,483; 123,976; 125,541
19	(T) 11	11	3	59,201;123,498; 123,792;
20	(T) 12	12	2	54,320; 124,361
21	(T) 13	13	1	9522
22	(T) 16	16	1	106,956
23	(AT) 5	5	2	20,342; 46,477
24	AT (7)	7	2	7285; 13,415
25	AT (9)	9	1	75,422
26	(TA) 5	5	3	19,311; 45,521; 45,967
27	TA (6)	6	3	30,631; 45,576; 82,952
28	(ATA) 4	4	2	64,698; 149,097
29	(TTC) 4	4	1	34,430
30	(AAT) 4	4	1	63,219
31	(TTA) 4	4	1	81,350
32	(TAT) 4	4	1	85,183
33	(TGA) 4	4	1	90,439
34	(TCT) 5	5	1	124,766
35	(ATC) 4	4	1	143,840
36	(ATAA) 3	3	1	45,662
37	(TAAA) 3	3	1	66,210
38	(AAAC) 3	3	1	67,218
39	(AAAT) 3	3	1	74,480
40	(AATA) 3	3	1	113,175
41	(AATC) 3	3	1	118,310
42	(AATT) 3	3	1	122,706

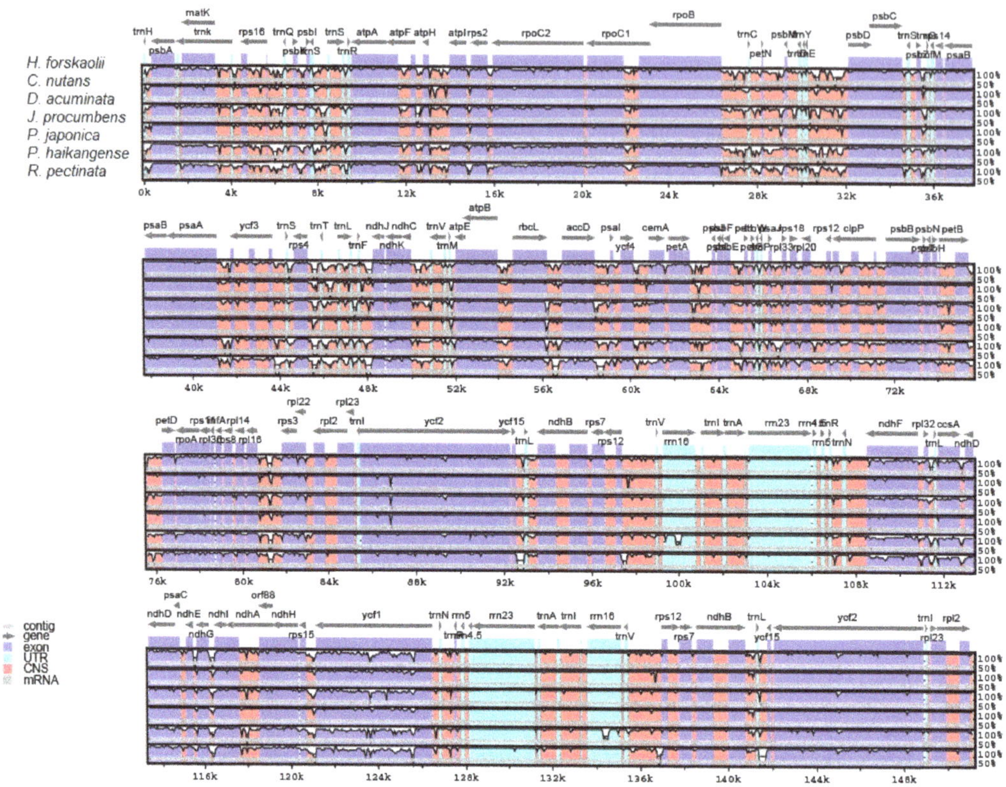

Figure 5. Variable regions in the cp genome of seven Justiceae species. The top arrow represents the direction of transcription; the colors blue and pink denote protein coding and conserved non-coding sequence, respectively; light green denotes tRNAs and rRNAs. The cp genome coordinates are shown by the *x*-axis, and the percentage identity ranges from 50% to 100% on the *y*-axis.

The structure and size of the chloroplast genome is often retained by angiosperms [53]; however, due to evolutionary processes, including genome contraction and expansion, there can be subtle variations in the size and location of the boundaries of inverted repeats and single copy regions [56,57]. We compared the JLB, JSB, JSA and JLA boundaries of the seven cp genomes of Justiceae and the results (Figure 6) showed some degree of similarity and variation among the compared cp genomes. The length of the seven cp genomes ranged from 149,627 bp (*R. pectinata)* to 152,849 bp (*P. haikangense*). Three species *J. flava*, *P. japonica* and *D. acuminata* have their JLB within the gene *rps19* with 102 bp overlapping into the IR. The JLB was bordered by *rps19* (LSC) and *rpl2* in *C. nutans* due to the contraction of the IR. In contrast, due to the IR expansion, the JLB of the *R. pectinata* was located within the *rpl22* while *P. haikangense* was between the *rpl22* (LSC) and *rps19* (IR). The JSB was found within *ndhF* with 2170–2184 bp overlapping into the IR except for the *C. nutans.* The *trnH* is located at the junction of the LSC/IRa border of all the compared cp genomes with the exception of *H. forskaolii*, which might be due to an expansion which leads to the loss of *rps19*. The cp genome of *C. nutans* varied with the other cp genomes by having the *ndhF* gene located in the SSC region. The *ycf1* pseudogene extended through the SSC and IRa with about 4300 bp in the SSC and 812 bp in the IRa in all the compared genomes except for *C. nutans*. The JSA and JSB had the most conserved borders among the compared cp genomes. The cp genome of *R. pectinata* was unique by having *rps3* in LSC/IRb border. The cp genome of *R. pectinata* had the smallest LSC region 81,979 bp while *P. haikangese* had the longest 83,878 bp. *H. forskaolii* was the only species that lost the *rps19*, which is located LSC/IR border.

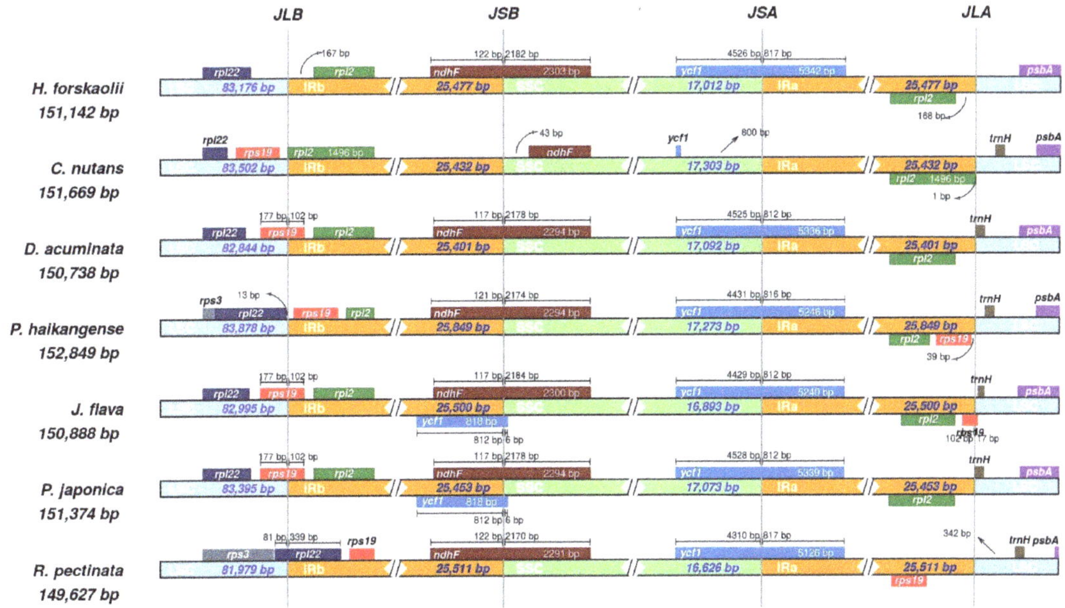

Figure 6. Structural variation in the junction of inverted repeat and single copy regions among the seven cp genomes of Justiceae. (JSA: Junction of the SSC and the IRA; JLB: Junction of the LSC and the IRB; JSB: Junction of the SSC and the IRB).

3.4. Divergence of Protein Coding Genes Sequence

To determine the genes that were undergoing selective pressure in the cp genome of *H. forskaolii*, DNAsp was used to calculate the dN/dS ratio, nonsynonymous (dN) and synonymous (dS) rates. The results showed that the dN/dS ratio was less than one in almost all of the paired genes except *atpF* and *clpP* (Figure 7), indicating that most of the genes were under negative selection with the exception of *clpP* and *atpF*. The synonymous (dS) values rangee from 0.0053 to 0.1628 in all the protein-coding genes (Figure 7). Twenty-nine genes, including *ycf15, ycf3, rps18, rps14, rpl36, rpl32, psbM, psbN, psbJ, psaI, petG, ndhJ, ndhC, infA* and *atpH* showed no nonsynonymous changes occurring in the plastome of *H. forskaolii*; comparable findings were reported for other cp genomes [58–60].

3.5. Identification of Sequence Divergence

Variable regions of the chloroplast genome are very useful in infering phylogenetic relationships and identification of species at the lowest taxonomic rank. These regions also play a vital role in providing information that helps in detecting differences between species and revealing the changes in the population structure [61,62]. The plastome sequence of *H. forskaolii* was found to be similar with that of related Justiceae species. The calculated pi values ranged from 0 to 0.79 (Figure 8), which shows slightly variation among the chloroplast genome and are relatively conserved. This pattern of variation was reported to occur in the plastomes of angiosperm [40]. Comparing the sequence divergence in the single copy regions and the inverted repeat region, the single copy region showed a higher variability than the invterted repeat. Highly variable protein-coding genes in the plastomes include *psbM, ndhH* and *ycf1*. Similarly, two intergenic spacers, *trnE-UUC-psbD* and *rps16-trnQ-UUG* were found to be highly variable. These results agree with the mVISTA divergence analysis and show that the regions could be used in the identification and authentication of Justiceae species.

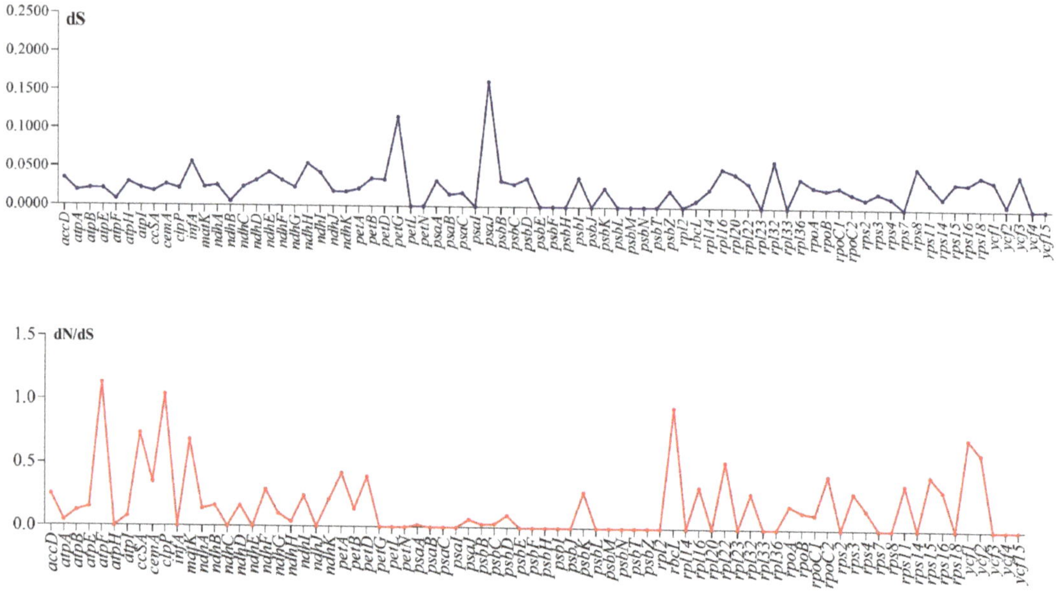

Figure 7. dN/dS ratio and synonymous values of the genes in the *H. forskaolii* chloroplast genome.

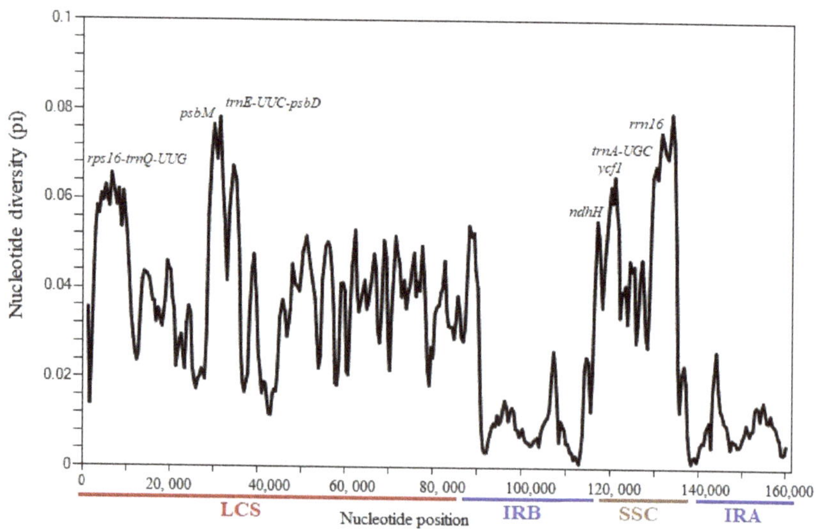

Figure 8. Sliding window analysis of nucleotide variability among the seven Justiceae species cp genomes (windowlength: 600 bp; step size: 200 bp).

3.6. Phylogenetic Analysis

The cp genome has been well utilized in inferring phylogenetic relationship due to its conserved nature and the presence of informative sites [63–65]. Various taxonomic complexes at different taxonomic levels have been resolved using phylogenetic trees reconstructed from cp genomes [66,67]. To reconstruct the phylogeny of Justiceae and infer the phylogenetic position of *H. forskaolii*, the cp genome sequences of 21 taxa were downloaded from Genbank. MAFFT version 7 was used to align the cp genome genome of *H. forskaolii* and all the downloaded cp genomes. Bayesian inference was used in reconstructing the phylogenetic tree. The results in (Figure 9), show the lineage of justiciod is strongly supported (1.0 PP) with all the twelve species belonging to the tribe Justiceae clustering in one clade (monophyletic). This is congruent with previous studies using molecular, *nrITS* and some of the chloroplast genes [0,17]. Withing the clade, there are six major sub clades, all with strongly supported posterior probability [PP] = 1.00. *H. forskaolii* forms a sister relationship with a sub clade containing *Peristrophe* and *Dicleptera* (Diclipterinae) with a strong support. A similar tree was reported by [17] and this confirmed that the species is a member of the Diclipterinae subtribe as reported by Bremekamp in his ealier classification. The core Diclipterianae (*Dicliptera, Hypoestes* and *Peristrophe*) is strongly supported in this study and members of this genera are reported to be united by some of the inflorescence part [68]. The sister relationship between *Peristrophe* and *Dicleptera* needs to be revisited by looking at their phylogenetic positions, though [69,70] suggested that the two taxa should be treated as different genera due to their differences in capsule dehiscence. The New World *Justicia* species forms a sister relationship with Diclipterinae with strong support. The phylogenetic tree clearly showed that *Justicia* is paraphyletic, as reported previously by [0,17], and there is a need to combine the phylogenetic approach with a morphological study to resolve this complexity.

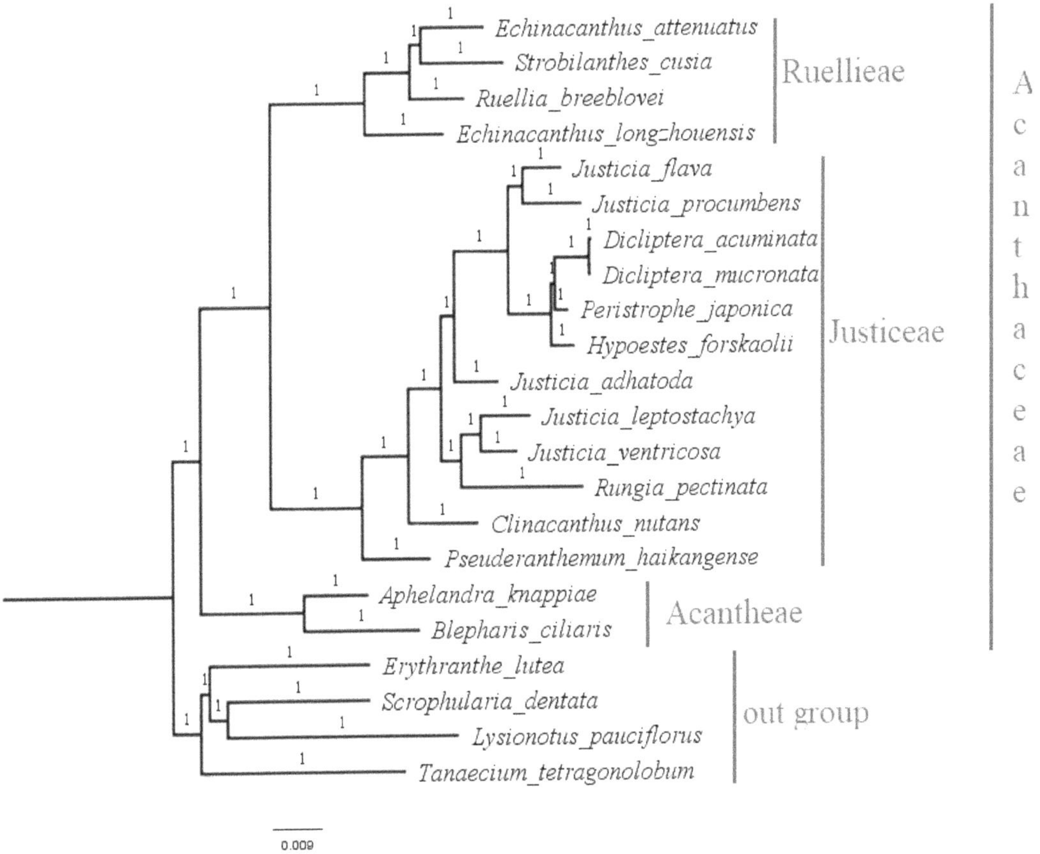

Figure 9. Bayesian inference phylogenetic tree reconstructed using the cp genome of 22 taxa. The branch nodes' numbers represent the posterior probabilities values (PP).

4. Conclusions

In this study, we have sequenced and reported on the cp genome of *H. forskaolii* to provide a valuable plastome genomic resources for the species. The plastomes have a typical gymonosperm cp genome structure they are comparable to another p genome of Acanthaceae. Simple sequence repeats used for evolutionary studies within the species were identified. The genome comparative analyses of seven Justiceae species revealed variable hotspot that could be used to develop a DNA barcode for the identification of the species. These hotspots will also be useful in phylogenetic relationship studies of the family Acanthaceae. The study has also revealed that only a few genes are under positive selection. The findings of the study have reported and confirmed the tribal position of major genera within Justiceae and has called for a further phylogenetic approach and morphological study to resolve the taxonomic complexities of the polyphyletic *Justicia* and the Justiceae.

Author Contributions: S.S.Y. collected the plant material, designed and performed the experiment, S.S.Y. and K.Y. analyzed the data and drafted the manuscript. All authors have read and agreed to the published version of the manuscript.

Funding: This research was funded by the University of Johannesburg postdoctoral funding scheme.

Data Availability Statement: The complete chloroplast genome sequence of *H. forskaolii* is deposited in the GenBank with I.D no: ON398071.

Acknowledgments: The authors acknowledge the University of Johannesburg for the postdoctoral fellowship position.

Conflicts of Interest: The authors declare no conflict of interest.

References

1. Ellis, B.W. *Taylor's Guide to Annuals: How to Select and Grow More Than 400 Annuals, Biennials, and Tender Perennials*; Houghton Mifflin Co.: Boston, MA, USA, 1999; pp. 1–448.
2. Almehdar, H.; Abdallah, H.M.; Osman, A.-M.M.; Abdel-Sattar, E.A. In Vitro cytotoxic screening of selected Saudi medicinal plants. *J. Nat. Med.* **2012**, *66*, 406–412. [CrossRef] [PubMed]
3. Shen, C.C.; Ni, C.L.; Huang, Y.L.; Huang, R.L.; Chen, C.C. Furanolabdane diterpenes from *Hypoestes* purpurea. *J. Nat. Prod.* **2004**, *67*, 1947–1949. [CrossRef] [PubMed]
4. Mothana, R.A.; Al-Musayeib, N.M.; Matheeussen, A.; Cos, P.; Maes, L. Assessment of the in vitro antiprotozoal and cytotoxic potential of 20 selected medicinal plants from the island of Soqotra. *Molecules* **2012**, *17*, 14349–14360. [CrossRef] [PubMed]
5. Balkwill, K.; Norris, F.G. Taxonomic studies in the Acanthaceae; The genus *Hypoestes* in southern Africa. *S. Afr. J. Bot.* **1985**, *51*, 133–144. [CrossRef]
6. Padysakova, E.; Bartos, M.; Tropek, R.; Janecek, S. Generalization versus specialization in pollination systems: Vistors, thieves, and pollinators of hypoestes aristate (Acanthaceae). *PLoS ONE* **2013**, *8*, e59299. [CrossRef]
7. Lindau, G. Acanthaceae. In *Die natürlichen Pflanzenfamilien, IV(3b)*; Engler, A., Prantl, K., Eds.; Engelmann: Leipzig, Germany, 1895; pp. 274–354.
8. Van Tieghem, P. Structure du pistil et de l'ovule du fruit et de la grane des Acathacees. *Ann. Sci. Nat. Serie 9. Botanique.* **1908**, *7*, 1–24.
9. Bremekamp, C.E.B. Notes on some acanthaceous genera of controversial position. *Acta Bot. Neerl.* **1955**, *4*, 644–655. [CrossRef]
10. Bremekamp, C.E.B. Delimitation and subdivision of the Acanthaceae. *Bull. Bot. Surv. India.* **1965**, *7*, 21–30.
11. Scotland, R.W.; Vollesen, K. Classification of Acanthaceae. *Kew Bull.* **2000**, *55*, 513–589. [CrossRef]
12. McDade, L.A.; Daniel, T.F.; Masta, S.E.; Riley, K.M. Phylogenetic relationships within the tribe Justicieae (Acanthaceae): Evidence from molecular sequences, morphology, and cytology. *Ann. Missouri Bot. Gard.* **2000**, *87*, 435–458. [CrossRef]
13. Kiel, C.A.; McDade, L.A.; Daniel, T.F.; Champluvier, D. Phylogenetic delimitation of Isoglossinae (Acanthaceae: Justicieae) and relationships among constituent genera. *Taxon* **2006**, *55*, 683–694. [CrossRef]
14. Daniel, T.F.; McDade, L.A.; Manktelow, M.; Kiel, C.A. The "Tetramerium Lineage" (Acanthaceae: Acanthoideae: Justicieae): Delimitation and intra-lineage relationships based on cp and nrITS sequence data. *Syst. Bot.* **2008**, *33*, 416–436. [CrossRef]
15. Kiel, C.A.; McDade, L.A. The Mirandea clade (Acanthaceae, Justicieae, Tetramerium Lineage): Phylogenetic signal from molecular data and micromorphology makes sense of taxonomic confusion caused by remarkable diversity of floral form. *Syst. Bot.* **2014**, *39*, 950–964. [CrossRef]
16. Côrtes, A.L.; Rapini, A.; Daniel, T.F. The Tetramerium lineage (Acanthaceae: Justicieae) does not support the Pleistocene Arc hypothesis for South American seasonally dry forests. *Amer. J. Bot.* **2015**, *102*, 992–1007. [CrossRef]

17. Kiel, C.A.; Daniel, F.T.; Darbyshire, I.; McDade, L.A. Unraveling relationships in the morphologically diverse and taxonomically challenging "justicioid" lineage (Acanthaceae: Justicieae). *Taxon* **2017**, *66*, 645–674. [CrossRef]
18. Neuhaus, H.; Emes, M. Nonphotosynthetic metabolism in plastids. *Ann. Rev. Plant Biol.* **2000**, *51*, 111–140. [CrossRef]
19. Grevich, J.J.; Daniell, H. Chloroplast Genetic Engineering: Recent Advances and Future Perspectives. *Crit. Rev. Plant Sci.* **2005**, *24*, 83–107. [CrossRef]
20. Wicke, S.; Schneeweiss, G.M.; Depamphilis, C.W.; Kai, F.M.; Quandt, D. The evolution of the plastid chromosome in land plants: Gene content, gene order, gene function. *Plant Mol. Biol.* **2011**, *76*, 273–297.
21. Shaw, J.; Shafer, H.L.; Leonard, O.R.; Kovach, M.J.; Schorr, M.; Morris, A.B. Chloroplast DNA sequence utility for the lowest phylogenetic and phylogeographic inferences in angiosperms: Thetortoise and the hare IV. *Am. J. Bot.* **2014**, *101*, 1987–2004. [CrossRef]
22. Guisinger, M.M.; Chumley, T.W.; Kuehl, J.V.; Boore, J.L.; Jansen, R.K. Implications of the plastid genome sequence of typha(Typhaceae, poales) for understanding genome evolution in poaceae. *J. Mol. Evol.* **2010**, *70*, 149–166. [CrossRef]
23. Yang, J.B.; Tang, M.; Li, H.T.; Zhang, Z.R.; Li, D.Z. Complete chloroplast genome of the genus Cymbidium: Lights into the species identification, phylogenetic implications and population genetic analysis. *BMC Evol. Biol.* **2013**, 13. [CrossRef] [PubMed]
24. Tillich, M.; Lehwark, P.; Pellizzer, T.; Ulbricht-Jones, E.S.; Fischer, A.; Bock, R.; Greiner, S. Geseq: Versatile and accurate annotation of organelle genomes. *Nucl. Acids Res.* **2017**, *45*, W6–W11. [CrossRef] [PubMed]
25. Dierckxsens, N.; Mardulyn, P.; Smits, G. NOVOPlasty: De novo assembly of organelle genomes from whole genome data. *Nucleic Acids Res.* **2016**, *45*, e18.
26. Lohse, M.; Drechsel, O.; Bock, R. OrganellarGenomeDRAW (OGDRAW): A tool for the easy generation of high-quality custom graphical maps of plastid and mitochondrial genomes. *Curr. Genet.* **2007**, *52*, 267–274. [CrossRef] [PubMed]
27. Kumar, S.; Stecher, G.; Tamura, K. MEGA7: Molecular Evolutionary Genetics Analysis version 7.0. *Mol. Biol. Evol.* **2015**, *30*, 2725–2729.
28. Thiel, T.; Michalek, W.; Varshney, R.; Graner, A. Exploiting EST databases for the development and characterization of gene-derived SSR-markers in barley (*Hordeum vulgare* L.). *Theor. Appl. Genet.* **2003**, *106*, 411–422. [CrossRef] [PubMed]
29. Kurtz, S.; Choudhuri, J.V.; Ohlebusch, E.; Schleiermacher, C.; Stoye, J.; Giegerich, R. Reputer: The manifold applications of repeat analysis on a genomic scale. *Nucleic Acids Res.* **2001**, *29*, 4633–4642. [CrossRef] [PubMed]
30. Mayor, C.; Brudno, M.; Schwartz, J.R.; Poliakov, A.; Rubin, E.M.; Frazer, K.A.; Pachter, L.S.; Dubchak, I. VISTA: Visualizing global DNA sequence alignments of arbitrary length. *Bioinformatics* **2000**, *16*, 1046–1047. [CrossRef]
31. Frazer, K.A.; Pachter, L.; Poliakov, A.; Rubin, E.M.; Dubchak, I. VISTA: Computational tools for comparative genomics. *Nucleic Acids Res.* **2004**, *32*, 273–279. [CrossRef]
32. Librado, P.; Rozas, J. DnaSP v6: A software for comprehensive analysis of DNA polymorphism data. *Bioinformatics* **2009**, *25*, 1451–1452. [CrossRef]
33. Katoh, K.; Standley, D.M. MAFFT multiple sequence alignment software version 7:improvements in performance and usability. *Mol. Biol.Evol.* **2013**, *30*, 772–780. [CrossRef] [PubMed]
34. Fredrik, R.; Maxim, T.; Paul, V.M.; Daniel, L.A.; Aaron, D.; Sebastian, H.; Bret, L.; Liang, L.; Mar, A.S.; John, P.H. MrBayes 3.2: Efficient Bayesian Phylogenetic Inference andModel Choice Across a Large Model Space. *Systematic* **2012**, *61*, 539–542.
35. Posada, D. jModelTest: Phylogenetic model averaging. *Mol. Biol. Evol.* **2008**, *25*, 1253–1259. [CrossRef] [PubMed]
36. Chen, H.; Shao, J.; Zhang, H.; Jiang, M.; Huang, L.; Zhang, Z.; Yang, D.; He, M.; Ronaghi, M.; Luo, X.; et al. Sequencing and analysis of *Strobilanthes cusia* (Nees) Kuntze chloroplast Genome revealed the rare simultaneous contraction and expansion of the inverted repeat region in Angiosperm. *Front. Plant Sci.* **2018**, *9*, 324. [CrossRef]
37. Chang, C.C.; Lin, H.C.; Lin, I.P.; Chow, T.Y.; Chen, H.H.; Chen, W.H.; Cheng, C.H.; Lin, C.Y.; Liu, S.M.; Chang, C.C.; et al. The chloroplast genome of *Phalaenopsis aphrodite* (Orchidaceae): Comparative analysis of evolutionary rate with that of grasses and its phylogenetic implications. *Mol. Biol. Evol.* **2006**, *23*, 279–291. [CrossRef]
38. Samaila, S.Y.; Shah, M.M. The complete chloroplast genome of *Lantana camara* L. (Verbenaceae). *Mitochondrial DNA part B* **2020**, *1*, 919.
39. Park, I.; Kim, W.J.; Yeo, S.-M.; Choi, G.; Kang, Y.-M.; Piao, R.; Moon, B.C. The complete chloroplast genome sequences of *Fritillaria ussuriensis* maxim. In addition, *Fritillaria cirrhosa* D. Don, and comparative analysis with other *Fritillaria* species. *Molecules* **2017**, *282*, 22.
40. Amenu, S.G.; Wei, N.; Wu, L.; Oyebanji, O.; Hu., G.; Zhou, Y.; Wang, Q. Phylogenomic and comparative analyses of Coffeeae alliance (Rubiaceae): Deep insights into phylogenetic relationships and plastome evolution. *BMC Plant Biol.* **2022**, *22*, 88. [CrossRef]
41. Campbell, W.H.; Gowri, G. Codon usage in higher plants, green algae, and cyanobacteria. *Plant Physiol.* **1990**, *92*, 1–11. [CrossRef]
42. Liu, Q.; Dou, S.; Ji, Z.; Xue, Q. Synonymous codon usage and gene function are strongly related in Oryza sativa. *Biosystems* **2005**, *80*, 123–131. [CrossRef]
43. Li, B.; Lin, F.R.; Huang, P.; Guo, W.Y.; Zheng, Y.Q. Complete chloroplast Genome sequence of *Decaisnea insignis*: Genome organization, Genomic resources and comparative analysis *Sci. Rep.* **2017**, *7*, 10073. [CrossRef] [PubMed]
44. Srivastava, D.; Shanker, A. Identification of simple sequence repeats in chloroplast genomes of Magnoliids through bioinformatics approach. *Interdiscip. Sci. Comput. Life Sci.* **2016**, *8*, 327–336. [CrossRef] [PubMed]

45. Zhou, J.; Chen, X.; Cui, Y.; Sun, W.; Li, Y.; Wang, Y. Molecular structure and phylogenetic analyses of complete chloroplast genomes of two *Aristolochia* medicinal species. *Int. J. Mol. Sci.* **2017**, *18*, 1839. [CrossRef] [PubMed]
46. Jiang, D.; Zhao, Z.; Zhang, T.; Zhong, W.; Liu, C.; Yuan, Q.; Huang, L. The chloroplast genome sequence of *Scutellaria baicalensis* provides insight into intraspecific and interspecific chloroplast genome diversity in *Scutellaria*. *Genes* **2017**, *8*, 227. [CrossRef] [PubMed]
47. Zhou, J.; Cui, Y.; Chen, X.; Li, Y.; Xu, Z.; Duan, B.; Li, Y.; Song, J.; Yao, H. Complete chloroplast genomes of *Papaver rhoeas* and *Papaver orientale*: Molecular structures, comparative analysis, and phylogenetic analysis. *Molecules* **2018**, *23*, 437. [CrossRef]
48. Li, Y.G.; Xu, W.Q.; Zou, W.T.; Jiang, D.Y.; Liu, X.H. Complete chloroplast genome sequences of two endangered Phoebe(Lauraceae) species. *Bot. Stud.* **2017**, *58*, 37. [CrossRef]
49. Greiner, S.; Wang, X.; Rauwolf, U.; Silber, M.V.; Mayer, K.; Meurer, J.; Haberer, G.; Herrmann, R.G. The complete nucleotide sequences of the five genetically distinct plastid genomes of Oenothera, subsection Oenothera: I. sequence evaluation and plastome evolution. *Nucleic Acids Res.* **2008**, *36*, 2366–2378. [CrossRef]
50. Song, Y.; Wang, S.; Ding, Y.; Xu, J.; Li, M.F.; Zhu, S.; Chen, N. Chloroplast Genomic Resource of Paris for Species Discrimination. *Sci. Rep.* **2017**, *7*, 3427. [CrossRef]
51. Bryan, G.J.; McNicol, J.W.; Meyer, R.C.; Ramsay, G.; De Jong, W.S. Polymorphic simple sequence repeat markers in chloroplast genomes of Solanaceous plants. *Theor. Appl. Genet.* **1999**, *99*, 859–867. [CrossRef]
52. Provan, J. Novel chloroplast microsatellites reveal cytoplasmic variation in Arabidopsis thaliana. *Mol. Ecol.* **2000**, *9*, 2183–2185. [CrossRef]
53. Ebert, D.; Peakall, R. Chloroplast simple sequence repeats (cpSSRs): Technical resources and recommendations for expanding cpSSR discovery and applications to a wide array of plant species. *Mol. Ecol. Resour.* **2009**, *9*, 673–690. [CrossRef] [PubMed]
54. Jacinta, N.M.; Xiang, D.; Jia-Xin, Y.; Elijah, M.M.; Vincent, O.W.; Millicent, A.O.; Josphat, K.S.; Paulm, M.M.; Guang-Wan, H. Complete chloroplast genome of Chlorophytum comosum abd Chlorophytum gallabatense: Geome structures, comparative and phylogenetic analysis. *Plants* **2020**.
55. Dhafer, A.A.; Samaila, S.Y.; Enaj, J.A.; Abidina, A. Complete chloroplast genome sequence of *Barleria prionitis*, Comparative chloroplast genomics and phylogenetic relationships among Acanthoideae. *BMC Genom.* **2020**, *21*, 393.
56. Philippe, H.; Delsuc, F.; Brinkmann, H.; Lartillot, N. Phylogenomics, Annual Review of Ecology. *Evol. Sys.* **2005**, *36*, 541–562. [CrossRef]
57. Raubeson, L.A.; Peery, R.; Chumley, T.W.; Dziubek, C.; Fourcade, H.M.; Boorem, J.L.; Jansen, R.K. Comparative chloroplast genomics: Analyses including new sequences from the angiosperms *Nupharadvena* and *Ranunculus macranthus*. *BMC Genom.* **2007**, *8*, 174–201. [CrossRef]
58. Zhou, T.; Chen, C.; Wei, Y.; Chang, Y.; Bai, G.; Li, Z.; Kanwal, Z.; Zhao, G. Comparative transcriptome and chloroplast genome analyses of two related Dipteronia Species. *Front. Plant Sci.* **2016**, *7*, 1512. [CrossRef]
59. Rousseau-Gueutin, M.; Bellot, S.; Martin, G.E.; Boutte, J.; Chelaifa, H.; Lima, O.; Michon-Coudouel, S.; Naquin, D.; Salmon, A.; Ainouche, K.; et al. The chloroplast genome of the hexaploid *Spartina maritima* (Poaceae, Chloridoideae): Comparative analyses and molecular dating. *Mol. Phylogenet. Evol.* **2015**, *93*, 516. [CrossRef]
60. Xu, J.-H.; Liu, Q.; Hu, W.; Want, T.; Xue, Q.; Messing, J. Dynamics of chloroplast genome in green plants. *Genomics* **2015**, *106*, 221–231. [CrossRef]
61. Hebert, P.D.N.; Ratnasingham, S.; de Waard, J.R. Barcoding animal life: Cytochrome C oxidase subunit 1 divergences among closely related species. *Proc. R. Soc. B Biol. Sci.* **2003**, *270*, S96–S99. [CrossRef]
62. Hebert, P.D.N.; Stoeckle, M.Y.; Zemlak, T.S.; Francis, C.M.; Charles, G. Identification of birds through DNA barcodes. *PLoS Biol.* **2004**, *2*, e312. [CrossRef]
63. Borsch, T.; Quandt, D. Mutational dynamics and phylogenetic utility of noncoding chloroplast DNA. *Plant Syst. Evol.* **2009**, *282*, 169–199.
64. Dong, W.P.; Liu, J.; Yu, J.; Wang, L.; Zhou, S.L. Highly variable chloroplast markers for evaluating plant phylogeny at low taxonomic levels and for DNA barcoding. *PLoS ONE* **2012**, *7*, e35071. [CrossRef] [PubMed]
65. Tong, W.; Kim, T.S.; Park, Y.J. Rice chloroplast genome variation architecture and phylogenetic dissection in diverse *Oryza* species assessed by whole-genome resequencing. *Rice* **2016**, *9*, 57. [CrossRef] [PubMed]
66. Dong, W.P.; Liu, H.; Xu, C.; Zuo, Y.J.; Chen, Z.J.; Zhou, S.L. A chloroplast genomic strategy for designing taxon specific DNA mini-barcodes: A case study on ginsengs. *BMC Genet.* **2014**, *15*, 138. [CrossRef] [PubMed]
67. Du, Y.P.; Bi, Y.; Yang, F.P.; Zhang, M.F.; Chen, X.Q.; Xue, J.; Zhang, X.H. Complete chloroplast genome sequences of *Lilium*: Insights into evolutionary dynamics and phylogenetic analyses. *Sci. Rep.* **2017**, *7*, 5751. [CrossRef]
68. Darbyshire, I.; Vollesen, K. The transfer of the genus Peristrophe to Dicliptera (Acanthaceae), with a new species described from eastern Africa. *Kew Bull.* **2007**, *62*, 119–128.
69. Balkwill, K.F. Taxonomic studies in the tribe Justicieae of the family Acanthaceae. [Unpublished]. Ph.D. Thesis, University of Natal, Pietermaritzburg, South Africa, 1985.
70. Balkwill, K.; Getiffe-Norris, F.; Balkwill, M.J. Taxonomic studies in the Acanthaceae: Dicliptera in southern Africa. *Kew Bull.* **1996**, *51*, 1–61. [CrossRef]

genes

MDPI

Article

Transcriptome Level Reveals the Triterpenoid Saponin Biosynthesis Pathway of *Bupleurum falcatum* L.

Yuchan Li [1], Jun Zhao [1], Hua Chen [1], Yanping Mao [1], Yuping Yang [1], Liang Feng [1], Chuanxin Mo [1], Lin Huang [1], Dabin Hou [1,*] and Ma Yu [1,2,*]

[1] School of Life Science and Engineering, Southwest University of Science and Technology, 59 Qinglong Road, Mianyang 621010, China

[2] Laboratory of Medicinal Plant Cultivation, Institute of Medicinal Plant Development (IMPLAD), Chinese Academy of Medical Sciences & Peking Union Medical College, Beijing 100193, China

* Correspondence: hdb@swust.edu.cn (D.H.); mayu0073@swust.edu.cn (M.Y.); Tel.: +86-816-608-9523 (D.H.); +86-151-962-78464 (M.Y.)

Abstract: *Bupleurum falcatum* L. is frequently used in traditional herbal medicine in Asia. Saikosaponins (SSs) are the main bioactive ingredients of *B. falcatum*, but the biosynthetic pathway of SSs is unclear, and the biosynthesis of species-specific phytometabolites is little known. Here we resolved the transcriptome profiles of *B. falcatum* to identify candidate genes that might be involved in the biosynthesis of SSs. By isoform sequencing (Iso-Seq) analyses of the whole plant, a total of 26.98 Gb of nucleotides were obtained and 124,188 unigenes were identified, and 81,594 unigenes were successfully annotated. A total of 1033 unigenes of 20 families related to the mevalonate (MVA) pathway and methylerythritol phosphate (MEP) pathway of the SS biosynthetic pathway were identified. The WGCNA (weighted gene co-expression network analysis) of these unigenes revealed that only the co-expression module of MEmagenta, which contained 343 unigenes, was highly correlated with the biosynthesis of SSs. Comparing differentially expressed gene analysis and the WGCNA indicated that 130 out of 343 genes of the MEmagenta module exhibited differential expression levels, and genes with the most "hubness" within this module were predicted. Manipulation of these genes might improve the biosynthesis of SSs.

Keywords: *Bupleurum falcatum*; transcriptome; saikosaponin biosynthetic pathway; WGCNA; gene expression

Citation: Li, Y.; Zhao, J.; Chen, H.; Mao, Y.; Yang, Y.; Feng, L.; Mo, C.; Huang, L.; Hou, D.; Yu, M. Transcriptome Level Reveals the Triterpenoid Saponin Biosynthesis Pathway of *Bupleurum falcatum* L. *Genes* **2022**, *13*, 2237. https://doi.org/10.3390/genes13122237

Academic Editors: Wajid Zaman and Hakim Manghwar

Received: 6 October 2022
Accepted: 25 November 2022
Published: 29 November 2022

1. Introduction

Radix Bupleuri (Chai Hu in Chinese) is the dried roots of the genus *Bupleurum*, Apiaceae family, which has been widely used in oriental traditional medicine for more than 2000 years, due to its remarkable therapeutic effect on fever, inflammation, influenza, hepatitis, malaria, and menopausal syndrome [1]. The principal bioactive constituents of *Radix Bupleuri* are its pentacyclic triterpenoid saponins, commonly known as saikosaponins (SSs), which are comprised of oleanane-type and ursane-type glycosides. Over 130 SSs have been isolated to date; according to their diverse skeleton structure, SSs are divided into 14 types [2]. SSs possess diverse pharmacological effects, especially SSa and SSd, which have immunomodulatory, anti-inflammatory, anticancer, antiviral, antibacterial, sedative, analgesic, and other pharmacological effects. Potent antioxidant, hepatoprotective, and cytotoxic effects were also reported [3,4].

Different genotypic varieties of *Bupleurum* plants differ significantly in the content and proportion of SSs [5,6], which are regulated by interior factors, e.g., growth stage, root structure, and environmental conditions, such as drought, light deficiency, and fertility [7–13]. SSs are mainly in the pericycle and primary phloem in young roots, while in mature roots, they were mainly concentrated in the secondary phloem and vascular cambium [14]. Therefore, combining cultivation techniques and manipulating gene expressions with regard to the biosynthesis of SSs could be a more operative approach to improving the SSs yield.

SSs are synthesized via the isoprenoid pathway, initially, IPP (isopentenyl pyrophosphate) and DMAPP (dimethylallyl diphosphate) are synthesized via the MVA (mevalonate) pathway and MEP (methylerythritol phosphate) pathway, respectively. Then the GPP (geranyl diphosphate) and DMAPP transform into farnesyl pyrophosphate (FPP), which then transforms into 2,3-oxidosqualene via two consecutive enzyme reactions. The β-amyrin synthase (β-AS) is responsible for cyclizing 2,3-oxidosqualene to generate β-amyrin. Finally, cytochrome P450s (CYPs) catalyze a series of hydroxylation/oxidation reactions and UDP-glycosyltransferases (UGTs) catalyze glycosylation reactions, so as to generate various saikosaponin monomers [15–17]. Several genes encoding SS biosynthesis enzymes were cloned and identified in *B. kaoi* (Bk), *B. chinense* (Bc), and *B. falcatum* (Bf), such as *BcSE1*, *BfSS1*, *CYP716Y1*, and *BkβAS* [18–21]. Nevertheless, the biosynthesis of SSs varies among species, and the underlying molecular regulatory mechanisms are still elusive. In the present study, weighted gene co-expression network analysis (WGCNA) was used to analyze the transcriptome and its relation with SSs content, so as to unearth the regulation networks of the SS biosynthesis route of *B. falcatum*.

2. Materials and Methods

2.1. Sampling of B. falcatum

A *B. falcatum* variety was obtained from a medicinal plant farm (altitude 600 m) in Rongxian in October 2013, which was authenticated by Professor Jianhe Wei (Institute of Medicinal Plant Development, Peking Union Medical College & Chinese Academy of Medical Sciences). The voucher specimen of this genotype (No. 0320130122) was stored in the herbarium of the School of Life Science and Engineering, Southwest University of Science & Technology, Mianyang, China.

Bf seeds were put on humid filter paper before germination. Sprouts were grown in modified Hoagland's nutrient media. The growth conditions were 24 ± 1 °C, a light/dark cycle of 12/12 h, and 55–65% relative humidity. The 5- and 15-day-old sprouts were subject to isoform sequencing (Iso-Seq), transcriptome scrutiny, and SSa/SSd quantifications. S1 was the whole fresh roots of 5-day-old seedlings; S2 was a 5 mm portion of the 15-day-old seedling root tips without the differentiation region; S3 was a 15-day-old seedling root with removed root tips as previously described (Supplementary Figure S1) [22].

2.2. Saikosaponin Extraction and HPLC (High-Performance Liquid Chromatography)

Samples S1–S3 with triplicates of each were freeze-dried for 72 h. The Waters HPLC system (Waters 1525 Binary HPLC Pump, Milford, MA, USA)) and an ASB-vensil C18 column (4.6 mm × 250 mm, 5 μm) were used to determine the SSa and SSd content. Their reference standards were purchased from the National Institutes for Food and Drug Control, Beijing, China. The determination methods and conditions previously described were utilized [23].

2.3. PacBio Iso-Seq Library Construction and Transcriptome Analyses

The root and leaf of the *B. falcatum seedling* were blended and subject to Iso-Seq sequencing library creation. The Iso-Seq analysis was performed as previously described [24]; the full-length transcript sequencing was conducted on the PacBio Sequel system (Pacific Biosciences, San Diego, CA, USA) following the isoform sequencing protocol, and the Clontech SMARTer PCR cDNA Synthesis Kit and BluePippin Size Selection System were utilized as described by the manufacturer PacBio (PN 100-092-800-03, San Diego, CA, USA).

The transcriptome sequencing of S1, S2, and S3 was performed on the HiSeq 2500 platform (Illumina, San Diego, CA, USA). Three replications were included in this study [25].

2.4. PacBio Sequencing Read Processing

The raw sequencing data were processed on the PacBio Sequel platform and SMRTLink5.1 software was utilized to divide the subread BAM files into non-CCS (circular consensus sequence) and CCS subreads. CCSs were then split into FL (full-length) and non-FL (nFL) reads after identifying the poly (A) tail signal and $5'/3'$ adaptors by pbclassify.py script. The FL transcripts were clustered by the ICE (iterative clustering for error correction) algorithm and then treated by arrow polishing. The additional nucleotide errors in consensus reads were proofread using the clean RNA-seq data through the software LoRDEC [26]. Finally, CD-HIT removed any redundancy in the corrected shared reads to obtain the final transcripts [27].

2.5. Functional Annotation of Genes

The full-length transcripts were detected via the ANGEL pipeline to obtain the protein-coding sequences (CDSs), amino acid sequences, and untranslated region sequences (UTR). Gene function was annotated by blasting CDSs against databases including NCBI nonredundant protein (Nr), Protein family (Pfam), euKaryotic Orthologous Groups (KOG), and Swiss-Prot, with the threshold $E \leq 10^{-10}$. A BLAST search was also performed against the Nucleotide sequence database (Nt) with the cutoff E-value 1×10^{-10} [28–31]. In addition, the KEGG (Kyoto Encyclopedia of Genes and Genomes) pathway and GO (Gene Ontology) functional enrichment analyses were implemented in the GOseq R package (3.6.2) and KOBAS software, respectively [32].

2.6. TF Prediction and lncRNA, SSR Analysis

CDSs of isoforms were aligned and assigned to different families based on transcription factors (TFs) prediction using iTAK predictive software [33]. The software PLEK and CNCI were adopted to assess the protein-coding potential of transcripts and distinguish protein-coding and non-coding sequences [34,35]. The CPC method uses biological sequence characteristics to determine the protein-coding likeliness of transcripts by comparing them with protein databases. Finally, an hmmscan homology search was performed via Pfam-scan databases to determine the LncRNA (long non-coding RNA). Simple sequence repeats (SSRs) were identified by MISA (http://pgrc.ipkgatersleben.de/misa/misa.html, accessed on 29 May 2020.), a method that identifies and localizes typical SSRs and composite ones intruded by bases.

2.7. Quantification of Gene Expression and Differentially Expressed Genes (DEGs) Analysis

The Bowtie2 program in RSEM removed superfluous sequences to acquire ultimate high-quality transcripts, which were the reference transcriptome. Then the levels of gene expression for each sample were estimated [36]. The clean reads were mapped to the reference transcriptome to obtain read counts compared to each gene in each sample. RSEM provides a method using the Bowtie2 program to calculate gene expression abundances via FPKM (fragments per kilobase per million mapped reads).

The gene readcounts were further analyzed in DESeq2 [37] to identify statistically significant DEGs among three root tissues. Additionally, p-values were fine-tuned with the Benjamini and Hochberg approach to control the FDR (false discovery rate) [36]. Genes with an adjusted p-value < 0.05 were designated as DEGs. Finally, the mean FPKM values of genes were changed to log2 ones for visualization with the heatmap package of R 3.6.2.

2.8. Selecting Candidate Genes of Saikosaponin Biosynthesis

Gene families related to saikosaponin synthesis pathways were selected to form Iso-Seq and transcriptome data, including AACT (acetyl-CoA acetyltransferase), HMGS (hydroxymethylglutaryl-CoA synthase), HMGR (3-hydroxy-3-methylglutaryl-coenzyme A reductase), MK (mevalonate kinase), PMK (phosphomevalonate kinase), MVD (mevalonate pyrophosphate decarboxylase), DXS (1-deoxy-D-xylulose-5-phosphate synthase), DXR (1-deoxy-D-xylulose 5-phosphate reductoisomerase), CMS (2-C-methyl-D-erythritol

4-phosphate cytidylyltransferase), CMK (4-diphosphocytidyl-2-C-methyl-D-erythritol kinase), HDS (4-hydroxy-3-methylbut-2-en-1-yl diphosphate synthase), MCS (2-C-methyl-D-erythritol 2,4-cyclodiphosphate synthase), GPS (geranylgeranyl pyrophosphate synthase), FPS (farnesyl diphosphate synthase), IDS (4-hydroxy-3-methylbut-2-enyl diphosphate reductase), IDI (isopentenyl-diphosphate Delta-isomerase), SS (squalene synthase), SE (squalene epoxidase), β-AS, CYP, and UGT. Herein a total of 1033 unigenes were selected in this work (Supplementary Table S1).

2.9. WGCNA

The 1033 candidate genes involved in the SS biosynthesis pathway were the input of R package WGCN, and a weighted co-expression network was built to reveal potential saponin biosynthesis modules of *B. falcatum*. The key parameters were set as follows: soft threshold power, 12; min module size, 20; merge cut height, 0.25. The ME (module eigengene, the first principal component of a module) value was computed for each module, so as to assess the relationship with SSs. The genes in the module which significantly related to saikosaponin content were analyzed in the DESeq package of R, and candidate genes with $P_{adj} < 0.05$ and log2 (fold change) > 2 selected by DESeq were defined as DEGs. The hub genes of the putative saikosaponin synthesis network were assigned based on the edge number and k_{ME} values [38]. The top score hub genes were computed and sorted with the MCC method implemented in the Cytoscape plugin cytoHbba [39]

2.10. Validation by qRT-PCR

From the hub genes, key enzyme genes that regulate the metabolic flux of MEP and MVA pathways, such as AACT, HMGR, and DXS, and β-AS, CYP, and UGT genes involved in the biosynthesis of SSs were selected. In qRT-PCR, gene-specific primers were used to validate the expression of 10 metabolic genes (Supplementary Table S2). The cDNA of S1, S2, and S3 were the same as those used for sequencing. The qRT-PCR reaction was conducted with a Green qPCR SuperMix kit (Transgen Biotech, Beijing, China) and CFX96 Real-time PCR system (Bio-Rad, Hercules, CA, USA), and the relative expression values were computed with the $2^{-\Delta\Delta}$Ct method for each gene, all reactions were carried out in biological triplicate [40,41].

3. Results

3.1. Saikosaponin a (SSa) and Saikosaponins d (SSd) Content

SSa and SSd were detected with HPLC in each region of the root tissue of *B. falcatum*. The result showed that no peaks of SSa and SSd were identified in the S1 and S2 samples, while they were significantly accumulated in S3 (Table 1). Based on this, the genes following the identical or opposite expression trend among the three samples are worth paying more attention to.

Table 1. Contents of SSa and SSd in the young root of Bf.

Sample	Ssa Content (μg/g)	Ssd Content (μg/g)
S1	0	0
S2	0	0
S3	444.56	312.29

3.2. Transcriptome Profile Analysis of B. falcatum

With the aim of profiling the whole picture of transcriptional dynamics during the young root development stage, and identifying genes participating in the SS biosynthesis of *B. falcatum*, the FL transcriptome was obtained by SMRT sequencing of the PacBio Sequel platform. We filtered the raw sequencing data to remove the adaptor and offline sequences less than 50 bp, and finally obtained 10,722,671 subreads based on 26.98G. The filtered subreads were 2517 bp in mean length. Through conditional screening, 485,498 CCSs were obtained (full passes, 1; quality, 0.80). In total, 438,691 FL non-chimeric reads (Flnc), with complete 5′ primers, 3′ primers, and poly-A tails, were acquired; the mean length of the Flnc was 3098 bp.

PacBio, the representative third-generation sequencing technology, possesses long read length superiority but is associated with the disadvantage of a higher single base error rate. Illumina data was therefore employed to correct it to decrease the error rate. After correction, 222,393 consensuses, 4061 N50, and 1962 N90 were obtained. Similar and redundant sequences were deleted with the software CD-HIT, resulting in 222,393 nonredundant transcripts.

The Bowtie2 program in RESM software was applied to align clean reads with unigenes. About 70.04% of clean reads were mapped to the *B. falcatum* reference transcriptome, with approximately 32 million clean reads mapped uniquely to the reference transcriptome assembly for each sample. In total, 124,188 unigenes were assembled.

3.3. Functional Annotation of Genes

In the functional annotation of unigenes, the databases Nr, Nt, Pfam, KOG, Swiss-Prot, GO, and KEGG were utilized. The Venn diagram demonstrated that 54,131 genes were concurrently annotated in all these databases (Supplementary Figure S2A). In total, 121,417 genes were annotated in ≥ 1 database, and 119,845 (96.5%) genes were interpreted by the NR database. Based on the sequence homology alignment, 107,815 (86.81%) unigenes were identified against *Daucus carota*; 1011 (0.81%) sequences showed momentous hits for *Vitis vinifera*.

A total of 118,312 genes (95.26%) genes were annotated by the KEGG database, and it revealed that 5915 (4.76%) genes were clustered in the signal transduction, 4564 in the signal transduction pathway, and 3885 (3.13%) in transport and catabolism (Supplementary Figure S2B). A total of 81,594 (65.70%) genes were assigned with GO terms, and the KOG database explained 78,450 (63.17%) genes (Supplementary Figure S2C,D).

3.4. Ts Prediction and SSR: LncRNA Analysis

A total of 7256 unigenes were classified into 69 TF families identified from the transcriptome of *B. falcatum* by iTAK software. Among them, the top ten TF families were SNF2, PHD, SET, C3H, WRKY, Jumonji, C2H2, bHLH, B3-ARF, and AP2/ERF-ERF. Moreover, 524 transcripts were predicted as other TFs. Six types of SSR, including dinucleotide, tetranucleotide, hexanucleotide, single nucleotide, pentanucleotide, and trinucleotide repeat sequences, were identified using MISA 1.0 software. LncRNAs are RNA molecules of longer than 200 bp without protein products. In addition, based on 222,393 PacBio Iso-Seq isoforms, CNCI, PLEK, CPC, and Pfam-scan, four tools were used to predict LncRNA, and the predicted number of lncRNAs were 26,648, 17,502, 5875, and 32,894, respectively. The length of lncRNAs was between 200 and 1000 bp, lower than that of mRNAs (1500–3000 bp).

3.5. Comparative Analysis of DEGs

To investigate transcript changes in *B. falcatum* between young root development stages, DEGs were filtered with $|\log2FoldChange| > 0$ & $P_{adj} < 0.05$. The DEGs numbers of three comparisons ("S1 vs. S2", "S2 vs. S3", and "S1 vs. S3") are shown in Figure 1A,B. Compared to undifferentiated root regions S1 and S2, 15,193 down-regulated and 10,004 up-regulated genes were identified in the taproot differentiation regions of 15-day-old sample S3, respectively. Only 34 or 91 genes were down- or up-regulated in comparison to "S1 vs. S2". Additionally, in total, 25,197 DEGs genes were identified in S3 in comparison to S1 and S2, and DEGs were annotated in the KEGG pathway (Figure 1C,D). Numerous DEGs involved in specialized metabolism were identified, such as those involved in sesquiterpenoid/triterpenoid production, diterpenoid synthesis, and steroid biosynthesis, among others.

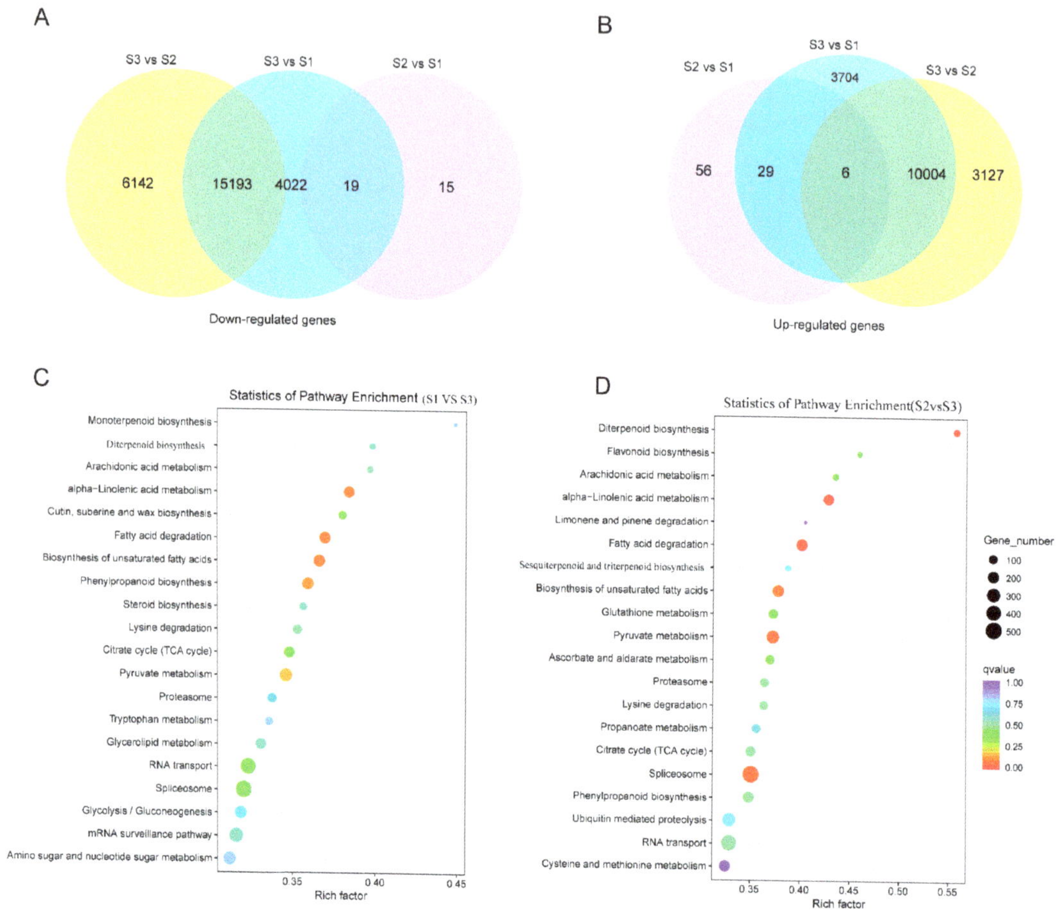

Figure 1. Differentially expressed genes in the seeding root of *B. falcatum* among three samples. (**A**,**B**) Venn diagram of all down-regulated genes and all up-regulated ones from the transcriptome. S1, the whole fresh roots of 5-day-old seedlings; S2, a 5 mm portion of root tips without the differentiation region of 15-day-old sprout; S3, differentiated root tissue of 15-day-old seedlings. (**C**,**D**) Pathway enrichment analysis of DEGs identified in "S1 vs. S3" (**C**), and "S2 vs. S3" (**D**), based on the KEGG database.

3.6. Gene Co-Expression Network Construction

By pairwise correlation analysis with gene expressions, WGCNA was employed to identify the candidate genes that were highly associated with SSs content in the *B. falcatum*. Ten gene modules of the co-expression network, labeled black, blue, brown, green, magenta, pink, purple, red, turquoise, and yellow, were identified. In each module, the number of target genes was between 25 and 343. At the *p*-value < 0.05 level, the correlation coefficients value among SSs content, different root region, and modules eigengenes revealed that the MEmengta module (343 unigenes) was positively significantly associated with SSa, SSd, and root region tissues (R = 0.95, 0.95, 0.87) (Figure 2).

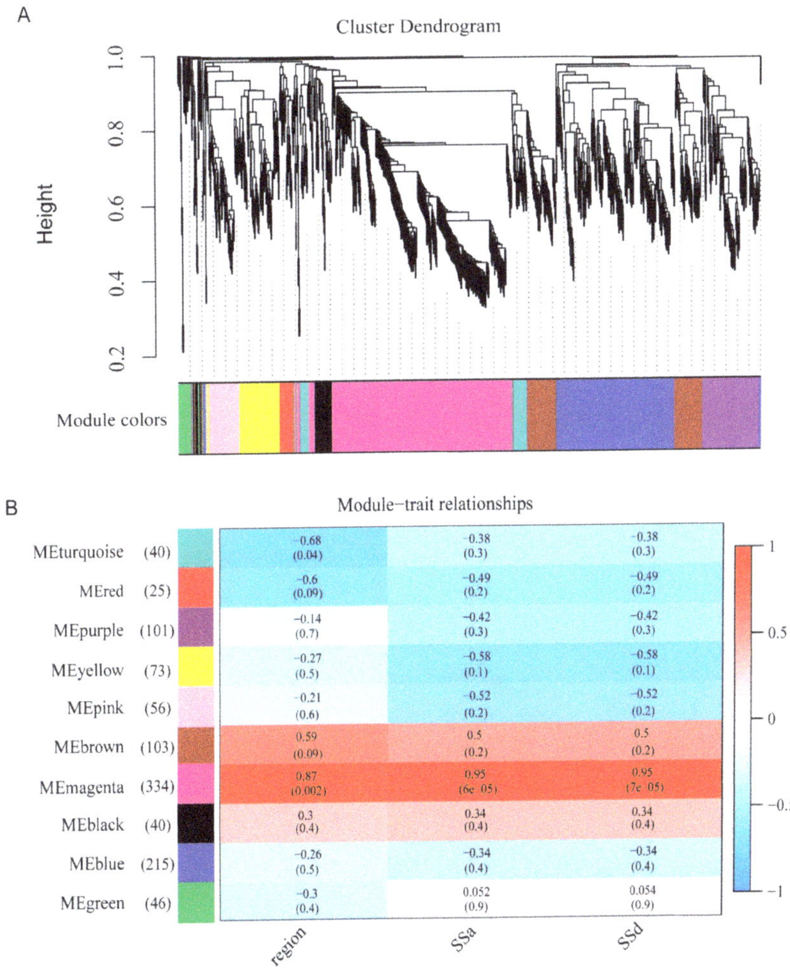

Figure 2. WGCNA. Hierarchical gene clustering was used to sort out the clustering tree of the co-expression network module according to the 1-tom matrix. Each module used a dissimilar color. (**A**) The hierarchical dendrogram displays co-expression modules revealed by WGCNA. One gene is represented by one leaf. Ten modules were found and assigned with different colors based on the calculation of eigengenes. (**B**) Correlation between traits (bottom) and modules (left), with numbers in left brackets showing the gene number of the module. Blue and red represent negative and positive correlations, respectively, and coefficient values are shown. The darker the color, the higher the correlation coefficient. The *p*-values are in brackets.

3.7. DEGs in MEmagenta Module Involved in the Saikosaponin Biosynthesis

To find genes participating in the differential metabolism of saikosaponins in the MEmagenta module, the expression levels and forms of all transcripts were analyzed in terms of the FPKM transcriptome data. A stringent threshold |log2 Fold change| >1 and $P_{adj} < 0.05$ screened out DEGs. In total, 130 unigenes closely related to saikosaponin biosynthesis showed significantly different expressions in different root tissue of *B. falcatum* (Supplementary Table S3). Among these DEGs, fifteen genes were divided into MVA pathways, including five AACT genes, one HMGS gene, six HMGR genes, and three PMK genes. For the MEP pathway, five gene families were identified, including DXS (6 unigenes),

DXR (3 unigenes), CMK (2 unigenes), HDS (7 unigenes), and IDS (1 unigenes). Additionally, we did not detect DEG in the gene families of MK, MVD, CMS, and MCS. These DEGs, including SS (1 unigenes), SE (1 unigene), β-AS (4), P450s (68), and UGT (23), may convert oleanane-type SSs (Figures 3 and 4).

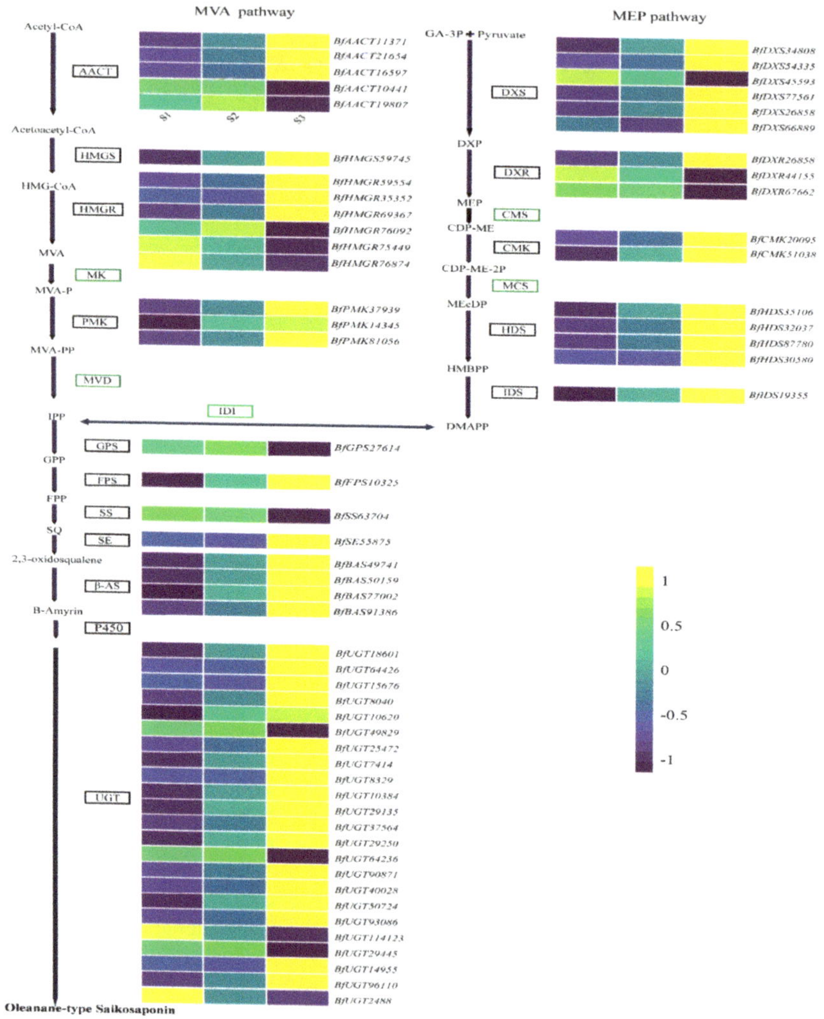

Figure 3. Differential expression of candidate genes of saikosaponin biosynthesis route among S1, S2, and S3. HMG-CoA: 3-hydroxy-3-methy-lglutaryl-COA; MVA-P: 5-phosphoevalonate; MVA-PP: 5-pyrophosphoevalonate; DXP: 1-deoxyxylulose 5-phosphateuvate; CDP-ME: methylerythritol cytidyl diphosphate; CDP-ME-2P: 4-diphosphocytidyl-2-C-methyl-D-erythritol; MEcDP: 2-C-methyl-D-erythritol-2,4-cyclodiphosphate; HMBPP: 4-hydroxy-3-methyl-butenyl-1-diphosphate.

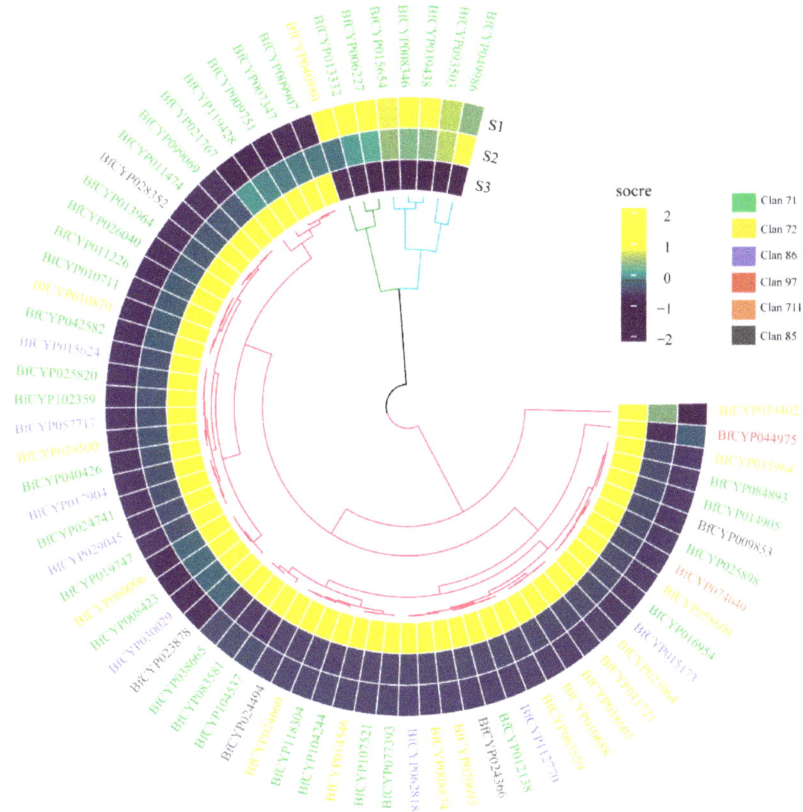

Figure 4. Gene expression profile and cluster analysis of candidate P450. These genes displayed meaningfully different expressions in S3 of *B. falcatum*.

3.8. Hub Gene Selection for the MEmagenta Module

Hub genes correlated with differential SSa and SSd levels in root tissues were identified based on the importance of genes in the MEmatenga module. The top 20 genes with the highest MCC values might connect more closely with other genes and play vital roles in the co-expression network (Supplementary Table S4). As shown in Figure 5, the hub gene included AACT (1 unigene), DXR (1 unigene), DXS (1 unigene), HMGR (3 unigenes), HDS (1 unigene), CMK (1 unigene), UGT (3 unigenes), P450 (7 unigenes), and β-AS (2 unigenes).

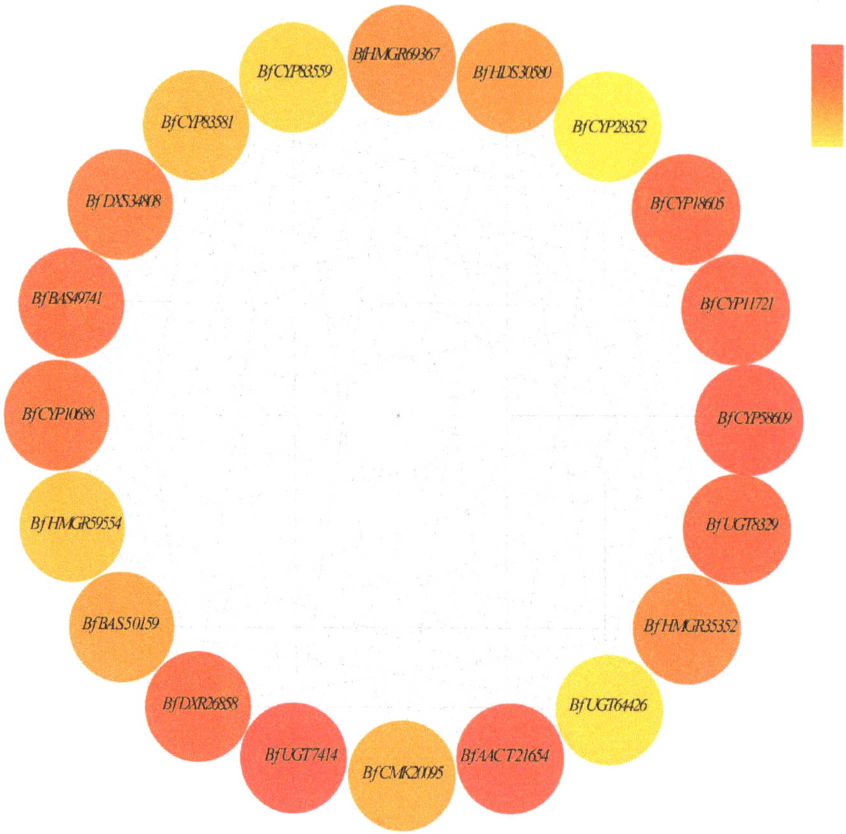

Figure 5. Hub genes of putative triterpenoid biosynthesis pathway. The MCC (maximal clique centrality) algorithm identified hub genes in PPI networks. The red node represents the high MCC score gene, and the yellow represents the low score gene.

3.9. *qRT-PCR*

qRT-PCR was conducted to confirm the transcriptome data and validate the hub genes expression level. The following genes were tested with triplicates, i.e., *BfAACT21654*, *BfHMGR35352*, *BfDXS34808*, *BfHDS30580*, *BfBAS49741*, *BfBAS50159*, *BfCYP18605*, *BfCYP58609*, *BfUGT8329* and *BfUGT7414*. The qRT-PCR showed that all genes had a higher transcript level in mature root tissue when compared with other parts of the root meristem. Therefore, SSs highly produced in mature root tissue may be caused by these highly expressed genes. These results agreed with transcriptome data (Figure 6).

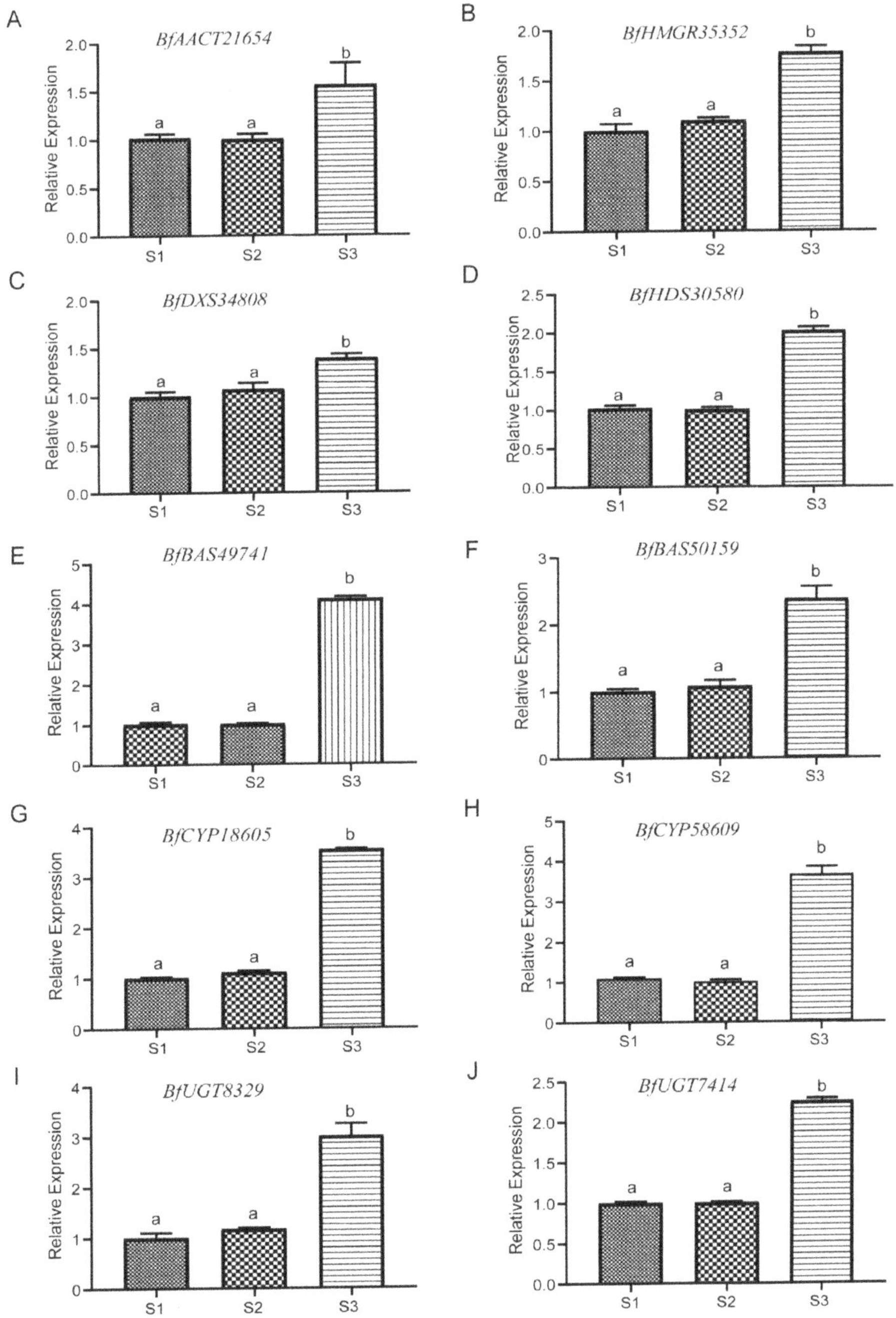

Figure 6. qRT-PCR verification of key genes of saikosaponin biosynthetic pathway. (**A–J**) The relative expression patterns of *BfAACT21654*, *BfHMGR35352*, *BfDXS34808*, *BfHDS30580*, *BfBAS49741*, *BfBAS50159*, *BfCYP18605*, *BfCYP58609*, *BfUGT8329* and *BfUGT7414*. The bar represents the standard error (SE). Data are presented as mean ± SE. Different letters represent different levels of significant difference.

4. Discussion

In the present study, we first reported the overall status of the full-length transcriptome profile of *B. falcatum* and identified 1033 unigenes of 20 families involving SSs synthesis pathways from Iso-Seq and transcriptome data. The WGCNA analysis of those 1033 unigenes revealed that only the co-expression module of MEmagenta, which contained 343 unigenes, was highly correlated with SSa and SSd. Additionally, 130 out of 343 genes of the MEmagenta module exhibited differential expression levels. We also identified 20 hub genes with the highest correlation in this module. Furthermore, real-time quantitative results also confirmed that these hub genes increased significantly in mature root tissue (S3).

MEP/MVA pathways participate in the biosynthesis of triterpenoid backbones in planta [42]. AACT catalyzes the formation of acetoacetyl-CoA via transferring an acetyl group from one acetyl-CoA to another, the first step of the MVA route; it plays a fundamental role in triterpenoid biosynthesis. For instance, overexpression of *GL-AACT* promoted the buildup of triterpenoids in *Ganoderma lucidum* [43]. As the first rate-limiting enzyme of the MVA route, HMGR regulates the terpenoid metabolism; for instance, Lee et al. [44] manipulated the HMGR gene by constructing biotin carboxyl carrier protein for co-expression in *Nicotiana benthamiana*, which increased the yield of sesquiterpenoids and triterpenoids. DXS is a limiting enzyme catalyzing the initial step of the MEP route in plant isoprenoid biosynthesis, just like HMGR in the MVA pathway [45]. In addition, reestablishing natural product pathways based on the plant as a chassis, deregulating crucial enzymes in the MVA or MEP pathway, and increasing the production of natural active substances are currently common approaches to increase the abundance of target metabolites, such as the production of taxadiene 5α-hydroxytaxadiene engineered by MEP pathway, which achieved 56.6 μg/g and 1.3 μg/g, respectively [46]. The enhanced production of artemisinic acid via the engineered MVA route attained an almost 500-fold yield [47]. In our study, five AACT genes, six HMGR genes, and six DXS genes were identified. Among them, three AACT genes, three HMGR genes, and five DXS genes showed significantly up-regulated expression in S3, with the exception of two AACT genes, three HMGR, and one DXS gene, which showed a conversely expressional tendency. Therefore, manipulating these genes with the same expression trend as SSs content in the root of *B. falcatum* could improve the availability of the precursor (2,3-oxidosqualene) for biosynthesizing triterpenes, especially hub genes, such as *BfAACT21654*, *BfDXS34808*, and *BfHMGR35352* (Figure 6A–C).

The C5 unit IPP generated by MEP/MVA routes can be transformed into the isomer DMAPP via IDI [48]. However, we failed to identify the expression disparity of the IDI gene. IPP/DMAPP are transformed into FPP and GPP via FPS and GPS, respectively. The *BfGPS27614* transcript expression decreased in S3 of *B. falcatum*, whereas the expression of the FPS gene *BfFPS10325* increased. A previous study revealed that when *BfSS1* was knocked down in transgenic Bf, the transcript buildup increased in downstream genes, e.g., SE and cycloartenol synthase. In this study, *BfSS63704* was down-regulated in S3, while *BfSE55875* was up-regulated, which showed a similar result to the study by Kim et al. (2011) [21].

β-AS is a member of the OSC (oxidosqualene cyclase) family, which acts in the initial step of the biosynthesis of SSs, and the resultant β-amyrin is the backbone of various SS monomers [49]. β-AS was an important branch point between specialized and primary metabolism and could regulate the metabolic accumulation of triterpenoid saponins. There is a positive correlation between β-AS activity and triterpene saponin content in plants. The down-regulation of β-AS gene expression by RNA interference in *Panax ginseng* hairy roots led to reduced β-amyrin and oleanane-type ginsenoside [50]. In *Medicago truncatula*, β-AS ectopic expression led to an increased yield of triterpene saponins [51]. Similarly, in our study, four β-AS genes were identified by WGCNA to be closely related to SSa and SSd. All these β-AS genes were highly expressed in S3, especially *BfBAS49741* and *BfBAS50159* (Figure 6E,F).

The cyclic backbone synthesized by β-AS is subject to site-specific oxidation via CYPs to yield diversified sapogenins, e.g., hydroxylation, carboxylation, and esterification [52,53].

CYP families of land plants were categorized into 11 clans, and the CYP71, CYP51 CYP72, and CYP85 clans are responsible for the biosynthesis of triterpenoids [54]. In addition, the CYP72A family forms the CYP72 clan, and the CYP716 family from the CYP85 clan is primarily involved in pentacyclic triterpenoid oleanolic acid-derived saponin biosynthesis [55,56]. For example, the *CYP716Y1* of Bf was responsible for the C-16α hydroxylation of oleanane triterpenes [18]. In the present study, we identified 68 CYP DEGs that are members of 19 subfamilies of 6 CYP clans (Supplementary Table S5). Among these genes, three *CYP716Y1* genes belonging to the CYP85 clan and nine unigenes belonging to the CYP72 clan showed significant up-regulation in S3. Furthermore, based on the hub gene analysis and RT-PCR results (Figures 5 and 6G,H), it was further confirmed that *BfCYP20695* and *BfCYP58609* may be critical in the biosynthesis of pentacyclic triterpenoid oleanolic acid-derived saponins.

Triterpenoids are further diversified via the scaffold modification of triterpenoids; UGTs catalyze the glycosylation of hydroxylated β-amyrin at C3 [15,57]. In transgenic *P. quinquefolius* hairy roots, *Pq3-O-UGT1* overexpression increased the level of protopanaxadiol group ginsenosides [58]. In this study, we identified 22 UGT genes, among these 18 unigenes were significantly up-regulated in S3, especially the hub gene *BfUGT8329*, while four genes exhibited a significant opposite trend. In this study, we identified 23 UGT genes, in addition to 5 genes that showed a down-regulation trend in S3, 18 other single genes showed significant up-regulation, especially the hub gene *BfUGT8329*.

5. Conclusions

In conclusion, the present study for the first time reports the full-length transcriptome profile of *B. falcatum*, and we performed a WGCNA and screened 129 DEGs involved in the SS biosynthetic pathway. Based on the up-regulated genes in S3 and identified hub genes, we constructed a hypothetical scheme of the SS biosynthesis pathway (Supplementary Figure S3). Finding the key enzyme genes of the biosynthesis route of the *Bupleurum* genus offers the option for acquiring appreciated triterpene saponins by synthetic biology. Consequently, it is reasonable to identify, clone, and functionally analyze the key enzymes of *B. falcatum* to promote biological research and commercial application of SSs. Moreover, all recognized key enzymes could be useful for diverse breeding studies of *B. falcatum* and other relevant species.

Supplementary Materials: The following supporting information can be downloaded at: https://www.mdpi.com/article/10.3390/genes13122237/s1, Table S1: Annotion of candidate genes for saikosaponin biosynthesis in *B. falcatum*; Table S2: Primer sequences used for qRT-PCR validation; Table S3: FPKM values of DEGs in MEmagenta module involved in the saikosaponin biosynthesis; Table S4: Top 20 in network CytoscapeInput-network.txt ranked by MCC method; Table S5: Summary of clan and subfamliy of P450 DEGs in MEmagenta module involved in the saikosaponin biosynthesis; Figure S1: Sampling positions of S1, S2 and S3 at different seedling stages; Figure S2: Functional annotation and classification of *B. falcatum* transcriptome; Figure S3: The hypothetical synthesis pathway for SSs in *B. falcatum*.

Author Contributions: Y.L. wrote the manuscript, J.Z. and H.C. performed the transcriptome analysis, Y.M. and Y.Y. seeded the seedling and prepared the plant materials, L.F. performed the validation by qRT-PCR, C.M. and L.H. performed HPLC analysis, M.Y. and D.H. funded the whole project and helped Y.L. to complete the manuscript. All authors have read and agreed to the published version of the manuscript.

Funding: This work was supported by the earmarked fund for CARS (CARS-21), the CAMS Innovation Fund for Medical Sciences (CIFMS) under grant number 2021-I2M-1-032, the Programs of Science and Technology Department of Sichuan Province (2019YFN0150, 2022NSFSC1639, 2022NSFSC1703), the Doctoral Research Funding of Southwest University of Science and Technology (19zx7117 and 21zx7116), and the Opening Project Fund of Key Laboratory of Rubber Biology and Genetic Resource Utilization, Ministry of Agriculture (RRL-KLOF202201).

Institutional Review Board Statement: Not applicable.

Informed Consent Statement: Not applicable.

Data Availability Statement: Data are contained within the article or supplementary material.

Conflicts of Interest: The work is presented in the manuscript with the consent of all authors. The authors declare that they have no conflict of interest.

References

1. Ashour, M.L.; Wink, M. Genus *Bupleurum*: A review of its phytochemistry, pharmacology and modes of action. *J. Pharm. Pharm.* **2011**, *63*, 305–321. [CrossRef] [PubMed]
2. Sui, C.; Han, W.J.; Zhu, C.R.; Wei, J.H. Recent Progress in Saikosaponin Biosynthesis in *Bupleurum*. *Curr. Pharm. Biotechnol.* **2021**, *22*, 329–340. [CrossRef]
3. Li, X.Q.; Song, Y.N.; Wang, S.J.; Rahman, K.; Zhu, J.Y.; Zhang, H. Saikosaponins: A review of pharmacological effects. *J. Asian Nat. Prod. Res.* **2018**, *20*, 399–411. [CrossRef]
4. Yuan, B.; Yang, R.; Ma, Y.; Zhou, S.; Zhang, X.; Liu, Y. A systematic review of the active saikosaponins and extracts isolated from Radix Bupleuri and their applications. *Pharm. Biol.* **2017**, *55*, 620–635. [CrossRef]
5. Liang, Z.; Oh, K.; Wang, Y.; Yi, T.; Chen, H.; Zhao, Z. Cell type-specific qualitative and quantitative analysis of saikosaponins in three *Bupleurum* species using laser microdissection and liquid chromatography-quadrupole/time of flight-mass spectrometry. *J. Pharm. Biomed. Anal.* **2014**, *97*, 157–165. [CrossRef]
6. Qin, X.; Dai, Y.; Liu, N.; Li, Z.; Liu, X.; Hu, J.; Choi, Y.; Verpoorte, R. Metabolic Fingerprinting by 1HNMR for Discrimination of the Two Species Used as Radix Bupleuri. *Planta Med.* **2012**, *78*, 926–933. [CrossRef]
7. Yu, M.; Chen, H.; Liu, Q.; Huang, J.; Semagn, K.; Liu, D.; Li, Y.C.; Yang, B.; He, Y.L.; Sui, C.; et al. Analysis of unigenes involved in lateral root development in *Bupleurum chinense* and *B. scorzonerifolium*. *Planta* **2021**, *253*, 128–139. [CrossRef]
8. Gong, J.; Liu, M.; Xu, S.; Jiang, Y.; Pan, Y.; Zhai, Z.; Luo, Q.; Yang, L.; Wang, Y. Effects of light deficiency on the accumulation of saikosaponins and the ecophysiological characteristics of wild *Bupleurum chinense* DC. *in China. Ind. Crops Prod.* **2017**, *99*, 179–188. [CrossRef]
9. Sun, J.L.; Duan, Z.J.; Zhang, Y.; Cao, S.S.; Tang, Z.H.; Abozeid, A. Metabolite Profiles Provide Insights into Underlying Mechanism in *Bupleurum* (Apiaceae) in Response to Three Levels of Phosphorus Fertilization. *Plants* **2022**, *11*, 752. [CrossRef]
10. Meng, J.; Chen, X.F.; Yang, W.Y.; Yao, R.Y.; Song, J.H.; Li, Z.F.; Yang, X.W. Determination of saikosaponin a and d in taproot and lateral root of *Bupleurum chinense* DC. *Chin. J. Pharma. Anal.* **2013**, *33*, 1218–1222.
11. Shon, T.K.; Agung, T.; Haryanto, D.; Yoshida, T. Variation and Distribution and Saikosaponin in *Bupleurum falcatum* L. *J. Fac. Agric. Kyushu Univ.* **1997**, *42*, 17–22. [CrossRef]
12. Zheng, Y.J.; Xia, P.G.; Zhao, H.G.; Zheng, J.F.; Chai, W.G.; Liang, Z.S.; Yan, K.J. Suitable soil moisture contents for water use efficiency and saponins accumulation in *Panax Notoginseng*. *Chin. Herb. Med.* **2021**, *13*, 267–273. [CrossRef] [PubMed]
13. Cao, P.; Wang, G.; Wei, X.; Chen, S.; Han, J. How to improve CHMs quality: Enlighten from CHMs ecological cultivation. *Chin. Herb. Med.* **2021**, *13*, 301–312. [CrossRef] [PubMed]
14. Tan, L.L.; Cai, X.; Zheng-Hai, H.U.; Ni, X.L. Localization and Dynamic Change of Saikosaponin in Root of *Bupleurum chinense*. *J. Integr. Plant Biol.* **2008**, *50*, 7. [CrossRef] [PubMed]
15. Seki, H.; Tamura, K.; Muranaka, T. P450s and UGTs: Key Players in the Structural Diversity of Triterpenoid Saponins. *Plant Cell Physiol.* **2015**, *56*, 1463–1471. [CrossRef] [PubMed]
16. Zheng, X.; Xu, H.; Ma, X.; Zhan, R.; Chen, W. Triterpenoid saponin biosynthetic pathway profiling and candidate gene mining of the Ilex asprella root using RNA-Seq. *Int. J. Mol. Sci.* **2014**, *15*, 5970–5987. [CrossRef] [PubMed]
17. Biswas, T.; Dwivedi, U.N. Plant triterpenoid saponins: Biosynthesis, in vitro production, and pharmacological relevance. *Protoplasma* **2019**, *256*, 1463–1486. [CrossRef]
18. Moses, T.; Pollier, J.; Almagro, L.; Buyst, D.; Van Montagu, M.; Pedreno, M.A.; Martins, J.C.; Thevelein, J.M.; Goossens, A. Combinatorial biosynthesis of sapogenins and saponins in *Saccharomyces cerevisiae* using a C-16alpha hydroxylase from *Bupleurum falcatum*. *Proc. Natl. Acad. Sci. USA* **2014**, *111*, 1634–1639. [CrossRef]
19. Lin, W.Y.; Peng, P.H.; Lin, T.Y. Cloning and characterization of β-amyrin synthase from *Bupleurum kaoi*. In Proceedings of the 8th International Congress of Plant Molecular Biology, Adelaide, Australia, 20–25 August 2005.
20. Gao, K.; Xu, J.S.; Sun, J.; Xu, Y.H.; Wei, J.H.; Sui, C.; Development, I.O.M.P.; Sciences, C.A.O.M. Molecular Cloning and Expression of Squalene Epoxidase from a Medicinal Plant, *Bupleurum chinense*. *Chin. Herb. Med.* **2016**, *8*, 67–74. [CrossRef]
21. Kim, Y.S.; Cho, J.H.; Park, S.; Han, J.Y.; Back, K.; Choi, Y.E. Gene regulation patterns in triterpene biosynthetic pathway driven by overexpression of squalene synthase and methyl jasmonate elicitation in *Bupleurum falcatum*. *Planta* **2011**, *233*, 343–355. [CrossRef]
22. Yu, M.; Chen, H.; Liu, S.; Li, Y.; Sui, C.; Hou, D.; Wei, J. Differential Expression of Genes Involved in Saikosaponin Biosynthesis Between *Bupleurum chinense* DC and *Bupleurum scorzonerifolium* Willd. *Front. Genet.* **2020**, *11*, 583245. [CrossRef] [PubMed]
23. Xu, J.; Wu, S.-R.; Xu, Y.-H.; Ge, Z.-Y.; Sui, C.; Wei, J.-H. Overexpression of BcbZIP134 negatively regulates the biosynthesis of saikosaponins. *Plant Cell Tissue Organ Cult. (PCTOC)* **2019**, *137*, 297–308. [CrossRef]
24. Wang, T.; Wang, H.; Cai, D.; Gao, Y.; Zhang, H.; Wang, Y.; Lin, C.; Ma, L.; Gu, L. Comprehensive profiling of rhizome-associated alternative splicing and alternative polyadenylation in moso bamboo (*Phyllostachys edulis*). *Plant J.* **2017**, *91*, 684–699. [CrossRef] [PubMed]

25. Yang, L.; Jin, Y.; Huang, W.; Sun, Q.; Liu, F.; Huang, X. Full-length transcriptome sequences of ephemeral plant Arabidopsis pumila provides insight into gene expression dynamics during continuous salt stress. *BMC Genom.* **2018**, *19*, 717. [CrossRef] [PubMed]
26. Salmela, L.; Rivals, E. LoRDEC: Accurate and efficient long read error correction. *Bioinformatics* **2014**, *30*, 3506–3514. [CrossRef]
27. Fu, L.; Niu, B.; Zhu, Z.; Wu, S.; Li, W. CD-HIT: Accelerated for clustering the next-generation sequencing data. *Bioinformatics* **2012**, *28*, 3150–3152. [CrossRef]
28. Li, W.Z.; Jaroszewski, L.; Adam, G. Tolerating some redundancy significantly speeds up clustering of large protein databases. *Bioinformatics* **2002**, *18*, 77–82. [CrossRef]
29. Tatusov, R.L.; Fedorova, N.D.; Jackson, J.D.; Jacobs, A.R.; Kiryutin, B.; Koonin, E.V.; Krylov, D.M.; Mazumder, R.; Mekhedov, S.L.; Nikolskaya, A.N. The COG database: An updated version includes eukaryotes. *BMC Bioinform.* **2003**, *4*, 41. [CrossRef]
30. Minoru, K.; Susumu, G.; Shuichi, K.; Yasushi, O.; Masahiro, H. The KEGG resource for deciphering the genome. *Nucleic Acids Res.* **2004**, *32*, 277–280.
31. Amos, B.; Rolf, A. The SWISS-PROT protein sequence database and its supplement TrEMBL in 2000. *Nucleic Acids Res.* **2000**, *28*, 45–48.
32. Young, M.; Wakefield, M.; Smyth, K.G.; Oshlack, A. Gene ontology analysis for RNA-seq: Accounting for selection bias. *Genome Biol.* **2010**, *11*, 14–25. [CrossRef]
33. Yi, Z.; Chen, J.; Sun, H.; Rosli, H.G.; Pombo, M.A.; Zhang, P.; Banf, M. iTAK: A program for genome-wide prediction and classification of plant transcription factors, transcriptional regulators, and protein kinases. *Mol. Plant* **2016**, *9*, 1667–1670.
34. Kong, L.; Yong, Z.; Ye, Z.Q.; Liu, X.Q.; Ge, G. CPC: Assess the protein-coding potential of transcripts using sequence features and support vector machine. *Nucleic Acids Res.* **2007**, *35*, 345–353. [CrossRef] [PubMed]
35. Bu, D.; Luo, H.; Huo, P.; Wang, Z.; Zhang, S.; He, Z.; Wu, Y.; Zhao, L.; Liu, J.; Guo, J.; et al. KOBAS-i: Intelligent prioritization and exploratory visualization of biological functions for gene enrichment analysis. *Nucleic Acids Res.* **2021**, *49*, 317–325. [CrossRef] [PubMed]
36. Bo, L.; Dewey, C.N. RSEM: Accurate transcript quantification from RNA-Seq data with or without a reference genome. *BMC Bioinform.* **2011**, *12*, 323–338.
37. Anders, S. Analysing RNA-Seq data with the DESeq package. *Mol. Biol.* **2010**, *43*, 1–17.
38. Xie, L.; Dong, C.; Shang, Q. Gene co-expression network analysis reveals pathways associated with graft healing by asymmetric profiling in tomato. *BMC Plant Biol.* **2019**, *19*, 373–384. [CrossRef]
39. Chin, C.H.; Chen, S.H.; Wu, H.H.; Ho, C.W.; Ko, M.T.; Lin, C.Y. *cytoHubba*: Identifying hub objects and sub-networks from complex interactome. *BMC Syst. Biol.* **2014**, *8*, 7. [CrossRef]
40. Kenneth, J.L.; Thomas, D.S. Analysis of relative gene expression data using real-time quantitative PCR and the 2-ΔΔCT method. *Methods* **2002**, *25*, 402–408.
41. Yu, M.; Liu, D.; Li, Y.C.; Sui, C.; Chen, G.D.; Tang, Z.K.; Yang, C.M.; Hou, D.B.; Wei, J.H. Validation of reference genes for expression analysis in three *Bupleurum* species. *Biotechnol. Biotechnol. Equip.* **2019**, *33*, 154–161. [CrossRef]
42. Stephen, G.H.; Richard, L. Terpenes, hormones and life: Isoprene rule revisited. *J. Endocrinol.* **2019**, *242*, 9–22.
43. Fang, X.; Shi, L.; Ren, A.; Jiang, A.L.; Wu, F.L.; Zhao, M.W. The cloning, characterization and functional analysis of a gene encoding an acetyl-CoA acetyltransferase involved in triterpene biosynthesis in *Ganoderma lucidum*. *Mycoscience* **2013**, *54*, 100–105. [CrossRef]
44. Lee, A.R.; Kwon, M.; Kang, M.K.; Kim, J.; Kim, S.U.; Ro, D.K. Increased sesqui- and triterpene production by co-expression of HMG-CoA reductase and biotin carboxyl carrier protein in tobacco (*Nicotiana benthamiana*). *Metab. Eng.* **2019**, *52*, 20–28. [CrossRef]
45. Estevez, J.M.; Cantero, A.; Reindl, A.; Reichler, S.; Leon, P. 1-Deoxy-D-xylulose-5-phosphate synthase, a limiting enzyme for plastidic isoprenoid biosynthesis in plants. *J. Biol. Chem.* **2001**, *276*, 22901–22909. [CrossRef]
46. Li, J.; Mutanda, I.; Wang, K.; Yang, L.; Wang, Y. Chloroplastic metabolic engineering coupled with isoprenoid pool enhancement for committed taxanes biosynthesis in *Nicotiana benthamiana*. *Nat. Commun.* **2019**, *10*, 4850–4861. [CrossRef] [PubMed]
47. Ro, D.K.; Paradise, E.M.; Ouellet, M.; Fisher, K.J.; Newman, K.L.; Ndungu, J.M.; Ho, K.A.; Eachus, R.A.; Ham, T.S.; Kirby, J. Production of the antimalarial drug precursor artemisinic acid in engineered yeast. *Nature* **2006**, *440*, 940–943. [CrossRef] [PubMed]
48. Xue, L.; He, Z.L.; Bi, X.C.; Xu, W.; Wei, T.; Wu, S.X.; Hu, S.N. Transcriptomic profiling reveals MEP pathway contributing to ginsenoside biosynthesis in *Panax ginseng*. *BMC Genom.* **2019**, *20*, 383–396. [CrossRef] [PubMed]
49. Gao, K.; Wu, S.R.; Wang, L.; Xu, Y.H.; Wei, J.H.; Sui, C. Cloning and analysis of β-amyrin synthase gene in *Bupleurum chinense*. *Genes Genom.* **2015**, *37*, 767–774. [CrossRef]
50. Zhao, C.; Xu, T.; Liang, Y.; Zhao, S.; Ren, L.; Wang, Q.; Dou, B. Functional analysis of β-amyrin synthase gene in ginsenoside biosynthesis by RNA interference. *Plant Cell Rep.* **2015**, *34*, 1307–1315. [CrossRef]
51. Confalonieri, M.; Cammareri, M.; Biazzi, E.; Pecchia, P.; Fevereiro, M.P.S.; Balestrazzi, A.; Tava, A.; Conicella, C. Enhanced triterpene saponin biosynthesis and root nodulation in transgenic barrel medic (*Medicago truncatula* Gaertn.) expressing a novel β-amyrin synthase (AsOXA1) gene. *Plant Biotechnol. J.* **2009**, *7*, 172–182. [PubMed]
52. Yu, J.K.; Zhang, D.B.; Yang, D.C. Biosynthesis and biotechnological production of ginsenosides. *Biotechnol. Adv. Int. Rev. J.* **2015**, *33*, 717–735.

53. Augustin, J.M.; Kuzina, V.; Andersen, S.B.; Bak, S. ChemInform Abstract: Molecular Activities, Biosynthesis and Evolution of Triterpenoid Saponins. *Phytochemistry* **2011**, *72*, 435–457. [CrossRef] [PubMed]
54. Shang, Y.; Huang, S.W. Multi-omics data-driven investigations of metabolic diversity of plant triterpenoids. *Plant J.* **2019**, *97*, 101–111. [CrossRef]
55. Biazzi, E.; Carelli, M.; Tava, A.; Abbruscato, P.; Losini, I.; Avato, P.; Scotti, C.; Calderini, O. CYP72A67 Catalyzes a Key Oxidative Step in Medicago truncatula Hemolytic Saponin Biosynthesis. *Mol. Plant* **2015**, *8*, 1493–1506. [CrossRef]
56. Ghosh, S. Triterpene Structural Diversification by Plant Cytochrome P450 Enzymes. *Front. Plant Sci.* **2017**, *8*, 1886–1900. [CrossRef] [PubMed]
57. Xiao, H.; Zhang, Y.; Wang, M. Discovery and Engineering of Cytochrome P450s for Terpenoid Biosynthesis. *Trends Biotechnol.* **2019**, *37*, 618–631. [CrossRef] [PubMed]
58. Lu, C.; Zhao, S.J.; Wang, X.S. Functional regulation of a UDP-glucosyltransferase gene (Pq3-O-UGT1) by RNA interference and overexpression in *Panax quinquefolius*. *Plant Cell Tissue Organ Cult.* **2017**, *129*, 445–456. [CrossRef]

genes

MDPI

Article

Interplay of Ecological Opportunities and Functional Traits Drives the Evolution and Diversification of Millettiod Legumes (Fabaceae)

Dongzhu Jiang †, Linzheng Liao †, Haitao Xing, Zhidan Chen, Xuemei Luo and Hong-Lei Li *

College of Landscape Architecture and Life Science/Institute of Special Plants, Chongqing University of Arts and Sciences, Chongqing 402168, China
* Correspondence: lihonglei215@163.com
† These authors contributed equally to this work.

Citation: Jiang, D.; Liao, L.; Xing, H.; Chen, Z.; Luo, X.; Li, H.-L. Interplay of Ecological Opportunities and Functional Traits Drives the Evolution and Diversification of Millettiod Legumes (Fabaceae). *Genes* **2022**, *13*, 2220. https://doi.org/10.3390/genes13122220

Academic Editors: Hakim Manghwar and Wajid Zaman

Received: 27 October 2022
Accepted: 22 November 2022
Published: 27 November 2022

Abstract: Understanding the striking diversity of the angiosperms is a paramount issue in biology and of interest to biologists. The Millettiod legumes is one of the most hyper-diverse groups of the legume family, containing many economically important medicine, furniture and craft species. In the present study, we explore how the interplay of past climate change, ecological opportunities and functional traits' evolution may have triggered diversification of the Millettiod legumes. Using a comprehensive species-level phylogeny from three plastid markers, we estimate divergence times, infer habit shifts, test the phylogenetic and temporal diversification heterogeneity, and reconstruct ancestral biogeographical ranges. We found that three dramatic accumulations of the Millettiod legumes occurred during the Miocene. The rapid diversification of the Millettiod legumes in the Miocene was driven by ecological opportunities created by the emergence of new niches and range expansion. Additionally, habit shifts and the switch between biomes might have facilitated the rapid diversification of the Millettiod legumes. The Millettiod legumes provide an excellent case for supporting the idea that the interplay of functional traits, biomes, and climatic and geographic factors drives evolutionary success.

Keywords: diversity; the Millettiod legumes; phylogeny; ecological opportunities; Miocene

1. Introduction

Angiosperms are the most prominent life form across global ecosystems, and exhibit a spectacular range of morphological and ecological diversity on the Earth [1]. However, the species diversity across the Tree of Life of the angiosperms is uneven. Understanding the striking diversity of the angiosperms is a paramount issue in biology and of interest to biologists [1–3]. Previous studies found that the rapid radiations of many angiosperm groups was driven by both biological (intrinsic) and environmental (extrinsic) factors [4–7]; however, the effects of functional traits, the biome, and climatic and geographic factors on plant diversification are largely unknown.

The Millettiod legumes is one of the most hyper-diverse groups in the legume family. This clade is composed of five tribes (Millettieae, Abreae, Phaseoleae, Desmodieae and Psoraleeae), including more than 2600 described species with a global distribution [8,9]. The Millettiod legumes contain many economically important species, which contribute enormously to the word's economy through food and vegetables, pharmaceuticals and medicines, furniture and crafts, chemicals and fertilizers, as well as horticulture [9]. The main dietary legumes include pigeon pea (*Cajanus cajan* Millsp.), soybean (*Glycine max* Merr.), cowpea (*Vigna unguiculata* Walp.), and bean (*Phaseolus*). Vetches (*Vicia amoena* Fisch.) and tropical kudzu (*Pueraria phaseoloides* Benth.) were important pastures used for forage. Species such as *Desmodium*, *Macoptilium* and *Centrosema* are used for the purpose of improving tropical pasture systems [10]. Showy-flowered species of *Mucuna* and *Strongylodon* are very popular in tropical gardens. Species of *Erythrina* are important woody tree legumes in forestry. Multipurpose shrubs, such as *Butea* and *Millettia*, are regional favorites as ornaments and for shade [9].

The Millettiod legumes show great diversity in their morphological and habit characteristics from ephemeral herbs and herbaceous climbers to shrubs, woody lianas and trees. The terminal taxa of the Millettiod legumes grow in various habitats such as succulent, grass, rainforest and temperate, which are in accordance with major zonobiomes [9,11]. The crown age of the Millettiod legumes is 44.1 Ma to 53 Ma. Global climate and tectomic changes since the Eocene may have triggered the diversification of many angiosperm lineages [12–16]. Moreover, the diversification of *Psoraleeae* is assumed to be a result of global climate change during the Pleistocene [17]. The evolution of phaseoloid legumes may be due to habit shifts and dispersal events along with the global climate changes since the Oligocene [18]. However, it remains puzzling and unexplored how the evolution of habits fostered diversification in the Millettiod legumes and how ecological forces have been regulating cladogenesis in different geographical areas.

A well-resolved phylogeny of Millettiod legumes is paramount for understanding the aspects of the temporal and biogeographical patterns of diversification. Some previous molecular phylogenetic studies have made progress in the delimitation of the Millettiod legumes based on nuclear ribosomal *ITS* or a few plastid loci [18–22]. These researches generated a well-resolved phylogenetic framework of the Millettiod legumes at tribal levels [3,23]. However, less than 40% of the generic diversity of the Millettiod legumes was sampled in previous studies. Therefore, a well-resolved phylogeny with dense taxon sampling at species level is needed to understand the evolutionary dynamics of the Millettiod legumes. Oyebanji et al. proved that plastome is an effective marker for resolving the evolutionary relationships of the Millettiod legumes [23]. In the past decades, plastome data of the Millettiod legumes increased rapidly along with previous studies [19–22,24–27]. The data accumulation of the Millettiod legumes provides an opportunity to reconstruct the phylogeny of this clade with comprehensive taxon sampling.

In the present study, we first reconstruct a species-level phylogeny for the phaseoloid legumes using three plastid loci with a more extensive generic sampling than in any previous studies. Based on the improved phylogenetic framework, we then explore how the interplay of past climate change, geographical expansion and habit shifts may have triggered diversification of the Millettiod legumes.

2. Material and Methods

2.1. Taxon and Nuclear Marker Sampling

Three plastid regions (*rbcL*, *matK* and *trnL-F*) were selected as DNA markers for the phylogenetic analyses of the Millettiod legumes. The combination of these three plastid regions has been proved to have high-efficiency in resolving intergeneric and interspecific relationships. To obtain dense sampling of the Millettiod legumes, all available nucleotide sequences of these three markers from GenBank were collected firstly. Then, sequences were filtered following the principles: (1) For each taxon, the same species and DNA sample across three plastid markers were kept in priority, but some composite accessions were necessary to represent genera. (2) When multiple sequences were available for each marker, the longest sequence was retained. (3) In the scenario of multiple sequences with the same length, one sequence was randomly selected. (4) In a genus, when one marker occurred in species different with other markers, the species with faster evolved marker were kept following the order: *trnL-F*, *rbcL*, *matK*. All the works above were completed by using a series of R scripts written by our group. Most of the DNA sequences have been used in previously published studies [19–22,24–27]. Species names and GenBank accession numbers are listed in Table S1.

2.2. Alignment and Phylogeny

For the two DNA coding genes (*rbcL* and *matK*), sequences were aligned directly in the program MUSCLE with the default settings [28]. Then the aligned matrix was manually adjusted using BioEdit version 5.0.9 (Hall, T.A., California, USA) [29]. For the plastid region *trnL-F*, sequences were aligned using a two-step strategy. First, sequences were divided into clusters according to taxonomic unit and aligned in the program MUSCLE under default high accuracy parameters with subsequently manual adjustment. Then, the matrix of each cluster was aligned using the profile–profile alignment algorithm in MUSCLE and final adjustments were made to the alignments using the refinement algorithm in MUSCLE. Alignment of each plastid region was manually trimmed for quality and maximum coverage. The final aligned matrix contains 429 OTUs.

To check the conflicts among three plastid regions, the initial phylogenetic analysis for each marker was carried out using maximum likelihood (ML) method implemented in the program RAxML version 7.6.6 (Stamatakis, A., Lausanne, Switzerland) [30]. After checking for significant conflict among different nodes (bootstrap support value exceeding 70%), all data from different markers were combined for further analyses. We used the ML method to construct the phylogeny of the Millettiod legumes based on the combined data. The ML analysis was performed using RAxML with the following options: three data partitions (*rbcL*, *matK* and *trnL-F*), GTR + I + G nucleotide substitution model, and gaps treated as missing data. A standard bootstrap analysis with 1000 bootstrap iterations was used to estimate the branch support. The program was run on the CIPRES network [31].

2.3. Molecular Dating

To date the branching events within the Millettiod legumes, a Bayesian MCMC approach as implemented in BEAST version 1.8.0 (Drummond, A.J., Auckland, Zealand) was used based on the combined dataset of 429 terminals [32]. We selected 7 calibration points from the result of Lavin et al. [33]. These calibration points could be confidently assigned to clades and nodes represented in our dataset (Table S2). A normal distribution was used for all 7 calibration points. The standard deviation was set to contain the lower and higher boundaries of the 95% highest posterior density values. MCMC searches were run for 500,000,000 generations, sampled every 5000 generations. Tracer v1.5 was used to monitor appropriate burn-in and the adequate effective sample sizes of the posterior distribution (>200). The maximum clade credibility tree was computed by TreeAnnotator v1.7.5 in BEAST software package [34]. BEAST analyses were performed in the CIPRES Web Portal 3.1 (Miller, M.A., California, USA) [31].

2.4. Trait Evolution Analyses

To clarify the trait evolutionary history of the Millettiod legumes, the habit and biome information were collected (Table S2). The habit information of the Millettiod legumes was collected from International Legume Database & Information Service (ILDIS). Two habit states were coded, herbaceous (including herbs and herbaceous climbing vines) vs. woody (including trees, woody climbers and shrubs) (Table S1). According to the geographic distributions and the biome patterns of Lewis et al. [9], five biome types were coded: (1) succulent (S), including semi-arid, non-fire-adapted, succulent-rich and grass-poor, dry tropical forest, thicket and bushland biome; (2) grass (G), comprising a fire-adapted, succulent-poor and grass-rich, seasonally dry tropical forest, woodland and savanna biome; (3) rainforest (R), comprising the equatorial tropics (wet forests) worldwide; (4) temperate north (TN), comprising the Mediterranean, warm and cold temperate regions of the North Hemisphere; and (5) temperate south (TS), comprising the area except for the first three biomes (Table S1). The parsimony method implemented in Mesquite v2.74 was used to reconstruct the trait evolution of the Millettiod legumes [35]. The maximum clade credibility tree obtained from the Bayesian analysis was used as input tree.

2.5. Diversification Analyses

To visualize the temporal dynamics of the diversification rates in the Millettiod legumes, we constructed the semilogarithmic lineage-through-time (LTT) plots using the R package APE v2.5-1 [36]. In total, 1000 dated phylogenies obtained from the Bayesian analyses were randomly selected and used to generate semilogarithmic LTT plots. To verify the rapid shifts in diversification rates at any specified time, the RC statistic was performed using the R package GEIGER v1.3-1 [37]. We used 11 models including pureBirth, bd, DDX, DDL, SPVAR, EXVAR, BOTHVAR, yule2rate, yule3rate, yule4rate and yule5rate to assess the significant rate changes. A diversification rate shift was confirmed by the criterion that more or fewer descendants than expected under the constant rate model happen in lineages.

2.6. Biogeographical Analyses

In order to infer the possible ancestral ranges of the Millettiod legumes, a Bayes-DIVA analysis [38] was conducted using the software package RASP [39]. Bayes-DIVA method is a popular method with the advantages of minimizing the phylogenetic uncertainties and generating credibility support values for alternative phylogenetic relationships [38,39]. We used 1000 trees from the BEAST output as a "trees file". The maximum clade credibility tree was used as a final representative tree. With the aim of predicating the ancestral areas of nodes deeper down into the tree, biogeographical analyses were run on continental spatial scale at species level. According to Lewies et al. [9] and the ILDIS database, nine geographic regions were coded: A, Africa; B, Asia; C, Australia; D, Indian Ocean area, including islands in the Indian Ocean region and Madagascar; E, Pacific, including islands in the Pacific; F, North America; H, Central America; I, Europe; K, South America (Table S1). The "maxareas" was constrained to 3 when constructing the ancestral areas of the Millettiod legumes.

3. Results

3.1. Phylogeny of the Millettiod Legumes

The aligned matrix of the combined three-marker dataset consists of 4728 characters with 2290 variable and 1690 parsimony-informative sites. The tree generated by the maximum likelihood (ML) analysis (Figure S1) was highly consistent with those retrieved from the Bayesian inference (BI) analysis (Figure S2), except for some weakly supported nodes (BS < 70%). The Millettiod legumes are strongly supported as monophyletic (BS 100%, PP 1.0). Within the Millettiod legumes, five major clades were identified: Basal millettioid (BS 94%, PP 1.0), Clitoriinae (BS 59%, PP 0.87), Disynstemon + (Abreae + (Diocleinae + core Millettieae)) (BS 81%, PP 1.0), *Platycyamus* + *Dewevrea* (BS 91%, PP 1.0), and the phaseoloid clade (BS 98%, PP 1.0). Within the phaseoloid clade, Desmodieae, Cajaninae and Phaseolinae are monophyletic with high support. Psoraleeae is monophyletic and nested in Glycininae.

3.2. Divergence-Time Estimates of the Millettiod Legumes

Based on our analyses, the Millettiod legumes diversified at 45.1 Ma (42.3–47.3 Ma, HPD) (Figures 1 and S3). The second lineage separated from the remaining Millettiod legumes at 44.5 Ma (40.1–45.6 Ma, HPD). The tribe Abreae split with its sister lineage at 34.3 Ma (32.2–38.2 Ma, HPD). The divergence time between core Millettieae and Diocleinae was estimated to be 27.3 Ma (26.4–35.4 Ma, HPD). The steam age of the phaseoloid clade was 37.3 Ma (36.1–41.4 Ma, HPD), and the crow age was 31.1 Ma (28.8–33.5 Ma, HPD). Desmodieae diverged from its close relatives at 24.8 Ma (23.7–30.0 Ma, HPD). Psoraleeae separated from *Glycine* at 11.8 Ma (8.8–14.4 Ma, HPD).

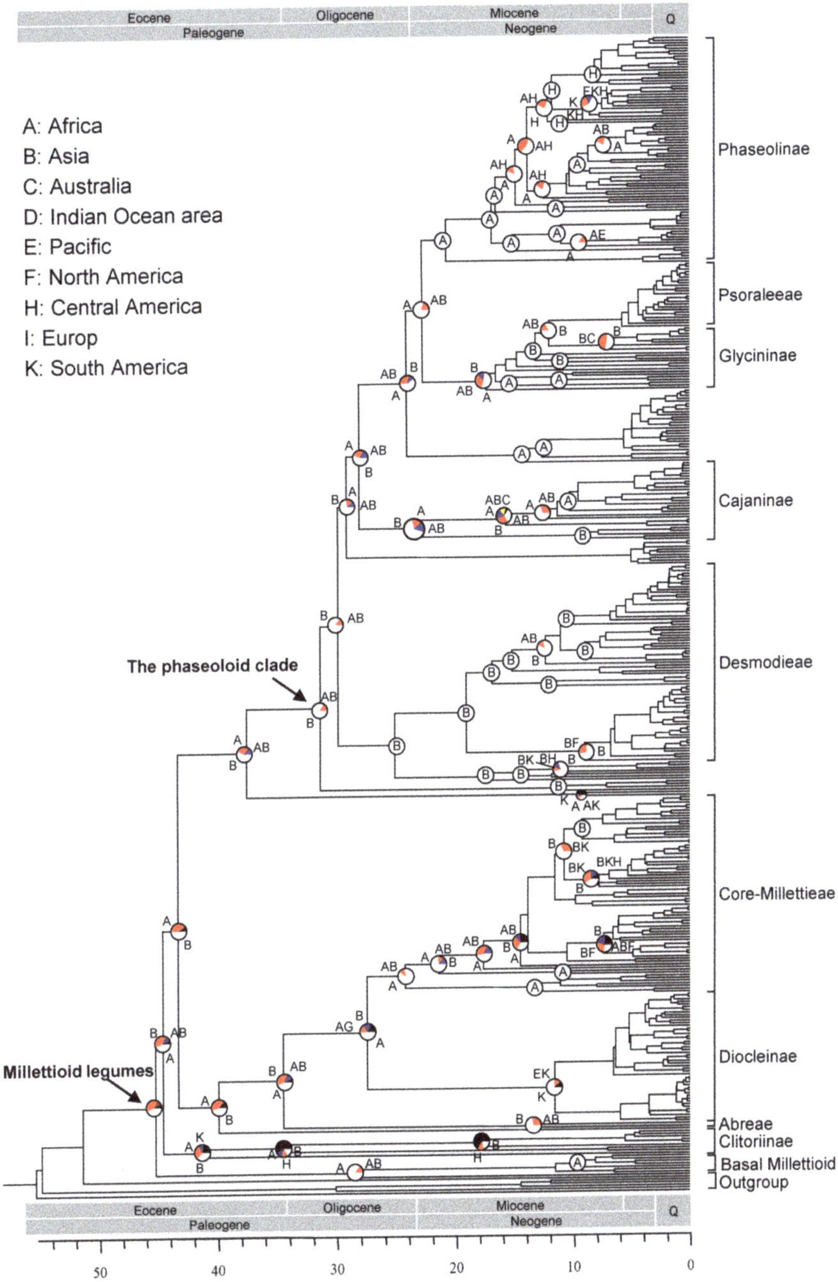

Figure 1. Combined chronogram and the summary of biogeographic analysis of the Millettiod legumes. Dating analysis was performed using BEAST software. The topology corresponds to the majority rule consensus tree of the Bayesian stationary sample. Large pie charts show the relative probabilities of alternative ancestral distributions obtained by statistical dispersal–vicariance analysis (white > red > blue > yellow); areas (frequencies < 0.1) are collectively given with black color.

3.3. *Trait Evolution*

Results of the ancestral habit state reconstruction of the Millettiod legumes are shown in Figure 2. The ancestral state of the Millettiod legumes was woody. Multiple transitions between woody and herbaceous occurred during the evolutionary history of the Millettiod legumes. Our results showed that switching frequency from woody to herbaceous was higher than that from herbaceous. The switching frequency from woody to herbaceous was 52, whereas the switching frequency from herbaceous to woody was 25. In our examining transitions analyses among biomes, the common ancestor of the Millettiod legumes was most likely distributed in a grass biome (Figure 3). The ancestor of core Millettieae similarly may have occupied a grass biome with subsequent shifts into the other biome types. The ancestor of the phaseoloid clade was also distributed in a grass biome. Multiple shifts between biome types in different clades were detected since the Miocene.

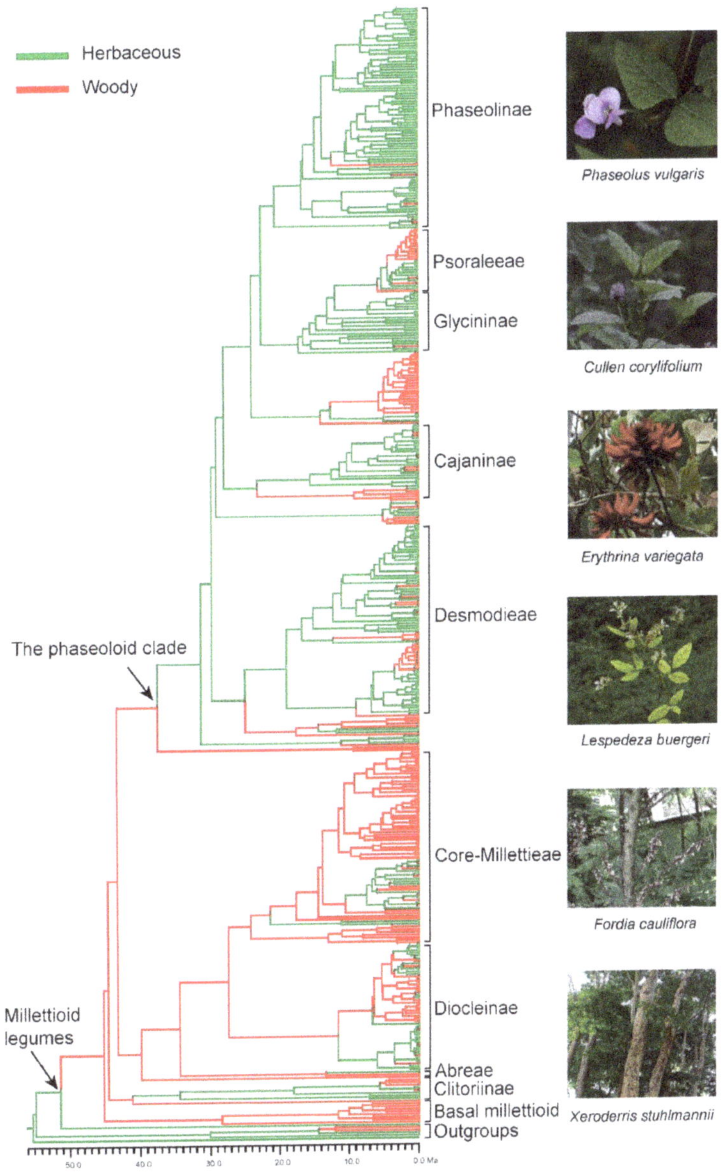

Figure 2. Habit shift analysis of the Millettiod legumes. Green represents herbaceous; red represents woody. Images are representative species from the Millettiod legumes.

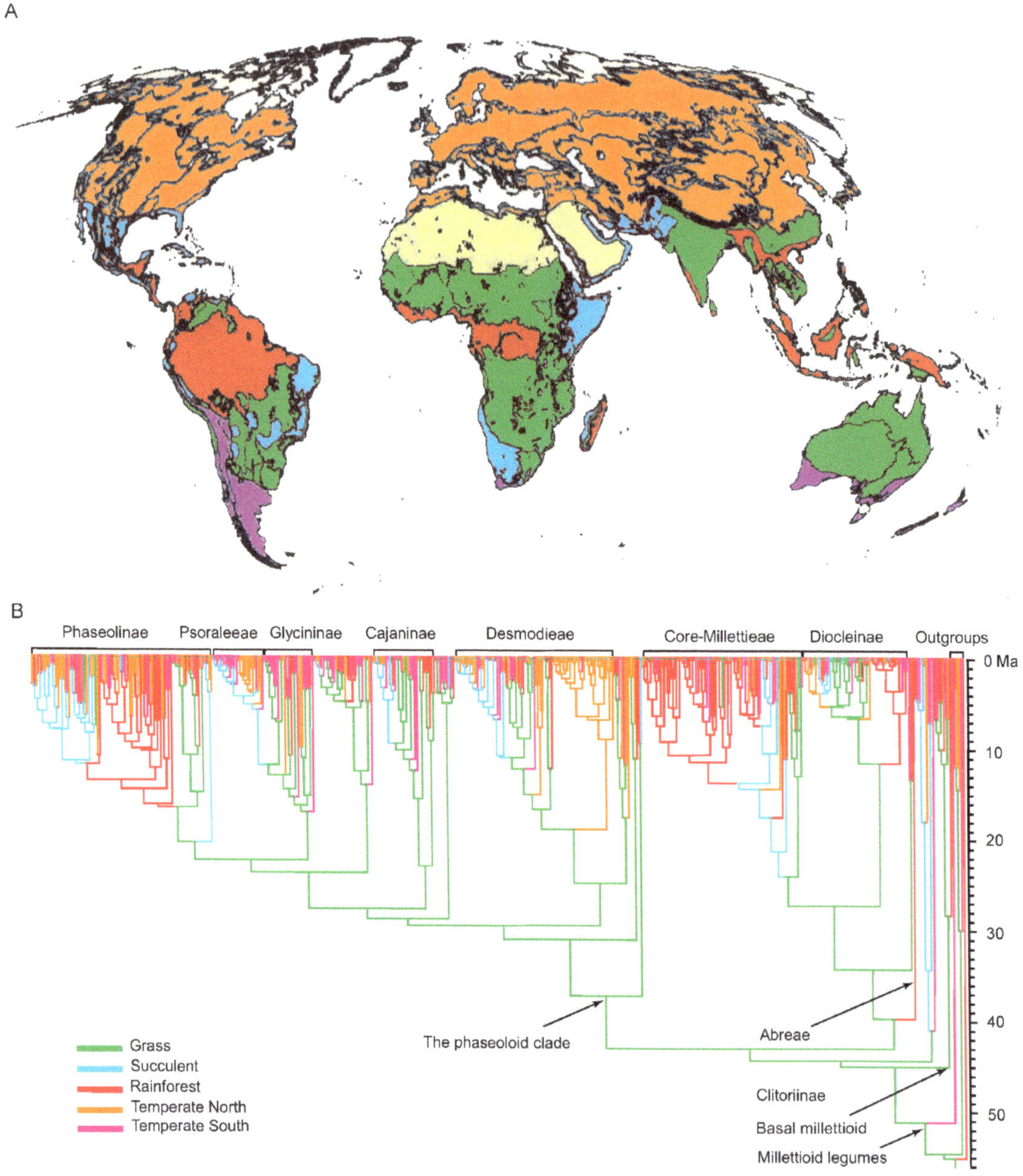

Figure 3. Biome evolutionary history of the Millettiod legumes. (**A**) World map of biome (edited from Lewis et al., 2005 [9]). (**B**) Ancestral biome state reconstruction of the Millettiod legumes using parsimony method.

3.4. Reconstruction of Historical Biogeography

The results of biogeographic reconstruction are summarized in Figures 1 and S4. The most recent common ancestor (MRCA) of the Millettiod legumes was equivocal, either in Africa (p = 0.44), Asia (p = 0.39) or the other regions (p = 0.17). Multiple colonization events occurred independently during the evolutionary history of the Millettiod legumes. The MRCA of core Millettieae was Africa (p = 0.88), and dispersed from Africa to Asia, North America, Central America and South America multiple times. The MRCA of the phaseoloid legumes occurred in Asia (p = 0.84), with a number of movements into Africa, Australia and Central America. Among all of the nine geographic regions, Africa and Asia harbored higher frequency of lineage emigration (185 and 204 times, respectively) than immigration (73 and 90 times, respectively). Conversely, the frequency of lineage emigration in Australia, the Indian Ocean area and Pacific harbor was lower with the numbers being 17, 10 and 17.5, respectively, whereas the frequencies of lineage immigration were 81, 87 and 69, respectively.

3.5. Diversification Rates

Our results of the semilogarithmic lineage-through-time (LTT) showed that the Millettiod legumes experienced a high diversification rate in the Miocene (Figure 4). When testing the significant rate changes based on 11 models, the AIC values of seven constant rate models, including pureBirth, bd, DDX, DDL, SPVAR, EXVAR and BOTHVAR, were found to be higher than the other four and were excluded (Table 1). Among the variable rate models, the AIC value of yule5rate was the lowest and was selected as the best fit model for the diversification rate analyses of the Millettiod legumes. Three rapid increases in diversification rate were found in the Millettiod legumes: the first was at 17.78 Ma with a rate change from 0.07 to 0.13 (sp Myr^{-1}); the second was at 11.57 Ma, with a rate change from 0.13 to 0.29 (sp Myr^{-1}); the third was at 8.46 Ma, with a rate change from 0.10 to 0.19 (sp Myr^{-1}).

Table 1. Results of diversification rate model-fitting test with parameters estimated under each model. "R" indicates the speciation rate, "St" indicates where the speciation rate shifts.

Model	AIC	Likelihood	R1	R2	R3	R4	R5	St1	St2	St3	St4
pureBirth	−1910.211	956.1055	0.16607	-	-	-	-	-	-	-	-
bd	−1928.547	966.2735	0.10794	-	-	-	-	-	-	-	-
DDX	−1930.799	967.3994	0.05211	-	-	-	-	-	-	-	-
DDL	−1908.208	956.1041	0.16608	-	-	-	-	-	-	-	-
SPVAR	−1926.028	966.0142	0.24839	-	-	-	-	-	-	-	-
EXVAR	−1926.544	966.2718	0.23662	-	-	-	-	-	-	-	-
BOTHVAR	−1924.329	966.1644	0.24951	-	-	-	-	-	-	-	-
yule2rate	−1929.337	967.6686	0.16650	0.10857	-	-	-	0.04	-	-	-
yule3rate	−1932.285	971.1423	0.07166	0.14114	0.19292	-	-	17.78	7.12	-	-
yule4rate	−1931.881	972.9404	0.09826	0.28943	0.09592	0.19	-	11.57	10.68	8.46	-
yule5rate	−1932.951	975.4753	0.07166	0.13342	0.28943	0.09592	0.19	17.78	11.57	10.68	8.46

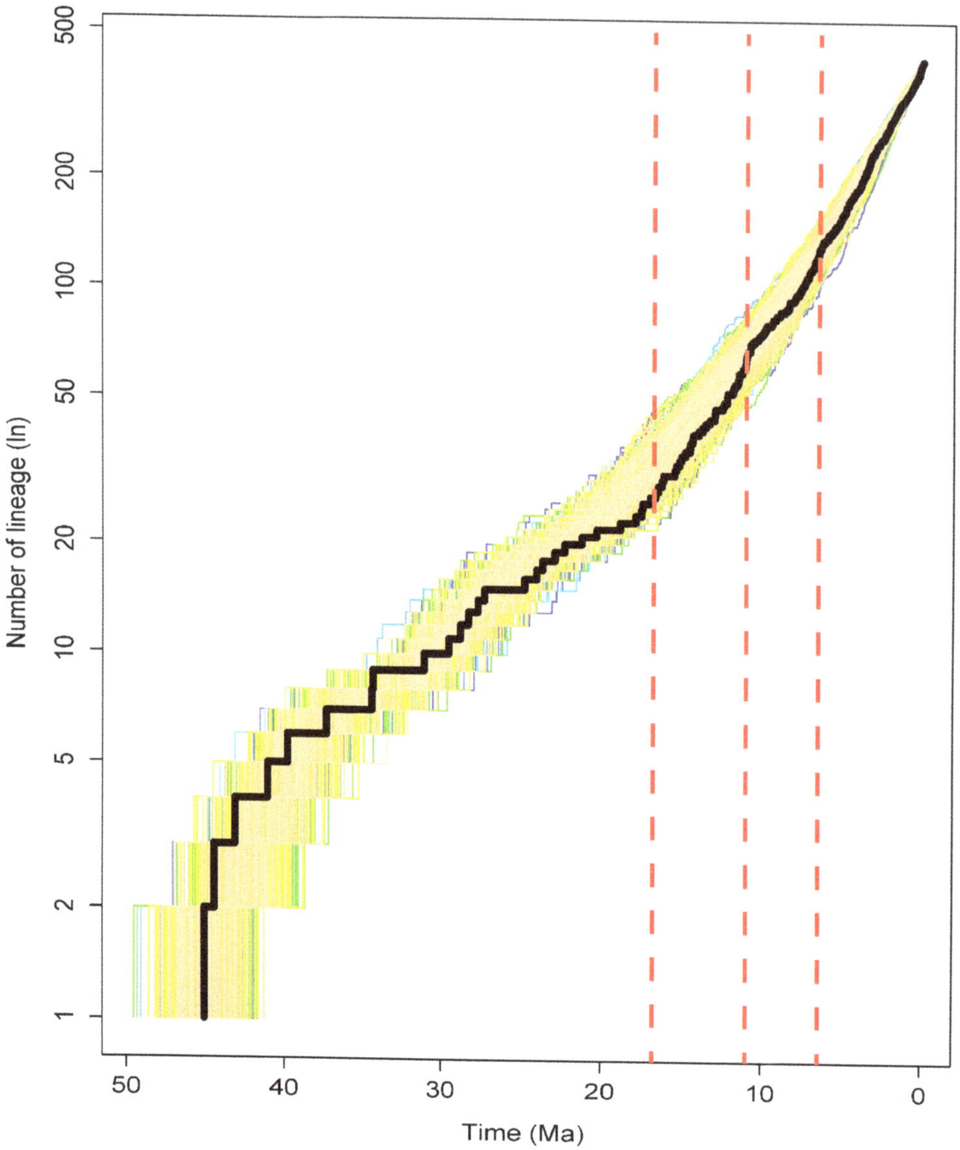

Figure 4. Lineage-through-time (LTT) plots of the Millettiod legumes. Dotted lines in red indicate the rapid increases in diversification rate.

4. Discussion

Reliable molecular dating is an essential precondition for revealing the diversification patterns of plant lineages. Phylogenetic analyses of the combined three-region DNA dataset generated a well-supported evolutionary framework of the Millettiod legumes (Figures 1, S1 and S2). The relationships among the main lineages of the Millettiod legumes were in accordance with previous studies [3,18,21–23]. Based on our time estimates, the crown age of the Millettiod legumes was 45.1 Ma (42.3–47.3 Ma, HPD; Figures 1 and S3), which corresponded to previous studies [1,33]. The crown age of the phaseoloid legumes was 31.1 Ma (28.8–33.5 Ma, HPD), which was consistent with the estimations of many previous studies [18,26,33]. Our result showed the crown age of Psoraleeae was 5.6 Ma (4.29–8.4 Ma, HPD), which was in accordance with the result of Egan and Crandall [17]. All these results indicated that the divergence times of the Millettiod legumes in the present study were credible.

Our biogeographic reconstruction and dating analyses indicated the Millettiod legumes occurred in Africa ($p = 0.44$) or Asia ($p = 0.39$) in the early Eocene. The LTT plot indicated that a dramatic burst of diversification of the Millettiod legumes occurred during the Miocene (c. 18–8 Ma; Figure 4). Three dramatic accumulations of the Millettiod legumes were detected at 17.8 Ma, 11.6 Ma and 8.5 Ma (Table 1). In accordance with the diversification patterns, high frequent migrations of the Millettiod legumes did not occur until the Miocene after its origin with the exception of some dispersal events into Central America and South America (clade Clitoriinae). Multiple colonization events seem to have occurred within a 10-million-year time window (c. 18–8 Ma, Figure 1). The ancestor of the Diocleinae experienced dispersal events into Central America, South America and the Pacific. There were at least three dispersal events into Central America and South America for the core Millettieae, and two such movements into the Indian Ocean area (Figures 1 and S4). Additionally, the Phaseoloids clade experienced multiple dispersal events from Africa to Central America. In the Miocene, the Qinghai–Tibetan Plateau experienced multiple stages of uplift with a rapid uplift at 10–7 Ma [40–44]. The African plate collided with the Eurasian at ca. 18–17 Ma [45]. In the southern hemisphere, the thrust of the South American plate against the Pacific plate led to the rise of the Andes mountains [46]. These geological events fragmented the ecosystems and subsequently opened many new niches [47,48], and forced the rapid diversification of many plant lineages [16,49–53]. Thus, our finding suggests that the rapid diversification of the Millettiod legumes in the Miocene was driven by ecological opportunities created by the emergence of new niches and range expansion.

In the Miocene, the global climate tended to be cooler, drier and more seasonal [54–56]. Both Asia and America experienced an arid period of marked aridity due to orogenetic changes [57–60]. The African plate also experienced an analogous arid period of aridity due to the closure of the Tethys Sea [45,61]. Our trait evolution analyses showed that a larger number of lineages in the 10-million-year time window (c. 18–8 Ma) were herbaceous and inhabited a grass biome. The grass biome generally has a unimodal rainfall pattern, which comprises a fire-adapted, succulent-poor and grass-rich, seasonally dry tropical forest, woodland and savanna [9,11]. The extensive aridity could have promoted the diversification of some groups inhabiting dry regions, such as *Bursera* [62] and the *Potentilla* [63]. Herbs could have produced higher per-year mutation rates on the basis of a shorter generation time, consequentially contributing to the adaptation of herbaceous lineages to dry climates [64–66]. Additionally, multiple shifts from the grass biome to other biomes, especially the rainforest biome, were found in the Miocene. The interdigitation of the grass and the rainforest biome favors a "refuge" interpretation of allopatric divergence, which was caused by the contraction and expansion of dry-adapted and wet vegetation [67]. Taxa adapted to different biomes are always pioneer species, which can take advantage most effectively of post-disturbance conditions [9]. Thus, the expansion of arid lands due to orogenetic and climatic changes increased the arid niche space, and the switch between biomes might have facilitated the rapid diversification of the Millettiod legumes during the Miocene.

Supplementary Materials: The following supporting information can be downloaded at: https://www.mdpi.com/article/10.3390/genes13122220/s1, Figure S1. Phylogenetic relationships of the phaseoloid legumes using maximum-likelihood (ML) method; Figure S2. Phylogenetic relationships of the phaseoloid legumes using Bayesian inference (BI) method; Figure S3. Chronogram for the Millettiod legumes obtained under a Bayesian relaxed clock; Figure S4. Biogeographic analysis of the Millettiod legumes by Statistical Dispersal-Vicariance Analysis (S-DIVA).; Table S1. Taxa, GenBank accession numbers, Life form code and Biome code for samples included in this study; Table S2. Calibrations used in this study. Node number followed Lavin et al., 2005.

Author Contributions: H.-L.L. conceived the study. D.J., L.L., H.X., Z.C. and X.L. performed the experiments. H.-L.L., D.J. and L.L. analyzed the data. H.-L.L., D.J., L.L. and Z.C. contributed reagents/materials/analysis tools. H.-L.L., D.J. and L.L. wrote the paper. H.-L.L. and D.J. edited the paper. All authors have read and agreed to the published version of the manuscript.

Funding: This research was supported by National Natural Science Foundation of China (No. 31500175 and 32270237), and the Foundation for High-level Talents of Chongqing University of Arts and Science (2017RTZ21, P2018TZ05).

Informed Consent Statement: Not applicable.

Acknowledgments: We would like to thank Miaoqin Xia, Xavier-ravi Baskaran and Zhiduan Chen for helping to read and revise the manuscript; Min Gong modified the Figures.

Conflicts of Interest: The authors declare no conflict of interest.

References

1. Lu, L.M.; Mao, L.F.; Yang, T.; Ye, J.F.; Liu, B.; Li, H.L.; Sun, M.; Miller, J.T.; Mathews, S.; Hu, H.H. Evolutionary history of the angiosperm flora of China. *Nature* **2018**, *554*, 234–238. [CrossRef]
2. Huang, W.; Zhang, L.; Columbus, J.T.; Hu, Y.; Zhao, Y.; Tang, L.; Guo, Z.; Chen, W.; McKain, M.; Bartlett, M. A well-supported nuclear phylogeny of Poaceae and implications for the evolution of C4 photosynthesis. *Mol. Plant* **2022**, *15*, 755–777. [CrossRef]
3. Zhao, Y.; Zhang, R.; Jiang, K.W.; Qi, J.; Hu, Y.; Guo, J.; Zhu, R.; Zhang, T.; Egan, A.N.; Yi, T.S. Nuclear phylotranscriptomics and phylogenomics support numerous polyploidization events and hypotheses for the evolution of rhizobial nitrogen-fixing symbiosis in Fabaceae. *Mol. Plant* **2021**, *14*, 748–773. [CrossRef] [PubMed]
4. Givnish, T.J.; Spalink, D.; Ames, M.; Lyon, S.P.; Hunter, S.J.; Zuluaga, A.; Iles, W.J.; Clements, M.A.; Arroyo, M.T.; Leebens-Mack, J. Orchid phylogenomics and multiple drivers of their extraordinary diversification. *Proc. R. Soc. B* **2015**, *282*, 20151553. [CrossRef]
5. Bouchena-Khelladi, Y.; Onstein, R.E.; Xing, Y.; Schwery, O.; Linder, H.P. On the complexity of triggering evolutionary radiations. *N. Phytol.* **2015**, *207*, 313–326. [CrossRef] [PubMed]
6. Hughes, C.E.; Atchison, G.W. The ubiquity of alpine plant radiations: From the Andes to the Hengduan Mountains. *N. Phytol.* **2015**, *207*, 275–282. [CrossRef] [PubMed]
7. Simões, M.; Breitkreuz, L.; Alvarado, M.; Baca, S.; Cooper, J.; Heins, L.; Herzog, K.; Lieberman, B. The evolving theory of evolutionary radiations. *Trends Ecol. Evol.* **2016**, *31*, 27–34. [CrossRef]
8. Schrire, B.D.; Lavin, M.; Barker, N.P.; Forest, F. Phylogeny of the tribe Indigofereae (Leguminosae Papilionoideae): Geographically structured more in succulent-rich and temperate settings than in grass-rich environments. *Am. J. Bot.* **2009**, *96*, 816–852. [CrossRef]
9. Lewis, G.; Schrire, B.; MacKinder, B.; Lock, M. *Legumes of the World*; Royal Botanic Gardens: Kew, UK, 2005.
10. Graham, P.H.; Vance, C.P. Legumes: Importance and constraints to greater use. *Plant Physiol.* **2003**, *131*, 872–877. [CrossRef]
11. Breckle, S.W. *Walter's Vegetation of the Earth: The Ecological Systems of the Geo Biosphere*; Springer: Berlin/Heidelberg, Germany, 2002.
12. Wade, B.; Pälike, H. Oligocene climate dynamics. *Paleoceanography* **2004**, *19*, PA4019. [CrossRef]
13. Palike, H.; Norris, R.D.; Herrle, J.O.; Wilson, P.A.; Coxall, H.K.; Lear, C.H.; Shackleton, N.J.; Tripati, A.K.; Wade, B.S. The heartbeat of the Oligocene climate system. *Science* **2006**, *314*, 1894–1898. [CrossRef]
14. Lu, H.; Guo, Z. Evolution of the monsoon and dry climate in East Asia during late Cenozoic: A review. *Sci. China Earth Sci.* **2014**, *57*, 70–79. [CrossRef]
15. Tada, R.; Zheng, H.; Clift, P.D. Evolution and variability of the Asian monsoon and its potential linkage with uplift of the Himalaya and Tibetan Plateau. *Prog. Earth Planet. Sci.* **2016**, *3*, 4. [CrossRef]
16. Zhao, J.L.; Gugger, P.F.; Xia, Y.M.; Li, Q.J. Ecological divergence of two closely related *Roscoea* species associated with late Quaternary climate change. *J. Biogeogr.* **2016**, *43*, 1990–2001. [CrossRef]
17. Egan, A.N.; Crandall, K.A. Divergence and diversification in North American Psoraleeae (Fabaceae) due to climate change. *BMC Biol.* **2008**, *6*, 55. [CrossRef]
18. Li, H.; Wang, W.; Lin, L.; Zhu, X.; Li, J.; Zhu, X.; Chen, Z. Diversification of the phaseoloid legumes: Effects of climate change, range expansion and habit shift. *Front. Plant Sci.* **2013**, *4*, 386. [CrossRef]
19. Kajita, T.; Ohashi, H.; Tateishi, Y.; Bailey, C.D.; Doyle, J.J. *RbcL* and legume phylogeny, with particular reference to Phaseoleae, Millettieae, and allies. *Syst. Bot.* **2001**, *26*, 515–536.
20. Cardoso, D.; de Queiroz, L.P.; Pennington, R.T.; de Lima, H.C.; Fonty, É.; Wojciechowski, M.F.; Lavin, M. Revisiting the phylogeny of papilionoid legumes: New insights from comprehensively sampled early branching lineages. *Am. J. Bot.* **2012**, *99*, 1991–2013. [CrossRef] [PubMed]
21. Azani, N.; Babineau, M.; Bailey, C.D.; Banks, H.; Barbosa, A.R.; Pinto, R.B.; Boatwright, J.S.; Borges, L.M.; Brown, G.K.; Bruneau, A.; et al. A new subfamily classification of the Leguminosae based on a taxonomically comprehensive phylogeny. *Taxon* **2017**, *66*, 44–77. [CrossRef]
22. Wojciechowski, M.F.; Lavin, M.; Sanderson, M.J. A phylogeny of legumes (Leguminosae) based on analysis of the plastid *matK* gene resolves many well-supported subclades within the family. *Am. J. Bot.* **2004**, *91*, 1846–1862. [CrossRef]

23. Oyebanji, O.; Zhang, R.; Chen, S.Y.; Yi, T.S. New insights into the plastome evolution of the Millettioid/Phaseoloid clade (Papilionoideae, Leguminosae). *Front. Plant Sci.* **2020**, *11*, 151. [CrossRef] [PubMed]
24. Vatanparast, M.; Powell, A.; Doyle, J.J.; Egan, A.N. Targeting legume loci: A comparison of three methods for target enrichment bait design in Leguminosae phylogenomics. *Appl. Plant Sci.* **2018**, *6*, e1036. [CrossRef] [PubMed]
25. Egan, A.N.; Vatanparast, M.; Cagle, W. Parsing polyphyletic *Pueraria*: Delimiting distinct evolutionary lineages through phylogeny. *Mol. Phylogenet. Evol.* **2016**, *104*, 44–59. [CrossRef] [PubMed]
26. Stefanović, S.; Pfeil, B.E.; Palmer, J.D.; Doyle, J.J. Relationships among phaseoloid legumes based on sequences from eight chloroplast regions. *Syst. Bot.* **2009**, *34*, 115–128. [CrossRef]
27. De Queiroz, L.P.; Pastore, J.F.B.; Cardoso, D.; Snak, C.; Lima, A.L.d.C.; Gagnon, E.; Vatanparast, M.; Holland, A.E.; Egan, A.N.A. Multilocus phylogenetic analysis reveals the monophyly of a recircumscribed papilionoid legume tribe Diocleae with well-supported generic relationships. *Mol. Phylogenet. Evol.* **2015**, *90*, 1–19. [CrossRef]
28. Edgar, R.C. MUSCLE: Multiple sequence alignment with high accuracy and high throughput. *Nucleic Acids Res.* **2004**, *32*, 1792–1797. [CrossRef]
29. Hall, T.A. In BioEdit: A user-friendly biological sequence alignment editor and analysis program for Windows 95/98/NT. *Nucleic Acids Symp. Ser.* **1999**, *41*, 95–98.
30. Stamatakis, A. RAxML-VI-HPC: Maximum likelihood-based phylogenetic analyses with thousands of taxa and mixed models. *Bioinformatics* **2006**, *22*, 2688–2690. [CrossRef]
31. Miller, M.A.; Pfeiffer, W.; Schwartz, T. Creating the CIPRES Science Gateway for inference of large phylogenetic trees. In Proceedings of the 2010 Gateway Computing Environments Workshop (GCE), New Orleans, LA, USA, 14 November 2010; pp. 1–8.
32. Drummond, A.J.; Rambaut, A. BEAST: Bayesian evolutionary analysis by sampling trees. *BMC Evol. Biol.* **2007**, *7*, 214. [CrossRef]
33. Lavin, M.; Herendeen, P.S.; Wojciechowski, M.F. Evolutionary rates analysis of Leguminosae implicates a rapid diversification of lineages during the tertiary. *Syst. Biol.* **2005**, *54*, 575–594. [CrossRef]
34. Drummond, A.J.; Suchard, M.A.; Xie, D.; Rambaut, A. Bayesian phylogenetics with BEAUti and the BEAST 1.7. *Mol. Biol. Evol.* **2012**, *29*, 1969–1973. [CrossRef] [PubMed]
35. Maddison, W.P. Mesquite: A modular system for evolutionary analysis. *Evolution* **2008**, *62*, 1103–1118.
36. Paradis, E.; Claude, J.; Strimmer, K.J.B. APE: Analyses of phylogenetics and evolution in R language. *Bioinformatics* **2004**, *20*, 289–290. [CrossRef] [PubMed]
37. Harmon, L.J.; Weir, J.T.; Brock, C.D.; Glor, R.E.; Challenger, W. GEIGER: Investigating evolutionary radiations. *Bioinformatics* **2008**, *24*, 129–131. [CrossRef] [PubMed]
38. Nylander, J.A.; Olsson, U.; Alström, P.; Sanmartin, I. Accounting for phylogenetic uncertainty in biogeography: A Bayesian approach to dispersal-vicariance analysis of the thrushes (Aves: *Turdus*). *Syst. Biol.* **2008**, *57*, 257–268. [CrossRef] [PubMed]
39. Yu, Y.; Harris, A.; He, X. *RASP (Reconstruct Ancestral State in Phylogenies) 1.1. Beta*; Sichuan University: Sichuan, China, 2011.
40. Zhisheng, A.; Kutzbach, J.E.; Prell, W.L.; Porter, S.C. Evolution of Asian monsoons and phased uplift of the Himalaya–Tibetan plateau since Late Miocene times. *Nature* **2001**, *411*, 62–66. [CrossRef]
41. Che, J.; Zhou, W.W.; Hu, J.S.; Yan, F.; Papenfuss, T.J.; Wake, D.B.; Zhang, Y.P. Spiny frogs (Paini) illuminate the history of the Himalayan region and Southeast Asia. *Proc. Natl. Acad. Sci. USA* **2010**, *107*, 13765–13770. [CrossRef]
42. Li, J.; Fang, X. Research on the uplift of the Qinghai-Xizang Plateau and environmental changes. *Chin. Sci. Bull.* **1998**, *43*, 1569–1574.
43. Liu, X.; Dong, B. Influence of the Tibetan Plateau uplift on the Asian monsoon-arid environment evolution. *Chin. Sci. Bull.* **2013**, *58*, 4277–4291. [CrossRef]
44. Shi, Y.; Li, J.; Li, B. Uplift of the Qinghai-Xizang (Tibetan) plateau and east Asia environmental change during late Cenozoic. *Acta Geogr. Sin.* **1999**, *54*, 20–28.
45. Pokorny, L.; Riina, R.; Mairal, M.; Meseguer, A.S.; Culshaw, V.; Cendoya, J.; Serrano, M.; Carbajal, R.; Ortiz, S.; Heuertz, M. Living on the edge: Timing of Rand Flora disjunctions congruent with ongoing aridification in Africa. *Front. Genet.* **2015**, *6*, 154. [CrossRef] [PubMed]
46. Hoorn, C.; Wesselingh, F.P.; Ter Steege, H.; Bermudez, M.; Mora, A.; Sevink, J.; Sanmartín, I.; Sanchez-Meseguer, A.; Anderson, C.; Figueiredo, J. Amazonia through time: Andean uplift, climate change, landscape evolution, and biodiversity. *Science* **2010**, *330*, 927–931. [CrossRef] [PubMed]
47. Prothero, D.R. The late eocene-oligocene extinctions. *Annu. Rev. Earth Planet Sci.* **1994**, *22*, 145–165. [CrossRef]
48. Crisp, M.D.; Cook, L.G. Explosive radiation or cryptic mass extinction? Interpreting signatures in molecular phylogenies. *Evolution* **2009**, *63*, 2257–2265. [CrossRef]
49. Favre, A.; Michalak, I.; Chen, C.H.; Wang, J.C.; Pringle, J.S.; Matuszak, S.; Sun, H.; Yuan, Y.M.; Struwe, L.; Muellner-Riehl, A.N. Out of Tibet: The spatiotemporal evolution of *Gentiana* (Gentianaceae). *J. Biogeogr.* **2016**, *43*, 1967–1978. [CrossRef]
50. Wen, J.; Zhang, J.Q.; Nie, Z.L.; Zhong, Y.; Sun, H. Evolutionary diversifications of plants on the Qinghai-Tibetan Plateau. *Front. Genet.* **2014**, *5*, 4. [CrossRef]
51. Bagheri, A.; Maassoumi, A.A.; Rahiminejad, M.R.; Brassac, J.; Blattner, F.R. Molecular phylogeny and divergence times of *Astragalus* section *Hymenostegis*: An analysis of a rapidly diversifying species group in Fabaceae. *Sci. Rep.* **2017**, *7*, 14033. [CrossRef]

52. Donkpegan, A.S.; Doucet, J.-L.; Hardy, O.J.; Heuertz, M.; Piñeiro, R. Miocene diversification in the Savannahs precedes tetraploid rainforest radiation in the african tree genus *Afzelia* (Detarioideae, Fabaceae). *Front. Plant Sci.* **2020**, *11*, 798. [CrossRef]
53. Oliveira, L.C.; Rodrigues, D.P.; Hopkins, H.C.F.; Lewis, G.P.; Hopkins, M.J.G. Phylogeny and historical biogeography of the pantropical genus *Parkia* (Leguminosae, Caesalpinioideae, mimosoid clade). *Mol. Phylogenet. Evol.* **2021**, *163*, 107219. [CrossRef]
54. Lu, H.; Wang, X.; Li, L. Aeolian sediment evidence that global cooling has driven late Cenozoic stepwise aridification in central Asia. *Geol. Soc. Spec. Publ.* **2010**, *342*, 29–44. [CrossRef]
55. Miao, Y.; Herrmann, M.; Wu, F.; Yan, X.; Yang, S. What controlled Mid-Late Miocene long-term aridification in Central Asia? Global cooling or Tibetan Plateau uplift: A review. *Earth Sci. Rev.* **2012**, *112*, 155–172. [CrossRef]
56. Mosbrugger, V.; Utescher, T.; Dilcher, D.L. Cenozoic continental climatic evolution of Central Europe. *Proc. Natl. Acad. Sci. USA* **2005**, *102*, 14964–14969. [CrossRef] [PubMed]
57. Graham, A. *Late Cretaceous and Cenozoic History of Latin American Vegetation and Terrestrial Environments*; Missouri Botanical Garden Press: St. Louis, MO, USA, 2010.
58. Guo, Z.; Ruddiman, W.F.; Hao, Q.Z.; Wu, H.; Qiao, Y.; Zhu, R.X.; Peng, S.; Wei, J.; Yuan, B.; Liu, T. Onset of Asian desertification by 22 Myr ago inferred from loess deposits in China. *Nature* **2002**, *416*, 159–163. [CrossRef]
59. Barbour, M.; Keeler-Wolf, T.; Schoenherr, A.A. *Terrestrial Vegetation of California*; University of California Press: Berkeley, CA, USA, 2007.
60. Ramstein, G.; Fluteau, F.; Besse, J.; Joussaume, S. Effect of orogeny, plate motion and land-sea distribution on Eurasian climate change over the past 30 million years. *Nature* **1997**, *386*, 788–795. [CrossRef]
61. Jacobs, B.F. Palaeobotanical studies from tropical Africa: Relevance to the evolution of forest, woodland and savannah biomes. *Philos. Trans. R. Soc. Lond. B Biol. Sci.* **2004**, *359*, 1573–1583. [CrossRef] [PubMed]
62. De-Nova, J.A.; Medina, R.; Montero, J.C.; Weeks, A.; Rosell, J.A.; Olson, M.E.; Eguiarte, L.E.; Magallón, S. Insights into the historical construction of species-rich Mesoamerican seasonally dry tropical forests: The diversification of *Bursera* (Burseraceae, Sapindales). *N. Phytol.* **2012**, *193*, 276–287. [CrossRef]
63. Töpel, M.; Antonelli, A.; Yesson, C.; Eriksen, B.; One, P. Past climate change and plant evolution in western North America: A case study in Rosaceae. *PLoS ONE* **2012**, *7*, e50358. [CrossRef] [PubMed]
64. Dodd, M.E.; Silvertown, J.; Chase, M.W. Phylogenetic analysis of trait evolution and species diversity variation among angiosperm families. *Evolution* **1999**, *53*, 732–744. [CrossRef]
65. Eriksson, O.; Bremer, B. Pollination systems, dispersal modes, life forms, and diversification rates in angiosperm families. *Evolution* **1992**, *46*, 258–266. [CrossRef]
66. Verdú, M. Age at maturity and diversification in woody angiosperms. *Evolution* **2002**, *56*, 1352–1361. [CrossRef]
67. Pennington, R.T.; Lavin, M.; Prado, D.E.; Pendry, C.A.; Pell, S.K.; Butterworth, C.A. Historical climate change and speciation: Neotropical seasonally dry forest plants show patterns of both Tertiary and Quaternary diversification. *Philos. Trans. R. Soc. Lond. B Biol. Sci.* **2004**, *359*, 515–538. [CrossRef] [PubMed]

genes

MDPI

Article

Characterization of the Evolutionary Pressure on *Anisodus tanguticus* Maxim. with Complete Chloroplast Genome Sequence

Dangwei Zhou [1,2,*,†], Furrukh Mehmood [3,4,†], Pengcheng Lin [1], Tingfeng Cheng [2], Huan Wang [2], Shenbo Shi [2], Jinkui Zhang [1], Jing Meng [1], Kun Zheng [1] and Péter Poczai [5,*]

1 The College of Pharmacy, Qinghai Nationalities University, Xining 810007, China
2 Key Laboratory of Adaptation and Evolution of Plateau Biota (AEPB), Northwest Institute of Plateau Biology, Chinese Academy of Sciences, Xining 810008, China
3 Department of Biochemistry, Faculty of Biological Sciences, Quaid-i-Azam University, Islamabad 45320, Pakistan
4 Department of Biochemistry, Faculty of Sciences, University of Sialkot, Daska Road, Punjab 51040, Pakistan
5 Faculty of Biological and Environmental Sciences, University of Helsinki, FI-00014 Helsinki, Finland
* Correspondence: dangweizhou@sina.com (D.Z.); pet.poczai@helsinki.fi (P.P.); Tel.: +86-17797146826 (D.Z.)
† These authors contributed equally to this work.

Abstract: *Anisodus tanguticus* Maxim. (Solanaceae), a traditional endangered Tibetan herb, is endemic to the Qinghai–Tibet Plateau. Here, we report the de novo assembled chloroplast (cp) genome sequences of *A. tanguticus* (155,765 bp). The cp contains a pair of inverted repeated (IRa and IRb) regions of 25,881 bp that are separated by a large single copy (LSC) region (86,516 bp) and a small single copy SSC (17,487 bp) region. A total of 132 functional genes were annotated in the cp genome, including 87 protein-coding genes, 37 tRNA genes, and 8 rRNA genes. Moreover, 199 simple sequence repeats (SSR) and 65 repeat structures were detected. Comparative plastome analyses revealed a conserved gene order and high similarity of protein-coding sequences. The *A. tanguticus* cp genome exhibits contraction and expansion, which differs from *Przewalskia tangutica* and other related Solanaceae species. We identified 30 highly polymorphic regions, mostly belonging to intergenic spacer regions (IGS), which may be suitable for the development of robust and cost-effective markers for inferring the phylogeny of the genus *Anisodus* and family Solanaceae. Analysis of the Ka/Ks ratios of the Hyoscyameae tribe revealed significant positive selection exerted on the cemA, rpoC2, and clpP genes, which suggests that protein metabolism may be an important strategy for *A. tanguticus* and other species in Hyoscyameae in adapting to the adverse environment on the Qinghai–Tibetan Plateau. Phylogenetic analysis revealed that *A. tanguticus* clustered closer with *Hyoscyamus niger* than *P. tangutica*. Our results provide reliable genetic information for future exploration of the taxonomy and phylogenetic evolution of the Hyoscyameae tribe and related species.

Keywords: *Anisodus tanguticus* Maxim.; chloroplast genome; comparative analysis; positive selection; phylogenetic relationship; hyoscyameae tribe

Citation: Zhou, D.; Mehmood, F.; Lin, P.; Cheng, T.; Wang, H.; Shi, S.; Zhang, J.; Meng, J.; Zheng, K.; Poczai, P. Characterization of the Evolutionary Pressure on *Anisodus tanguticus* Maxim. with Complete Chloroplast Genome Sequence. *Genes* **2022**, *13*, 2125. https://doi.org/10.3390/genes13112125

Academic Editors: Wajid Zaman and Hakim Manghwar

Received: 3 October 2022
Accepted: 7 November 2022
Published: 15 November 2022

1. Introduction

Chloroplasts (cp) in plant cells are photosynthetic organelles that play a critical role in the production of essential energy and metabolites [1]. Cps contain a 120–160 kb genome with typical conserved quadripartite regions, including an LSC region, an SSC region, two identical copies of IR regions, and approximately 110–130 encoded genes distributed in a circular genome [2,3]. As most plant species exhibit maternal inheritance in their cps, the cp genome has undergone less recombination and evolution [3–5]. Therefore, cp sequences are considered a vital resource of molecular genetic markers to analyze the profiling of gene distribution and their molecular phylogenetic relationships [6]. Furthermore, because of the high level of expression, the absence of post-transcriptional gene silencing and site-specific transformation [7], cps are useful in the agriculture and pharmaceutical industries. In recent decades, rapid advances in next-generation sequencing technologies along with reduced costs have enabled a better understanding of the phylogenetic status and the cp transformation technique in traditional herbs using cp genome information. *A. tanguticus* is a rare and endangered Solanaceae species found on the Qinghai–Tibetan Plateau in grasslands or meadows (2200–4200 m) [8]. According to traditional Tibetan medicine, this species includes an abundance of hyoscyamine, atropine, and total alkaloids [9–11] and is widely used for the treatment of asthma, epilepsy, and inflammation [12,13]. Despite the development of a domesticated plant [11], the main medicinal resource of *A. tanguticus* still relies on wild plant exploration and collection, which has led to the species becoming endangered [14]. Furthermore, as populations are self-incompatible and interspecific hybridization occurs [15], the population structure and genetic diversity of *A. tanguticus* in the Qinghai–Tibetan region should be evaluated. DNA barcoding is thought to be an effective approach for medicinal component identification and quality control of produced goods in addition to protecting consumers [16,17]. Previous phylogenetic studies of Solanaceae plants have been based mostly on a few cp genes or ITS sequences [18,19]. There is currently limited cp genome information of *A. tanguticus*. Here, we assembled and de novo sequenced the whole cp genome of *A. tanguticus* with Illumina sequencing platforms. Additionally, cp genome annotation and structure analysis were performed and several oligonucleotide repeats, SSR, and substitutions were identified as mutation hotspots for future DNA barcoding. Further insights on *A. tanguticus* plastome evolution and its phylogenetic relationships have been revealed through comparative studies with reported cp genomes of Solanaceae species.

2. Materials and Methods

2.1. DNA Extraction

Fresh young leaves of *A. tanguticus* were collected from Haibei station (37.48° N, 101.2° E; Alt. 3200 m), in Qinghai province. According to the manufacturer's instructions, a DNeasy Plant MiniKit (QIAGEN, Düsseldorf, Germany) was utilized to isolate DNA from the fresh young leaves. Spectrophotometry and electrophoresis on a 1% (w/v) agarose gel were used to evaluate the purity of the DNA. DNA of sufficient integrity and purity DNA was used for library construction and sequencing with an IlluminaHiseq2500 (San Diego, CA, USA).

2.2. Plastome Assembly and Annotation

Trimmomatic V0.36 was used to remove adapters and poor-quality reads from the raw reads [20]. Furthermore, clean reads were then mapped to the database, which was constructed from all cp genome sequences reported in the NCBI based on coverage depth and similarity. The mapped reads were then assembled into contigs [21] using SOAPdenovo2. The reads were filtered with careful error correction and different k-mers (55, 87, and 121) [22]. The scaffold of the cp genome was constructed with SSPACE [23] and GapCloser was used to fill the gaps [21]. PCR amplification and Sanger sequencing were used to validate the assembly by verifying the four boundaries between single copy (SC) and inverted repeat (IR) regions of the assembled sequences. Dual Organellar GenoMe Annotator (DOGMA, http://dogma.ccbb.utexas.edu/, accessed on 15 March 2021), CPGAVAS, and manual corrections were used to annotate the full cp genomes [24,25]. The software tRNAscan-SE was used for the prediction of tRNA genes. Circular cp genome map was drawn by Organellar Genome DRAW (OGDRAW) V1.2 [26]. The correct and complete cp genome was deposited in the National Center for Biotechnology Information (NCBI) under accession number MW246825 (*A. tanguticus*). Web-based REPuter (https://bibiserv.cebitec.uni-bielefeld.de/reputer/ accessed on 15 March 2021) was used for the analysis of repeat sequences, including forward, palindrome, reverse, and complement repeats. Sequence identity was set to >90% with a minimum repeat size of 30 base pairs. Micro Satellite Identification Tool (MISA) [27] was used to identify simple sequence repeats (SSRs), with minimum repeats of mono-, di-, tri-, tetra-, penta-, and hexa-nucleotides set to 10, 5, 4, 3, and 3, respectively. Using the tool CodonW1.4.2 (http://downloads.fyxm.net/CodonW-76666.html accessed on 15 March 2021), only 53 protein-coding genes with lengths more than 300 bp were chosen for analysis of synonymous codon usage. The relative synonymous codon usage (RSCU) and overall codon usage were both analyzed.

2.3. Comparative Analysis of cp Genomes

The plastome of *A. tanguticus* was compared with the cp genomes of *Prewalskia tangutica, Scopolia parviflora, H. niger, Datura stramonium,* and *Nicotiana sylvestris* in Solanaceae using Mauve software to identify evolutionary events such as gene loss, duplication, and rearrangement [28]. As a reference, the annotation of N. sylvestris was selected.

2.4. Molecular Evolution Analysis

The protein-coding genes of *A. tanguticus* were evaluated for synonymous (Ks) and non-synonymous (Ka) substitution rates in comparison with the six closely related Solanaceae species (*P. tangutica, S. parviflora, H. niger, D. stramonium,* and *N. tabacum*). Between the compared species, the corresponding functional protein-coding gene was separately aligned using MAFFT [29]. The Ka/Ks value was then determined using the NG method of computation, created by Nei and Gojobori and implemented in KaKs calculator 2.0 [30], with the settings genetic code Table S1 (bacterial and plant plastid code). In the results, some genes had Ka/Ks values of "NA", which meant they were not relevant. When Ks = 0, this indicates no substitutions in the alignment, or 100% match. Concatenated sequences of 67 common protein-coding genes among the studied species were used in a phylogenetic analysis based on the complete cp genomes. The two sets of sequences were aligned by MAFFT [29], and the alignments were then adjusted by the Gblocks program [31]. Phylogenetic trees were created by the RAxMLversion 8.0 program using the GTRGAMMA model by the maximum likelihood (ML) approach [32]. Each branch's bootstrap analysis was calculated using 1000 replications.

3. Results and Discussion

3.1. Chloroplast Genome Features of *A. tanguticus*

The cp genome of *A. tanguticus* is 155,765 nucleotides and has a quadripartite architecture. The genome consists of a pair of IRs (25,881 bp), a large single copy (LSC) region (86,516 bp), and a small single copy (SSC) region (17,487 bp) (Figure 1). These results are similar to other species of Solanaceae (Table 1). The cp genome's total GC content was 37.63% and the GC content of the IR regions was 42.87%, which is higher than those of the LSC region (35.65%) and SSC region (31.93%). Similar results have also been observed in other species [32,33], which may be due to the four rRNAs and tRNAs [34,35]. The *A. tanguticus* plastome has 132 unique genes, including 87 protein-coding genes and 37 tRNA and eight rRNA genes. The cp genome contained 131 genes altogether, including 21 functional gene duplications in IR areas (Table 1). The plastome of *A. tanguticus* (87) has more protein-coding genes than *P. tangutica* (85) and *H. niger* (80) (Table 1). Among these, eight protein-coding genes (atpF, ndhA, ndhB, petD, rpoC1, rpl16, rpl2, and rps16) and six tRNAs had one intron, while three genes (ycf3, rps12, and clp2) contained two introns (Table 2). Nineteen genes, including four ribosomal (rrn4.5, rrn5, rrn16, and rrn23), seven trn (trn-UGC, trnI-CAU, trnI-GAU, trnL-CAA, trnN-GUU, trnR-ACG, and trnV-GAC), ndhB, rps7, rps12, rpl2, rpl23, ycf1, ycf2, and ycf15 had two repeats in the *A. tanguticus* plastome (Table 2). The largest intron (2502 bp) was found in the trnK-UUU gene, and it contained matK. These results are consistent with the mustard family (Brassicaceae) cp genome [35].

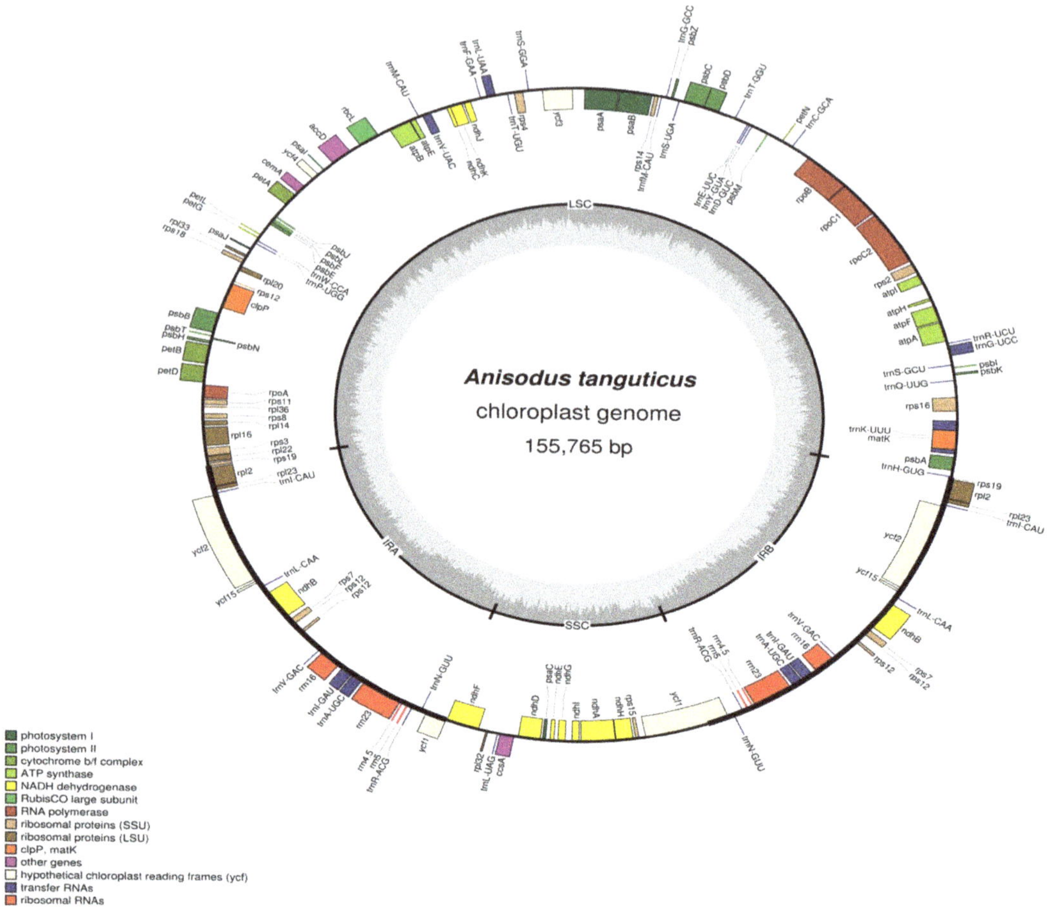

Figure 1. Gene map of the *A. tanguticus* chloroplast genome. (Genes shown outside the outer circle are transcribed clockwise and those insides are transcribed counterclockwise. The colored bars indicated different functional groups).

Table 1. Comparison analyses of *A. tanguticus* with six close species.

Genome Feature	*A. tanguticus*	*P. tangutica*	*S. parviflora*	*H. niger*	*D. stramonium*	*N. tabacum*	*S. tuberosum*
Genome Size (bp)	155,765	155,569	156,193	155,720	155,871	155, 943	155,298
LSC (bp)	86,516	86,707	86,364	86,105	86,301	86, 686	85,749
SSC (bp)	17,487	18,288	25,905	17,863	18,366	18, 573	18,373
IR (bp)	25,881	25,287	25,876	25,876	25,602	25,342	25,595
GC content (%)	37.63	37.6	37.6	37.6	37.9	37.9	37.9
Total number of genes	132	138	131	118	134	156	130
Protein-coding gene	87	85	86	80	88	111	81
tRNA	37	44	37	30	38	37	37
rRNA	8	8	8	4	8	8	8

Table 2. List of genes in the cp genome of *A. tanguticus*.

Category for Genes	Groups of Genes	Name of Genes
Photosynthesis	Subunits of photosystem I	psaA, psaB, psaC, psaI, psaJ
	Subunits of photosystem II	psbA, psbB, psbC, psbD, psbE, psbF, psbH psbI, psbJ, psbK, psbL, psbM, psbN, psbT, psbZ
	Subunits of cytochrome b/f complex	petA, petB, petD *, petG, petL, petN
	Large subunit of Rubisco	rbcL
	Subunits of ATP synthase	atpA, atpB, atpE, atpF *, atpH, atpI
	Subunits of NADH-dehydrogenase	ndhA *, ndhB [a]*, ndhC, ndhD, ndhE, ndhF, ndhG, ndhHndhI, ndhJ, ndhK
Self-replication	Ribosomal RNA genes	rrn16 [a], rrn23 [a], rrn5 [a], rrn4.5 [a]
	Transfer RNA genes	*trnA-UGC* *[a], *trnC-GCA, trnD-GUC, trnE-UUC* *[a], *trnF-GAA, trnfM-CAU, trnG-GCC, trnG-UCC* *, *trnH-GUG, trnI-CAU* [a], *trnI-GAU* [a]*, *trnK-UUU* *, *trnL-CAA* [a], *trnL-UAG, trnL-UAA* *, *trnM-CAU, trnN-GUU* [a], *trnP-UGG, trnQ-UUG, trnR-ACG* [a], *trnR-UCU, trnS-GCU, trnS-GGA, trnS-UGA, trnY-GUA, trnT-GGU, trnT-UGU, trnV-GAC* [a], *trnV-UAC* *, *trnW-CCA*
	Small subunit of ribosome	rps2, rps3, rps4, rps7 [a], rps8, rps11, rps12 [a]**, rps14, rps15, rps16, rps18, rps19
	Large subunit of ribosome	rpl2 [a]*, rpl14, rpl16 *, rpl20, rpl22, rpl23 [a], rpl32, rpl33, rpl36
	DNA-dependent RNA polymerase	rpoA, rpoB, rpoC1 *, rpoC2
Other genes	Maturase	matK
	Envelope membrane protein	cemA
	Subunit of acetyl-CoA	accD
	C-type cytochrome synthesis gene	ccsA
	Protease	clpP **
Genes of unknown function	Conserved Open reading frames	ycf1 [a], ycf2 [a], ycf3 **, ycf4, ycf15 [a]

Note: * means the gene contained one intron; ** means the gene contained two introns; [a] indicates the number of the repeat unit is 2.

3.2. Codon Usage Analysis

RSCU frequency plays an important role in reflecting mutation bias during evolution [36]. RSCU values > 1 show the use of a codon more frequently than expected and vice versa [37]. The *A. tanguticus* plastome has a total of 80,910 bp of protein-coding genes, which encodes 26,970 codons. Among these, 2867 (6%) encoded leucine, which is the most abundant amino acid, whereas only 483 (1.0%) encoded tyrosine, which is the least prevalent amino acid in the *A. tanguticus* cp genome. According to the codon-anticodon recognition patterns, 29 tRNAs had codons that matched all 20 amino acids (Figure 2, Supplementary Materials, Table S1). AT content at the first, second, and third codon positions covered 54.3%, 61.9%, and 67.8%, respectively. This phenomenon was also higher in other plants [38,39]. There were 31 codons that exhibited clearly biased usage (RSCU > 1). Except for two codons (UUG for Leu and AUG for Met), all biased codon usage bases were A/U at the third position. Similarly, the codon for Trp did not exhibit biased usage (RSCU = 1) (Figure 2, Supplementary Materials, Table S1). This result is consistent with the *N. officinale* cp genome [35]. However, unlike these species, ATG showed the highest RSCU (1.9936) value of all codons in the *A. tanguticus* cp genome. The significant evolutionary character of the genome is evident in the codon usage bias [37,38]. Our results suggest that the *A. tanguticus* cp genome has experienced a different evolution process.

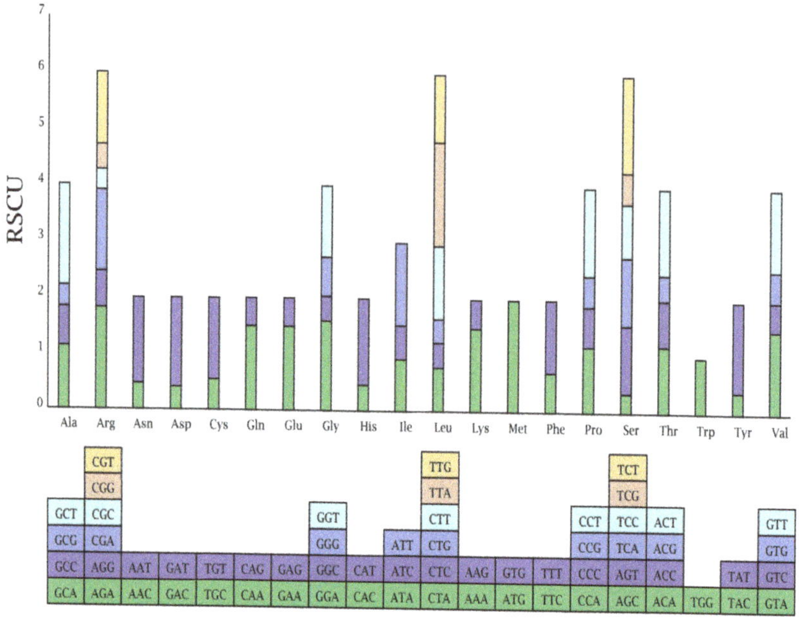

Figure 2. Codon content of 20 amino acids and the stop codon of 81 coding genes of the *A. tanguticus*.

3.3. Repeat Sequence and SSR Analysis

Repeat sequences are a constant feature of illegitimate recombination and slipped-strand mispairing, and they can be utilized in plant phylogenetic studies [39,40]. The *A. tanguticus* cp genome contains 39 repeats, of which 18 are direct repeat sequences, 18 are palindromic sequences, and three are reverse repeat sequences. There were no complementary repeat sequences (Figure 3A). Analyses of oligonucleotide repeats revealed that all sequences were ≥30 bp, and direct and palindromic repeats covered 46% of the total dispersed repeat sequences in the cp genome. Interestingly, there was one palindromic sequence of 25,881 bp (Figure 3A). The number of forward and palindromic sequences was close among the different types of repeats, although they still differed among the six species (Figure 3B). Further analysis of repeat length among the four species of Hyoscyameae tribe revealed that the length of 40–49 bp in forward and palindromic sequences was greater than the other two species (Figure 3C,D), and the length of 50–59 bp may have played an important role in *H. niger* cp genome rearrangement (Figure 3C,D). However, the length of 30–40 bp for forward and palindromic sequences did not differ significantly among six species. Surprisingly, the number of reverse sequences with lengths of 30–39 bp was substantially greater in the cp genomes of *A. tanguticus* and *P. tangutica*, and there were no reverse sequences in the cp genomes of *H. niger* and *D. stramonium* (Figure 3E). Previous research has discovered long repetitions of >30 bp, which may contribute to cp genome rearrangement and increase population genetic diversity [41,42]. Our findings revealed that forward and palindromic sequences with lengths of 40–49 bp may play an essential role in the Hyoscyameae tribe, while reverse sequences with lengths of 30–39 bp are more abundant in *A. tanguticus* and *P. tangutica* than in the Hyoscyameae tribe. This may suggest that the reverse sequences of *A. tanguticus* and *P. tangutica* may play a more important role in cp genome rearrangement to adapt to the alpine environment on the Qinghai–Tibet Plateau. SSRs were 1–6 bp in the cp genome and were widely dispersed in intergenic, intron, or coding regions. SSRs had a high mutation rate and diverse copy number, which confers their important value in phylogenetic and population analyses [43,44]. Using MISA, 199 SSRs were identified in the *A. tanguticus* plastome (Supplementary Materials, Table S2). Among these, there were 120 mononucleotide, eight dinucleotide, 72 trinucleotide, seven tetranucleotide, and one pentanucleotide repeats. In addition, 60.3% mononucleotide repeats were A/T, which is consistent with results from *Physalis* species [45]. There were 61 trinucleotide SSRs in the *A. tangutica* cp genome, which is similar to four species of Hyoscyameae (Figure 3F) and more than that observed in four *Physalis* species [45]. Interestingly, dinucleotide SSRs were considerably lower than that of *P. tangutica*, *S. parviflora*, and *H. niger*, which contained 43, 43, and 42 dinucleotide SSRs, respectively. Although there was one pentanucleotide repeat in *A. tanguticus*, *S. parviflora*, and *H. niger*, there were three such repeats in the *P. tangutica* cp genome (Figure 3F). Therefore, these SSR differences may be helpful for phylogenetic and evolutionary studies. Furthermore, the LSC regions had 63.3% SSR markers, which was considerably greater than that of the SSC and IR regions. Similarly, nearly all SSRs in the *A. tanguticus* cp genome were comprised of A/T, which added 62.3% to the genome's base composition.

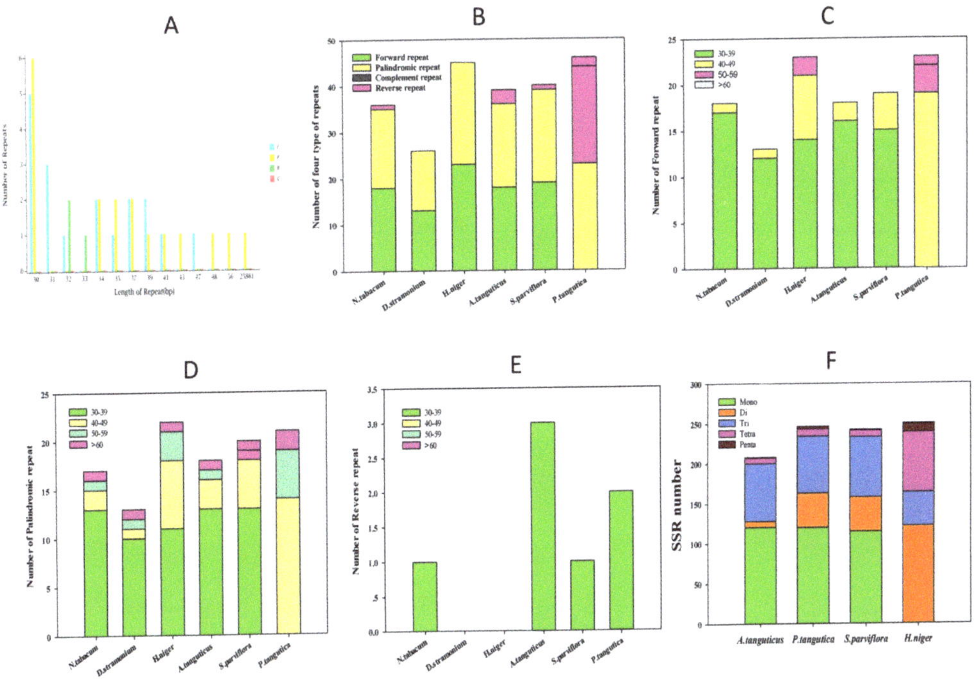

Figure 3. Repeat sequences and SSR analysis of *A. tanguticus* and related species of Solanaceae. (**A**) Profiling of repeat sequences of *A. tanguticus*; (**B**) number of four types of repeat of *A. tanguticus* and five related species; (**C**) comparison of a forward repeat of *A. tanguticus* and five related species; (**D**) comparison of a palindromic repeat of *A. tanguticus* and five related species; (**E**) comparison of a reverse repeat of *A. tanguticus* and five related species; (**F**) SSR number of *A. tanguticus* and three related species.

3.4. IR Contraction and Expansion

The expansion and contraction of the IR boundaries were supposed to be an evolutionary trait of plastomes, and contraction and expansion at the IR region borders were believed to explain plastome differences [46,47]. To characterize the variation at the IR boundary, we selected the nine phylogenetically related Solanaceae species (*P. peruviana*, *Iochroma ellipticum*, *Capsicum frutescens*, *Solanum lycopersicum*, *D. stramonium*, *N. sylvestris*, *H. niger*, *S. parviflora*, and *P. tangutica*). Genomic structure and gene order are conserved in these species, especially in the Hyoscyamus group (Figure 4). The IRa/SSC border for the *A. tanguticus* cp genome was found in the 3′ region of the whole ycf1 gene. This is similar to *H. niger*, *P. tangutica*, and the other species except for *S. parviflora*. Similarly, except for *S. parviflora* and *N. sylvestris*, *A. tanguticus* and the other seven species contained the ycf1 pseudogene in the IRb region. This suggests that the ycf1 pseudogene in the IRb was lost in the evolution of the two species. In the cp genomes of *A. tanguticus* and the Solanaceae species mentioned above, the rps19 gene is located within the LSC/IRb border, except *P. tangutica*, where rps19 exists only in the LSC region. In some species, the 3′ region truncated rps19 pseudogene is in the IRa/LSC border [48]. Similarly, rps19 only appeared in *A. tanguticus* and *S. parviflora* in the IRa region. Our results indicate that the rps19 gene duplication is lost in the IRa/LSC border in the cp genomes of eight Solanaceae species (Figure 4). In the *P. tangutica* cp genome, there was also no rpl22 gene in the LSC region and rpl2 expanded to the IRb/LSC border (Figure 4). Similarly, the rpl2 gene is duplicated in *A. tanguticus* and five other species, while one duplication is lost in *P. tangutica*, *H. niger*, *I. ellipticum*, and *P. peruviana*. In general, our results showed that the IR boundary expansion and contraction differed from the Solanaceae species.

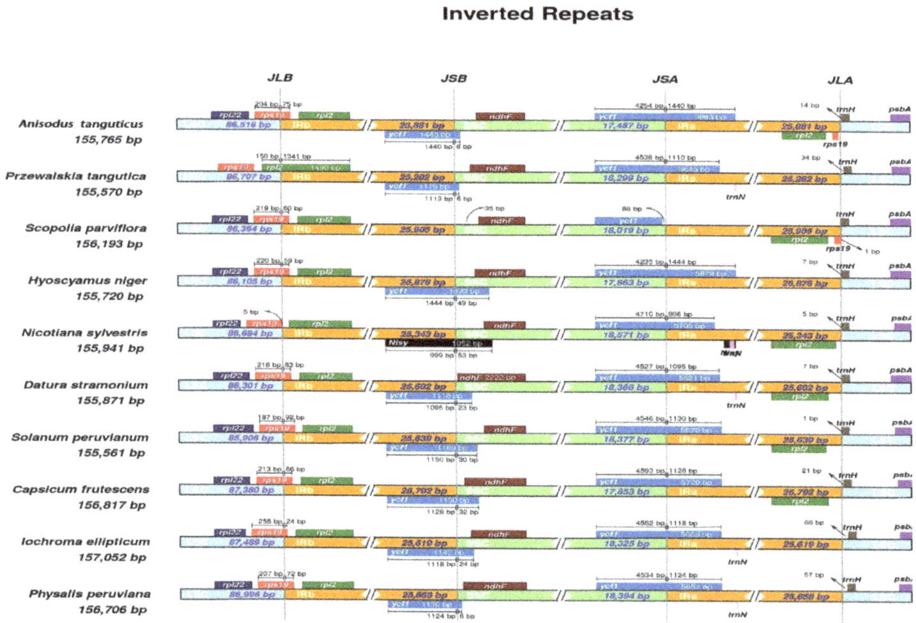

Figure 4. Comparison of the border position of LSC, SSC, and LR regions among *A. tanguticus* and nine species' chloroplast genomes in Solanaceae. Genes are represented by boxes above (negative strand) and below (positive strand) the proportionate line. The size of each gene and their relative position at the junctions are shown in base pairs (bp).

3.5. Comparative Chloroplast Genome Analysis

The whole cp genome of *A. tanguticus* was compared with five other species (*P. tangutica*, *H. niger*, *S. parviflora*, *N. tabacum*, and *D. stramonium*) using the Mauve program [28]. All cp genomes appeared as locally collinear blocks (Figure 5). The coding proteins and the tRNA and rRNA collinearly blocks are relatively conserved in structures and regions. However, gene loss occurred in the six species, such as 150,000–155,000, in which the *P. tangutica* and *H. niger* cp genomes lost two genes compared with the other four species. In addition, there was no gene rearrangement in Solanaceae genomes (Figure 5); this result is consistent with the other species [48,49]. The CD regions in the cp genomes of these six species were analyzed with MAFFT software and VCFtools was used to calculate the nucleotide variability (π) of each gene (Figure 6, Supplementary Materials, Table S3).

The mean Pi value was 0.006936 among the six species. Similarly, IR regions showed less variability than that of the LSC and SSC regions. Among 113 genes, 26 genes showed no change in the six species, whereas 30 mutation hotspot regions were discovered (Pi > 0.01). These highly divergent proteins included matK, clpP, rps16, rps12, and ycf1, which had the highest Pi value (0.04246) (Figure 6, Supplementary Materials, Table S3). These regions might be going through rapid nucleotide substitution at the species level, suggesting the possible use of molecular markers for phylogenetic studies and plant identification.

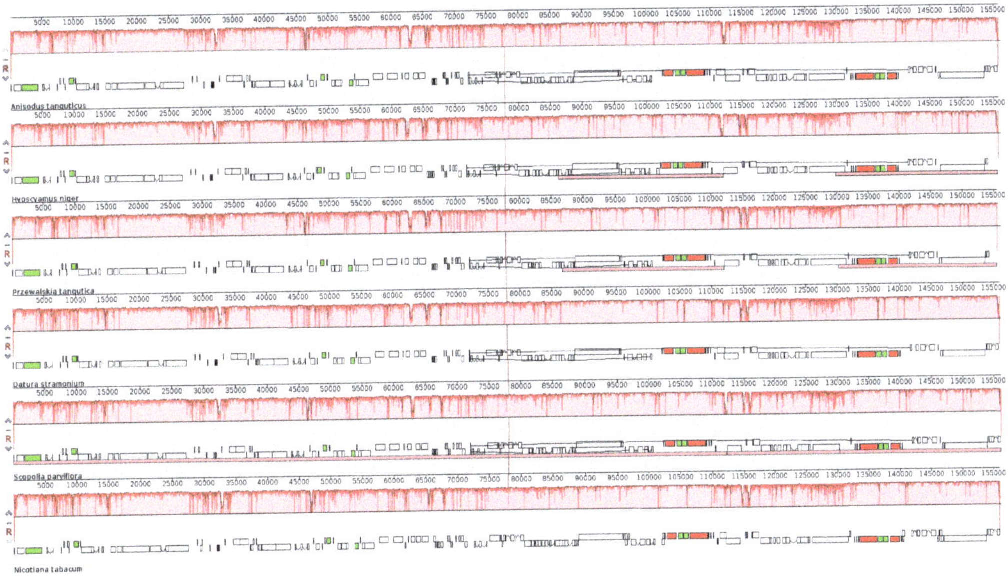

Figure 5. Sequence alignment of four chloroplast genomes in Solanaceae with *A. tanguticus* by Mauve software. Within each of the alignments, local collinear blocks are presented by blocks of the same color connected by lines. Annotations of rRNA, protein-coding, and tRNA genes are shown in red, white, and green boxes, respectively.

3.6. Synonymous (Ks) and Non-Synonymous (Ka) Substitution Rate Analysis

Synonymous (Ks) and non-synonymous (Ka) were parameters for assessing the rate of gene divergence in a molecular evolution study [30]. To characterize the molecular evolution pattern, we calculated the Ka/Ks ratios of the *A. tanguticus* cp genome and the results were compared with those of five closely related species (*P. tangutica*, *H. niger*, *S. parviflora*, *D. stramonium*, and *N. tabacum*) (Figure 7; Supplementary Materials, Table S4). Of the 86 protein-coding genes, most had Ka/Ks values < 1. The average ratio was 0.24–0.32 for the five comparison groups. This shows that these species were mostly the outcome of vigorous purifying selection. In the Hyoscyameae tribe, cemA, rpoC2, and clpP exhibited positive selection in *S. parviflora* and *P. tangutica*. In a previous study, the clpP gene showed positive selection in various plant lineages, such as Oenothera [50], and cemA and rpoC2 also appeared to have undergone adaptation evolution in Gossypium and Populus, respectively [51,52]. Our results are consistent with these observations. In a recent study, clpP was shown to degrade misfolded proteins and protected cps from abiotic and biotic stress [53]; loss of this protein may lead to inappropriate structural changes in cps [54]. cemA proteins play an important role in protein sorting signals [55] and rpoC2 is an important subunit of PEP, which is required for transcription of photosynthesis genes and is involved in cp biogenesis [56]. In Nicotiana, positive selection acting on the ATP synthase and NAD(P)H dehydrogenase encoding genes was believed to be an adaptation to novel ecological conditions [57]. Here, our results indicate that protein metabolism may be an important adaptation strategy for *A. tanguticus* and other species in Hyoscyameae. As an endemic species on the Qinghai–Tibetan Plateau, *A. tanguticus* is often exposed to low temperatures, high UV-B radiation, and other abiotic stresses. Degrading misfolded proteins and a sorting signal system may help maintain plasmid function. Therefore, our results suggest that these genes are also important for Hyoscyameae tribe species evolution and adaptation to adverse environments.

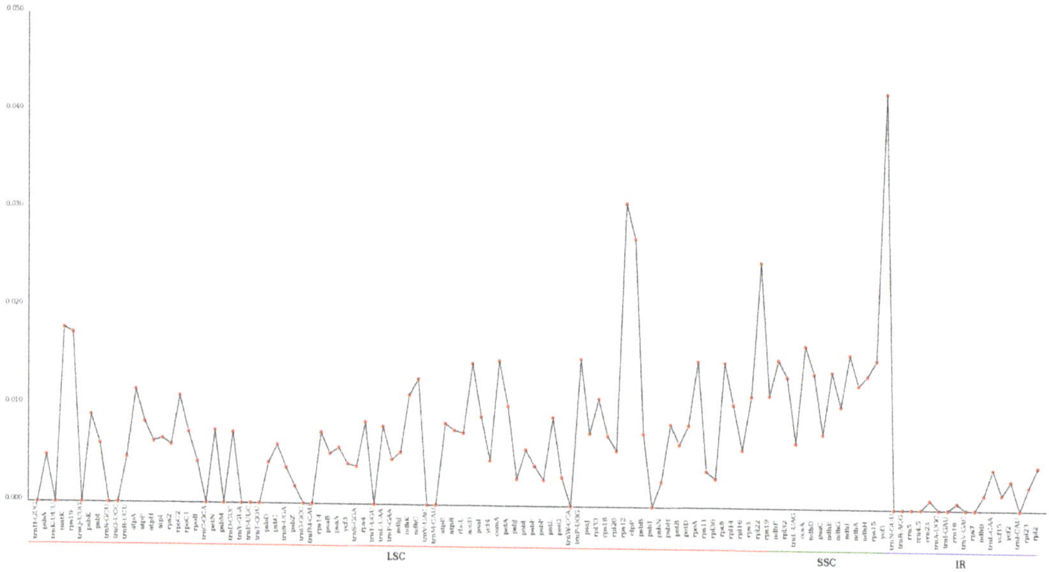

Figure 6. Comparison of nucleotide variability (Pi) among *A. tanguticus* and related species.

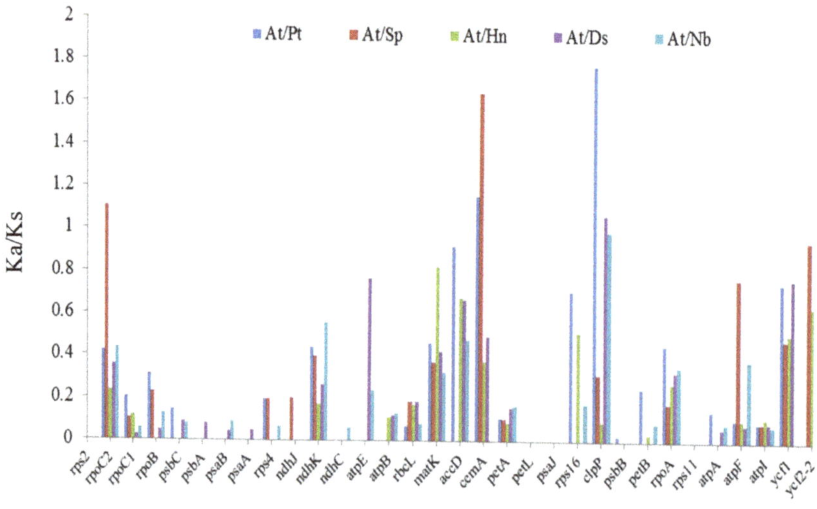

Figure 7. The Ka/Ks ratios of protein-coding genes of *A. tanguticus* and related species.

3.7. Phylogenetic Analysis of the *A. tanguticus* cp Genome

The tribe Hyoscyameae is distributed throughout Eurasia and is considered as a monophyly in Solanaceae [18,19]. However, its genera relationship within this tribe remains unclear. Cp genomes provide whole phylogenetic information, which is suitable for establishing phylogenetic relationships in plants. Whole cp genomes and protein-coding genes of 21 species were utilized to create an ML phylogenetic tree to understand the phylogenetic position of *A. tanguticus* and to further elucidate the evolutionary relationship in the Solanaceae (Supplementary Materials, Table S5). Most nodes were strongly supported by 100% bootstrap value (Figure 8A). Moreover, *A. tanguticus* was clustered with three species (*P. tangutica*, *H. niger*, and *S. parviflora*) with 100% bootstrap support. The three Hyoscyameae tribe species were clustered together more closely. *D. stramonium*, *N. sylvestris*, and other species were clustered outside this branch with 100% bootstrap support. These findings are consistent with previous studies [18,19,57,58]. In a Hyoscyameae tribe study, *A. tanguticus* grouped with *H. niger* when several chloroplast markers were combined [18,19]. Our results are consistent with this observation. However, when whole genome data are used, *P. tangutica* and *S. parviflora* exhibited a closer relationship in the Hyoscyameae tribe (Figure 8B). This may be due to a data information shortage, such as in *Physochlaina*. Since *A. tanguticus* is an endemic species that grows on the Qinghai–Tibetan Plateau, sequencing, and characterization of additional cp genomes of Hyoscyameae tribe species may provide further insight on the phylogenetic relationships and evolution among these species.

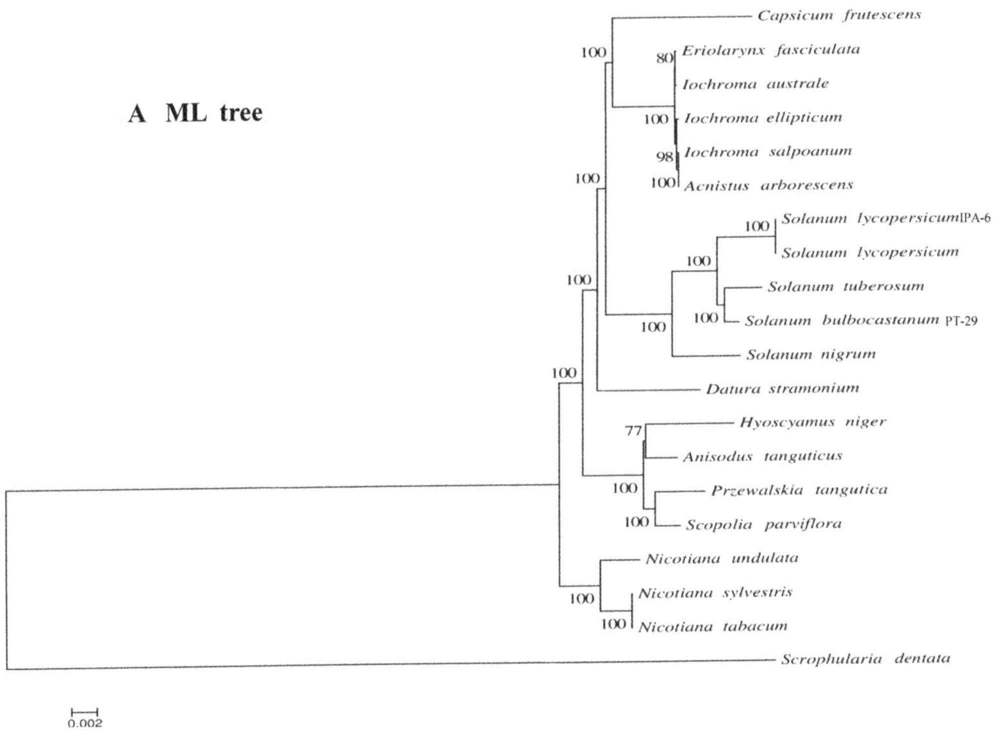

Figure 8. *Cont.*

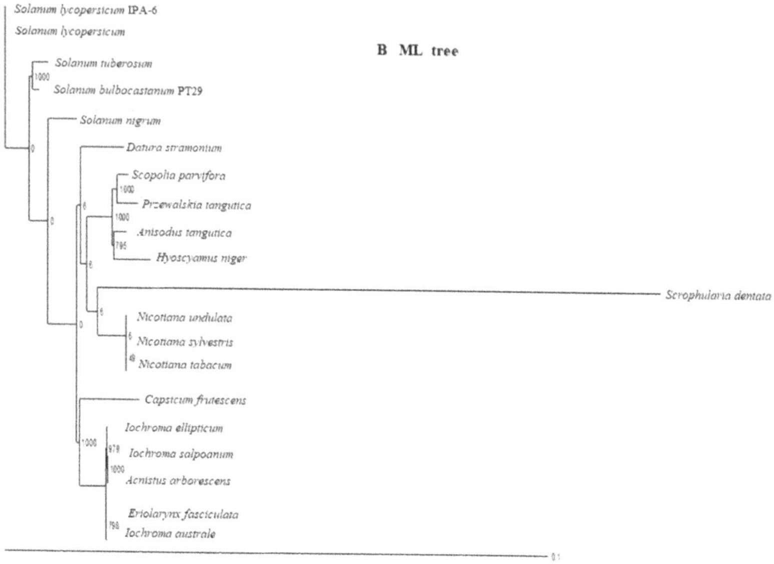

Figure 8. The phylogenetic tree base on 20 cp chloroplast species was constructed with ML methods. Bootstrap values were shown at the nodes. ((**A**). ML tree of coding genes; (**B**). ML tree of whole cp genomes).

4. Conclusions

High throughput sequencing was used to obtain the whole cp genome of *A. tanguticus*. The cp genome is 152,714 bp and encodes 126 functional genes. Analysis of repeat sequences revealed that forward and palindromic repeats of 40–49 bp were more common in the Hyoscyameae tribe, and 30–39 bp reverse repeats may play a more significant role in *A. tanguticus* and *P. tangutica* cp genome rearrangement and in adapting to extreme conditions on the Qinghai–Tibetan Plateau. The *A. tanguticus* cp genome contained a total of 199 SSRs and can be utilized for the development of molecular markers for phylogenetic and population studies. The genome structure and gene organization were similar to three previously reported species in the Hyoscyameae tribe, while gene contraction and expansion were clearly different in the *A. tanguticus* cp genome. Evolutionary analysis indicated that clpP, cemA, and ropC2 showed positive selection, which suggests that protein metabolism may be an important adaptation strategy for *A. tanguticus* and other species in Hyoscyameae on the Qinghai–Tibetan Plateau. Phylogenetic analysis indicated that *A. tanguticus* has a closer relationship with *P. tangutica*. These results provide significant information for forthcoming studies on phylogenetic and evolutionary adaptation in *A. tanguticus* and its related species.

Supplementary Materials: The following supporting information can be downloaded at: https://www.mdpi.com/article/10.3390/genes13112125/s1, Table S1: The codon usage and codon-anticodon recognition pattern for the A. tanguticus cp genome; Table S2a: Comparison of repeat sequences on A. tanguticus and five related species; Table S2b: Distribution of SSRs in the A. tanguticus cp genome; Table S3: Comparison of nucleotide variability (Pi) among A. tanguticus and related species; Table S4: The Ka/Ks ratios of protein-coding genes of A. tanguticus and related species; Table S5 The chloroplast genome of plant species for construction phylogenetic tree.

Author Contributions: D.Z. designed and managed the project; P.L., T.C. and H.W. performed the molecular experiment; F.M., S.S., J.Z., performed the bioinformation analysis; J.M. and K.Z. provided the sequencing data; D.Z., F.M. and P.P. wrote and revised the manuscript. All authors have read and agreed to the published version of the manuscript.

Funding: This work was supported by grants from the Training Qualified People Plan "Hope of West China" (Y229151211), the Foundation of Science in Qinghai (2021-ZJ-734), Construction Project for Innovation Plat form of Qinghai province, and the Innovation Group on Medicinal Resource Utilization (2021XJP102) and Institute fund of medicinal technology.

Institutional Review Board Statement: Not applicable.

Informed Consent Statement: Not applicable.

Data Availability Statement: Data are available from the corresponding author.

Acknowledgments: We acknowledge Nanjing Genepioneer Biotechnologies Inc. for sequencing of *A. tanguticus* and appreciate Chuanbei Jiang, Peipei Cao, Xiaolong Luo and Mingxing Yang for assistance in data analysis.

Conflicts of Interest: The authors declare no conflict of interest.

Abbreviations

chloroplast genome: cP; inverted repeated: IR; large single copy: LSC; small single copy: SSC; simple sequence repeats: SSR; intergenic spacer regions: IGS; relative synonymous codon usage: RSCU; maximum likelihood: ML; locally collinear blocks: LCBs; synonymous Ks; non-synonymous Ka.

References

1. Rodríguez-Ezpeleta, N.; Brinkmann, H.; Burey, S.C.; Roure, B.; Burger, G.; Löffelhardt, W.; Bohnert, H.J.; Philippe, H.; Lang, B.F. Monophyly of primary photosynthetic eukaryotes: Green plants, red algae, and glaucophytes. *Curr. Biol.* **2005**, *15*, 1325–1330. [CrossRef] [PubMed]
2. Palmer, J.D. Comparative organization of chloroplast genomes. *Annu. Rev. Genet.* **1985**, *19*, 325–354. [CrossRef] [PubMed]
3. Smith, D.R. Mutation rates in plastid genomes: They are lower than you might think. *Genome Biol. Evol.* **2015**, *7*, 1227–1234. [CrossRef] [PubMed]
4. Wolfe, K.H.; Li, W.H.; Sharp, P.M. Rates of nucleotide substitution vary greatly among plant mitochondrial, chloroplast, and nuclear DNAs. *Proc. Natl. Acad. Sci. USA* **1987**, *84*, 9054–9058. [CrossRef] [PubMed]
5. Duchene, D.; Bromham, L. Rates of molecular evolution and diversification in plants: Chloroplast substitution rates correlate with species-richness in the Proteaceae. *BMC Evol. Biol.* **2013**, *13*, 65. [CrossRef] [PubMed]
6. Shaw, J.; Lickey, E.B.; Schilling, E.E.; Small, R.L. Comparison of whole chloroplast genome sequences to choose noncoding regions for phylogenetic studies in angiosperms: The tortoise and the hare III. *Am. J. Bot.* **2007**, *94*, 275–288. [CrossRef] [PubMed]
7. Maliga, P. Plastid transformation in higher plants. *Annu. Rev. Plant Biol.* **2004**, *55*, 289–313. [CrossRef]
8. Liu, S.W. *Flora of Qinghai*; Qinghai People's Publishing House: Xining, China, 1999.
9. Yang, Y.C. *Tibetan Medcine*; Qinghai People's Publishing House: Xining, China, 1990.
10. Xiao, P.G.; Xia, G.C.; He, L.Y. The occurrence of some important tropane alkaloids in Chinese solanaceous plants. *J. Integr. Plant Biol.* **1973**, *15*, 187–194.4.
11. Wang, H.; Zhang, X.F.; Chen, G.C.; Li, T.C.; Zhou, G.Y.; Shen, J.W. Comparative study of contents of four tropane alkaloids in cultural and wild Anisodus tanguticus. *Acta Bot. Boreal.-Occid. Sin.* **2005**, *25*, 575–577.
12. Spinks, A.; Wasiak, J.; Villanueva, E.; Bernath, V. Scopolamine (hyoscine) for preventing and treating motion sickness. *Cochrane Database Syst. Rev.* **2007**, *18*, CD002851.
13. Lei, T.X.; Cai, X.J.; Wang, H.; Li, S.L.; Shen, J.W.; Zhou, D.W. Progress on molecular mechanism of tropane alkaloids synthesis and plant bioengineering research. *Acta Bot. Boreal. Occid. Sin.* **2016**, *36*, 0204–0214.
14. Zheng, W.; Wang, L.Y.; Meng, L.H.; Liu, J.Q. Genetic variation in the endangered Anisodus tanguticus (Solanaceae), an alpine perennial endemic to the Qinghai-Tibetan Plateau. *Genetica* **2008**, *132*, 123–129. [CrossRef] [PubMed]
15. Duan, Y.W.; Zhang, T.F.; Liu, J.Q. Pollination biology of Anisodus tanguticus (Solanaceae). *Biodivers. Sci.* **2007**, *15*, 584–591.
16. Li, M.; Cao, H.; But, P.P.H.; Shaw, P.C. Identification of herbal medicinal materials using DNA barcodes. *J. Syst. Evol.* **2011**, *49*, 271–283. [CrossRef]
17. Raclariu, A.C.; Heinrich, M.; Ichim, M.C.; de Boer, H. Benefits and limitations of DNA barcoding and metabarcoding in herbal product authentication. *Phytochem. Anal.* **2018**, *29*, 123–128. [CrossRef] [PubMed]
18. Tu, T.Y. Phylogeny and Biogeography of the tribes Nolaneae, Hyoscyameae and Mandragoreae of Solanaceae. Ph.D. Thesis, Kunming Institute of Botany, Chinese Academy of Sciences, Kunming, China, 2008.
19. Sanchez-Puerta, M.V.; Abbona, C.C. The Chloroplast genome of Hyoscyamus niger and a phylogenetic study of the tribe Hyoscyameae (Solanaceae). *PLoS ONE* **2014**, *9*, e98353. [CrossRef]

20. Bolger, A.M.; Lohse, M.; Usadel, B. Trimmomatic: A flexible trimmer for Illumina sequence data. *Bioinformatics* **2014**, *30*, 2114–2120. [CrossRef]
21. Luo, R.; Liu, B.; Xie, Y.; Li, Z.; Huang, W.; Yuan, J.; He, G.; Chen, Y.; Qi, P.; Liu, Y. SOAPdenovo2: An empirically improved memory-efficient short-read de novo assembler. *Gigascience* **2012**, *1*, 18. [CrossRef]
22. Bankevich, A.; Nurk, S.; Antipov, D.; Gurevich, A.A.; Dvorkin, M.; Kulikov, A.S.; Lesin, V.M.; Nikolenko, S.I.; Pham, S.; Prjibelski, A.D. SPAdes: A new genome assembly algorithm and its applications to single-cell sequencing. *J. Comput. Biol.* **2012**, *19*, 455–477. [CrossRef]
23. Boetzer, M.; Henkel, C.V.; Jansen, H.J.; Butler, D.; Pirovano, W. Scaffolding pre-assembled contigs using SSPACE. *Bioinformatics* **2011**, *27*, 578–579. [CrossRef]
24. Wyman, S.K.; Jansen, R.K.; Boore, J.L. Automatic annotation of organellar genomes with DOGMA. *Bioinformatics* **2004**, *20*, 3252–3255. [CrossRef] [PubMed]
25. Liu, C.; Shi, L.; Zhu, Y.; Chen, H.; Zhang, J.; Lin, X.; Guan, X. CPGAVAS, an integrated web server for the annotation, visualization, analysis, and GenBank submission of completely sequenced chloroplast genome sequences. *BMC Genom.* **2012**, *13*, 715. [CrossRef] [PubMed]
26. Lohse, M.; Drechsel, O.; Bock, R. OrganellarGenomeDRAW (OGDRAW): A tool for the easy generation of high-quality custom graphical maps of plastid and mitochondrial genomes. *Curr. Genet.* **2007**, *52*, 7–274. [CrossRef] [PubMed]
27. Thiel, T.; Michalek, W.; Varshney, R.; Graner, A. Exploiting EST databases for the development and characterization of gene-derived SSR-markers in barley (*Hordeum vulgare* L.). *Theor. Appl. Genet.* **2003**, *106*, 411–422. [CrossRef]
28. Darling, A.C.E.; Mau, B.; Blattner, F.R.; Perna, N.T. Mauve: Multiple alignment of conserved genomic sequence with rearrangements. *Genome Res.* **2004**, *14*, 1394–1403. [CrossRef] [PubMed]
29. Katoh, K.; Standley, D.M. MAFFT multiple sequence alignment software version 7: Improvements in performance and usability. *Mol. Biol. Evol.* **2013**, *30*, 772–780. [CrossRef]
30. Wang, D.; Zhang, Y.; Zhang, Z.; Zhu, J.; Yu, J. KaKs_calculator 2.0: A toolkit incorporating gamma-series methods and sliding window strategies. *Genom. Proteom. Bioinform.* **2010**, *8*, 77–80. [CrossRef]
31. Castresana, J. Selection of conserved blocks from multiple alignments for their use in phylogenetic analysis. *Mol. Biol. Evol.* **2000**, *17*, 540–552. [CrossRef]
32. Stamatakis, A. RAxML version 8: A tool for phylogenetic analysis and post-analysis of large phylogenies. *Bioinformatics* **2014**, *30*, 1312–1313. [CrossRef]
33. Zhang, X.; Rong, C.X.; Qin, L.; Mo, C.Y.; Fan, L.; Yan, J.; Zhang, M.R. Complete chloroplast genome sequence of Malus hupehensis: Genome Structure, comparative analysis, and phylogenetic relationships. *Molecules* **2018**, *23*, 2917. [CrossRef]
34. Shen, X.; Wu, M.; Liao, B.; Liu, Z.; Bai, R.; Xiao, S.; Li, X.; Zhang, B.; Xu, J.; Chen, S. Complete chloroplast genome sequence and phylogenetic analysis of the medicinal plant Artemisia annua. *Molecules* **2017**, *22*, 1330. [CrossRef] [PubMed]
35. Yan, C.; Du, J.C.; Gao, L.; Li, Y.; Hou, X.L. The complete chloroplast genome sequence of watercress (Nasturtium officinale R. Br.): Genome organization, adaptive evolution and phylogenetic relationships in Cardamineae. *Gene* **2019**, *699*, 24–36. [CrossRef] [PubMed]
36. Liu, Q.; Xue, Q. Comparative studies on codon usage pattern of chloroplasts and their host nuclear genes in four plant species. *J. Genet.* **2005**, *84*, 55–62. [PubMed]
37. Sharp, P.M.; Li, W.H. The codon adaptation index-a measure of directional synonymous codon usage bias, and its potential applications. *Nucleic Acids Res.* **1987**, *15*, 1281–1295. [CrossRef]
38. Redwan, R.M.; Saidin, A.; Kumar, S.V. Complete chloroplast genome sequence of MD-2 pineapple and its comparative analysis among nine other plants from the subclass Commelinidae. *BMC Plant Biol.* **2015**, *15*, 196. [CrossRef]
39. Cavaliersmith, T. Chloroplast evolution: Secondary symbiogenesis and multiple losses. *Curr. Biol.* **2002**, *12*, R62–R64. [CrossRef]
40. Wang, W.; Yu, H.; Wang, J.; Lei, W.; Gao, J.; Qiu, X.; Wang, J. The complete chloroplast genome sequences of the medicinal plant Forsythia suspensa (Oleaceae). *Int. J. Mol. Sci.* **2017**, *18*, 2288. [CrossRef]
41. Nie, X.; Lv, S.; Zhang, Y.; Du, X.; Wang, L.; Biradar, S.S.; Tan, X.; Wan, F.; Weining, S. Complete chloroplast genome sequence of a major invasive species, Croftonweed (Ageratina adenophora). *PLoS ONE* **2012**, *7*, e36869. [CrossRef] [PubMed]
42. Timme, R.E.; Kuehl, J.V.; Boore, J.L.; Jansen, R.K. A comparative analysis of the Lactuca and Helianthus (Asteraceae) plastid genomes: Identification of divergent regions and categorization of shared repeats. *Am. J. Bot.* **2007**, *94*, 302–312. [CrossRef]
43. Huang, H.; Shi, C.; Liu, Y.; Mao, S.Y.; Gao, L.Z. Thirteen Camellia chloroplast genome sequences determined by high-throughput sequencing: Genome structure and phylogenetic relationships. *BMC Evol. Biol.* **2014**, *14*, 151. [CrossRef]
44. Zhao, Y.B.; Yin, J.L.; Guo, H.Y.; Zhang, Y.Y.; Xiao, W.; Sun, C.; Wu, J.Y.; Qu, X.B.; Yu, J.; Wang, X.M.; et al. The complete chloroplast genome provides insight into the evolution and polymorphism of Panax ginseng. *Front. Plant Sci.* **2015**, *5*, 696. [CrossRef] [PubMed]
45. Feng, S.G.; Zheng, K.X.; Jiao, K.L.; Cai, Y.C.; Chen, C.L.; Mao, Y.Y.; Wang, L.Y.; Zhan, X.R.; Ying, Q.C.; Wang, H.Z. Complete chloroplast genomes of four Physalis species (Solanaceae): Lights into genome structure, comparative analysis, and phylogenetic relationships. *BMC Plant Biol.* **2020**, *20*, 242. [CrossRef] [PubMed]
46. Raubeson, L.A.; Peery, R.; Chumley, T.W.; Dziubek, C.; Fourcade, H.M.; Boore, J.L.; Jansen, R.K. Comparative chloroplast genomics: Analyses including new sequences from the angiosperms Nuphar advena and Ranunculus macranthus. *BMC Genom.* **2007**, *8*, 174. [CrossRef] [PubMed]

47. Wang, R.J.; Cheng, C.L.; Chang, C.C.; Wu, C.L.; Su, T.M.; Chaw, S.M. Dynamics and evolution of the inverted repeat-large single copy junctions in the chloroplast genomes of monocots. *BMC Evol. Biol.* **2008**, *8*, 36. [CrossRef]
48. Yang, J.; Yue, M.; Niu, C.; Ma, X.F.; Li, Z.H. Comparative analysis of the complete chloroplast genome of four endangered herbals of Notopterygium. *Genes* **2017**, *8*, 124. [CrossRef]
49. Zhao, K.H.; Li, L.Q.; Lu, Y.Z.; Yang, J.B.; Zhang, Z.R.; Zhao, F.Y.; Quan, H.; Ma, X.J.; Liao, Z.H.; Lan, X.Z. Characterization and comparative analysis of two Rheum complete chloroplast genomes. *BioMed. Res. Int.* **2020**, *2020*, 6490164. [CrossRef] [PubMed]
50. Oxelman, B.; Erixon, P. Whole-gene positive selection, elevated synonymous substitution rates, duplication, and indel evolution of the chloroplast clpP1 gene. *PLoS ONE* **2008**, *3*, e1386.
51. Zong, D.; Zhou, A.P.; Zhang, Y.; Zou, X.L.; Li, D.; Duan, A.N.; He, C.Z. Characterization of the complete chloroplast genomes of five Populus species from the western Sichuan plateau, southwest China: Comparative and phylogenetic analyses. *Peer J.* **2019**, *7*, e6386.
52. Wu, Y.; Liu, F.; Yang, D.G.; Li, W.; Zhou, X.J.; Pei, X.Y.; Liu, Y.G.; He, K.L.; Zhang, W.S.; Ren, Z.Y.; et al. Comparative chloroplast genomics of Gossypium species: Insights into repeat sequence variations and phylogeny. *Front. Plant Sci.* **2018**, *9*, 376. [CrossRef]
53. Ali, M.D.; Baek, K.H. Protective roles of cytosolic and plastidal proteasomes on abiotic stress and pathogene invasion. *Plants* **2020**, *9*, 832. [CrossRef]
54. Ali, M.D.; Baek, K.H. Co-Suppression of NbClpC1 and NbClpC2, encoding Clp protease chaperons, elicits significant changes in the metabolic profile of Nicotiana benthamiana. *Plants* **2020**, *9*, 259. [CrossRef] [PubMed]
55. Zoschke, R.; Barkan, A. Genome-wide analysis of thylakoid-bound ribosomes in maize reveals principles of cotranslational targeting to the thylakoid membrane. *Proc. Natl. Acad. Sci. USA* **2015**, *112*, E1678–E1687. [CrossRef] [PubMed]
56. Börner, T.; Yu, A.; Yan, O.; Zubo, A.; Victor, V.; Kusnetsov, V.V. Chloroplast RNA polymerases: Role in chloroplast biogenesis. *Biochim. Biophys. Acta* **2015**, *1874*, 761–769. [CrossRef] [PubMed]
57. Mehmood, F.; Abdullah.; Ubaid, Z.; Shahzadi, I.; Ahmed, I.; Waheed, M.T.; Poczai, P.; Mirza, B. Plastid genomics of Nicotiana (Solanaceae): Insights into molecular evolution, positive selection and the origin of the maternal genome of Aztec tobacco (Nicotiana rustica). *Peer J.* **2020**, *8*, e9552. [CrossRef]
58. Zhang, D.; Fan, L.Q.; Zhang, L.; Wang, T.J.; Guo, X.Y.; Hu, Q.J. Characterization of the complete chloroplast genome of the endangered Przewalskia tangutica Maxim. *Conserv. Genet. Resour.* **2017**, *9*, 409–413. [CrossRef]

genes

MDPI

Article

Potential of In Vitro Culture of *Scutellaria baicalensis* in the Formation of Genetic Variation Confirmed by ScoT Markers

Jacek Gawroński and Magdalena Dyduch-Siemińska *

Department of Genetics and Horticultural Plant Breeding, Institute of Plant Genetics, Breeding and Biotechnology, University of Life Sciences in Lublin, Akademicka 15 Street, 20-950 Lublin, Poland
* Correspondence: magdalena.dyduch@up.lublin.pl

Abstract: The in vitro culture technique can be used for micropropagation of medicinal plants as well as for creating genotypes with an improved profile of phytochemical compounds. For this purpose, somaclonal variability may be used for the induction of genetic diversity among regenerants. The paper presents a protocol for obtaining *Scutellaria baicalensis* regenerants by indirect organogenesis and the assessment of their genetic variability with the use of start codon-targeted markers. The most intense process of indirect shoot organogenesis was observed on Murashige and Skoog medium supplemented with kinetin and 6-Benzylaminopurine (0.5 mg $\times$ dm^{-3} each)—7.4 shoot per explant on average. The callogenesis process occurred on the medium supplemented with TDZ, while the medium supplemented with GA_3 allowed for direct shoot organogenesis and was used for the micropropagation of regenerants. In the analysis of plantlets obtained by indirect organogenesis, 11 ScoT markers generated a total of 130 amplicons, 45 of which were polymorphic. This analysis showed genetic diversity of regenerants in relation to the donor plant as well as within them, with mean similarity among the analyzed genotypes at the level of 0.90. This study confirms that the use of in vitro cultures allows for the possibility to generate genetic variability in *Scutellaria baicalensis*, which can be effectively revealed with the use of the SCoT marker.

Keywords: indirect organogenesis; genetic diversity; medicinal plant; molecular marker; plant tissue culture; somaclonal variation

Citation: Gawroński, J.; Dyduch-Siemińska, M. Potential of In Vitro Culture of *Scutellaria baicalensis* in the Formation of Genetic Variation Confirmed by ScoT Markers. *Genes* **2022**, *13*, 2114. https://doi.org/10.3390/genes13112114

Academic Editors: Wajid Zaman and Hakim Manghwar

Received: 30 September 2022
Accepted: 11 November 2022
Published: 14 November 2022

1. Introduction

Scutellaria baicalensis Georgi. (common name—Baikal skullcap) is a widespread medicinal plant of the family *Lamiaceae*. It belongs to the genus *Scutellaria*, subfamily *Scutellarioideae*. The species is diploid (2n = 18) and belongs to entomophilous cross-pollinated plants [1,2]. Baikal skullcap is one of the 50 essential herbs of traditional Chinese medicine that have been used since ancient times to treat inflammation, high blood pressure, bacterial and viral infections, and to improve the overall health of the body. *S. baicalensis* contains many active substances, the main source of which is the root and, to a lesser extent, the aerial parts of the plant [3]. Baicalin, baicalein, vogonoside and vogonin are the most popular flavonoids, and their content in *S. baicalensis* root is as high as 20%, of which approximately 12–17% constitute baicalin and 3–4% constitute vogonoside [4]. In addition to flavonoids, *S. baicalensis* contains volatile oils, which are mainly responsible for the scent and sweet taste, and additionally exhibit antibacterial activity against Gram-positive and Gram-negative bacteria [5]. *S. baicalensis* also contains over 100 diterpenoids, mainly in the aerial part, which exhibit antibacterial and antiviral activity [6]. An important aspect in the cultivation of Baikal skullcap is the appropriate selection of conditions that have a significant impact on the content of active ingredients in the plant. In vitro cultures are an alternative to conventional methods of vegetative reproduction. Culturing plant material in vitro can induce or reveal variation between cells, tissues and organs, thereby creating variability within cultures or between regenerated plants (somaclonal variation). Some or all of the somaclons may differ phenotypically and genetically from the donor plants from which the culture was derived. Variability of this type, which usually occurs spontaneously and is largely uncontrolled, may be the result of genetic and epigenetic changes occurring in cells in vivo as well as during in vitro culture [7]. In vitro culture alone or in combination with mutagenesis induced by physicochemical and biological factors can be used to produce plants with increased genetic variability and mutants as a potential source of new commercial varieties. For this reason, plants regenerated from tissue, organ, callus, or protoplast cultures and through somatic embryogenesis may show variability at the phenotypic and genotypic level [8]. This type of variability can be studied using molecular markers. In 2009, Collard and Mackill [9] developed a new alternative and repeatable technique based on start codon-targeted (SCoT) markers. These markers are based on DNA and target the conserved ATG translation start codons. The SCoT system has been successfully used to assess genetic diversity or identify varieties and map quantitative trait loci in different species, generating high polymorphism and repeatability of the results.

The main aim of the study was: (i) to obtain regenerants of *S. baicalensis* through indirect organogenesis; and (ii) the assessment of genetic variation in plants regenerated in vitro using SCoT markers. The study also optimized the procedure for obtaining *S. baicalensis* regenerants and their multiplication in vitro, which will enable access to ready protocols for efficient propagation of the analyzed species.

2. Materials and Methods

In order to assess the possibility of inducing genetic variation during in vitro culture of the tested species, a two-step procedure for obtaining regenerants was applied.

2.1. Procedure for Obtaining Regenerants

2.1.1. Callus Induction and Shoot Regeneration

At this stage, through the use of nutrient supplementation with different types and concentrations of plant growth regulators (PGRs), it was determined which ones had the greatest effect on the process of indirect shoot organogenesis, i.e., callus tissue formation and shoot regeneration. The regenerated shoots could potentially show variation at the DNA level due to the genetic instability of the callus tissue. These shoots were obtained in such a way that the mature *S. baicalensis* plants grown at the Experimental Farm of the Department of Vegetable and Medicinal Plants of the University of Life Sciences in Lublin (51°14′53″ N, 2°34′13″ E) were used as a source material for the experiment, i.e., donor plants (DP—Figure 2). Directly after harvest, the explants were transported to the in vitro culture laboratory at the Institute of Plant Genetics, Breeding and Biotechnology, University of Life Sciences in Lublin. Healthy shoots with 10–12 nodes were harvested from field-grown plants and washed under tap water for an hour. Shoots were subsequently rinsed for 15 min in distilled water with a drop of Tween-20 on a stirrer. Shoots were sterilized with 75% ethanol for 1 min and 1% sodium hypochlorite for 5 min. Shoots were then washed three times (5 min each) in sterile distilled water and cut into fragments to obtain nodal explants (Figure 1). Explants were transferred to jars containing approximately 15 mL Murashige and Skoog (MS) [10] medium with different types and concentrations of plant growth regulators (PGRs). The type and concentration of PGs used are listed in Table 1. The medium composition was as follows: Murashige and Skoog basal medium supplemented with sucrose (30.0 g dm^{-3}), thiamine (0.4 g × dm^{-3}), pyridoxine (0.5 mg × dm^{-3}), nicotinic acid (0.5 mg × dm^{-3}), inositol (100 dm^{-3}), PGRs and agar-agar (8.0 g × dm^{-3}). (Sigma-Aldrich—St. Louis, MO, USA). For the experiment, the pH of the medium was adjusted to 5.8 with 1 M NaOH and 1 M HCl before autoclaving at 121 °C for 20 min. After initiation, the explants were cultured under 40 μmol m^{-2} s^{-1} light provided by cool white fluorescent tubes with a 16-h photoperiod and a temperature of 21 °C ± 2 °C. The experiment was carried out in triplicate. One replicate consisted of three jars of five explants each. Observations of the number of propagated shoots, their average length, average number of nodes and callus formation were carried out after 42 days of the culture.

Table 1. Phytohormones used in the media, their combinations and the observed effect on the regeneration process of *S. baicaliensis*.

Medium Number	Plant Growth Regulator Concentration and Combination (mg × dm^{-3})					Mean Number of Shoot per Explant	Mean Shoot Length (cm)	Mean Number of Nodes per Shoot	Calli Formation
	KIN	BAP	TDZ	GA_3	2iP				
1.	0.5	0.5	-	-	-	7.4 a * (I)	1.7 b	3.0 b	+++
2.	-	-	0.5	-	-	-	-	-	+++
3.	-	-	-	0.5	-	2.1 b (D)	6.3 a	4.4 a	-
4.	-	-	-	-	0.5	6.0 ab (I)	5.5 a	3.5 ab	−/+
5.	-	-	-	0.5	0.5	4.8 ab (I)	4.8 a	3.2 ab	+

*—means in columns followed by the same letter do not differ significantly at 5% level of probability, D—direct organogenesis, I—indirect organogenesis, +++ very high degree of callus tissue formation, ++ high degree of callus tissue formation, + low degree of callus tissue formation, - no callus tissue formation; KIN-kinetin; BAP-6-Benzylaminopurine; TDZ-thidiazuron; GA_3-gibberellic acid; 2iP-2-isopentenyladenine.

2.1.2. Shoot Multiplication of Regenerants

The hypothesis concerning the genetic diversity of regenerants obtained in the first stage was verified in the second stage by conducting direct shoot organogenesis, and the obtained plants were intended for DNA analysis. To this end, 10 randomly selected shoots regenerated on MS medium supplemented with 0.5 mg × dm^{-3} BAP (6-Benzylaminopurine) and kinetin 0.5 mg × dm^{-3} were cut into nodal fragments and used as secondary explants. Three nodal explants were collected from the selected regenerant for individual multiplication. Explants were placed on an MS medium supplemented with GA_3. The number of propagated shoots, their average length, the average number of nodes and root system development stage were assessed after 6 weeks of the culture. After this time, the plants were ready for the process of acclimatization.

2.2. *Acclimatization*

In vitro-rooted plantlets were removed from the jar. Their roots were washed with running tap water to remove the medium. Regenerants were inserted into plastic vessels filled with soil and perlite in a 2:1 ratio and kept in a growth chamber at 23–25 °C and 16-h photoperiod for 20 days. During this time, the high relative humidity in the vessel was gradually reduced by removing the cover. The plants were subsequently transferred to a greenhouse for further acclimatization.

2.3. *Molecular Assays*

2.3.1. DNA Extraction

Genomic DNA was isolated from a fragment of a shoot with leaves of the mother plant and regenerated plants after the multiplication process was completed (Section 2.2). DNA was extracted following the CTAB method described by Doyle and Doyle [11]. The DNA concentration was determined using a Nanodrop spectrophotometer (Thermo Scientific). All test samples were diluted to a final concentration of 25 ng μL^{-1}.

2.3.2. SCoT Analysis

To perform the genetic analysis of *S. baicalensis*, the PCR was optimized for the SCoT markers tested. The optimization involved determining the optimal magnesium ion concentration, which affects polymerase performance. Thus, primers that initiated stable amplification of clearly separated bands could be selected. Eleven 18-base primers selected from 20 arbitrary primers were used for PCR amplification. DNA amplification of SCoT markers was carried out in a final volume of 10 μL containing 0.5 U of Taq DNA Polymerase (Fermentas), 0.8 μL of oligonucleotide primer (0.8 μM), 1 μM dNTPs, 1 × PCR buffer with 1.5 mM $MgCl_2$, and 25 ng of genomic DNA as a template. Amplification was performed in a gradient thermal cycler (Biometra GmbH) with the following reaction conditions: initial predenaturation at 94 °C for 3 min, followed by 35 denaturation cycles at 94 °C for 1 min, annealing at 50 °C for 1 min, and extension at 72 °C for 2 min. The final extension was carried out for 5 min at 72 °C with the holding temperature of 4 °C. To verify reproducibility, the primers were tested twice on the same sample.

PCR products were electrophoresced in 1.5% agarose gels stained with ethidium bromide at constant voltage (3 V cm^{-1}) until bromophenol blue/loading dye migrated to the other end of the gel. The gel was visualized in a UV transilluminator and photographed using a GeneSnap ver. 7.09 (SynGene) gel documentation system. NZYDNA Ladder III (NZYTech) was used to establish the molecular weight of the products. Among obtained SCoT products, only reproducible and clear fragments were scored from the photographs. Bands detected in analyzed genotypes and scored as present (1) or absent (0) were considered polymorphic profiles, while specific bands were restricted to a specific individual. Indistinct or weak bands were excluded from the analysis.

2.4. Statistical Analysis

Statistical analysis of the results of shoot multiplication of regenerants was carried out using ANOVA, and the significance of differences between mean values was calculated using Duncan's multiple range tests performed at $p < 0.05$. The similarity coefficient between the studied genotypes in the SCoT analysis was assessed according to the Dice formula [12]. A cluster analysis was conducted using the UPGMA (unweighted pair-group method with arithmetic mean) distance method implemented in the PAST software [13].

3. Results and Discussion

3.1. Callus Induction and Shoot Regeneration

Phytohormones applied at the regeneration stage caused different directions of culture development (Table 1, Figure 1).

The process of direct shoot organogenesis was observed—medium 3, indirect shoot organogenesis (intensive callus formation with intensive shoot production, as well as the formation of a small amount of callus with simultaneous shoot formation)—medium 1 and 5, moreover formation of only callus tissue—medium 2. Since the prerequisite for obtaining genetic differentiation among regenerants is their formation via callus tissue, it should be pointed out that its formation was most intense on the medium containing KIN + BAP (0.5 mg × dm^{-3} each) and the medium with TDZ (0.5 mg × dm^{-3}) addition. In contrast, small amounts of callus were observed on media with a combination of GA_3 + 2iP (0.5 mg × dm^{-3} each) and with 2iP only (0.5 mg × dm^{-3}). With respect to the medium containing TDZ (0.5 mg × dm^{-3}), intensive formation of callus tissue was observed, on which subsequently a large number of shoot buds formed; unfortunately, they were not capable of regenerating shoots. Despite the statement of Guo et al. [14] that TDZ induced shoot regeneration in many plant species of the genus *Scutellaria*, (Table 1), such an effect was not recorded. Similarly, in a study of Ozdemir et al. [15], TDZ effectively stimulated the formation of callus tissue on various types of explants derived from seedlings of this species. However, on some of them, although shoot buds were clearly visible on the explants, they did not show any shoot regeneration. In a study by Stojakowska et al. [16], nodal explants on TDZ-containing media produced caulogeniccalli, and those obtained at a concentration of 0.5 μM TDZ developed numerous shoot buds, but only a few were capable of regenerating shoots. In contrast, Zhang et al. [17] reported that calli induced by 0.3 mg × dm^{-3} TDZ produced shoots directly, while those obtained with a higher TDZ concentration required a change of medium composition to initiate the process. In turn, Gharari et al. [18] indicated that by increasing TDZ concentrations (0.5 mg × dm^{-3}), the response rate of explants for shoot induction increased. Therefore, in connection with the aim of the research, it is necessary to indicate that potential genetic variability should be sought among the regenerants obtained on medium containing the combination of KIN + BAP (0.5 mg × dm^{-3} each). Nevertheless, it should be noted that, as shown by other authors, the process of indirect shoot organogenesis can be carried out using various combinations of phytohormones. According to Gharari et al. [19], stem explants of

S. araxensis allowed efficient shoot organogenesis through the callus stage, and obtaining up to 18 new shoots per explant on medium containing 0.5 mg × dm^{-3} BAP in addition to 0.5 mg × dm^{-3} IBA. Hwang et al. [20] reported that callus induction in *S. baicalensis* occurred in the presence of 1-Naphthaleneacetic acid (NAA) 1 mg × dm^{-3} plus BAP mg × dm^{-3}, while Trivedi et al. [21] showed that relatively high concentrations of TDZ (4 mg × dm^{-3}) stimulated this process, and subsequently NAA 1 mg × dm^{-3} was sufficient for shoot regeneration. In contrast, according to a study by Ozdemir et al. [22], in *S. orientalis* subsp. *bicolor*, there was the possibility of callus regeneration in the presence of BAP alone or in combination with NAA.

Figure 1. In vitro regeneration of *Scutellaria baicalensis* on Murashige and Skoog (MS) media. (**1a**)—primary explants used to establish the culture; (**1b**)—Donor plants in in vivo condition; (**1c**)—Indirect organogenesis from nodal explants MS medium supplemented with KIN + BAP (0.5 mg × dm^{-3} each); (**1d**)—Cultures on the media according to Table 1 after 42 days. (**1e**)—Plantlet on the media according to Table 1 after 42 days. (**1f**)—Plantlet acclimatization in pot containing soil and perlite in a 2:1 ratio.

Considering the number of shoots produced by the explant, the medium containing KIN + BAP (0.5 mg × dm^{-3} each) proved to be most optimal, and an average of 7.4 new shoots were obtained on this substrate. Slightly lower values, although statistically not significantly different from the above, were recorded on medium enriched with 2iP (0.5 mg × dm^{-3})—6.0 shoots per explant, and the medium with GA_3 and 2iP (0.5 mg × dm^{-3} each)—4.8 shoots per explant. In contrast, only 2.1 shoots per explant were obtained on substrate with GA_3. The latter is generally known for its effect in the direction of shoot elongation [19], and only caused the development of two axillary buds present on the explant, with the length of the resulting shoots being greater compared to those obtained on the other media. Stem elongation in the presence of GA_3 also resulted in the development of an average of one internode more on the shoot compared to the other PGRs applied. Nevertheless, when using the regenerant cultures obtained at this stage for further multiplication using nodal explants, the highest culture efficiency can be obtained using KIN + BAP medium (7.4 × 3.0 = 22.2 nodal segments) and 2iP (6.0 × 3.5 = 21 nodal segments). This stage of culture could be summarized by the statement that the PGRs used in the medium induced a differential response of the explants (same genotype), resulting in the possibility of genetic variability occurrence among regenerants, particularly those obtained on medium 1.

3.2. Shoot Multiplication of Regenerants

Due to the fact that the process of direct regeneration from axillary buds was observed on the medium with GA_3 addition, it was applied in the second stage of culture, which required limiting the possibility of variability, and as reported by Ngezahayo and Liu [23], axillary bud proliferation was the most frequently used, and also considered the most suitable to guarantee the genetic stability of the regenerated plants. Since a genetically diverse group of genotypes is likely to be evaluated at this stage, it should be assumed that they would show a slightly different morphogenetic response to the applied medium (the same for all genotypes). The data in Table 2 and Figure 2 clearly show the differentiation of the regenerants at the phenotypic level expressed by the coefficient of variation, whose value was the highest for the mean shoot number, while slightly lower for the other two traits—mean shoot length and mean number of nodes per shoot. This especially reflected the variability in the number of shoots, which ranged from 2.0 to 9.4. Variability regarding this trait was also observed in other species of the genus. The average number of shoots in *S. alpina* and *S. altissima* was 5.4 and 2.0, respectively, and was obtained on a medium with identical PGR composition [24]. The observed differences could be due to the fact that the studied species reacted differently to the applied media. On the other hand, the identical number of shoots (an average of 4.5) obtained on MS medium supplemented with meta-Topolin + NAA for both *S. barbata* and *S. racemosa*, in this case indicated a similar response of each species [25]. The interactions between genotype, a type and PGR concentration in the assessment of morphogenetic capacity, were described previously by Dyduch-Siemińska [26]. Comparing the mean value of shoot length and mean number of nodes per shoot of the analyzed regenerants in relation to the donor plant cultured on the same medium (Table 1—medium 3), genotypes R5 and R9 could be considered the most distinct characterized by advantageous morphogenetic potential, while R7 was particularly unfavorable. Regenerants after this stage of culture were fully prepared for the acclimatization process and after its completion (Figure 1f), they were planted at the Experimental Farm of the Department of Vegetable and Medicinal Plants to evaluate their phytochemical properties.

Table 2. Evaluation of the morphogenetic capacity of regenerants derived from medium number 1 regenerated on medium number 3.

Plant Number	Mean Number of Shoot	Mean Shoot Length	Mean Number of Nodes per Shoot	Root System Development Stage [1]
R1	3.4 bc *	4.1 d	5.7 de	1
R2	4.6 b	6.0 c	8.4 c	3
R3	2.0 c	7.8 bc	7.0 cd	1
R4	3.4 bc	7.7 bc	6.7 cd	3
R5	2.0 c	13.3 a	16.5 a	2
R6	7.4 a	2.8 e	4.6 e	2
R7	9.4 a	3.2 de	4.0 e	3
R8	8.6 a	5.0 cd	5.4 de	2
R9	2.4 bc	8.7 b	10.5 b	3
R10	3.6 bc	8.4 b	6.9 cd	2
Mean	4.6	6.7	7.5	-
Coefficient of variation (%)	56.8	44.8	45.2	-

*—means in columns followed by the same letter do not differ significantly at 5% level of probability, [1]—Root system development stage: 1—only a few short roots, 2—a few roots up to 2 cm long, 3—several roots longer than 2 cm.

Figure 2. Phenotypic diversity of regenerants (from R-1 to R-10) on medium number 3 after 6 weeks of the culture.

3.3. *Molecular Assays*

Since DNA methylation, amplification, activation of transposable elements, polyploidy, changes in chromosome number or DNA sequence can be affected by in vitro culture conditions and consequently lead to genetic changes in regenerated plants [27,28], it is necessary to monitor the occurrence of somaclonal variations during in vitro propagation and assess the genetic constitution and variability of the plants regenerated in vitro. The genetic diversity was successfully detected in the genus *Scutellaria* and mainly concerned the assessment of the genetic diversity of the study's wild and cultivated populations using RAPD markers [29–31], ISSR markers [32,33], and SSR (simple sequence repeat) markers [34]. The use of the aforementioned markers to assess the variability of Baikal skullcap genotypes obtained using a tissue culture has been reported sporadically in the literature, but SCoT-type markers were used for the first time in this study.

Eleven SCoT primers were analyzed in this study, and their sequences are listed in Table 3. The visualized amplification products after electrophoresis are shown in the images (Figure 3). All primers produced a total of 130 polymorphic bands, averaging more than 11 bands per primer. The percentage of polymorphism obtained ranged from 20 to 45%, averaging 34.86% for the eleven primers mentioned above. Previous studies in medicinal plants [35,36] used RAPD and ISSR markers and indicated the presence of somaclonal variation in callus tissue, as well as between the regenerants obtained from it [37]. Specific products generated by eight primers (SCoT 4, SCoT 14, SCoT 19, SCoT 24, SCoT 25, SCoT 35, SCoT 46 and SCoT 50) were also detected. A total of 11 products were obtained, with an average of 0.73 bands for all primers amplifying these products. All analyzed primers generated monomorphic products. Primer SCoT 46 produced the highest number of bands of this type, and SCoT 35 produced the lowest. The size of the obtained products ranged between 310 and 9800 bp.

Based on the polymorphism identified using the SCoT technique, the genetic similarity between the studied genotypes was determined according to Dice's formula after Nei and Li [12]. Table 4 shows the values of genetic similarity between the genotypes. The similarity of individual regenerants to the donor plant ranged from 0.82 to 0.86. Regenerates R3 and R8 showed the highest similarity to the donor plant (0.86), while R7 showed the lowest, at the level of 0.82. The highest similarity was found for regenerant R3 and R4, as well as R8 and R9, amounting to 0.96. A similarity below 0.9 was observed between regenerant

R1 and R2; R2–R8; R2–R9. The mean similarity among the analyzed genotypes was 0.90, which indicated rather minor changes within the studied group of regenerants. The level of polymorphism in the range of 10–15% indicates low genetic variability for the SSR, RAPD and ISSR markers [35,36,38]. Etminan et al. [39] and Shahlaei et al. [40] reported that the verification of the results with SCoT markers could lead to different conclusions due to the higher informativeness of this marker compared to other marker systems. The variability observed in our study was due to the use of more sensitive markers, such as SCoT-type markers. Therefore, the utilization of this marker system in the identification of genetic variation in regenerants grown in in vitro cultures allows the detection of more subtle changes at the DNA level. The results are also visualized in the form of a dendrogram (Figure 4). It was observed that the studied genotypes could be assigned to two groups, which consisted of one single genotype (DP) and a cluster consisting of the remaining 10 regenerants. The presented method of grouping the studied genotypes clearly indicates the distinctiveness of DP in comparison to all in vitro regenerated plants. This confirms the existence of genetic variation within the analyzed genotypes, resulting from the process of indirect organogenesis used to obtain them. The reasons for this differentiation may result from in vitro culture conditions affecting different regions of the genome, including the types and level of PGRs use, which can stimulate rapid and disordered cell proliferation or the accumulation of somatic mutation during the tissue culture period [41–43].

Table 3. Assessment of the number and type of products generated with the primers used in the study.

Primer	Primer Sequence 5′-3′	Number of Products				% of Polymorphism	Size Range (bp)
		Total	Polymorphic	Specific	Monomorphic		
SCoT 4	CAACAATGGCTACCACCT	10	2	1	7	20.00	600–6000
SCoT 9	CAACAATGGCTACCAGCA	9	4	0	5	44.44	1200–5800
SCoT 14	ACGACATGGCGACCACGC	13	3	2	8	23.08	420–3200
SCoT 19	ACCATGGCTACCACCGGC	15	5	2	8	33.33	310–9300
SCoT 24	CACCATGGCTACCACCAT	11	5	1	5	45.45	520–9000
SCoT 25	ACCATGGCTACCACCGGG	10	3	1	6	30.00	370–6000
SCoT 31	CCATGGCTACCACCGCCT	12	5	0	7	41.67	470–8000
SCoT 35	CATGGCTACCACCGGCCC	8	3	1	4	37.50	380–9800
SCoT 36	GCAACAATGGCTACCACC	13	5	0	8	38.46	480–5500
SCoT 46	ACAATGGCTACCACTGAG	15	4	1	10	26.67	500–8200
SCoT 50	ACAATGGCTACCACTGGG	14	6	2	6	42.86	380–5000
	Mean per primer	11.82	4.09	1.00	6.73	34.86	-
	Total	130	45	11	74	-	310–9800

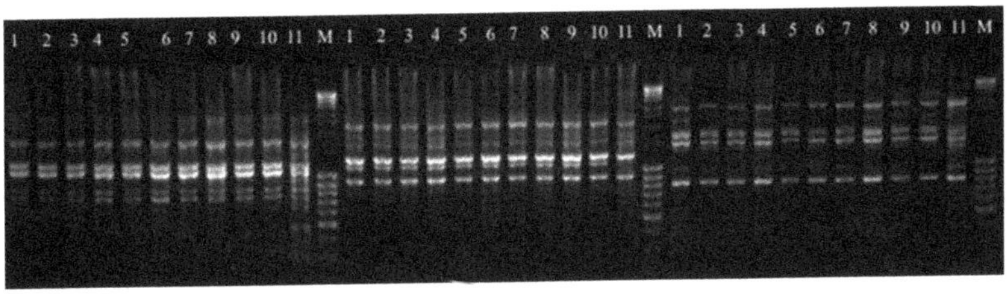

Figure 3. SCoT profile of regenerants 1–10 (from R-1to R-10 respectively) and DP—11 for primer 19, 24 and 25 respectively. M—NZYDNA Ladder III.

Table 4. Matrix of genetic similarity between the studied genotypes obtained on the basis of SCoT markers.

	R1	R2	R3	R4	R5	R6	R7	R8	R9	R10	DP
R1	1.00	0.86	0.94	0.94	0.92	0.90	0.90	0.92	0.92	0.92	0.85
R2		1.00	0.90	0.90	0.93	0.93	0.92	0.89	0.89	0.90	0.83
R3			1.00	0.96	0.93	0.91	0.94	0.95	0.93	0.93	0.86
R4				1.00	0.92	0.92	0.91	0.93	0.92	0.90	0.84
R5					1.00	0.92	0.92	0.92	0.92	0.92	0.83
R6						1.00	0.94	0.91	0.91	0.90	0.83
R7							1.00	0.95	0.93	0.93	0.82
R8								1.00	0.96	0.94	0.86
R9									1.00	0.95	0.84
R10										1.00	0.83
DP											1.00

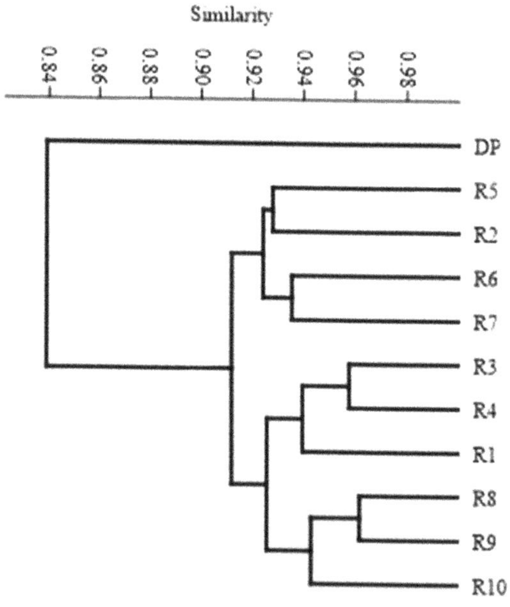

Figure 4. UPGMA dendrogram of DP and regenerants (from R-1 to R-10) based on SCoT marker analysis.

4. Conclusions

In the literature, a number of works can be found indicating the possibility of regenerating shoots from nodal segments in various *Scutellaria* species, however, none of them analyzed the impact of regeneration through the callus stage (indirect organogenesis) on the variability within the obtained regenerants, and this has been presented in this study both at the phenotypic level and in detail at the genotypic level. The authenticated protocol presented in this work enables the micropropagation of *S. baicalensis*. The application of a protocol using indirect organogenesis generates somaclonal variation among the regenerants. The effective revealing of this variability at the DNA level was possible thanks to the precise SCoT marker system, which was presented for the first time for the studied species. This provides a basis for further biochemical studies of the obtained regenerants aimed at isolating genotypes with higher secondary metabolite content of *S. baicalensis*.

Author Contributions: Conceptualization, J.G. and M.D.-S.; methodology, J.G.; software, J.G.; validation, J.G. and M.D.-S.; formal analysis, J.G.; investigation, M.D.-S.; resources, J.G.; data curation, J.G.; writing—original draft preparation, J.G.; writing—review and editing, M.D.-S.; visualization, M.D.-S.; supervision, J.G.; project administration, M.D.-S.; funding acquisition, J.G. All authors have read and agreed to the published version of the manuscript.

Funding: The work was funded from the statutory activity of the University of Life Sciences in Lublin UP/RGH/6/022.

Institutional Review Board Statement: Not applicable.

Informed Consent Statement: Not applicable.

Data Availability Statement: Not applicable.

Conflicts of Interest: The authors declare no conflict of interest.

References

1. Xu, Z.; Gao, R.; Pu, X.; Xu, R.; Wang, J.; Zheng, S.; Zeng, Y.; Chen, J.; He, C.; Song, J. Comparative Genome Analysis of *Scutellaria baicalensis* and *Scutellaria barbata* Reveals the Evolution of Active Flavonoid Biosynthesis. *Genom. Proteom. Bioinform.* **2000**, *18*, 230–240. [CrossRef]
2. Zhang, H.; Gao, Z.; He, D.; Wang, W. Pollination biology characters of *Scutellaria baicalensis*. *Acta Hortic. Sin.* **2011**, *38*, 2209–2214.
3. Ciocarlan, N. Scutellaria baicalensis Georgi in the national botanical garden (institute), Republic of Moldova. In *Main, Rare and Non-Traditional Plant Types—From Research to Development (Agricultural and Biological Sciences)*; National Academy of Agrarian Sciences of Ukraine: Georgia, Ukraine, 2021; Volume 4, pp. 123–127.
4. Jarosławska, A.; Sokół-Łętowska, A.; Oszmiański, I.J. Use of natural polyphenols to stabilization of sunflower oil. *Food* **2003**, *2*, 77–86.
5. Zhao, T.; Tang, H.; Xie, L.; Zheng, Y.; Ma, Z.; Sun, Q.; Li, I.X. *Scutellaria baicalensis* Georgi. (Lamiaceae): A review of itstraditional uses, botany, phytochemistry, pharmacology and toxicology. *JPP* **2019**, *71*, 1353–1369. [CrossRef]
6. Song, J.-W.; Long, J.-Y.; Xie, L.; Zhang, L.-L.; Xie, Q.-X.; Chen, H.-J.; Deng, M.; Li, X.-F. Applications, phytochemistry, pharmacological effects, pharmacokinetics, toxicity of *Scutellaria baicalensis* Georgi. and its probably potential therapeutic effects on COVID-19: A review. *Chin. Med.* **2020**, *15*, 102. [CrossRef] [PubMed]
7. Krishna, H.; Alizadeh, M.; Singh, D.; Singh, U.; Chauhan, N.; Eftekhari, M.; Kishan, R. Somaclonal variations and their applications in horticultural crops improvement. *3 Biotech* **2016**, *6*, 54. [CrossRef] [PubMed]
8. Orbović, V.; Ćalović, M.; Viloria, Z.; Nielsen, B.; Gmitter, F.; Castle, W.; Grosser, J. Analysis of genetic variability in various tissue culture-derived lemon plant populations using RAPD and flow cytometry. *Euphytica* **2000**, *161*, 329–335. [CrossRef]
9. Collard, B.C.Y.; Mackill, D.J. Start Codon Targeted (SCoT) Polymorphism: A Simple, Novel DNA Marker Technique for Generating Gene-Targeted Markers in Plants. *Plant Mol. Biol. Rep.* **2009**, *27*, 86–93. [CrossRef]
10. Murashige, T.; Skoog, F.A. Revised medium for rapid growth and bioassays with tobacco tissue culture. *Physiol. Plant.* **1962**, *15*, 473–497. [CrossRef]
11. Doyle, J.J.; Doyle, J.L. A rapid DNA isolation procedure for small quantities of fresh leaf tissue. *Phytochem. Bull.* **1987**, *9*, 11–15.
12. Nei, M.; Li, W. Mathematical model for studying genetic variation in terms of restriction endonucleases. *Proc. Natl. Acad. Sci. USA* **1979**, *76*, 5269–5273. [CrossRef] [PubMed]
13. Hammer, Ø.; Harper, D.A.T.; Ryan, P.D. Past: Paleontological statistics software package for education and data analysis. *Palaeontol. Elect.* **2001**, *4*, 9.
14. Guo, B.; Abbasi, B.H.; Zeb, A.; Xu, L.L.; Wei, Y.H. Thidiazuron: A multi-dimensional plant growth regulator. *Afr. J. Biotechnol.* **2011**, *10*, 45. [CrossRef]
15. Ozdemir, F.; Yildirim, M.; Kahriz, P. Micropropagation of Endemic *Scutellaria orientalis* L. Subsp. Bicolor Using Modified MS Medium & TDZ. *Emir. J. Food Agric.* **2015**, *27*, 818–824. [CrossRef]
16. Stojakowska, A.; Malarz, J.; Kohlmünzer, J. Micropropagation of *Scutellaria baicalensis* Georgi. *Acta Soc. Bot. Pol.* **1999**, *68*, 103–107. [CrossRef]
17. Zhang, C.G.; Li, W.; Mao, Y.F.; Zhao, D.L.; Dong, W.; Guo, G.Q. Endogenous Hormonal Levels in *Scutellaria baicalensis* Calli Induced by Thidiazuron. *Russ. J. Plant Physiol.* **2005**, *52*, 345–351. [CrossRef]
18. Gharari, Z.; Bagheri, K.; Sharafi, A.; Danafar, H. Thidiazuron induced efficient in vitro organogenesis and regeneration of *Scutellaria bornmuelleri*: An important medicinal plant. *Vitr. Cell. Dev. Biol.-Plant* **2019**, *55*, 133–138. [CrossRef]
19. Gharari, Z.; Bagheri, K.; Sharafi, A. High-frequency adventitious shoot organogenesis from in vitro stem explants of *Scutellaria araxensis* Grossh. *BioTechnologia* **2022**, *103*, 143–151. [CrossRef]
20. Hwang, I.-T.; Lee, J.-J.; Lee, J.; Paik, S.-W.; Kim, Y.-H. Production of Baicalin, Baicalein, and Wogonin on Plant Tissue Culture of *Scutellaria baicalensis*. *Korean J. Plant Resour.* **2015**, *28*, 526–532. [CrossRef]
21. Trivedi, M.; Trivedi, R.K.; Guag, Z.C.; Guo, G.; Zheng, G. Hormone-Induced Indirect Regeneration Protocol for *Scutellaria baicalensis* Georgi (Huang-qin). *J. Crop Improv.* **2011**, *25*, 550–559. [CrossRef]

22. Ozdemir, F.A.; Kilic, O.; Atalan, E. In vitro callus propagation and antibacterial activities of callus an edible endemic and medicinal plant *Scutellaria orientalis* L. subsp. bicolor. *Prog. Nutr.* **2016**, *18*, 81–86.
23. Ngezahayo, F.; Liu, B.Y. Axillary Bud Proliferation Approach for Plant Biodiversity Conservation and Restoration. *Int. J. Biodivers.* **2014**, *2014*, 727025. [CrossRef]
24. Zakaria, I.A.; Sherman, S.; Vaidya, B.; Joshee, N. In vitro propagation and synseed mediated short-term conservation of *Scutellaria alpine* L. and *Scutellaria altissima* L. *Plant Cell Cult. Micropropag.* **2020**, *16*, e163. [CrossRef]
25. Brearley, T.; Vaidya, B.; Joshee, N. Cytokinin, Carbon Source, and Acclimatization Requirements for In Vitro Propagation of *Scutellaria barbata* D. Don and *Scutellaria racemose* Pers. *Am. J. Plant Sci.* **2014**, *5*, 3662–3672. [CrossRef]
26. Dyduch-Siemińska, M. A fast and effective protocol for obtaining genetically diverse stevia (*Stevia rebaudiana* Bertoni) regenerants through indirect organogenesis. *Agron. Sci.* **2021**, *76*, 47–62. [CrossRef]
27. Bairu, M.W.; Aremu, A.O.; Van Staden, J. Somaclonal variation in plants: Causes and detection methods. *Plant Growth Regul.* **2011**, *63*, 147–173. [CrossRef]
28. Govindaraju, S.; Arulselvi, P.I. Effect of cytokinin combined elicitors (l-phenylalanine, salicylic acid and chitosan) on in vitro propagation, secondary metabolites and molecular characterization of medicinal herb—*Coleus aromaticus* Benth (L.). *J. Saudi Soc. Agric. Sci.* **2018**, *17*, 435–444. [CrossRef]
29. Su, S.; He, C.M.; Li, L.C.; Chen, K.J.; Zhou, T.S. Genetic characterization and phytochemical analysis of wild and cultivated populations of *Scutellaria baicalensis*. *Chem. Biodivers.* **2008**, *5*, 1353–1363. [CrossRef]
30. Hosokawa, K.; Minami, M.; Kawahara, K.; Nakamura, I.; Shibata, T. Discrimination among three species of medicinal *Scutellaria* plants using RAPD markers. *Planta Med.* **2000**, *66*, 270–272. [CrossRef]
31. Shao, A.J.; Li, X.; Huang, L.Q.; Lin, S.F.; Chen, J. RAPD analysis of *Scutellaria baicalensis* from different germplasms. *Zhongguo Zhong Yao Za Zhi* **2006**, *31*, 452–455.
32. Bai, C.; Wen, M.; Zhang, L.; Li, G. Genetic diversity and sampling strategy of *Scutellaria baicalensis* germplasm resources based on ISSR. *Genet. Resour. Crop Evol.* **2013**, *60*, 1673–1685. [CrossRef]
33. Muraseva, D.S.; Guseva, A.A. ISSR primer screening for analysis of genetic diversity among *Scutellaria tuvensis* (Lamiaceae) populations. *BIO Web Conf.* **2021**, *38*, 00082. [CrossRef]
34. Yuan, Y.; Long, P.; Jiang, C.; Li, M.; Huang, L. Development and characterization of simple sequence repeat (SSR) markers based on a full-length cDNA library of *Scutellaria baicalensis*. *Genomics* **2015**, *105*, 61–67. [CrossRef] [PubMed]
35. Mahmood, T.; Jameel, A.; Abbasi, B.H.; Manir, F.; Naqvi, S.M.S. In vitro callogenesis and detection of somaclonal variations in *Plantago ovata* L. *J. Crop Sci. Biotechnol.* **2012**, *15*, 289–295. [CrossRef]
36. Ghorbanpour, M.; Khadivi-Khub, A. Somaclonal variation in callus samples of *Plantago major* using inter-simple sequence repeat marker. *Caryologia* **2015**, *68*, 19–24. [CrossRef]
37. Vitamvas, J.; Viehmannova, I.; Cepkova, P.H.; Mrhalova, H.; Eliasova, K. Assessment of somaclonal variation in indirect morphogenesis-derived plants of *Arracacia xanthorrhiza*. *Pesqui. Agropecuária Bras.* **2019**, *54*, e00301. [CrossRef]
38. Chávez-Cortazar, A.; Mata-Rosas, M.; Oyama, K.; Samain, M.S.; Quesada, M. Induction of somatic embryogenesis and evaluation of genetic stability in regenerated plants of *Magnolia dealbata*. *Biol. Plant.* **2020**, *64*, 224–233. [CrossRef]
39. Etminan, A.; Pour-Aboughadareh, A.; Noori, A.; Ahmadi-Rad, A.; Shooshtari, L.; Mahdavian, Z.; Yousefiazar-Khanian, M. Genetic relationships and diversity among wild *Salvia* accessions revealed by ISSR and SCoT markers. *Biotechnol. Biotechnol. Equip.* **2018**, *32*, 610–617. [CrossRef]
40. Shahlaei, A.; Torabi, S.; Khosroshahli, M. Efficacy of SCoT and ISSR marekers in assessment of tomato (*Lycopersicum esculentum* Mill.) genetic diversity. *Int. J. Biosci.* **2014**, *5*, 14–22.
41. Rathore, M.S.; Chikara, J.; Mastan, S.G.; Rahman, H.; Anand, K.G.V.; Shekhawat, N.S. Assessment of genetic stability and instability of tissue culture-propagated plantlets of *Aloe vera* L. by RAPD and ISSR markers. *Appl. Biochem. Biotechnol.* **2011**, *165*, 1356–1365. [CrossRef]
42. Rahmani, M.S.; Pijut, P.M.; Shabanian, N.; Nasri, M. Genetic fidelity assessment of in vitro-regenerated plants of *Albizia julibrissin* using SCoT and IRAP fingerprinting. *Vitr. Cell. Dev. Biol.-Plant.* **2015**, *51*, 407–419. [CrossRef]
43. Soares, D.M.M.; Sattler, M.C.; Ferreira, M.F.D.S.; Praça-Fontes, M.M. Assessment of genetic stability in three generations of in vitro propagated *Jatropha curcas* L. plantlets using ISSR markers. *Trop. Plant Biol.* **2016**, *9*, 229–238. [CrossRef]

genes

MDPI

Article

Population Diversity Analysis Provide Insights into Provenance Identification of *Dendrobium catenatum*

Xin-Yi Wu [1,2], Ting-Zhang Li [1,2], Fang Zheng [1], Jian-Bing Chen [1], Yue-Hong Yan [1] and Jiu-Xiang Huang [2,*]

1 Key Laboratory of National Forestry and Grassland Administration for Orchid Conservation and Utilization, The Orchid Conservation and Research Center of Shenzhen and the National Orchid Conservation Center of China, Shenzhen Key Laboratory for Orchid Conservation and Utilization, Shenzhen 518114, China
2 South China Limestone Plants Research Center, College of Forestry and Landscape Architecture, South China Agricultural University, Guangzhou 510642, China
* Correspondence: jxhuang@scau.edu.cn

Abstract: *Dendrobium catenatum* (*Dendrobium officinale*) is a valuable genuine herb. The source of this species is difficult to be identified by traditional methods including morphology, spectroscopy, and chromatography. We used the restriction site-associated DNA sequencing (RAD-seq) approach to perform the high-throughput sequencing of 24 *D. catenatum* provenances. In this study, 371.18 Gb clean data were obtained, and 655,057 high-quality SNPs were selected after their filtration. We used phylogenetic tree, genetic structure, and principal component analyses to examine the genetic diversities and genetic relationships of the 109 accessions. We found that *D. catenatum* could be divided into two groups, and each group was closely related to the distribution of the sampling sites. At the population level, the average nucleotide diversity (π) of the *D. catenatum* population mutation parameters was 0.1584 and the expected heterozygosity (H_E) was 0.1575. The GXLPTP07 accessions showed the highest genetic diversity in terms of the private allele number, observed heterozygosity, and nucleotide diversity. The Mantel test showed a significant positive correlation between the genetic and geographic distances among the overall distribution. A genetic information database of *D. catenatum* was established, which confirmed that RAD-seq technology has the potential to be applied in the identification of medicinal *Dendrobium* of different origins.

Keywords: *Dendrobium catenatum*; genetic diversity; provenance identification; genuine regional drugs; RAD-seq; SNP

Citation: Wu, X.-Y.; Li, T.-Z.; Zheng, F.; Chen, J.-B.; Yan, Y.-H.; Huang, J.-X. Population Diversity Analysis Provide Insights into Provenance Identification of *Dendrobium catenatum*. *Genes* **2022**, *13*, 2093. https://doi.org/10.3390/genes13112093

Academic Editors: Wajid Zaman and Hakim Manghwar

Received: 11 October 2022
Accepted: 5 November 2022
Published: 10 November 2022

1. Introduction

The genus *Dendrobium* Swartz (1799) is one of the largest genera in Orchidaceae [1], and it consists of approximately 1450 species [2]. *Dendrobium* is highly valued for its medicinal, ornamental, scientific and economics uses, but its medicinal value has attracted the most attention. There are nearly 80 species of *Dendrobium* in China. Many *Dendrobium* species have been extensively used in Traditional Chinese Medicine (TCM), including *D. catenatum* Lindl (*D. officinale* Kimura and Migo) and *D. huoshanense* C.Z. Tang and S.J. Cheng [3,4]. Many studies have shown that *Dendrobium* polysaccharide has antitumor effects [5], immune enhancement [6,7], antioxidation [8,9], anti-inflammatory [10,11], and hypoglycemic effects [12,13]. There is also research showing that *Dendrobium* can improve learning and memory [14]. The medicinal *Dendrobium* industry has developed rapidly in recent years, and some growers have carried out cross-breeding for medicinal *Dendrobium* in pursuit of increasing their yield and profits, which has caused confusion in the market regarding the *Dendrobium* varieties. There are great differences in the pharmaceutical components between *Dendrobium* species [15], which affect the efficacy or curative effect of them. A large number of medicinal *Dendrobium* products on the market have lost their active ingredients, and variable quality and effective identification methods are lacking. Traditional morphological identification, microscopic identification, and the physical and chemical identification methods can only be used for the interspecies identification of medicinal *Dendrobium* [16–18]. Obviously, it is difficult for conventional techniques to identify the provenance of medicinal *Dendrobium* as they are similar in their morphological and anatomical characteristics and even in their chemical components. One of the most important factors for the modernization of Chinese medicines is the quality of the traditional Chinese medicines, and the key support for it is the study of its genuineness [19]. A variety of genuine medicinal materials had been confirmed that the effective components of different producing areas are very different [20–22]. *D. catenatum* is a geo-authentic Chinese medicinal material, and so its identification of provenance is very important.

For identification purposes, researchers have established a full-sequence database that is based on the rDNA ITS region of the *Dendrobium* variety "Fengdou" [23]. The phylogenetic tree constructed from *matK* and *rbcL* data could distinguish five types of medicinal *Dendrobium* [24]. For the purpose of the swift and precise identification of thirteen wild and cultivated *Dendrobium* species belonging to two sections *Formosae* and *Chrysotoxae*, the researchers designed the rDNA ITS region sequence analysis [25]. In recent years, the molecular identification of *Dendrobium* plants has also been studied. Inter-simple sequence repeats (ISSR) molecular fingerprinting markers had been employed to authenticate eight populations of *D. officinale* using 10 primers [26]. Two genuine population had been authenticated based on the SNPs of the rDNA ITS region [27]. However, microsatellite studies have typically used a low number of markers (<25), which increases the danger of underestimating the genetic structure due to a lack of polymorphic markers [28]. When one is using a low number of genetic markers, larger sample sizes are required to accurately estimate the allele frequencies and diversity [29], which might be difficult to achieve, especially for *Dendrobium* as it is an endangered species that has been severely damaged. The SNPs produce more precise estimates of the population-level diversity, and they use a higher power to identify the groups in clusters than the methods of using microsatellites do [30]. To evaluate and quantify the potential marker-specific biases, researchers have compared the microsatellite variation with genome-wide SNPs. They concluded that a few thousand random SNPs are sufficient to accurately estimate the genome-wide diversity and to distinguish between the populations with different levels of genetic variation [31].

Currently, with the development of high-throughput sequencing technology, SNPs are increasingly used in species identification and population genetic research [32]. Compared with traditional molecular markers, the SNPs that are obtained by high-throughput sequencing can provide a large amount of accurate and reliable information for a genetic and evolutionary analysis [33]. Reduced representation genome sequencing has been widely used in the field of population research in areas such as population genetic analysis [34], marker development [35], genetic map construction and whole-genome association analysis [36,37]. These methods, which consist of restriction site-associated DNA sequencing (RAD-seq), have been successfully applied to the study of the population structure, genetic distance, and genetic diversity of many species [38–41]. Currently, high-quality *D. catenatum* genome data have been obtained and published [42]. Therefore, the RAD-seq of this species of medicinal *Dendrobium* can provide a large number of reliable SNPs to distinguish the provenances of them.

To protect the wild population of *D. catenatum* and meet the increasing market demand, the following protective measures are recommended: (a) in situ conservation; (b) building an ex situ conservation base; (c) establishing a provenance database to strictly control the use of wild populations while ensuring the reliability and detectability of their provenance. For the creation of management and conservation strategies, information on the genetic diversity and population structure are crucial. To better understand the genetic diversity, genetic organization, and divergence of the wild populations of *D. catenatum*, we produced and analyzed the SNPs for these populations using RAD-seq. Accurately identifying the provenances of *D. catenatum* based on RAD-seq will help growers to choose high-quality provenances for artificial cultivation and propagation. This research not only has important significance for the evolution, molecular breeding and biogeography of

Dendrobium, but it also provides important enlightenment for the identification methods of other medicinal plants.

2. Materials and Methods

2.1. Sample Collection and DNA Extraction

The plant materials that were sequenced in this study were collected from the National Orchid Conservation & Research Center of Shenzhen, and the specimens were deposited in the National Orchid Conservation Center herbarium (NOCC) (Table 1). All of the plant materials were collected with the permission of this institution, and all of the samples were identified by Prof. Zhong-Jian Liu. We complied with the IUCN Policy Statement on Research Involving Species at Risk of Extinction and the Convention on the Trade in Endangered Species of Wild Fauna. Detailed information about localities and samples are given in Table 1 and Figure 1. Specifically, we collected 109 samples were from 24 wild populations of *D. catenatum*, and 10 samples were from 2 wild populations of *D. huoshanense*. We selected 3–5 individuals from each population and sampled their leaves. The leaves were dried in sealed plastic bags that were filled with silica gel until the DNA extraction was performed.

Table 1. In this study, the population locations and voucher information of *D. catenatum* and *D. huoshanense*.

Sample Name	Location	Specimen Code	Population Code	Individual Number
D. catenatum	Guangze, Fujian Province	Z.J. Liu 10572	FJGZTP01	5
	Liancheng, Fujian Province	Z.J. Liu 10583	FJLCTP02	4
	Pingjiang, Hunan Province	Z.J. Liu 10580	HNPJTP04	5
	Yiyang, Hunan Province	Z.J. Liu 10582	HNYYTP05	5
	Yuanling, Hunan Province	Z.J. Liu 10578	HNYLTP06	5
	Xinning, Hunan Province	Z.J. Liu 9506	HNXNTP36	3
	Lipu, Guangxi Province	Z.J. Liu 10574	GXLPTP07	5
	Gongcheng, Guangxi Province	Z.J. Liu 9497	GXGCTP09	4
	Xing'an, Guangxi Province	Z.J. Liu 9518	GXXATP10	5
	Lishui, Zhengjiang Province	Z.J. Liu 9520	ZJLSTP12	5
	Wuyi, Zhengjiang Province	Z.J. Liu 9511	ZJWYTP32	5
	Quzhou, Zhengjiang Province	Z.J. Liu 9510	ZJQZTP13	5
	Yanshan, Jiangxi Province	Z.J. Liu 10576	JXQSTP14	5
	Xiushui, Jiangxi Province	Z.J. Liu 10573	JXXSTP15	4
	Huichang, Jiangxi Province	Z.J. Liu 10581	JXHCTP16	5
	Xingguo, Jiangxi Province	Z.J. Liu 9517	JXXGTP17	4
	Nanxiong, Guangdong Province	Z.J. Liu 10584	GDNXTP19	5
	Heyuan, Guangdong Province	Z.J. Liu 10585	GDHYTP20	4
	Shaoguan, Guangdong Province	Z.J. Liu 9504	GDSGTP22	5
	Qujing, Yunnan Province	Z.J. Liu 10579	YNQJTP24	4
	Wangzishan, Yunnan Province	Z.J. Liu 7462	WZSTP34	4
	Xianning, Hubei Province	Z.J. Liu 9514	HBXNTP25	5
	Sandu, Guizhou Province	Z.J. Liu 11147	GZSDTP29	4
	Dushan, Guizhou Province	Z.J. Liu 11148	GZDSTP30	4
D. huoshanense	Huangshan, Anhui Province	Z.J. Liu 9508	AHHSHS26	5
	Longhushan, Jiangxi Province	Z.J. Liu 9500	JXLHSHS28	5

The total genomic DNA was extracted using a Plant Genomic DNA kit (Tiangen, Beijing, China) according to the manufacturer's protocol. For all of the samples, the DNA was quantified using a Qubit spectrophotometer (Invitrogen, Carlsbad, CA, USA).

2.2 RAD Library Development and Sequencing

The RAD sequencing libraries were generated using the VAHTS Universal DNA Library Prep Kit for Illumina (Vazyme Biotech Co., Ltd., Nanjing, China, ND604) following the manufacturer's recommendations. In brief, the RAD-seq reduced representation libraries were prepared following the digestion procedure using the *Hae* III (New England

Biolabs, Ipswich, MA, USA) enzyme, which was followed by a barcode ligation, a DNA purification and a selective DNA amplification and a size selection. Pair-end sequencing with a read length of 150 bp was performed to produce approximately 3 Gb of raw data for each sample using the Illumina HiSeq 4000 platform (Illumina, San Diego, CA, USA) at Novogene (Beijing, China).

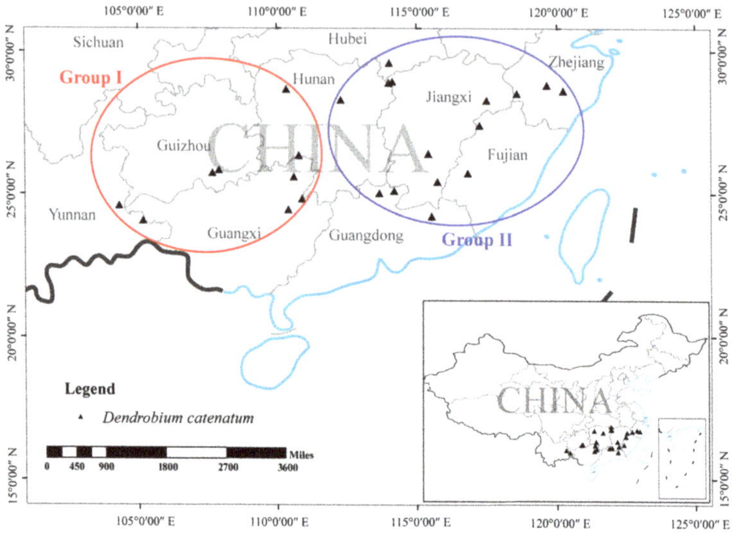

Figure 1. Sampling locations of *D. catenatum* species used in this study.

2.3. SNP Calling

The raw Illumina paired-end reads were filtered using a personally designed program "filter-G 30.0-adapter 1-poly_A 1-Q 20,0.5-mean 25-insert $ insertSize-Q20 90" to discard the reads with adapter sequences, poly_A tails, a poor base quality (Q < 20), and those that were less than 25 bp in length. The cleaned reads were aligned to the *D. catenatum* genome (NCBI accession number NC_037361.1) using BWA (version 0.7.12) [43] default parameters, and they were sorted using Samtools (version 1.9), and following this, they were de-duplicated using the Picard tool [44]. In order to reduce the impact of mapping bias, we further excluded the sites with extraordinarily high or extremely low coverage. The RealignerTargetCreator and IndelRealigner modules from GATK (version 3.8) [45] were used to improve the local alignments around the indels. The resulting alignment files were subjected to genotyping using the GATK UnifiedGenotyper at each reference locus. Finally, hard filtering was applied to the raw variant set using GATK recommended parameters, "QD < 2.0 | | MQ < 40.0 | | FS > 60.0 | | SOR > 3.0 | | MQRankSum < −12.5 | | ReadPosRankSum < −8.0". We further filtered the dataset using VCFtools (version 0.1.13) [46] to ensure that we had high-quality SNPs, removing the SNPs with a missing rate > 20% and minor allele frequencies < 0.02.

2.4. Phylogenetic Tree Construction

The maximum likelihood (ML) analysis was performed using IQtree (version 2.0.3) [47] with 1000 bootstrap [48] replicates, and the settings were as described. The results were graphically visualized and edited in FigTree (version 1.4.2).

2.5. Population Structure Analyses

Genetic structure was investigated by a PCA. A PCA was conducted using GATK, and the program Admixture [49] was used to infer the genotype structure. A population number (K) ranging from 2 to 8 was assumed, and the CV scores were used to determine the best-fit K value.

2.6. Genetic Diversity and Differentiation

A variety of genetic diversity indices were calculated using the Stacks' populations program, including the private allele number (A_P), nucleotide diversity (π), heterozygosity (H_O and H_E), and inbreeding coefficient (F_{IS}) [50]. Sliding windows that were 10 kb in size were used to calculate Tajima's D by VCFtools (version 0.1.13) [46]. To analyze the pairwise population differentiation between the wild germplasms, the F_{ST} values were also computed using the PopGenome package in R [51]. The pairwise geographic distances from longitude and latitude were identified using the R package geosphere. We used the Mantel test for the associations between the F_{ST} and geographic distance. PopLDdecay (version 3.31) was used to calculate the linkage disequilibrium between the SNP pairs within a 500 kb window [52]. The linkage disequilibrium decay was measured the distance at which the Pearson's correlation efficient (r^2) dropped to half of the maximum.

3. Results

3.1. Sequence Data Quality

For the 109 sequenced samples, 403.61 Gb of raw data with an average of 3.70 Gb per individual were generated, ranging from 1.74 to 5.29 Gb. After filtering the sequence data, a total of 371.18 Gb of clean data (1.61 Gb to 4.89 Gb for each individual, with an average of 3.37 Gb) was maintained, presenting an average effective rate of 91.97% (Supplementary Figure S1). In short, the sequencing data were of high quality and could be used for a subsequent analysis. Finally, 655,057 high-quality SNPs were selected after the filtration was performed. Nearly half of the SNPs in the population level exhibited base transitions, and the total transition-to-transversion (ts/tv) ratio was 1.36.

3.2. Genetic Diversity

When we were analyzing the variant positions for all of the polymorphic loci at the germplasm level, the observed heterozygosity (H_O), expected heterozygosity (H_E), nucleotide diversity (π) and Wright's F-statistic (F_{IS}) of the wild *D. catenatum* were 0.0992, 0.1575, 0.1584, and 0.3836, respectively (Table 1). The number of private alleles (A_P) in the populations ranged from 1934 (FJLCTP02) to 9655 (GZDSTP30). The observed heterozygosity at the population level ranged from 0.0724 (GXXATP10) to 0.1516 (GZDSTP30); the expected heterozygosity for each population ranged from 0.0833 (FJLCTP02) to 0.1312 (GXLPTP07); the nucleotide diversity for each population ranged from 0.0961 (GDSGTP22) to 0.1484 (GXLPTP07); the inbreeding coefficient in each population ranged from −0.0434 (GZDSTP30) to 0.1135 (GXLPTP07).

Tajima's D value was a locus-based indicator of the intraspecific polymorphism. Tajima's D values were positive and statistically different from zero for all of the populations (Table 2). It is therefore possible to reject the null hypothesis of neutral evolution. In addition, many intermediate-frequency alleles were found in the populations, which may be a consequence of the bottleneck effects, the population structures, or the selection for balancing.

3.3. Phylogenetic Tree

In the maximum likelihood (ML) phylogenetic tree that was derived from our finalized RAD-seq matrix, *D. huoshanense* was the outgroup (Figure 2). There were high bootstrap values on most of the branches of the phylogenetic tree, which indicated a strong reliability of the tree. All of the samples were separated into two clades based on the ML tree (Group I and Group II). Group I mainly contained *D. catenatum* from the Yungui Plateau and the western part of the Nanling Mountains, while Group II mainly contained populations from the Yandangshan Mountains, Wuyishan Mountains, and the eastern part of the Nanling Mountains.

Table 2. The statistical values of genetic diversity within the populations from variants and all of the positions data. (A_P, private allele number; H_O, observed heterozygosity; H_E, expected heterozygosity; π, nucleotide diversity; F_{IS}, inbreeding coefficient of an individual relative to the subpopulation.)

Taxon	Population	A_P	H_O	H_E	π	F_{IS}	Tajima's D
D. catenatum			0.0992	0.1575	0.1584	0.3836	
	FJGZTP01	3254	0.0786	0.0969	0.1165	0.0725	0.6534
	FJLCTP02	1934	0.0729	0.0833	0.099	0.0496	0.3964
	HNPJTP04	2871	0.0895	0.1131	0.1297	0.0824	0.4221
	HNYYTP05	4263	0.0818	0.1082	0.1223	0.084	0.4411
	HNYLTP06	4689	0.0976	0.1196	0.1371	0.081	0.5556
	GXLPTP07	5512	0.0952	0.1312	0.1484	0.1135	0.4684
	GXGCTP09	2679	0.0965	0.1041	0.1355	0.066	0.7907
	GXXATP10	2940	0.0724	0.0891	0.1074	0.0628	0.8556
	ZJLSTP12	2355	0.0994	0.1182	0.1341	0.0737	0.3298
	ZJQZTP13	3084	0.1123	0.1105	0.1245	0.0256	0.3962
	JXQSTP14	3844	0.123	0.1115	0.1243	0.0053	0.3384
	JXXSTP15	2552	0.0783	0.107	0.1257	0.0922	0.3767
	JXHCTP16	3417	0.0778	0.1037	0.1188	0.0866	0.3404
	JXXGTP17	4215	0.0919	0.096	0.1134	0.039	0.5287
	GDNXTP19	3515	0.0848	0.1022	0.1154	0.0659	0.3163
	GDHYTP20	5221	0.1019	0.1155	0.133	0.0638	0.2443
	GDSGTP22	3845	0.1078	0.0862	0.0961	−0.0208	0.4748
	YNQJTP24	3933	0.1144	0.1153	0.1343	0.039	0.4327
	HBXNTP25	2757	0.0979	0.1057	0.1185	0.0426	0.5005
	GZSDTP29	4461	0.0975	0.1147	0.1326	0.0696	0.4221
	GZDSTP30	9655	0.1516	0.1121	0.1288	−0.0434	0.4646
	ZJWYTP32	2932	0.0885	0.106	0.1191	0.0664	0.2539
	WZSTP34	4820	0.125	0.1165	0.1342	0.0208	0.4762
	HNXNTP36	2967	0.1034	0.0994	0.1218	0.0332	0.4315

In the Group I, the YNQJTP24 accessions were the sister group to the WZSTP34 accessions, which both had 100% bootstrap values. The samples from Guizhou Dushan (GZDSTP30), Guangxi Xing'an (GXXATP10) and Hunan Yiyang (HNYYTP06) are clustered into a branch. The samples from the provenances GZSDTP29, GXGCTP09, and GXLCTP07 were divided into two branches. In the Group II, for most of the provenances of *D. catenatum*, the individuals of the same provenance are clustered together with high support values, with the exception of the individuals of three populations from Zhejiang Wuyi (ZJWYTP32), Zhejiang Lishui (ZJLSTP12), and Zhejiang Quzhou (ZJQZTP13). The samples from Zhejiang were gathered into a large branch, and the sister branch was composed of FJGZTP01, FJLCTP02, and JXQSTP14. The populations from Hubei Xinning (HBXNTP25), Jiangxi Xiushui (JXXSTP15), Hunan Pingjiang (HNPJTP04), and Hunan Yiyang (HNYYTP05) are clustered into a large branch, and the other three provenances being independently clustered into a small branch, with the exception of JXXSTP15. The samples from Guangdong Shaoguan (GDXGTP22) and Guangdong Heyuan (GDHYTP20) are clustered into a branch. At the same time, the distribution of the *D. catenatum* samples in the phylogenetic tree are closely related to their geographical locations. The base of the phylogenetic tree is represented by the population that can be located in the west.

The geographical distribution of the *Dendrobium* samples from the west to the east is roughly consistent with their distribution from the base to the top of the phylogenetic tree.

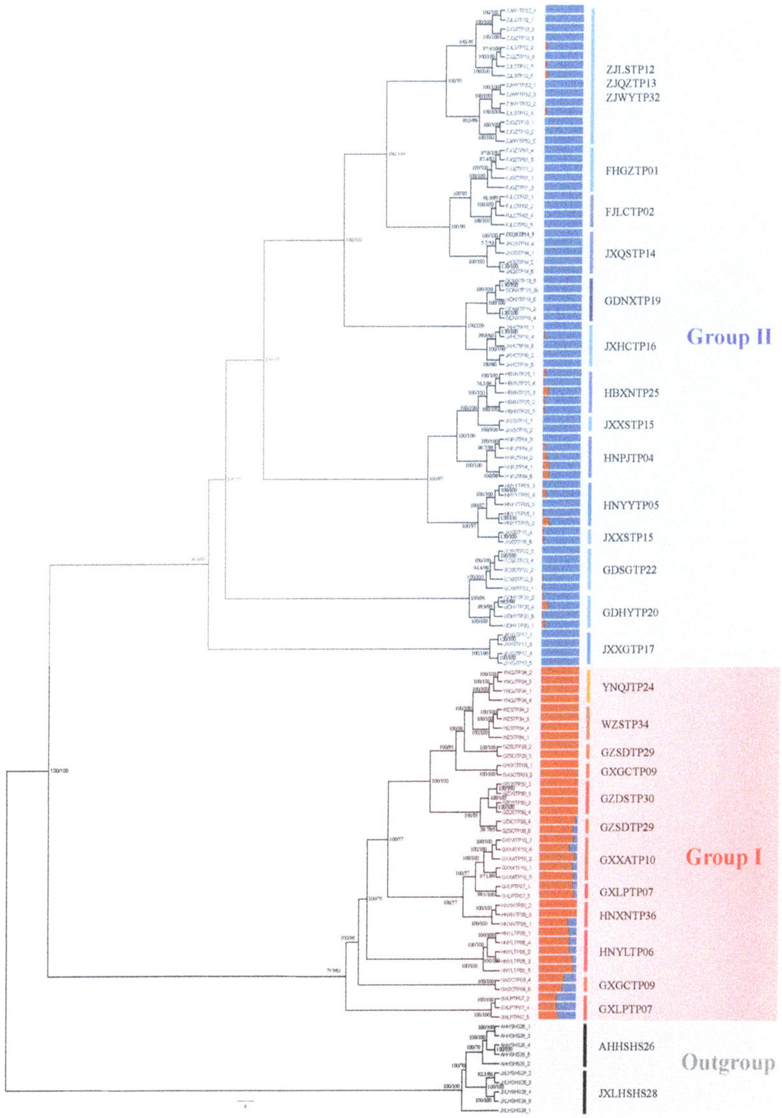

Figure 2. ML phylogenetic tree of the *D. catenatum* accessions and model-based clustering with K from 2. Numbers near the nodes are bootstrap percentages.

3.4. Population Structure

Based on the ML tree (Figure 2), we calculated the population structure with K values that range from two to eight (Figure 3). In the clustering analysis using the Admixture software, K = 2 was the most likely genetic cluster number, because its cross-validation error (CV) was the lowest. However, the CV values were very close when K = 3 and K = 4. According to the principal component analysis (Figure 4), when the K value equaled two, 109 materials were not clearly assigned to the two groups, which was inconsistent with the results of the phylogenetic tree analysis. Due to the differences in the population structure and phylogenetic tree results, it was necessary to analyze the population structure under different K values.

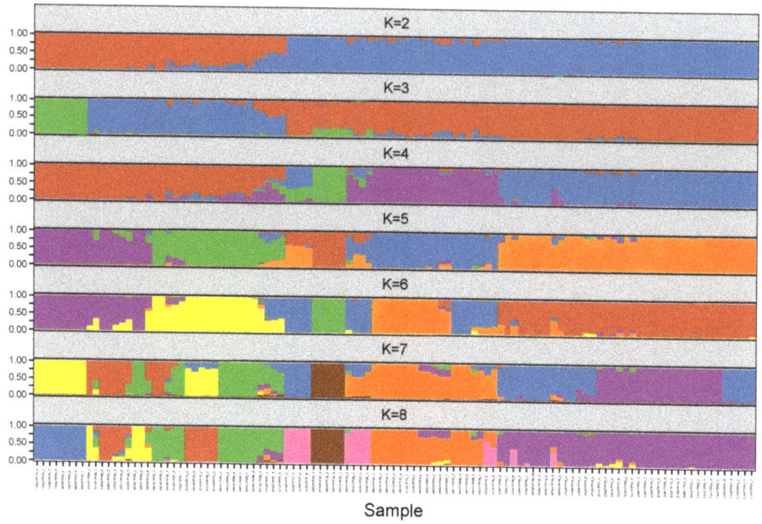

Figure 3. Genetic structure of cultivated 109 *D. catenatum* for K = 2–8 based on the Admixture software (K = 2 with cross validation error is 0.18580, while K = 3 with ross validation error is 0.19141).

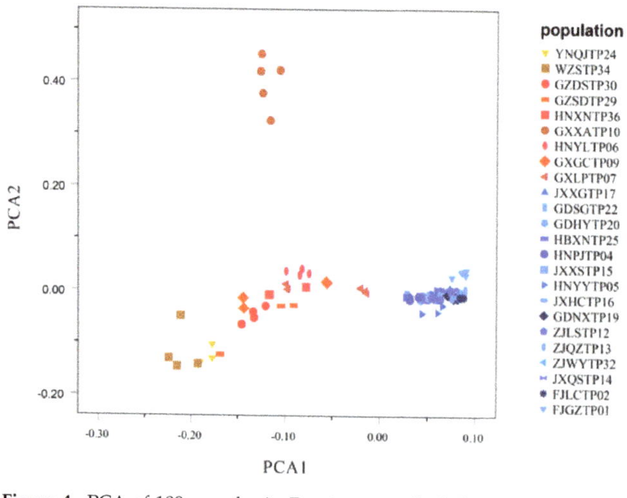

Figure 4. PCA of 109 samples in *D. catenatum*. Red shapes indicate Group I, while blue shapes indicate Group II.

When K = 2, the Group I and Group II accessions could be separated. When K = 3, the Yunnan populations in Group I were distinguished, which was inconsistent with K = 4. In conjunction with the phylogenetic tree and structure analysis, these results clearly classified *D. catenatum* into two groups.

3.5. Genetic Relationships

A principal component analysis (PCA) was performed with 109 individuals. According to the first component, the accessions were divided into two groups: Group I and Group II. Overall, GXXATP10 was clustered separately, and other *D. catenatum* accessions could be divided into two groups. Among them, the distribution of Group II was more concentrated (Figure 4).

The linkage disequilibrium (LD) decay, which was measured by the physical distance, at which the pairwise correlation coefficient dropped to half of its maximum value, occurred at 48.2 kb in Group I (r^2 = 0.277) and 13.6 kb in Group II (r^2 = 0.242) (Figure 5), respectively. Group II had the higher linkage disequilibrium decay rate.

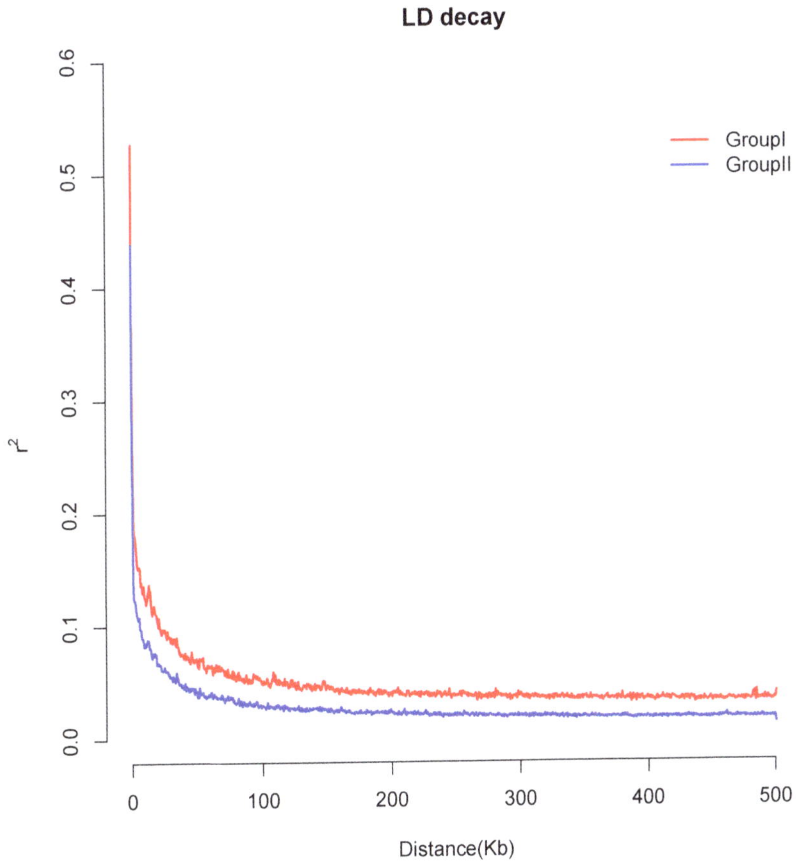

Figure 5. Linkage disequilibrium decay patterns of different *D. catenatum* group.

3.6. Genetic Differentiation

The pairwise F_{ST} values between the wild populations of *D. catenatum* varied from 0.0533 (HNYLTP06 with GXGCTP09) to 0.4755 (FJLCTP02 with GZDSTP30), with 110 of the 276 population pairs having F_{ST} values that are greater than 0.25 (Figure 6). It became clear that, GXXATP10 had the highest genetic differentiation from the other 24 populations which was consistent with the PCA results (Figure 4). GXLPTP07 was in the western part of the Nanling Mountains, and it obtained the lowest mean pairwise F_{ST} value, 0.1504; low values represent the genetic flow between the populations and show a greater genetic difference between the accessions within a population than it does between the populations. The pairwise genetic distance between the locations F_{ST} was highly correlated with the geographic distance (Figure 7); the level of significance was $p < 0.05$.

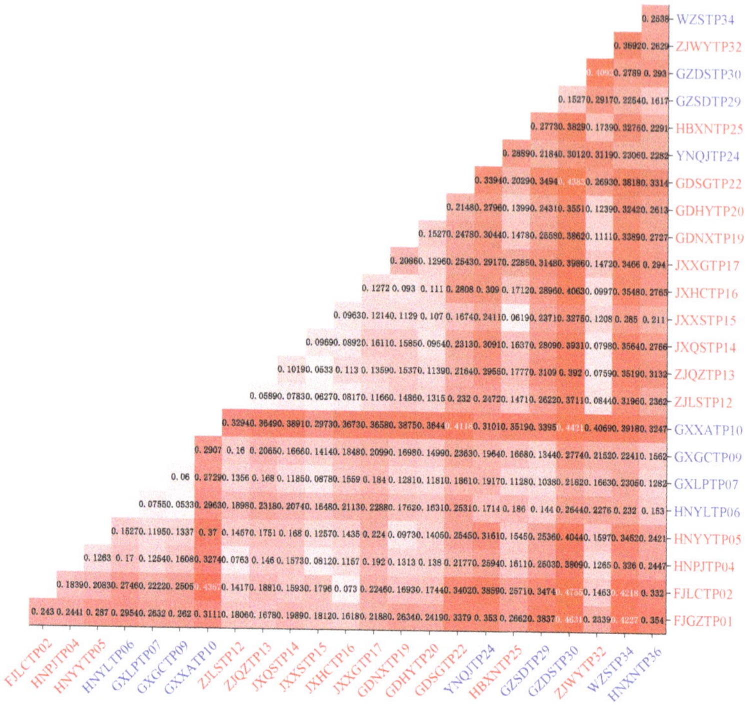

Figure 6. Genetic distance (F_{ST}) for *D. catenatum*. The samples from Group I were in red font, and the samples from Group II were in blue font.

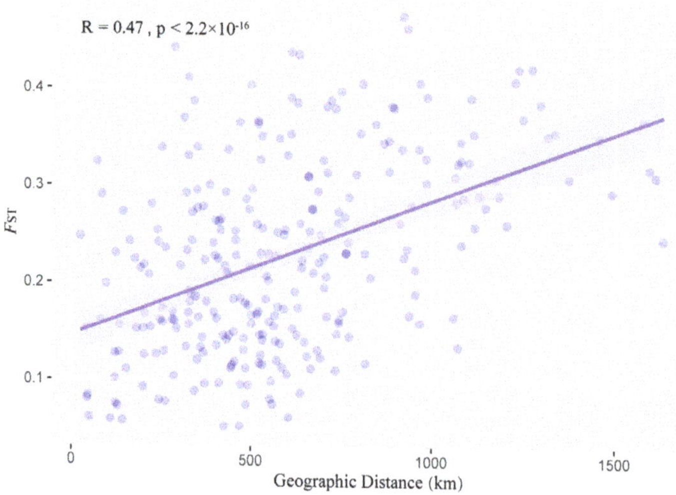

Figure 7. Relationships between the pairwise genetic distance (F_{ST}) and geographic distance with blue bars and shading represent the 95% highest posterior density.

4. Discussion

Genomic data provide a novel perspective for studying the genetic diversity and method of identification of medicinal *Dendrobium* species. In the past, scholars have been authenticated eight wild populations of *D. officinale* by ISSR [26]. Because of the low number of microsatellite markers, marker ascertainment bias, and there being a high variance in the microsatellite-derived estimates, the genetic differentiation among the populations (F_{ST}) estimates of microsatellites were significantly larger than those from the SNPs [31]. The simple PCR methods are simple and quite effective and cheap. However, their weakness is obvious at the same time. Firstly, these data are not reproducible enough, and it is difficult to share and reference them, and they can only be analyzed and used in the same experimental project. Secondly, these markers cannot be mapped to the genome for further analysis and applications. The data that are obtained by RAD not only have the advantages of being convenient and highly reproducible, but also, they can be accurately located in the genome position for more applications. For example, the candidate genes with known functions can be identified by annotating the outlier loci through the selection elimination analysis [53,54]. At the same time, the SNPs that are obtained by RAD-seq can provide more accurate and reliable information for a genetic and evolutionary analysis. Researchers utilized anadromous pike (*Esox lucius*) to assess the microsatellite and RAD-seq results of a study of population differentiation and genetic structure, and they discovered that the full RAD-seq dataset can provide the most accurate detection of the finer-scaled genetic structuring [55]. The nuclear microsatellites and the RAD-seq data for a threatened freshwater fish species were compared by other researchers. The results showed that RAD-seq more clearly and consistently identified the hierarchical phylogenetic structure [56].

In this research, a whole-genome restriction enzyme digestion of 109 accessions from 24 provenances was performed to obtain accurate variation information. Except for GXGCTP09, the clean data size of each sample in the other 23 populations was at least 2 times the *D. catenatum* genome size (1.11 Gb), and some individual populations reached 4 times this (Supplementary Figure S1). The average sequencing volume of each sample was 3.37 Gb. The amount of sequencing data was sufficient to meet the requirements of the subsequent analysis, thus ensuring the accuracy of population genetic analysis. In terms of the result of the RAD-seq, we detected 655,057 SNPs, which exceeded the number of genetic variations that were detected by amplified fragment length polymorphism (AFLP) and the random amplified polymorphic DNA (RAPD) markers [57,58]. These data not only can be used in this experimental analysis, but they are also convenient for researchers for other studies.

The number of transitions was predicted to be much larger than the number of transversions due to the biased mutational processes within the plant genomes. This nucleotide mutation pattern is also observed in other plants, such as peanuts [59], maize [60], *Amorphophallus paeoniifolius* [39], *Arabidopsis*, and apricots [61,62]. Consistent with this prediction, the ts/tv ratio of the *D. catenatum* populations was 1.36, indicating that there was a strong transition bias. This value was higher than the result (1.34) that was reported by Zhang et al. [42], and lower than that for *D. huoshanense* (1.47) [63].

Based on the ML tree, the principal component analysis, the genetic structure analysis, and the population differentiation analysis divided the 24 wild populations of *D. catenatum* into two groups. The two groups where the provenances of the wild *D. catenatum* are arranged from the west to the east are very consistent with the geographical distribution of the samples. In another study, Ding et al. used RAPD to study eight wild *D. catenatum* populations [58]. Among them, Shaowu and Shunchang in Fujian were clustered into a small branch, which was the sister branch of Jiangxi Nanfeng, and Tian'e in Guangxi and Guangnan in Yunnan were clustered into a small group. Our results in this study are consistent with the previous findings. In a systematic geographical study of *D. catenatum* and four related taxa, the regional evolution of *D. catenatum* was divided into six populations [64]. They were the South Yungui Plateau, the East Yungui Plateau, Nanling Mountain, Wuyishan Mountain, Dabieshan Mountain, and Yandang Mountain. In this study, Group I contained the Yungui Plateau and the western part of the Nanling Mountains, while Group II contained the Yandangshan Mountains, Wuyishan Mountains, and the eastern part of the Nanling Mountains. This is largely consistent with the division that had been made previously.

Genetic diversity forms during the evolution of a species, and it often plays a key role in the gradual evolution and long-term survival of the species in a changeable environment. The average nucleotide diversity (π) of the *D. catenatum* population mutation parameters was 0.1584 at the population level (Table 2). *D. catenatum* had a smaller mean nucleotide variation than other crops did such as *A. paeoniifolius* (π = 0.3592) [39]. At the population level, the expected heterozygosity (H_E) was 0.1575, which was very close to the previous results (0.1477) [57]. Overall, the observed heterozygosity and the expected heterozygosity were relatively low.

Among all of the wild populations, GXLPTP07 had the most significant private allele number (AP) of 5221, and this indicated the presence of substantial genetic variation, which could guide genetic improvement in the future. The genetic diversity indices (H_E, A_P, and π) of the GXLPTP07 accessions were higher than those of the accessions in the other ecological groups (Table 2). We speculate that the GXLPTP07 was closer to the origin center of Group I in terms of genetic distance. In the principal component analysis, GXXATP10 was clustered separately. Combined with the analysis of genetic diversity, GXXATP10 had the highest Tajima's D values, indicating that the population underwent a balanced selection or a sudden contraction. The results of the H_O, H_E, and inbreeding coefficient (F_{IS}) show that $H_O < H_E$, $F_{IS} > 0$, which may have resulted from the heterozygote loss and inbreeding, and the selection pressure is high (Figure 3, Table 2). Based on PC1, we speculate that GXXATP10 was a descendant that was formed by the intersection of the East Yungui Plateau population and the western part of the Nanling Mountains population in the past, or a nonnative species that had been introduced from another place.

The Mantel test discovered a significant positive relationship between the genetic and geographical distances in the total distribution ($r = 0.47$, $p < 2.2 \times 10^{-16}$), suggesting that isolation by distance plays an important role in genome-wide variation. Combined with the results of F_{ST}, the phylogenetic tree and the Linkage disequilibrium analysis speculated that the Group I population is closer to the origin center in terms of its genetic distance, which supports that it originated in the Nanling Mountains and the Yungui Plateau before migrating eastward [64].

In the wild, *Dendrobium* species are either epiphytic or lithophytic. Based on this habit, there is no need to compete for the niche of most terrestrial plants. However, people over-excavate and conduct commercial trade for profit, thus causing serious damage to the wild resources [65]. This leads to a decline in the genetic differentiation among the *Dendrobium* species [66].

5. Conclusions

In this study, a genetic information database of *D. catenatum* was established, which confirmed that RAD-seq technology has the potential to be applied in the identification of medicinal *Dendrobium* of different origins. The level of genetic diversity in the population of wild *D. catenatum* was relatively low. The 24 wild populations were divided into two groups, and the Group I population is closer to the origin center in terms of its genetic distance. The isolation by distance plays an important role in genome-wide variation. This study provides information for the development of identification technology for medicinal Dendrobium species in order to provide a new perspective for the research of genuine regional drugs. Genetic diversity research provides a theoretical basis for the protection of *Dendrobium*, and it also lays a solid foundation for the breeding of fine *Dendrobium* provenances. Overall, our study provides abundant genomic resources for wild *Dendrobium* species and makes important contributions to its genetic improvement and molecular breeding.

Supplementary Materials: The following supporting information can be downloaded at: https://www.mdpi.com/article/10.3390/genes13112093/s1, Figure S1: Average clean data size per sample in 24 populations.

Author Contributions: Conceptualization, X.-Y.W. and J.-X.H.; data curation, X.-Y.W., T.-Z.L. and F.Z.; writing—original draft, X.-Y.W.; writing—review and editing, Y.-H.Y. and J.-X.H.; supervision, J.-B.C. All authors have read and agreed to the published version of the manuscript.

Funding: This work was supported by Department of Wildlife Conservation, National Forestry and Grassland Administration, China (15 June 2018).

Institutional Review Board Statement: Not applicable.

Informed Consent Statement: Not applicable.

Data Availability Statement: All genomic data generated in this study are deposited in NCBI database (https://www.ncbi.nlm.nih.gov/) (BioProject accession: PRJNA774562). These data will remain private until the related manuscript has been accepted. All other data generated in this manuscript are available from the corresponding author upon reasonable request.

Acknowledgments: We would like to thank Zhong-Jian Liu at the Key Laboratory of National Forestry and Grassland Administration for Orchid Conservation and Utilization at College of Landscape Architecture, Fujian Agriculture and Forestry University, for identifying the samples in the present study and other help.

Conflicts of Interest: The authors declare no conflict of interest.

References

1. Cribb, P.; Govaerts, R. Just how many orchids are there? In Proceedings of the 18th World Orchid Conference, Dijon, France, 11–20 March 2005; Rayanl-Roques, A., Roguenant, A., Prat, D., Eds.; Naturalia: Dijon, France, 2005; pp. 161–172.
2. Pridgeon, A.M.; Cribb, P.J.; Chase, M.W.; Rasmussen, F.N. *Genera Orchidacearum Volume 6: Epidendroideae (Part 3)*; OUP: Oxford, UK, 2014; ISBN 978-0-19-100389-9.
3. Chinese Pharmacopoeia Commission. *Pharmacopoeia of the People's Republic of China*; Chinese Medicine Science and technology Publishing House: Beijing, China, 2020.
4. Bao, X.S.; Shun, Q.S.; Chen, L.Z. *The Medicinal Plants of Dendrobium (Shi-Hu) in China*; Shanghai Medical University Press, Fudan University Press: Shanghai, China, 2001; pp. 56–60.
5. He, T.G.; Yang, L.T.; Li, Y.R.; Wang, C.Q.; Huang, D.P. Physicochemical properties and antitumor activity of polysaccharide dcpp1a-1 from suspension-cultured protocorms of Dendrobium candidum. *Nat. Prod. Res. Dev.* **2007**, 578–583. [CrossRef]
6. Liu, X.-F.; Zhu, J.; Ge, S.-Y.; Xia, L.-J.; Yang, H.-Y.; Qian, Y.-T.; Ren, F.-Z. Orally Administered Dendrobium Officinale and Its Polysaccharides Enhance Immune Functions in BALB/c Mice. *Nat. Prod. Commun.* **2011**, *6*, 867–870. [CrossRef]
7. Huang, K.; Li, Y.; Tao, S.; Wei, G.; Huang, Y.; Chen, D.; Wu, C. Purification, Characterization and Biological Activity of Polysaccharides from Dendrobium Officinale. *Molecules* **2016**, *21*, 701. [CrossRef]
8. Zha, X.Q.; Luo, J.P.; Luo, S.Z.; Jiang, S.T. Structure Identification of a New Immunostimulating Polysaccharide from the Stems of Dendrobium Huoshanense. *Carbohydr. Polym.* **2007**, *69*, 86–93. [CrossRef]
9. Zhang, J.; Guo, Y.; Si, J.-P.; Sun, X.-B.; Sun, G.-B.; Liu, J.-J. A Polysaccharide of Dendrobium Officinale Ameliorates H_2O_2-Induced Apoptosis in H9c2 Cardiomyocytes via PI3K/AKT and MAPK Pathways. *Int. J. Biol. Macromol.* **2017**, *104*, 1–10. [CrossRef] [PubMed]
10. Tian, C.-C.; Zha, X.-Q.; Luo, J.-P. A Polysaccharide from Dendrobium Huoshanense Prevents Hepatic Inflammatory Response Caused by Carbon Tetrachloride. *Biotechnol. Biotechnol. Equip.* **2015**, *29*, 132–138. [CrossRef] [PubMed]
11. Zeng, Q.; Ko, C.-H.; Siu, W.-S.; Li, K.-K.; Wong, C.-W.; Han, X.-Q.; Yang, L.; Lau, C.B.-S.; Hu, J.-M.; Leung, P.-C. Inhibitory Effect of Different Dendrobium Species on LPS-Induced Inflammation in Macrophages via Suppression of MAPK Pathways. *Chin. J. Nat. Med.* **2018**, *16*, 481–489. [CrossRef]
12. Zhang, Q.; Li, J.; Luo, M.; Xie, G.-Y.; Zeng, W.; Wu, Y.; Zhu, Y.; Yang, X.; Guo, A.-Y. Systematic Transcriptome and Regulatory Network Analyses Reveal the Hypoglycemic Mechanism of Dendrobium Fimbriatum. *Mol. Ther.-Nucleic Acids* **2020**, *19*, 1–14. [CrossRef]
13. Pan, L.-H.; Li, X.-F.; Wang, M.-N.; Zha, X.-Q.; Yang, X.-F.; Liu, Z.-J.; Luo, Y.-B.; Luo, J.-P. Comparison of Hypoglycemic and Antioxidative Effects of Polysaccharides from Four Different Dendrobium Species. *Int. J. Biol. Macromol.* **2014**, *64*, 420–427. [CrossRef]
14. Si, J.-P.; Wang, Q.; Liu, Z.-J.; Liu, J.-J.; Luo, Y.-B. Breakthrough in key science and technologies in *Dendrobium catenatum* industry. *China J. Chin. Mater.* **2017**, *42*, 2223–2227. [CrossRef]
15. Lam, Y.; Ng, T.B.; Yao, R.M.; Shi, J.; Xu, K.; Sze, S.C.W.; Zhang, K.Y. Evaluation of Chemical Constituents and Important Mechanism of Pharmacological Biology in Dendrobium Plants. *Evid.-Based Complement. Altern. Med.* **2015**, *2015*, e841752. [CrossRef]
16. Ma, G.; Guo, Y.; Xu, G.; Xu, L.; An, D. Microscopic Identification Studies on Dendrobium Stems By Clustering Analysis. *J. China Pharm. Univ.* **1996**, *27*, 208–210.

17. Zha, X.-Q.; Pan, L.-H.; Luo, J.-P.; Wang, J.-H.; Wei, P.; Bansal, V. Enzymatic Fingerprints of Polysaccharides of Dendrobium Officinale and Their Application in Identification of Dendrobium Species. *J. Nat. Med.* **2012**, *66*, 525–534. [CrossRef] [PubMed]
18. Wei, G.; Shun, Q.; Huang, Y.; Yang, M.; Chen, L.; Li, M.; Ye, J.; Wang, H.; Jiang, R.; He, Z. Comparative study of hplc characteristic spectrum of Dendrobium officinale from three provenance sources. *Tradit. Chin. Drug Res. Clin. Pharmacol.* **2014**, *25*, 467–471.
19. Huang, L.; Zhang, X.; Chen, S. Research Progress in Quality Ecology of Genuine Regional Drugs. *Mod. Tradit. Chin. Med. Mater. Med.-World Sci. Technol.* **2019**, *21*, 844–853. [CrossRef]
20. Xia, B.; Zhou, Y.; Tan, H.S.; Ding, L.S.; Xu, H.X. Advanced Ultra-Performance Liquid Chromatography–Photodiode Array–Quadrupole Time-of-Flight Mass Spectrometric Methods for Simultaneous Screening and Quantification of Triterpenoids in Poria Cocos. *Food Chem.* **2014**, *152*, 237–244. [CrossRef]
21. Tan, H.-S.; Hu, D.-D.; Song, J.-Z.; Xu, Y.; Cai, S.-F.; Chen, Q.-L.; Meng, Q.-W.; Li, S.-L.; Chen, S.-L.; Mao, Q.; et al. Distinguishing Radix Angelica Sinensis from Different Regions by HS-SFME/GC–MS. *Food Chem.* **2015**, *186*, 200–206. [CrossRef]
22. Zhang, X.; Gu, C.; Ahmad, B.; Huang, L. Optimization of Extract Method for *Cynomorium Songaricum* Rupr. by Response Surface Methodology. *J. Anal. Methods Chem.* **2017**, *2017*, 6153802. [CrossRef]
23. Ding, X.; Wang, Z.; Xu, H.; Xu, L.; Zhou, K. Database establishment of the whole rDNA ITS region of Dendrobium species of "fengdou" and authentication by analysis of their sequences. *Acta Pharm. Sin. B.* **2002**, *37*, 567–573.
24. Asahina, H.; Shinozaki, J.; Masuda, K.; Morimitsu, Y.; Satake, M. Identification of Medicinal Dendrobium Species by Phylogenetic Analyses Using *matK* and *rbcL* Sequences. *J. Nat. Med.* **2010**, *64*, 133–138. [CrossRef]
25. Liu, H.; Fang, C.; Zhang, T.; Guo, L.; Ye, Q. Molecular Authentication and Differentiation of Dendrobium Species by rDNA ITS Region Sequence Analysis. *AMB Expr.* **2019**, *9*, 53. [CrossRef] [PubMed]
26. Shen, J.; Ding, X.; Liu, D.; Ding, G.; He, J.; Li, X.; Tang, F.; Chu, B. Intersimple Sequence Repeats (ISSR) Molecular Fingerprinting Markers for Authenticating Populations of Dendrobium Officinale Kimura et Migo. *Biol. Pharm. Bull.* **2006**, *29*, 420–422. [CrossRef] [PubMed]
27. Ding, G.; Xu, G.; Zhang, W.; Lu, S.; Li, X.; Gu, S.; Ding, X.-Y. Preliminary Geoherbalism Study of Dendrobium Officinale Food by DNA Molecular Markers. *Eur. Food Res. Technol.* **2008**, *227*, 1283–1286. [CrossRef]
28. Hodel, R.G.J.; Chen, S.; Payton, A.C.; McDaniel, S.F.; Soltis, P.; Soltis, D.E. Adding Loci Improves Phylogeographic Resolution in Red Mangroves despite Increased Missing Data: Comparing Microsatellites and RAD-Seq and Investigating Loci Filtering. *Sci. Rep.* **2017**, *7*, 17598. [CrossRef] [PubMed]
29. Hale, M.L.; Burg, T.M.; Steeves, T.E. Sampling for Microsatellite-Based Population Genetic Studies: 25 to 30 Individuals per Population Is Enough to Accurately Estimate Allele Frequencies. *PLoS ONE* **2012**, *7*, e45170. [CrossRef]
30. Zimmerman, S.J.; Aldridge, C.L.; Oyler-McCance, S.J. An Empirical Comparison of Population Genetic Analyses Using Microsatellite and SNP Data for a Species of Conservation Concern. *BMC Genom.* **2020**, *21*, 382. [CrossRef]
31. Fischer, M.C.; Rellstab, C.; Leuzinger, M.; Roumet, M.; Gugerli, F.; Shimizu, K.K.; Holderegger, R.; Widmer, A. Estimating Genomic Diversity and Population Differentiation—An Empirical Comparison of Microsatellite and SNP Variation in Arabidopsis Halleri. *BMC Genom.* **2017**, *18*, 69. [CrossRef]
32. Shendure, J.; Ji, H. Next-Generation DNA Sequencing. *Nat. Biotechnol.* **2008**, *26*, 1135–1145. [CrossRef]
33. Ganal, M.W.; Altmann, T.; Roder, M.S. SNP Identification in Crop Plants. *Curr. Opin. Plant Biol.* **2009**, *12*, 211–217. [CrossRef]
34. Ma, B.; Liao, L.; Peng, Q.; Fang, T.; Zhou, H.; Korban, S.S.; Han, Y. Reduced Representation Genome Sequencing Reveals Patterns of Genetic Diversity and Selection in Apple. *J. Integr. Plant Biol.* **2017**, *59*, 190–204. [CrossRef]
35. Davey, J.W.; Hohenlohe, P.A.; Etter, P.D.; Boone, J.Q.; Catchen, J.M.; Blaxter, M.L. Genome-Wide Genetic Marker Discovery and Genotyping Using next-Generation Sequencing. *Nat. Rev. Genet.* **2011**, *12*, 499–510. [CrossRef] [PubMed]
36. Andolfatto, P.; Davison, D.; Erezyilmaz, D.; Hu, T.T.; Mast, J.; Sunayama-Morita, T.; Stern, D.L. Multiplexed Shotgun Genotyping for Rapid and Efficient Genetic Mapping. *Genome Res.* **2011**, *21*, 610–617. [CrossRef] [PubMed]
37. Liu, N.; Li, M.; Hu, X.; Ma, Q.; Mu, Y.; Tan, Z.; Xia, Q.; Zhang, G.; Nian, H. Construction of High-Density Genetic Map and QTL Mapping of Yield-Related and Two Quality Traits in Soybean RILs Population by RAD-Sequencing. *BMC Genom.* **2017**, *18*, 466. [CrossRef] [PubMed]
38. Valdisser, P.A.M.R.; Pappas, G.J.; de Menezes, I.P.P.; Müller, B.S.F.; Pereira, W.J.; Narciso, M.G.; Brondani, C.; Souza, T.L.P.O.; Borba, T.C.O.; Vianello, R.P. SNP Discovery in Common Bean by Restriction-Associated DNA (RAD) Sequencing for Genetic Diversity and Population Structure Analysis. *Mol. Genet. Genom.* **2016**, *291*, 1277–1291. [CrossRef] [PubMed]
39. Gao, Y.; Yin, S.; Wu, L.; Dai, D.; Wang, H.; Liu, C.; Tang, L. Genetic Diversity and Structure of Wild and Cultivated Amorphophallus Paeoniifolius Populations in Southwestern China as Revealed by RAD-Seq. *Sci. Rep.* **2017**, *7*, 14183. [CrossRef]
40. Hirase, S.; Tezuka, A.; Nagano, A.J.; Kikuchi, K.; Iwasaki, W. Genetic Isolation by Distance in the Yellowfin Goby Populations Revealed by RAD Sequencing. *Ichthyol. Res.* **2020**, *67*, 98–104. [CrossRef]
41. Martin Cerezo, M.L.; Kucka, M.; Zub, K.; Chan, Y.F.; Bryk, J. Population Structure of Apodemus Flavicollis and Comparison to Apodemus Sylvaticus in Northern Poland Based on RAD-Seq. *BMC Genom.* **2020**, *21*, 241. [CrossRef]
42. Zhang, G.-Q.; Xu, Q.; Bian, C.; Tsai, W.-C.; Yeh, C.-M.; Liu, K.-W.; Yoshida, K.; Zhang, L.-S.; Chang, S.-B.; Chen, F.; et al. The *Dendrobium Catenatum* Lindl. Genome Sequence Provides Insights into Polysaccharide Synthase, Floral Development and Adaptive Evolution. *Sci. Rep.* **2016**, *6*, 19029. [CrossRef]
43. Li, H.; Durbin, R. Fast and Accurate Short Read Alignment with Burrows–Wheeler Transform. *Bioinformatics* **2009**, *25*, 1754–1760. [CrossRef]

44. McKenna, A.; Hanna, M.; Banks, E.; Sivachenko, A.; Cibulskis, K.; Kernytsky, A.; Garimella, K.; Altshuler, D.; Gabriel, S.; Daly, M.; et al. The Genome Analysis Toolkit: A MapReduce Framework for Analyzing next-Generation DNA Sequencing Data. *Genome Res.* **2010**, *20*, 1297–1303. [CrossRef]
45. Van der Auwera, G.A.; Carneiro, M.O.; Hartl, C.; Poplin, R.; del Angel, G.; Levy-Moonshine, A.; Jordan, T.; Shakir, K.; Roazen, D.; Thibault, J.; et al. From FastQ Data to High-Confidence Variant Calls: The Genome Analysis Toolkit Best Practices Pipeline. *Curr. Protoc. Bioinform.* **2013**, *43*, 11.10.1–11.10.33. [CrossRef]
46. Danecek, P.; Auton, A.; Abecasis, G.; Albers, C.A.; Banks, E.; DePristo, M.A.; Cornelis, A.; Handsaker, R.E.; Lunter, G.; Marth, G.T.; et al. The Variant Call Format and VCFtools. *Bioinformatics* **2011**, *27*, 2156–2158. [CrossRef] [PubMed]
47. Nguyen, L.-T.; Schmidt, H.A.; von Haeseler, A.; Minh, B.Q. IQ-TREE: A Fast and Effective Stochastic Algorithm for Estimating Maximum-Likelihood Phylogenies. *Mol. Biol. Evol.* **2015**, *32*, 268–274. [CrossRef] [PubMed]
48. Minh, B.Q.; Nguyen, M.A.T.; von Haeseler, A. Ultrafast Approximation for Phylogenetic Bootstrap. *Mol. Biol. Evol.* **2013**, *30*, 1188–1195. [CrossRef]
49. Alexander, D.H.; Novembre, J.; Lange, K. Fast Model-Based Estimation of Ancestry in Unrelated Individuals. *Genome Res.* **2009**, *19*, 1655–1664. [CrossRef]
50. Catchen, J.; Hohenlohe, P.A.; Hohenlohe, P.A.; Bassham, S.; Amores, A.; Cresko, W.A. Stacks: An Analysis Tool Set for Population Genomics. *Mol. Ecol.* **2013**, *22*, 3124–3140. [CrossRef]
51. Pfeifer, B.; Ulrich Wittelsbürger; Ramos-Onsins, S.E.; Lercher, M.J. PopGenome: An Efficient Swiss Army Knife for Population Genomic Analyses in R. *Mol. Biol. Evol.* **2014**, *31*, 1929–1936. [CrossRef]
52. Chi, Z.; Shan-Shan, D.; Jun-Yang, X.; Wei-Ming, H.; Tie-Lin, Y. PopLDdecay: A Fast and Effective Tool for Linkage Disequilibrium Decay Analysis Based on Variant Call Format Files. *Bioinformatics* **2018**, *35*, 1786–1788. [CrossRef]
53. Bay, R.A.; Harrigan, R.J.; Underwood, V.L.; Gibbs, H.L.; Smith, T.B.; Ruegg, K. Genomic Signals of Selection Predict Climate-Driven Population Declines in a Migratory Bird. *Science* **2018**, *359*, 83–86. [CrossRef]
54. Xie, H.; Hou, J.; Fu, N.; Wei, M.; Li, Y.; Yu, K.; Song, H.; Li, S.; Liu, J. Identification of QTL Related to Anther Color and Hull Color by RAD Sequencing in a RIL Population of Setaria Italica. *BMC Genom.* **2021**, *22*, 556. [CrossRef]
55. Sunde, J.; Yıldırım, Y.; Tibblin, P.; Forsman, A. Comparing the Performance of Microsatellites and RADseq in Population Genetic Studies: Analysis of Data for Pike (Esox Lucius) and a Synthesis of Previous Studies. *Front. Genet.* **2020**, *11*, 218. [CrossRef] [PubMed]
56. Bohling, J.; Small, M.; Von Bargen, J.; Louden, A.; DeHaan, P. Comparing Inferences Derived from Microsatellite and RADseq Datasets: A Case Study Involving Threatened Bull Trout. *Conserv. Genet.* **2019**, *20*, 329–342. [CrossRef]
57. Li, X.; Ding, X.; Chu, B.; Zhou, Q.; Ding, G.; Gu, S. Genetic Diversity Analysis and Conservation of the Endangered Chinese Endemic Herb Dendrobium Officinale Kimura et Migo (Orchidaceae) Based on AFLP. *Genetica* **2008**, *133*, 159–166. [CrossRef]
58. Ding, G.; Ding, X.; Shen, J.; Tang, F.; Liu, D.; He, J.; Li, X.; Chu, B. Genetic diversity and molecular authentication of wild populations of Dendrobium officinale by RAPD. *Yao Xue Xue Bao* **2005**, *40*, 1028–1032. [PubMed]
59. Gupta, S.K.; Baek, J.; Carrasquilla-Garcia, N.; Penmetsa, R.V. Genome-Wide Polymorphism Detection in Peanut Using next-Generation Restriction-Site-Associated DNA (RAD) Sequencing. *Mol. Breed.* **2015**, *35*, 145. [CrossRef]
60. Jiao, Y.; Zhao, H.; Ren, L.; Song, W.; Zeng, B.; Guo, J.; Wang, B.; Liu, Z.; Chen, J.; Li, W.; et al. Corrigendum: Genome-Wide Genetic Changes during Modern Breeding of Maize. *Nat. Genet.* **2014**, *46*, 1039–1040. [CrossRef]
61. Dillon, M.M.; Sung, W.; Lynch, M.; Cooper, V.S. The Rate and Molecular Spectrum of Spontaneous Mutations in the GC-Rich Multichromosome Genome of Burkholderia Cenocepacia. *Genetics* **2015**, *200*, 935–946. [CrossRef]
62. Li, W.; Liu, L.; Wang, Y.; Zhang, Q.; Fan, G.; Zhang, S.; Wang, Y.; Liao, K. Genetic Diversity, Population Structakure, and Relationships of Apricot (Prunus) Based on Restriction Site-Associated DNA Sequencing. *Hortic. Res.* **2020**, *7*, 69. [CrossRef]
63. Ye, M.; Wang, X.; Zhou, Y.; Huang, S.; Liu, A. Genetic Diversity and Population Structure of Cultivated Dendrobium Huoshanense (C.Z. Tang et S.J. Cheng) Using SNP Markers Generated from GBS Analysis. *Pak. J. Bot* **2021**, *53*, 1683–1690. [CrossRef]
64. Hou, B.; Luo, J.; Zhang, Y.; Niu, Z.; Xue, Q.; Ding, X. Iteration Expansion and Regional Evolution: Phylogeography of Dendrobium Officinale and Four Related Taxa in Southern China. *Sci. Rep.* **2017**, *7*, 43525. [CrossRef]
65. Hinsley, A.; de Boer, H.J.; Fay, M.F.; Gale, S.W.; Gardiner, L.M.; Gunasekara, R.S.; Kumar, P.; Masters, S.; Metusala, D.; Roberts, D.L.; et al. A Review of the Trade in Orchids and Its Implications for Conservation. *Bot. J. Linn. Soc.* **2018**, *186*, 435–455. [CrossRef]
66. Niu, Z.; Hou, Z.; Wang, M.; Ye, M.; Zhang, B.; Xue, Q.; Liu, W.; Ding, X. A Comparative Plastomics Approach Reveals Available Molecular Markers for the Phylogeographic Study of Dendrobium Huoshanense, an Endangered Orchid with Extremely Small Populations. *Ecol. Evol.* **2020**, *10*, 5332–5342. [CrossRef] [PubMed]

genes

MDPI

Article

Molecular Characterization of *Tinospora cordifolia* (Willd.) Miers Using Novel g-SSR Markers and Their Comparison with EST-SSR and SCoT Markers for Genetic Diversity Study

Ritu Paliwal [1,2], Rakesh Singh [1,*], Debjani Roy Choudhury [1], Gunjan Tiwari [1,3], Ashok Kumar [4], K. C. Bhat [5] and Rita Singh [2]

[1] Division of Genomic Resources, ICAR-National Bureau of Plant Genetic Resources, New Delhi 110012, India
[2] Department of Biotechnology, School of Engineering and Technology, Sharda University, Greater Noida 201306, India
[3] Central Institute of Medicinal and Aromatic Planys, Lucknow 226015, India
[4] Division of Germplasm Evaluation, ICAR-National Bureau of Plant Genetic Resources, New Delhi 110012, India
[5] Division of Plant Exploration and Germplasm Collection, ICAR-National Bureau of Plant Genetic Resources, New Delhi 110012, India
* Correspondence: rakesh.singh2@icar.gov.in

Citation: Paliwal, R.; Singh, R.; Choudhury, D.R.; Tiwari, G.; Kumar, A.; Bhat, K.C.; Singh, R. Molecular Characterization of *Tinospora cordifolia* (Willd.) Miers Using Novel g-SSR Markers and Their Comparison with EST-SSR and SCoT Markers for Genetic Diversity Study. *Genes* **2022**, *13*, 2042. https://doi.org/10.3390/genes13112042

Academic Editors: Wajid Zaman and Hakim Manghwar

Received: 14 September 2022
Accepted: 24 October 2022
Published: 5 November 2022

Abstract: In the present study, novel genomic-SSR (g-SSR) markers generated in our laboratory were used to characterize *Tinospora cordifolia* and related species. The g-SSR marker was also compared with EST-SSR and SCoT markers used earlier in our laboratory to assess the genetic diversity of *T. cordifolia*. A total of 26 accessions of *T. cordifolia* and 1 accession each of *Tinospora rumphii* and *Tinospora sinensis* were characterized using 65 novel g-SSR markers. A total of 125 alleles were detected with 49 polymorphic g-SSR markers. The number of alleles per locus varied from 1–4 with a mean value of 2.55 alleles per locus. Mean PIC, gene diversity, and heterozygosity were estimated to be 0.33, 0.41, and 0.65, respectively. The two species, namely *T. rumphii* and *T. sinensis*, showed cross-species transferability of g-SSRs developed in *T. cordifolia*. The success rate of cross-species transferability in *T. rumphii* was 95.3% and 93.8% in *T. sinensis*, proving the usefulness of this marker in genetic diversity studies of related species. The *Tinospora* accessions were also used for molecular characterization using SCoT and EST-SSR markers and compared for genetic diversity and cross-species transferability. The PIC, gene diversity, heterozygosity, and principal coordinate analysis showed that g-SSR is the better maker for a genetic diversity study of *T. cordifolia*. Additionally, high cross-species transferability of g-SSRs was found (95.3% and 93.8%) compared to EST-SSRs (68.8% and 67.7%) in *T. rumphii* and *T. sinensis*, respectively.

Keywords: *Tinospora cordifolia*; genetic diversity; g-SSR; EST-SSR; SCoT markers; principal coordinate analysis (PCoA); cross-species transferability

1. Introduction

Genetic diversity plays an important role in the stability of a species because it provides the required adaptation to the existing biotic and abiotic environmental conditions and enables change in the genetic composition to cope with the changes in the environment. There are various methods for evaluating genetic diversity in organisms including phenotypic, isozyme, biochemical, and molecular markers [1–3]. With the development of molecular biology, molecular markers have attained a special position. They are more authentic, less time-consuming, and allow the detailed analysis and evaluation of genetic diversity in plants. Among molecular markers, microsatellite markers are more reliable for a genetic diversity study because these markers are evenly distributed in the whole genome with tandem repeats of short nucleotides (1–6 bp) [4,5]. The SSRs are located in both the coding and non-coding regions of the genome and contribute a high level of polymorphism. They are multi-allelic, highly reproducible, and co-dominant in nature [6,7]. Hence, SSRs are highly advantageous, easy to assay, and have different applications like the construction of linkage maps, diversity assessment, and marker-assisted selection [8]. Start codon targeted markers (SCoT) are gene-targeted markers located in the translational start codon. This technique designs single primers from short, conserved regions flanking the ATG start codon. They are better markers than RAPD and ISSR and have gained importance in the study of genetic diversity analysis [9]. EST-SSR (expressed sequence tags microsatellite markers) has high intraspecific polymorphism, a co-dominant nature, and high reproducibility. They originate from genomic coding regions and directly reflect the diversity present in the genes. EST SSRs have been used to evaluate the genetic diversity of coconut [10] and *Dendrobium officinale* [11] and to construct core germplasm collection in olives [12].

Tinospora cordifolia (Willd.) Miers. is a diploid (2n = 22), large, glabrous, deciduous, climbing shrub that belongs to the family Menispermaceae. *T. cordifolia* usually requires fast-growing species like Jatropha, Moringa, and Neem (*Azadirachta indica*) for support to grow. *T. cordifolia* growing with Neem is known by the name "Neem Giloy", and as it has many of the chemical properties of Neem and Giloe, it shows better therapeutic applications [13]. After the COVID-19 pandemic, many reports have shown the importance of Giloe in overcoming drug-induced liver injuries [14,15]. *T. Cordifolia* is well-known plant of deciduous and dry forests, dispersed in the tropical regions and above sea level from Kumaon to Assam, extending through West Bengal, Bihar, Deccan, Konkan, Karnataka, and Kerala. It is also found in Bangladesh, Sri Lanka, and China [16–19].

Due to the presence of enormous medicinal properties, this plant has been over-exploited by pharmaceutical companies and folk people for traditional remedies which led to the necessity of exploring more genetic information about this plant which would be useful in conservation. The literature survey shows that very little molecular characterization work has been performed so far on *T. cordifolia*. There are only a few reports available regarding the genetic diversity and phylogenetic study of *T. Cordifolia* using isozyme [20], ITS and cpDNA markers [21,22], SCoT markers [23], RAPD and ISSR markers [24–27], and SSR markers [28,29]. SSRs markers are excellent for linkage analysis, agronomic trait selection, and varietal identification. The present study has been done on the molecular characterization of *T. Cordifolia* using novel genomic SSR (g-SSR) markers developed in our laboratory [28] and is reported for the first time for the molecular characterization of *T. cordifolia*. Here, we are also reporting a comparison of the newly developed g-SSR with the EST-SSR markers developed in our laboratory [30] and SCoT markers [23] in deciphering the genetic diversity and population structure present in genus *Tinospora* along with cross-species transferability.

2. Materials and Methods

2.1. Plant Material and Genomic DNA Extraction

A total of 28 accessions which includes *T. cordifolia* (26 accessions), *T. rumphii* (one accession), and *T. sinensis* (one accession) cultivated and maintained at Issapur farms of ICAR-NBPGR, New Delhi, India, were selected (Table 1). Young leaves of these accessions were collected for genomic DNA extraction by using the modified CTAB method [31]. To remove phenolic compounds, 2% polyvinyl pyrrolidone (PVP) was added to the DNA extraction buffer. The quality of the DNA was checked on 0.8% agarose gel, and the quantity of DNA was estimated by a NanoDrop™ 1000 spectrophotometer (Thermo Fisher, Waltham, MA, USA). After quantification, the working concentration of each accession was made to 10 ng/µL and stored at 4 °C.

Table 1. Details of *Tinospora* accessions used in genetic diversity study.

S. No.	IC Number	Status	Source	District	State
1	IC-281970	Wild	Disturbed wild	Outer Delhi	Delhi
2	IC-281966	Wild	Disturbed wild	Central Delhi	Delhi
3	IC-281965	Wild	Disturbed wild	Central Delhi	Delhi
4	IC-281969	Wild	Disturbed wild	West Delhi	Delhi
5	IC-281952	Wild	Disturbed wild	Central Delhi	Delhi
6	IC-281973	Wild	Disturbed wild	West Delhi	Delhi
7	IC-417323	Wild	Natural wild	Balrampur	Uttar Pradesh
8	IC-471321	Wild	Disturbed wild	Balrampur	Uttar Pradesh
9	IC-281955	Wild	Natural wild	Central Delhi	Delhi
10	IC-417328	Wild	Natural wild	Faizabad	Uttar Pradesh
11	IC-281960	Wild	Disturbed wild	Outer Delhi	Delhi
12	IC-281963	Wild	Disturbed wild	Central Delhi	Delhi
13	IC-417327	Wild	Natural wild	Basti	Uttar Pradesh
14	IC-281967	Wild	Disturbed wild	West Delhi	Delhi
15	IC-281954	Wild	Disturbed wild	Central Delhi	Delhi
16	IC-281962	Wild	Disturbed wild	North Delhi	Delhi
17	IC-281961	Wild	Disturbed wild	North Delhi	Delhi
18	IC-281972	Wild	Disturbed wild	Outer Delhi	Delhi
19	IC-281968	Wild	Disturbed wild	West Delhi	Delhi
20	IC-281959	Wild	Disturbed wild	West Delhi	Delhi
21	IC-281958	Wild	Disturbed wild	North Delhi	Delhi
22	IC-281957	Wild	Disturbed wild	Central Delhi	Delhi
23	IC-281971	Wild	Disturbed wild	Outer Delhi	Delhi
24	IC-281953	Wild	Garden	Central Delhi	Delhi
25	KCOP/41	Wild	Garden	Sirsa	Haryana
26	KCOP/18	Wild	Disturbed wild	Mahendragarh	Haryana
27	KCB/38 [a]	Wild	Disturbed wild	North Cachar Hills	Assam
28	KC/NS/GD-42/14 [b]	Wild	Disturbed wild	East Siang	Arunachal Pradesh

IC-Indigenous collections, [a]-Tinospora rumphii, [b]-Tinospora sinensis.

2.2. *g-SSR PCR Amplification*

Sixty-five g-SSR markers were used for the genetic diversity analysis of twenty-eight *Tinospora* accessions. All markers were generated by the "SSR enrichment method" in our laboratory [23] and primers were synthesized from Macrogen Inc. Out of 65 g-SSR primers, only 49 primer pairs were able to show good scorable bands on gel (Table S1). PCR reaction was performed in the total volume of 10 μL containing 3 μL genomic DNA (10 ng/μL), 1 μL of 10X buffer, 0.8 μL of 25 mM MgCl2, 0.3 μL of 10 mM dNTPs, 0.2 μL of each primer (10 nmol), 0.2 μL of Taq DNA polymerase enzyme (Fermentas, Life Sciences, City, Waltham, MS, USA), and 4.3 μL distilled water. PCR amplification was performed in a thermocycler using the following PCR cycle: initial denaturation at 94 °C for 4 min followed by 36 cycles of denaturation at 94 °C for the 30 s, annealing temperature (Ta °C) (Table S1) for 45 s, extension at 72 °C for 1 min, and final extension at 72 °C for 10 min. All amplified products were mixed with gel-loading dye (bromophenol blue, xylene cyanol, and sucrose) and then analyzed on 4% Metaphor agarose (Lonza, Basel, Switzerland) gel for 4 h at a constant supply of 120 V. Gel pictures were recorded using a Gel Documentation System (α Imager®, South San Francisco, CA, USA).

2.3. *EST-SSR Amplification*

A total of 96 EST-SSRs were used for the genetic diversity analysis of two additional *Tinospora* accessions (KCOP/41 and KCOP/18), one accession of *T. rumphii* (KCB/38), and one accession of *T. sinensis* (KC/NS/GD-42/14). The PCR amplification conditions were the same as Singh et al., 2016 [30]. All amplified products were mixed with gel-loading dye (bromophenol blue, xylene cyanol, and sucrose) and then analyzed on 4% Metaphor agarose (Lonza, Bend, OR, USA) gel for 4 h at a constant supply of 120 V. Gel pictures were recorded using a Gel Documentation System (α Imager®, USA).

2.4. SCoT Marker Amplification

Nineteen SCoT markers were used for the genetic diversity analysis of five additional *T. cordifolia* accession (IC-281970, IC-471321, IC-281968, KCOP/41, KCOP/18), one accession of *T. rumphii* (KCB/38), and one accession of *T. sinensis* (KC/NS/GD-42/14). The PCR amplification conditions were the same as Paliwal et al., 2013 [23]. All amplified products were mixed with gel-loading dye (bromophenol blue, xylene cyanol, and sucrose) and then analyzed on 1% agarose (Lonza, USA) gel for 4 h at a constant supply of 80 V. Gel pictures were recorded using a Gel Documentation System (α Imager®, USA).

2.5. Data Analysis

PCR products were scored visually on all 28 accessions of *Tinospora*. The band size of the amplified products was determined by comparison with size markers using a 50 bp DNA ladder (Fermentas, Life Sciences, USA). The genetic diversity assessment was done based on parameters like the number of alleles per locus, the major allele frequency, heterozygosity, gene diversity, and the polymorphic Information Content (PIC) for each locus using Power Marker V3.25 [32]. Genetic distances among all *Tinospora* accessions were calculated using Power Marker V3.25, and a Phylogenetic tree was constructed using a distance matrix in Power Marker and visualized using Figtree v 1.4.3 [33]. Populations were defined according to the states mentioned in Table 1. Delhi, Uttar Pradesh, and Haryana were considered three different populations, whereas Assam and Arunachal Pradesh were considered the fourth population (NE-North east). The model-based program STRUCTURE 2.3.4 [34] was used to infer the number of populations. The membership of each genotype was run for a range of genetic clusters from the value of K = 1 to K = 10. Each run was implemented over a burn-in period of 100,000 and 100,000 Monte Carlo Markov Chain replicates, the number of iterations being three for each population. Each K was then plotted to find the maximum peak of ΔK [35]. The online available "Structure harvester" program (http://taylor0.biology.ucla.edu (accessed on 15 July 2022)) was used to determine the final number of populations. The genetic relationship among *Tinospora* accessions was also analyzed by Principal Coordinate Analysis (PCoA) and Analysis of Molecular Variance (AMOVA) using GenAlEx V6.5 [36].

The genetic diversity of g-SSR markers was also compared with EST-SSR markers and SCoT markers for all 28 accessions. Hence, data scored across 28 accessions of *Tinospora* using 49 polymorphic g-SSRs, 80 polymorphic EST-SSRs [30], and 19 SCoT [23] markers were compared using different genetic parameters, i.e., the number of alleles per locus, heterozygosity, major allele frequency, polymorphic Information Content (PIC), and gene diversity for each locus using Power Marker 3.5 [32]. The genetic relatedness among 28 *Tinospora* accessions was also determined by the model-based program STRUCTURE and principal coordinate analysis (PCoA) generated by GenAlEx V6.5 [34] with 49 g-SSR, 80 EST-SSR, and 19 SCoT markers. Cross-transferability ability of g-SSRs and EST-SSRs was also compared in *T. rumphii* and *T. sinensis*.

3. Results

3.1. Diversity Assessment in Tinospora Species Using g-SSR

Sixty-five g-SSR markers were used to characterize twenty-eight accessions of *Tinospora* (26 accessions of *T. cordifolia*, 1 accession of *T. rumphii*, and 1 accession of *T. sinensis*). All 65 g-SSR markers were amplified by 26 *T. cordifolia* accessions, whereas, in the case of *T. rumphii*, 62 markers and 61 markers for *T. sinensis* were amplified. This shows that the cross-species transferability success rate was 95.3% in *T. rumphii* and 93.8% in *T. sinensis*. Out of these 65 markers used in the present study, 16 were monomorphic, whereas 49 g-SSR markers showed polymorphism. Only 49 polymorphic markers were used for the further study, which amplified a total of 125 alleles. The number of alleles per locus varied from 1 to 4 with a mean value of 2.55 alleles per locus. The maximum number of alleles (4) was amplified with primers TcgSSR15 and TcgSSR17, whereas 24 primers amplified 3 alleles and 22 primers amplified 2 alleles (Table S2). The *PIC* values ranged from 0.03

(TcgSSR-30 and TcgSSR-38) to 0.54 (TcgSSR-17) with a mean value of 0.33 (Table S2). The variability at each g-SSR locus was estimated in terms of major allele frequency which ranges from 0.46 (TcgSSR-17) to 1.00 (TcgSSR-43) with a mean value of 0.65 (Table S2). The heterozygosity varied from 0.04 to 1.00 with a mean value of 0.65. The gene diversity value for 49 polymorphic markers varied from 0.04 to 0.62 with a mean value of 0.41. TcgSSR-30 and TcgSSR-38 showed the minimum value (0.04), while TcgSSR17 (0.62) showed the maximum value of gene diversity (Table S2).

3.2. Genetic Relatedness Analysis

Cluster analysis based on 49 g-SSR markers grouped 28 *Tinospora* accessions into 3 major clusters (Figure 1a). In cluster 1, nineteen accessions were grouped, in cluster 2 two accessions were grouped, and in cluster 3 seven accessions were grouped. In cluster 1, twelve from Delhi, three from Uttar Pradesh, two from Haryana, and one accession each from Assam (*T. rumphii*) and Arunachal Pradesh (*T. sinensis*) were present. *T. sinensis and T. rumphii* showed more similarity with two accessions of *T. cordifolia* (KCPO/18 and KCOP/41) from Haryana state. The accessions used in the present study were collected from three states (Delhi, Haryana, and Uttar Pradesh); however, location-specific grouping was not observed. The two species *T. rumphii* (KCB/38) and *T. sinensis* (KC/NS/GD-42/14) in cluster 1 showed that these two species share more similarities to each other than *T. cordifolia*. Phylogenetic analysis with EST-SSR markers (Figure 1b) showed three clusters. In clusters 1, 2, and 3, ten, fifteen, and three accessions were grouped, respectively. Seven accessions from Delhi and three accessions from Uttar Pradesh were present in cluster 1. Ten accessions from Delhi, two accessions from Haryana, one accession from Uttar Pradesh, one accession from Assam, and one accession from Arunachal Pradesh were present in cluster 2. Three accessions from Delhi were present in cluster 3. Cluster analysis with EST-SSR markers revealed that accessions KCOP/41 (Haryana), KCOP/18 (Haryana), KCB/38 (*T. rumphii*), KC/NS/GD-42/14 (*T. sinensis*), and IC-281970 from Delhi got grouped r (Cluster 2). Phylogenetic analysis with SCoT markers showed one cluster and two ungrouped accessions. Eighteen accessions from Delhi, four accessions from Uttar Pradesh, two accessions from Haryana, and one accession each from Assam and Arunachal Pradesh were present in cluster 1. The ungrouped accessions (IC-281958 and IC-281953) were from Delhi. With SCoT markers, (Figure 1c) the accessions KCOP/41 (Haryana), KCOP/18 (Haryana), KCB/38 (*T. rumphii*), KC/NS/GD-42/14 (*T. sinensis*), and IC-281970 (Delhi) grouped (Cluster 1). The comparison of the phylogenetic tree with all three-marker systems showed that *Tinospora* accessions KCOP/41 (Haryana), KCOP/18 (Haryana), KCB/38 (*T. rumphii*), S/GD-42/14 (*T. sinensis*), and IC-281970 share more similarity.

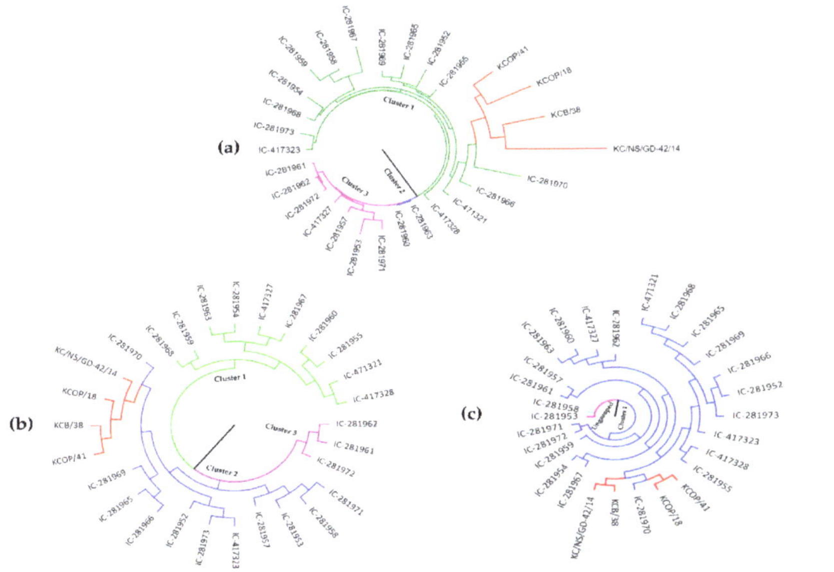

Figure 1. Genetic relationships study among 28 *Tinospora* accessions based on (**a**) g-SSR, (**b**) EST-SSR, and (**c**) SCoT markers.

3.3. Population Structure Analysis

Delta k was at the maximum at K = 2, and this was considered as the number of populations for 28 *Tinospora* accessions with g-SSR markers. In the structure analysis, different populations were categorized as pure or admixed. For categorization purposes, accessions with a score higher than 0.80 were considered pure, and those with a score lower than 0.80 as admixed (Figure 2a). Here, in population 1, there were 23 accessions categorized as pure, while in population 2 there were four accessions of which two were from other species T. *rumphii* (KCB/38) and *T. sinensis* (KC/NS/GD-42/14) along with two accessions of *T. cordifolia* (KCOP/41 and KCOP/18) from Haryana state. All these were pure except one *T. cordifolia* accession IC-281970 which was admixed (The numbers in the bar plot correspond to the serial no. of accessions given in Table 1). Delta k was the maximum at K = 3 for EST-SSR markers (Figure 2b). Population 1 had 23 accessions, population 2 had 2 accessions, and population 3 had 3 accessions. Population 1 had 21 pure and 2 admixed accessions. The two admixed were *T. cordifolia* accession IC-281968 from Delhi and IC-471321 accession from Uttar Pradesh. In population 2, *T. cordifolia* accession IC-281970 was pure and *T. sinensis* (KC/NS/GD-42/14) was considered admixed. In population 3, there were three accessions, KCOP/41, KCOP/18, and KCB/38 (*T. rumphii*), and all were considered pure. The population structure with SCoT markers revealed three populations (Figure 2c). Population 1 with five accessions showed accession KCB/38 (*T. rumphii*) and KC/NS/GD-42/14 (*T. sinensis*) were considered pure and accession KCOP/41, KCOP/18, and IC-281970 were in the admixed category. Population 2 had ten accessions with eight pure and two admixed. Population 3 had 13 accessions with 11 pure and 2 admixed. With EST-SSR markers, the *T. sinensis* and *T. rumphii* were separated into two different populations which were not observed with the g-SSR and SCoT markers.

(a)

(b)

(c)

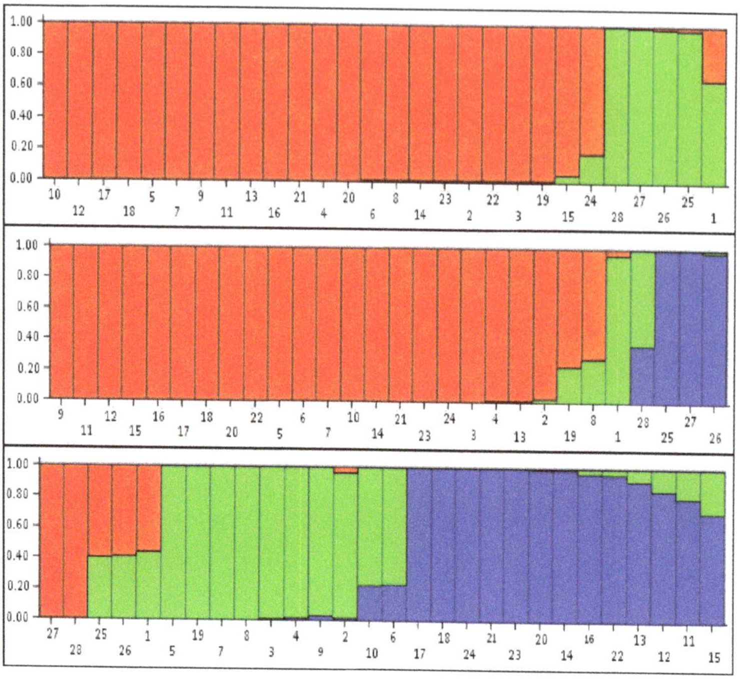

Figure 2. STRUCTURE bar plots of (**a**) g SSR (K = 2), (**b**) EST-SSR (K = 3), and (**c**) SCoT (K = 3) of 28 accessions of *Tinospora*.

3.4. Principal Coordinate and AMOVA Analysis

The principal coordinate analysis (PCoA) was performed with twenty-eight accessions considering the states as populations with all three marker systems. With g-SSR markers, axis-1 explained 16.97%, axis-2 explained 10.95%, and axis-3 explained 9.44% variation with the total percentage variation contributed by the first three axes being 37.37% (Table S4). This shows that a large diversity exists in the *Tinospora* species. The grouping pattern in the PCoA plot showed that *T. rumphii* (KCB/38) from Assam and *T. sinensis* (KC/NS/GD-42/14) from Arunachal Pradesh (NE (red), Figure 3a) were distinct from the other accessions of *T. cordifolia* and were more similar to each other. Among *T. cordifolia* accessions, KCOP/41 and KCOP/18 from Haryana (blue) and IC-281970 from Delhi were distinct, whereas other accessions showed more genetic relatedness to each other. The principal coordinate analysis (PCoA) performed with EST-SSR markers showed that axis-1 explained 26.84%, axis-2 explained 15.51%, and axis-3 explained 8.79% variation with the total percentage variation contributed by the first three axes being 51.13%, showing moderate genetic diversity among the accessions (Table S4). The distribution pattern in the PCoA plot showed *T. rumphii* (KCB/38) from Assam and *T. sinensis* (KC/NS/GD-42/14) from Arunachal Pradesh (NE (red) Figure 3b) being distinct and more similar to each other. Two *T. cordifolia* accessions KCOP/41 and KCOP/18 from Haryana (blue) were also found to be distinct. This result was similar to the PCoA plot pattern obtained by g-SSR markers, but unlike the pattern obtained in cluster analysis where all the above-mentioned accessions were grouped (Figure 1b). Principal coordinate analysis (PCoA) with SCoT markers axis 1 explained 26.65%, axis 2 explained 15.09%, and axis 3 explained 10.88% variation with total percentage variation contributed by the first three axes being 52.62% (Table S4), showing moderate genetic diversity among the accessions. The PCoA plot (Figure 3c) showed *T. rumphii* (KCB/38) from Assam and *T. sinensis* (KC/NS/GD-42/14) from Arunachal Pradesh (NE (red), Figure 3c) being mixed with other *Tinospora* accessions. Here, intermixing among the individual accessions was observed.

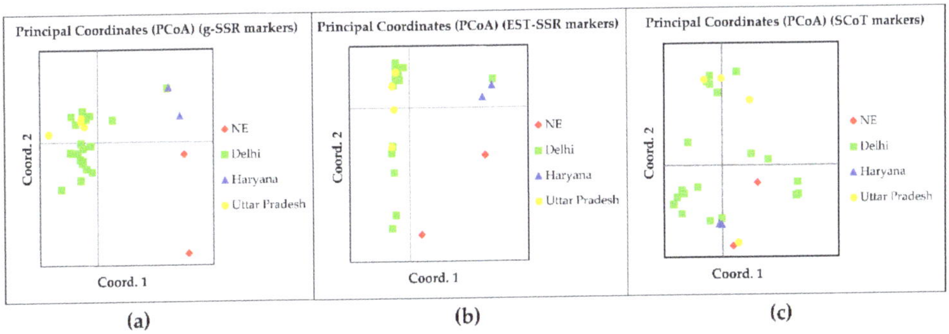

Figure 3. Principal Coordinate Analysis (PCoA) of 28 *Tinospora* accessions based on (**a**) g-SSR (**b**) EST-SSR (**c**) SCoT markers.

AMOVA study was performed to estimate population differentiation among all *Tinospora* accessions. g-SSR data showed that 7% variations were present among the population, no variation among individuals, and 93% variations within the individuals (Figure 4a). Whereas, EST-SSR markers showed a 10% variation among the population, no variation among individuals, and 90% variation within individuals (Figure 4b). With SCoT markers there was a 4% variation among the population, 20% variation among individuals, and 76% variation within individuals (Figure 4c). AMOVA analysis indicates that EST-SSR markers are better markers to differentiate the *Tinospora* populations followed by the g-SSR marker system.

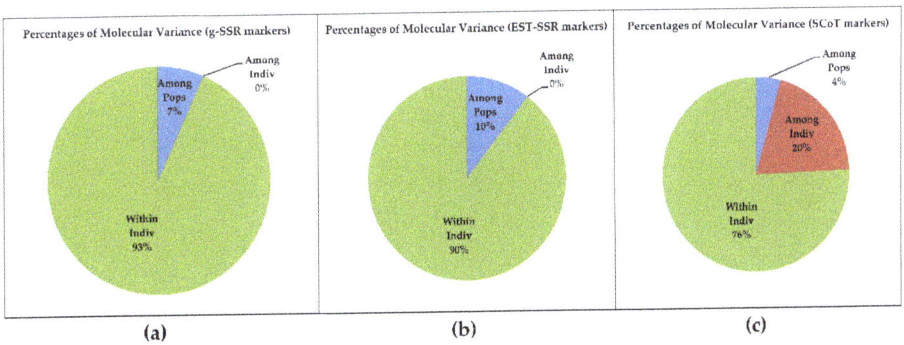

Figure 4. Analysis of Molecular variance (AMOVA) of 28 accessions of *Tinospora* based on (**a**) g-SSR (**b**) EST-SSR (**c**) SCoT markers.

3.5. Comparison of g-SSR, EST-SSR and SCoT Markers in Genetic Diversity Study

A comparison of 49 g-SSR, 80 EST-SSR, and 19 SCoT markers used for molecular characterization of 28 *Tinospora* accessions were made based on genetic diversity parameters, PCoA, AMOVA, and cross-species transferability level. The average number of bands produced per primer was compared, and it was found that in the case of g-SSRs, the average number of bands produced per primer was 2.55 while it was 2.30 and 6.0 for EST-SSRs and SCoT markers, respectively. The average major allele frequency estimated for g-SSRs, EST-SSRs, and SCoT were 0.65, 0.70, and 0.83, respectively (Tables S4 and S5 and Figure 5a). The heterozygosity varied from 0.04 to 1.00 with a mean value of 0.65 for g-SSRs and it ranged from 0.04 to 1.00 with a mean value of 0.55 for EST-SSR markers (Tables S2–S4 and Figure 5a). The gene diversity varied from 0.04 to 0.62 with a mean value of 0.41 for g-SSRs, it varied from 0.04 to 0.59 with a mean value of 0.37 for EST-SSR markers, while in SCoT markers it varied from 0.04 to 0.39 with a mean value of 0.24 (Tables S2–S4 and Figure 5a). The PIC values ranged from 0.04 to 0.54 with a mean value of 0.33 for g-SSRs, it ranged from 0.04 to 0.52 with a mean value of 0.30 for EST-SSR markers, while it varied from 0.04 to 0.30 with a mean value of 0.20 for SCoT markers (Tables S2–S4 and Figure 5a). Similarly, average per cent polymorphism was 75.4% in g-SSRs, while EST-SSRs and SCoT markers showed 83.3% and 84% average per cent polymorphic, respectively (Figure 5b). The cross-transferability success rate was found to be 95.3% in *T. rumphii* and 93.8% in *T. sinensis* with g-SSRs, while with EST-SSRs it was slightly lower, that is, 68.8% in *T. rumphii* and 67.7% in *T. sinensis* (Figure 6). PCoA analysis by g-SSR and EST-SSR presents *T. rumphii* and *T. sinensis* in separate coordinates with more similarity with each other, while the SCoT marker shows these species in the same coordinates but with more similarity with the accessions of Delhi and UP (Figure 3). AMOVA analysis shows that g-SSR gave a maximum variance of 93% within individuals, and SCoT markers gave a minimum variance of 76% within individuals. Among the population, the maximum variance was shown by EST-SSR markers, i.e., 10%, and the minimum by SCoT markers, i.e., 4%. There is no variance among individuals shown by g-SSR and EST-SSR. In contrast, SCoT markers show a 20% variance among individuals (Figure 4). The principal coordinate analysis based on 49 g-SSR data showed that the total percentage of variation explained by the first three axes was 37.4%, whereas analysis based on 80 EST-SSRs data explained 51.1% of the total, and, for analysis based on 19 SCoT markers, the total percentage variation explained by first three axes was 52.6% (Table S5 and Figure 5b).

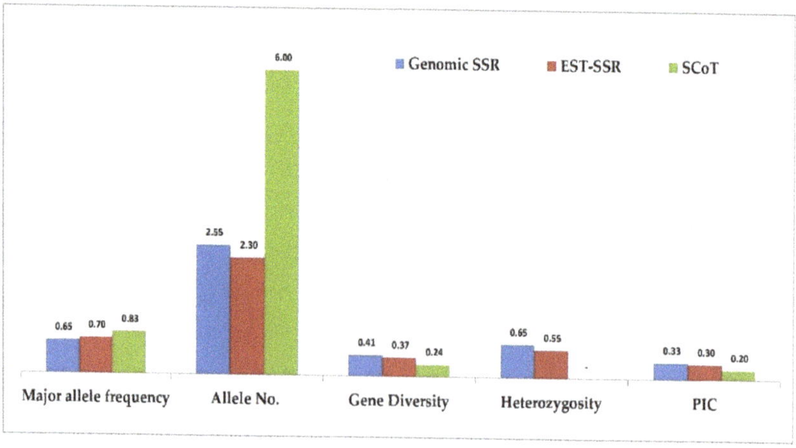

(a)

Figure 5. *Cont.*

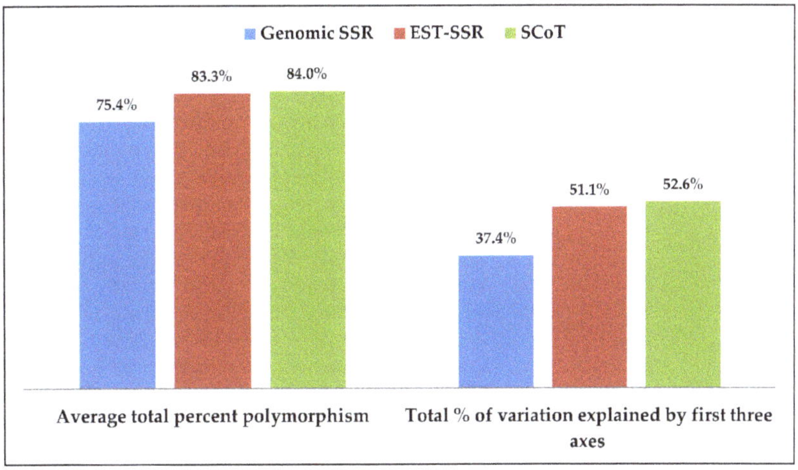

(b)

Figure 5. (**a**) Comparison of g-SSR, EST-SSR, and SCoT markers on diversity parameters of 28 *Tinospora* accessions. (**b**) Comparison of percent polymorphism between g-SSR, EST-SSR, and SCoT markers of 28 *Tinospora* accessions.

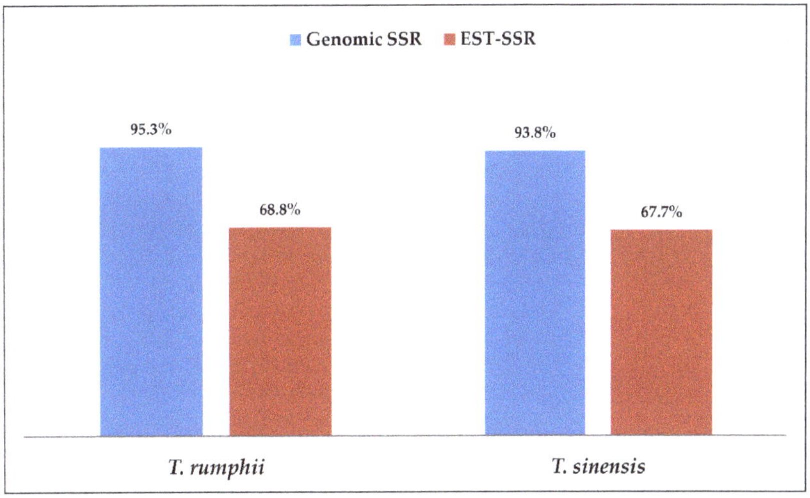

Figure 6. Comparison of cross-species transferability rate in *Tinospora rumphii* and *Tinospora sinensis* by using g-SSR and EST-SSR markers.

4. Discussion

T. cordifolia is a plant of immense therapeutic value; therefore, it needs to be conserved. It is necessary to know about the total genetic diversity present in *T. cordifolia* from all over the country for its proper management and conservation. As there are very few reports available on the genetic diversity study of *T. cordifolia*, the present study attempts to characterize *T. cordifolia* using g-SSR markers. Twenty-six accessions of *T. cordifolia*, 1 accession of *T. rumphii*, and 1 accession of *T. sinensis* were characterized with 49 g-SSR markers. In our study, the number of alleles per locus varied from 1 to 4 with a mean value of 2.55 alleles per locus with g-SSR markers. In the case of *Paris polyphylla* [37], a medicinal plant, the number of alleles per locus ranged from 2 to 5. Similarly, average allele numbers have been reported in jute 2.2 [38], groundnut 2.5 [39], and sunflower 2.67 [40] using SSR markers, which is very similar to our report. According to Botstein et al. [41], the PIC value represents diversity within accessions and the degree of polymorphism at each locus. The Polymorphic Information Content (*PIC*) in *Tinospora* varied from 0.03 to 0.53 with a mean value of 0.33, which is less than the mean *PIC* observed in other medicinal plants. Cheng et al. [42] observed a PIC value greater than 0.7 in *Paeonia lactiflora*, a *PIC* value of 0.76 was observed in *Astragalus mongholicus* [43], and a value of more than 0.7 was seen in *Smilax brasiliensis* [44]. However, there are reports that showed low PIC values such as in the case of Groundnut (0.25) [39] and *Moringa* (0.15) [45]. The gene diversity varied from 0.04 to 0.62 with a mean value of 0.41. This value is more significant than the gene diversity reported in *Moringa* (0.18) [45], but higher gene diversity has been reported in the common bean (0.6) [46]. Heterozygosity ranged from 0.04 to 1.0 with a mean value of 0.66. A similar heterozygosity value was reported in *Astragalus mongholicus* ranging from 0.36 to 1.0 [47].

The principal coordinate analysis (PCoA) showed that the total percentage variation contributed by the first three axes was 37.40%, which shows a high diversity among the accessions. In the common bean *Phaseolus vulgaris*, the PCoA also showed a separation of 53% along coordinate 1 and 13% along coordinate 2 [46]. In the grass pea, the first two principal components explained 43.42% and 29.17% of the molecular variance, respectively [43]. Population Structure analysis also showed differentiation among *T. cordifolia*, *T. sinensis*, and *T. rumphii* study based on g-SSR markers. Similar studies were also reported in medicinal plants like *Andrographis paniculate* [48]. The genomic SSR markers developed in *T. cordifolia* were found transferable in the species *Tinospora rumphii* and *Tinospora sinensis*. Similarly, a study has been done on *Camellia oleifera* where SSRs markers showed a 90.4% transferability rate in *Camellia chekangoleosa* and 78.8% in *Camellia japonica* [49]. In another study on a tea plant (*Camellia sinensis*), a 100% SSR transferability rate was reported in cultivated *Camellia assamica* and *Camellia assamica* subsp. *Lasiocalyx* [50]. In peppermint (*Mentha piperita*), an 87% transferability rate of *Mentha arvensis* and 37% in *Mentha citrate* [51] were reported, and the soybean (*Glycine max*) showed a success rate of up to 85% in *Glycine clandestine* [52]. These studies show that there were different transferability rates based on the related species.

To compare g-SSR with EST-SSR markers which were earlier studied with 24 accessions, four new accessions were added using 80 EST-SSR markers. The revised analysis showed that the number of alleles generated was 2.30 alleles per locus. The cross-transferability success rate was 68.8% in *T. rumphii* and 67.7% in *T. sinensis*, which was lower than the transferability reported by the g-SSR makers (95.3% and 93.8%). Three populations (K = 3) were obtained with EST-SSR markers, and a similar population number was reported in *Monochasma savatieri* with EST-SSR markers [53]. The principal coordinate analysis (PCoA) with EST-SSR markers gave a total percentage variation contributed by the first three axes of 51.13%, showing good diversity among the accessions. The same type of percent variation has also been reported in the medicinal plant *Monochasma savatieri* using EST-SSR markers [53]. An AMOVA-based study in Tinospora by EST-SSRs showed 90% variation within individuals, which is higher than the medicinal plant *Panax vietnamensis* (63.17%) using EST-SSR markers [54]. It is observed that EST-SSR markers have been able to differentiate the *Tinospora* populations at a moderate level in comparison to g-SSR markers. This may be due to the generation of EST-SSR markers from more conserved and expressed gene sequences compared to genomic sequences which are present in the whole genome and have less or no selection pressure than expressed regions.

For comparison with g-SSR and SCoT markers, seven accessions were added using 19 ScoT markers. ScoT markers showed high polymorphism (84%) with a high average number of alleles per primer (6.0) reported. A study of the *Hedera helix* also showed high polymorphism (95.78%) with a low allele number (1.34) [55]. The PIC value with SCoT was 0.20, similar to that found in the *Hedera helix* [55]. The population structure analysis showed that SCoT markers could not differentiate the two species of *Tinospora* from *T. cordifolia* accessions from Delhi, UP states. Similar studies with SCoT markers showed pure and admixture individuals in medicinal plants like *Dendrobium* and *Crepidium* with three (K = 3) and two (K = 2) populations, respectively [56,57]. AMOVA analysis showed a 4% variation among the population which was observed at 34% in *Clerodendrum serratum* [58]. When gene diversity parameters were compared with *g-SSR, EST-SSR*, and *SCoT markers* on 28 *Tinospora* accessions, it was found that the major allele frequency is the lowest with g-SSRs (0.65) and the maximum in SCoT markers (0.83). The number of alleles per locus found was the maximum (6.0) with SCoT followed by g-SSRs (2.55), while it was the minimum with EST-SSR (2.30). The other parameters like heterozygosity (mean value of 0.65 for g-SSR and 0.55 for EST-SSR), gene diversity (mean value of 0.41 for g-SSR, 0.37 for EST-SSR, and 0.24 for SCoT), and *PIC* value (mean value of 0.35 for g-SSR, 0.30 for EST-SSR, and 0.20 for SCoT) showed that g-SSR markers were highly polymorphic compared to EST-SSRs and SCoT markers. In a similar type of study, high PIC was reported with EST-SSR (0.70) in comparison to g-SSR (0.63) in sugarcane [59], which is contradictory to our findings.

5. Conclusions

The genetic diversity assessment among *T. cordifolia* accessions with novel g-SSR markers revealed high genetic variation among accessions, and its comparison with EST-SSR and SCoT markers-based PIC, gene diversity, heterozygosity, and principal coordinate analysis supported the same conclusion when these parameters were compared among three markers systems (g-SSR, EST-SSR, and SCoT). In addition, the two species of genus *Tinospora* used in the present study, namely *T. rumphii* and *T. sinensis*, showed higher cross-species transferability of g-SSRs compared to EST-SSRs, which shows the usefulness of g-SSR markers in the genetic diversity study of *T. cordifolia* and related species. Thus, g-SSR markers developed in our laboratory were found to be highly polymorphic and powerful tools for genetic diversity analyses of the *Tinospora* germplasms.

Supplementary Materials: The following supporting information can be downloaded at: https://www.mdpi.com/article/10.3390/genes13112042/s1. Table S1: Details of g-SSR primers used for characterization 28 Tinospora accessions. Table S2: Major allele frequency, Allele number, Gene diversity, Heterozygosity and PIC generated by 49 g-SSR in 28 *Tinospora accessions*. Table S3: Major allele frequency, Allele number, Gene diversity, Heterozygosity and PIC generated by 80 EST-SSR in 28 *Tinospora* accessions. Table S4: Major allele frequency, Allele number, Gene diversity, Heterozygosity and PIC generated by 19 SCoT markers in 28 *Tinospora* accessions. Table S5: Percentage of variation explained by the first 3 axes among the *Tinospora* accessions by g-SSR, EST-SSR and SCoT markers.

Author Contributions: R.S. (Rakesh Singh) conceived and designed the experiments; R.P. performed the experiments; R.P., D.R.C. and G.T. analyzed the data; R.S. (Rakesh Singh), A.K. and K.C.B. contributed reagents/materials/analysis tools; R.S. (Rakesh Singh), R.P. and R.S. (Rita Singh) contributed to the writing of the manuscript. All authors have read and agreed to the published version of the manuscript.

Funding: The research work was fully supported by the ICAR-National Bureau of Plant Genetic Resources, New Delhi.

Institutional Review Board Statement: Not applicable.

Informed Consent Statement: Not applicable.

Data Availability Statement: Not applicable.

Acknowledgments: We are thankful to the Director, NBPGR, New Delhi, who provided facilities for this work. Financial support was granted by the Indian Council of Agricultural Research (ICAR), India and an SRF fellowship received by Ritu Paliwal from the Council of Scientific and Industrial Research (CSIR), India.

Conflicts of Interest: There is no conflict of interest in this manuscript.

References

1. Glaszmann, J.C.; Fautret, A.; Noyer, J.L.; Feldmann, P.; Lanaud, C. Biochemical Genetic-Markers in Sugarcane. *Theor. Appl. Genet.* **1989**, *78*, 537–543. [CrossRef] [PubMed]
2. Kumar, P.; Gupta, V.K.; Misra, A.K.; Modi, D.R.; Pandey, B.K. Potential of molecular markers in plant biotechnology. *Plant Omics J.* **2009**, *2*, 141–162.
3. Lister, C.; Dean, C. Recombinant inbred lines for mapping RFLP and phenotypic markers in *Arabidopsis thaliana*. *Plant J.* **1993**, *4*, 745–750. [CrossRef]
4. Litt, M.; Luty, J.A. A hypervariable microsatellite revealed by in vitro amplification of a dinucleotide repeat within the cardiac muscle actin gene. *Am. J. Hum. Genet.* **1989**, *44*, 397–401. [PubMed]
5. Tautz, D.; Renz, M. Simple sequences are ubiquitous repetitive components of eukaryotic genomes. *Nucleic Acids Res.* **1984**, *12*, 4127–4138. [CrossRef] [PubMed]
6. Gur-Arie, R.; Cohen, C.J.; Eitan, Y.; Shelef, L.; Hallerman, E.M.; Kashi, Y. Simple sequence repeats in Escherichia coli: Abundance, distribution, composition, and polymorphism. *Genome Res.* **2000**, *10*, 62–71. [CrossRef] [PubMed]
7. Zane, L.; Bargelloni, L.; Patarnello, T. Strategies for microsatellite isolation: A review. *Mol. Ecol.* **2002**, *11*, 1–6. [CrossRef]
8. Wang, J.M.; Yang, J.M.; Zhu, J.H.; Jia, Q.J.; Tao, Y.Z. Assessment of genetic diversity by simple sequence repeat markers among forty elite varieties in the germplasm for malting barley breeding. *J. Zhejiang Univ. Sci. B.* **2010**, *11*, 792–800. [CrossRef]
9. Zhang, Y.; Yan, H.; Jiang, X.; Wang, X.; Huang, L.; Xu, B.; Zhang, X.; Zhang, L. Genetic variation, population structure and linkage disequilibrium in Switchgrass with ISSR, SCoT and EST-SSR markers. *Hereditas* **2016**, *153*, 4–15. [CrossRef]
10. Preethi, P.; Rahman, S.; Naganeeswaran, S.; Sabana, A.A.; Gangaraj, K.P.; Jerard, B.A.; Niral, V.; Rajesh, M.K. Development of EST-SSR markers for genetic diversity analysis in coconut (*Cocos nucifera* L.). *Mol. Biol.* **2020**, *47*, 9385–9397. [CrossRef]
11. Xie, J.; Zuo, J.; Huang, Y.; Li, C.; Chen, Y. The and germplasm collection for cultivated *Dendrobium officinale* K. Kimura & Migo individuals revealed by EST-SSR markers. *Genet. Resour. Crop. Evol.* **2020**, *67*, 1209–1219. [CrossRef]
12. Dervishi, A.; Jakše, J.; Ismaili, H.; Javornik, B.; Štajner, N. Genetic structure and core collection of olive germplasm from albania revealed by microsatellite markers. *Genes* **2021**, *12*, 256. [CrossRef] [PubMed]
13. Chaudhari, S.; Shaikh, N. Gaduchi-the best ayurvedic herb. *Pharma Innov. -J.* **2013**, *2*, 97–102.
14. Nagral, A.; Adhyaru, K.; Rudra, O.S.; Gharat, A.; Bhandare, S. Herbal immune booster-induced liver injury in the COVID-19 pandemic-a case series. *J. Clin. Exp. Hepatol.* **2021**, *11*, 732–738. [CrossRef] [PubMed]
15. Devarbhavi, H. Response to Tinospora cordifolia (giloy)-induced liver injury during the COVID-19 pandemic—Multicenter nationwide study from India. *Hepatol. Commun.* **2022**, *6*, 2995–2996. [CrossRef]
16. Raghu, A.V.; Geetha, S.P.; Martin, G.; Balachandran, I.; Ravindran, P.N. In vitro propagation through mature nodes of *Tinospora cordifolia* (Willd.) Hook. F. & Thoms.: An important ayurvedic medicinal plant. *Vitr. Cell Dev. Biol.* **2006**, *42*, 584–588. [CrossRef]
17. Sinha, K.; Mishra, N.P.; Singh, J.; Khanuja, S.P.S. *Tinospora cordifolia* (Guduchi), a reservoir plant for therapeutic applications: A review. *Indian J. Tradit. Know.* **2004**, *3*, 257–270.
18. Sharma, V.; Pandey, D. Beneficial effects of *Tinospora cordifolia* on blood profiles in male mice exposed to lead. *Toxicol. Int.* **2010**, *17*, 8–11. [CrossRef]
19. Spandana, U.; Ali, S.L.; Nirmala, T.; Santhi, M.; Babu, S.D.S. A Review on *Tinospora cordifolia*. *Int. J. Pharm. Sci. Rev. Res.* **2013**, *4*, 61–68.
20. Kalpesh, I.; Mohan, J.S.S. Assessment of genetic diversity in the medicinal climber *Tinospora cordifolia* (Willd.) Miers (Menispermaceae) from Gujarat, India. *Afr. J. Biotechnol.* **2009**, *8*, 6499–6505.
21. Ahmad, S.M.; Verma, V.; Qazi, P.H.; Ganaie, M.M.; Bakshi, S.K.; Qazi, G.N. Molecular phylogeny in Indian Tinospora species by DNA based molecular markers. *Plant Syst. Evol.* **2006**, *256*, 75–87. [CrossRef]
22. Ahmad, S.M.; Verma, V.; Qazi, P.H.; Hoot, S.B. Phylogenetic patterns and genetic diversity of Indian Tinospora species based on chloroplast sequence data and cytochrome P450 polymorphisms. *Plant Syst. Evol.* **2009**, *281*, 87–96. [CrossRef]
23. Paliwal, R.; Singh, R.; Singh, A.K.; Kumar, S.; Kumar, A.; Majumdar, R.S. Molecular characterization of Giloe (*Tinospora cordifolia* (Willd.) Miers ex Hook F & Thoms.) accessions using Start Codon Targeted (ScoT) markers. *Int. J. Med. Arom. Plants.* **2013**, *3*, 413–422.
24. Rout, G.R. Identification of *Tinospora cordifolia* (Willd.) Miers ex Hook F & Thoms using RAPD markers. *Z. Nat. C* **2006**, *61*, 118–122. [CrossRef]
25. Malik, A.; Arya, A.; Kaushik, V.; Sindhu, A. Studies on genetic variability of Tinospora cordifolia collected from different agroclimatic zones of Haryana using RAPD markers. *Res. Crop.* **2019**, *1*, 20.
26. Rana, V.; Thakur, K.; Sood, R.; Sharma, V.; Sharma, T.R. Genetic diversity analysis of *Tinospora cordifolia* germplasm collected from north-western Himalayan region of India. *J. Genet.* **2012**, *9*, 99–103. [CrossRef]
27. Shinde, V.M.; Dhalwal, K. DNA Fingerprinting of *Tinospora cordifolia* using RAPD analysis. *J. Glob. Pharma Technol.* **2010**, *2*, 38–42.
28. Paliwal, R.; Kumar, R.; Choudhury, D.R.; Singh, A.K.; Kumar, S.; Kumar, A.; Bhatt, K.C.; Singh, R.; Mahato, A.K.; Singh, N.K.; et al. Development of genomic simple sequence repeats (g-SSR) markers in Tinospora cordifolia and their application in diversity analyses. *Plant Gene* **2016**, *5*, 118–125. [CrossRef]
29. Lade, S.; Pande, V.; Rana, T.S.; Yadav, H.K. Estimation of genetic diversity and population structure in *Tinospora cordifolia* using SSR markers. *3 Biotech* **2020**, *10*, 310. [CrossRef]

30. Singh, R.; Kumar, R.; Mahato, A.K.; Paliwal, R.; Singh, A.K.; Kumar, S.; Marla, S.S.; Kumar, A.; Singh, N.K. De novo transcriptome sequencing facilitates genomic resource generation in *Tinospora cordifolia*. *Funct. Integr. Genom.* **2016**, *16*, 581–591. [CrossRef]
31. Doyle, J.J.; Doyle, J.L. Isolation of plant DNA from fresh tissue. *Focus* **1990**, *12*, 13–15.
32. Liu, K.; Muse, S.V. PowerMarker: An integrated analysis environment for genetic marker analysis. *Bioinformatics* **2005**, *21*, 2128–2129. [CrossRef] [PubMed]
33. Rambaut, A. FigTreev1.4.2. A Graphical Viewer of Phylogenetic Trees. 2014. Available online: http://tree.bio.ed.ac.uk/software/figtree/ (accessed on 2 September 2022).
34. Pritchard, J.K.; Stephens, M.; Donnelly, P. Inference of population structure using multilocus genotype data. *Genetics* **2000**, *155*, 945–959. [CrossRef] [PubMed]
35. Evanno, G.; Regnaut, S.; Goudet, J. Detecting the number of clusters of individuals using the software STRUCTURE: A simulation study. *Mol. Ecol.* **2005**, *14*, 2611–2620. [CrossRef] [PubMed]
36. Peakall, R.O.; Smouse, P.E. GENALEX 6: Genetic analysis in Excel. Population genetic software for teaching and research. *Mol. Ecol. Notes.* **2006**, *6*, 288–295. [CrossRef]
37. Zheng, J.Y.; Wang, H.; Chen, X.X.; Wang, P.; Gao, P.; Li, X.N.; Zhu, G.P. Microsatellite markers for assessing genetic diversity of the medicinal plant Paris polyphylla var. chinensis (Trilliaceae). *Genet. Mol. Res.* **2012**, *11*, 1975–1980. [CrossRef] [PubMed]
38. Das, M.; Banerjee, S.; Dhariwal, R.; Vyas, S.; Mir, R.R.; Topdar, N.; Kundu, A.; Khurana, J.P.; Tyagi, A.K.; Sarkar, D.; et al. Development of SSR markers and construction of a linkage map in jute. *J. Genet.* **2012**, *91*, 21–31. [CrossRef]
39. Gautami, B.; Ravi, K.; Narasu, M.L.; Hoisington, D.A.; Varshney, R.K. Novel set of groundnuts SSR markers for germplasm analysis and interspecific transferability. *Int. J. Integr. Biol.* **2009**, *7*, 100–106.
40. Hvarleva, T.; Bakalova, A.; Chepinski, I.; Hristova-Cherbadji, M.; Hristov, M.; Atanasov, A. Characterization of Bulgarian sunflower cultivars and inbred lines with microsatellite markers. *Biotechnol. Biotechnol. Equip.* **2007**, *21*, 408–412. [CrossRef]
41. Botstein, D.; White, R.L.; Skolnick, M.; Davis, R. Construction of a genetic linkage map in man using restriction fragment length polymorphisms. *Am. J. Hum. Genet.* **1980**, *32*, 314–331.
42. Cheng, Y.; Kim, C.-H.; Shin, D.-I.; Kim, S.-M.; Koo, H.M.; Park, Y. Development of simple sequence repeat (SSR) markers to study diversity in the herbaceous peony (*Paeonia lactiflora*). *J. Med. Plant Res.* **2011**, *5*, 6744–6751. [CrossRef]
43. Wang, P.C.; Zhao, L.L.; Mo, B.T.; Zhang, Y.; Chen, J.; Wang, L.B. Genetic diversity of *Ceratoides arborescens*, a species endemic to China, detected by inter-simple sequence repeat (ISSR). *Genet. Mol. Res.* **2015**, *14*, 5658–5666. [CrossRef] [PubMed]
44. Martins, A.R.; Abreu, A.G.; Bajay, M.M.; Villela, P.M.; Batista, C.E.; Monteiro, M.; Alves-Pereira, A.; Figueira, G.M.; Pinheiro, J.B.; Appezzato-da-Glória, B.; et al. Development and Characterization of Microsatellite markers for the Medicinal Plant *Smilax brasiliensis* (Smilacaceae) and related species. *Appl. Plant Sci.* **2013**, *1*, 1200507. [CrossRef]
45. Kumar Ganesan, S.; Singh, R.; Choudhury, D.R.; Bharadwaj, J.; Gupta, V.; Singode, A. Genetic diversity and population structure study of drumstick (Moringa oleifera Lam.) using morphological and SSR markers. *Ind. Crop. Prod.* **2014**, *60*, 316–325. [CrossRef]
46. Kwak, M.; Gepts, P. Structure of genetic diversity in the two major gene pools of common bean (*Phaseolus vulgaris* L., Fabaceae). *Theor. Appl. Genet.* **2009**, *118*, 979–992. [CrossRef] [PubMed]
47. Wang, A.; Wujisguleng, W.; Liu, Y.; Liu, Y.; Long, C. Isolation and characterization of polymorphic microsatellite loci for the valuable medicinal plant *Astragalus Mongholicus*. *Open J. Genet.* **2013**, *3*, 89–92. [CrossRef]
48. Kumar, R.; Kumar, C.; Paliwal, R.; Choudhury, D.R.; Singh, I.; Kumar, A.; Kumari, A.; Singh, R. Development of novel genomic simple sequence repeat (g-SSR) markers and their validation for genetic diversity analyses in Kalmegh [Andrographis paniculata (Burm. F.) Nees]. *Plants* **2020**, *9*, 1734. [CrossRef]
49. Jia, B.G.; Lin, Q.; Feng, Y.Z.; Hu, X.Y.; Tan, X.F.; Shao, F.G.; Zhang, L. Development and cross-species transferability of unigene-derived microsatellite markers in an edible oil woody plant, Camellia oleifera (Theaceae). *Genet. Mol. Res.* **2015**, *14*, 6906–6916. [CrossRef]
50. Sharma, H.; Kumar, R.; Sharma, V.; Kumar, V.; Bhardwaj, P.; Ahuja, P.S.; Sharma, R.K. Identification and cross-species transferability of 112 novel unigene-derived microsatellite markers in tea (Camellia sinensis). *Am. J. Bot.* **2011**, *98*, e1–e6. [CrossRef]
51. Kumar, B.; Kumar, U.; Yadav, H.K. Identification of EST-SSRs and molecular diversity analysis in Mentha piperita. *Crop. J.* **2015**, *3*, 335–342. [CrossRef]
52. Peakall, R.; Gilmore, S.; Keys, W.; Morgante, M.; Rafalski, A. Cross-species amplification of soybean (Glycine max) simple sequence repeats (SSRs) within the genus and other legume genera: Implications for the transferability of SSRs in plants. *Mol. Biol. Evol.* **1998**, *15*, 1275–1287. [CrossRef] [PubMed]
53. Yang, W.; Bai, Z.; Wang, F.; Zou, M.; Wang, X.; Xie, J.; Zhang, F. Analysis of the genetic diversity and population structure of Monochasma savatieri Franch. ex Maxim using novel EST-SSR markers. *BMC Genom.* **2022**, *23*, 597. [CrossRef] [PubMed]
54. Vu, D.D.; Shah, S.N.M.; Pham, M.P.; Bui, V.T.; Nguyen, M.T.; Nguyen, T.P.T. De novo assembly and Transcriptome characterization of an endemic species of Vietnam, Panax vietnamensis Ha et Grushv., including the development of EST-SSR markers for population genetics. *BMC Plant Biol.* **2020**, *20*, 358. [CrossRef]
55. Li, H.; Wang, Y.; Iqbal, R. SCoT molecular markers and population differentiation in *Hedera helix* L. *Genetika* **2021**, *53*, 739–756. [CrossRef]
56. Tikendra, L.; Potshangbam, A.M.; Amom, T.; Dey, A.; Nongdam, P. Understanding the genetic diversity and population structure of Dendrobium chrysotoxum Lindl.-An endangered medicinal orchid and implication for its conservation. *S. Afr. J. Bot.* **2021**, *138*, 364–376. [CrossRef]

57. Thakur, J.; Dwivedi, M.D.; Singh, N.; Uniyal, P.L.; Goel, S.; Pandey, A.K. Applicability of Start Codon Targeted (SCoT) and Inter Simple Sequence Repeat (ISSR) markers in assessing genetic diversity in Crepidium acuminatum (D. Don) Szlach. *J. Appl. Res. Med. Aromat. Plants* **2021**, *23*, 100310. [CrossRef]
58. Apana, N.; Amom, T.; Tikendra, L.; Potshangbam, A.M.; Dey, A.; Nongdam, P. Genetic diversity and population structure of *Clerodendrum serratum* (L.) Moon using CBDP, iPBS and SCoT markers. *J. Appl. Res. Med. Aromat. Plants* **2021**, *25*, 100349. [CrossRef]
59. Parthiban, S.; Govindaraj, P.; Senthilkumar, S. Comparison of relative efficiency of genomic SSR and EST-SSR markers in estimating genetic diversity in sugarcane. *3 Biotech* **2018**, *8*, 144. [CrossRef]

genes

MDPI

Review

Glucomannan in *Dendrobium catenatum*: Bioactivities, Biosynthesis and Perspective

Luyan Qi [1,†], Yan Shi [1,†], Cong Li [1,2], Jingjing Liu [1,2], Sun-Li Chong [1], Kean-Jin Lim [1], Jinping Si [1,2], Zhigang Han [2,*] and Donghong Chen [1,*]

1 State Key Laboratory of Subtropical Silviculture, Zhejiang A&F University, Lin'an, Hangzhou 311300, China
2 National Innovation Alliance of *Dendrobium catenatum* Industry, Engineering Technology Research Center of *Dendrobium catenatum* of National Forestry and Grassland Administration, Hangzhou 311300, China
* Correspondence: hanzg@zafu.edu.cn (Z.H.); donghong.chen@zafu.edu.cn (D.C.)
† These authors contributed equally to this work.

Abstract: *Dendrobium catenatum* is a classical and precious dual-use plant for both medicine and food in China. It was first recorded in *Shen Nong's Herbal Classic*, and has the traditional functions of nourishing yin, antipyresis, tonifying the stomach, and promoting fluid production. The stem is its medicinal part and is rich in active polysaccharide glucomannan. As an excellent dietary fiber, glucomannan has been experimentally confirmed to be involved in anti-cancer, enhancing immunity, lowering blood sugar and blood lipids, etc. Here, the status quo of the *D. catenatum* industry, the structure, bioactivities, biosynthesis pathway and key genes of glucomannan are systematically described to provide a crucial foundation and theoretical basis for understanding the value of *D. catenatum* and the potential application of glucomannan in crop biofortification.

Keywords: *Dendrobium catenatum*; glucomannan; biosynthetic pathway; structure; bioactivity; dietary fiber; hidden hunger; biofortification

Citation: Qi, L.; Shi, Y.; Li, C.; Liu, J.; Chong, S.-L.; Lim, K.-J.; Si, J.; Han, Z.; Chen, D. Glucomannan in *Dendrobium catenatum*: Bioactivities, Biosynthesis and Perspective. *Genes* **2022**, *13*, 1957. https://doi.org/10.3390/genes13111957

Academic Editors: Wajid Zaman and Hakim Manghwar

Received: 17 September 2022
Accepted: 24 October 2022
Published: 27 October 2022

1. Introduction

Dendrobium catenatum (also named *D. officinale*) is a rare and precious Traditional Chinese Medicinal (TCM) plant, and it was first recorded in the earliest works of Chinese Medicine, *Shen Nong's Herbal Classic*, written in the Eastern Han Dynasty nearly 2300 years ago. The main medicinal part of *D. catenatum* is the stem, which has the effects of nourishing *yin*, antipyresis, tonifying the stomach, and promoting fluid [1]. Wild *D. catenatum* is on the verge of extinction, and it was listed as a third-class protected species in the *Regulations on the Protection and Management of Wild Medicinal Resources* issued by the State Council of China in 1987 and as an endangered plant by the *China Plant Red Data Book* in 1992. Subsequently, *D. catenatum* was listed as critically endangered on the International Union for Conservation of Nature (IUCN) Red list (Critically Endangered A4c ver 3.1), and as a second-class national protected endangered plant on the *List of Rare and Endangered Plants of the People's Republic of China* in 2009 [2]. In the 2010 edition of the *Pharmacopoeia of the People's Republic of China*, *D. catenatum* (*D. officinale*) was separated from the "*Dendrobium*" item to form a single item. In 2018, *D. catenatum* was listed as the top one of the new "Zhe-Ba-Wei" Chinese Medicinal Materials by Zhejiang province. In 2019, *D. catenatum* was included in the catalogue of pilot work of food and drug material management by the National Health Commission of the People's Republic of China, which indicates that *D. catenatum* is officially recognized as an edible material. In 2021, as a second-level protected plant, *D. catenatum* was further included in the new version of the *List of National Key Protected Wild Plants* jointly issued by the National Forestry and Grassland Administration and the Ministry of Agriculture and Rural Affairs of the People's Republic of China.

As a traditional dual-purpose plant for both food and medicine, *D. catenatum* research and industry have experienced three developmental stages (I~III). (I) The initiation stage.

Since the 1970s, preliminary studies on the improvement of both seedling propagation and artificial cultivation technologies have been carried out. *D. catenatum* raw materials were obtained from wild resources by unrecoverable over-exploitation. (II) The blooming stage. In the beginning of the 2000s, with breakthroughs in key technologies for seed production, tissue culture, and planting substrates of *D. catenatum*, the artificial-sheltered cultivation mode (Figure 1A) with high yield was dramatically developed, resulting in a sharp rise of a ten-billion-level industry [3]. (III) The plateau stage. From the 2020s onward, simulated wild cultivation mode with *D. catenatum* directly planted on the rock and trunk (Figure 1B,C) is popular due to its high TCM quality, contributing to breaking through the bottleneck and further transformation and upgradation of the *D. catenatum* industry. Therefore, *D. catenatum* has progressed from its natural wild state (endangered) to facility cultivation (high-yield), then to imitation wild cultivation (high-quality), leading to species conservation and industrial rise with combined ecological and economic benefits.

Modern studies have shown that the main active components of *D. catenatum* are polysaccharides, alkaloids, phenols, terpenes, and flavonoids [4]. Among them, polysaccharide is closely related to its pharmacological activity and is an important index for the evaluation of *D. catenatum's quality*. The main active polysaccharide in *D. catenatum* stem is glucomannan, which has antioxidant and mild immunostimulatory activity to protect macrophages from hydrogen peroxide (H_2O_2)-induced oxidative damage [5]. Glucomannan, as an excellent soluble dietary fiber, has the virtues of regulating the intestines and stomach, improving immunity, lowering blood sugar and blood fat, fighting cancer, and losing weight [6,7]. In addition, mannose as the main glucomannan component has been identified to inhibit cancer cell growth by interfering with cellular glucose metabolism [8].

Figure 1. Epiphytic cultivation model of *D. catenatum*. (**A**). Facility-aided cultivation; (**B**) Rock-dependent eco-cultivation; (**C**) Trunk-dependent eco-cultivation; (**D**) Stereo cultivation.

Hidden hunger refers to malnutrition due to nutritional imbalance in the human body, even having taken in enough energy. More than two billion people worldwide have suffered from hidden hunger, which is highly correlated with the status that three staple cereals (rice, maize, and wheat) provide 60% of the world's food intake [9]. Recently, rediscovery and utilization of a nutrient-rich variety of "forgotten crops" in Asia offers a viable and promising approach to eliminate hidden hunger [10]. Glucomannan belongs to soluble hemicellulose and is an excellent source of soluble dietary fiber as the seventh essential nutrient in the human body. Therefore, incorporating *D. catenatum* as a rich source of glucomannan into the daily diet contributes to promoting nutritional diversification and alleviating hidden hunger.

2. Feature and Structure of Glucomannan

Unlike other soluble fibers, glucomannan is characterized by high viscosity [11]. As a hemicellulose polysaccharide, glucomannan is ubiquitous in the plant cell wall. Moreover, glucomannan exists as energy storage substance in Araceae, Liliaceae, and Iridaceae and Orchidaceae [12]. It was reported that glucomannan contributes to plant tolerance to the lack of water as a compatible solute and the succulence in *Aloe vera* [13]. Storage glucomannans in distinct species display different structures. The mannose:glucose (Man:Glc)ratio is 1.6:1.0 in glucomannan from *Amorphophallus konjac* and 3.0:1.0 in glucomannan from *Orchis mascula* [14,15]. The Man:Glc ratios of three glucomannans (ASP-4N, ASP-6N and ASP-8N) in *Aloe* leaves are 19.13:1, 8.97:1, and 2.96:1, respectively [16]. In addition, glucomannan can also be obtained from microorganisms, such as the cell walls of bacteria, yeast, or fungus [17,18]. Natural glucomannan is composed of D-glucose and D-mannose linked by a β-1,4-glucopyranoside bond to form polymer heteropolysaccharides in a certain molar ratio [19]. Moreover, the C3 position of D-mannose on the main chain can be linked to polysaccharides in the form of β-1,3-glycosidic bond or β-1,6-glycosidic bond [20].

Glucomannan is the main polysaccharide active component in *D. catenatum* (Figure 2, Table 1), and its derivatives also contain other monosaccharides, such as galacturonic acid, glucuronic acid, and galactose [21]. The contents of mannose and glucose in total polysaccharides from *D. catenatum* are 120.60 mg·g^{-1} and 71.23 mg·g^{-1}, respectively [22]. Several glucomannans from *D. catenatum* have been reported to exhibit different Man:Glc molar ratios [5,23,24], and contain abundant O-acetyl groups [25–27]. Acetylation can increase the activity and health benefits of glucomannan [28].

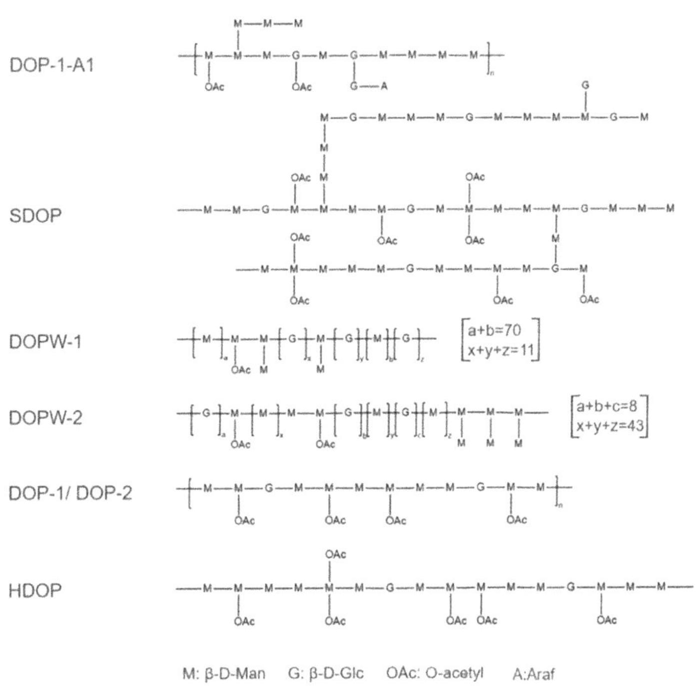

Figure 2. Compositions of different glucomannans in *D. catenatum* ([27,29,30]).

Table 1. Polysaccharides isolated from *D. catenatum*: monosaccharide compositions, molecular weights, structure unit/backbone chain, and bioactivities.

Name	Molecular Weights (Mw, kDa)	Monosaccharide Compositions	Bioactivities	References
DOP2	4699	Man:Glc = 7.64:1.00	Unknown	[23]
DOP3	5480	Man:Glc = 4.50:1.00	Unknown	[23]
DOP4	5408	Man:Glc = 3.57:1.00	Unknown	[23]
DOP-1-A1	130	Man:Glc = 40.2:8.4	Unknown	[25]
SDOP	1660	Man:Glc = 4.9:1.0	Unknown	[26]
DOPW-1	389.98	Man:Glc = 10.75:1.00	Unknown	[29]
DOPW-2	374.11	Man:Glc = 8.82:1.00	Unknown	[29]
DOP-1	389.98	Man:Glc = 5.18:1.00	Immunomodulatory activity	[27]
DOP-2	374.11	Man:Glc = 4.78:1.00	Immunomodulatory activity	[27]
DOP-W3-b	15.43	Man:Glc = 4.5:1.0	Immunomodulatory activity	[31]
DOP-I-1	730	Man:Glc = 5.8:1.0	Immunomodulatory activity	[32]
DOPa	810	Man:Glc = 5.6:1.0	Immunomodulatory activity	[32]
DOPb	670	Man:Glc = 5.9:1.0	Immunomodulatory activity	[32]
DOPA-1	394	Man:Glc = 5.8:1.0	Immunomodulatory activity	[5]
DOPA-2	362	Man:Glc = 4.5:1.0	Immunomodulatory activity	[5]
DWDOP1	1341	Man: Glc = 6.79:1.00	Unknown	[33]
FWDOP1	1415	Man: Glc = 7.46:1.00	Anti-tumor activity	[33]
DOPA-1	229	Man:Glc:Gal = 1.00:0.42:0.27	Anti-tumor activity	[34]
DOP1-DES	298	Man:Glc = 2.2:1.0	Unknown	[35]
DOP2-DES	30	Man:Glc = 3.7:1.0	Unknown	[35]
LDOP-1	91.8	Man:Gal:Glc:Gal:Ara = 2.0:1.7:1.3:1.6:0.7	Anti-inflammatory activity	[30]
DLP-1	1380	Man:Glc = 71.69:22.89	Immunomodulatory activity	
DCP	221	Man:Glc:Gal = 69.5:30.2:0.3	Immunomodulatory activity	[36]

DOP/DOPA/DWDOP/FWDOP/LDOP, *Dendrobium officinale* polysaccharide; SDOP, single *Dendrobium officinale* polysaccharide; DOPW, water-soluble *Dendrobium officinale* polysaccharide fraction; DCP, *D. catenatum* polysaccharides; Man:Glc, mannose:glucose.

3. Applications of Glucomannan

3.1. Medical Applications

D. catenatum has multiple traditional functions and has been developed into various medicines, such as Mailuoning injection (Jinling Pharmaceutical Co., Nanjing, China), Tongsaimai tablet (Jiangsu Kangyuan Sunshine Pharmaceutical Co., Nanjing, China), and Dendrobium nightlight pill (Tong Ren Tang, Beijing, China). Modern medicine has shown that *D. catenatum* polysaccharides have diverse bioactivities, including immunomodulatory, anti-tumor, gastro-protective, hypoglycemic, anti-inflammatory, hepatoprotective, and vasodilating effects [37]. In this review, major emphasis will be placed on the gastrointestinal protection, immunomodulatory, anti-cancer, hypoglycemic, and hypolipidemic functions of *D. catenatum* glucomannan (Figure 3).

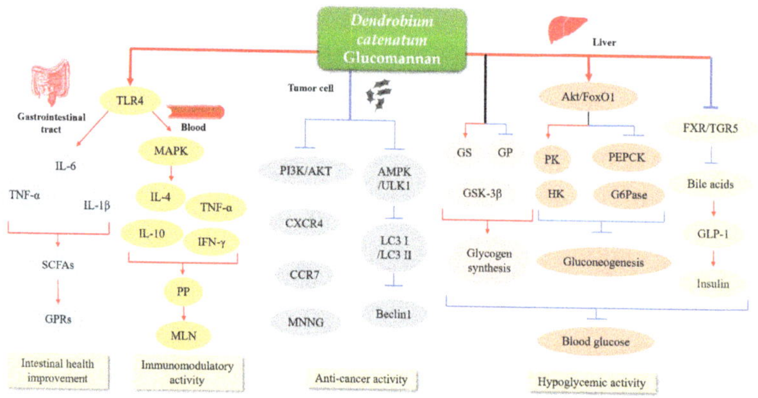

Figure 3. The effects of glucomannan from *D. catenatum* stem. DOP, *Dendrobium officinale* polysaccharide; TLR4, toll-like receptor4; IL-4, Interleukin-4; IL-6, Interleukin-6; IL-10, Interleukin-10;

IL-1β, Interleukin-1β; TNF, tumor necrosis factor; PI3K/AKT, phosphatidylinositol 3 kinase/protein kinase B; CXCR4, chemokine receptor4; CCR7, CC chemokine receptor 7 CC chemokine receptor 7; AMPK, adenosine monophosphate-activated protein kinase; ULK-1, UNC-51 like autophagy activating kinase 1; LC3, light chain 3; SCFAs, short-chain fatty acids; GPRs, g-protein-coupled receptors; MAPK, mitogen-activated protein kinase; PP, Peyer's patche; MLN, mesenteric lymph node; MNNG, 1-methyl-2-nitro-1-nitroguanidine; GS, glycogen synthase; GP, glycogen phosphorylase; GSK-3β, glycogen synthase kinase 3β; PK, pyruvate kinase; HK, hexokinase; PEPCK, phosphoenolpyruvate carboxykinase; G6Pase, glucose-6-phosphatase; GLP-1, glucagon-like peptide-1.

3.1.1. Intestinal Health Improvement

It has been reported that glucomannan can modulate mouse cecal and fecal microbiota with favorable prebiotic effects [38]. The unique properties of glucomannan hydrolysate make it valuable as a prebiotic in a wide range of food, feed, and pharmaceutical products [39]. Prebiotics promote specific changes in the gastrointestinal microbiota [40], promote the growth of probiotics such as Lactobacillus and Bifidobacterium, inhibit the proliferation of harmful bacteria, reduce inflammation, improve the integrity of the intestinal mucosa, promote nutrient absorption [41], and control the blood glucose level of patients with type 2 diabetes [42]. Glucomannan selectively stimulates the production of beneficial gut microflora such as probiotics and benefits the treatment of functional gastrointestinal disorders related to abdominal pain in mice [43–45]. The combined laxative of glucomannan-probiotic promotes defecation in constipated rats [46]. Mannose-oligosaccharides, the oxidative degradation product of glucomannan, are potential prebiotics, which can affect the growth and species abundance of fecal microbiota and regulate the balance of intestinal flora [47,48]. For instance, *D. catenatum* polysaccharide DOP can restore the diversity of intestinal flora and regulate the abundance of intestinal flora by inhibiting the overexpression of pro-inflammatory cytokines (TNF-α, IL-6, and IL-1β), restoring the level of short-chain fatty acids (SCFAs), activating G-protein-coupled receptors (GPRs), and regulating the intestinal flora to alleviate the symptoms of colitis in mice [49].

Glucomannan also plays a direct role in protecting the intestinal epithelium. *D. catenatum* polysaccharide DOP is not easily digested and absorbed in the human body but is degraded into SCFAs by the gut microbiota in the large intestine, therefore improving intestinal health [22,50]. In addition, *D. catenatum* glucomannan can reduce intestinal epithelial injury and regulate intestinal mucosal immunity by keeping a balanced ratio of pro- and anti-inflammatory cytokines and regulating the expressions of toll-like receptors (i.e., TLR-2, TLR-4, TLR-6, and TLR-9) important for recognizing pathogen-associated molecular patterns derived from various microbes in mice [51].

3.1.2. Immunomodulatory Activity

Glucomannan, one of the natural bioactive ingredients, can be used as an ideal immunomodulatory agent. Glucomannan can reduce brain inflammation, improve hippocampal neuron damage, maintain hippocampal cognitive function, and play an anti-epileptic role in epileptic rats [52]. *D. catenatum* polysaccharide DOPW3-B can improve the intestinal mucosal immune activity by increasing at Peyer's patches the levels of interferon-γ (IFN-γ) and interleukin-4 (IL-4), two key effector cytokines for the differentiation of T helper types 1 and 2 with a positive effect on mesenteric lymph nodes [31,53]. Glucomannan can increase the expression of several cytokines that are important for immune homeostasis (e.g., TNF-α, IL-β and IL-10) [54]. Meanwhile, glucomannan plays a pivotal role in regulating the activation and proliferation of macrophages [55], and promoting phagocytosis [29]. Finally, the pretreatment of *D. officinale* with organic solvents enhances the immunostimulatory activity of polysaccharides and affects the mannose/glucose ratio of polysaccharides, which plays an important role in immunostimulation [56].

3.1.3. Anti-Cancer Activity

In a zebrafish xenograft model, *D. catenatum* polysaccharide DopW-1 inhibits the proliferation of HT-29 cells by the apoptosis pathway and has an anti-tumor effect on colorectal cancer [57]. It is known that hyper-activation of the phosphatidylinositol 3 kinase/protein kinase B (PI3K/AKT) signaling pathway in human cancers can promote the proliferation and survival of tumor cells [58]. Glucomannan can block the PI3K/AKT pathway, thereby promoting the apoptotic rate and reducing the proliferation ability of tumor cells [59]. In parallel, glucomannan can also inhibit the expression of major chemokine receptors, chemokine receptor 4 (*CXCR4*) and CC chemokine receptor 7 (*CCR7*), found in a wide range of tumor cells, thus reducing the dissemination ability of tumor cells [60]. In addition to directly interfering with tumor cells, glucomannan also acts in an indirect manner. The *D. catenatum* polysaccharide DOPa-3 significantly inhibits the formation and growth of colon tumors and alleviates colon injuries [61]. In rats, DOPA-4 effectively inhibits precancerous lesions of gastric cancer induced by 150μg/mL MNNG [62]. *D. catenatum* polysaccharides can reduce oxidative stress level, inhibit stress-induced activation of adenosine monophosphate-activated protein kinase (AMPK)/UNC-51 like autophagy activating kinase 1 (ULK1) pathway and the expression of light chain 3 (LC3) I and LC3 II proteins, reduce Beclin1 expression, and then reduce hypoxia/reoxygenation induced astrocyte autophagy, reduce human astrocyte apoptosis, and promote astrocyte survival [63].

3.1.4. Hypoglycemic Activity

Glucomannan is suitable as a dietary fiber supplement for the treatment of being overweight, hyperlipidemia, and diabetes. *D. catenatum* glucomannan DOP can alleviate hyperglycemia in high fat diet (HFD)/streptozocin (STZ)-induced diabetic mice through promoting the synthesis of liver glycogen and inhibiting the degradation of liver glycogen [64].

DOP treatment can reduce the level of bile acid in diabetic rats, reduce the binding of bile acid to the nuclear receptor FXR or the membrane receptor TGR5, increase the level of glucagon-like peptide-1 (GLP-1), and improve glucose and lipid metabolism and insulin sensitivity [65]. DOP promotes glycogen synthesis by regulating the expression of glycogen synthase kinase 3β (GSK-3β) and glycogen synthase (GS) in the liver or glucose transporter 4 (GLUT4) in muscle. Glucose levels are reduced by regulating the activity of glucose metabolism enzymes in the liver, including pyruvate kinase (PK), hexokinase (HK), and phosphoenolpyruvate carboxykinase (PEPCK) [66]. On the other hand, DOP can delay diabetic cataract by decreasing the level of serum malondialdehyde (MDA), increasing the activity of superoxide dismutase (SOD) and enhancing its antioxidant capacity [49]. Glucomannan AABP-2B, isolated and purified from *Anemarrhena asphodeloides*, demonstrates its hypoglycemic effect by inhibiting α-glucosidase activity and activating irS-1/PI3K/Akt signaling pathway in insulin-resistant cells [67]. Effects of glucomannan on insulin sensitivity contribute to weight loss, and taking as little as 4 g of glucomannan per day can promote weight loss [68].

3.2. Daily Application

3.2.1. Cosmetics

D. catenatum extracts are used as cosmetic raw materials to effectively solve the skin problems caused by *yin* deficiency and fire hyperactivity due to their rich polysaccharides, flavonoids, and other nutrients that are absorbed through the skin and have good moisturizing, anti-aging, and anti-wrinkle effects [69]. The aqueous extract of *D. catenatum* can resist drying damage to epidermal cells, increase cell vitality, and improve skin moisture content [70]. *D. catenatum* polysaccharide DSP has strong antioxidant activity, can scavenge DPPH free radicals, and has a better inhibitory effect on lipid peroxidation and oxidative damage of red blood cells [71,72]. Macromolecular polysaccharide as the main moisturizing component in *D. catenatum* is a kind of polyhydroxyl polymer, whose polar groups can form hydrogen bonds with water molecules to bind water. Meanwhile, polysaccharide can form a uniform film on the skin surface to prevent water loss [73]. *Dendrobium huoshanense* polysaccharide has certain hygroscopic and moisturizing properties with a higher moisture retention rate than glycerin, is non-irritating to skin, and serves as a natural moisturizing agent [74]. So far, many related skin care products are popular in the market, such as moistening and skin brightening masks with *D. catenatum* (SENYU, Jinhua, China), *Dendrobium* emulsion (MISS QUEEN, Ningbo, China), and *D. catenatum* skin corset firming lotion (BOTANIERA, Hangzhou, China).

3.2.2. Food and Functional Food

Glucomannan has wide application prospects in the food industry and can be used as a food additive, meal substitute food, and health care products, such as *Tiepifengdou* capsules, oral liquid, decoction pieces, yogurt [75], teabags [76,77], beverages, and noodles [78], and has been applied to a number of patents [79]. Due to its large molecular weight and strong water binding ability, glucomannan has excellent hydrophilicity, gelation, emulsification, film formation, thickening, and other unique functional properties [80]. With alkali treatment, the acetyl group in glucomannan is removed to promote the formation of intramolecular and intermolecular hydrogen bonds and get a gel with excellent stability [81]. Glucomannan is a licensed food additive, which is used as a stabilizing, thickening, and gellating agent [82]. For example, in yogurt and other drinks, glucomannan can be used as a thickening agent to increase flavor and nutrition [82]. In sausages, hams, and other meat products, glucomannan can replace fat, reduce fat content, increase viscosity, and water retention, to improve the texture and flavor of meat [83]. In starch products, the addition of glucomannan affects the paste characteristics, rheological properties, and texture of the starch system [84]. In addition, glucomannan can be used for food preservation, such as glucomannan film, which has good stability and food applicability and can be used for fruit and vegetable coating preservation and flavor microcapsule production [85].

4. Glucomannan Biosynthesis Pathway in Plant

In the biosynthesis pathway of glucomannan (Figure 4), the photosynthates of leaves are transported to the *D. catenatum* stem in the form of sucrose, which is then decomposed into glucose (Glc), UDP glucose (UDP-Glc), and fructose (Fru) under the action of sucrose synthase (SUS) and invertase (INV), and then, under the action of hexokinase (HXK) and fructokinase (FRK), Glc-6-P is further catalyzed by phosphoglucomutase (PGM) to produce glucose-1-phosphate (Glc-1-P). At the same time, Fru-6-P is catalyzed by phosphate mannose isomerase (PMI) to produce mannose-6-phosphate (Man-6-P), which is converted to mannose-1-phosphate (Man-1-P) by phosphomannomutase (PMM). GDP mannose pyrophosphorylase (GMP) catalyzes the production of GDP mannose (GDP-Man). Glc-1-P is converted to GDP-glucose (GDP-Glc) by GDP-Glc pyrophosphorylase (GGP) and Glc-1-P is also converted to ADP-glucose (ADP-Glc) by ADP-Glc pyrophosphorylase (AGP). Subsequently, GDP-Man, GDP-Glu, or ADP-Glu are each transported into the Golgi apparatus by specific transporters, and they are used as substrates of cellulose-like synthases A/D (CSLA/D) to synthesize glucomannan. Additionally, GDP-Man is also used for the synthesis of vitamin C/ascorbic acid (AsA) [86]. Glucomannan synthesized in the Golgi matrix might have two roles. On the one hand, it is localized in the cell wall through vesicle transport and functions as a structural polysaccharide [87]. On the other hand, glucomannan also acts as a storage polysaccharide in some plants, such as *D. catenatum*, *Konjac*, and *A. vera*. In *Konjac*, glucomannan was discovered to accumulate in the egg-shaped idioblast within the parenchyma [12].

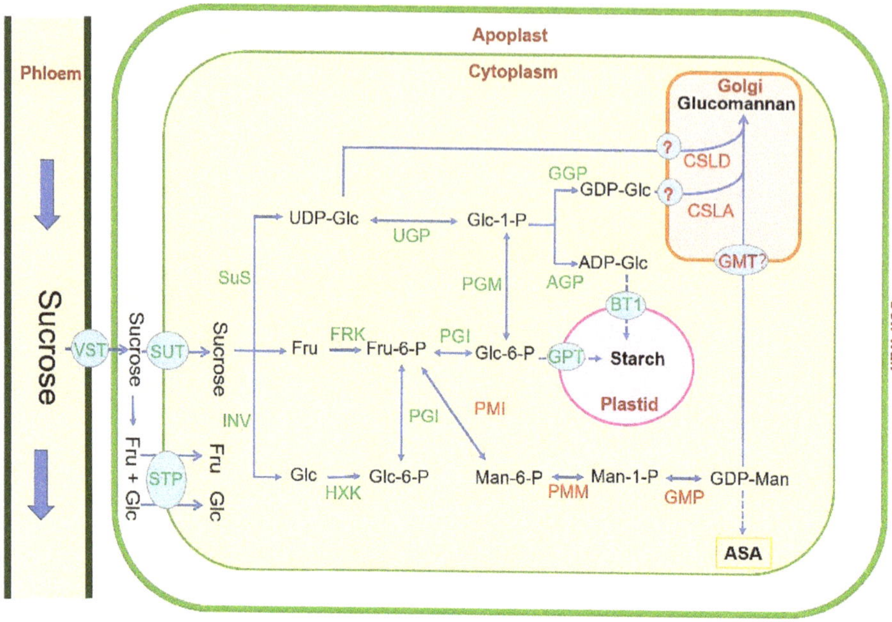

Figure 4. Putative glucomannan biosynthetic pathway in *D. catenatum* (modified from [41,44,45,88] Glc, glucose; Fru, fructose; Fru-6-P, fructose-6-phosphate; Glc-1-P, glucosophosphate-1-P; Glc-6-P, glucosophosphate-6-P; Man, mannose; Man-6-P, mannose-6-phosphate; Man-1-P, mannose-1-phosphate; AsA, ascorbic acid; SUT, sucrose transporter; STP, sugar transporter proteins; SuS, sucrose synthase; INV, invertase; FRK, fructokinase; HXK, hexokinase; PGI, phosphoglucose isomerase; UGP, UDP-Glc pyrophosphorylase; AGP, ADP-Glc pyrophosphorylase; GGP, GDP-Glc pyrophosphorylase; PGM, phosphoglucomutase; GPT, glucose-6-phosphate transporter; BT1, brittle-1 protein, an ADP-Glc transporter; PMI, phosphate mannose isomerase; PMM, phosphomannomutase; GMP, GDP-mannose pyrophosphorylase; GMT, GDP-mannose transporter; CSLA, cellulose-like synthase A; CSLD, cellulose-like synthase D.

5. Research Progresses of Key Glucomannan Biosynthetic Genes in Plant

The pathway of glucomannan synthesis involves some specific key enzymes: phosphate mannose isomerase (PMI), phosphomannomutase (PMM), GDP-mannose pyrophosphorylase (GMP), and cellulose-like synthase A/D (CSLA/D). PMI, PMM, and GMP can provide precursors for the synthesis of GDP-Man, which serves as not only the glycosyl donor and metabolic intermediate widely existing in various organisms but also the substrate of the biosynthesis of glucomannan and ascorbic acid (AsA) [86]. In *D. catenatum*, the expression levels of *DcPMI*, *DcPMM*, *DcGMP2*, and *DcCSLA4/8/13* in the stem are significantly higher than those in other tissues (Figure 5), indicating their important roles for glucomannan accumulation in the *D. catenatum* stem.

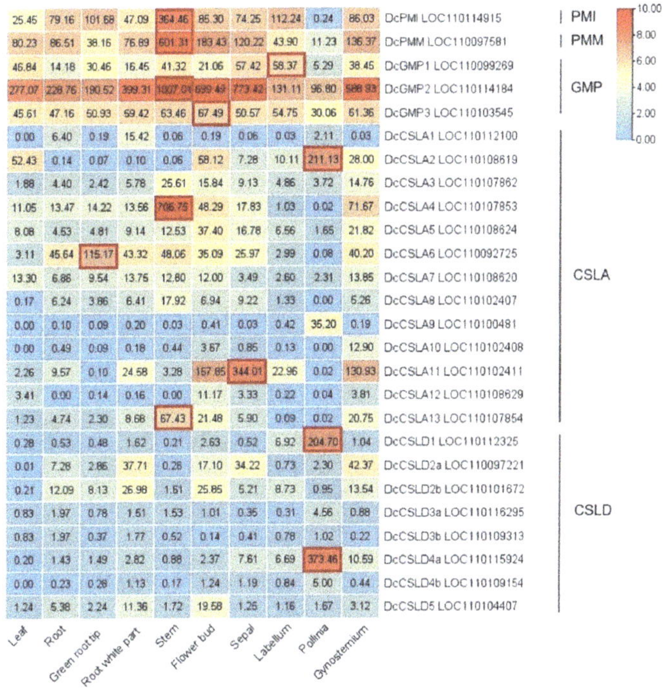

Figure 5. Expression profile of glucomannan pathway genes in tissues and organs of *D. catenatum*. The raw RNA-seq reads of different tissues and organs in *D. catenatum* are derived from the NCBI database (www.ncbi.nlm.nih.gov, accessed on 16 September 2022), including the leaf (SRR4431601), the root (SRR5722140), the green root tip (SRR4431599), the white part of the root (SRR4431598), the stem (SRR4431600), the flower bud (SRR4431603), the sepal (SRR4431597), the labellum (SRR4431602), the pollinia (SRR5722145), and the gynostemium (SRR4431596). A heatmap has been generated via TBtools software [89]. The color scale represents log2 of FPKM expression values; green and red indicate a low and high level of gene expression, respectively.

5.1. Phosphate Mannose Isomerase (PMI): Fru-6-P ↔ Man-6-P

Phosphate mannose isomerase (PMI) catalyzes the reversible conversion between Fru-6-P to Man-6-P in eukaryotes and prokaryotes and is a key enzyme during GDP-Man production [90], required for the first step of the mannose/L-galactose pathway during AsA biosynthesis in plants [91]. In Arabidopsis, there are two PMI1 isozymes, but PMI1, rather than PMI2, is involved in AsA biosynthesis. PMI1 has constitutive expression in both vegetative and reproductive organs under normal growth conditions, whereas PMI2 has no expression in any organs under light. Continuous light can induce PMI1 expression and an increased AsA level in leaves, whereas long-term darkness can induce PMI2 expression and decrease the AsA level. The diurnal expression pattern of PMI1 is in parallel with the total PMI activity and the AsA content in leaves. Moreover, knockdown of PMI1 results in a substantial decrease in the total AsA content of leaves, whereas knockout of PMI2 does not affect the total AsA levels in leaves [92]. In addition, AsA plays an essential role in scavenging reactive oxygen species (ROS) produced during cell metabolism and stress, and increase in the AsA level contributes to an improvement in abiotic stress tolerance [93]. Overexpression of *BcPMI2* from non-heading Chinese cabbage (*Brassica campestris* ssp. *chinensis* Makino) improves AsA content and tolerance to oxidative and salt stress under NaCl or H_2O_2 treatment, although transgenic tobacco has lower contents of AsA and soluble sugar than WT under normal conditions [94].

To date, PMI-like genes are mainly used in the mannose selection-based plant transformation system (plant PMI/Man system), which is non-antibiotic and environmentally friendly [95]. Most plant cells are sensitive to mannose and fail to utilize mannose as a sole carbon source because mannose can be taken up and phosphorylated to Man-6-P by endogenous hexokinase, but Man-6-P is not further utilized in the absence of enough PMI activity, subsequently leading to the accumulation of Man-6-P, inhibition of phosphoglucose isomerase, glycolysis blocking, and cell growth arrest [96]. However, plant cells with high PMI activity through genetic modification can metabolize Man-6-P and enter into glycolysis for ATP production and normal growth [97]. As a new safe positive selectable marker gene, *PMI* has been successfully used in the genetic selection of many plants, such as rice, maize, wheat, sorghum, sugarcane, onion, tomato, potato, cassava, grape, apple, cucumber, sugar beet, papaya, Arabidopsis, cabbage, and so on [98–100]. The *Escherichia coli PMI* (*EcPMI*) gene was first used in the plant PMI/Man system, which is the most popular one [101]. In 2009, the first genetically modified maize MIR604 via the EcPMI/Man selection system was approved for food and feed use, import, and processing in the European Union (European Food Safety Authority, 2009). *Saccharomyces cerevisiae PMI* (*ScPMI*) can also be used as a selectable marker in rice transformation [98]. However, non-plant type PMIs can still raise public safety concerns, and plant type PMIs may offer a superior alternative [102]. Hu et al. first proved that plant *PMI* genes from *Chlorella variabilis* and *Oryza sativa* can also be used as selectable markers to obtain transgenic plants exhibiting an accumulation of PMI transcripts and enhancement of PMI activity [100]. Through the selection system based on the green microalga *Chlorococcum* sp. *PMI (ChlPMI)*, a polycistronic gene cluster containing *crtB*, *HpBHY*, *CrBKT*, and *SlLCYB* is transformed into tomato, resulting in the production of high astaxanthin content [103].

5.2. Phosphomannomutase (PMM): Man-6-P ↔ Man-1-P

Phosphomannomutase (PMM) catalyzes the interconversion between Man-6-P and Man-1-P and provides precursors for the synthesis of GDP-Man. GDP-Man is not only used for the synthesis of glucomannan but is also an indispensable intermediate in the AsA biosynthesis pathway. The PMM enzyme isolated from the *cinnamon* seed has the activity of converting D-glucose produced by photosynthesis or glycolysis into a mannose component of plant storage polysaccharide [104]. At present, there is much research on the catalytic role of PMM in the AsA synthesis pathway. In tobacco, decreased expression of PMM via the VIGS technique resulted in a reduction of AsA content in leaves, while overexpression of *PMM* increases AsA content in leaves [105]. Transgenic tobacco plants overexpressing the acerola (*Malpighia glabra*) *PMM* gene show around a 2-fold increase in AsA content compared with WT, which correlates with the level of PMM transcripts and the corresponding enzymatic activities [106]. Overexpression of rice PMM in transgenic rice increased AsA content in seeds by 25–50% [107]. Overexpressing *D. catenatum PMM* gene in Arabidopsis, resulting in a significant increase in the contents of both AsA and polysaccharide [108].

5.3. GDP-Mannose Pyrophosphorylase (GMP): Man-1-P ↔ GDP-Man

GDP-mannose pyrophosphorylase (GMP) (also named Vitamin C Defective1, VTC1) catalyzes the conversion of Man-1-P and GTP to GDP-Man and pyrophosphate. GMP plays an important role in maintaining the AsA level and redox balance in plants. Compared with WT, the Arabidopsis *vtc1-1* mutant accumulated only 25% of leaf AsA content, but had no effect on the AsA redox state [109]. Arabidopsis KONJAC1 (KJC1) and KJC2 interact with VTC1 to stimulate GMP activity to affect the accumulation of AsA and glucomannan, and VTC1 mutants cause severe dwarfism [110]. The acerola plant has very high AsA levels, consistent with that of the *M. glabra MgGMP* gene promoter, which has higher activity than 35S and Arabidopsis GMP promoters. Transgenic tobacco plants containing the *MgGMP* gene and its original promoter, which displayed about 2-fold increased levels of AsA [111]. Transgenic tomatoes overexpressing a yeast-derived GMP exhibit up to a

31 and 17-fold increase in GMP activity in leaves and green fruit, respectively. The AsA levels increase by up to 70% in leaves, 50% in green fruit, and 35% in red fruit, especially in photosynthesizing organs [112]. The overexpression of tomato *SlGMP3* increases total AsA levels and tolerance to oxidative stress in tomatoes and high or low temperature stress in tobacco, whereas knockdown of SlGMP3 in tomatoes leads to substantially decreased AsA content and a defective phenotype with lesions and senescence due to failing to instantly detoxify ROS [113,114]. Overexpression of tomato GMP in potatoes can improve the content of AsA and dehydroascorbate (DHA) under low temperature stress and enhance the cold tolerance of potatoes [115]. Transgenic tobacco expressing *Pogonatherum paniceum* GMP has a high germination rate and high AsA content under drought and salt stress, and has strong salt tolerance and drought resistance [116]. Overexpression of *D. catenatum* GMP in Arabidopsis results in increased mannose content in water-soluble polysaccharides and enhances salt stress tolerance [117], consistent with salt hypersensitivity in the Arabidopsis *vtc-1* mutant [118].

In addition, the GMPase activity level can regulate the sensitivity of Arabidopsis to ammonium [119]. The *vtc1-1* mutant exhibits stunted root growth with an elevated of NH_4^+ efflux at the elongation zone and inhibition of cell elongation in the presence of NH_4^+ [120], but this NH_4^+-hypersensitive phenotype in mutant is independent of AsA-deficiency. In fact, the *GMP* gene disruption can also cause N-glycosylation disorder of proteins associated with the destruction of hormone homeostasis and the increase of nitric oxide content under high NH_4^+ conditions, consequently leading to conditional sensitivity to ammonium ions [121]. However, impaired GDP-mannose biosynthesis and defective N-glycosylation are required for but are not the primary causes of conditional NH_4^+ sensitivity in *vtc1-1*, whereas pH alterations in the presence of NH_4^+ associated with lost N for assimilation and alkalinization of the cytosol account for the drastic root growth defect in WT and *vtc1-1* [122].

5.4. Cellulose-like Synthase A/D (CSLA/D): GDP-Man + UDP/GDP-Glu → Glucomannan

The cellulose synthase gene superfamily consists of the CesA family and 10 Csl families: CslA~CslH, CslJ, and CslM [123,124]. The 30 Arabidopsis CSL proteins is clustered into six families: CslA, CslB, CslC, CslD, CslE, and CslG [125], and the 33 rice CSL proteins are also divided into six families: CslA, CslC, CslD, CslE, CslF, and CslH [126], but only the CslA, CslC, CslD, and CslE families are shared by both. CSL proteins are usually located in the Golgi apparatus, which mediates the synthesis of hemicellulose and then transports it to the cell wall [127]. The CSLA/D families are mainly responsible for the skeleton synthesis of mannan and glucomannan [128,129]. So far, nine *CSLA* and six *CSLD* genes have been identified in Arabidopsis, while 10 *CSLA* and four *CSLD* genes have been identified in rice [130,131], and 13 *CSLA* [132] and eight *CSLD* members have been identified in *D. catenatum* [21].

Heterologous expression assays showed CSLA proteins from a variety of species catalyze the biosynthesis of the β-1,4-mannan or glucomannan backbone in vitro, such as CtManS in guar (*Cyamopsis tetragonoloba*) [133], AtCSLA2/7/9 in Arabidopsis [128], OsCSLA1 in rice [134], PtCslA1 in *Populus trichocarpa* [135], AkCSLA3 in *A. konjac* [136], and TfManS in fenugreek (*Trigonella foenum-graecum*) [137]. In Arabidopsis, the *csla9* mutant exhibits significantly decreased glucomannan, and the *csla2 csla3 csla9* triple mutant is absent of detectable glucomannan in inflorescence stems, although these mutants have no alteration in stem development or strength [138]. The *clsa7* mutant is embryo lethal due to defective embryogenesis with evidently delayed development, abnormal cell patterning, and an arrested globular stage, associated with reduced nuclei proliferation and failed cellularization in the endosperm [139]. Overexpression of AtCSLA2, AtCSLA7, and AtCSLA9 leads to elevated glucomannan content in stems; AtCSLA9 overexpression can rescue the embryo lethality of *csla7*, indicating their functional redundancy [138]. Therefore, AtCSLA2, AtCSLA3, and AtCSLA9 are necessary for glucomannan synthesis in the stem, while AtCSLA7 is necessary for glucomannan synthesis in the embryo [138]. Heterologous expression of the O-fucosyltransferase family member AtMSR1 can enhance the glucomannan synthesis capability of AtCSLA2 and AkCSLA3, possibly via affecting enzymatic activity by protein glycosylation [140,141]. The mannosyl level in stem glucomannans is decreased by around 40% in Arabidopsis *msr1* single mutant and by more than 50% in *msr1 msr2* double mutant [142]. Transcription factors ANAC041, bZIP1, and MYB46 can directly bind the promoter of AtCLSA9 and regulate its expression [143]. Furthermore, overexpression of MYB46 leads to a significant increase in mannan content. In *D. catenatum*, overexpression of DoCSLA6 in *Arabidopsis* promotes mannan biosynthesis [144].

CSLD family members are required for the growth of stem and root, tip growth of root hair and pollen tube, and female gametophyte development and fertility. In Arabidopsis, the *csld2* and *csld3* mutants exhibit root hair bursts [145–148]; the *csld5* mutant has reduced stem growth [149]; the *csld1* and *csld4* mutants are defective in male transmission and pollen tube production; the *csld2 csld5*, *csld3*, and *csld5* and *csld2*, *csld3*, and *csld5* mutants display dwarfism and severely reduced viability [129]; and *csld2* and *csld3* show synergid cell degeneration during megagametogenesis and reduced pollen tube penetration during fertilization [150]. In rice, the *oscsld4* (*nd1*) mutant is dwarfed and its culm possesses a decreased content of xylan and cellulose but an increased amount of homogalacturonan, whereas disruption of the Arabidopsis *AtCSLD5* gene results in decreased xylan and homogalacturonan synthase activities in *Arabidopsis* [151]. It seems that CSLD proteins are not only limited in (gluco)mannan biosynthesis. At first, researchers tended to believe that CSLD proteins have major functions in mannan synthesis because microsomes isolated from tobacco (*Nicotiana benthamiana*) leaves heterologously expressing AtCSLD5 or co-expressing AtCSLD2/3 have elevated mannan synthase activity, specifically using GDP-Man as an activated nucleotide-sugar donor [152]. However, genetic rescue assays with CSLD-CESA chimeric proteins and in vitro biochemical reconstitution suggested that CSLD3 prefers to function as a UDP-Glu-dependent β-1,4-glucan synthase and forms a protein complex displaying a similar ultrastructural feature to the CESA6-forming complex [153,154]. Overexpression of cotton *GhCSLD3* in Arabidopsis enhances primary cell wall synthesis and restores the defects of the *atceas6* mutant, including significantly reduced cellulose content, a defected cell wall, and a lower dry mass, indicating that GhCSLD3 and AtCESA6 may play a similar role in cellulose or cellulose-like polysaccharide synthesis [155]. Furthermore, the CSLD family is the most similar of the *CSL* gene families to the CESA family at the amino acid level [125] and, together with the CSLF family (accounting for mixed-linkage glucan synthesis), displays the closest relationship with the CESA family at the phylogeny level [156]. Therefore, the CSLD family might function as both a (gluco)mannan synthase and a β-1,4-glucan synthase.

6. Summary and Perspective

Food production concerns a new hot topic from meeting the need of "eating fully" to "eating healthily" [7,157]. In modern life, dietary structure and habits have changed, and the source of staple food is mostly limited to several starch crops, including rice, wheat, maize, and potatoes, leading to nutrition imbalance and hidden hunger [158]. Furthermore, severe malnutrition will hinder human growth and development, accompanied by mental disorders and diseases [159]. More than 7% of chronic diseases in contemporary society are caused by hidden hunger. The FAO proposed *Zero Hunger* as the 2nd Sustainable Development Goal to eliminate all forms of hunger by 2030, and biodiversity for food and agriculture is indispensable to achieve the *Zero Hunger* program [160]. Dietary diversity and a higher intake of dietary fiber can reduce the risk of chronic diseases such as diabetes, obesity, and cardiovascular disease. Intriguingly, *D. catenatum* as a traditional dual-purpose plant for both food and medicine in China, is able to supple glucomannan as an excellent dietary fiber and has great potential to fight hidden hunger.

The efficiency of *D. catenatum* is largely dependent on the active polysaccharide glucomannan, which has multiple medicinal effects such as anti-tumor, lowering blood lipids, and preventing diabetes. To date, most of the studies on the glucomannan biosynthetic pathway have focused on its role as a cell wall structural component. PMI, PMM, and GMP sequentially catalyze the production of GDP-mannose, thus providing precursors for the synthesis of glucomannan, and CSLA/D is required for the synthesis of the glucomannan skeleton. However, little is known about the subcellular localization, transportation, and regulation of storage glucomannan, which may be clarified by the comprehensive application of newly developed technologies including immunolocalization, high throughput omics, and gene editing, consequently contributing to creating a glucomannan biofortified crop through synthetic biology (Figure 6). For instance, a stack of four identified committed genes, including *manC* and *manB* from *Escherichia coli BL21(DE3), manA and pgi* from *Bacillus subtilis*, are introduced into food-grade *B. subtilis* to obtain mannan [161]. Four anthocyanin synthetic genes, *sZmPSY1*, *sPaCrtI*, *sCrBKT*, and *sHpBHY*, are introduced into rice endosperm to produce astaxanthin-rich rice with high antioxidant activity [162]. Therefore, it is promising that glucomannan biofortified crops in the future will contribute to reducing hidden hunger and chronic diseases and promoting human health.

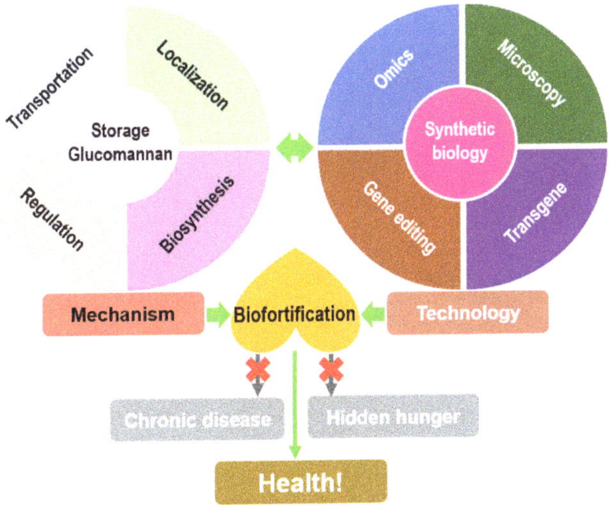

Figure 6. Biofortification strategy of glucomannan based on the in-depth dissection of molecular mechanisms and combined application of new-developed technologies.

Author Contributions: Drafting the manuscript: L.Q. and Y.S.; Gathering data: Y.S.; Contributing to the section of introduction: C.L.; Contributing to the section of feature and structure of glucomannan: J.L.; Contributing to the section of summary and perspective: J.S.; Conceiving the framework: D.C. and Z.H.; Revising the manuscript: S.-L.C. and K.-J.L. All authors have read and agreed to the published version of the manuscript.

Funding: This work was funded by National Natural Science Foundation of China (31870310), and the Major Science and Technology Projects of Breeding New Varieties of Agriculture in Zhejiang Province (2021C02074).

Institutional Review Board Statement: Not applicable.

Informed Consent Statement: Not applicable.

Data Availability Statement: The data used to support the findings of this study is available.

Acknowledgments: We appreciate Alexandre Berr (Institut de biologie moléculaire des plantes, France) for language revision.

Conflicts of Interest: The authors declare no conflict of interest.

References

1. Si, J.; Zhang, Y.; Luo, Y.; Liu, J.; Liu, Z. Herbal textual research on relationship between chinese medicine "shihu" (*Dendrobium* spp.) and "tiepi shihu" (*D. catenatum*). *China J. Chin. Mater. Med.* **2017**, *42*, 2001–2005.
2. Cheng, J.; Dang, P.-P.; Zhao, Z.; Yuan, L.-C.; Zhou, Z.-H.; Wolf, D.; Luo, Y.-B. An assessment of the chinese medicinal Dendrobium industry: Supply, demand and sustainability. *J. Ethnopharmacol.* **2019**, *229*, 81–88. [CrossRef]
3. Si, J.; Wang, Q.; Liu, Z.; Liu, J.; Luo, Y. Breakthrough in key science and technologies in *Dendrobium catenatum* Industry. *China J. Chin. Mater. Med.* **2017**, *42*, 2223–2227.
4. Jiao, C.; Song, C.; Zheng, S.; Zhu, Y.; Jin, Q.; Cai, Y.; Lin, Y. Metabolic profiling of Dendrobium officinale in response to precursors and methyl jasmonate. *Int. J. Mol. Sci.* **2018**, *19*, 728. [CrossRef]
5. Huang, K.; Li, Y.; Tao, S.; Wei, G.; Huang, Y.; Chen, D.; Wu, C. Purification, characterization and biological activity of polysaccharides from *Dendrobium officinale*. *Molecules* **2016**, *21*, 701. [CrossRef]
6. Keithley, J.; Swanson, B. Glucomannan and obesity: A critical review. *Altern. Ther. Health Med.* **2005**, *11*, 30–34.
7. Chen, D.; Han, Z.; Si, J. Huangjing (*Polygonati rhizoma*) is an emerging crop with great potential to fight chronic and hidden hunger. *Sci. China Life Sci.* **2021**, *64*, 1564–1566. [CrossRef]
8. Gonzalez, P.S.; O'Prey, J.; Cardaci, S.; Barthet, V.J.A.; Sakamaki, J.; Beaumatin, F.; Roseweir, A.; Gay, D.M.; Mackay, G.; Malviya, G.; et al. Mannose impairs tumour growth and enhances chemotherapy. *Nature* **2018**, *563*, 719–723. [CrossRef]
9. Food and Agriculture Organization of the United Nations. *Future Smart Food: Rediscovering Hidden Treasures of Neglected and Underutilized Species for Zero Hunger in Asia*; Li, X., Siddique, K.H.M., Eds.; UN: Rome, Italy, 2018; ISBN 978-92-1-047392-7.
10. Siddique, K.H.M.; Li, X.; Gruber, K. Rediscovering Asia's forgotten crops to fight chronic and hidden hunger. *Nat. Plants* **2021**, *7*, 116–122. [CrossRef]
11. McCarty, M.F. Glucomannan minimizes the postprandial insulin surge: A potential adjuvant for hepatothermic therapy. *Med. Hypotheses* **2002**, *58*, 487–490. [CrossRef]
12. Chua, M.; Hocking, T.J.; Chan, K.; Baldwin, T.C. Temporal and spatial regulation of glucomannan deposition and mobilization in corms of *Amorphophallus konjac* (Araceae). *Am. J. Bot.* **2013**, *100*, 337–345. [CrossRef] [PubMed]
13. Silva, H.; Sagardia, S.; Seguel, O.; Torres, C.; Tapia, C.; Franck, N.; Cardemil, L. Effect of water availability on growth and water use efficiency for biomass and gel production in Aloe Vera (*Aloe barbadensis* M.). *Ind. Crops Prod.* **2010**, *31*, 20–27. [CrossRef]
14. Chen, J.; Li, J.; Li, B. Identification of molecular driving forces involved in the gelation of konjac glucomannan: Effect of degree of deacetylation on hydrophobic association. *Carbohydr. Polym.* **2011**, *86*, 865–871. [CrossRef]
15. Cescutti, P.; Campa, C.; Delben, F.; Rizzo, R. Structure of the oligomers obtained by enzymatic hydrolysis of the glucomannan produced by the plant Amorphophallus konjac. *Carbohydr. Res.* **2002**, *337*, 2505–2511. [CrossRef]
16. Shi, X.-D.; Nie, S.-P.; Yin, J.-Y.; Que, Z.-Q.; Zhang, L.-J.; Huang, X.-J. Polysaccharide from leaf skin of Aloe barbadensis Miller: Part I. Extraction, fractionation, physicochemical properties and structural characterization. *Food Hydrocoll.* **2017**, *73*, 176–183. [CrossRef]
17. Pereira, J.H.; Chen, Z.; McAndrew, R.P.; Sapra, R.; Chhabra, S.R.; Sale, K.L.; Simmons, B.A.; Adams, P.D. Biochemical characterization and crystal structure of endoglucanase Cel5A from the hyperthermophilic Thermotoga maritima. *J. Struct. Biol.* **2010**, *172*, 372–379. [CrossRef]
18. Tester, R.; Al-Ghazzewi, F. Glucomannans and nutrition. *Food Hydrocoll.* **2017**, *68*, 246–254. [CrossRef]
19. Zhang, C.; Chen, J.; Yang, F. Konjac glucomannan, a promising polysaccharide for OCDDS. *Carbohydr. Polym.* **2014**, *104*, 175–181. [CrossRef]
20. Katsuraya, K.; Okuyama, K.; Hatanaka, K.; Oshima, R.; Sato, T.; Matsuzaki, K. Constitution of konjac glucomannan: Chemical analysis and 13C NMR spectroscopy. *Carbohydr. Polym.* **2003**, *53*, 183–189. [CrossRef]
21. Xi, H.; Li, Q.; Chen, X.; Liu, C.; Zhao, Y.; Yao, J.; Chen, D.; Liu, J.; Si, J.; Zhang, L. Genome-wide identification of cellulose-like synthase D gene family in *Dendrobium catenatum*. *Biotechnol. Biotechnol. Equip.* **2021**, *35*, 1163–1176. [CrossRef]
22. Li, L.; Yao, H.; Li, X.; Zhang, Q.; Wu, X.; Wong, T.; Zheng, H.; Fung, H.; Yang, B.; Ma, D.; et al. Destiny of *Dendrobium officinale* Polysaccharide after Oral Administration: Indigestible and Nonabsorbing, Ends in Modulating Gut Microbiota. *J. Agric. Food Chem.* **2019**, *67*, 5968–5977. [CrossRef] [PubMed]
23. Luo, Q.L.; Tang, Z.H.; Zhang, X.F.; Wang, L.S.; Lin, C.W.; Luo, X. Isolation, Purification and Chemical Composition Analysis of Polysaccharides from *Dendrobium officinale*. *J. Guangxi Univ. (Nat. Sci. Ed.)* **2016**, *41*, 2060–2066.
24. Hsieh, Y.S.-Y.; Chien, C.; Liao, S.K.-S.; Liao, S.-F.; Hung, W.-T.; Yang, W.-B.; Lin, C.-C.; Cheng, T.-J.R.; Chang, C.-C.; Fang, J.-M.; et al. Structure and bioactivity of the polysaccharides in medicinal plant *Dendrobium huoshanense*. *Bioorg. Med. Chem.* **2008**, *16*, 6054–6068. [CrossRef]
25. Hua, Y.; Zhang, M.; Fu, C.; Chen, Z.; Chan, G.Y.S. Structural characterization of a 2-O-acetylglucomannan from *Dendrobium officinale* stem. *Carbohydr. Res.* **2004**, *339*, 2219–2224. [CrossRef] [PubMed]
26. Gao, Y.; Hu, X.; Wang, Y.; Jiang, Z.; Zhang, H.; Zhang, M.; Hu, P. Primary structural analysis of polysaccharides from *Dendrobium officinale*. *Chem. J. Univ.* **2018**, *39*, 934–940.
27. Kuang, M.-T.; Li, J.-Y.; Yang, X.-B.; Yang, L.; Xu, J.-Y.; Yan, S.; Lv, Y.-F.; Ren, F.-C.; Hu, J.-M.; Zhou, J. Structural characterization and hypoglycemic effect *via* stimulating glucagon-like peptide-1 secretion of two polysaccharides from *Dendrobium officinale*. *Carbohydr. Polym.* **2020**, *241*, 116326. [CrossRef]

28. Li, M.; Feng, G.; Wang, H.; Yang, R.; Xu, Z.; Sun, Y.-M. Deacetylated konjac glucomannan is less effective in reducing dietary-induced hyperlipidemia and hepatic steatosis in C57BL/6 mice. *J. Agric. Food Chem.* **2017**, *65*, 1556–1565. [CrossRef]
29. Tao, S.; Lei, Z.; Huang, K.; Li, Y.; Ren, Z.; Zhang, X.; Wei, G.; Chen, H. Structural characterization and immunomodulatory activity of two novel polysaccharides derived from the stem of *Dendrobium officinale* Kimura et Migo. *J. Funct. Foods* **2019**, *57*, 121–134. [CrossRef]
30. Xing, X.; Cui, S.W.; Nie, S.; Phillips, G.O.; Goff, H.D.; Wang, Q. Study on *Dendrobium officinale* o-acetyl-glucomannan (dendronan®): Part II. fine structures of o-acetylated residues. *Carbohydr. Polym.* **2015**, *117*, 422–433. [CrossRef]
31. Xie, S.-Z.; Liu, B.; Zhang, D.-D.; Zha, X.-Q.; Pan, L.-H.; Luo, J.-P. Intestinal immunomodulating activity and structural characterization of a new polysaccharide from stems of *Dendrobium officinale*. *Food Funct.* **2016**, *7*, 2789–2799. [CrossRef]
32. Wei, W.; Feng, L.; Bao, W.-R.; Ma, D.-L.; Leung, C.-H.; Nie, S.-P.; Han, Q.-B. Structure characterization and immunomodulating effects of polysaccharides isolated from *Dendrobium officinale*. *J. Agric. Food Chem.* **2016**, *64*, 881–889. [CrossRef] [PubMed]
33. Yu, W.; Ren, Z.; Zhang, X.; Xing, S.; Tao, S.; Liu, C.; Wei, G.; Yuan, Y.; Lei, Z. Structural characterization of polysaccharides from *Dendrobium officinale* and their effects on apoptosis of hela cell line. *Molecules* **2018**, *23*, 2484. [CrossRef] [PubMed]
34. Wei, Y.; Wang, L.; Wang, D.; Wang, D.; Wen, C.; Han, B.; Ouyang, Z. Characterization and anti-tumor activity of a polysaccharide isolated from *Dendrobium officinale* grown in the Huoshan County. *Chin. Med.* **2018**, *13*, 47. [CrossRef]
35. Liang, J.; Chen, S.; Hu, Y.; Yang, Y.; Yuan, J.; Wu, Y.; Li, S.; Lin, J.; He, L.; Hou, S.; et al. Protective roles and mechanisms of *Dendrobium officinal* polysaccharides on secondary liver injury in acute colitis. *Int. J. Biol. Macromol.* **2018**, *107*, 2201–2210. [CrossRef]
36. Liu, J.; Yu, L.; Wang, C.; Zhang, Y.; Xi, H.; Si, J.; Zhang, L.; Yan, J. Preparation, structural features and in vitro immunostimulatory activity of a glucomannan from fresh *Dendrobium catenatum* stems. *Front. Nutr.* **2022**, *8*, 823803. [CrossRef] [PubMed]
37. Chen, W.; Wu, J.; Li, X.; Lu, J.; Wu, W.; Sun, Y.; Zhu, B.; Qin, L. Isolation, structural properties, bioactivities of polysaccharides from *Dendrobium officinale* Kimura et. Migo: A review. *Int. J. Biol. Macromol.* **2021**, *184*, 1000–1013. [CrossRef]
38. Chen, H.-L.; Fan, Y.-H.; Chen, M.-E.; Chan, Y. Unhydrolyzed and hydrolyzed konjac glucomannans modulated cecal and fecal microflora in Balb/c mice. *Nutrition* **2005**, *21*, 1059–1064. [CrossRef] [PubMed]
39. Behera, S.S.; Ray, R.C. Konjac glucomannan, a promising polysaccharide of *Amorphophallus konjac* K. Koch in health care. *Int. J. Biol. Macromol.* **2016**, *92*, 942–956. [CrossRef]
40. Holscher, H.D. Dietary fiber and prebiotics and the gastrointestinal microbiota. *Gut Microbes* **2017**, *8*, 172–184. [CrossRef]
41. Nakamura, Y.K.; Omaye, S.T. Metabolic diseases and pro- and prebiotics: Mechanistic insights. *Nutr. Metab.* **2012**, *9*, 60. [CrossRef]
42. Colantonio, A.G.; Werner, S.L.; Brown, M. The Effects of prebiotics and substances with prebiotic properties on metabolic and inflammatory biomarkers in individuals with type 2 diabetes mellitus: A systematic review. *J. Acad. Nutr. Diet.* **2020**, *120*, 587–607.e2. [CrossRef] [PubMed]
43. Tanabe, K.; Nakamura, S.; Moriyama-Hashiguchi, M.; Kitajima, M.; Ejima, H.; Imori, C.; Oku, T. Dietary fructooligosaccharide and glucomannan alter gut microbiota and improve bone metabolism in senescence-accelerated mouse. *J. Agric. Food Chem.* **2019**, *67*, 867–874. [CrossRef] [PubMed]
44. Connolly, M.L.; Lovegrove, J.A.; Tuohy, K.M. Konjac glucomannan hydrolysate beneficially modulates bacterial composition and activity within the faecal microbiota. *J. Funct. Foods* **2010**, *2*, 219–224. [CrossRef]
45. Al-Ghazzewi, F.H.; Khanna, S.; Tester, R.F.; Piggott, J. The potential use of hydrolysed konjac glucomannan as a prebiotic. *J. Sci. Food Agric.* **2007**, *87*, 1758–1766. [CrossRef]
46. Lu, Y.; Zhang, J.; Zhang, Z.; Liang, X.; Liu, T.; Yi, H.; Gong, P.; Wang, L.; Yang, W.; Zhang, X.; et al. Konjac glucomannan with probiotics acts as a combination laxative to relieve constipation in mice by increasing short-chain fatty acid metabolism and 5-hydroxytryptamine hormone release. *Nutrition* **2021**, *84*, 111112. [CrossRef]
47. Pongsapipatana, N.; Charoenwattanasatien, R.; Pramanpol, N.; Nguyen, T.-H.; Haltrich, D.; Nitisinprasert, S.; Keawsompong, S. Crystallization, structural characterization and kinetic analysis of a GH26 β-mannanase from *Klebsiella oxytoca* KUB-CW2-3. *Acta Crystallogr. Sect. Struct. Biol.* **2021**, *77*, 1425–1436. [CrossRef]
48. Li, J.; Jiao, G.; Sun, Y.; Chen, J.; Zhong, Y.; Yan, L.; Jiang, D.; Ma, Y.; Xia, L. Modification of starch composition, structure and properties through editing of *TaSBEIIa* in both winter and spring wheat varieties by CRISPR/Cas9. *Plant Biotechnol. J.* **2021**, *19*, 937–951. [CrossRef]
49. Zhang, Y.; Wu, Z.; Liu, J.; Zheng, Z.; Li, Q.; Wang, H.; Chen, Z.; Wang, K. Identification of the core active structure of a *Dendrobium officinale* polysaccharide and its protective effect against dextran sulfate sodium-induced colitis *via* alleviating gut microbiota dysbiosis. *Food Res. Int.* **2020**, *137*, 109641. [CrossRef]
50. Rooks, M.G.; Garrett, W.S. Gut Microbiota, Metabolites and host immunity. *Nat. Rev. Immunol.* **2016**, *16*, 341–352. [CrossRef]
51. Zhang, L.-J.; Huang, X.-J.; Shi, X.-D.; Chen, H.-H.; Cui, S.W.; Nie, S.-P. Protective effect of three glucomannans from different plants against DSS induced colitis in female BALB/c mice. *Food Funct.* **2019**, *10*, 1928–1939. [CrossRef]
52. Zhang, K.; Zhou, X.; Wang, J.; Zhou, Y.; Qi, W.; Chen, H.; Nie, S.; Xie, M. *Dendrobium officinale* polysaccharide triggers mitochondrial disorder to induce colon cancer cell death via ROS-AMPK-autophagy pathway. *Carbohydr. Polym.* **2021**, *264*, 118018. [CrossRef] [PubMed]
53. Zhu, J. T helper 2 (Th2) cell differentiation, type 2 innate lymphoid cell (ILC2) development and regulation of interleukin-4 (IL-4) and IL-13 production. *Cytokine* **2015**, *75*, 14–24. [CrossRef] [PubMed]

54. Zhong, C.; Tian, W.; Chen, H.; Yang, Y.; Xu, Y.; Chen, Y.; Chen, P.; Zhu, S.; Li, P.; Du, B. Structural characterization and immunoregulatory activity of polysaccharides from *Dendrobium officinale* leaves. *J. Food Biochem.* **2022**, *46*, e14023. [CrossRef] [PubMed]
55. Gurusmatika, S.; Nishi, K.; Harmayani, E.; Pranoto, Y.; Sugahara, T. Immunomodulatory activity of octenyl succinic anhydride modified porang (*Amorphophallus oncophyllus*) glucomannan on mouse macrophage-like J774.1 cells and mouse primary peritoneal macrophages. *Molecules* **2017**, *22*, 1187. [CrossRef] [PubMed]
56. Jo, K.; Kim, S.; Yu, K.; Chung, Y.B.; Kim, W.J.; Suh, H.J.; Kim, H. Changes in the component sugar and immunostimulating activity of polysaccharides isolated from *Dendrobium officinale* in the pretreatments. *J. Sci. Food Agric.* **2022**, *102*, 3021–3028. [CrossRef] [PubMed]
57. Tao, S.; Ren, Z.; Yang, Z.; Duan, S.; Wan, Z.; Huang, J.; Liu, C.; Wei, G. Effects of different molecular weight polysaccharides from *Dendrobium officinale* kimura & migo on human colorectal cancer and transcriptome analysis of differentially expressed genes. *Front. Pharmacol.* **2021**, *12*, 704486. [CrossRef]
58. Porta, C.; Paglino, C.; Mosca, A. Targeting PI3K/Akt/mTOR signaling in cancer. *Front. Oncol.* **2014**, *4*, 00064. [CrossRef]
59. Guanen, Q.; Junjie, S.; Baolin, W.; Chaoyang, W.; Yajuan, Y.; Jing, L.; Junpeng, L.; Gaili, N.; Zhongping, W.; Jun, W. MiR-214 promotes cell meastasis and inhibites apoptosis of esophageal squamous cell carcinoma via PI3K/AKT/mTOR signaling pathway. *Biomed. Pharmacother.* **2018**, *105*, 350–361. [CrossRef]
60. Wu, C.; Qiu, S.; Liu, P.; Ge, Y.; Gao, X. *Rhizoma Amorphophalli* inhibits TNBC cell proliferation, migration, invasion and metastasis through the PI3K/Akt/mTOR pathway. *J. Ethnopharmacol.* **2018**, *211*, 89–100. [CrossRef]
61. Liang, J.; Li, H.; Chen, J.; He, L.; Du, X.; Zhou, L.; Xiong, Q.; Lai, X.; Yang, Y.; Huang, S.; et al. *Dendrobium officinale* polysaccharides alleviate colon tumorigenesis via restoring intestinal barrier function and enhancing anti-tumor immune response. *Pharmacol. Res.* **2019**, *148*, 104417. [CrossRef]
62. Zhao, Y.; Li, B.; Wang, G.; Ge, S.; Lan, X.; Xu, G.; Liu, H. *Dendrobium officinale* polysaccharides inhibit 1-methyl-2-nitro-1-nitrosoguanidine induced precancerous lesions of gastric cancer in rats through regulating wnt/β-catenin pathway and altering serum endogenous metabolites. *Molecules* **2019**, *24*, 2660. [CrossRef] [PubMed]
63. Han, H.; Liu, W.; Chen, F.; Li, N. Effects of *Dendrobium officinale* polysaccharide on AMPK/ULK1 Pathway Related Autophagy in Astrocytes Induced by Hypoxia/Reoxygenation. *China J. Mod. Appl. Pharm.* **2021**, *38*, 2101–2115.
64. Liu, Y.; Yang, L.; Zhang, Y.; Liu, X.; Wu, Z.; Gilbert, R.G.; Deng, B.; Wang, K. *Dendrobium officinale* polysaccharide ameliorates diabetic hepatic glucose metabolism *via* glucagon-mediated signaling pathways and modifying liver-glycogen structure. *J. Ethnopharmacol.* **2020**, *248*, 112308. [CrossRef] [PubMed]
65. Chen, H.; Nie, Q.; Hu, J.; Huang, X.; Huang, W.; Nie, S. Metabolism amelioration of *Dendrobium officinale* polysaccharide on type II diabetic rats. *Food Hydrocoll.* **2020**, *102*, 105582. [CrossRef]
66. Wang, K.; Wang, H.; Liu, Y.; Shui, W.; Wang, J.; Cao, P.; Wang, H.; You, R.; Zhang, Y. *Dendrobium officinale* polysaccharide attenuates type 2 diabetes mellitus via the regulation of PI3K/Akt-mediated glycogen synthesis and glucose metabolism. *J. Funct. Foods* **2018**, *40*, 261–271. [CrossRef]
67. Chen, J.; Wan, L.; Zheng, Q.; Lan, M.; Zhang, X.; Li, Y.; Li, B.; Li, L. Structural characterization and *in vitro* hypoglycaemic activity of glucomannan from *Anemarrhena asphodeloides* bunge. *Food Funct.* **2022**, *13*, 1797–1807. [CrossRef]
68. Walsh, D.E.; Yaghoubian, V.; Behforooz, A. Effect of glucomannan on obese patients: A clinical study. *Int. J. Obes.* **1984**, *8*, 289–293.
69. Li, Q.; Xie, C.; Li, X.; Wang, X. Chemical constituents of *Dendrobium officinale* and their development and utilization in cosmetics. *China Surfactant Deterg. Cosmet.* **2017**, *47*, 109–113.
70. Chen, M.; Sun, Y.; Zhao, Y. Study on moisturizing properties of *Dendrobium officinale* extract. *J Shanghai Uni Tradit. Chin Med.* **2015**, *29*, 70–73. [CrossRef]
71. Bao, S.; Zha, X.; Hao, J.; Luo, J. Study on antioxidant activity of polysaccharide from *Dendrobium officinale* with different molecular weight in vitro. *Food Sci* **2009**, *30*, 123–127.
72. Luo, Q.; Tang, Z.; Zhang, X.; Zhong, Y.; Yao, S.; Wang, L.; Lin, C.; Luo, X. Chemical properties and antioxidant activity of a water-soluble polysaccharide from *Dendrobium officinale*. *Int. J. Biol. Macromol.* **2016**, *89*, 219–227. [CrossRef] [PubMed]
73. Huang, X.; Han, Z.; Zhang, J. Effects of fermentation on active constituents of *Dendrobium officinale* and its application in cosmetics. *China Surfactant Deterg.* **2021**, *44*, 46–50.
74. Gu, F.; Jiang, X.; Chen, Y.; Han, B.; Chen, N.; Wei, C. Study on hygroscopic and moisturizing properties and skin irritation of polysaccharides from *Dendrobium huoshanense*. *Nat. Prod. Res. Dev.* **2018**, *30*, 1701–1705. [CrossRef]
75. Jiang, W.; Zhou, M.; Li, C.; Zhang, Z.; He, S. Development of functional yoghurt of *Dendrobium officinale*. *Fujian Agric. Sci. Technol.* **2021**, *51*, 19–23. [CrossRef]
76. Meng, Y.; Lu, H.; Yang, S.; Zhang, Z.; Chen, L.; Liu, B.; Wang, L. Preparation technology and function of *Dendrobium officinale* mixed flower tea. *Food Ferment. Ind.* **2021**, *47*, 170–179. [CrossRef]
77. Luo, M.; Xie, W. Development of *Dendrobium officinale* leaf health tea bag. *Food Ind.* **2021**, *42*, 38–43.
78. Tang, W.; Xia, J.; Chen, Y. Effects of different cutting methods on active components and antioxidant activity of *Dendrobium officinale* leaf tea. *Food Sci. Technol.* **2021**, *46*, 74–82. [CrossRef]
79. Chen, S.; Yan, M.; Lv, G.; Liu, X. Development status and progress of *Dendrobium officinale* health food. *Chin. J. Pharm.* **2013**, *48*, 1625–1628.

80. Tan, Y.; Liu, X.; Yuan, F. Structure, properties and application of *Konjac* Glucomannan in Food. *China Condiment* **2019**, *44*, 168–174+178.
81. Baianu, I.C.; Ozu, E.M. Gelling mechanisms of glucomannan polysaccharides and their interactions with proteins. *ACS* **2002**, *8*, 298–305.
82. Xue, H.; Wu, D.; Xu, Q.; Zhu, Y.; Cheng, C. Application and research progress of *Konjac* Glucomannan in yogurt. *Packag. Food Mach.* **2021**, *39*, 58–62.
83. Chen, J.; Zhang, K.; Du, J.; Hu, Y.; Wang, L.; Wang, C.; Ni, X.; Jiang, F. Effects of konjac glucomannan and its derivatives on the physical properties of poultry reconstituted ham. *Food Sci.* **2010**, *31*, 36–39.
84. Zhao, D.; Zhou, Y.; Liu, H.; Liang, J.; Cheng, Y.; Nirasawa, S. Effects of dough mixing time before adding konjac glucomannan on the quality of noodles. *J. Food Sci. Technol.* **2017**, *54*, 3837–3846. [CrossRef] [PubMed]
85. Huang, Y.; Zhang, Y.; Xu, X.; Zhong, G. Optimization of konjac glucomannan edible film formulation. *Sci. Technol. Food Ind.* **2016**, *37*, 330–336. [CrossRef]
86. Gilbert, L.; Alhagdow, M.; Nunes-Nesi, A.; Quemener, B.; Guillon, F.; Bouchet, B.; Faurobert, M.; Gouble, B.; Page, D.; Garcia, V.; et al. GDP-d-mannose 3,5-epimerase (GME) plays a key role at the intersection of ascorbate and non-cellulosic cell-wall biosynthesis in tomato. *Plant J.* **2009**, *60*, 499–508. [CrossRef]
87. Joët, T.; Laffargue, A.; Salmona, J.; Doulbeau, S.; Descroix, F.; Bertrand, B.; Lashermes, P.; Dussert, S. Regulation of galactomannan biosynthesis in coffee seeds. *J. Exp. Bot.* **2014**, *65*, 323–337. [CrossRef]
88. Manzoor, S.; Wani, S.M.; Ahmad Mir, S.; Rizwan, D. Role of probiotics and prebiotics in mitigation of different diseases. *Nutrition* **2022**, *96*, 111602. [CrossRef]
89. Chen, C.; Chen, H.; Zhang, Y.; Thomas, H.R.; Frank, M.H.; He, Y.; Xia, R. TBtools: An integrative toolkit developed for interactive analyses of big biological data. *Mol. Plant* **2020**, *13*, 1194–1202. [CrossRef]
90. Roux, C.; Gresh, N.; Perera, L.E.; Piquemal, J.-P.; Salmon, L. Binding of 5-phospho-D-arabinonohydroxamate and 5-phospho-D-arabinonate inhibitors to zinc phosphomannose isomerase fromCandida albicans studied by polarizable molecular mechanics and quantum mechanics. *J. Comput. Chem.* **2007**, *28*, 938–957. [CrossRef]
91. Wheeler, G.L.; Jones, M.A.; Smirnoff, N. The biosynthetic pathway of vitamin C in higher plants. *Natre* **1998**, *393*, 365–369. [CrossRef]
92. Maruta, T.; Yonemitsu, M.; Yabuta, Y.; Tamoi, M.; Ishikawa, T.; Shigeoka, S. Arabidopsis Phosphomannose isomerase 1, but not phosphomannose isomerase 2, is essential for ascorbic acid biosynthesis. *J. Biol. Chem.* **2008**, *283*, 28842–28851. [CrossRef] [PubMed]
93. Xiao, M.; Li, Z.; Zhu, L.; Wang, J.; Zhang, B.; Zheng, F.; Zhao, B.; Zhang, H.; Wang, Y.; Zhang, Z. The multiple roles of ascorbate in the abiotic stress response of plants: Antioxidant, cofactor, and regulator. *Front. Plant Sci.* **2021**, *12*, 598173. [CrossRef] [PubMed]
94. Wang, X.; Zhang, S.; Hu, D.; Zhao, X.; Li, Y.; Liu, T.; Wang, J.; Hou, X.; Li, Y. BcPMI2, isolated from non-heading Chinese cabbage encoding phosphomannose isomerase, improves stress tolerance in transgenic tobacco. *Mol. Biol. Rep.* **2014**, *41*, 2207–2216. [CrossRef] [PubMed]
95. Zhu, Y.J.; Agbayani, R.; McCafferty, H.; Albert, H.H.; Moore, P.H. Effective selection of transgenic papaya plants with the PMI/Man selection system. *Plant Cell Rep.* **2005**, *24*, 426–432. [CrossRef]
96. He, Z.; Fu, Y.; Si, H.; Hu, G.; Zhang, S.; Yu, Y.; Sun, Z. Phosphomannose-isomerase (*PMI*) gene as a selectable marker for rice transformation via *Agrobacterium*. *Plant Sci.* **2004**, *166*, 17–22. [CrossRef]
97. Fujiki, Y.; Yoshikawa, Y.; Sato, T.; Inada, N.; Ito, M.; Nishida, I.; Watanabe, A. Dark-inducible genes from *Arabidopsis thaliana* are associated with leaf senescence and repressed by sugars. *Physiol. Plant.* **2001**, *111*, 345–352. [CrossRef]
98. Wang, T.; Liu, L.; Tang, Y.; Zhang, X.; Zhang, M.; Zheng, Y.; Zhang, F. Using the phosphomannose isomerase (*PMI*) gene from saccharomyces cerevisiae for selection in rice transformation. *J. Integr. Agric.* **2012**, *11*, 1391–1398. [CrossRef]
99. Duan, Y.; Zhai, C.; Li, H.; Li, J.; Mei, W.; Gui, H.; Ni, D.; Song, F.; Li, L.; Zhang, W.; et al. An efficient and high-throughput protocol for Agrobacterium-mediated transformation based on phosphomannose isomerase positive selection in Japonica rice (*Oryza sativa* L.). *Plant Cell Rep.* **2012**, *31*, 1611–1624. [CrossRef]
100. Hu, L.; Li, H.; Qin, R.; Xu, R.; Li, J.; Li, L.; Wei, P.; Yang, J. Plant phosphomannose isomerase as a selectable marker for rice transformation. *Sci. Rep.* **2016**, *6*, 25921. [CrossRef]
101. Joersbo, M.; Donaldson, I.; Kreiberg, J.; Petersen, S.G.; Brunstedt, J.; Okkels, F.T. Analysis of mannose selection used for transformation of sugar beet. *Mol. Breed.* **1998**, *4*, 111–117. [CrossRef]
102. Miki, B.; McHugh, S. Selectable marker genes in transgenic plants: Applications, alternatives and biosafety. *J. Biotechnol.* **2004**, *107*, 193–232. [CrossRef] [PubMed]
103. Lin, Y.; Huang, J. Characterization of an algal phosphomannose isomerase gene and its application as a selectable marker for genetic manipulation of tomato. *Plant Divers.* **2021**, *43*, 63–70. [CrossRef] [PubMed]
104. Small, D.M.; Matheson, N.K. Phosphomannomutase and phosphoglucomutase in developing Cassia corymbosa seeds. *Phytochemistry* **1979**, *18*, 1147–1150. [CrossRef]
105. Qian, W.; Yu, C.; Qin, H.; Liu, X.; Zhang, A.; Johansen, I.E.; Wang, D. Molecular and functional analysis of phosphomannomutase (PMM) from higher plants and genetic evidence for the involvement of PMM in ascorbic acid biosynthesis in Arabidopsis and Nicotiana benthamiana: Functional analysis of plant phosphomannomutase. *Plant J.* **2007**, *49*, 399–413. [CrossRef] [PubMed]

106. Badejo, A.A.; Fujikawa, Y.; Esaka, M. Gene expression of ascorbic acid biosynthesis related enzymes of the Smirnoff-Wheeler pathway in acerola (*Malpighia glabra*). *J. Plant Physiol.* **2009**, *166*, 652–660. [CrossRef] [PubMed]
107. Gao, L.; Xia, Z.; Zhang, J.; Wang, D.; Zhai, W. Transgenosis of the phosphomannomutase transgene increases vitamin C content in rice. *Chin. J. Rice Sci.* **2016**, *30*, 441–446.
108. He, C.; Zeng, S.; Teixeira da Silva, J.A.; Yu, Z.; Tan, J.; Duan, J. Molecular cloning and functional analysis of the phosphomannomutase (*PMM*) gene from *Dendrobium officinale* and evidence for the involvement of an abiotic stress response during germination. *Protoplasma* **2017**, *254*, 1693–1704. [CrossRef]
109. Veljovic-Jovanovic, S.D.; Pignocchi, C.; Noctor, G.; Foyer, C.H. Low ascorbic acid in the *vtc-1* mutant of Arabidopsis is associated with decreased growth and intracellular redistribution of the antioxidant system. *Plant Physiol.* **2001**, *127*, 426–435. [CrossRef]
110. Sawake, S.; Tajima, N.; Mortimer, J.C.; Lao, J.; Ishikawa, T.; Yu, X.; Yamanashi, Y.; Yoshimi, Y.; Kawai-Yamada, M.; Dupree, P.; et al. Konjac1 and 2 are key factors for gdp-mannose generation and affect l-ascorbic acid and glucomannan biosynthesis in Arabidopsis. *Plant Cell* **2015**, *27*, 3397–3409. [CrossRef]
111. Badejo, A.A.; Tanaka, N.; Esaka, M. Analysis of GDP-D-mannose pyrophosphorylase gene promoter from Acerola (Malpighia glabra) and increase in ascorbate content of transgenic tobacco expressing the Acerola gene. *Plant Cell Physiol.* **2008**, *49*, 126–132. [CrossRef]
112. Cronje, C.; George, G.M.; Fernie, A.R.; Bekker, J.; Kossmann, J.; Bauer, R. Manipulation of l-ascorbic acid biosynthesis pathways in Solanum lycopersicum: Elevated GDP-mannose pyrophosphorylase activity enhances l-ascorbate levels in red fruit. *Planta* **2012**, *235*, 553–564. [CrossRef] [PubMed]
113. Wang, H.; Yu, C.; Zhu, Z.; Yu, X. Overexpression in tobacco of a tomato *gmpase* gene improves tolerance to both low and high temperature stress by enhancing antioxidation capacity. *Plant Cell Rep.* **2011**, *31*, 2068–2075. [CrossRef] [PubMed]
114. Zhang, C.; Ouyang, B.; Yang, C.; Zhang, X.; Liu, H.; Zhang, Y.; Zhang, J.; Li, H.; Ye, Z. Reducing AsA leads to leaf lesion and defence response in knock-down of the AsA biosynthetic enzyme GDP-D-mannose pyrophosphorylase gene in tomato plant. *PLoS ONE* **2013**, *8*, e61987. [CrossRef]
115. Li, C.; Zhang, L.; Shi, Q.; Li, Q.; Guo, X.; Li, X.; Yu, X. Effect of Tomato GMPase overexpression on tolerance of potato plants to temperature stress. *Scientia. Agric. Sin.* **2011**, *38*, 692–700.
116. Ai, T.; Liao, X.; Li, R.; Fan, L.; Luo, F.; Xu, Y.; Wang, S. GDP-D-mannose pyrophosphorylase from *Pogonatherum paniceum* enhances salinity and drought tolerance of transgenic tobacco. *Z. Für Nat. C* **2016**, *71*, 243–252. [CrossRef]
117. He, C.; Yu, Z.; Teixeira da Silva, J.A.; Zhang, J.; Liu, X.; Wang, X.; Zhang, X.; Zeng, S.; Wu, K.; Tan, J.; et al. DoGMP1 from *Dendrobium officinale* contributes to mannose content of water-soluble polysaccharides and plays a role in salt stress response. *Sci. Rep.* **2017**, *7*, 41010. [CrossRef]
118. Huang, C.; He, W.; Guo, J.; Chang, X.; Su, P.; Zhang, L. Increased sensitivity to salt stress in an ascorbate-deficient Arabidopsis mutant. *J. Exp. Bot.* **2005**, *56*, 3041–3049. [CrossRef]
119. Qin, C.; Qian, W.; Wang, W.; Wu, Y.; Yu, C.; Jiang, X.; Wang, D.; Wu, P. GDP-mannose pyrophosphorylase is a genetic determinant of ammonium sensitivity in *Arabidopsis thaliana*. *Proc. Natl. Acad. Sci. USA* **2008**, *105*, 18308–18313. [CrossRef]
120. Li, Q.; Li, B.-H.; Kronzucker, H.J.; Shi, W.-M. Root growth inhibition by NH4+ in Arabidopsis is mediated by the root tip and is linked to NH4+ efflux and GMPase activity: Root growth inhibition by ammonium. *Plant Cell Environ.* **2010**, *33*, 1529–1542. [CrossRef]
121. Barth, C.; Gouzd, Z.A.; Steele, H.P.; Imperio, R.M. A mutation in GDP-mannose pyrophosphorylase causes conditional hypersensitivity to ammonium, resulting in Arabidopsis root growth inhibition, altered ammonium metabolism, and hormone homeostasis. *J. Exp. Bot.* **2010**, *61*, 379–394. [CrossRef]
122. Kempinski, C.F.; Haffar, R.; Barth, C. Toward the mechanism of NH4+ sensitivity mediated by Arabidopsis GDP-mannose pyrophosphorylase: NH4+ sensitivity mediated by GMPase. *Plant Cell Environ.* **2011**, *34*, 847–858. [CrossRef] [PubMed]
123. Yin, Y.; Huang, J.; Xu, Y. The cellulose synthase superfamily in fully sequenced plants and algae. *BMC Plant Biol.* **2009**, *9*, 99. [CrossRef] [PubMed]
124. Little, A.; Schwerdt, J.G.; Shirley, N.J.; Khor, S.F.; Neumann, K.; O'Donovan, L.A.; Lahnstein, J.; Collins, H.M.; Henderson, M.; Fincher, G.B.; et al. Revised Phylogeny of the *Cellulose Synthase* Gene Superfamily: Insights into Cell Wall Evolution. *Plant Physiol.* **2018**, *177*, 1124–1141. [CrossRef] [PubMed]
125. Richmond, T.A.; Somerville, C.R. The cellulose synthase superfamily. *Plant Physiol.* **2000**, *124*, 495–498. [CrossRef]
126. Keegstra, K.; Walton, J. β-glucans–brewer's bane, dietician's delight. *Science* **2006**, *311*, 1872–1873. [CrossRef]
127. Davis, J.; Brandizzi, F.; Liepman, A.H.; Keegstra, K. Arabidopsis mannan synthase CSLA9 and glucan synthase CSLC4 have opposite orientations in the Golgi membrane: Hemicellulosic glycan synthase topology. *Plant J.* **2010**, *64*, 1028–1037. [CrossRef]
128. Liepman, A.H.; Wilkerson, C.G.; Keegstra, K. Expression of cellulose synthase-like (*Csl*) genes in insect cells reveals that *CslA* family members encode mannan synthases. *Proc. Natl. Acad. Sci. USA* **2005**, *102*, 2221–2226. [CrossRef]
129. Yin, L.; Verhertbruggen, Y.; Oikawa, A.; Manisseri, C.; Knierim, B.; Prak, L.; Jensen, J.K.; Knox, J.P.; Auer, M.; Willats, W.G.T.; et al. The Cooperative activities of CSLD2, CSLD3, and CSLD5 Are required for normal arabidopsis development. *Mol. Plant* **2011**, *4*, 1024–1037. [CrossRef]
130. Wang, X.; Cnops, G.; Vanderhaeghen, R.; Block, S.D.; Montagu, M.V.; Lijsebettens, M.V. *AtCSLD3*, a cellulose synthase-like gene important for root hair growth in Arabidopsis. *Plant Physiol.* **2001**, *126*, 575–586. [CrossRef]
131. Hazen, S.P.; Scott-Craig, J.S.; Walton, J.D. Cellulose synthase-like genes of rice. *Plant Physiol.* **2002**, *128*, 336–340. [CrossRef]

132. Gao, Y.; Chen, X.; Chen, D.; Liu, J.; Si, J. Genome-wide identification and expression analysis of *CSLA* gene family of *Dendrobium catenatum*. *China J. Chin Mater. Med.* **2020**, *45*, 3120–3127.
133. Dhugga, K.S.; Barreiro, R.; Whitten, B.; Stecca, K.; Hazebroek, J.; Randhawa, G.S.; Dolan, M.; Kinney, A.J.; Tomes, D.; Nichols, S.; et al. Guar Seed ß-Mannan Synthase Is a Member of the Cellulose Synthase Super Gene Family. *Science* **2004**, *303*, 363–366. [CrossRef] [PubMed]
134. Liepman, A.H.; Nairn, C.J.; Willats, W.G.T.; Sørensen, I.; Roberts, A.W.; Keegstra, K. Functional genomic analysis supports conservation of function among cellulose synthase-like A gene family members and suggests diverse roles of mannans in plants. *Plant Physiol.* **2007**, *143*, 1881–1893. [CrossRef] [PubMed]
135. Suzuki, S.; Li, L.; Sun, Y.-H.; Chiang, V.L. The Cellulose Synthase Gene superfamily and biochemical functions of xylem-specific cellulose synthase-like genes in *Populus trichocarpa*. *Plant Physiol.* **2006**, *142*, 1233–1245. [CrossRef]
136. Gille, S.; Cheng, K.; Skinner, M.E.; Liepman, A.H.; Wilkerson, C.G.; Pauly, M. Deep sequencing of voodoo lily (*Amorphophallus konjac*): An approach to identify relevant genes involved in the synthesis of the hemicellulose glucomannan. *Planta* **2011**, *234*, 515–526. [CrossRef]
137. Wang, Y.; Alonso, A.P.; Wilkerson, C.G.; Keegstra, K. Deep EST profiling of developing fenugreek endosperm to investigate galactomannan biosynthesis and its regulation. *Plant Mol. Biol.* **2012**, *79*, 243–258. [CrossRef]
138. Goubet, F.; Barton, C.J.; Mortimer, J.C.; Yu, X.; Zhang, Z.; Miles, G.P.; Richens, J.; Liepman, A.H.; Seffen, K.; Dupree, P. Cell wall glucomannan in Arabidopsis is synthesised by CSLA glycosyltransferases, and influences the progression of embryogenesis. *Plant J.* **2009**, *60*, 527–538. [CrossRef]
139. Goubet, F.; Misrahi, A.; Park, S.K.; Zhang, Z.; Twell, D.; Dupree, P. AtCSLA7, a cellulose synthase-like putative glycosyltransferase, is important for pollen tube growth and embryogenesis in Arabidopsis. *Plant Physiol.* **2003**, *131*, 547–557. [CrossRef]
140. Voiniciuc, C.; Dama, M.; Gawenda, N.; Stritt, F.; Pauly, M. Mechanistic insights from plant heteromannan synthesis in yeast. *Proc. Natl. Acad. Sci. USA* **2019**, *116*, 522–527. [CrossRef]
141. Robert, M.; Waldhauer, J.; Stritt, F.; Yang, B.; Pauly, M.; Voiniciuc, C. Modular biosynthesis of plant hemicellulose and its impact on yeast cells. *Biotechnol. Biofuels* **2021**, *14*, 140. [CrossRef]
142. Wang, Y.; Mortimer, J.C.; Davis, J.; Dupree, P.; Keegstra, K. Identification of an additional protein involved in mannan biosynthesis. *Plant J.* **2013**, *73*, 105–117. [CrossRef] [PubMed]
143. Kim, W.-C.; Reca, I.-B.; Kim, Y.; Park, S.; Thomashow, M.F.; Keegstra, K.; Han, K.-H. Transcription factors that directly regulate the expression of CSLA9 encoding mannan synthase in Arabidopsis thaliana. *Plant Mol. Biol.* **2014**, *84*, 577–587. [CrossRef] [PubMed]
144. He, C.; Wu, K.; Zhang, J.; Liu, X.; Zeng, S.; Yu, Z.; Zhang, X.; Teixeira da Silva, J.A.; Deng, R.; Tan, J.; et al. Cytochemical localization of polysaccharides in *Dendrobium officinale* and the involvement of DoCSLA6 in the synthesis of mannan polysaccharides. *Front. Plant Sci.* **2017**, *8*, 00173. [CrossRef] [PubMed]
145. Bernal, A.J.; Yoo, C.-M.; Mutwil, M.; Jensen, J.K.; Hou, G.; Blaukopf, C.; Sørensen, I.; Blancaflor, E.B.; Scheller, H.V.; Willats, W.G.T. Functional analysis of the cellulose synthase-like genes *CSLD1*, *CSLD2*, and *CSLD4* in tip-growing Arabidopsis cells. *Plant Physiol.* **2008**, *148*, 1238–1253. [CrossRef] [PubMed]
146. Schnall, J.A.; Quatrano, R.S. Abscisic acid elicits the water-stress response in root hairs of *Arabidopsis thaliana*. *Plant Physiol.* **1992**, *100*, 216–218. [CrossRef] [PubMed]
147. Favery, B.; Ryan, E.; Foreman, J.; Linstead, P.; Boudonck, K.; Steer, M.; Shaw, P.; Dolan, L. *KOJAK* encodes a cellulose synthase-like protein required for root hair cell morphogenesis in *Arabidopsis*. *Genes Dev.* **2001**, *15*, 79–89. [CrossRef]
148. Galway, M.E.; Eng, R.C.; Schiefelbein, J.W.; Wasteneys, G.O. Root hair-specific disruption of cellulose and xyloglucan in AtCSLD3 mutants, and factors affecting the post-rupture resumption of mutant root hair growth. *Planta* **2011**, *233*, 985–999. [CrossRef]
149. Bernal, A.J.; Jensen, J.K.; Harholt, J.; Sørensen, S.; Moller, I.; Blaukopf, C.; Johansen, B.; de Lotto, R.; Pauly, M.; Scheller, H.V.; et al. Disruption of *ATCSLD5* results in reduced growth, reduced xylan and homogalacturonan synthase activity and altered xylan occurrence in Arabidopsis. *Plant J.* **2007**, *52*, 791–802. [CrossRef]
150. Yoo, C.-M.; Quan, L.; Blancaflor, E.B. Divergence and redundancy in CSLD2 and CSLD3 function during *Arabidopsis Thaliana* root hair and female gametophyte development. *Front. Plant Sci.* **2012**, *3*, 00111. [CrossRef]
151. Li, M.; Xiong, G.; Li, R.; Cui, J.; Tang, D.; Zhang, B.; Pauly, M.; Cheng, Z.; Zhou, Y. Rice cellulose synthase-like D4 is essential for normal cell-wall biosynthesis and plant growth: Rice cellulose synthase-like D4 is essential for normal cell-wall. *Plant J.* **2009**, *60*, 1055–1069. [CrossRef]
152. Verhertbruggen, Y.; Yin, L.; Oikawa, A.; Scheller, H.V. Mannan synthase activity in the CSLD family. *Plant Signal. Behav.* **2011**, *6*, 1620–1623. [CrossRef] [PubMed]
153. Park, S.; Szumlanski, A.L.; Gu, F.; Guo, F.; Nielsen, E. A role for CSLD3 during cell-wall synthesis in apical plasma membranes of tip-growing root-hair cells. *Nat. Cell Biol.* **2011**, *13*, 973–980. [CrossRef] [PubMed]
154. Yang, J.; Bak, G.; Burgin, T.; Barnes, W.J.; Mayes, H.B.; Peña, M.J.; Urbanowicz, B.R.; Nielsen, E. Biochemical and genetic analysis identify CSLD3 as a beta-1,4-glucan synthase that functions during plant cell wall synthesis. *Plant Cell* **2020**, *32*, 1749–1767. [CrossRef] [PubMed]
155. Hu, H.; Zhang, R.; Tang, Y.; Peng, C.; Wu, L.; Feng, S.; Chen, P.; Wang, Y.; Du, X.; Peng, L. Cotton CSLD3 restores cell elongation and cell wall integrity mainly by enhancing primary cellulose production in the Arabidopsis cesa6 mutant. *Plant Mol. Biol.* **2019**, *101*, 389–401. [CrossRef] [PubMed]

156. Yin, Y.; Johns, M.A.; Cao, H.; Rupani, M. A survey of plant and algal genomes and transcriptomes reveals new insights into the evolution and function of the cellulose synthase superfamily. *BMC Genom.* **2014**, *15*, 260. [CrossRef]
157. Si, J.; Zhu, Y. Polygonati Rhizome—A new high-quality crop with great potential and not occupying farmland. *Sci. Sin. Vitae.* **2021**, *51*, 1477–1484. [CrossRef]
158. Lowe, N.M. The global challenge of hidden hunger: Perspectives from the field. *Proc. Nutr. Soc.* **2021**, *80*, 283–289. [CrossRef]
159. Akhtar, S. Malnutrition in South Asia—A Critical Reappraisal. *Crit. Rev. Food Sci. Nutr.* **2016**, *56*, 2320–2330. [CrossRef]
160. Li, X.; Yadav, R.; Siddique, K.H.M. Neglected and underutilized crop species: The key to improving dietary diversity and fighting hunger and malnutrition in Asia and the Pacific. *Front. Nutr.* **2020**, *7*, 593711. [CrossRef]
161. Jin, P.; Liang, Z.; Li, H.; Chen, C.; Xue, Y.; Du, Q. Biosynthesis of low-molecular-weight mannan using metabolically engineered Bacillus subtilis 168. *Carbohydr. Polym.* **2021**, *251*, 117115. [CrossRef]
162. Zhu, Q.; Zeng, D.; Yu, S.; Cui, C.; Li, J.; Li, H.; Chen, J.; Zhang, R.; Zhao, X.; Chen, L.; et al. From golden rice to aSTARice: Bioengineering astaxanthin biosynthesis in rice endosperm. *Mol. Plant* **2018**, *11*, 1440–1448. [CrossRef] [PubMed]

genes

Article

Phylogenetic Analysis Based on DNA Barcoding and Genetic Diversity Assessment of *Morinda officinalis* How in Vietnam Inferred by Microsatellites

Thanh Pham [1,*], Quynh Thi Nguyen [1], Duc Minh Tran [2], Hoi Nguyen [2], Hung Thai Le [2], Que Thi Hong Hoang [2], Yen Thi Van [2] and Thang Nam Tran [2,*]

1 Department of Biology, University of Education, Hue University, 34 Le Loi, Hue 530000, Vietnam
2 Faculty of Forestry, University of Agriculture and Forestry, Hue University, 102 Phung Hung, Hue 530000, Vietnam
* Correspondence: phamthanh@hueuni.edu.vn (T.P.); tnthang@hueuni.edu.vn (T.N.T.)

Abstract: *Morinda officinalis* How is well-known as a valuable medicinal plant found in some regions of Vietnam. This species is mainly used for treating male impotence, irregular menstruation, and rheumatoid arthritis. This study aimed to identify the species of and genetic diversity in three *M. officinalis* populations: one each in Quang Binh (QB), Thua Thien Hue (TTH), and Quang Nam (QN). In this study, four DNA barcoding markers (*ITS1*, *ITS2*, *matK*, and *rbcL*) were used to identify the species and 22 microsatellite markers were applied for population structure and diversity analyses. The results showed that the sequences of gene regions studied in *M. officinalis* had a high similarity (>95%) to the *ITS1*, *ITS2*, *matK*, and *rbcL* sequences of *M. officinalis* on BLAST. Of the four DNA barcoding markers used, *ITS1* and *ITS2* showed higher efficiency in DNA amplification of *M. officinalis*. From this study, 27 GenBank codes were published on BLAST. The results also revealed high levels of genetic diversity in populations. The average observed and expected heterozygosity values were $H_O = 0.513$ and $H_E = 0.612$, respectively. The average F_{ST} value was 0.206. Analysis of molecular variance (AMOVA) showed 70% variation within populations and 30% among populations. The population structure of *M. officinalis* inferred in STRUCTURE revealed that the optimum number of genetic groups for the admixture model was K = 2. These findings provided vital background information for future studies in the conservation of *M. officinalis* in both ex situ and in situ plans.

Keywords: conservation; genetic variability; *M. officinalis*; population structure

Citation: Pham, T.; Nguyen, Q.T.; Tran, D.M.; Nguyen, H.; Le, H.T.; Hoang, Q.T.H.; Van, Y.T.; Tran, T.N. Phylogenetic Analysis Based on DNA Barcoding and Genetic Diversity Assessment of *Morinda officinalis* How in Vietnam Inferred by Microsatellites. *Genes* **2022**, *13*, 1938. https://doi.org/10.3390/genes13111938

Academic Editors: Wajid Zaman and Hakim Manghwar

Received: 29 September 2022
Accepted: 21 October 2022
Published: 25 October 2022

1. Introduction

Morinda officinalis is a perennial vine mainly distributed in tropical and subtropical regions [1]. In Vietnam, *M. officinalis* is found in the wild in provinces such as Cao Bang, Lao Cai, Ha Giang, Quang Binh, Thua Thien Hue, and Quang Tri [2–4].

M. officinalis contains various bioactive components and has been used for decades as a tonic and an antirheumatic medicinal herb in some Asian countries [2,5,6]. The root of *M. officinalis* has long been used as a tonic or nutrient supplement for alleviating diseases such as depression, Alzheimer's disease, impotence, osteoporosis, and rheumatoid arthritis [6].

In Vietnam, due to the rapid increase in the demand for medicinal herbs, this plant has been over-exploited, leading to the depletion of raw materials. Additionally, because *M. officinalis* has slow growth and poor regeneration, its natural population has significantly shrunk and become endangered [2,4]. With the natural population reduced by at least 50%, this species has recently been classified as an endangered precious medicinal plant species [7]. Therefore, it is necessary to study the genetic diversity and structure of the natural population in order to conserve it effectively.

Nowadays, phylogenetic studies are necessary to conserve rare medicinal plants. In addition to representing the relationships among species in the tree of life, phylogenetic studies provide a framework for interdisciplinary investigations in taxonomy, evolutionary biology, biogeography, ecology, and conservation [8]. More recently, phylogenetic approaches based on molecular data have also proven to be an indispensable tool for genome comparisons. These approaches are used to identify genes and regulatory elements, interpret modern and ancient individual genomes, and reconstruct ancestral genomes [9], providing conservationists with background information to make conservation policies efficiently.

DNA barcoding is a universally used and reliable method of identifying plant species and has become a major focus in the fields of biodiversity and conservation. This molecular technique is not influenced by external factors or development stage, and DNA can be easily isolated from all tissues, providing an important basis for species identification at the genetic level [10]. Recently, *rbcL* and *matK* plastid coding genes were recommended as barcodes for plant species and have become the most used markers in flowering plants, as the rbcL region is highly suitable for amplification and sequencing [10,11]. Meanwhile, the nuclear *ITS* region includes both the *ITS1* and *ITS2* regions, with relatively strong discrimination power, that serve as complementary barcodes to matK and rbcL in plants. The four markers (*ITS1*, *ITS2*, *MatK*, and *rbcL*) used in this study have also proven effective in identifying medicinal species [12–14].

Currently, molecular markers help detect variations or polymorphisms that exist among individuals in the population for specific regions of DNA [15]. Among commonly used molecular markers such as AFLP (amplified fragment length polymorphism), RAPD (random amplified polymorphic), SSR (simple sequence repeat), and ISSR (inter simple sequence repeat), simple sequence repeat (SSR) markers are useful tools for research in plant genetics, breeding, and identification of individuals and species due to the allelic sequence diversity. SSRs are widely spread in the genome and have high codominant inheritance, polymorphism, and multiallelic variation [16–19].

Studies on molecular markers have been carried out on *M. officinalis* [5,20]. Liao et al. in particular developed an SSR marker dataset to serve in further research related to this plant.

The SSR marker has the advantage of using only a tiny amount of DNA. In addition, this method involves a more straightforward, faster, and more cost-effective technique than other methods.

In this study, before evaluating the genetic diversity and population structure of *M. officinalis* by using SSR markers, we used DNA barcoding to identify this species in three different provinces in central Vietnam (Quang Binh, Thua Thien Hue, and Quang Nam). The current study aimed to pave the way for protecting wild *M. officinalis* populations.

2. Materials and Methods

2.1. Plant Materials

The leaves of 37 *M. officinalis* trees were randomly collected from 3 populations in the central provinces of Quang Binh, Thua Thien Hue, and Quang Nam in Vietnam. In the field, the samples were placed in plastic bags containing silica gel. Next, the samples were transferred to the Genetic Laboratory of the Biology Department at Hue University of Education and stored at −20 °C until DNA extraction. Sampling locations were recorded using a global positioning system (GPS) (Figure 1, Table S1).

2.2. DNA Extraction

Genomic DNA was extracted from 100 mg of young leaf samples using a Plant Genomic DNA Extraction Kit (Bioteke Corporation, Beijing, China) according to the manufacturer's protocol. The quality of the extracted DNA was further estimated via 0.8% agarose gel electrophoresis. Safe dye (Phu Sa Biochem, Can Tho, Vietnam) was applied for DNA gel stain. Gel imaging was performed using a UV gel imaging system (Major Science, Saratoga, CA, USA). The isolated DNA was then stored at −20 °C until further analysis.

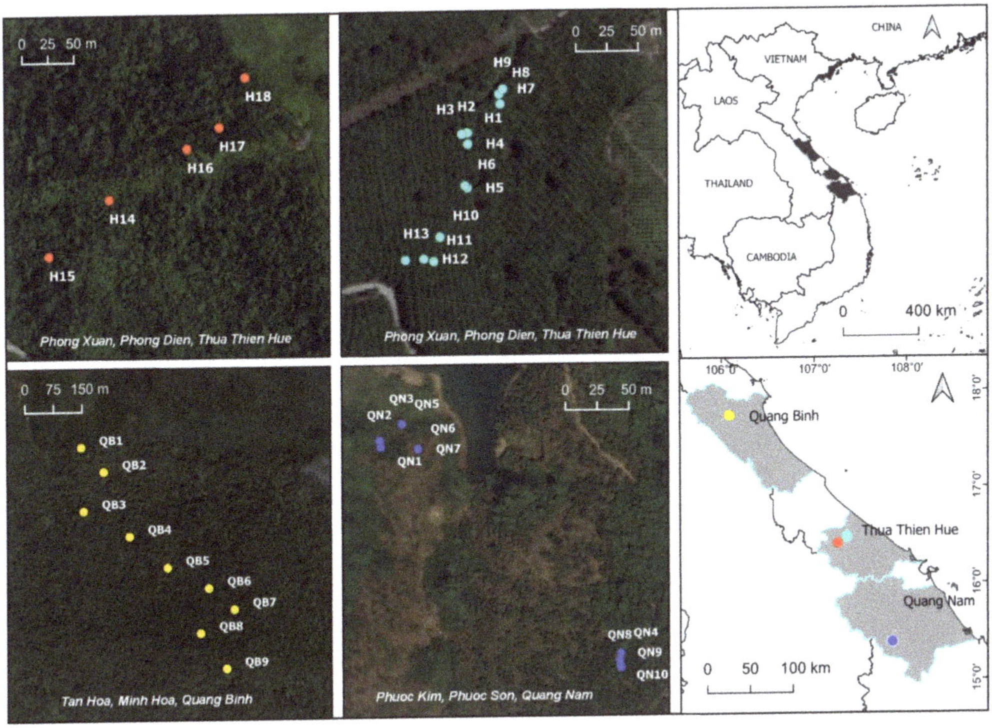

Figure 1. Map showing collection sites of three populations of *M. officinalis* in central Vietnam.

2.3. DNA Barcoding Amplification and Sequencing

A polymerase chain reaction (PCR) was performed using standard universal plant DNA barcoding primers (Table 1), with 25 μL of the reaction mixture containing 2.0 μL of template DNA (50 ng), 12.5 μL of 2X Taq Master Mix, 0.5 μL of each primer (10 pmol) (Table 2), and 9.5 μL of deionized water. PCR amplification was carried out with an Aeris™ PCR Aeris Thermal Cycler (21 Changi South Street 1, Singapore). The PCR conditions followed those laid out in the previous publication [8].

Table 1. The universal primers for DNA barcoding used in this study.

Locus	Primer Name	Sequences (5′–3′)	Reference
ITS1	*5a fwd*	CCTTATCATTTAGAGCAAGGAG	
	4 rev	TCCTCCGCTTATTGATATGC	
ITS2	*S2F*	ATGCGATACTTGGTGTGAAT	
	S3R	GACGCTTCTCCAGACTACAAT	[12]
rbcL	*1f*	ATGTCACCACAAACAGAAAC	
	724r	TCGCATGTACCTGCAGTAGC	
matK	*390F*	CGATCTATTCATTCAATATTTC	
	1326R	TCTAGCACACGAAAGTCGAAGT	

Table 2. Efficiency of PCR amplification of four potential barcodes in three populations of *M. officinalis*.

Population Codes	Number of Amplified Samples	PCR Efficiency of *ITS1* (%)	PCR Efficiency of *ITS2* (%)	PCR Efficiency of *MatK* (%)	PCR Efficiency of *rbcL* (%)
QB	3	100.0	100.0	0.0	33.3
TTH	3	100.0	100.0	100.0	66.5
QN	3	100.0	100.0	33.3	66.5
Total	9	100	100	44.44	55.56

To check the presence or absence of bands, amplified PCR products were electrophoresed using 0.8% agarose gel (1 × TAE buffer and 5 µL/mL safe dye). Gel imaging was carried out using a UV gel imaging system (Major Science, Saratoga, CA, USA). Band size of amplified products was determined using a Thermo Scientific GeneRuler 100 bp DNA Ladder (Thermo Fisher Scientific, Waltham, MA, USA).

The PCR products were sent to First Base Laboratories Sdn. Bhd (Taman Serdang Perdana, Seri Kembangan, Selangor, Malaysia) for purification and sequencing service, using the same primers as those used for the PCR.

2.4. Microsatellite Amplification

In all, 37 genomic DNA samples were used in this study. A polymerase chain reaction (PCR) was performed in 25 µL of reaction mixture that contained 2.0 µL of template DNA, 12.5 µL of 2X Taq Master Mix, 0.5 µL of each primer, and 9.5 µL of deionized water.

Liao et al. [20] have described the 22 microsatellite loci used to generate data for the current study (Table S2).

PCR amplification was carried out with an Aeris™ PCR Aeris Thermal Cycler as follows: Initial denaturation was carried out at 95 °C for 3 min. This was followed by 35 cycles of 45 s each at 94 °C for denaturation, 45 s of alignment at the annealing temperature (50–52 °C) for each primer pair, and 45 s of alignment at 72 °C for extension. Finally, 10 min of alignment at 72 °C for the final cycle completed the extension of any remaining products. The samples were kept at 4 °C until they were analyzed.

The amplification products were separated on 8% polyacrylamide gels in 1 × TAE buffer using the Mini Vertical Gel Electrophoresis Apparatus (Major Science, Saratoga, CA, USA) and visualized with safe dye (Phu Sa company). A UV gel imaging system (Major Science, Saratoga, CA, USA) was used to verify the presence of amplified fragments.

2.5. Data Analysis

2.5.1. DNA Barcoding

To obtain the sequence of each region (*ITS1, ITS2, matK*, and *rbcL*), the forward and reverse sequences were aligned using BioEdit version 7.2.5 software [21]. In searching for the similarities between those sequences and the sequences deposited in the GenBank database, the sequences of this study were analyzed using the BLAST (Basic Local Alignment Search Tool) program at http://www.ncbi.nlm.nih.gov/BLAST (accessed on 20 June 2022) [22] (Table S3).

The alignment was then exported to Molecular Evolutionary Genetics Analysis (MEGA-X) software for phylogenetic analysis [23]. The maximum likelihood trees were constructed for *ITS1, ITS2, matK*, and *rbcL* data separately using the Kimura 2-parameter model [24] with 1000 bootstrap replicates for node supports.

2.5.2. Genetic Diversity

Size of bands was detected by GelAnalyzer 19.1 software (www.gelanalyzer.com (accessed on 4 January 2022), by Istvan Lazar Jr. and Istvan Lazar Sr).

To determine the level of genetic variation within a population, the following genetic diversity parameters were calculated: mean number of alleles per locus (A), number of unique alleles, the effective number of alleles (Ne), average observed heterozygosity (Ho), average expected heterozygosity (He), and fixation index (F_{IS}). All calculations were performed in GenAIEx v.6.5 [25].

A genetic distance matrix of pairwise F_{ST} values was also used to perform a hierarchical analysis of molecular variance (AMOVA) in GenAIEx v.6.5 [25]. Significance levels were determined using 999 permutations. AMOVA was used to estimate and partition the total variances at two hierarchy levels: within populations and among populations.

Principal coordinate analysis (PCoA) based on the codominant genotypic distance for 37 studied samples of 3 populations of *M. officinalis* was carried out in GenAIEx [25].

To determine the optimal value of the genetic clusters (K), a Bayesian analysis of the population structure was performed with STRUCTURE v.2.3.4 (https://web.stanford.edu/group/pritchardlab/structure.html (accessed on 28 July 2022) [26]. Once the admixture model was set with a correlated allele frequency and ancestry models, 10 separate runs of the number of clusters (K) in the dataset were performed from 1 to 10 for each K value at 500,000 Markov Chain Monte Carlo (MCMC) repetitions and a 100,000 burn-in period. The optimal value of K was detected using Structure Harvester [27] based on the ΔK value by Evanno et al. [28].

3. Results

3.1. DNA Barcoding

For this study's DNA barcoding amplification process, we selected three samples per population, using four different primers to identify species. The results showed that *ITS1* and *ITS2* markers gave fragments of molecular weight as expected and bands of PCR products were clear for further research in all samples of the three populations, whereas some samples that used *matK* and *rbcL* markers for amplification had negative results and therefore their sequences were not determined.

A total of 27 assembled sequences were obtained and deposited in the GenBank: nine assembled sequences of *ITS1* amplicon, nine assembled sequences of *ITS2* amplicon, four assembled sequences of *matK* amplicon, and five assembled sequences of *rbcL* amplicon. The BLAST tool was used to identify the plant for each sequence, and the closest species in the GenBank was obtained. The results showed that all markers used and plants deposited in this study belonged to the genus *Morinda*. *Gynochthodes officinalis* (synonym of *M. officinalis*) was the most closely related species, with a higher extent than other species of the *Morinda* genus.

In our study, *ITS1* and *ITS2* were the most effective in terms of the amplification process (100%) (Table 2).

In the case of ITS1 region sequences (~179 bp), through use of the *ITS1* pair of primers (*ITS 5a* forward and 4 reverse) and the BLAST tool for identification, the closest species was found to be *M. officinalis*, at 98.88% similarity (Table 3). The phylogenetic tree for those sequences was constructed with the closest species (Figure 2). In the case of the *ITS2* region sequences (~235 bp), through use of the *ITS2* pair of primers (*ITS S2F* and *ITS S3R*), the closest species was found to be *Gynochthodes officinalis* (synonym of *M. officinalis*), at 100% similarity (Table 4). The phylogenetic tree for those sequences was constructed with the closest species (Figure 3).

Table 3. *M. officinalis* plants in this study and the percentage of similarity to the closest species in the GenBank as per *ITS1* gene sequence.

No.	Species Description	Scientific Name	Aligned *ITS1* Gene Sequence (bp)	Coverage (%) E	Similarity (%)	Accession Numbers
1	H01-*ITS1-M.officinalis*	*M. officinalis*	179 (86bp–264bp)	100	100	ON819584.1 *
2	H03-*ITS1-M. officinalis*	*M. officinalis*	179 (92bp–270bp)	100	100	ON819588.1 *
3	*Morinda officinalis*	*M. officinalis*	179 (93bp–271bp)	100	98.88	AB715224.1
4	*Morinda officinalis* h15	*M. officinalis*	179 (11bp–189bp)	100	98.88	FJ915618.1
5	*Morinda officinalis* h4	*M. officinalis*	179 (11bp–189bp)	100	98.32	FJ915607.1
6	*Morinda officinalis* voucher	*M. officinalis*	179 (19bp–197bp)	100	97.21	JF977048.1
7	*Gynochthodes officinalis*	*M. officinalis*	179 (73bp–251bp)	100	97.21	MH714873.1
8	QB01-*ITS1-M. officinalis*	*M. officinalis*	179 (85bp–263bp)	100	97.21	ON819596.1 *
9	QN03-*ITS1-M. officinalis*	*M. officinalis*	179 (83bp–261bp)	100	97.21	ON819592.1*
10	QN02-*ITS1-M. officinalis*	*M. officinalis*	179 (83bp–261bp)	100	97.21	ON819591.1 *
11	QB03-*ITS1-M. officinalis*	*M. officinalis*	179 (86bp–264bp)	100	96.65	ON819598.1 *
12	QB02-*ITS1-M. officinalis*	*M. officinalis*	179 (85bp–263bp)	100	96.65	ON819597.1 *
13	H02-*ITS1-M. officinalis*	*M. officinalis*	179 (83bp–261bp)	100	95.53	ON819586.1 *
14	*Gynochthodes officinalis*	*M. officinalis*	179 (73bp–251pb)	100	97.21	MH714872.1

Note: "*" is the sample in this study.

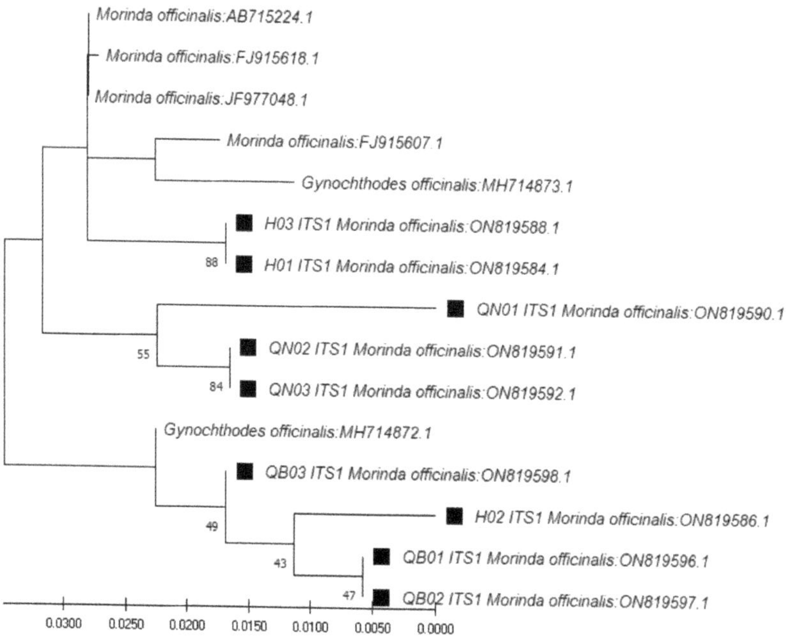

Figure 2. Phylogenetic tree by the maximum likelihood of similarity between *M. officinalis* and the closest species as per *ITS1* region sequences. The samples of our study are marked by black squares next to their names.

Table 4. *M. officinalis* plants in this study and the percentage of similarity to the closest species in the GenBank as per *ITS2* gene sequences.

No.	Species Description	Scientific Name	Aligned *ITS2* Gene Sequence (bp)	Coverage (%) E	Similarity (%)	Accession Numbers
1	H01-*ITS2-M.officinalis*	*M. officinalis*	235 (108bp–342bp)	100	100	ON819585.1 *
2	*Gynochthodes officinalis*	*M. officinalis*	235 (53bp–287bp)	100	100	MG730033.1
3	*Gynochthodes officinalis* B7	*M. officinalis*	235 (102bp–336bp)	100	100	MF375766.1
4	*Gynochthodes officinalis* B5	*M. officinalis*	235 (102bp–336bp)	100	100	MF375765.1
5	*Gynochthodes officinalis* B3	*M. officinalis*	235 (105bp–339bp)	100	100	MF375764.1
6	QB01-*ITS2-M.officinalis*	*M. officinalis*	235 (108bp–342bp)	100	99.15	ON819599.1 *
7	QN02-*ITS2-M.officinalis*	*M. officinalis*	235 (106bp–340bp)	100	99.15	ON819594.1 *
8	QB03-*ITS2-M.officinalis*	*M. officinalis*	235 (108bp–342bp)	100	99.15	ON819601.1 *
9	QB02-*ITS2-M.officinalis*	*M. officinalis*	235 (108bp–342bp)	100	99.15	ON819600.1 *
10	QN01-*ITS2-M.officinalis*	*G. officinalis*	235 (110bp–344bp)	100	100	ON819593.1 *
11	QN03-*ITS2-M.officinalis*	*M. officinalis*	235 (106bp–340bp)	100	99.15	ON819595.1 *
12	H03_*ITS2-M.officinalis*	*M. officinalis*	235 (75bp–309bp)	100	99.15	ON819589.1 *
13	H02-*ITS2-M.officinalis*	*M. officinalis*	235 (71bp–305bp)	100	99.15	ON819587.1 *

Note: "*" is the sample in this study.

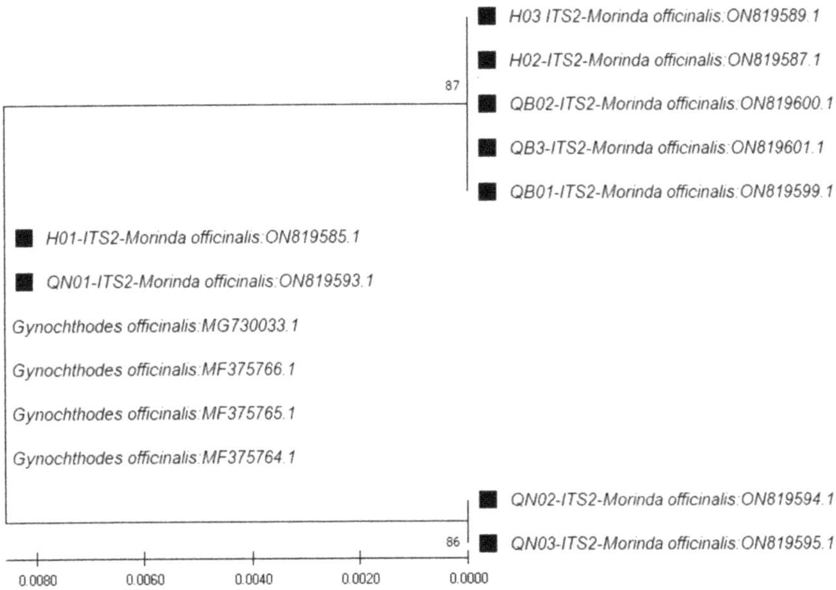

Figure 3. Phylogenetic tree by the maximum likelihood of similarity between *M. officinalis* and the closest species as per *ITS2* region sequences. The samples of our study are marked by black squares next to their names.

In the case of the *matK* region sequences (ranging from 861 bp to 899 bp), through use of the *matK* pair of primers (*390F* and *1326R*), the closest species was found to be *M. officinalis*, at 99.89% similarity (Table 5), followed by *Gynochthodes parvifolia*, at 99.78% similarity. The phylogenetic tree for these sequences was constructed with the closest species (Figure 4).

Table 5. *M. officinalis* plants in this study and the percentage of similarity with the closest species in the GenBank as per *matK* gene sequences.

No.	Species Description	Scientific Name	Aligned Sequence (bp)	Coverage (%) E	Similarity (%)	Accession Numbers
1	*Gynochthodes officinalis _H01_matK*	*M. officinalis*	899	100	100	ON926564.1 *
2	*Morinda officinalis*	*M. officinalis*	897	99.00	99.89	KR869730.1
3	*Gynochthodes officinalis_QN02_matK*	*M. officinalis*	897	99.00	99.89	ON926566.1 *
4	*Gynochthodes parvifolia*	*G. parvifolia*	897	99.00	99.78	NC_054151.1
5	*Gynochthodes cochinchinensis*	*M. officinalis*	897	99.00	99.78	NC_053818.1
6	*Gynochthodes officinalis_QN03_matK*	*M. officinalis*	897	99.00	99.67	ON926567.1 *
7	*Morinda officinalis* isolate h7_*matK*	*M. officinalis*	864	96.00	99.88	GQ130298.1 *
8	*Morinda officinalis* h2 _*matK*	*M. officinalis*	864	96.00	99.88	GQ130293.1
9	*Morinda officinalis* h1 _*matK*	*M. officinalis*	864	96.00	99.88	GQ130292.1
10	*Morinda officinalis* h6_*matK*	*M. officinalis*	864	96.00	99.77	GQ130297.1
11	*Morinda officinalis* h5	*M. officinalis*	864	96.00	99.77	GQ130296.1
12	*Gynochthodes officinalis_QN01_matK*	*M. officinalis*	861	95.00	99.77	ON926565.1 *

Note: "*" is the sample in this study.

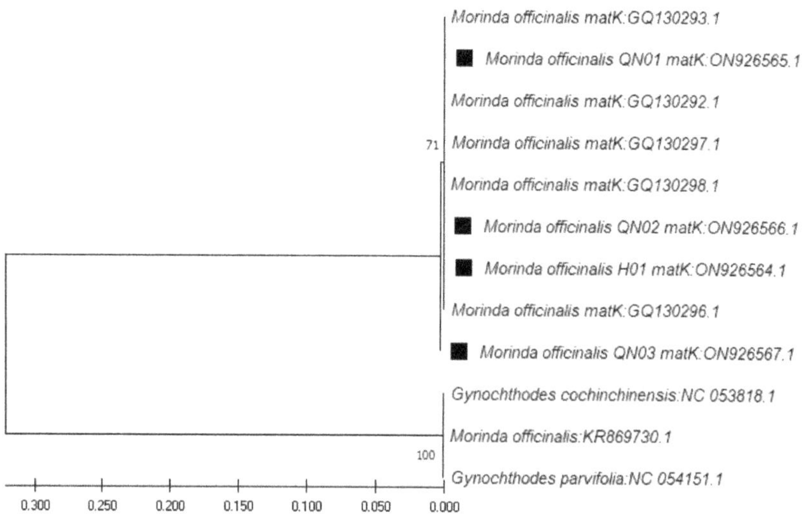

Figure 4. Phylogenetic tree by maximum likelihood of similarity between *M. officinalis* and the closest species as per *matK* region sequences. The samples of our study are marked by black squares next to their names.

In use of the *rbcL* gene, the BLAST result on NCBI showed that the nucleotide sequences obtained were highly similar to those of species of *Morinda officinalis* (accession numbers KR869730.1, NC_053818.1, and GQ436556.1), *G. parvifolia* (accession number NC_054151.1), and *Morinda* sp. SH-2010 (accession numbers AB586541.1 and AB586543.1), with similarity ranging from 99.31% to 100% (Table 6). The phylogenetic tree for those sequences was constructed with the closest species (Figure 5).

Table 6. *M. officinalis* plants in this study and percentage of similarity with the closest species in the GenBank base as per *rbcL* gene sequences.

No.	Species Description	Scientific Name	Aligned Sequence (bp)	Coverage (%) E	Similarity (%)	Accession Numbers
1	*Gynochthodes officinalis_H01_rbcL*	*M. officinalis*	738	100.00	100	ON926568.1 *
2	*Gynochthodes officinalis_QB02_rbcL*	*M. officinalis*	738	99.00	99.73	ON926572.1 *
3	*Morinda officinalis*	*M. officinalis*	737	99.00	99.46	KR869730.1
4	*Gynochthodes cochinchinensis*	*M. officinalis*	737	99.00	99.46	NC_053818.1
5	*Gynochthodes parvifolia*	*G. parvifolia*	737	99.00	99.32	NC_054151.1
6	*Gynochthodes officinalis_H02_rbcL*	*M. officinalis*	713	96.00	100.00	ON926569.1 *
7	*Morinda* sp.	*Morinda* sp.	737	97.00	99.31	AB586541.1
8	*Morinda* sp.	*Morinda* sp.	721	97.00	99.31	AB586543.1
9	*Morinda officinalis*	*M. officinalis*	703	95.00	99.86	GQ436556.1
10	*Gynochthodes officinalis_QN03_rbcL*	*M. officinalis*	670	90.00	99.85	ON926571.1 *
11	*Gynochthodes officinalis_QN01_ rbcL*	*M. officinalis*	670	90.00	99.55	ON926570.1 *

Note: "*" is the sample in this study.

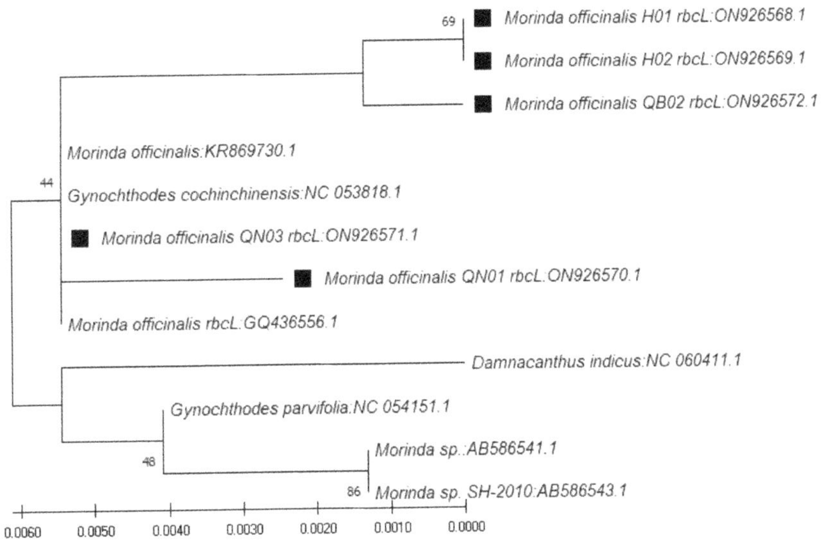

Figure 5. Phylogenetic tree by maximum likelihood of similarity between *M. officinalis* and the closest species as per *rbcL* region sequences. The samples of our study are marked by black squares next to their names.

3.2. Nucleotide Components

In terms of the occurrence of each type of nucleotide in the *ITS1* and *ITS2* gene region, cysteine (C) accounted for the highest proportion, ranging from 35.3% to 38.9%, followed by guanidine (G), which ranged from 31.1% to 34.0%. Timin (uracil) was observed with the lowest percentage (ranging from 10.0% to 14.5%) (Table 7). For matK and rbcL, however, the percentage of Timin (uracil) was higher than that of the remaining nucleotides (ranging from 28.1% to 36.9%) (Table 8).

Table 7. Nucleotide components of the *ITS1* and *ITS2* gene regions of nine *M. officinalis* samples.

Sam.	Compute Nucleotide Composition (%)											
	ITS1						*ITS2*					
	T(U)	C	A	G	C + G	Total	T(U)	C	A	G	C + G	Total
H01	12.3	35.8	21.2	30.7	66.5	179	14.5	35.7	15.7	34.0	69.7	235
H02	11.2	36.9	20.1	31.8	68.7	179	14.0	36.2	16.2	33.6	69.8	235
H03	12.3	35.8	21.2	30.7	66.5	179	14.0	36.2	16.2	33.6	69.8	235
QN01	10.0	38.9	20.0	31.1	70	180	14.5	35.7	15.7	34.0	69.3	235
QN02	11.7	35.8	20.7	31.8	66.8	179	14.9	35.3	16.2	33.6	68.9	235
QN03	12.3	36.3	20.1	31.3	67.6	179	14.9	35.3	16.2	33.6	68.9	235
QB1	11.7	36.3	20.7	31.3	67.6	179	14.0	36.2	16.2	33.6	69.8	235
QB2	12.3	36.3	20.1	31.3	67.6	179	14.0	36.2	16.2	33.6	69.8	235
QB3	12.3	36.3	20.1	31.3	67.6	179	14.0	36.2	16.2	33.6	69.8	235
Avg.	11.8	36.5	20.5	31.3	68.0	179.1	14.3	35.9	16.1	33.7	69.6	235

Table 8. Nucleotide components of the *matk* and *rbcL* gene regions of four *M. officinalis* samples.

Sam.	Compute Nucleotide Composition (%)											
	matK						*rbcL*					
	T(U)	C	A	G	C + G	Total	T(U)	C	A	G	C + G	Total
H01	36.5	17.8	28.8	16.9	34.7	899	28.6	20.3	28.3	22.8	43.1	738
H02	-	-	-	-	-	-	28.1	20.6	28.6	22.7	43.3	713
QN01	36.9	18.2	27.8	17.1	35.3	861	28.5	20.7	27.5	23.3	44.0	670
QN02	36.5	17.8	28.7	17.1	34.9	897	-	-	-	-	-	-
QN03	36.5	18.1	28.4	17.1	35.2	897	28.8	20.7	27.5	23.0	43.7	670
QB2	-	-	-	-	-	-	28.5	20.5	28.2	22.8	43.3	737
Avg.	36.6	18.0	28.4	17.0	35	888.5	28.5	20.6	28.0	22.9	43.5	705.6

The (G + C) percentage was the highest, with an average of 69.6% in the *ITS2* gene region (Tables 7 and 8).

3.3. Genetic Diversity and Population Structure

A total of 140 different alleles were detected from 22 microsatellite markers in *M. officinalis* in the three surveyed populations, with sizes ranging from 106 bp to 328 bp.

In this study, of all 22 loci, the locus MO05 showed the highest number of alleles in all populations (Na = 9), while the locus MO90 showed the lowest number of alleles, with Na = 2.333. The F_{IS} values of the loci ranged between −0.287 and 0.965, with an average of 0.216. Half of the loci (MO02, MO04, MO05, MO12, MO26, MO30, M039, MO41, MO60, MO63, and MO96) showed positive F_{IS} values (Table 9). The F_{IT} values of the loci ranged between −0.086 and 0.978.

All the loci showed positive F_{ST} values ranging between 0.127 and 0.554, with an average of 0.262. This result indicated a relatively high genetic differentiation at the locus level in all populations (Table 9).

Table 10 presents the genetic diversity of *M. officinalis* populations. The allele numbers (Na) ranged from 2 to 8 in the QB population, from 3 to 12 in the TTH population, and from 0 to 10 in the QN population. Specifically, the TTH population had the highest number of alleles, with an average of 6.455, whereas the population with the lowest number of alleles (3.955) was the QB population (Table 10). The average percentage of polymorphic loci was 98.48%, of which two populations, QB and TTH, reached the absolute value (100%).

Table 9. Genetic diversity characteristics of 22 microsatellite loci of *M. officinalis*.

Locus	Na	Ne	Ho	He	F_{IS}	F_{IT}	F_{ST}	Nm
MO02	4.667	3.494	0.893	0.698	−0.279	−0.086	0.151	1.405
MO04	4.333	2.902	0.652	0.633	−0.030	0.222	0.244	0.774
MO05	9.000	4.280	0.796	0.733	−0.086	0.109	0.179	1.144
MO12	5.667	3.820	0.815	0.727	−0.121	0.021	0.127	1.718
MO19	7.667	5.277	0.619	0.782	0.209	0.322	0.144	1.491
MO26	6.000	4.061	0.889	0.739	−0.204	−0.015	0.157	1.343
MO30	7.333	5.525	0.963	0.783	−0.230	−0.044	0.152	1.400
MO38	4.000	2.628	0.144	0.574	0.749	0.828	0.318	0.537
MO39	4.000	3.421	0.785	0.669	−0.174	0.071	0.209	0.948
MO41	3.000	2.184	0.689	0.539	−0.278	0.174	0.354	0.457
MO43	2.667	2.142	0.019	0.531	0.965	0.978	0.370	0.425
MO47	3.333	2.091	0.304	0.466	0.349	0.586	0.364	0.437
MO53	3.333	1.892	0.167	0.393	0.576	0.698	0.287	0.621
MO57	7.667	4.660	0.456	0.747	0.391	0.488	0.160	1.311
MO60	5.667	3.501	0.381	0.704	0.458	0.556	0.180	1.139
MO61	5.333	2.863	0.337	0.579	0.418	0.556	0.237	0.805
MO63	4.000	2.333	0.726	0.564	−0.287	0.053	0.265	0.695
MO88	5.000	2.917	0.433	0.652	0.336	0.461	0.189	1.074
MO89	5.333	2.540	0.144	0.425	0.660	0.818	0.463	0.290
MO90	2.333	1.764	0.085	0.396	0.785	0.892	0.498	0.252
MO94	3.333	1.834	0.093	0.405	0.771	0.898	0.554	0.202
MO96	4.667	4.076	0.907	0.736	−0.233	−0.038	0.159	1.327
Mean ± SE					0.216 ± 0.090	0.389 ± 0.077	0.262 ± 0.027	0.900 ± 0.098

Table 10. Genetic diversity estimates of three populations of *M. officinalis*.

Pop.	N	Na	Np	Ne	P%	Ho	He	F_{ST}
QB	9	3.955	1.818	2.567	100.00	0.556	0.554	0.085
TTH	18	6.455	4.364	3.786	100.00	0.394	0.683	0.451
QN	10	4.364	2.318	3.221	95.45	0.591	0.601	0.076
Mean ± SE		4.924 ± 0.296		3.191 ± 0.197	98.48 ± 1.52	0.513 ± 0.047	0.612 ±0.023	0.206 ±0.067

The highest total number of private alleles (Np = 4.364) was observed in the TTH population, followed by the QN population (Np = 2.318) and the QB population (Np = 1.818). Similarly, of the three populations, the effective allele numbers of the TTH population were the highest at 3.786, compared to 3.221 and 2.567 for the QN and QB populations, respectively.

Observed heterozygosity (Ho) and expected heterozygosity (He) averaged 0.513 and 0.612, respectively. The QN population had the highest expected heterozygosity (0.591) of the three populations, whereas the value of the observed heterozygosity of that population was 0.601. The fixation index (F_{ST}) was positive for all populations (F_{ST} = 0.206).

The results of the AMOVA pointed to higher levels of variation within populations (70%) than between populations (30%) (Table 11), indicating a relatively low genetic divergence among the three studied populations (TTH, QB, QN). However, within populations, individuals were found to have high genetic differentiation.

Table 11. Analysis of molecular variance (AMOVA) for different groups of *M. officinalis* How.

Source	df	SS	MS	Est. Var.	%
Among populations	2	158.697	79.348	3.089	30%
Within populations	71	512.222	7.214	7.214	70%
Total	73	670.919		10.303	100%
Stat	Value	P(rand >= data)			
Fst	0.300	0.001			

An admixture model was performed to determine the group population of the 37 individuals of *M. officinalis* based on Bayesian analysis using the STRUCTURE program. The population structure of *M. officinalis* inferred in STRUCTURE revealed that the optimum number of genetic groups for the admixture model was K = 2, with a delta K value of 74.084117. After a PCA analysis, the three populations were classified into two distinct clusters. K = 2, thus, can be considered to be the most suitable number of clusters for the structure of the populations in this study (Figures 6 and 7). The QB and QN populations were in one cluster (green cluster), whereas the TTH population was in the remaining cluster (red cluster) (Figure 7).

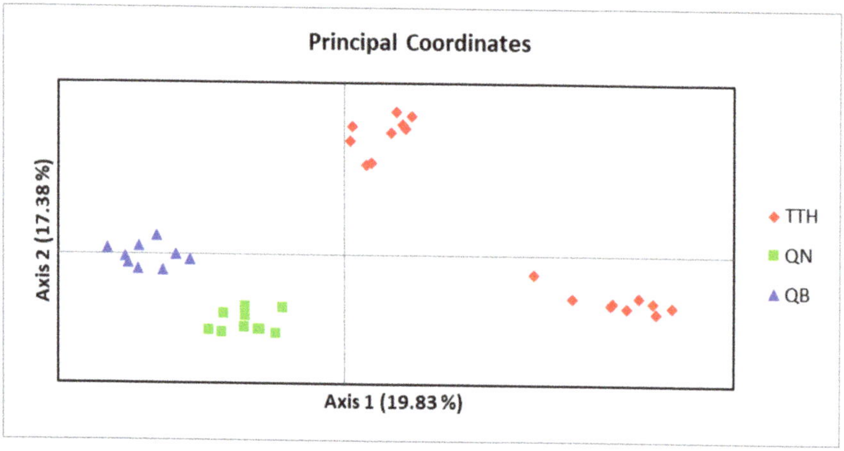

Figure 6. Principal coordinate analysis (PCoA) based on codominant genotypic distance for 37 studied samples of three populations of *M. officinalis* (TTH: TTH population; QN: QN population; QB: QB population).

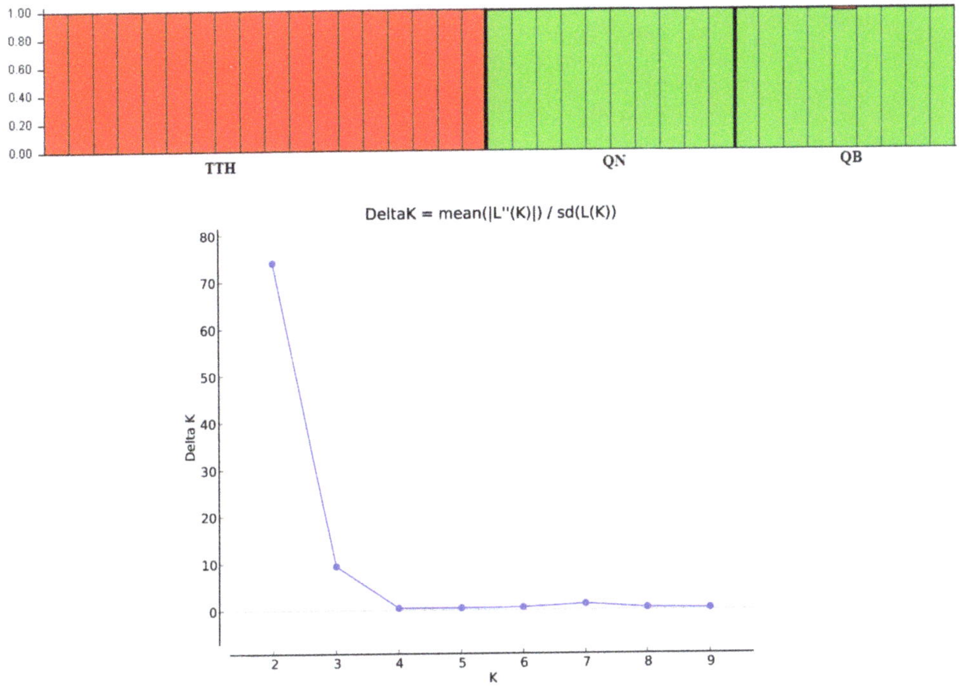

Figure 7. Delta K distribution graph and bar plot of the admixture assignment for three *M. officinalis* populations to clusters (K = 2; highest ΔK value = 74.084117) based on Bayesian analysis.

4. Discussion

DNA barcoding is a novel approach for identifying and classifying species based on the nucleotide diversity of conserved sequences. Recently, many studies have indicated that DNA markers *ITS1*, *ITS2*, *matK*, and *rbcL* are highly effective in identifying medicinal plants at species and genus levels. Therefore, this study was conducted to investigate the efficiency of those markers in barcoding *M. officinalis* plant species for the first time to identify the best marker for this valuable plant.

In previous studies, the *matK* gene and (*ITS*) region genetic markers have been proven highly efficient in distinguishing plant species and therefore identified as potential candidates for barcoding plants [29]. This study demonstrated that the *ITS1* and *ITS2* regions of *M. officinalis* were amplified more effectively compared to the other two markers (*matK* and *rbcL*).

Sequence analysis using the above four primers showed that the samples studied in this experiment belonged to the genus *Morinda* and the most closely related species was *Gynochthodes officinalis* (synonym of *M. officinalis*). In the present study, the *ITS2* marker showed the highest efficiency in terms of amplification and species identification. The higher capacity of *ITS2* in terms of phylogenetic reconstruction has been proven in previous studies [12,13,29].

The origin and ecology of a species are often expressed through its genetic diversity [30]. Many previous studies have proven a positive correlation between population genetic diversity, population size, and geographic distribution range [30,31]. Species with broad distribution and a large population size generally maintain high genetic diversity in comparison to species with narrow distribution and small population size.

Heterozygosity can reflect the genetic variation in natural populations and is considered a measure of genetic diversity. The higher the heterozygosity in a population is, the more the genetic variability in it is. *M. officinalis* had high genetic diversity (Ho = 0.513 and

He = 0.612) in this study compared to in Luo's publication for the same species (Ho = 0.3436 and He = 0.2881) [5].

F_{ST} is considered one of the measures of genetic variation in populations. An F_{ST} value of higher than 0.15 can indicate significant differentiation in a population [32]. Therefore, remarkable diversity was observed in the three populations in this study (Table 10). The F_{ST} value ranged from 0.076 (QN population) to 0.451 (TTH population), with an average F_{ST} value of 0.206. This result was higher than that in the study of *P. vietnamensis* (F_{ST} = 0.13) [19] and lower than those in the studies of *Cinnamomum balansae* (F_{ST} = 0.601) [26] and *Pulsatilla patens* (L.) (F_{ST} = 0.22) [33] and (F_{ST} = 0.13) using microsatellite markers.

The results of the AMOVA also revealed that 30% of the total variation was found among subpopulations, while the rest (70%) was within subpopulations. These results indicated a relatively low genetic differentiation among the three populations and a high genetic divergence within those populations. Genetic variation among populations was highly influenced by gene flow, selection, genetic drift, and other factors.

A Bayesian analysis in STRUCTURE showed two different groups of genetically mixed individuals of *M. officinalis*. Even though the populations were separated by large geographical distances, the majority of individuals from the QB and QN populations shared the same ancestral origin.

5. Conclusions

The study proved the *ITS1* region and the *ITS2* region as reliable markers in the barcoding of *M. officinalis*. In addition, the results confirmed that *M. officinalis* had low genetic differentiation among the three studied populations and high genetic divergence within those three populations. The three populations of *M. officinalis* were divided into genetic groups on the basis of the codominant genotypic distance. Our study of *M. officinalis* in terms of genetic diversity and population structure is intended to make a large contribution to the conservation of this species.

Supplementary Materials: The following supporting information can be downloaded at https://www.mdpi.com/article/10.3390/genes13111938/s1: Table S1. Geographical characteristics of the 37 samples of *M. officinalis* used in this study; Table S2. The primers used for PCR and sequencing in this study; Table S3. The samples were deposited in the GenBank and their accession numbers.

Author Contributions: T.P.: conceptualization, methodology, investigation, writing—original draft, revision. Q.T.N.: methodology, investigation, writing—original draft, revision. D.M.T., H.N., H.T.L., Q.T.H.H. and Y.T.V.: collecting samples; T.N.T.: conceptualization, writing—review, revision and editing, project administration. All authors have read and agreed to the published version of the manuscript.

Funding: The Ministry of Education and Training under grant number B2021-DHH-18.

Institutional Review Board Statement: Not applicable.

Informed Consent Statement: Not applicable.

Data Availability Statement: The datasets generated during and/or analyzed during the current study are available from the corresponding author on reasonable request.

Acknowledgments: This research was funded by the Ministry of Education and Training under grant number B2021-DHH-18 and the Strong Research Group Program of Hue University.

Conflicts of Interest: The authors declare no conflict of interest.

References

1. Bin, W.Y.; Guo, W.J.; Jian, Z.C.; Ting, H.; Ping, Q.L.; Zhong, W.J.; Yan, Z.Q.; Wu, Y.-B.; Zheng, C.-J.; Han, T.; et al. Quantitative and chemical profiles analysis of the root of *Morinda officinalis* based on reversed-phase high performance liquid chromatography combined with chemometrics methods. *J. Med. Plants Res.* **2013**, *7*, 2249–2258. [CrossRef]
2. Hieu, N.C.; Hong, T.L.; Oanh, N.L.; Nguyen, D.T.; Hoa, N.T. Effects of different plant densities and fertilizers on root rot disease of indian mulberry (*Morinda officinalis* How.) in Thai Nguyen. *TNU J. Sci. Technol.* **2019**, *202*, 199–204. (In Vietnamese)
3. The, H.T.; Thao, N.T.P.; Thao, N.T.; Thuy, N.T. In vitro Propagation of *Morinda officinalis* How. *J. Sci. Devel.* **2013**, *11*, 285–292. (In Vietnamese)

4. Hieu, H.D.; Ha, C.T.T.; Ngoc, P.B.; Nhan, D.L.; Huong, N.T.T.; Ha, C.H. Study on genetic diversity in *Morinda officinalis* F. C. How populations in Quang Ninh using ISSR marker. *Acad. J. Biol.* **2016**, *38*, 89–95. [CrossRef]
5. Luo, Z.; Chen, Z.; Liu, M.; Yang, L.; Zhao, Z.; Yang, D.; Ding, P. Phenotypic, chemical component and molecular assessment of genetic diversity and population structure of *Morinda officinalis* germplasm. *BMC Genom.* **2022**, *23*, 605. [CrossRef]
6. Zhang, J.H.; Xin, H.L.; Xu, Y.M.; Shen, Y.; He, Y.Q.; Lin, B.; Song, H.T.; Yang, H.Y.; Qin, L.P.; Zhang, Q.Y.; et al. *Morinda officinalis* How.—A comprehensive review of traditional uses, phytochemistry and pharmacology. *J. Ethnopharmacol.* **2018**, *213*, 230–255. [CrossRef]
7. Thịnh, N.Q.; Linh, P.T.; Trang, N.T.H.; Huy, Đ.Q. Study of conserving the medicinal plant resources in Phu Luong district, Thai Nguyen province, Vietnam. *TNU J. Sci. Technol.* **2019**, *194*, 47–52. (In Vietnamese)
8. Zaman, W.; Ye, J.F.; Ahmad, M.; Saqib, S.; Shinwari, Z.K.; Chen, Z. Phylogenetic exploration of traditional Chinese medicinal plants: A case study on Lamiaceae (angiosperms). *Pak. J. Bot.* **2022**, *54*. [CrossRef]
9. Yang, Z.; Rannala, B. Molecular phylogenetics: Principles and practice. *Nat. Rev. Genet.* **2012**, *13*, 303–314. [CrossRef]
10. Al-Juhani, W.S.; Khalik, K.N. Identification and molecular study of medicinal *Plectranthus* species (Lamiaceae) from Saudi Arabia using plastid DNA regions and ITS2 of the nrDNA gene. *J. King Saud Univ. Sci.* **2021**, *33*, 101452. [CrossRef]
11. Li, M.; Cao, H.; But, P.P.; Shaw, P.C. Identification of herbal medicinal materials using DNA barcodes. *J. Syst. Evol.* **2011**, *49*, 271–283. [CrossRef]
12. Chen, S.; Yao, H.; Han, J.; Liu, C.; Song, J.; Shi, L.; Zhu, Y.; Ma, X.; Gao, T.; Pang, X.; et al. Validation of the ITS2 region as a novel DNA barcode for identifying medicinal plant species. *PLoS ONE* **2010**, *5*, e8613. [CrossRef]
13. Wang, X.; Zheng, S.; Liu, Y.; Han, J. Chinese Herbal Medicines (CHM) Original article ITS2, a Better DNA Barcode than ITS in Identification of Species in *Artemisia* L. *Chin. Herb. Med.* **2016**, *8*, 352–358. [CrossRef]
14. Wattoo, J.I.; Saleem, M.Z.; Shahzad, M.S.; Arif, A.; Hameed, A.; Saleem, M.A. DNA Barcoding: Amplification and sequence analysis of rbcl and matK genome regions in three divergent plant species. *Int. J. Adv. Life Sci.* **2016**, *4*, 3–7.
15. Avinash, M.; Anurag, K.S.; Gaur, R.K. Molecular markers: Tool for genetic analysis. *Anim. Biotechnol.* **2014**, *1*, 289–305.
16. Kalia, R.K.; Rai, M.K.; Kalia, S.; Singh, R.; Dhawan, A.K. Microsatellite markers: An overview of the recent progress in plants. *Euphytica* **2011**, *177*, 309–334. [CrossRef]
17. Li, Y.C.; Korol, A.B.; Fahima, T.; Nevo, E. Microsatellites within genes: Structure, function, and evolution. *Mol. Biol. Evol.* **2004**, *21*, 991–1007. [CrossRef]
18. Vieira, M.L.C.; Santini, L.; Diniz, A.L.; Munhoz, C.D.F. Microsatellite markers: What they mean and why they are so useful. *Genet. Mol. Biol.* **2016**, *39*, 312–328. [CrossRef]
19. Vu, D.D.; Shah, S.N.M.; Pham, M.P.; Bui, V.T.; Nguyen, M.T.; Nguyen, T.P.T. De novo assembly and Transcriptome characterization of an endemic species of Vietnam, Panax vietnamensis Ha et Grushv., including the development of EST-SSR markers for population genetics. *BMC Plant Biol.* **2020**, *20*, 358. [CrossRef]
20. Liao, B.; Lee, S.; Meng, K.; Yin, Q.; Huang, C.; Fan, Q.; Liao, W.; Chen, S. Characterization and novel Est-SSR marker development of an important Chinese medicinal plant, *Morinda officinalis* How (Rubiaceae) Boyong. *Biotechnol. Biotechnol. Equip.* **2019**, *33*, 1311–1318. [CrossRef]
21. Hall, T.A. BIOEDIT: A user-friendly biological sequence alignment editor and analysis program for Windows 95/98/NT. *Nucleic Acids Symp. Ser.* **1999**, *41*, 95–98.
22. Altschul, S.F.; Madden, T.L.; Schäffer, A.A.; Zhang, J.; Zhang, Z.; Miller, W.; Lipman, D.J. Gapped BLAST and PSI-BLAST: A new generation of protein database search programs. *Nucleic Acids Res.* **1997**, *25*, 3389–3402. [CrossRef] [PubMed]
23. Kumar, S.; Stecher, G.; Li, M.; Knyaz, C.; Tamura, K. MEGA X: Molecular Evolutionary Genetics Analysis across Computing Platforms. *Mol. Biol. Evol.* **2018**, *35*, 1547–1549. [CrossRef] [PubMed]
24. Kimura, M. A simple method for estimating evolutionary rates of base substitutions through comparative studies of nucleotide sequences. *J. Mol. Evol.* **1980**, *16*, 111–120. [CrossRef] [PubMed]
25. Peakall, R.; Smouse, P.E. GenALEx 6.5: Genetic analysis in Excel. Population genetic software for teaching and research—An update. *Bioinformatics* **2012**, *28*, 2537–2539. [CrossRef] [PubMed]
26. Pritchard, J.K.; Stephens, M.; Donnelly, P. Inference of population structure using multilocus genotype data. *Genetics* **2000**, *155*, 945–959. [CrossRef]
27. Earl, D.A.; VonHoldt, B.M. STRUCTURE HARVESTER: A website and program for visualizing STRUCTURE output and implementing the Evanno method. *Conserv. Genet. Resour.* **2012**, *4*, 359–361. [CrossRef]
28. Evanno, G.; Regnaut, S.; Goudet, J. Detecting the number of clusters of individuals using the software STRUCTURE: A simulation study. *Mol. Ecol.* **2005**, *14*, 2611–2620. [CrossRef]
29. Yu, J.; Wu, X.; Liu, C.; Newmaster, S.; Ragupathy, S.; Kress, W.J. Progress in the use of DNA barcodes in the identification and classification of medicinal plants. *Ecotoxicol. Environ. Saf.* **2021**, *208*, 111691. [CrossRef]
30. Cui, B.; Vu, D.D.; Vu, D.C.; Bui, T.T.X.; Rahman, S.U.; Pham, M.P.; Nguyen, M.T.; Shah, S.N.M.; Tran, V.H. Genetic diversity and population structure of Cinnamomum balansae Lecomte inferred by microsatellites. *Open Life Sci.* **2022**, *17*, 323–332. [CrossRef]
31. Vandewoestijne, S.; Schtickzelle, N.; Baguette, M. Positive correlation between genetic diversity and fitness in a large, well-connected metapopulation. *BMC Biol.* **2008**, *6*, 46. [CrossRef]

32. Luo, Z.; Brock, J.; Dyer, J.M.; Kutchan, T.; Schachtman, D.; Augustin, M.; Ge, Y.; Fahlgren, N.; Abdel-Haleem, H. Genetic diversity and population structure of a Camelina sativa spring panel. *Front. Plant Sci.* **2019**, *10*, 184. [CrossRef]
33. Szczecińska, M.; Sramko, G.; Wołosz, K.; Sawicki, J. Genetic diversity and population structure of the rare and endangered plant species *Pulsatilla patens* (L.) Mill in East Central Europe. *PLoS ONE* **2016**, *11*, e0151730. [CrossRef] [PubMed]

genes

MDPI

Article

Systematic Identification and Validation of Suitable Reference Genes for the Normalization of Gene Expression in *Prunella vulgaris* under Different Organs and Spike Development Stages

Hui Zheng [1], Hongguang Zhao [2], Xuemin Zhang [3], Zongsuo Liang [4,*] and Qiuling He [1,*]

1 Key Laboratory of Plant Secondary Metabolism and Regulation of Zhejiang Province, College of Life Science and Medicine, Zhejiang Sci-Tech University, Hangzhou 310018, China
2 Tasly Botanical Pharmaceutical Co., Ltd., Shangluo 726000, China
3 Tasly R&D Institute, Tasly Holding Group Co., Ltd., Tianjin 300410, China
4 Shaoxing Academy of Biomedicine, Zhejiang Sci-Tech University, Shaoxing 312000, China
* Correspondence: liangzs@zstu.edu.cn (Z.L.); qlhe@zstu.edu.cn (Q.H.); Tel.: +86-186-6810-7267 (Q.H.)

Abstract: The quantitative real-time PCR (qRT-PCR) is an efficient and sensitive method for determining gene expression levels, but the accuracy of the results substantially depends on the stability of the reference gene (RG). Therefore, choosing an appropriate reference gene is a critical step in normalizing qRT-PCR data. *Prunella vulgaris* L. is a traditional Chinese medicine herb widely used in China. Its main medicinal part is the fruiting spike which is termed Spica Prunellae. However, thus far, few studies have been conducted on the mechanism of Spica Prunellae development. Meanwhile, no reliable RGs have been reported in *P. vulgaris*. The expression levels of 14 candidate RGs were analyzed in this study in various organs and at different stages of Spica Prunellae development. Four statistical algorithms (Delta Ct, BestKeeper, NormFinder, and geNorm) were utilized to identify the RGs' stability, and an integrated stability rating was generated via the RefFinder website online. The final ranking results revealed that *eIF-2* was the most stable RG, whereas *VAB2* was the least suitable as an RG. Furthermore, *eIF-2* + *Histon3.3* was identified as the best RG combination in different periods and the total samples. Finally, the expressions of the *PvTAT* and *Pv4CL2* genes related to the regulation of rosmarinic acid synthesis in different organs were used to verify the stable and unstable RGs. The stable RGs in *P. vulgaris* were originally identified and verified in this work. This achievement provides strong support for obtaining a reliable qPCR analysis and lays the foundation for in-depth research on the developmental mechanism of Spica Prunellae.

Keywords: *Prunella vulgaris* L.; reference gene; data normalization; RT-qPCR; Spica Prunellae; medicinal herb

Citation: Zheng, H.; Zhao, H.; Zhang, X.; Liang, Z.; He, Q. Systematic Identification and Validation of Suitable Reference Genes for the Normalization of Gene Expression in *Prunella vulgaris* under Different Organs and Spike Development Stages. *Genes* **2022**, *13*, 1947. https://doi.org/10.3390/genes13111947

Academic Editors: Wajid Zaman, Hakim Manghwar and Qinghu Ma

Received: 15 September 2022
Accepted: 24 October 2022
Published: 25 October 2022

1. Introduction

Gene expression refers to the amount of mRNA expression of a particular gene in a tissue or cell at a specific time. At present, the analysis methods of plant gene expression include Southern hybridization, Northern hybridization, in situ hybridization, traditional PCR, and qRT-PCR [1]. Among them, the qRT-PCR completes an expression analysis by detecting real-time fluorescent signal changes during the entire PCR reaction, and it has received a lot of attention in molecular biology research because of its excellent accuracy, sensitivity, specificity, and cost-effectiveness [2,3]. However, a number of variables, such as RNA integrity, purity, quality, reverse transcription efficiency, primer performance, normalization, and a host of other elements, might affect how accurate the data on expression are [4,5]. In order to control variables and ensure the accuracy of results, selecting one or more stable RGs is crucial, especially for samples with large fluctuations of expression levels [6]. Using inappropriate RGs to normalize the expression of additional genes in an RT-qPCR analysis may lead to misunderstandings about the expression level of the target gene [2,7,8]. Housekeeping genes are essential to maintain the fundamental biological processes of cells and can be expressed stably in cells, so they are often used as internal RGs [1,9]. The expression levels of ideal RGs in plants will not change significantly with the changes in external conditions, samples, and treatment methods [10]. However, research has demonstrated that the expression of RGs is not always steady under certain experimental conditions [11]. That is, given specific experimental circumstances, the stable expression of any housekeeping gene is only consistent in a restricted range of cells [11]. Thus, choosing the appropriate RGs is crucial when using a qPCR to evaluate the gene expression level of plants under particular circumstances.

The expression of the internal RG is relatively constant across tissues and cells, so it is often used as a reference when detecting the change in gene expression levels [12]. Its role is to correct for the experimental errors that exist in the amount and process of sample loading, so as to ensure the accuracy of the experimental results [13]. Theoretically, RGs are genes that are constantly expressed in all types of tissues and cells under all experimental conditions. However, studies have demonstrated that the expression level of conventionally employed RGs is variable based on different experimental conditions and materials [11]. As a result, RGs tend to vary amongst different materials [6]. Housekeeping genes are essential for the maintenance of basic cellular activities and are often involved in processes such as cytoskeleton building, vesicle transport, glycolysis, protein synthesis, and protein degradation [14]. Traditionally, glyceraldehyde-3-phosphate dehydrogenase (*GAPDH*), β-actin (*ACT*), 18S ribosomal RNA (*18S rRNA*), 28S ribosomal RNA (*28S rRNA*), ubiquitin (*UBQ*), cyclophilin (*CYP*), tubulin β (*TUB*), elongation factor 1-α (*EF1A*), protein phosphatase 2A regulatory subunit (*PP2A*), and tubulin α (*TUA*) are genes that play a housekeeping role in cells and are often used as RGs for the standardization of qPCR data [15,16]. Previous studies on screening endogenous genes have focused on model plants, major food crops, and some cash crops. Nevertheless, as the therapeutic potential of medicinal plants is becoming more widely recognized, it is essential to investigate their molecular mechanisms and identify the suitable RGs under various experimental settings. In this process, many new genes have also been determined as appropriate RGs. Currently, some medicinal plants have performed the study of internal RG selection and validation, including *Angelica sinensis* [17], *Desmodium styracifolium Merr* [18], *Rubus* [19], *Schima superba* [1], and *Isatis indigotica fortune* [20].

An expression stability analysis of each candidate gene is the focus of internal RG screening. Currently, there are five major statistical algorithms for identifying the optimal RGs which can be stably expressed in specific conditions: geNorm [21], NormFinder [22], BestKeeper [23], ΔCt method [24], and RefFinder [25]. The geNorm algorithm computes the M value of the stability of each RG to select the most stable endogenous gene, and the smaller the M value, the more stable the RG is, and vice versa. geNorm also calculates the pairwise variation V value of the RGs and determines the optimal number of endogenous genes depending on the value of V_n/V_{n+1}. The optimal number of endogenous genes is n when V_n/V_{n+1} is <0.15, which is the default value for V. There should be n + 1 endogenous genes if $V_n/V_{n+1} > 0.15$. The NormFinder algorithm will filter the most appropriate RGs according to the stability value, and the most appropriate RG will be the one with the smallest stability value. The NormFinder program can evaluate the most stable endogenous genes not only within groups but also between groups. The BestKeeper software was used to calculate each gene's correlation coefficient (r), standard deviation (SD), and coefficient of variation (CV). The internal RG is more stable when r is higher, while SD and CV are lower, and vice versa. Furthermore, the internal RG exhibits unstable expression when SD > 1. This procedure can evaluate the stability of the RGs and the degrees of expressiveness of different target genes. The ΔCt method is performed to make a two-by-two comparison by calculating the SD of each pair of candidate internal RGs and the mean SD value of each gene to finally identify the more stable internal RGs. RefFinder is a web tool for the comprehensive ranking of candidate gene stability based on the geNorm, NormFinder, BestKeeper, and Delta Ct algorithms. Each evaluation software has its merits, and the results will be more accurate when used together. Their help identified the optimal RGs for many species, such as *D. styracifolium* [18], *Punica granatum* L. [26], and *Oryza sativa* L. [27].

Flower bud differentiation is the process by which the growing point of a plant stem changes from meristematic leaves and axillary buds to inflorescences or flowers. Flower bud differentiation is an important stage that determines the development of Spica Prunellae, and it is a complex process that arises in response to the integration of signals from the external environment and internal factors [28]. The Spica Prunellae develops once a year, usually starting from the end of April and ending in the beginning of June, with a total differentiation period of nearly 40 days.

P. vulgaris is a low-growing perennial herbaceous plant widely distributed across northeastern Asia, such as in China, Japan, and Korea. Among them, Henan, Anhui, Jiangsu, Hunan, and other provinces are the main producing areas in China [29]. *P. vulgaris* has a wide range of medicinal values, and the whole plant can be used as medicine. The principal medicinal part of the plant is the dried mature spike, which has been in use for thousands of years in China. Different organs (root, stem, leaf, and spike) of *P. vulgaris* are sold at different prices in the market, among which the spike has the highest medicinal value and economic benefits. Recent reports on *P. vulgaris* have mostly concentrated on its pharmacological properties and clinical applications. Modern pharmacological and biological studies showed that *P. vulgaris.* exhibits numerous functions, including antibacterial, anticancer [30], anti-inflammatory [31], antiviral [32], immune-enhancing [33], antioxidant [34], antiproliferative [35], antihyperlipidemic [36], and free radical scavenging activities [37]. However, the molecular mechanism of its flower spike development is still unclear. With the development of molecular biology, the research to reveal the molecular mechanism of Spica Prunellae development has become increasingly intense. At the same time, different materials typically contain various stable genes that may be expressed at different expression levels at different periods of growth and development. Therefore, screening appropriate internal RGs in different development periods and organs in *P. vulgaris* is highly important. However, so far, there are no reports on this aspect of endogenous gene screening in *P. vulgaris*. In total, 14 commonly used candidate RGs: Actin (*ACT7, ACT12, ACT1*), *18S rRNA*, Translation initiation factor (*eIF-3, eIF4A-III, eIF-2*), Histone (*His3.3*), α-tubulin (*TUA6*), Cyclophilin (*CYP38*), Protein phosphatase 2Asubunit (*PP2A-2, PP2A-3, PP2A- 4*), and Homeodomain transcription factor (*VAB2*) were selected for this study based on the analysis of our previous transcriptome datasets in *P. vulgaris*. The expression levels of candidate RGs in different organs (root, stem, leaf, and spike) of *P. vulgaris* and the Spica Prunellae at different developmental stages (heading stage, early flowering stage, full-flowering stage, and ripening stage) were detected by the qPCR technique (Figure 1). Four alternative statistical algorithms—Delta CT, geNorm, NormFinder, and BestKeeper—were employed to evaluate the stability of the fourteen genes' expression levels. The online software RefFinder was also used to carry out a thorough ranking of these RGs' stability under each unique experimental scenario. The expression of *PvTAT* and *Pv4CL2*, genes regulating the synthesis of rosmarinic acid in *P. vulgaris*, were studied throughout several organs to further verify the reliability of the screened RGs. These findings will hopefully provide an important reference for further investigation into the molecular mechanism of Spica Prunellae development.

Figure 1. Intact plants and four distinct developmental stages of Spica Prunellae. P1: bolting stage; P2: early flowering stage; P3: complete flowering stage; P4: mature stage.

2. Materials and Methods

2.1. Plant Materials

P. vulgaris seeds obtained from Shangluo Tasly Pharmaceutical Co., Ltd., Shaanxi Province, were selected for this study. Plants were cultivated in the artificial climate chamber of Zhejiang Sci-Tech University (120°35′ E; 30°31′ N), Zhejiang, China. The cultivation conditions were programmed for a 16 h day (25 °C) and 8 h night (20 °C) cycle with 60% relative humidity. According to the established phenological period of *P. vulgaris*, Spica Prunellae at different growth stages (bolting stage, early flowering stage, complete flowering stage, and mature stage) and different organs (root, stem, leaf, and spike) of *P. vulgaris* in the complete flowering stage were collected [38]. The gathered samples were washed with distilled water, dried with absorbent papers quickly, liquid nitrogen-snap frozen, then kept at −80 °C for subsequent use. Three biological replicates were adopted in the stability analysis for each sample and time point, and five biological replicates were applied to the validation.

2.2. Total RNA Extraction and cDNA Synthesis

Total RNAs were extracted from samples using the RNAprep Pure Plant Plus Kit DP441 (Tiangen Biotech, Beijing, China) according to the manufacturer's protocol and then treated with DNase I (Tiangen Biotech) to eliminate contaminating DNA. The concentration and purity of RNA were measured using a NanoDrop™ 2000 spectrophotometer (Thermo, Waltham, MA, USA), and the integrity of the RNA was confirmed by 1.5% (*w*/*v*) agarose gel electrophoresis. For the purpose of the cDNA synthesis, only RNA samples with content greater than 150 ng/μL, an A260/A280 ratio ranging from 1.8 to 2.2, and an A260/A230 ratio greater than 1.8 were required. The reverse transcription kits TaKaRa RR036A (TaKaRa

Bio Inc., Dalian, China) were utilized to create the first-strand cDNA from 1 μg of RNA for 20 μL of reaction with oligo dT primers. To perform the qPCR, the acquired cDNA was subsequently diluted using RNase Free ddH_2O.

2.3. Selection and Validation of Candidate RGs and the Design of qPCR Primers

In total, 25 candidate RGs were selected after a review of the research literature on plant RGs. The transcriptome sequences of *P. vulgaris* were retrieved from GenBank (accession number: SRR7873856). To ensure the accuracy and reliability of the screened RGs, the CDS sequence of Arabidopsis was obtained from GenBank and used as the query sequence in BlASTn to locate the homologous sequence from *P. vulgaris*. Finally, 14 candidate RG sequences (*ACT7, ACT12, ACT1, 18S rRNA, eIF-3, eIF4A-III, eIF-2, His3.3, TUA6, CYP38, PP2A-2, PP2A-3, PP2A-4*, and *VAB2*) with the highest homology, the e-values all less than 1×10^{-5}, and the alignment rates ranged from 78.01% to 95.58% were acquired. The quantitative primers of the candidate RGs were designed by Primer-BLAST (https://www.ncbi.nlm.nih.gov/tools/primer-blast/, accessed on 27 November 2021) with a manual inspection. The parameters were as follows: primer length 18–25 bp, amplification product size 150–200 bp, GC content 45–65%, melting temperature 58–60 °C, no hairpin structure, homodimer, and heterodimer (Table 1). The designed primers were synthesized by Youkang Biotechnology Co., Ltd. (Hangzhou, Zhejiang, China). A melting curve analysis was used to evaluate the specificity of each primer pair (Figure S2), and the size of each amplicon was determined by electrophoresis on a 1.5% agarose gel (Figure S1). A ten-fold dilution of cDNA was used to assess the primer efficiency, and the amplification efficiency (E) and correlation coefficient (R^2) of the primers were then computed using the established standard curve (Figure S3). The slope of the curve was used to determine the E of the primers based on the theoretical formula $E = [10^{(-1/slope)} - 1] \times 100\%$ [39,40].

Table 1. Description of the fourteen candidate reference genes.

Gene Abbreviation	Gene ID	Gene Name	Arabidopsis Homolog Locus	Primer Sequences (5′–3′) (Forward/Reverse)	Tm (°C)	Length (bp)	Efficiency	R^2
ACT7	SRR7873856.1.862141	Actin	AT5G09810	GTTACGAGCTTCCCGATGGA GATCCACCACTGAGCACGAT	59.54 59.82	193	105.56%	0.9989
ACT12	SRR7873856.1.883268	Actin	AT3G46520	ACGGGTATCGTGCTTGACTC GAACAATTTCCGCTCGGCAG	59.83 60.18	182	108.36%	0.9957
ACT1	SRR7873856.1.2050798	Actin	AT2G37620	GTCGGGACTGTGTGACTGAC CCCGGAAGAGCACCTAACTC	60.23 59.82	193	107.67%	0.9983
18S rRNA	SRR7873856.1.1835427	18S ribosomal RNA	AT3G41768	GACGGAGGTAGGGTTCGATT CACCAGACTTGCCTCCAATG	58.89 58.83	197	108.49%	0.9933
eIF-3	SRR7873856.1.968612	Translation initiation factor	AT4G20980	GGCTCTTGAGTCGCTCCAAT GCGAATCGTCGGTGTTCAAG	60.11 59.91	195	107.04%	0.9993
eIF4A-III	SRR7873856.1.1836036	Translation initiation factor	AT3G19760	CCACCTTTTGCCTCCAACAC GGTACCGGGAAAACCTCCAT	59.61 59.38	191	106.06%	0.9971
eIF-2	SRR7873856.1.18172	Translation initiation factor	AT1G76720	TTTTGGGAGAGCGGACACAA AGCTGCCTTGGAGACTGAAA	59.82 59.23	196	102.96%	0.9927
His3.3	SRR7873856.1.2135040	Histone	AT4G40030	CACAAGGTAGGCCTCTGCTG AAGAAGCCCACAGATACCGC	60.39 60.11	193	105.94%	0.9945
TUA6	SRR7873856.1.1877380	α-tubulin	AT4G14960	TCCACCCACTCCCTTCTTGA TTCATCCACGTTCAGGCTCC	60.10 60.04	193	105.31%	0.9960
CYP38	SRR7873856.1.1584051	Cyclophilin	AT3G01480	CGCTCGAGAGGGTCGATAAC GCCTGCTACCACTTGACTGA	60.04 59.68	188	102.00%	0.9921
PP2A-2	SRR7873856.1.2107443	Protein phosphatase 2Asubunit	AT1G10430	TTTAGATCAGAGGTGCGCGG AAATTGCTCTCGCGCCTGAT	60.18 60.75	185	106.41%	0.9945
PP2A-3	SRR7873856.1.2002889	Protein phosphatase 2A subunit	AT2G42500	CCAGCACCTCGAGGGAGATA TTCCGACTGCACTGGTTGAA	60.47 59.82	182	103.48%	0.9971
PP2A-4	SRR7873856.1.256172	Protein phosphatase 2A subunit	AT3G58500	GTGGCTTTGAAAGTGCGCTA TGATTCAACCAAGGCGGTCA	59.41 59.89	184	100.20%	0.9936
VAB2	SRR7873856.1.1180932	Homeodomain transcription factor	AT3G05020	CTCGGAATTGTCGTCAGGCT ATCGTTGGCCGTTCAGGAAA	60.11 60.25	197	98.37%	0.9943

2.4. Quantitative Real-Time PCR (qRT-PCR) Analyses

The qPCR reactions were carried out using TB Green Premix Ex Taq II (TaKaRa Bio Inc., Dalian, China) in 96-well plates with an ABI 7500 Real-Time PCR System (Applied Biosystems, Waltham, MA, USA). Each PCR reaction mixture (final volume of 20 μL) consisted of TB Green Premix Ex Taq II (5 μL), each forward, and reverse primers with a concentration of 10 μM (0.4 μL), RNase-free dH_2O (2.6 μL), ROX II (0.1 μL), and 1.5 μL of the cDNA template (diluted ten times with RNase-free ddH_2O). All of the RGs in every run were kept in NTC (no template control). All experiments were maintained in three technical replicates and three to five biological replicates. The PCR amplification cycles' conditions: 95 °C for 10 min, followed by 40 cycles of 95 °C for 5 s, and 58 °C for 30 s. A melt curve analysis was performed at 95 °C for 15 s, 60 °C for 1 min, and 95 °C for 15 s at the end of the PCR run. Ct values were recorded and taken for further analyses.

2.5. Analysis of Expression Stability of Candidate RGs

Utilizing Microsoft Excel 2019, a descriptive statistical analysis was carried out to assess the expression stability of potential RGs. Four different software programs (Delta Ct, NormFinder, geNorm, and BestKeeper) based on the experimental design and manufacturers' instructions were used to ascertain the expression stability of 14 candidate RGs in various organs and experimental periods.

The raw Ct values obtained by qRT-PCR were transformed into relative expression levels for each RG using the formula $2^{-\Delta CT}$ [41] (ΔCT = each corresponding Ct value—the lowest Ct value for that gene in different samples), which was subsequently employed for an additional analysis in geNorm and NormFinder. These values were imported into geNorm to calculate the variable M that gauges the consistency of gene expression. The M values were inversely correlated with the stability of the genes, meaning that the expression of this RG in the identified genes was the most stable when the M value was the lowest [21]. geNorm can also determine the paired variation V value of the normalization factor using the geometric mean from expression levels of the most stable RGs, and the value of V_n/V_{n+1} can be used to calculate the optimal number of internal RGs (with a cut-off value of $V_{n+1} < 0.15$) [42].

NormFinder is a mathematical model describing the expressed stable values of RGs by computing the variance, both within-group variance and between-group [22]. Moreover, the stability value (SV) determines the ideal gene or set of genes for standardization.

The analysis of the Delta Ct and BestKeeper software employed raw Ct data. The Delta Ct has the ability to determine the standard deviations (SD) value of the Ct value of each selected RG. Lower RG stability is associated with greater SD values [41]. The BestKeeper software algorithm utilizes the coefficient of variance (CV), and the SD of the Ct value. The value of the correlation coefficient between the candidate gene determines the stability of the RGs [23]. Genes with higher stability typically have lower SD (lower than 1) and CV values.

In addition to these four algorithms, RefFinder (an online tool) was also used to generate an overall comprehensive ranking of RGs. The geometric mean was calculated for the final overall ranking of the genes by the RefFinder algorithm, which combined ranks from the Delta Ct, geNorm, NormFinder, and BestKeeper.

2.6. RGs Validation

To further confirm the dependability of the selected RGs, the expression patterns of *PvTAT* and *Pv4CL2*, genes involved in the controlling of the synthesis of rosmarinic acid in *P. vulgaris*, were analyzed throughout different organs. The relative expression levels of the target genes *PvTAT* and *Pv4CL2* were computed according to the formula $2^{-\Delta\Delta Ct}$ [43].

2.7. Data Analysis

Delta Ct, geNorm, NormFinder, and BestKeeper were used for the data analysis. Figures were created utilizing GraphPad Prism 6 (GraphPad Software, Inc., La Jolla, CA, USA). Microsoft Excel 2019 and SPSS 21.0 software were used for the statistical analysis.

3. Results

3.1. Selection of RGs and Analysis of Primer Amplification Specificity and Efficiency

Based on the previous studies, 14 genes were identified as potential RGs by mining the transcriptome data of *P. vulgaris* (GenBank accession: SRR7873856). The details of gene symbols, gene IDs, gene names, Arabidopsis homologous gene numbers, primer sequences, annealing temperature (°C), amplification length (bp), PCR efficiency (E), and correlation coefficient (R^2) for the 14 candidate genes were shown in Table 1. Specific qPCR primers were created on the basis of the sequences of transcriptome data, and the specificity of the primers was evaluated by both the gel electrophoresis and melting curves. The results showed that all the internal RG primers were single bands in an agarose gel electrophoresis analysis, and the amplification products were consistent with the expected fragment size and without any dimers or non-specific amplification appearing (Figure S1). The identical outcomes for all of the primer sets with single amplification peaks were confirmed by a qRT-PCR melting curve analysis (Figure S2). The amplification efficiency of 14 pairs of primers varied between 98.37% and 108.49%, and all the correlation coefficients (R^2) values were above 0.99 (Figure S3). Briefly, the outcomes suggested that the primers of these 14 genes were thoughtfully created with reasonable specificity and showed excellent amplification ability, suggesting that these primers were suitable for subsequent experiments.

3.2. Expression Profiles of Candidate RGs

To evaluate the expression stability of 14 candidate internal RGs under different conditions, the mean cycle threshold values (Cps) were used to determine their transcript abundances, ranging from 5.81 to 36.74. The Cp value was inversely proportional to the transcription level of the gene. In other words, the expression abundance increased as the Cp value decreased. According to Figure 2, the mean Cp values of 14 RGs ranged from 7 to 35, with the bulk falling between 21 and 24. Across all samples, the lowest and highest Ct values appeared in *18S rRNA* and *ACT1* with 5.81–10.77 and 32.58–36.74, indicating that they were the most highly and lowly expressed genes, respectively. Among the 14 genes, *His3.3* (21.62–22.90) was expressed in high abundance with the small range of variation, while the expression levels of *VAB2* (19.53–25.90) were the most variable (Supplementary Table S1). Overall, the expression profiles of the endogenous genes were displayed through Ct values combined with the box plot, allowing us to have a glimpse of the gene stability. Our initial findings demonstrated that the expression levels of *His3.3*, *eIF-2*, *eIF4A-3*, *PP2A-2*, *ACT7*, and *eIF-3* were relatively stable and showed high expression levels, while the expression of *18S rRNA* and *VAB2* varied greatly.

3.3. Expression Stability Estimation of Candidate RGs by Five Bioinformatic Programs

For further proof that the RGs were stable, Spica Prunellae at different developmental stages (bolting stage, early flowering stage, full flowering stage, maturity stage) and various organs (leaf, stem, root, spike) at the full flowering stage were collected as samples.

All the samples were tested to assess the stability of the potential RGs by four algorithms: BestKeeper, geNorm, NormFinder, and Delta Ct. Finally, the RefFinder tool (https://blooge.cn/RefFinder/, accessed on 3 March 2022) was utilized to comprehensively rank the expression stability of all potential RGs to screen out the most suitable ones for subsequent research.

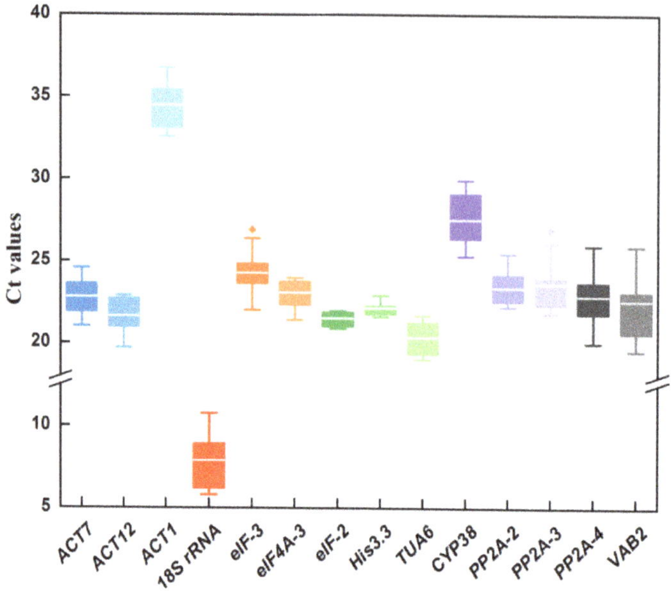

Figure 2. Expression levels of 14 candidate reference genes in all *P. vulgaris* samples. The box chart indicates the interquartile range. The outer box shows the 25th to 75th percentile, and the inner box indicates the average. Lines across the boxes depict the medians and whiskers represent the maximum and minimum values. The dots represent the outlier.

3.3.1. geNorm Analysis

The average expression stability measurement (M) value and paired variation (V) value calculated by the geNorm software algorithm were used to rank the gene expression stability and determine the optimal number of RGs for accurate normalization. The M value was negatively correlated with gene stability. The gene expression stability of the RGs increased with decreasing M value. Meanwhile, if the M value of a gene was greater than 1.5, it indicated that it was not suitable as an RG [21,44]. The results revealed that all the candidate internal RGs had different stability levels under different conditions (Figure 3). The M value of *eIF-2* was the lowest, whereas the M value of *ACT1* was the highest, indicating that *eIF-2* was the gene with the highest level of stability across all samples, while the *ACT1* was the least stable one. The optimal internal RG in the different periods was also *eIF-2*, while the most erratic gene was *eIF-3*. Notably, the steadiest RG was *TUA6* in an organ subset, while *ACT1* seemed to be less stable than other candidate RGs.

In the qRT-PCR, an RG may lead to biased results, while combining internal RGs can yield more precise and trustworthy gene expression data [1]. geNorm uses the relationship between the V and the cutoff value of 0.153336 to determine whether additional RGs need to be added to the RG set [17]. Ideally, the V_n/V_{n+1} value should be less than 0.153336, i.e., the introduction of a new gene will not significantly affect the normalization [42]. Otherwise, n + 1 RGs need to be introduced. As shown in Figure 4, V_2/V_3 values were below 0.153336 for different organs, and different periods as well as all samples combined, demonstrating that adding a third internal RG had no discernible impact on the calibration of the qPCR data. Therefore, the optimal RG set was *eIF-2* + *His3.3* for both total samples and different periods, while the most suitable set of internal RGs was *TUA6* + *PP2A-3* in an organ subset.

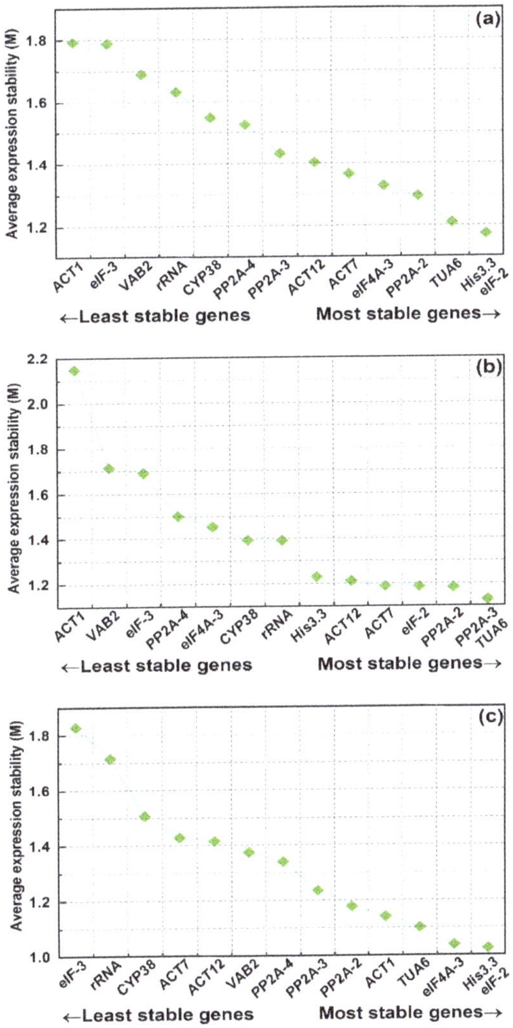

Figure 3. Gene expression stability and ranking of fourteen reference genes based on geNorm. The expression stability value (M) for each gene was obtained and graphed. The lower the M value, the more stable the expression. The most stable genes are on the right and the least stable genes are on the left. (**a**) Total samples: all root, stem, leaf, and spike samples of *P. vulgaris*; (**b**) different organs: root, stem, leaf, and spike samples during the full flowering stage; (**c**) different periods: Spica Prunellae from four stages: bolting stage, early flowering stage, full flowering stage, maturity stage.

3.3.2. NormFinder Analysis

The average pairwise variance of a gene in comparison to other RGs was calculated by the NormFinder to directly assess the stability of RGs [22]. Similar to geNorm, the NormFinder analysis also uses the $2^{-\Delta Ct}$ (ΔCT = each corresponding Ct value—the lowest Ct value for that gene) method [9]. In general, the lower the stability value (SV), the higher the expression stability of the corresponding gene. It can be seen from Figure 5 that the highest-ranked gene was not completely consistent under different experimental conditions. *eIF-2* and *eIF4A-3* were the most stable genes in all samples and different periods, respectively. Among the different organs, *TUA6* had the lowest stability value of 0.349. The results indicated that *TUA6* had the best stability among the 14 internal RGs in organs' subsets, while *ACT1* (1.367) showed the highest variability. This agrees with the conclusions of the geNorm analysis. Interestingly, the worst-performing gene across all samples and different periods was *eIF-3*.

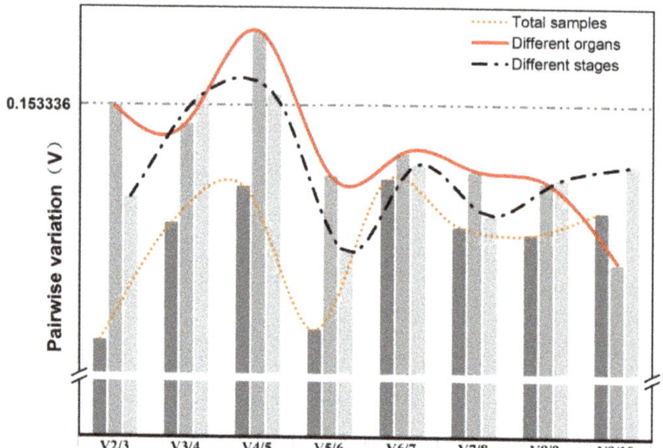

	V2/3	V3/4	V4/5	V5/6	V6/7	V7/8	V8/9	V9/10
Total samples	0.089	0.121	0.131	0.092	0.133	0.12	0.118	0.124
Different organs	0.153	0.148	0.173	0.134	0.14	0.135	0.132	0.11
Diefferent stages	0.128	0.156	0.156	0.113	0.137	0.123	0.133	0.137

Figure 4. Pairwise variation (V) of the 14 candidate reference genes calculated by geNorm to determine the optimal numbers of reference genes for accurate normalization. Different conditions are included and marked in square frames with different colors and the threshold used was 0.153336.

3.3.3. BestKeeper Analysis

The BestKeeper software assesses the stability of internal RGs by directly calculating their coefficient of variation (CV), standard deviation (SD), and correlation coefficient (r) on the basis of each gene's raw Ct value. It is worth noting that the most stable RG must have the highest r value, the lowest SD, and CV values [23]. It indicates that the expression of the RG is unstable and cannot be used for normalization when SD > 1 [23]. As shown in Table 2, the number of genes satisfying SD < 1 varied under different experimental conditions. Only *His3.3*, *eIF-2*, and *eIF4A-3* showed SD < 1 in organs' subsets, which complied with the demands of internal RGs. Unfortunately, this was inconsistent with the findings of the geNorm and NormFinder analyses. *His3.3* (1.50 ± 0.33) and *eIF-2* (1.60 ± 0.34) were the highest-ranked among all samples and different periods, respectively. Interestingly, the BestKeeper analysis showed that the expression of *18S rRNA* was the least stable under different conditions and could not be used for normalization.

3.3.4. Delta Ct Analysis

A Delta Ct analysis was used to calculate the mean standard deviation (SD) of each gene to determine the stability of the internal RGs. Higher genetic stability is associated with smaller SD [41]. With average SD values of just 0.374, *eIF-2* was the gene with the highest level of stability in organs' subsets. The most stable RG was *His3.3* in both the total samples and the different periods, while the least stable gene was *VAB2*.

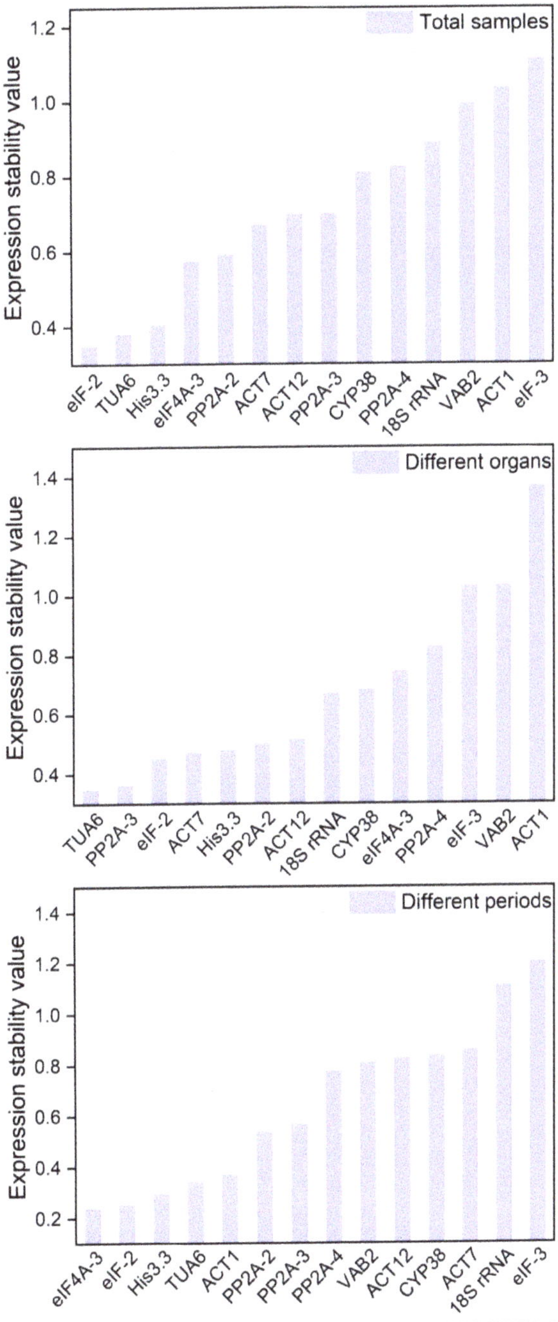

Figure 5. Expression stability of 14 reference genes in *P. vulgaris* analyzed by NormFinder.

Table 2. Stability analysis of 14 candidate reference genes based on BestKeeper software.

Rank	Total Samples				Different Organs				Different Periods			
	Gene	CV $\pm$ SD	r	*p*-Value	Gene	CV $\pm$ SD	r	*p*-Value	Gene	CV $\pm$ SD	r	*p*-Value
1	*His3.3*	1.50 $\pm$ 0.33	0.644	0.005	*His3.3*	1.42 $\pm$ 0.32	0.672	0.033	*eIF-2*	1.60 $\pm$ 0.34	0.910	0.001
2	*eIF-2*	1.76 $\pm$ 0.38	0.926	0.001	*eIF-2*	1.79 $\pm$ 0.39	0.937	0.001	*His3.3*	1.93 $\pm$ 0.43	0.799	0.010
3	*eIF4A-3*	2.85 $\pm$ 0.66	0.663	0.004	*eIF4A-3*	3.05 $\pm$ 0.71	0.573	0.083	*PP2A-2*	1.95 $\pm$ 0.45	0.208	0.593
4	*PP2A-2*	3.30 $\pm$ 0.77	0.489	0.046	*CYP38*	4.34 $\pm$ 1.18	0.861	0.001	*eIF4A-3*	2.23 $\pm$ 0.51	0.836	0.005
5	*ACT1*	3.32 $\pm$ 1.15	0.321	0.210	*eIF-3*	4.41 $\pm$ 1.08	0.306	0.389	*ACT7*	2.56 $\pm$ 0.59	0.513	0.810
6	*eIF-3*	3.77 $\pm$ 0.91	0.028	0.914	*ACT1*	4.46 $\pm$ 1.54	0.208	0.565	*ACT1*	2.88 $\pm$ 1.01	0.950	0.001
7	*ACT7*	3.97 $\pm$ 0.91	0.527	0.030	*PP2A-2*	4.52 $\pm$ 1.07	0.810	0.004	*ACT12*	3.24 $\pm$ 0.71	0.108	0.780
8	*CYP38*	4.66 $\pm$ 1.28	0.699	0.002	*PP2A-3*	4.62 $\pm$ 1.06	0.934	0.001	*eIF-3*	3.48 $\pm$ 0.83	0.534	0.055
9	*ACT12*	4.71 $\pm$ 1.02	0.454	0.068	*ACT7*	4.71 $\pm$ 1.07	0.864	0.001	*PP2A-3*	4.52 $\pm$ 1.12	0.924	0.001
10	*PP2A-3*	4.93 $\pm$ 1.17	0.907	0.001	*TUA6*	4.99 $\pm$ 1.02	0.808	0.005	*TUA6*	4.59 $\pm$ 0.94	0.889	0.001
11	*TUA6*	4.94 $\pm$ 1.01	0.837	0.001	*ACT12*	5.79 $\pm$ 1.25	0.745	0.013	*CYP38*	5.08 $\pm$ 1.44	0.709	0.032
12	*PP2A-4*	6.51 $\pm$ 1.49	0.998	0.001	*PP2A-4*	7.43 $\pm$ 1.67	0.992	0.001	*PP2A-4*	6.81 $\pm$ 1.62	0.995	0.001
13	*VAB2*	7.43 $\pm$ 1.68	0.995	0.001	*VAB2*	8.99 $\pm$ 1.99	0.995	0.001	*VAB2*	7.01 $\pm$ 1.67	0.999	0.001
14	*18S rRNA*	19.52 $\pm$ 1.54	0.895	0.001	*18S rRNA*	15.67 $\pm$ 1.27	0.944	0.001	*18S rRNA*	24.88 $\pm$ 2.04	0.931	0.001

3.3.5. RefFinder Analysis

An online analysis tool called RefFinder determined the geometric mean of each RG depending on the results of four statistical methods, Delta Ct, geNorm, NormFinder, and BestKeeper, to comprehensively rank the 14 candidate internal RGs. The smaller the geometric mean, the more consistent the expression of genes. The research found that *eIF-2* had the lowest indices in total data and different periods and was the most stable RG (Table 3), while *VAB2* and *18S rRNA* were the least stable genes under these two conditions, respectively. In contrast, *TUA6* performed the best, and *ACT1* the worst, in different organs. The differences in results between different software are understandable and acceptable due to the different assessment methods. Overall, in descending order, the stability of these 14 candidate RGs was: *eIF-2*, *His3.3*, *PP2A-2*, *TUA6*, *eIF4A-3*, *ACT7*, *ACT12*, *PP2A-3*, *PP2A-4*, *CYP38*, *eIF-3*, *18S rRNA*, *ACT1*, and *VAB2*.

Table 3. Comprehensive evaluation of expression stability of fourteen candidate reference genes in *P. vulgaris* using the RefFinder algorithm.

Group	Rank	Delta Ct	BestKeeper	NormFinder	geNorm	RefFinder	
		Gene	Gene	Gene	Gene	Gene	SV
Total samples	1	*His3.3*	*His3.3*	*eIF-2*	*eIF-2*	*eIF-2*	1.190
	2	*eIF-2*	*eIF-2*	*TUA6*	*His3.3*	*His3.3*	1.570
	3	*eIF4A-3*	*eIF4A-3*	*His3.3*	*TUA6*	*PP2A-2*	3.940
	4	*PP2A-2*	*PP2A-2*	*eIF4A-3*	*PP2A-2*	*TUA6*	3.980
	5	*ACT7*	*ACT1*	*PP2A-2*	*eIF4A-3*	*eIF4A-3*	4.530
	6	*TUA6*	*eIF-3*	*ACT7*	*ACT7*	*ACT7*	5.180
	7	*ACT12*	*ACT7*	*ACT12*	*ACT12*	*ACT12*	6.650
	8	*eIF-3*	*CYP38*	*PP2A-3*	*PP2A-3*	*PP2A-3*	8.460
	9	*ACT1*	*ACT12*	*CYP38*	*PP2A-4*	*PP2A-4*	9.930
	10	*PP2A-3*	*PP2A-3*	*PP2A-4*	*CYP38*	*CYP38*	9.970
	11	*CYP38*	*TUA6*	*18S rRNA*	*18S rRNA*	*eIF-3*	10.920
	12	*18S rRNA*	*PP2A-4*	*VAB2*	*VAB2*	*18S rRNA*	11.470
	13	*PP2A-4*	*VAB2*	*ACT1*	*eIF-3*	*ACT1*	12.310
	14	*VAB2*	*18S rRNA*	*eIF-3*	*ACT1*	*VAB2*	12.470

Table 3. *Cont.*

Group	Rank	Delta Ct	BestKeeper	NormFinder	geNorm	RefFinder	
		Gene	Gene	Gene	Gene	Gene	SV
Different organs	1	*eIF-2*	*His3.3*	*TUA6*	*TUA6*	*TUA6*	1.410
	2	*His3.3*	*eIF-2*	*PP2A-3*	*PP2A-3*	*PP2A-2*	3.220
	3	*PP2A-2*	*eIF4A-3*	*eIF-2*	*PP2A-2*	*eIF-2*	3.310
	4	*eIF4A-3*	*CYP38*	*ACT7*	*eIF-2*	*PP2A-3*	3.440
	5	*ACT7*	*eIF-3*	*His3.3*	*ACT7*	*His3.3*	3.810
	6	*ACT12*	*ACT1*	*PP2A-2*	*ACT12*	*ACT7*	4.860
	7	*eIF-3*	*PP2A-2*	*ACT12*	*His3.3*	*ACT12*	5.960
	8	*TUA6*	*PP2A-3*	*18S rRNA*	*18S rRNA*	*eIF4A-3*	7.000
	9	*ACT1*	*ACT7*	*CYP38*	*CYP38*	*18S rRNA*	8.920
	10	*PP2A-3*	*TUA6*	*eIF4A-3*	*eIF4A-3*	*CYP38*	9.240
	11	*CYP38*	*ACT12*	*PP2A-4*	*PP2A-4*	*eIF-3*	10.610
	12	*PP2A-4*	*PP2A-4*	*eIF-3*	*eIF-3*	*PP2A-4*	11.720
	13	*VAB2*	*VAB2*	*VAB2*	*VAB2*	*VAB2*	13.240
	14	*18S rRNA*	*18S rRNA*	*ACT1*	*ACT1*	*ACT1*	13.470
Different periods	1	*His3.3*	*eIF-2*	*eIF4A-3*	*eIF-2*	*eIF-2*	1.190
	2	*eIF-2*	*His3.3*	*eIF-2*	*His3.3*	*His3.3*	1.860
	3	*eIF4A-3*	*PP2A-2*	*His3.3*	*eIF4A-3*	*eIF4A-3*	2.450
	4	*TUA6*	*eIF4A-3*	*TUA6*	*TUA6*	*PP2A-2*	4.560
	5	*PP2A-2*	*ACT7*	*ACT1*	*ACT1*	*TUA6*	5.470
	6	*ACT7*	*ACT1*	*PP2A-2*	*PP2A-2*	*ACT1*	6.510
	7	*eIF-3*	*ACT12*	*PP2A-3*	*PP2A-3*	*ACT7*	7.580
	8	*ACT12*	*eIF-3*	*PP2A-4*	*PP2A-4*	*ACT12*	7.750
	9	*PP2A-3*	*PP2A-3*	*VAB2*	*VAB2*	*PP2A-3*	8.150
	10	*CYP38*	*TUA6*	*ACT12*	*ACT12*	*PP2A-4*	9.360
	11	*18S rRNA*	*CYP38*	*CYP38*	*ACT7*	*VAB2*	10.370
	12	*ACT1*	*PP2A-4*	*ACT7*	*CYP38*	*CYP38*	11.490
	13	*PP2A-4*	*VAB2*	*18S rRNA*	*18S rRNA*	*eIF-3*	11.770
	14	*VAB2*	*18S rRNA*	*eIF-3*	*eIF-3*	*18S rRNA*	13.240

3.4. RGs' Validation

The previous study reported that the specific pattern of expression of *PvTAT* and *Pv4CL2*, genes involved in the control of rosmarinic acid production in *P. vulgaris*, was significantly distinct in different organs [45,46]. To confirm the accuracy of the identified internal RGs, the expression profiles of *PvTAT* and *Pv4CL2* in different organs of *P. vulgaris* were standardized by the two most reliable candidate RGs (*eIF-2* and *His3.3* alone or in combination) and the most unstable internal RGs (*VAB2*) from the above analysis. As shown in Figure 6, the expression levels of *PvTAT* and *Pv4CL2* normalized with *eIF-2*, *His3.3*, and the combination of *eIF-2+His3.3* showed perfect consistency in different organs. *PvTAT* was expressed at the highest level in roots ($p < 0.05$), then came the leaves, stems, and spikes. *Pv4CL2* had the highest significant expression in the roots ($p < 0.05$) and decreased from stems to leaves to spikes. In contrast, the expression level of *PvTAT* and *Pv4CL2* normalized with the worst internal reference gene *VAB2* and showed significant fluctuations. The relative expression level of *PvTAT* and *Pv4CL2* in the donor stems was significantly overestimated ($p < 0.05$), while the relative expression level in roots was obviously underestimated with incompatible expression patterns. In summary, the combination of *eIF-2* and *His3.3* should be the optimum RG set for the normalization of the qRT-PCR in diverse organs.

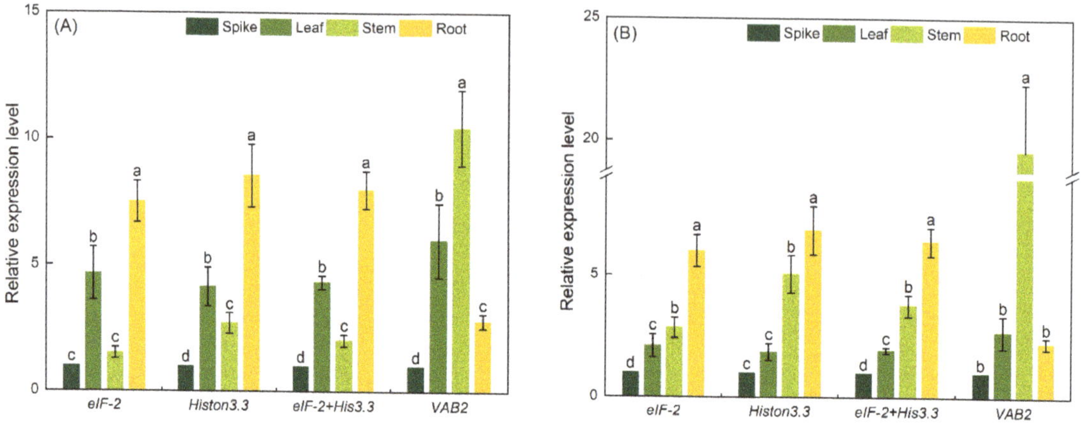

Figure 6. Validation of the identified reference genes. Relative expression of (**A**) *PvTAT* and (**B**) *Pv4CL2* in different organs of *P. vulgaris* that normalized by the most stable reference genes *eIF-2*, *His3.3*, *eIF-2* + *His3.3*, and the least stable reference gene *VAB2*. The different letters signified significant differences at $p < 0.05$ by LSD test.

4. Discussion

In recent years, studying specific gene expression and regulatory mechanisms in various species has become an emerging hotspot [47]. The qPCR is a broadly accepted technique for gene expression analysis in molecular biology [48]. Due to the advantages of high sensitivity, high throughput, good reproducibility, and high specificity, the RT-qPCR technique has now become an important tool for gene expression studies in many laboratories [49]. Nevertheless, the accuracy of quantitative experimental results may be significantly impacted by variables somewhat like primer specificity, sample variation, and RNA quality [5]. During a gene expression analysis, internal RGs are often used to normalize the data to minimize or rectify errors generated during the quantification of target RGs. Consequently, to obtain accurate quantitative results, the selecting of appropriate internal RGs in a variety of experimental circumstances for a specific species is critical [7]. Internal RGs are genes that can be expressed as relatively stable in an organism under a particular condition. Commonly used RGs such as *ACT*, *UBQ*, *18S rRNA*, and *TUA* are usually genes that contribute to the preservation of cell structure or basic biochemical metabolic activities of the organism, and the expression stability of these genes is less affected by environmental factors. The ideal housekeeping genes must maintain relatively constant expression levels under different physiological conditions. Unfortunately, gene expression stability is influenced by many factors and varies in various organs or tissues, or at various growth and development phases. In other words, screening for stably expressed RGs under specific experimental conditions is crucial for analyzing qPCR results. *P. vulgaris* is a traditional Chinese medicine that was discovered and used as a medicinal herb thousands of years ago and has a high medicinal value. The medicinal part of *P. vulgaris* is mainly the dried and mature flower spike, which mainly contains rosmarinic acid, caffeic acid, ursolic acid, and other effective substances [29]. Nowadays, the research hotspots of *P. vulgaris* mainly focus on pharmacology and clinical application, and it is found to have therapeutic effects on various diseases including cancer [50,51]. Regrettably, the limited production of the active medicinal ingredients of *P. vulgaris* cannot meet the increasing market demand. Therefore, it is significant to exploit the genetic resources related to the biosynthesis of secondary metabolites and elucidate the molecular mechanism of Spica Prunellae development. Unfortunately, because of the absence of genetic information, the study on the molecular mechanism of *P. vulgaris* has been progressing slowly. Additionally, research on suitable RGs of *P. vulgaris* has also not been reported comprehensively.

In this research, 14 frequently employed candidate RGs (*18S rRNA*, *ACT1*, *ACT7*, *ACT12*, *eIF-3*, *eIF4A-3*, *eIF-2*, *His3.3*, *TUA6*, *CYP38*, *PP2A-2 PP2A-3*, *PP2A-4*, and *VAB2*) were screened based on *P. vulgaris* transcriptome data (GenBank accession: SRR7873856). Considering the differences in gene expression levels of the internal RGs in different organs and different developmental stages of the fruiting spike, it is necessary to combine more statistical tools to further analyze the expression stability of these candidate RGs under various circumstances. The different screening principles and focus of bioinformatics programs may lead to differences in the analysis results [52]. To get around the drawbacks of utilizing just a single analysis program, this study combined four commonly used Excel-based statistical algorithms, namely Delta Ct, geNorm, BestKeeper, and NormFinder, to assess the stability of internal RGs to improve the reliability of experimental data. Finally, the online software RefFinder was then used to rank each candidate internal RG in terms of the comprehensive index. It has been previously reported that in some cases, BestKeeper has a larger difference in results compared to other algorithms. For example, among all the samples of *D. styracifolium*, *SF3B* ranked first in the BestKeeper algorithm, while ranking low in other software [18]. In *I. indigotica*, BestKeeper ranked *ACT* as a medium, while geNorm and NormFinder methods ranked it as the most stable gene [20]. The same phenomenon also occurred in our study. *TUA6* ranked 11th in the BestKeeper analysis results of all samples, while it ranked 6th in Delta Ct, 2nd in NormFinder, 3rd in geNorm, and 4th in RefFinder. The geNorm analysis showed that among 14 genes in different organs of *P. vulgaris*, *TUA6* was the most suitable as RGs. This was consistent with NormFinder, but not the results of Delta Ct and BestKeeper. *eIF-2* and *His3.3* had lower SD and CV values and showed better stability in Delta Ct and BestKeeper analyses, respectively. Variations in the rankings of these algorithms among other species have also been reported [1,6,9,14,19,20], which may be because of the various underlying principles of the algorithms and the different screening focus used by the four analytical software, resulting in inconsistent results obtained even under the same conditions.

Some prior research has indicated that using a combination of RGs is preferable to using a single RG. RefFinder, an online tool integrating four algorithms, was used to comprehensively evaluate the candidate RGs. In addition, considering that the use of a single RG for normalization could produce inaccurate quantitative results, geNorm was used to calculate the V value to determine the optimal RG combination. It is noteworthy that the ideal V value must be less than the cutoff value of 0.153336 [17]. As shown in Figure 3, the pairwise variation values V_2/V_3 in different organs, different developmental periods, and all samples were smaller than the cutoff value. The results indicated that under these conditions, gene expression normalization can be accomplished without using the third internal RG. It was shown that the optimal combination of RGs in different organs was *TUA6* and *PP2A-2* among the 14 candidate RGs, while the optimal combination in different developmental periods and all samples was *eIF-2* and *His3.3*. All the results indicated the need to select the appropriate RGs according to the specific experimental conditions. Notably, the stability of *eIF-2* and *His3.3* genes was significantly better than that of other RGs under different conditions, while *ACT1* and *VAB2* were almost always ranked at the bottom of the stability list. So far, many prior studies have selected *eIF-2* and *His3.3* as internal RGs for standardization, such as *Nitraria tangutorum* [9] and *Ipomoea batatas* L. [53], etc.

As RGs, candidate genes with consistent expression levels, including *UBQ*, *ACT*, *18S rRNA*, and *GAPDH*, are commonly utilized in many studies [1,19,20]. However, the expression levels of these RGs vary among species and instances. RGs should be selected based on specific experiments. Even if a gene was shown to be expressed in a stable manner in numerous closely related species, it is crucial to confirm and validate the stability of the RG before applying it to gene expression normalization.

PvTAT is a key enzyme gene in the tyrosine branch of the rosmarinic acid metabolic pathway [45], which is intimately associated with the regulation of rosmarinic acid biosynthesis in *P. vulgaris*. *Pv4CL2* is one of the key enzymes involved in the phenylpropanoid pathway and also in the accumulation of rosmarinic acid [46]. The highest transcript abundances of *PvTAT* were found in roots and leaves, while the relatively lower transcript level was found in the stems. Furthermore, *PvTAT* showed the lowest expression levels in spikes [45]. The transcription of *Pv4CL2* was highest in roots and lower in stems, leaves, and spikes [46]. To confirm the accuracy of the experimental results even further, the expression patterns of the target genes *PvTAT* and *Pv4CL2* in different organs were detected and identified utilizing the two most stable genes, *eIF-2* and *His3.3*, and an unstable gene, *VAB2*, as RGs. The results showed that the expression patterns were almost identical, when *eIF-2*, *His3.3*, and *eIF-2* + *His3.3* were used to calibrate the expression of the target genes. However, when *VAB2* was used for normalization, its normalized expression patterns were incompatible with *eIF-2* + *His3.3* (Figure 6). These data again demonstrate that a reliable RG is a prerequisite for the correct normalization of target gene expression levels. Using inappropriate internal RGs can lead to the misestimation of target gene expression levels, resulting in the misinterpretation of experimental results. In conclusion, it is evident from this study that *eIF-2* + *His3.3* had the best stability in different developmental periods of spikes and total data, and they also had good stability in different organs, which has been verified. Therefore, these two RGs can be used as the optimal RG combinations of *P. vulgaris* for subsequent experimental studies.

5. Conclusions

The goal of this work was to identify the most suitable RGs for the qRT-PCR of mRNAs in different organs (root, stem, leaf, and spike) and Spica Prunellae developmental stages (bolting stage, early flowering stage, full flowering stage, and mature stage). Five popular algorithms (Delta Ct, geNorm, NormFinder, BestKeeper, and RefFinfer) were used to assess the expression stability of 14 typical candidate RGs in *P. vulgaris*. The results showed that the optimal RG and RG sets were *eIF-2* and *eIF-2* + *His3.3* in both different developmental stages and all samples, while *TUA6* and *TUA6* + *PP2A-2* were the most suitable RG and RG combinations in different organs. Therefore, in an RT-qPCR examination of this plant, they can be employed as RGs to normalize data depending on your demands. Furthermore, *VAB2*, *ACT1*, and *18S rRNA* showed low expression stability under different experimental conditions. The results of this study can help to accurately quantify the expression levels of target genes in *P. vulgaris* and provide guidance for obtaining reliable qPCR results subsequently, and also establish the groundwork for future research on the developmental mechanism of *P. vulgaris* flower spikes in the future.

Supplementary Materials: The following supporting information can be downloaded at: https://www.mdpi.com/article/10.3390/genes13111947/s1, Figure S1: qPCR amplification specificity of 14 candidate reference genes. Amplification fragments were separated by 1.5% agarose gel electrophoresis; Figure S2: Melt curves for the 14 candidate reference genes; Figure S3: Standard curves of the 14 candidate reference genes; Table S1: Cq values of 14 candidate reference genes in all *P. vulgaris* samples.

Author Contributions: Conceptualization, Q.H. and Z.L.; methodology, Q.H.; software, H.Z. (Hui Zheng) and Q.H.; validation, H.Z. (Hui Zheng); data curation, H.Z. (Hui Zheng); writing—original draft preparation, H.Z. (Hui Zheng); writing—review and editing, Q.H.; supervision, Q.H. and H.Z. (Hongguang Zhao); funding acquisition, X.Z. and Z.L. All authors have read and agreed to the published version of the manuscript.

Funding: This research was funded by the Natural Science Foundation of Zhejiang Province, grant number LGN21H280004, and was funded by the Zhejiang Sci-Tech University and Tasly Group Modern Chinese Medicine Resources Development Key Technology Research Center, grant number TZYHT201903250246.

Institutional Review Board Statement: Not applicable.

Informed Consent Statement: Not applicable.

Data Availability Statement: The data presented in this study are available in Supplementary Materials.

Conflicts of Interest: The authors declare no conflict of interest.

References

1. Yang, Z.; Zhang, R.; Zhou, Z. Identification and Validation of Reference Genes for Gene Expression Analysis in Schima superba. *Genes* **2021**, *12*, 732. [CrossRef]
2. Xu, L.; Xu, H.; Cao, Y.; Yang, P.; Feng, Y.; Tang, Y.; Yuan, S.; Ming, J. Validation of Reference Genes for Quantitative Real-Time PCR during Bicolor Tepal Development in Asiatic Hybrid Lilies (*Lilium* spp.). *Front. Plant Sci.* **2017**, *8*, 669. [CrossRef]
3. Hossain, M.S.; Ahmed, R.; Haque, M.S.; Alam, M.M.; Islam, M.S. Identification and validation of reference genes for real-time quantitative RT-PCR analysis in jute. *BMC Mol. Biol.* **2019**, *20*, 13. [CrossRef]
4. Deng, Y.; Li, Y.; Sun, H. Selection of reference genes for RT-qPCR normalization in blueberry (*Vaccinium corymbosum* × *angustifolium*) under various abiotic stresses. *FEBS Open Bio* **2020**, *10*, 1418–1435. [CrossRef]
5. Kumar, D.; Das, P.K.; Sarmah, B.K. Reference gene validation for normalization of RT-qPCR assay associated with germination and survival of rice under hypoxic condition. *J. Appl. Genet.* **2018**, *59*, 419–430. [CrossRef]
6. Qian, J.; Gao, Y.; Wáng, Y.; Wu, Y.; Wāng, Y.; Zhao, Y.; Chen, H.; Bao, D.; Xu, J.; Bian, X. Selection and Evaluation of Appropriate Reference Genes for RT-qPCR Normalization of Volvariella volvacea Gene Expression under Different Conditions. *BioMed Res. Int.* **2018**, *2018*, 6125706. [CrossRef]
7. Bustin, S.A.; Benes, V.; Nolan, T.; Pfaffl, M.W. Quantitative real-time RT-PCR—A perspective. *J. Mol. Endocrinol.* **2005**, *34*, 597–601. [CrossRef]
8. Zhang, J.; Xie, W.; Yu, X.; Zhang, Z.; Zhao, Y.; Wang, N.; Wang, Y. Selection of Suitable Reference Genes for RT-qPCR Gene Expression Analysis in Siberian Wild Rye (*Elymus sibiricus*) under Different Experimental Conditions. *Genes* **2019**, *10*, 451. [CrossRef]
9. Wang, B.; Duan, H.; Chong, P.; Su, S.; Shan, L.; Yi, D.; Wang, L.; Li, Y. Systematic selection and validation of suitable reference genes for quantitative real-time PCR normalization studies of gene expression in Nitraria tangutorum. *Sci. Rep.* **2020**, *10*, 15891. [CrossRef]
10. Lin, Y.; Zhang, A.; Yang, S.; Huang, L. Reference gene selection for real-time quantitative PCR normalization in Hemarthria compressa and Hemarthria altissima leaf tissue. *Mol. Biol. Rep.* **2019**, *46*, 4763–4769. [CrossRef]
11. Chen, C.; Yuan, M.; Song, J.; Liu, Y.; Xia, Z.; Yuan, Y.; Wang, W.; Xie, Q.; Guan, X.; Chen, Q.; et al. Genome-wide identification and testing of superior reference genes for transcript normalization during analyses of flesh development in Asian pear cultivars. *Sci. Hortic.* **2020**, *271*, 109459. [CrossRef]
12. Vera Hernández, F.P.; Martínez Núñez, M.; Ruiz Rivas, M.; Vázquez Portillo, R.E.; Bibbins Martínez, M.D.; Luna Suárez, S.; Rosas Cárdenas, F.F. Reference genes for RT-qPCR normalisation in different tissues, developmental stages and stress conditions of amaranth. *Plant Biol.* **2018**, *20*, 713–721. [CrossRef]
13. Lucho, S.R.; do Amaral, M.N.; Benitez, L.C.; Milech, C.; Kleinowski, A.M.; Bianchi, V.J.; Braga, E.J.B. Validation of reference genes for RT-qPCR studies in Stevia rebaudiana in response to elicitor agents. *Physiol. Mol. Biol. Plants Int. J. Funct. Plant Biol.* **2018**, *24*, 767–779. [CrossRef]
14. Dixit, S.; Jangid, V.K.; Grover, A. Evaluation of suitable reference genes in Brassica juncea and its wild relative Camelina sativa for qRT-PCR analysis under various stress conditions. *PLoS ONE* **2019**, *14*, e0222530. [CrossRef]
15. Jia, D.-H.; Wang, B.; Li, X.-L.; Tan, W.; Gan, B.-C.; Peng, W.-H. Validation of reference genes for quantitative gene expression analysis in Auricularia cornea. *J. Microbiol. Methods* **2019**, *163*, 105658. [CrossRef]
16. Mo, J.; Xu, J.; Jin, W.; Yang, L.; Yin, T.; Shi, J. Identification of Reference Genes for Quantitative Gene Expression Studies in Pinus massoniana and Its Introgression Hybrid. *Forests* **2019**, *10*, 787. [CrossRef]
17. He, Y.; Zhong, Y.; Bao, Z.; Wang, W.; Xu, X.; Gai, Y.; Wu, J. Evaluation of Angelica decursiva reference genes under various stimuli for RT-qPCR data normalization. *Sci. Rep.* **2021**, *11*, 18993. [CrossRef]
18. Wang, Z.; Yu, F.; Shi, D.; Wang, Y.; Xu, F.; Zeng, S. Selection and validation of reference genes for RT-qPCR analysis in Desmodium styracifolium Merr. *3 Biotech* **2021**, *11*, 403. [CrossRef]
19. Wu, Y.; Zhang, C.; Yang, H.; Lyu, L.; Li, W.; Wu, W. Selection and Validation of Candidate Reference Genes for Gene Expression Analysis by RT-qPCR in Rubus. *Int. J. Mol. Sci.* **2021**, *22*, 10533. [CrossRef]
20. Qu, R.; Miao, Y.; Cui, Y.; Cao, Y.; Zhou, Y.; Tang, X.; Yang, J.; Wang, F. Selection of reference genes for the quantitative real-time PCR normalization of gene expression in Isatis indigotica fortune. *BMC Mol. Biol.* **2019**, *20*, 9. [CrossRef]
21. Vandesompele, J.; De Preter, K.; Pattyn, F.; Poppe, B.; Van Roy, N.; De Paepe, A.; Speleman, F. Accurate normalization of real-time quantitative RT-PCR data by geometric averaging of multiple internal control genes. *Genome Biol.* **2002**, *3*, Research0034. [CrossRef]
22. Andersen, C.L.; Jensen, J.L.; Ørntoft, T.F. Normalization of real-time quantitative reverse transcription-PCR data: A model-based variance estimation approach to identify genes suited for normalization, applied to bladder and colon cancer data sets. *Cancer Res.* **2004**, *64*, 5245–5250. [CrossRef]

23. Pfaffl, M.W.; Tichopad, A.; Prgomet, C.; Neuvians, T.P. Determination of stable housekeeping genes, differentially regulated target genes and sample integrity: BestKeeper—Excel-based tool using pair-wise correlations. *Biotechnol. Lett.* **2004**, *26*, 509–515. [CrossRef] [PubMed]
24. Schmittgen, T.D.; Livak, K.J. Analyzing real-time PCR data by the comparative C(T) method. *Nat. Protoc.* **2008**, *3*, 1101–1108. [CrossRef]
25. Xie, F.; Xiao, P.; Chen, D.; Xu, L.; Zhang, B. miRDeepFinder: A miRNA analysis tool for deep sequencing of plant small RNAs. *Plant Mol. Biol.* **2012**, *80*, 75–84. [CrossRef]
26. Doddaraju, P.; Kumar, P.; Dashyal, M.S.; Girigowda, M. Identification of suitable reference genes for expression studies in pomegranate under different biotic and abiotic stress conditions. *Mol. Biol. Rep.* **2021**, *48*, 3935–3943. [CrossRef]
27. Joseph, J.T.; Poolakkalody, N.J.; Shah, J.M. Plant reference genes for development and stress response studies. *J. Biosci.* **2018**, *43*, 173–187. [CrossRef]
28. Fan, L.; Chen, M.; Dong, B.; Wang, N.; Yu, Q.; Wang, X.; Xuan, L.; Wang, Y.; Zhang, S.; Shen, Y. Transcriptomic Analysis of Flower Bud Differentiation in Magnolia sinostellata. *Genes* **2018**, *9*, 212. [CrossRef]
29. Bai, Y.; Xia, B.; Xie, W.; Zhou, Y.; Xie, J.; Li, H.; Liao, D.; Lin, L.; Li, C. Phytochemistry and pharmacological activities of the genus Prunella. *Food Chem.* **2016**, *204*, 483–496. [CrossRef]
30. Song, Y.-G.; Kang, L.; Tian, S.; Cui, L.-L.; Li, Y.; Bai, M.; Fang, X.-Y.; Cao, L.-H.; Coleman, K.; Miao, M.-S. Study on the anti-hepatocarcinoma effect and molecular mechanism of *Prunella vulgaris* total flavonoids. *J. Ethnopharmacol.* **2021**, *273*, 113891. [CrossRef]
31. Fang, X.; Yu, M.M.; Yuen, W.H.; Zee, S.Y.; Chang, R.C. Immune modulatory effects of *Prunella vulgaris* L. on monocytes/macrophages. *Int. J. Mol. Med.* **2005**, *16*, 1109–1116. [CrossRef] [PubMed]
32. Brindley, M.A.; Widrlechner, M.P.; McCoy, J.A.; Murphy, P.; Hauck, C.; Rizshsky, L.; Nikolau, B.; Maury, W. Inhibition of lentivirus replication by aqueous extracts of *Prunella vulgaris*. *Virol. J.* **2009**, *6*, 8. [CrossRef] [PubMed]
33. Li, C.; Huang, Q.; Fu, X.; Yue, X.J.; Liu, R.H.; You, L.J. Characterization, antioxidant and immunomodulatory activities of polysaccharides from *Prunella vulgaris* Linn. *Int. J. Biol. Macromol.* **2015**, *75*, 298–305. [CrossRef] [PubMed]
34. Hwang, Y.J.; Lee, E.J.; Kim, H.R.; Hwang, K.A. In vitro antioxidant and anticancer effects of solvent fractions from *Prunella vulgaris* var. lilacina. *BMC Complement. Altern. Med.* **2013**, *13*, 310. [CrossRef]
35. Li, C.; Huang, Q.; Xiao, J.; Fu, X.; You, L.; Liu, R.H. Preparation of *Prunella vulgaris* polysaccharide-zinc complex and its antiproliferative activity in HepG2 cells. *Int. J. Biol. Macromol.* **2016**, *91*, 671–679. [CrossRef]
36. Raafat, K.; Wurglics, M.; Schubert-Zsilavecz, M. *Prunella vulgaris* L. active components and their hypoglycemic and antinociceptive effects in alloxan-induced diabetic mice. *Biomed. Pharmacother.* **2016**, *84*, 1008–1018. [CrossRef]
37. Wang, S.J.; Wang, X.H.; Dai, Y.Y.; Ma, M.H.; Rahman, K.; Nian, H.; Zhang, H. *Prunella vulgaris*: A Comprehensive Review of Chemical Constituents, Pharmacological Effects and Clinical Applications. *Curr. Pharm. Des.* **2019**, *25*, 359–369. [CrossRef]
38. Guo, Q. Study on Biological Characters and Active Components of SP6 Generation of *Prunella vulgaris* L. Master's Thesis, Northwest A&F University, Xianyang, China, 2019.
39. Radonić, A.; Thulke, S.; Mackay, I.M.; Landt, O.; Siegert, W.; Nitsche, A. Guideline to reference gene selection for quantitative real-time PCR. *Biochem. Biophys. Res. Commun.* **2004**, *313*, 856–862. [CrossRef]
40. Li, J.; Han, X.; Wang, C.; Qi, W.; Zhang, W.; Tang, L.; Zhao, X. Validation of Suitable Reference Genes for RT-qPCR Data in Achyranthes bidentata Blume under Different Experimental Conditions. *Front. Plant Sci.* **2017**, *8*, 776. [CrossRef]
41. Luo, M.; Gao, Z.; Li, H.; Li, Q.; Zhang, C.; Xu, W.; Song, S.; Ma, C.; Wang, S. Selection of reference genes for miRNA qRT-PCR under abiotic stress in grapevine. *Sci. Rep.* **2018**, *8*, 4444. [CrossRef]
42. Cheng, T.; Zhu, F.; Sheng, J.; Zhao, L.; Zhou, F.; Hu, Z.; Diao, Y.; Jin, S. Selection of suitable reference genes for quantitive real-time PCR normalization in *Miscanthus lutarioriparia*. *Mol. Biol. Rep.* **2019**, *46*, 4545–4553. [CrossRef] [PubMed]
43. Livak, K.J.; Schmittgen, T.D. Analysis of relative gene expression data using real-time quantitative PCR and the 2(-Delta Delta C(T)) Method. *Methods* **2001**, *25*, 402–408. [CrossRef] [PubMed]
44. Xiao, X.; Ma, J.; Wang, J.; Wu, X.; Li, P.; Yao, Y. Validation of suitable reference genes for gene expression analysis in the halophyte Salicornia europaea by real-time quantitative PCR. *Front. Plant Sci.* **2014**, *5*, 788. [CrossRef] [PubMed]
45. Ru, M.; Wang, K.; Bai, Z.; Peng, L.; He, S.; Wang, Y.; Liang, Z. A tyrosine aminotransferase involved in rosmarinic acid biosynthesis in *Prunella vulgaris* L. *Sci. Rep.* **2017**, *7*, 4892. [CrossRef]
46. Kim, Y.B.; Shin, Y.; Tuan, P.A.; Li, X.; Park, Y.; Park, N.I.; Park, S.U. Molecular cloning and characterization of genes involved in rosmarinic acid biosynthesis from *Prunella vulgaris*. *Biol. Pharm. Bull.* **2014**, *37*, 1221–1227. [CrossRef] [PubMed]
47. Fei, X.; Shi, Q.; Yang, T.; Fei, Z.; Wei, A. Expression Stabilities of Ten Candidate Reference Genes for RT-qPCR in *Zanthoxylum bungeanum* Maxim. *Molecules* **2018**, *23*, 802. [CrossRef]
48. Archer, M.; Xu, J. Current Practices for Reference Gene Selection in RT-qPCR of Aspergillus: Outlook and Recommendations for the Future. *Genes* **2021**, *12*, 960. [CrossRef]
49. Zhao, X.; Yang, H.; Chen, M.; Song, X.; Yu, C.; Zhao, Y.; Wu, Y. Reference Gene Selection for Quantitative Real-Time PCR of Mycelia from Lentinula edodes under High-Temperature Stress. *BioMed Res. Int.* **2018**, *2018*, 1670328. [CrossRef]
50. Zhang, X.; Shen, T.; Zhou, X.; Tang, X.; Gao, R.; Xu, L.; Wang, L.; Zhou, Z.; Lin, J.; Hu, Y. Network pharmacology based virtual screening of active constituents of *Prunella vulgaris* L. and the molecular mechanism against breast cancer. *Sci. Rep.* **2020**, *10*, 15730. [CrossRef]

51. Guo, Q.; Qu, H.; Zhang, H.; Zhong, X. *Prunella vulgaris* L. Attenuates Experimental Autoimmune Thyroiditis by Inhibiting HMGB1/TLR9 Signaling. *Drug Des. Dev. Ther.* **2021**, *15*, 4559–4574. [CrossRef]
52. Die, J.V.; Román, B.; Nadal, S.; González-Verdejo, C.I. Evaluation of candidate reference genes for expression studies in *Pisum sativum* under different experimental conditions. *Planta* **2010**, *232*, 145–153. [CrossRef] [PubMed]
53. Yu, J.; Su, Y.; Sun, J.; Liu, J.; Li, Z.; Zhang, B. Selection of stable reference genes for gene expression analysis in sweet potato (*Ipomoea batatas* L.). *Mol. Cell. Probes* **2020**, *53*, 101610. [CrossRef] [PubMed]

genes

MDPI

Article

Integrated Metabolite and Transcriptome Profiling-Mediated Gene Mining of *Sida cordifolia* Reveals Medicinally Important Genes

Deepthi Padmanabhan [1], Purushothaman Natarajan [1,2] and Senthilkumar Palanisamy [1,*]

[1] Department of Genetic Engineering, School of Bioengineering, SRM, Institute of Science and Technology, Kattankulathur 603203, India
[2] Department of Biology, West Virginia State University, Institute, WV 25112-1000, USA
* Correspondence: mpsenthilkumar@gmail.com; Tel.: +91-9443086561

Abstract: *Sida cordifolia* is a medicinal shrub that is conventionally used in the Indian system of medicine;however, the genes contributing to its medicinal properties have been minimally explored, thus limiting its application. High-throughputsequencing and Liquid Chromatography with tandem mass spectrometry(LC-MS/MS) technologies were applied to unravel the medicinally important bioactive compounds. As a result, transcriptomic sequencing generated more than 12 GB of clean data, and 187,215 transcripts were obtained by de novoassembly. These transcripts were broadly classified into 20 classes, based on the gene ontology classification, and 6551 unigenes were annotated using Kyoto Encyclopedia of Genes and Genomes (KEGG) database with more than 142 unigenes involved in the biosynthesis of secondary metabolites. LC-MS/MS analysis of three tissues of *Sida cordifolia* revealed that acacetin and procyanidin are some important metabolites identified thatcontribute to its medicinal value. Several key enzymes witha crucial role in phenylpropanoid and flavonoid biosynthetic pathways were identified, especially phenylalanine ammonia lyase, which might be an important rate-limiting enzyme. Real-Time Quantitative Reverse Transcription Polymerase chain reaction (qRT-PCR) analysis revealed enzymes, such as Phenylalanine ammonia lyase (PAL), Cinnamyl alcohol dehydrogenase 1 (CAD), Cinnamoyl-CoA reductase 1 (CF1) and Trans cinnamate 4-monooxygenase(TCM), which were predominantly expressed in root compared to leaf and stem tissue. The study provides a speculative insight for the screening of active metabolites and metabolic engineering in *Sida cordifolia*.

Keywords: gene expression; medicinal plants; acacetin; transcriptomics; biosynthesis; flavonoids

Citation: Padmanabhan, D.; Natarajan, P.; Palanisamy, S. Integrated Metabolite and Transcriptome Profiling-Mediated Gene Mining of *Sida cordifolia* Reveals Medicinally Important Genes. *Genes* **2022**, *13*, 1909. https://doi.org/10.3390/genes13101909

Academic Editors: Wajid Zaman and Hakim Manghwar

Received: 24 September 2022
Accepted: 17 October 2022
Published: 20 October 2022

1. Introduction

Plants have been used as an immortal source of medicine forages. Globally, herbal medicines have played an important role in human health to treat chronic and acute conditions without any toxic effect. They are usually used as therapeutics to health conditions including diabetes mellitus, wounds, cancer, heart diseases, tuberculosis, hypertension, etc. These herbs are used in medicinal fields, as they are rich in bioactive phytocompounds, such as flavonoids, alkaloids, terpenoids, polyphenols, and tannins possessing various pharmacological properties [1]. A traditional sub shrub, *Sida cordifolia*, belonging to the family *Malvaceae*, is widely spread across countries like India, Brazil, and Africa and is extensively used in the ayurvedic system of medicine. It has laid the foundation to perform the transcriptomic approach to elucidate the various biosynthetic pathways responsible for its medicinal properties [2]. Some important constituents isolated from various extracts of *Sida cordifolia* are 1, 2, 3, 9-Terta hydropyrrolo[21-b]-quinazolin-3-yl-amine, ephedrine, vasicine, pseudo-ephedrine, vasicinone, hypaphorine, vasicinol, stigmasterol, and sterculic acid [3–7]. Pharmacological properties exhibited by different extracts of *Sida cordifolia* are antimicrobial, anti-inflammatory, analgesic, anti-ulcer, nephroprotective, anti-diabetic, hepatoprotective, anticancer, and central nervous system depressant activity [8]. The inevitable identification of genes responsible for the production of various secondary metabolites by *S.cordifolia* is achieved using the next-generation sequencing technology to unravel the novel transcripts from medicinal plants with respect to secondary metabolite biosynthesis and gene expression analysis [9]. The technology is employed to generate functional data for non-model plants and EST sequences for the annotation of genes and analysis of targeted discovery. The de novo sequencing prompts large volumes of information that can be used to identify various molecular markers, novel gene identification, or discovery and polymorphism [10–13]. Transcriptomics is the study of collection of all the transcripts available in the specific tissue andhas been used to study the structure and function of the gene at the molecular levels.The study of small molecular weight metabolites that exist in all cells to maintain their growth and function is referred to as metabolomics. The metabolite profiling reflects the overall biochemical and physiological conditions of the plant for the survival in that particular environment [14]. LC-MS analysis is preferable for evaluating large groups of secondary metabolites like phenolic compounds, flavonoids, and alkaloids [14]. Though the plant has been extensively used in herbal formulations, the biosynthetic pathways responsible for the synthesis of secondary metabolites that possess the pharmacological properties have yet to be unveiled. Thus, in this study, we aimed to annotate the transcripts, identify the putative transcripts that are involved in the major biosynthetic pathways attributing to the medicinal properties of the plant using transcriptomics, and analyze the different metabolites present in them by metabolomics.

2. Materials and Methods

2.1. Sample Collection and RNA Extraction

Root, stem, and leaf tissues were collected from a healthy plant near Potheri in Tamil Nadu, and identification was taxonomically done in SRMIST, IIISM. The RNA was extracted using TRizol®Reagent followed by phase-separation using Chloroform. After centrifugation at 12,000 rpm for 15 min at 4 °C. To the aqueous phase, 0.7 volume of isopropanol was added and centrifuged at 12,000 rpm for 10 min at 4 °C, followed by ethanol, which was to the pellet obtained from the previous step. Then, nuclease-free water was addedto the air-dried pellet. To remove the DNA contamination, the RNA was treated with DNase. A further purification of the total RNA was done using Qiagen RNeasy®MinElute clean-up kit. The integrity of RNA was analyzed using the Agilent Bioanalyzer with RIN value range of 7.4 to 9.2, a Thermo Scientific Nanodrop Lite spectrophotometer was used to check the quality of RNA, and the samples were visualized in the agarose gel electrophoresis [15,16].

2.2. RNA Library Preparation and Illumina Sequencing

Purified RNA was used for cDNA preparation. The poly A tail present in the mRNA was refined from the total RNA using magnetic beads attached to poly-T oligo, and a fragmentation buffer was used for fragmenting the mRNAs into smaller fragments. The first strand synthesis of these shorter fragments was carried out using Invitrogen's Superscript II reverse transcriptase, and for the second strand synthesis, RNase H and DNA polymerase I were used. Single addition of dATP was done to the fragmented cDNA and then connected with adapters followed by a further selection of templates for PCR amplification. NextSeq 500 was used for sequencing, and the raw data in the FastQ format were obtained (Illumina Inc., San Diego, CA, USA) [15,16].

2.3. *Assembly of Transcripts*

The raw data obtained from sequencing were subjected to quality control and reads with (Phred score $\geq$ 30), were removed by the software FastQCv.0.11.9, and the adapter sequences were trimmed by Cutadapt software. TRINITY v2.14.0 de novo assembler was used for the re-construction of transcriptome data with three software modules: the Inchworm module assembles the transcripts by k-mer; the Chrysalis module clusters the contigs, thereby constructing a *de Bruijn* graph for the contigs; and the Butterfly module analyzes the contigs based on the graphs created and lists the isoforms. CD-HIT v4.8.1 program was used to reduce the sequence redundancy and increase the analysis performance [17,18].

2.4. *Gene Ontology Classification and Functional Annotation*

The de novo assembled transcripts of *Sida cordifolia* were collated against different databases including protein non-redundant database (nr) from NCBI with the BLASTX tool, and an E-value not greater than 1E-05 was considered as a significant match. Further Omics Box tools were used for gene ontology (GO) classification and annotation with enzyme codes (EC). The pathway mapping was done by KEGG Automated Annotation Server for retrieving the pathway maps of Kyoto Encyclopedia of Genes and Genomes (KEGG) [19].

2.5. *SSR Identification*

The microsatellites were identified using the Krait software, which offers an ultrafast and user-friendly graphic interface for investigating genome-wide microsatellites. The software also identifies VNTRS from a large genome size, locates the SSRs in the gene coding region, and statistically analyzes and plots the graphs [20]. Simple sequence repeats are short repeat DNA sequences with 1–6 bp in length, which have high polymorphism and can be used as a tool in genetic mapping, population genetics, and phylogenetic analysis [21].

2.6. *Transcript Quantification*

The SALMON quantification tool was used to quantify the transcript abundance from the sequenced data andcombines the dual-phase parallel interface algorithm. The tool quantifies the transcripts based on the GC content, quasi mapping was conducted for accuracy, and fast mapping of genes was conducted to study the expression levels [22].

2.7. *Reverse Transcription PCR Validation*

The validation assembled transcriptome data of *Sida cordifolia* from the root tissue by reverse transcription PCR was carried out. The full-length transcripts were found using the BLAST tool to find the complete transcript to design the primers. Then, the PRIMER BLAST tool from NCBI was used to design the primers, and in silico PCR was conducted, where Actin was considered as a housekeeping gene. The gene expression analysis was normalized using reverse transcription PCR, and gradient PCR was conductedto optimize the annealing temperatures from 52 °C to 60 °C. PCR amplification was carried out by the following parameters: 95 °C for 5 min, 95 °C for 30 s, 57–59 °C for 30 s, 72 °C for 30 s, and 72 °C for 5 min for 40 cycles.

2.8. *Validation of Gene Expression Analysis*

S.cordifolia transcriptome data were validated by quantitative real-time PCR with QuantStudio 5 (Thermo Scientific, Wilmington, DE, USA) PCR machine and the QuantiNovaSYBRGreen PCR (Qiagen Inc., GmbH, Germany) kit. Actin was used as an internal reference, and a negative control reaction was set up in all experiments where it was used. The relative gene expression levels were analyzed and calculated using the$2^{-\Delta\Delta Ct}$ method, which represents the cycle threshold of the target gene with the housekeeping gene Actin. qRT-PCR analysis was carried out with three replicates [23].

2.9. Extraction of Metabolites from *Sida cordifolia*

The leaf, root, and stem tissues of *Sida cordifolia* were collected, dried using a microwave oven, and pulverized using an electric blender. Then, 10 g of powdered plant tissues were transferred to a Schott bottle, and 100 mL of 99.9% methanol was added to it and macerated for three days. The solvent was evaporated using a rotary evaporator. The samples were centrifuged at 3000 rpm, and the supernatant was transferred into a fresh tube and stored at room temperature until further analysis.

2.10. LC-MS/MS Analysis

The secondary metabolites were analyzed using Shimadzu LC-MS/MS-8040 (QQQ) (liquid chromatography–mass spectrometry, Triple Quadrupole, North America) with a scan range of 50–1000 m/z in both positive and negative modes. Mobile phase A: 5 mM ammonium formateand 0.1% formic acid in water; phase B: 5 mM ammonium formateand 0.1% formic acid in methanol with a flow rate 0.6 ml/min an isocratic A—20% and B—80% were used. The column Union was used at a temperature 40 °C with a flow rate of 500 µL/min. The data acquisition was done by the software LabSolutions™ LCMS, and the compounds were identified from free-source online tools, such as METLIN, PUBCHEM, and KEGG databases, along with some previously published articles.

3. Results

3.1. RNA Sequencing Analysis of Raw and Processed Data

The root transcriptome of *Sida cordifolia* was obtained with 59,484,771 raw reads. From the raw reads, adapter sequences were trimmed to reduce the redundancy, and 59,484,597 clean reads were obtained further with a GC content of 40.52%. The raw reads were submitted to NCBI Sequence Read Achieve (SRA) and an accession number PRJNA841821 was obtained. Overall, 187,215 transcripts were assembled using a de novo assembler TRINITY.The average length of bases was 1035.19, and the median contig length was 717. The contig N50 value was found to be 1622 (Table 1).

Table 1. Summary of Illumina paired-end sequencing and de novo assembly of *Sida cordifolia*.

Particulars	Numbers
Number of raw reads	59,484,771
Number of clean reads	59,484,597
No.of bases (after processing)	50,962,206
Mean Phred score	30
Total transcripts	187,215
Total length (bases)	193,803,049
Average length (bases)	1035.19
Median Contig length	717
GC content (%)	40.53
Contig N50 (Transcripts)	1622

3.2. Functional Annotation of Unigenes

A similarity search using BLASTX was conductedagainst the nr database from NCBI; with a total of 187,215 unigenes assembled, a total of 54,375 unigenes were non-annotated and 132,840 unigenes were annotated (Figure 1). Due to inadequate information of the *Sida cordifolia* genome, some annotated genes were classified into predicted, uncharacterized, or hypothetical proteins.The results from BLASTX were transferred to OMICS BOX to annotate further. The similarity search of assembled transcripts showed high similarity with *Gossypium hirsutum, Gossypium raimondii, Gossypium arboretum, Duriozibethinus, Theobroma cacao, Herrania umbratica*, and *Quercus suber* and the least similarity with *Stipa magnifica* and *Stipa borysthenica* (Supplementary Figure S1).

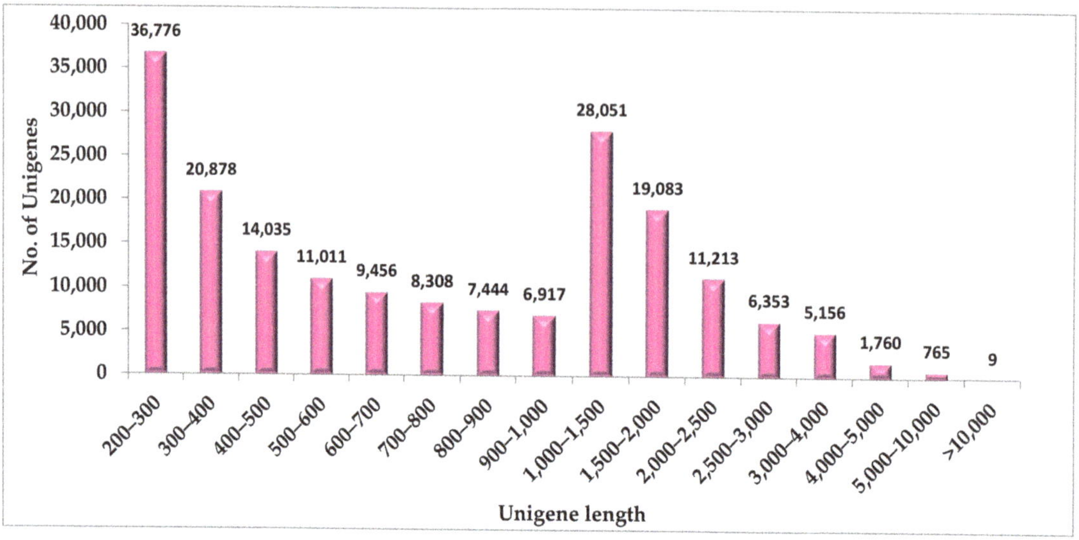

Figure 1. Unigene length distribution of *S.cordifolia* root transcriptome.

3.3. Functional Classification of Unigenes

Using the OMICS BOX software, the assembled unigenes were annotated onto GO classification into three different categories viz. molecular function, cellular process, and biological process.The unigenes were categorized into 60 subcategories. In the cellular component, 9467 unigenes were classified into 20 classes. This category includes intracellular anatomical structure with high unigenes in the organelle, cytoplasm, and membrane and the leastunigenes in the nucleoplasm, supramolecular complex, and external encapsulating structure. In the molecular function, a total of 19,593 unigenes were classified into 20 different classes with high unigenes in organic cyclic compound binding, heterocyclic compound binding, and ion binding and low unigenesin isomerase activity, protein-containing complex binding, and carbohydrate binding. In the biological process, 25,315 unigenes were grouped into 20 classes with the maximum number of unigenes in organic substance metabolic process, cellular metabolic process, and primary metabolic process and the minimum number in signal transduction, response to chemicals, and vesicle-mediated transport (Supplementary Figure S2).

3.4. Biological Pathway Analysis

In total, 6551 unigenes were annotated using the KEGG database and assigned to 150 pathway maps. A total of 6409 unigenes were found to be annotated in metabolic pathways onto different categories, which include nucleotide metabolism (2539 unigenes), carbohydrate metabolism (347 unigenes), metabolism of co-factors and vitamins (1333 unigenes), energy metabolism (823 unigenes), biosynthesis of antibiotics (413 unigenes), xenobiotics biodegradation and metabolism (135 unigenes), lipid metabolism (267 unigenes), and amino acid metabolism (477 unigenes) (Supplementary Figure S3). Further, 142 unigenes were classified into biosynthesis of secondary metabolites, which was further subdivided into different categories: sesquiterpenoid and triterpenoid biosynthesis (57 unigenes), terpenoid backbone biosynthesis (22 unigenes), ubiquinone and other terpenoid-quinone biosynthesis (26 unigenes), phenylpropanoid biosynthesis (11 unigenes), isoquinoline alkaloid biosynthesis (8 unigenes), and flavonoid biosynthesis (6 unigenes) (Figure 2).

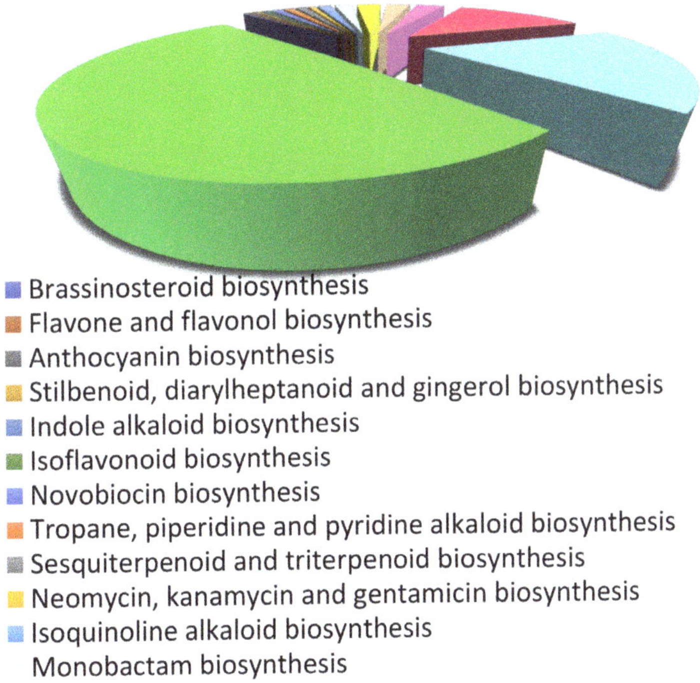

Figure 2. KEGG pathway analysis based on secondary metabolite biosynthesis in root transcriptome of *Sida cordifolia*.

3.5. *Untargeted Metabolic Profiling of Sida cordifolia*

Transcriptomic analysis of metabolic pathways and the validation of key metabolites require further confirmation by the identification of metabolites. The metabolomic analysis of samples was completed by the untargeted metabolomic method. There were a total of 298 different metabolites in leaf, stem, and root tissues of *S.cordifolia* in both the positive and negative mode. Flavonoids were the major group of secondary metabolites found in the metabolome analysis of *S.cordifolia* where Naringin, Cinnamic acid, Cinnamaldehyde, Caffeic acid, Kaempferol derivatives, and Quercetin were predominantly present in root, stem, and leaf tissues. Some tissue-specific metabolites, such as Rosmarinic acid, Caffeic acid and its derivatives, Kaempferol derivatives, Rutin, Quercitrin, and Gallic acid were identified in root tissue. Apigenin, Gallagic acid, Quercetin and its derivates, Caffeic acid and its derivatives, and Kaempferol and its derivatives were found in leaf tissue. In stem tissue, Ferulic acid, Gallic acid, Quinic acid derivative, and Malic acid were found (Table 2). In some metabolic pathways, such as, flavonoid biosynthesis, isoflavonoid biosynthesis, flavones and flavonol biosynthesis, phenylalanine, tyrosine and tryptophan biosynthesis, and ubiquonone biosynthesis, intermediate metabolites could be identified in the mass spectra. The results revealed that the transcriptomic and metabolic analysis of the transcripts that catalyzes the enzymes is consistent with their metabolites that belong to the pathway.

Table 2. Identification of various secondary metabolites from root, stem, and leaves tissue of *Sida cordifolia* by LC-MS/MS.

S.No	*m/z*	Compound Name	Pathway Involved in	Expressed in Root (R), Leaf (L), Stem (S)	Medicinal Properties	Reference
1	149.9, 148.9	Cinnamic acid	Flavonoid pathway	R, L, S	Antioxidant, antimicrobial, anticancer, neuroprotective, anti-inflammatroy, antidiabetic	[24]
2	315	Hydroxytyrosol 4-O-glucoside	Phenylpropanoid pathway	L	Anti-inflammatory, antioxidant, neuroprotective, immunomodulatory effects	[25]
3	133.0135, 115.009, 114.9900, 89.0200	Malic acid	TCA cycle	R, S	Antimicrobial activity	[26]
4	283.0617 177.0812 171.05 1510	Acacetin	Flavone pathway	R	Anticancer, anti-inflammatory	[27]
5	359.1372 285.0421 257.0316	Rosmarinic acid	Flavonoid pathway	R, S	Antimicrobial, immunomodulatory, antidiabetic, anti-allergic, anti-inflammatory, hepato and renal protectant	[28]
6	285.0396, 257.0512, 229.0516	Luteolin	Flavonoid pathway	R	Cardioprotective, antioxidant, antiviral, anti-inflammatory, antimicrobial, neuroprotective	[29]
7	489.1477 341.1138 179.0572 147.0313	Caffeic acid monohexoside derivative	Phenylpropanoid pathway	R, S	Antioxidant, anti-inflammatory, antineoplastic	[30]
8	327.1123 179.0571 147.0290	Caffeic acid derivative	Phenylpropanoid pathway	R, S, L	Antioxidant, anti-inflammatory, antineoplastic	[30]
9	577.2495 341.1110 179.0535	Caffeic acid-O- hexoside derivative	Phenylpropanoid pathway	R, S	Antioxidant, anti-inflammatory, antineoplastic	[30]
10	342	Caffeoyl glucose	Phenylpropanoid pathway	L	Antioxidant, anti-inflammatory, antineoplastic	[30]
11	311	Caffeoyl tartaric acid	Phenylpropanoid pathway	L	Antioxidant, anti-inflammatory, antineoplastic	[30]

Table 2. *Cont.*

S.No	*m*/*z*	Compound Name	Pathway Involved in	Expressed in Root (R), Leaf (L), Stem (S)	Medicinal Properties	Reference
12	179	Caffeic acid	Phenylpropanoid pathway	L, S	Antioxidant, anti-inflammatory, antineoplastic	[30]
13	567.228 405.1753 179.0515	Dihydrosinapyl caffeoyl hexoside	Phenylpropanoid pathway	R	Antitumor, neuroprotectant	[30]
14	461.1678 389.0100 385.1501 315.1100	Kaempferol-3-O-Glucuronide	Flavonoid pathway	R	Treats Alzheimer's disease, Parkinson's disease, ischemia stroke, epilepsy, major depressive disorder, anxiety disorders, neuropathic pain, and glioblastoma	[31]
15	417	Kaempferol-3-O-arabinoside	Flavone pathway	R, S	Treats Alzheimer's disease, Parkinson's disease, ischemia stroke, epilepsy, major depressive disorder, anxiety disorders, neuropathic pain, and glioblastoma	[31]
16	293	Kaempferol-3-O-(p-coumaroyl)-glucoside	Flavone pathway	S	Treats Alzheimer's disease, Parkinson's disease, ischemia stroke, epilepsy, major depressive disorder, anxiety disorders, neuropathic pain, and glioblastoma	[31]
17	301	Quercetin	Flavonoid pathway	R, L	Anticancer, antiviral, antiprotozoal, and antimicrobial effects;treatment of allergic, metabolic, and inflammatory disorders; eye and cardiovascular diseases; and arthritis	[32]
18	301 151 179	Quercetin hexoside	Flavonoid pathway	L	Anticancer, antiviral, antiprotozoal, and antimicrobial effects; treatment of allergic, metabolic, and inflammatory disorders; eye and cardiovascular diseases; and arthritis	[32]
19	610	Quercetin 3-O-glucoside	Flavonoid pathway	L	Anticancer, antiviral, antiprotozoal, and antimicrobial effects; treatment of allergic, metabolic, and inflammatory disorders; eye and cardiovascular diseases; and arthritis	[32]

Table 2. *Cont.*

S.No	*m/z*	Compound Name	Pathway Involved in	Expressed in Root (R), Leaf (L), Stem (S)	Medicinal Properties	Reference
20	610	Quercetin 3-O-rutinoside	Flavonoid pathway	L	Anticancer, antiviral, antiprotozoal, and antimicrobial effects; treatment of allergic, metabolic, and inflammatory disorders; eye and cardiovascular diseases; and arthritis	[32]
21	577.2699 417.1785	Procyanidin B2	Flavonoid pathway	R	Antioxidant, antibacterial, antitumor, anti-inflammatory, treat neurogenerative disorder	[33]
22	407, 425, 451, 289	Procyanidin dimer 1	Flavonoid pathway	L	Antioxidant, antibacterial, antitumor, anti-inflammatory, treat neurogenerative disorder	[33]
23	419.1364 337.1521 294.8212	Dihydroxymethoxy-glucopyranosylstilbene	Stilbene pathway	R	Anticancer, antitumor	[34]
24	429.17	Hecogenin	Saponin pathway	R	Anti-inflammatory, antioxidant, antifungal, hypotensive, anti-hyperalgesic, and anti-nociceptive effects	[35]
25	431.1631 413.3001 397.120, 201.0110	Kaempferol-3-O-rhamnoside	Flavonoid pathway	R	Anticancer	[36]
26	595.260, 448.181, 431.1930	Saponin cinnamoyl derivative	Saponin pathway	R	Fungicidal, antimicrobial, antiviral, anti-inflammatory, anticancer, antioxidant, and immunomodulatory effects	[37]
27	577.250, 431.2069	Kaempferol dirhamnoside	Flavonoid pathway	R	Antimicrobial, antioxidant	[38]
28	609.148, 449.160, 301.0412	Rutin	Flavonoid pathway	R	Antioxidant, anticonvulsant, anti-alzheimer, antidepressant, analgesic, antiarthritic	[39]

Table 2. *Cont.*

S.No	*m/z*	Compound Name	Pathway Involved in	Expressed in Root (R), Leaf (L), Stem (S)	Medicinal Properties	Reference
29	581.199, 419.350, 273.2141	Naringin	Flavonoid pathway	R, S	Antioxidant, anti-inflammatory, anticancer, anicarcenogenic, anti-ulcer	[39]
30	827	Naringin 6′-malonate	Flavonoid pathway	L	Antioxidant, anti-inflammatory, anticancer, anicarcenogenic, anti-ulcer	[39]
31	133.101, 131.08, 117.0665 115.050, 91.0212	Cinnamaldehyde	Phenylpropanoid pathway	R	Antifungal, antidiabetic,	[40]
32	197.117, 179.121, 135.110, 133.0211	Loliolide	Terpene pathway	R	Antioxidant, antifungal, antibacterial, antidiabetic, anticancer, antiviral, antituberculosis, anti-melanogenic, anti-inflammatory, and anti-aging	[41]
33	245, 179, 205	Catechin	Flavanol pathway	L	UV protectant, antimicrobial, antiallergenic antiinflammatory, antiviral, anticancer	[42]
34	353	Chlorogenic acid	Phenylpropanoid pathway	L, S	Antioxidant activity, antibacterial, hepatoprotective, cardioprotective, anti-inflammatory, antipyretic, neuroprotective, anti-obesity, antiviral, antimicrobial, anti-hypertension	[43]
35	355	Ferulic acid 4-O-glucoside	Flavonoid pathway	L	Vasodilator	[44]
36	339	3-p-Coumaroylquinic acid	Phenylpropanoid pathway	L	Antioxidant, antidiabetic, anticancer activity, antimicrobial, antiviral, anti-aging, protective, antinociceptive, and analgesic	[45]
37	191	Quinic acid derivative	Phenylpropanoid pathway	S	Antioxidant, antidiabetic, anticancer activity, antimicrobial, antiviral, anti-aging, protective, antinociceptive, and analgesic	[45]
38	315	Protocatechuic acid 4-O-glucoside	Tannin pathway	L	Antioxidant, anti-inflammatory, antihyperglyemic, antimicrobial,	[46]

Table 2. *Cont.*

S.No	*m/z*	Compound Name	Pathway Involved in	Expressed in Root (R), Leaf (L), Stem (S)	Medicinal Properties	Reference
39	465	Dihydromyricetin 3-O-rhamnoside	Flavonoid pathway	L	Antioxidant	[47]
40	273	3′,4′,7-Trihydroxyisoflavanone	Flavanone pathway	L	Inhibitor of UV radiation	[48]
41	605	Gallagic acid	Tannin pathway	L	Antioxidant, anti-inflamatory, antiplasmodial activity	[49]
42	593	Apigenin	Flavone pathway	L	Antioxidant, anti-inflammatory, anti-amyloidogenic, neuroprotective, and cognition-enhancing substance	[50]
43	193	Gallic acid	Flavonoid pathway	S	Antioxidant, anti-inflammatory, and antineoplastic	[51]

3.6. KEGG Enrichment Analysis of Transcripts Involved in Various Secondary Metabolite Biosynthesis

3.6.1. TranscriptsInvolvedin Phenylpropanoid Biosynthesis Pathway

The compounds derived from this pathway are initiated by the amino acid phenylalanine which serves as a precursor. Precursor phenylalanine, which is converted into Cinnamoyl-CoA, p-Coumaryl-CoA, p-Coumarylquinic acid, Caffeoylquinic acid, Caffeoyl-CoA, Feruloyl-CoA and Sinapoyl-CA. Eleven genes were intricate in the biosynthesis pathway. The root transcriptome analysis revealed the presence of the genes Phenylalanine ammonia lyase (EC: 4.3.1.24, EC: 4.3.1.25), 4-Coumarate CoA ligase (EC: 6.2.1.12), trans-cinnamate 4-monooxygenase (EC: 1.14.14.91), Cinnamoyl-CoA reductase 1 (EC: 1.2.1.44), and Cinnamyl alcohol dehydrogenase (EC: 1.1.1.195) (Supplementary Figure S4).

3.6.2. Transcripts Involved in Flavonoid Biosynthesis Pathway

The precursor for flavonoid biosynthesis is phenylalanine. The pathway is elucidated by the conversion of phenylalanine to p-Coumaryl-CoA, and the rate-limiting enzyme is Chalcone synthase. The genes identified from the transcriptome data are trans-cinnamate 4-monooxygenase (EC: 1.14.14.91), Chalcone-flavonone isomerase (EC: 5.5.1.6), and naringenin, 2-oxoglutarate 3-dioxygenase (EC: 1.14.11.9) (Supplementary Figure S5).

3.7. SSR Identification

A total of 187,215 transcripts were analyzed, and 36197 perfect SSRs were identified with a total length of 626,976 bp, and the average length of SSRs was 17.33. Relative abundance and density were found to be 186.77 and 3235.12, respectively. The most abundant repeat motifswere mono nucleotide (12,898;35.6%), followed by tri-nucleotide (12,732;35.2%), di-nucleotide (6868;19%), tetra-nucleotide (2526;6.98%), penta-nucleotide (653;1.8%), and hexa-nucleotide (520;1.44%). SSRs with 5tandem repeats were most common in *Sida cordifolia,* followed by 12 repeats and 7tandem repeats. Among mono-nucleotide repeat motifs, Awashighly abundant (35%).In tri-nucleotide repeats, the highest frequency was observed in AAG (11.51%) followed by AAT (4.36%) (Supplementary Figure S6)

3.8. Gene Expression Analysis of Sida cordifolia

The qRT-PCR analysis was done to validate the transcriptome date and the expression analysis of the selected genes. The selected transcripts included Phenylalanine ammonia lyase (EC: 4.3.1.24, 4.3.1.25), Trans-cinnamate4-monooxygenase (EC: 1.14.14.91), Chalcone-flavonone isomerase (EC: 5.5.1.6), and Cinnamyl alcohol dehydrogenase 1(EC: 1.1.1.219) (Table 3). The primer sequences used for the analysis are depicted in the Supplementary Table S1. The analysis revealed varied expression patterns of the selected genes, where all the genes were upregulated in the root tissue rather than the stem and leaf. Actin was used as a housekeeping gene. The results obtained showed significant agreement with the transcriptome data (Figure 3).

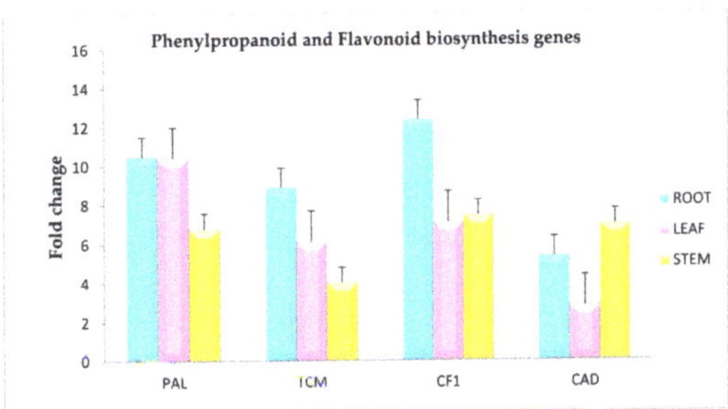

Figure 3. Validation of gene expression of phenylpropanoid and flavonoid biosynthesis genes by real-time PCR from *Sida cordifolia.*

Table 3. Major genes involved in phenylpropanoid and flavonoid biosynthesis pathways identified from *Sida cordifolia* root transcriptome.

Gene Name	EC Number	Unigene ID	Unigene Length	TPM Value	No. of Unigenes
Phenylalanine Ammonia Lyase (PAL)	4.3.1.24 4.3.1.25	TRINITY_DN21246_c0_g1_i1	358	518.921	12
		TRINITY_DN21246_c0_g1_i2	313	288.181	
		TRINITY_DN21246_c2_g2_i2	2069	170.628	
		TRINITY_DN21246_c2_g2_i3	364	123.988	
		TRINITY_DN22573_c0_g1_i1	1135	57.6983	
		TRINITY_DN22573_c0_g1_i2	983	213.701	
		TRINITY_DN21246_c0_g1_i1	358	518.921	
		TRINITY_DN21246_c0_g1_i2	313	288.181	
		TRINITY_DN21246_c2_g2_i2	2069	170.628	
		TRINITY_DN21246_c2_g2_i3	364	123.988	
		TRINITY_DN22573_c0_g1_i1	1135	57.6983	
		TRINITY_DN22573_c0_g1_i2	983	213.701	
Trans cinnamate 4-monooxygenase	1.14.14.91	TRINITY_DN19528_c1_g1_i1	2073	682.03	2
		TRINITY_DN19528_c1_g1_i7	271	8.5076	
Cinnamoyl-CoA reductase 1	1.2.1.44	TRINITY_DN16088_c0_g1_i9	3601	1.132281	1
Cinnamyl alcohol dehydrogenase 1	1.1.1.195	TRINITY_DN18521_c0_g1_i1	1921	17.6824	6
		TRINITY_DN18521_c0_g1_i2	1837	32.1333	
		TRINITY_DN18521_c0_g1_i3	1450	52.8276	
		TRINITY_DN18521_c0_g2_i1	1379	4.49343	
		TRINITY_DN18521_c0_g1_i8	1724	37.4843	
		TRINITY_DN18521_c0_g1_i10	2145	44.9757	
4-coumarate–CoA ligase 2	6.2.1.12	TRINITY_DN19412_c1_g1_i1	2201	16.54958	1

4. Discussion

4.1. De Novo Assembly, Functional Classification, and Annotation

RNA sequencing is a high-throughput sequencing method thathas been an integral part of metabolome research in non-model species with relatively high rapidity. This sequencing technology is effective to study the annotation and to elucidate various transcripts responsible for the biosynthetic pathways, gene expression analysis, and distribution of secondary metabolites in different tissues [52].The Illumina sequencing of the root tissue of *Sida cordifolia* revealed 59,484,771 reads, which were further processed to remove the adapter and redundant sequences. The de novo assembler TRINITY was used to assemble the transcripts, and a total of 187,215 numbers of transcripts were identified.

Functional annotation and classification comprise a vital step thatprovides us with the homology alignment. Gene ontology classification was done to annotate the transcripts into three major categories such as cellular function, molecular function, and biological process. In our study, out of 187,215 unigenes, a total of 54,375 unigenes were non-annotated, and 132,840 unigenes were annotated against different databases. *Sida cordifolia* showed high similarity with the genus *Gossypium* and the least similarity with *Stipa* genus. These results can be because of the scarcity of reference genomic resources for *Sida genus*.

4.2. KEGG Pathway Analysis and Identification of Candidate Genes Involved in Secondary Metabolite Biosynthesis

Biological pathway analysis against the KEGG database was conductedto identify the various pathways involved in the transcriptome data with a total of 150 pathway maps, out of which a higher number of transcripts were annotated into metabolism pathway than the secondary metabolite biosynthesis pathway maps. The phenylpropanoid biosynthesis pathway is the key biosynthetic pathway responsible for the production of various secondary metabolites. The pathway is initiated by catalyzing phenylalanine to cinnamate and ammonia by the enzyme Phenylalanine ammonialyase (PAL). PAL is the key regulatory enzyme involved in the pathway that is found ubiquitously in plants [53]. The enzyme serves as a precursor for the flavonoid and lignin biosynthetic pathways [54]. A total of 12 transcripts were identified that are responsible for synthesis of PAL gene. Trans-cinnamic acid is also a precursor for the flavonoid and biosynthetic pathways. The increased activity of the enzyme PAL in turn increases the production of phenylpropanoid products, which vary with different stress stimuli, developmental stage, and tissue and cell differentiation. The enzyme is delineated to be stimulated by infection, radiations, drastic change in temperature, or drought stress [55–58]. Cinnamoyl-CoA reductase 1 involved in the monolignol pathway engenders the conversion of p-coumaroyl-, feuloyl-, and sinapoyl-CoA to p-coumaraldehyde, coniferaldehyde, and sinapaldehyde [59,60].

Flavonoids are naturally occurring secondary metabolites that are rich in antioxidant, antibacterial, anti-inflammatory, antifungal, and antidiabetic activity and act as an anticancer agent. On the basis of a number of hydroxyl groups, flavonoids are grouped into flavones, flavonols, flavones, anthocyanidins, and isolflavones. Flavonoids are synthesized from the phenylpropanoid biosynthesis pathway with the key enzymes involved in them. The enzymes that are involved in the first step of flavonoids biosynthesis are 4-Coumaroyl-CoA ligase, 4-Coumarate-3-Hydroxylase, and Phenylalanine ammonialyase. The genes involved in flavonoid biosynthesis pathway are flavone synthase, Dihydroflavonol-4-Reductase, Chalcone synthase, and Chalcone isomerase [61], wherethe enzyme 4-coumarate-CoA ligase catalyzes the activation of 4-coumarate, which occurs in multiple isoenzymes forms and exhibits substrate affinity with the metabolic function. It has a pivotal role in the biosynthesis of secondary metabolites from general phenylpropanoid metabolism. Other secondary metabolite biosynthesis pathways identified are ubiquinone, triterpenoid, sesquiterpenoid and isoquinoline alkaloid biosynthesis pathways.

A flavonoid—quercetin—and its derivatives are extensively used due to their bioactive effects as they possess many pharmacological properties such as anti-arthritic, cardiovascular, anticancer, anti-Alzheimer's, antimicrobial, and wound-healing effects[62–64]. Studies suggest that flavonoids are found to be potential inhibitors of coronaviruses against the SARS-CoV-2 infection by binding to the targets that promotes the replication and entry of viruses [65,66]. The KEGG pathway analysis revealed that the flavonoids and its derivatives are derived from the phenylpropanoid biosynthesis pathway from L-tyrosine with intermediate metabolites, such as Naringenin and Kaempferol. Phenolic compounds function by reducing the reactive oxygen species and stabbing the lipid peroxidation, thus acting as a potent antioxidant agent. They also play a vital role in prevention and treatment of chronic illnesses like neurodegenerative disorders. The integrated LC-MS/MS and transcriptomic analysis revealed that acacetin and procyanidin metabolites were some important secondary metabolites that could be responsible for the pharmacological properties exhibited by the plant, which is used to treat various inflammations and bronchodilator activity and has a cardiovascular protecting nature.

4.3. Simple Sequence Repeats Analysis

Molecular markers reveal the genetic relationship between the species as they are unaffected by the environment and have easy detection and high heritability [67,68]. These markers are widely used in plant improvement and genetic conservation [69]. A total of 36,197 SSRs were identified with highest number of mononucleotide motifs compared to di-, tri-, tetra-, penta-, and hexa-nucleotide motifs. SSRs with five tandem repeats were most common in *Sida cordifolia* with highly abundant A (35%) mono-nucleotide repeats and in tri-nucleotide repeats, and the highest frequency was observed in AAG (11.51%) followed by AAT (4.36%). The SSRs were discovered to help plant genetics for molecular reproduction as well as identification of candidate molecular markers for understanding the genetic variations in the *Sida* family.

5. Conclusions

To expedite molecular level studies in *Sida cordifolia* and the characterization of the root transcriptome to identify the unitranscripts responsible for the biosynthesis of secondary metabolites, since the root tissue of the plant contributesto several medicinal properties of the plant. The result suggests that flavonoids are the major secondary metabolites that are responsible for the medicinal properties exhibited by the plant. The assembled transcripts were validated using reverse transcription PCR method, and the expression levels of the key genes from the phenylpropanoid and flavonoid pathways wereobtainedusing qRT-PCR. The secondary metabolites identified from the metabolomic studies lead the way for understanding the molecular mechanism of pharmacological properties exhibited by *S.cordifolia*. This is the first study that integrates the molecular basis and metabolomics of this plant.

Supplementary Materials: The following supporting information can be downloaded at: https://www.mdpi.com/article/10.3390/genes13101909/s1, Figure S1: Top hit species distribution of *Sida cordifolia*. Figure S2: Gene ontology classification of*Sida cordifolia*. Figure S3: KEGG pathway analysis based on the metabolism pathways in *Sida cordifolia*. Figure S4: KEGG pathway analysis of phenylpropanoid biosynthesis. Figure S5: KEGG pathway analysis of flavonoid biosynthesis. Figure S6: SSR identification from root transcriptome of*Sida cordifolia*. Table S1: List of genes selected to validate gene expression and their primer sequences.

Author Contributions: Conceptualization, writing—review and editing, supervision, S.P.; software, P.N.; methodology, validation, writing—original draft preparation, D.P. All authors have read and agreed to the published version of the manuscript.

Funding: This research received no external funding.

Institutional Review Board Statement: Not applicable.

Informed Consent Statement: Not applicable.

Data Availability Statement: The datasets used in the study was deposited in the SRA under the SRA accession number PRJNA841821.

Acknowledgments: All authors acknowledge the RNA-Seq support by SRM-DBT platform and High-Performance Computing Cluster Facility (Genome Server) at SRM Institute of Science and Technology (SRMIST).

Conflicts of Interest: The authors declare no conflict of interest.

References

1. Prasathkumar, M.; Anisha, S.; Dhrisya, C.; Becky, R.; Sadhasivam, S. Therapeutic and Pharmacological Efficacy of Selective Indian Medicinal Plants—A Review. *Phytomed. Plus* **2021**, *1*, 100029. [CrossRef]
2. Ahmed, H.; Juraimi, A.S.; Swamy, M.K.; Ahmad-Hamdani, M.S.; Omar, D.; Rafii, M.Y.; Sinniah, U.R.; Akhtar, M.S. Botany, Chemistry, and Pharmaceutical Significance of Sida cordifolia: A Traditional Medicinal Plant. In *Anticancer Plants: Properties and Application*; Springer: Singapore, 2018; Volume 1, pp. 517–537.
3. Ghosal, S.; Chauhan, R.B.P.; Mehta, R. Alkaloids of Sida Cordifolia. *Phytochemistry* **1975**, *14*, 830–832. [CrossRef]
4. Ghosh, S.; Dutta, A. Chemical examination of *Sida cordifolia* Linn. *J. Indian Chem. Soc.* **1930**, *7*, 825–829.
5. Tamura, S.; Kaneko, M.; Shiomi, A.; Yang, G.M.; Yamaura, T.; Murakami, N. Unprecedented NES Non-Antagonistic Inhibitor for Nuclear Export of Rev from Sida Cordifolia. *Bioorg. Med. Chem. Lett.* **2010**, *20*, 1837–1839. [CrossRef]
6. Aminah, N.S.; Laili, E.R.; Rafi, M.; Rochman, A.; Insanu, M.; Tun, K.N.W. Secondary metabolite compounds from Sida genus and their bioactivity. *Heliyon* **2021**, *7*, 2405–8440. [CrossRef]
7. Momin, M.A.; Bellah, S.F.; Rahman, S.M.; Rahman, A.A.; Murshid, G.M.; Emran, T.B. Phytopharmacological evaluation of ethanol extract of Sida cordifolia L. roots. *Asian Pac. J. Trop. Biomed.* **2014**, *4*, 18–24. [CrossRef]
8. Silva, R.L.; Melo, G.B.D.; Melo, V.A.D.; Antoniolli, Â.R.; Michellone, P.R.T.; Zucoloto, S.; Castro e Silva, O.D. Effect of the aqueous extract of *Sida cordifolia* on liver regeneration after partial hepatectomy. *Acta Cir. Bras.* **2006**, *21*, 37–39. [CrossRef]
9. Köse, M.A.; Çetinsağ, N.; Gürcan, K. De Novo Transcriptome Assembly and SSR Marker Development in Apricot (Prunus Armeniaca). *Turk. J. Agric. For.* **2017**, *41*, 305–315. [CrossRef]
10. Fan, J.; Lou, Y.; Shi, H.; Chen, L.; Cao, L. Transcriptomic Analysis of Dark-Induced Senescence in Bermudagrass (CynodonDactylon). *Plants* **2019**, *8*, 614. [CrossRef]

11. Wu, T.; Luo, S.; Wang, R.; Zhong, Y.; Xu, X.; Lin, Y.; He, X.; Sun, B.; Huang, H. The First Illumina-Based de Novo Transcriptome Sequencing and Analysis of Pumpkin (Cucurbita MoschataDuch.) and SSR Marker Development. *Mol. Breed* **2014**, *34*, 1437–1447. [CrossRef]
12. Zhang, Y.; Yang, L.; Yang, J.; Hu, H.; Wei, G.; Cui, J.; Xu, J. Transcriptome and Metabolome Analyses Reveal Differences in Terpenoid and Flavonoid Biosynthesis in Cryptomeria fortunei Needles Across Different Seasons. *Front. Plant Sci.* **2022**, *13*, 862746. [CrossRef]
13. Fang, H.; Qi, X.; Li, Y.; Yu, X.; Xu, D.; Liang, C.; Li, W.; Liu, X. De Novo Transcriptomic Analysis of Light-Induced Flavonoid Pathway, Transcription Factors in the Flower Buds of. *Trees* **2020**, *34*, 267–283. [CrossRef]
14. Khan, I.; Rahman, H.; Abd El-Salam, N.M.; Tawab, A.; Hussain, A.; Khan, T.A.; Khan, U.A.; Qasim, M.; Adnan, M.; Azizullah, A.; et al. *Punica granatum* peel extracts: HPLC fractionation and LC MS analysis to quest compounds having activity against multidrug resistant bacteria. *BMC Complement Altern. Med.* **2017**, *17*, 247. [CrossRef]
15. Lateef, A.; Prabhudas, S.K.; Natarajan, P. RNA Sequencing and de Novo Assembly of Solanum Trilobatum Leaf Transcriptome to Identify Putative Transcripts for Major Metabolic Pathways. *Sci. Rep.* **2018**, *8*, 15375. [CrossRef]
16. Joshi, N.A.; Fass, J.N. Sickle: A Sliding-Window, Adaptive, Quality-Based Trimming Tool for FastQ files (Version 1.33) [Software]. 2011. Available online: https://github.com/najoshi/sickle (accessed on 23 September 2022).
17. Martin, M. Cutadapt removes adapter sequences from high-throughput sequencing reads. *EMBnet J.* **2011**, *17*, 10–12. [CrossRef]
18. Li, W.; Godzik, A. CD-HIT: A fast program for clustering and comparing large sets of protein or nucleotide sequences. *Bioin formatics* **2006**, *22*, 1658–1659. [CrossRef]
19. Chen, L.; Zhang, Y.-H.; Wang, S.; Zhang, Y.; Huang, T.; Cai, Y.-D. Prediction and Analysis of Essential Genes Using the Enrichments of Gene Ontology and KEGG Pathways. *PLoS ONE* **2017**, *12*, e0184129. [CrossRef]
20. Du, L.; Zhang, C.; Liu, Q.; Zhang, X.; Yue, B.; Hancock, J. Krait: An Ultrafast Tool for Genome-Wide Survey of Microsatellites and Primer Design. *Bioinformatics* **2018**, *34*, 681–683. [CrossRef]
21. Song, X.; Yang, Q.; Bai, Y.; Gong, K.; Wu, T.; Yu, T.; Pei, Q.; Duan, W.; Huang, Z.; Wang, Z.; et al. Comprehensive analysis of SSRs and database construction using all complete gene-coding sequences in major horticultural and representative plants. *Hortic Res.* **2021**, *8*, 122. [CrossRef]
22. Patro, R.; Duggal, G.; Love, M.I.; Irizarry, R.A.; Kingsford, C. Salmon Provides Fast and Bias-Aware Quantification of Transcript Expression. *Nat. Methods* **2017**, *14*, 417–419. [CrossRef]
23. Rao, X.; Huang, X.; Zhou, Z.; Lin, X. An improvement of the 2 (– delta delta CT) methods for quantitative real-time poly merase chain reaction data analysis. *Biostat. Bioinform. Biomath.* **2013**, *3*, 71–85.
24. Zhang, J.; Chen, W.; Ju, W.; Liu, F.; Tan, H. Determination of plasma concentration of cinnamic acid by LC-MS-MS and study of pharmacokinetics in healthy volunteers after mailuoning injection. *Zhongguo Zhong Yao Za Zhi Zhongguo Zhongyao Zazhi China J. Chin. Mater. Med.* **2010**, *35*, 1887–1889.
25. Romero, C.; Brenes, M.; García, P.; Garrido, A. Hydroxytyrosol 4-beta-D-glucoside, an important phenolic compound in olive fruits and derived products. *J. Agric. Food Chem.* **2002**, *50*, 3835–3839. [CrossRef]
26. Al Kadhi., O.; Melchini, A.; Mithen, R.; Saha, S. Development of a LC-MS/MS Method for the Simultaneous Detection of Tricarboxylic Acid Cycle Intermediates in a Range of Biological Matrices. *J. Anal. Methods Chem.* **2017**, *2017*, 5391832. [CrossRef]
27. Li, Y.; Guang, C.; Zhao, N.; Feng, X.; Qiu, F. LC–MS/MS Method for Simultaneous Determination of Linarin and Its Metabolites in Rat Plasma and Liver Tissue Samples: Application to Pharmacokinetic and Liver Tissue Distribution Study After Oral Administration of Linarin. *Molecules* **2019**, *24*, 3342. [CrossRef]
28. Solomon, K.S.; Amoah.; Maique, W.; Biavatti, A. LC-MS validated method for quantification of rosmarinic acid and sesquiterpene lactones in *Hedyosmum brasiliense*. *J. Chromatogr. Sci.* **2018**, *56*, 812–818.
29. Cheruvu, H.S.; Yadav, N.K.; Valicherla, G.R.; Arya, R.K.; Hussain, Z.; Sharma, C.; Arya, K.R.; Singh, R.K.; Datta, D.; Gayen, J.R. LC-MS/MS method for the simultaneous quantification of luteolin, wedelolactone and apigenin in mice plasma using hansen solubility parameters for liquid-liquid extraction: Application to pharmacokinetics of Eclipta alba chloroform fraction. *J. Chromatogr. B Analyt. Technol. Biomed Life Sci.* **2018**, *1801*, 76–86. [CrossRef]
30. Espíndola, K.M.M.; Ferreira, R.G.; Narvaez, L.E.M.; Rosario, A.C.R.S.; de Silva, A.H.M.; Silva, A.G.B.; Vieira, A.P.O.; Monteiro, M.C. Chemical and Pharmacological Aspects of Caffeic Acid and Its Activity in Hepatocarcinoma. *Front. Oncol.* **2019**, *9*, 541. [CrossRef]
31. Silva dos Santos, J.; Goncalves Cirino, J.P.; de Oliveira Carvalho, P.; Ortega, M.M. The Pharmacological Action of Kaempferol in Central Nervous System Diseases: A Review. *Front. Pharmacol.* **2021**, *11*, 565700. [CrossRef]
32. Batiha, G.E.-S.; Beshbishy, A.M.; Ikram, M.; Mulla, Z.S.; El-Hack, M.E.A.; Taha, A.E.; Algammal, A.M.; Elewa, Y.H.A. The Pharmacological Activity, Biochemical Properties, and Pharmacokinetics of the Major Natural Polyphenolic Flavonoid: Quercetin. *Foods* **2020**, *9*, 374. [CrossRef]
33. Rue, E.A.; Rush, M.D.; van Breemen, R.B. Procyanidins: A comprehensive review encompassing structure elucidation via mass spectrometry. *Phytochem. Rev.* **2018**, *17*, 1–16. [CrossRef] [PubMed]
34. El Sayed, A.M.; Basam, S.M.; Marzouk, H.S.; El-Hawary, S. LC–MS/MS and GC–MS profiling as well as the antimicrobial effect of leaves of selected Yucca species introduced to Egypt. *Sci. Rep.* **2020**, *10*, 17778. [CrossRef] [PubMed]
35. Cruz, M.S.; Barroso, S.C.; Navoni, J.A.; Teles, M.M.R.S.; Barbosa-Filho, J.M.; Rocha, H.A.D.O.; do Amaral, V.S. Effect of Hecogenin on DNA Instability. *Toxicol. Rep.* **2016**, *3*, 539–543. [CrossRef] [PubMed]

36. Diantini, A.; Subarnas, A.; Lestari, K.; Halimah, E.; Susilawati, Y.; Supriyatna, S.; Julaeha, E.; Achmad, T.H.; Suradji, E.W.; Yamazaki, C.; et al. Kaempferol-3-O-Rhamnoside Isolated from the Leaves of *SchimaWallichiiKorth.* Inhibits MCF-7 Breast Cancer Cell Proliferation through Activation of the Caspase Cascade Pathway. *Oncol. Lett.* **2012**, *3*, 1069–1072. [CrossRef]
37. Juang, Y.P.; Liang, P.H. Biological and Pharmacological Effects of Synthetic Saponins. *Molecules* **2020**, *27*, 4974. [CrossRef]
38. Tatsimo, S.J.N.; Tamokou, J.D.D.; Havyarimana, L.; Csupor, D.; Forgo, P.; Hohmann, J.; Kuiate, J.-R.; Tane, P. Antimicrobial and antioxidant activity of kaempferol rhamnoside derivatives from *Bryophyllumpinnatum*. *BMC Res. Notes* **2012**, *5*, 158. [CrossRef]
39. He, J.; Feng, Y.; Ouyang, H.Z.; Yu, B.; Chang, Y.X.; Pan, G.X.; Dong, G.Y.; Wang, T.; Gao, X.M. A sensitive LC-MS/MS method for simultaneous determination of six flavonoids in rat plasma: Application to a pharmacokinetic study of total flavonoids from mulberry leaves. *J. Pharm. Biomed Anal.* **2013**, *84*, 189–195.
40. Sharma, U.K.; Sharma, A.K.; Gupta, A.; Kumar, R.; Pandey, A.; Pandey, A.K. Pharmacological activities of cinnamaldehyde and eugenol: Antioxidant, cytotoxic and anti-leishmanial studies. *Cell Mol. Biol.* **2017**, *63*, 73–78. [CrossRef]
41. Grabarczyk, M.; Wińska, K.; Mączka, W.; Potaniec, B.; Anioł, M. Loliolide-the most ubiquitous lactone. *Acta Univers. Lodz. Folia Biol.Oecol.* **2015**, *11*, 1–8. [CrossRef]
42. Bae, J.; Kim, N.; Shin, Y.; Kim, S.-Y.; Kim, Y.-J. Activity of catechins and their applications. *Biomed. Dermatol.* **2020**, *4*, 8.
43. Naveed, M.; Hejazi, V.; Abbas, M.; Kamboh, A.A.; Khan, G.J.; Shumzaid, M.; Ahmad, F.; Babazadeh, D.; Fang, X.F.; Modarresi-Ghazani, F.; et al. Chlorogenic acid (CGA): A pharmacological review and call for further research. *Biomed. Pharmacother.* **2018**, *97*, 67–74. [CrossRef]
44. Serreli, G.; Sayec, M.L.; Thou, E.; Lacour, C.; Diotallevi, C.; Dhunna, M.A.; Deiana, M.; Spencer, J.P.E.; Corona, G. Ferulic Acid Derivatives and Avenanthramides Modulate Endothelial Function through Maintenance of Nitric Oxide Balance in HUVEC Cells. *Nutrients* **2021**, *12*, 2026. [CrossRef]
45. Benali, T.; Bakrim, S.; Ghchime, R.; Benkhaira, N.; Omari, N.E.; Balahbib, A.; Taha, D.; Zengin, G.; Hasan, M.M.; Bibi, S.; et al. Pharmacological insights into the multifaceted biological properties of quinic acid. *Biotechnol. Genet. Eng. Rev.* **2022**, *19*, 1–30. [CrossRef]
46. Semaming, Y.; Pannengpetch, P.; Chattipakorn, S.C.; Chattipakorn, N. Pharmacological properties of protocatechuic Acid and its potential roles as complementary medicine. *Evid. Based Complement Altern. Med. Ecam* **2015**, 593902. [CrossRef]
47. Hong, Y.; Wang, Z.; Barrow, C.J.; Dunshea, F.R.; Suleria, H.A.R. High-Throughput Screening and Characterization of Phenolic Compounds in Stone Fruits Waste by LC-ESI-QTOF-MS/MS and Their Potential Antioxidant Activities. *Antioxidants.* **2021**, *10*, 234. [CrossRef]
48. Lee, D.E.; Lee, K.W.; Byun, S.; Jung, S.K.; Song, N.; Lim, S.H.; Heo, Y.S.; Kim, J.E.; Kang, N.J.; Kim, B.Y.; et al. 7,3′,4′-Trihydroxyisoflavone, a metabolite of the soy isoflavone daidzein, suppresses ultraviolet B-induced skin cancer by targeting Cot and MKK4. *J. Biol. Chem.* **2011**, *22*, 14246–14256. [CrossRef]
49. Reddy, M.K.; Gupta, S.K.; Jacob, M.R.; Khan, S.I.; Ferreira, D. Antioxidant, antimalarial and antimicrobial activities of tannin-rich fractions, ellagitannins and phenolic acids from *Punica granatum* L. *Planta Med.* **2007**, *73*, 461–467. [CrossRef]
50. Salehi, B.; Venditti, A.; Sharifi-Rad, M.; Kręgiel, D.; Sharifi-Rad, J.; Durazzo, A.; Lucarini, M.; Santini, A.; Souto, E.B.; Novellino, E.; et al. The Therapeutic Potential of Apigenin. *Int. J. Mol. Sci.* **2019**, *15*, 1305. [CrossRef]
51. Kahkeshani, N.; Farzaei, F.; Fotouhi, M.; Alavi, S.S.; Bahramsoltani, R.; Naseri, R.; Momtaz, S.; Abbasabadi, Z.; Rahimi, R.; Farzaei, M.H.; et al. Pharmacological effects of gallic acid in health and diseases: A mechanistic review. *Iran. J. Basic Med. Sci.* **2019**, *22*, 225–237.
52. Zhou, L.; Sun, J.; Zhang, T.; Tang, Y.; Liu, J.; Gao, C.; Zhai, Y.; Guo, Y.; Feng, L.; Zhang, X.; et al. Comparative Transcriptome Analyses of Different *Rheum officinale* Tissues Reveal Differentially Expressed Genes Associated with Anthraquinone, Catechin, and Gallic Acid Biosynthesis. *Genes* **2022**, *13*, 1592. [CrossRef]
53. Kong, J.Q. Phenylalanine Ammonia-Lyase, a Key Component Used for Phenylpropanoids Production by Metabolic Engineering. *RSC Adv.* **2015**, *5*, 62587–62603. [CrossRef]
54. Chun, H.J.; Bae, D.; Cho, H.M.; Lee, S.H.; Jin, B.J.; Yun, D.-J.; Hong, Y.-S.; Kim, M.C. Lignin biosynthesis genes play critical roles in the adaptation of *Arabidopsis* plants to high-salt stress. *Plant Signal. Behav.* **2019**, *14*, 1625697. [CrossRef]
55. Yamada, S.; Nabe, K.; Izuo, N.; Nakamichi, K.; Chibata, I. Production of l-Phenylalanine from trans-Cinnamic Acid with Rhodotorula glutinis Containing l-Phenylalanine Ammonia-Lyase Activity. *Appl. Environ. Microbiol.* **1981**, *42*, 773–778. [CrossRef]
56. Winkel-Shirley, B. Flavonoid Biosynthesis. A Colorful Model for Genetics, Biochemistry, Cell Biology, and Biotechnology. *Plant Physiol.* **2001**, *126*, 485–493. [CrossRef]
57. Hou, L.; Wang, L.; Wu, X.; Gao, W.; Zhang, J.X.; Huang, C.Y. Expression patterns of two *pal* genes of *Pleurotus ostreatus* across developmental stages and under heat stress. *BMC Microbiol.* **2019**, *19*, 231. [CrossRef]
58. Bagal, U.R.; Leebens-Mack, J.H.; Lorenz, W.W.; Dean, J.F. The phenylalanine ammonia lyase (PAL) gene family shows a gymnosperm-specific lineage. *BMC Genom.* **2012**, *13* (Suppl. 3), S1. [CrossRef]
59. Umezawa, T. The cinnamate/monolignol pathway. *Phytochem. Rev.* **2010**, *9*, 1–17. [CrossRef]
60. Kawasaki, T.; Koita, H.; Nakatsubo, T.; Hasegawa, K.; Wakabayashi, K.; Takahashi, H.; Umemura, K.; Umezawa, T.; Shimamoto, K. Cinnamoyl-CoA reductase, a key enzyme in lignin biosynthesis, is an effector of small GTPase Rac in defense signaling in rice. *Proc. Natl. Acad. Sci. USA* **2006**, *103*, 230–235. [CrossRef]
61. Iwashina, T. The Structure and Distribution of the Flavonoids in Plants. *J. Plant Res.* **2000**, *113*, 287–299. [CrossRef]

62. Falcone Ferreyra, M.L.; Rius, S.P.; Casati, P. Flavonoids: Biosynthesis, biological functions, and biotechnological applications. *Front. Plant Sci.* **2012**, *28*, 222. [CrossRef]
63. Azeem, M.; Hanif, M.; Mahmood, K.; Ameer, N.; Chughtai, F.R.S.; Abid, U. An insight into anticancer, antioxidant, antimicrobial, antidiabetic and anti-inflammatory effects of quercetin: A review. *Polym. Bull.* **2022**, 1–22. [CrossRef] [PubMed]
64. David, A.V.A.; Arulmoli, R.; Parasuraman, S. Overviews of Biological Importance of Quercetin: A Bioactive Flavonoid. *Pharmacogn. Rev.* **2016**, *10*, 84–89.
65. Saeedi-Boroujeni, A.; Mahmoudian-Sani, M.R. Anti-inflammatory potential of Quercetin in COVID-19 treatment. *J. Inflamm.* **2021**, *18*, 3. [CrossRef] [PubMed]
66. Cherrak, S.A.; Merzouk, H.; Mokhtari-Soulimane, N. Potential bioactive glycosylated flavonoids as SARS-CoV-2 main protease inhibitors: A molecular docking and simulation studies. *PLoS ONE* **2020**, *15*, e0240653. [CrossRef]
67. Huang, C.J.; Chu, F.H.; Huang, Y.S.; Tu, Y.C.; Hung, Y.M.; Tseng, Y.H.; Pu, C.E.; Hsu, C.T.; Chao, C.H.; Chou, Y.S.; et al. SSR individual identification system construction and population genetics analysis for *Chamaecyparis formosensis*. *Sci. Rep.* **2022**, *12*, 4126.
68. Yasodha, R.; Vasudeva, R.; Balakrishnan, S.; Sakthi, A.R.; Abel, N.; Binai, N.; Rajashekar, B.; Bachpai, V.K.W.; Pillai, C.; Dev, S.A. Draft Genome of a High Value Tropical Timber Tree, Teak (*Tectona Grandis* L. f): Insights into SSR Diversity, Phylogeny and Conservation. *DNA Res.* **2018**, *25*, 409–419. [CrossRef]
69. Padmanabhan, D.; Lateef, A.; Natarajan, P.; Palanisamy, S. *De novo* transcriptome analysis of *Justicia adhatoda* reveals candidate genes involved in major biosynthetic pathway. *Mol. Biol. Rep.* **2022**, 1–8. [CrossRef]

genes

MDPI

Article

Establishment of an Efficient In Vitro Propagation of *Cnidium officinale* Makino and Selection of Superior Clones through Flow Cytometric Assessment of DNA Content

Hyung-Eun Kim [1,2], Jong-Eun Han [1], Hyoshin Lee [3], Hosakatte Niranjana Murthy [1,4], Hyuk-Joon Kwon [5], Gun-Myung Lee [5] and So-Young Park [1,*]

1 Department of Horticultural Science, Chungbuk National University, Cheongju 28644, Korea
2 Korea Disease Control and Prevention Agency, Osong-eup, Cheongju 28159, Korea
3 Department of Forest Genetic Resources, National Institute of Forest Science, 39 Onjeong-ro, Suwon 16631, Korea
4 Department of Botany, Karnatak University, Dharwad 580003, India
5 Food Science R&D Center, Kolmar BNH Co., Seocho-gu, Seoul 30003, Korea
* Correspondence: soypark7@cbnu.ac.kr; Tel.: +82-432-612-531

Abstract: *Cnidium officinale* is a valuable medicinal plant cultivated in Asia for its rhizomes. This study reports the in vitro regeneration of *Cnidium officinale* plants and the induction of rhizomes from microshoots. The rhizomatous buds of *Cnidium officinale* induced multiple shoots on Murashige and Skoog (MS) medium supplemented with 0.5 mg L^{-1} BA, which led to the regeneration of plants within four weeks of culture. After four weeks of culture, the plants were assessed for fresh weight, the number of leaves, the number of roots, and the length of roots to compare the performance of the different clones. The clones with good growth characteristics were selected with the aid of a flow cytometric analysis of 2C nuclear DNA content. The plants bearing high DNA values showed better growth characteristics. Various factors, namely, sucrose concentration (30, 50, 70, and 90 g L^{-1}), ABA (0, 0.5, 1.0, and 2.0 mg L^{-1}), the synergistic effects of BA (1.0 mg L^{-1}) + NAA (0.5 mg L^{-1}) and BA (1.0 mg L^{-1}) + NAA (0.5 mg L^{-1}) + ABA (1.0 mg L^{-1}) with or without activated charcoal (1 g L^{-1}), and light and dark incubation were tested on rhizome formation from microshoots. The results of the above experiments suggest that MS medium supplemented with 50 g L^{-1} sucrose, 1.0 mg L^{-1} ABA, and 1 g L^{-1} AC is good for the induction of rhizomes from the shoots of *Cnidium officinale*. Plantlets with rhizomes were successfully transferred to pots, and they showed 100% survival.

Keywords: flow cytometry; medicinal plant; multiple shoots; plant regeneration; rhizomes

Citation: Kim, H.-E.; Han, J.-E.; Lee, H.; Murthy, H.N.; Kwon, H.-J.; Lee, G.-M.; Park, S.-Y. Establishment of an Efficient In Vitro Propagation of *Cnidium officinale* Makino and Selection of Superior Clones through Flow Cytometric Assessment of DNA Content. *Genes* **2022**, *13*, 1815. https://doi.org/10.3390/genes13101815

Academic Editors: Hakim Manghwar and Wajid Zaman

Received: 19 September 2022
Accepted: 6 October 2022
Published: 8 October 2022

1. Introduction

Cnidium officinale Makino is an important herbaceous medicinal plant that belongs to the Apiaceae family. It is widely cultivated in China, Japan, and Korea for its rhizomes. Dried rhizomes are used in Chinese, Japanese, and Korean medicine as a tonic to improve blood circulation, overcome the problems of inflammation, and treat women's menstrual problems [1]. The rhizomes of *Cnidium officinale* have been reported to possess several bioactive compounds, including polysaccharides [2–4] and volatile alkylphthalide derivatives [5,6]. Rhizome extracts and isolated compounds have been reported to exhibit various pharmacological activities, including antiangiogenic [7], anticancer [8–10], antidiabetic [11], anti-inflammatory [12,13], immunomodulatory [2], analgesic [14], and antimicrobial [15,16] effects. Yang et al. [17] demonstrated that the ethanolic rhizome extracts of *Cnidium officinale* are usable in the treatment of ischemic injury and, thus, that rhizomes could be used as recovery medicine after plastic surgery, for the disturbance of peripheral blood circulation, and for overcoming postmenopausal problems. Kim [18] and Cha [19] showed the positive effects of *Cnidium officinale* extracts on the repression of melanogenesis and in decreasing hyperpigmentation, as well as its use as a cosmetic ingredient.

Cnidium officinale does not set seeds; therefore, the multiplication of this plant only occurs through the division of rhizomes, and there is a high risk of virus dissemination affecting yield and quality during cultivation [1]. Therefore, in vitro propagation methods have been developed for this species using shoot tip culture and the induction of somatic embryogenesis using inflorescence and flowers [1,20,21]. However, plant regeneration efficacy is low and inconsistent. Further, *Cnidium officinale* is a rhizomatous plant, and the in vitro induction of rhizomes from regenerated shoots is desirable. In vitro rhizome induction has been reported in many bulbous plants, such as *Kaempferia galanga*, *Kaempfria rotunda* [22], *Curcuma longa* [23], and *Bambosa bambos* var. *gigantea* [24]. The in vitro production of rhizomes has commercial potential for the large-scale generation of pathogen-free seed material for propagation and utilization by the pharmaceutical industry. Further, in vitro rhizome production is also useful for storage, transport, and germplasm preservation. Therefore, in the present study, an efficient plant regeneration protocol was developed in *Cnidium officinale* using rhizome bud explants. The regenerated plants/clones were assessed for their DNA content using the flow cytometric method. Additionally, the DNA content of the regenerated clones was correlated with growth characteristics. In addition, a reliable method for in vitro rhizome induction from regenerated shoots was established.

2. Materials and Methods

2.1. Preparation of Plant Material and Induction of Microshoots from Rhizome Buds

The rhizomes of *Cnidium officinale* Makino growing in the greenhouse of Chungbuk National University, Korea, were used in the experiment. For the establishment of the in vitro culture of *Cnidium officinale*, buds (~1.5 cm long) collected from the rhizome were washed in running tap water for 1 h, and the bacteria and fungi were primarily removed by immersing them in a sterilizer containing 1000 ppm Physan 20 (Maril, Tustin, CA, USA) for 20 s. Subsequently, surface sterilization was conducted for 20 s with 70% ethanol and for 10 min in a solution with added Tween 20, surfactant (ThermoFisher, Seoul, Korea), and 2% NaOCl. The explants were thoroughly washed five to six times with sterile double-distilled water, and they were cultured on Murashige and Skoog [25] medium with 0.5 mg L^{-1} benzyl adenine (BA), 30 g L^{-1} sucrose, and 2.4 g L^{-1} Gelrite (Duchefa Biochimie, Haarlem, the Netherlands). The cultures were incubated at 25 ± 1 °C under a 16 h photoperiod under the white fluorescent light of 35 $\mu\ m^{-2}\ s^{-1}$. Subculture was conducted at four week intervals. The cultured explants developed shoots, simultaneously developing roots at the base of the shoots. Plantlets were maintained in MS medium with 0.5 mg L^{-1} BA through repeated subculturing. We compared the regeneration rate of ten different clones, which were available in our germplasm collection, and calculated the shoot proliferation rate in subsequent subcultures for up to three cycles.

2.2. Flow Cytometry

Flow cytometry was performed to determine the amount of 2C DNA in the in vitro-grown *Cnidium officinale* plants. Briefly, a fresh staining solution was prepared on the day of analysis by mixing the staining buffer with PI solution and RNase solution. Leaf discs (0.5 cm^2) were prepared from the leaves of *Cnidium officinale* and *Nicotiana tabacum* cv. xanthi (internal standard) plants and placed in a Petri dish containing 200 μL of extraction buffer. Nuclei were isolated from the leaf discs using a CyStain PI Absolute P Kit (Partec, Görlitz, Germany), according to the manufacturer's instructions. The solution was filtered through a 50 μm nylon mesh. Then, 800 μL of staining solution was added to the filtered solution in order to label the DNA with a fluorescent dye. The samples were incubated in the staining solution on a stationary surface in the dark for 20 min and then analyzed using CytoFLEX (Beckman Coulter Inc., Fullerton, CA, USA). At least 5000 nuclei were analyzed per sample.

2.3. *Assessment of in Vitro Growth of Regenerated Plants*

Ten plants regenerated in vitro from each clone of *Cnidium officinale* were selected, and they were re-cultured on regeneration medium (MS + 0.5 mg L^{-1} benzyl adenine, 30 g L^{-1} sucrose, and 2.4 g L^{-1} Gelrite) and maintained four weeks. After four weeks of culture, the plants were assessed for their plant height, their fresh weight, the number of leaves, and the number and length of roots. Subsequently, the plants were re-cultured for another passage of a four-week cycle, and the number of plantlets in each cycle and the shoot proliferation rate (shoot proliferation rate = the total number of shoots after subculture vs. the total number of shoots before subculture) were estimated.

2.4. *In Vitro Induction of Rhizome*

The microshoots (2 cm in length) obtained after the four passages of re-culturing were harvested and transferred to MS media supplemented with 30, 50, 70, and 90 g L^{-1} sucrose, or 0, 0.5, 1.0, and 2.0 mg L^{-1} abscisic acid (ABA) and 50 g L^{-1} sucrose for rhizome induction. Similarly, the effect of 0.5 mg L^{-1} naphthalene acetic acid (NAA) in combination with 1.0 mg L^{-1} BA alone, or in combination with 1.0 mg L^{-1} ABA with or without 1 g L^{-1} activated charcoal (AC) was also tested on the rhizome induction. In these treatments, the medium was supplied with 50 g L^{-1} sucrose. Depending on the experiment, the cultures were either kept in continuous dark conditions or in 12 h light (35 µmol m^{-2} s^{-1}) and 8 h dark conditions at 25 ± 1 °C temperature. The cultures were maintained for up to eight weeks, and data were collected.

2.5. *Acclimatization*

One day before acclimatization, the lid of the culture vessel was opened by ~20% for adaptation to the ex vitro conditions. Then, the plants with rhizomes regenerated in vitro were harvested from the vessel, carefully washed with distilled water, and planted in wet Growfoam (Horticubes, Smithers-Oasis, Kent, OH, USA). The plants in Growfoam were placed in a plastic container (25 × 25 × 3 cm), and the water was changed every week. The plants were fertilized with 1.0 g L^{-1} Hyponex solution once per month. The transplanted plants showed 100% survival. After 3 months of growing ex vitro, the plants were transplanted and cultivated in plastic pots (diameter 10 cm) filled with an artificial soil mixture (peat moss 1:1 perlite and vermiculite 1). The potted plants were fertilized with 1.0 g·L^{-1} Hyponex solution (N20:P20:K20) once per month.

2.6. *Statistical Analysis*

A one-way analysis of variance (ANOVA) was performed to determine significant differences between the results of shoot/plant regeneration, flow cytometric data, the assessment of growth characteristics, and the in vitro rhizome induction experiments. The statistical significance of the differences between the mean values was then assessed using Duncan's multiple range test at $p < 0.05$. All statistical analyses were performed using SAS 9.4 software (SAS Institute Inc., Cary, NC, USA).

3. Results

3.1. *Establishment of Culture*

The *Cnidium officinale* rhizome buds (Figure 1a) were cultured on MS medium, with 0.5 mg L^{-1} BA involved in swelling one week after inoculation, and produced multiple shoots in subsequent weeks (Figure 1b). The roots simultaneously formed at the basal shoots on the same medium, and plantlets with well-developed shoots and roots were obtained after 4 weeks of incubation (Figure 1c). The shoot proliferation rate varied from 1.5 to 3.4 times across the different clones. Among the ten clones used (clone nos. 1, 2, 5, 6, 8, 11, 14, 15, 22, and 25), clones 5 and 6 performed well and regenerated 16–18 shoots per explant after the third cycle (Figure 2).

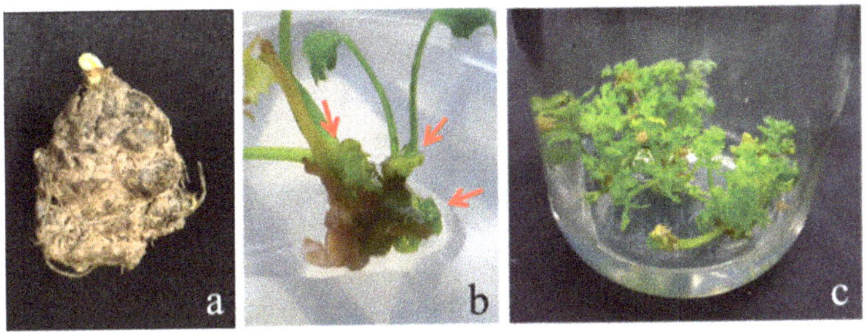

Figure 1. In vitro regeneration of *Cnidium officinale* plants on MS medium with 0.5 mg·L^{-1} benzyl adenine (BA) and 30 g·L^{-1} sucrose: (**a**) rhizome, (**b**) shoot development (arrows denote multiple shoots), and (**c**) plantlets.

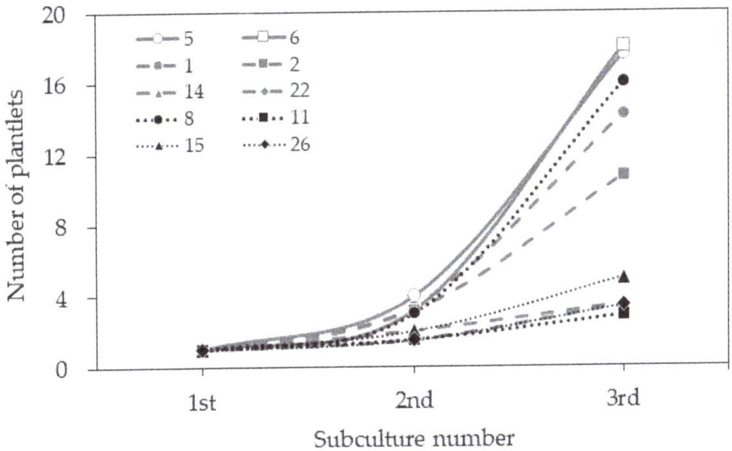

Figure 2. The number of plants regenerated from different clones of *Cnidium officinale* on MS medium with 5 mg·L^{-1} benzyl adenine (BA) and 30 g·L^{-1} sucrose over three 4-week passages of subculture.

3.2. Flow Cytometric Analysis of DNA Content in Different Clones of Cnidium officinale *and Selection of Superior Clones with the Highest Regeneration Potential*

According to the chromosome count database (CCDB, http://ccdb.tau.ac.il/ (accessed on 19 September 2022) [26], the chromosome number of *Cnidium officinale* populations/clones (germplasm) is highly variable, and the chromosome numbers of 2n = 22, 33, 44 have been reported for this species. The chromosome count of the available *Cnidium officinale* germplasm/clones was not known; therefore, we carried out a flow cytometric analysis of the 2C DNA content of the different clones used in the current experiment and selected the superior clones for further regeneration experiments and utilization. The flow cytometric analysis data of the *Cnidium officinale* clones used in the current experiment are presented in Figure 3 and Table 1. The DNA values varied across the different clones, ranging from 7.68 to 8.60 pg/2C (Table 1). Based on such data, the clones were classified into clones with a high DNA content (HD, clones 5 and 6), clones with a moderate DNA content (MD, clones 1, 2, 14, and 22), clones with a low DNA content (LD, clones 8, 11, 15, and 26). The clones with HD, MD, and LD were compared based on their regeneration potential on MS medium with 0.5 mg L^{-1} BA by using rhizome bud cultures. The clones with a higher DNA content (clones 5 and 6) showed the highest regeneration potential in terms of the regeneration rate (Figure 2).

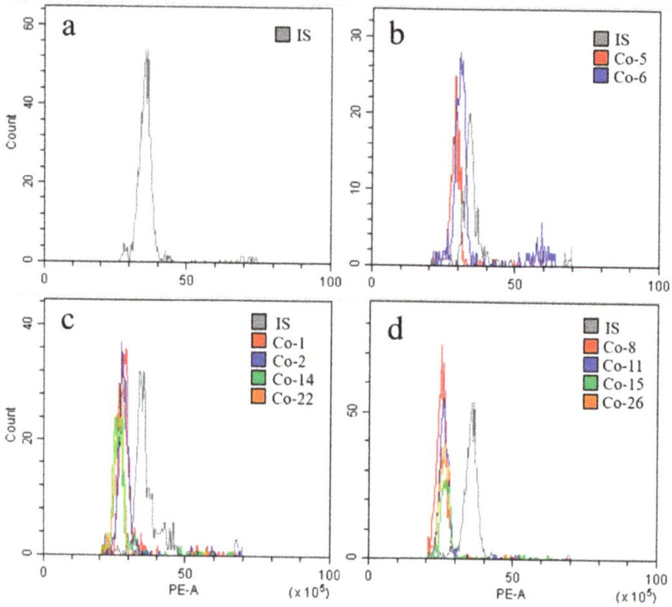

Figure 3. Analysis of 2C DNA content in different clones of *Cnidium officinale* plants using flow cytometry: histograms showing the DNA content of *Nicotiana tabacum* cv. xanthi for internal standard (IS) (**a**); clones with high DNA content (HD) (**b**); clones with moderate DNA content (MD) (**c**); and clones with low DNA content (LD) (**d**).

Table 1. DNA content of ten in vitro-grown clones of *Cnidium officinale*.

Group	Clone	Median	CV	DNA Content [z] (pg/2C)
HD [y]	5	2,988,019.06	5.9%	8.60 ± 0.15a [x]
HD	6	2,961,047.90	6.0%	8.52 ± 0.20ab
MD	1	2,816,142.06	5.8%	8.10 ± 0.10abc
MD	2	2,869,026.08	5.8%	8.26 ± 0.41abc
MD	14	2,779,891.16	6.0%	8.00 ± 0.24abc
MD	22	2,683,071.82	6.8%	7.72 ± 0.28bc
LD	8	2,634,563.66	6.8%	7.58 ± 0.23c
LD	11	2,602,247.30	6.3%	7.49 ± 0.20c
LD	15	2,638,274.26	6.7%	7.59 ± 0.38c
LD	26	2,636,395.84	6.1%	7.68 ± 0.18c

[z] Internal standard for calculating DNA content is *Nicotiana tabacum* cv. xanthi (10.07 pg/2C). [y] HD: high DNA content, MD: moderate DNA content, LD: low DNA content. [x] Different letters indicate significantly different values at $p < 0.05$ according to Duncan's multiple range test ($n = 5$).

3.3. Comparison of Growth Characteristics of Different Clones

The microshoots of several clones with known 2C DNA values were subcultured on MS medium with 0.5 mg L^{-1} BA and maintained for four weeks in culture to assess their growth characteristics and performance during in vitro propagation; the data are presented in Table 2. In terms of overall growth performance, i.e., fresh weight, the number of leaves, and the number of roots, the clones with higher DNA values performed well, and clone 5 was the best performer. A correlation analysis of the several clones with different DNA amounts with growth characteristics revealed that there was a positive correlation between the clones with the highest DNA content and propagation rate, the number of leaves, and the number of roots (Figure 4).

Table 2. Growth characteristics of ten in vitro-grown clones of *Cnidium officinale*.

Clone	Plant Height (mm)	Fresh Weight (mg/Plantlet)	No. of Leaves (ea/Plantlet)	No. of Roots (ea/Plantlet)	Root Length (mm)
1	24.5 ± 0.9bc [z]	168.0 ± 12.1ab	2.4 ± 0.3	3.6 ± 0.4bc	11.1 ± 1.3a
2	18.9 ± 0.7ef	160.0 ± 18.5ab	2.2 ± 0.2	2.4 ± 0.3c	4.6 ± 0.3d
5	21.4 ± 1.2cde	230.0 ± 23.3a	2.6 ± 0.5	8.3 ± 0.3a	6.4 ± 0.9cd
6	22.1 ± 1.0bcde	198.0 ± 18.6ab	2.8 ± 0.2	8.3 ± 1.0a	10.3 ± 0.6ab
8	19.9 ± 0.9def	206.0 ± 14.6a	2.1 ± 0.2	2.3 ± 0.6c	5.9 ± 1.7cd
11	17.6 ± 0.8f	202.0 ± 13.4a	2.4 ± 0.2	3.1 ± 0.9bc	6.3 ± 0.5cd
14	25.3 ± 1.6ab	204.0 ± 23.4a	2.2 ± 0.3	3.3 ± 0.5bc	7.2 ± 1.3cd
15	20.8 ± 1.2def	210.0 ± 29.4a	2.3 ± 0.2	3.6 ± 0.5bc	8.0 ± 0.7bc
22	27.9 ± 1.2a	208.0 ± 15.3a	2.4 ± 0.3	2.4 ± 0.3c	8.2 ± 1.0abc
26	22.4 ± 1.3bcd	134.0 ± 15.1b	2.0 ± 0.1	5.3 ± 0.8b	7.9 ± 1.2bc

[z] Different letters indicate significantly different values at $p < 0.05$ according to Duncan's multiple range test ($n = 5$).

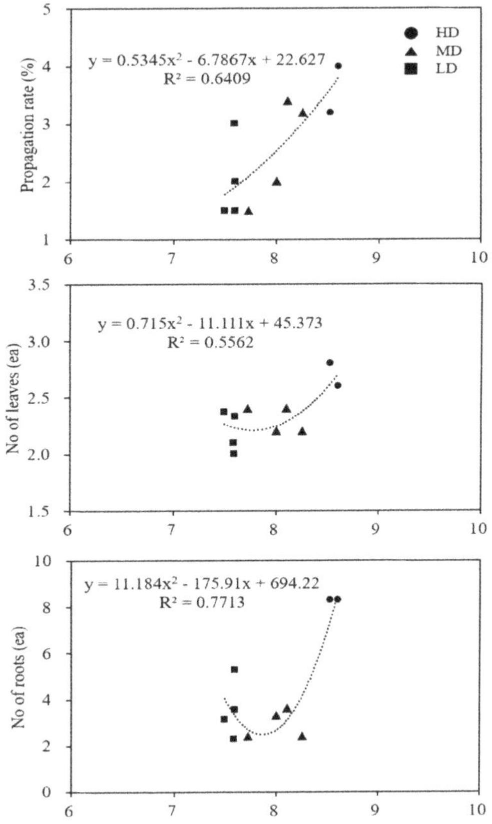

Figure 4. Correlation between DNA content and growth characteristics of ten clones of *Cnidium officinale*: HD, cones with high DNA content; MD, clones with moderate DNA content; and LD, clones with low DNA content.

3.4. In Vitro Induction of Rhizome

The results obtained for rhizome formation on media with different concentrations of sucrose and ABA, as well as for the light and dark incubation of the cultures, are presented in Figure 5. After two weeks of incubation, rhizome induction was evident at the base of the cultured shoots. Initially, the swelling of the shoot bases was observed, followed by the appearance of small rhizomes at the base within four weeks of incubation. Subsequently, the rhizomes were involved in the shoots' growth, and roots simultaneously developed in the nodal regions of the rhizomes. Rhizome induction from the shoots was influenced by the levels of sucrose in the medium. The medium supplemented with 30 g L^{-1} to 50 g L^{-1} sucrose favored rhizome development (Figure 5). However, with a further increase in the concentration of sucrose from 50 g L^{-1} onwards (70 and 90 g L^{-1}), there was a marked decrease in the percentage of rhizome formation and the rooting rate. Rhizomes formed on the medium in the presence of ABA, and the 0.5 mg L^{-1} and 1.0 mg L^{-1} ABA treatments were better in terms of the percentage of rhizome formation, the number of rhizomes forming shoots, and the rooting rate. A comparison of the light and dark treatments revealed that light incubation is stimulatory in inducing rhizomes from the shoots (Figure 5).

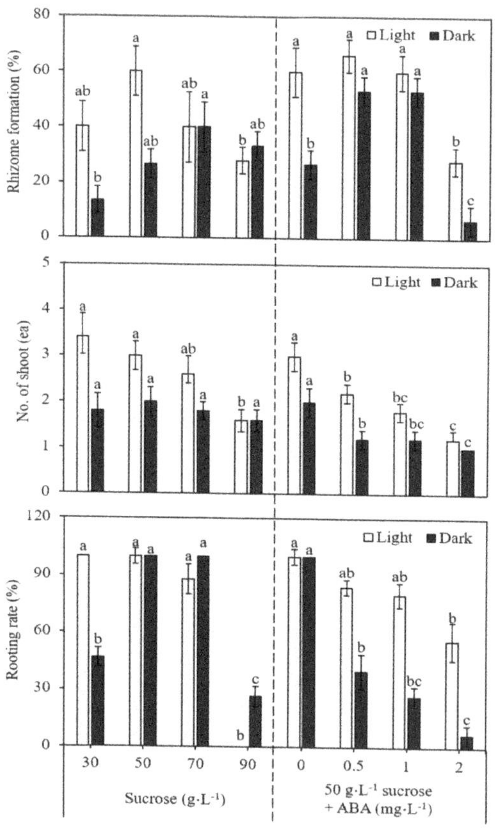

Figure 5. Effect of sucrose and ABA on rhizome induction of *Cnidium officinale* and under the same treatment (light vs. dark). Different letters in each bar significantly differ from each other as per DMRT ($p < 0.05$).

In another set of experiments, we compared the effect of ABA (1.0 mg L^{-1}) and the interaction of BA (1.0 mg L^{-1}) + NAA (0.5 mg L^{-1}) and BA (1.0 mg L^{-1}) + NAA (0.5 mg L^{-1}) + ABA (1.0 mg L^{-1}) on rhizome induction for the in vitro-cultured shoots. At the same time, we also tested the effect of the light and dark treatments, along with hormonal combinations, with and without the addition of activated charcoal (1 g L^{-1}), and the results are presented in Figure 6. Among these different treatments, the cultures that received light treatment performed better than those that received dark treatment, and again, the ABA treatment was far better than the BA + NAA and BA + NAA + ABA treatments. The results of the above experiments suggest that MS medium supplemented with 50 g L^{-1} sucrose, 1.0 mg L^{-1} ABA, and 1 g L^{-1} AC is good for the induction of rhizomes from the shoots of *Cnidium officinale* (Figure 7a,b).

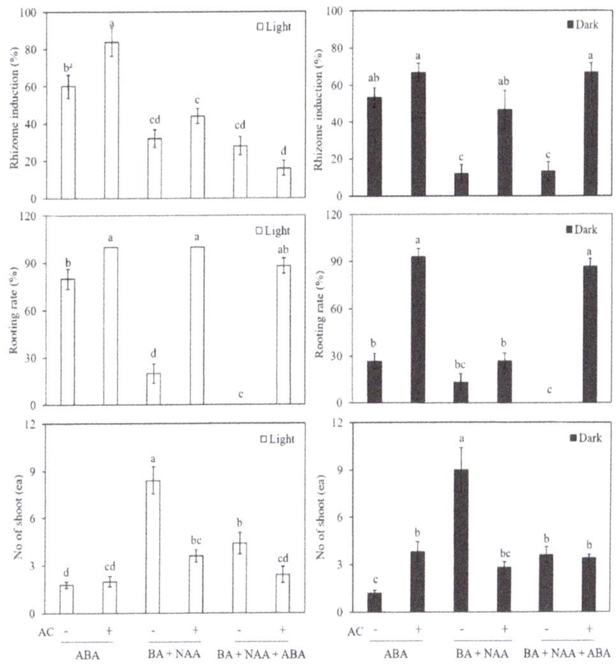

Figure 6. Effect of ABA, BA, and NAA with or without activated charcoal (AC) and under light or dark conditions on rhizome induction of *Cnidium officinale*. Different letters in each bar significantly differ from each as per DMRT ($p < 0.05$).

Figure 7. Rhizomes of *Cnidium officinale* developed with the in vitro-grown plantlets (**a**,**b**); (**c**) four-week-old plants growing with Growfoam; and (**d**) six-month-old plants growing in pots containing vermiculite.

13. Tran, H.N.K.; Cao, T.Q.; Kim, J.A.; Youn, U.Y.; Kim, S.; Woo, M.H.; Min, B.S. Anti-inflammatory activity of compounds from the rhizome of *Cnidium officinale*. *Arch. Pharm. Res.* **2018**, *41*, 977–985. [CrossRef]
14. Lim, E.Y.; Kim, J.G.; Lee, J.; Lee, C.; Shim, J.; Kim, Y.T. Analgesic effect of *Cnidium officinale* extracts on postoperative, neuropathic, and menopausal pain in rat models. *Evid.-Based Complement. Altern. Med.* **2019**, *2019*, 9698727. [CrossRef] [PubMed]
15. Lee, M.J.; Shim, Y.S.; An, S.Y.; Kang, M.K. Surface characterization, biocompatibility and antifungal efficacy of a denature-lining material containing *Cnidium officinale* extracts. *Molecules* **2021**, *26*, 1440. [CrossRef]
16. Lee, M.J.; Kang, M.K. Analysis of the antimicrobial, cytotoxic and antioxidant activities of *Cnidium officinale* extracts. *Plants* **2020**, *9*, 988. [CrossRef]
17. Yang, H.; Jung, D.H.; Lee, H.W. Therapeutic effect of *Cnidium officinale* Makino extract on ovariectomized hand-limb ischemic mice. *Int. Med. Res.* **2019**, *8*, 107–115.
18. Kim, Y.J. Inhibition effect of *Cnidium officinale* Makino extracts on MMP1 expression in human dermal fibroblasts. *Asian J. Beauty Cosmetol.* **2018**, *16*, 131–138. [CrossRef]
19. Cha, H.J. *Cnidium officinale* Makino extracts inhibit α-MSH-induced melanogenesis in B16F10 mouse melanoma cells. *Asian J. Beauty Cosmetol.* **2018**, *16*, 122–130. [CrossRef]
20. Cho, D.Y.; Lee, E.K.; Soh, W.Y. Cotyledon structure and germinability if somatic embryos formed from inflorescence explants of *Cnidium officinale* M. *Korean J. Plant Tissue Cult.* **2000**, *27*, 137–142.
21. Pant, B.; Kohda, H.; Namera, A. Clonal propagation of *Cnidium officinale* by shoot tip culture. *Planta Med.* **1996**, *62*, 281–283. [CrossRef] [PubMed]
22. Chirangini, P.; Sinha, S.K.; Sharma, G.J. In vitro propagation and mirorhizome induction in *Kaempferia galanga* Linn. and *K. rotunda* Linn. *Ind. J. Biotechnol.* **2005**, *4*, 404–408.
23. Nayak, S.; Naik, P.K. Factors affecting in vitro mirorhizome formation and growth in *Curcuma longa* L. and improved field performance of micropropagated plants. *ScienceAsia* **2006**, *32*, 31–37. [CrossRef]
24. Kapoor, P.; Usha Rao, I. In vitro rhizome induction and plantlet formation from multiple shoots in *Bambusa bambos* var. *gigantea* Bennet and Gaur by using growth regulators and sucrose. *Plant Cell Tiss. Organ Cult.* **2006**, *85*, 211–217. [CrossRef]
25. Murashige, T.; Skoog, F. A revised medium for rapid growth and bioassays with tobacco tissue cultures. *Physiol. Planta.* **1962**, *15*, 473–497. [CrossRef]
26. Rice, A.; Glick, L.; Abadi, S.; Einhorn, M.; Kopelaman, N.M.; Salman-Minkov, A.; Mayzel, J.; Mayrose, I. The Chromosome Counts Database (CCDB)—A community resource of plant chromosome numbers. *New Phytol.* **2015**, *206*, 19–26. [CrossRef] [PubMed]
27. Adil, M.; Kang, D.I.; Jeong, B.R. Data on recurrent somatic embryogenesis and in vitro micropropagtion of *Cnidium officinale* Makino. *Data Br.* **2018**, *19*, 2311–2314. [CrossRef] [PubMed]
28. Sliwinska, E.; Thiem, B. Genome size stability in six medicinal plants species propagated in vitro. *Biol. Plant.* **2007**, *51*, 556–558. [CrossRef]
29. Yan, J.; Zhang, J.; Sun, K.; Chang, D.; Bai, S.; Shen, Y.; Huang, L.; Zhang, J.; Zhang, Y.; Dong, Y. Ploidy level and DNA content of *Erianthus arundinaceus* as determined by flow cytometry and the association with biological characteristics. *PLoS ONE* **2016**, *11*, e0151948. [CrossRef] [PubMed]
30. Zafer, N.; Mujib, A.; Aali, M.; Tonk, D.; Gulzar, B.; Malik, M.; Sayeed, R.; Mamgain, J. Genome size analysis of field grown and tissue culture regenerated *Rauvolfia serpentina* (L.) by flow cytometry: Histology and scanning electron microscopic study for in vitro morphogenesis. *Ind. Corps Prod.* **2019**, *128*, 545–555. [CrossRef]
31. Sharma, T.R.; Singh, B.M. In vitro microrhizome production in *Zingiber officinale* Rosc. *Plant Cell Rep.* **1995**, *15*, 274–277. [CrossRef] [PubMed]
32. Gopal, J.; Minocha, J.L.; Dhaliwal, H.S. Microtuerization in potato (*Solanum tuberosum* L.). *Plant Cell Rep.* **1998**, *17*, 794–798. [CrossRef] [PubMed]
33. Yang, M.; Zhu, L.; Pan, C.; Xu, L.; Liu, Y.; Ke, W.; Yang, P. Transcriptomic analysis of the regulation of rhizome formation in temperate and tropical lotus (*Nelumbo nucifera*). *Sci. Rep.* **2015**, *5*, 13059. [CrossRef] [PubMed]
34. Sakamura, F.; Ogihara, K.; Suga, T.; Taniguchi, K.; Tanaka, R. Volatile constituents of *Zingiber officinalae* rhizomes produced by in vitro shoot tip culture. *Phytochemistry* **1986**, *25*, 1333–1335. [CrossRef]
35. Shimasaki, K.; Uemoto, S. Rhizome induction and plantlet regeneration of *Cymbidium goeringii* from flower bud cultures in vitro. *Plant Cell Tiss. Organ Cult.* **1991**, *25*, 49–52. [CrossRef]

genes

MDPI

Article

Identification and Registration for High-Yielding Strain through ST and MLT of *Curcuma caesia* Roxb. (Jor Lab KH-2): A High-Value Medicinal Plant

Mohan Lal *, Sunita Munda, Twahira Begum, Tanmita Gupta, Manabi Paw, Sanjoy Kumar Chanda and Himangshu Lekhak

ARD Division, CSIR-North East Institute of Science and Technology (NEIST), Jorhat 785006, Assam, India
* Correspondence: mohan@neist.res.in

Abstract: (1) Background: *Curcuma caesia* Roxb. is a high valued crop which is extensively used in pharmaceuticals, flavour and fragrances. *C. caesia* is recognised as an endangered species due to its extensive collection from the wild through human intervention. Therefore, to prevent the species from extinction, it is very necessary to conserve and cultivate this plant species for the sustainable availability of the raw material. (2) Methods: In the present plant breeding programme, a multi-year study was performed for the identification of superior genotypes which will help in conservation. To fulfil this objective, a total of 135 accessions of *C. caesia* were collected from different regions of India and were set up for experimental selection trial for three years (2016–2018). After proper evaluation of the genotypes based on six agronomical traits, five high-yielding genotypes were identified which underwent multilocation trial for two years (2019 and 2020). The stability analysis using the Eberhart–Russell method, AMMI and GGE biplot were used to study the consistency of the genotypes in varied environments compared with the check variety. (3) Results: Analysis of variance indicated significant genotype and environment interaction for the yield traits, i.e., dry rhizome recovery, rhizome yield and essential oil yield. The coefficient of variation (CV) was highest for tillers per plant (21.76) and lowest for the plant height (4.93). All the results clearly demonstrated Jor Lab KH-2 as the highest yielding and stable genotype in varied environments compared with the check variety and other selected genotypes. (4) Conclusions: This genotype was then submitted to ICAR-NBPGR, New Delhi, for germplasm registration and received its confirmation vide registration number INGR 21159. This genotype will greatly benefit the breeders and will also help in the conservation of this endangered species. This is the first report on the identification and registration of a high-yielding variety of *C. caesia*.

Keywords: *Curcuma caesia*; AMMI; GGE; high essential oil; MLT; regression

Citation: Lal, M.; Munda, S.; Begum, T.; Gupta, T.; Paw, M.; Chanda, S.K.; Lekhak, H. Identification and Registration for High-Yielding Strain through ST and MLT of *Curcuma caesia* Roxb. (Jor Lab KH-2): A High-Value Medicinal Plant. *Genes* **2022**, *13*, 1807. https://doi.org/10.3390/genes13101807

Academic Editors: Wajid Zaman and Hakim Manghwar

Received: 20 September 2022
Accepted: 1 October 2022
Published: 6 October 2022

1. Introduction

Curcuma caesia Roxb. commonly known as kala haldi, is named after the ginger family and is the largest family of the order Zingiberales [1]. There are around 1300 species in it, grouped into 53 genera [2]. The genus *Curcuma* is classified under the Zingiberaceae family which consists of more than 80 species, some of which have been used in traditional systems of medicine (Ayurveda, Siddha, Unani) for a long time [3], one of which is *C. caesia*. It is widely distributed in Thailand, Malaysia, Nepal, China, Bangladesh and India. It is native to central and northeast India [4]. It is endangered to Asia's south-eastern region and rarely found in the Himalayan foothills, Sikkim's North Hill Forest, and the East Godavari Pappi Hills [5–7].

The rhizome of the perennial herb *C. caesia* is bluish-black in colour, hence it is called as black turmeric [7]. The plant has 42 chromosomes and is a diploid species [4,8]. It is usually erect and grows between 0.5 and 1.0 m in height with a short stem. The plant bears pale yellow-coloured, long, tubular flowers, with reddish borders, which are smaller than bracts. It is identified by the presence of a large underground-ovoid bluish-black rhizome.

Black turmeric produces a cluster of 10–20 leaves, which are characterised by the presence of a deep violet streak running throughout the leaf lamina. The rhizome is usually 2–6 cm in diameter and has a camphoraceous odour [8]. The rhizome is highly important because of its medicinal properties [9]. The ethnobotanical uses include application of *C. caesia* leaf paste to a scorpion or snake bite, and also the dried powder is mixed with *Andrographis paniculata* and applied to insect and snake bites [10]. Dried leaves and rhizomes are used to treat toothache, wounds, fertility, impotency, vomiting, piles, leprosy, cancer, and allergies [11]. The fresh rhizomes are utilised in leprosy, cancer, epilepsy, vomiting, menstrual disorder, anthelminthic aphrodisiac, and gonorrhoeal discharge [12]. The rhizome of *C. caesia* is mashed into a paste to be used in rheumatic arthritis [13]; also, it is applied for rapid healing after cuts and injuries to control bleeding [14]. The *C. caesia* rhizome is also used to treat tonsillitis [15]. *C. caesia* roots are powdered and taken orally with water during gastric disorder [16]. The rhizome is used as a tonic for brain and heart disease and is also effective against asthma, bronchitis, piles, leukoderma [17], allergic eruptions, epileptic seizures, enlargement of the spleen, tuberculous glands of the neck and tumours [18]. The plasters made from the leaves of the crop are used to treat adenitis, lymphangitis, and furunculosis [8].

The rhizome's essential oil contains a number of biological benefits [6,19]. It possesses many medicinal properties such as resistance to fungal infection [20], anti-asthmatic activity, acts as relaxant for smooth muscle [21], antioxidant activity [19,22], broncho dilating activity [5], locomotors depressant, CNS depressant and anxiolytic activity [23], anti-bacterial, anthelmintic activity [24], and anti-ulcer activity [25]. The strong odour of the rhizome is due to the presence of the essential oil which is rich in eucalyptol, camphor, and starch, etc. [7]. It contains flavour and fragrances due to the presence of bioactive secondary metabolites which also correlates with the medicinal uses and many important useful pharmaceutical products [26]. The essential oil of *C. caesia* rhizomes contains 30 components, out of which curcumene, bornyl acetate, borneol, elemene, 1, 8-cineole, ar-curcumene, (Z)-ocimene, ar-turmerone and camphor represents the major components constituting 97.48% of the essential oil [27]. Other major components are linalool followed by ocimene, 1- ar-curcumene, zingiberol, 1, 8-cineole and borneol [13]. Eucalyptol (28.55%), camphor (21.73%) and epicurzerenone (19.62%) identified as major components in the rhizome essential oil of *C. caesia* [19]. The bioactive components from the rhizomes are responsible for anti-coagulant, hypoglycaemia, wound healing, anti-inflammatory, anti-oxidative properties, and also exhibits the free radical scavenging property [1,28,29].

Due to widespread habitat destruction caused by human activities, including over utilization of black turmeric for traditional medicine, industrialisation, and urbanisation, *C. caesia* is currently regarded as being vulnerable [6,30]. Therefore, the need of the hour is to identify or develop high-yielding essential oil and rhizome varieties to maintain the sustainable cultivation of these high-value medicinal plant species. The phenotyping of germplasm could be used for the development of new varieties, to enhance the quality parameters in terms of yield, to match genotypes to locations, to help choose the best offspring and parents for breeding and to select more suitable genotypes. The measurement of quantitative or qualitative values for biochemical, physiological, morphological and performance-related aspects is referred to as "plant phenotyping". These play a vital role in determining quality, growth, and stress resistance traits by serving as observable intermediaries between the environment and gene(s) expression [31]. Plant breeding through selection trial (ST) and multilocation trial (MLT) is the most widely accepted programme for the development or identification of superior variety. Multilocation trial is the most essential criteria of the varietal development which determines the ability of the genotype to perform consistently in different environments, confirming its stability. Because of environmental influences or interactions between the genotype and the environment, the selection in some contexts necessitates the presence of effective selection features [32]. The primary attribute that significantly responds to the two sources of variation is the yield [33]. The Eberhart–Russell method and AMMI model are the most widely accepted analysis for confirmation of stability [34,35]. Additionally, nowadays, GGE biplot analysis is also used by the breeder for the evaluation of stability by ranking the environments and genotypes, making it superior to AMMI biplot analysis [36]. A genotype is considered to be superior if it consistently maintains its high-yielding characteristic across different environments [35,37]. Therefore, the identification of this superior variety will help in the conservation of this species for future prospects and will expand the possibilities for the commercial production of camphor, which is the major component in this crop, and offer the pharmaceutical, flavour and fragrances industries new sources of raw materials. The identified strain Jor Lab KH-2 of *C. caesia* was found to possess high rhizome yield with an average of 124 cm plant height, 49 cm leaf length, 26% dry rhizome recovery, 0.80% essential oil in fresh rhizome weight basis and 26.18 tones/ha rhizome yield. The essential oil of this species has very high commercial values and contains antimicrobial, anti-inflammatory activities and much less genotoxicity [7]. This is the first report on the identification and registration of the superior germplasm of *C. caesia* confirmed through stability analysis which showed consistent and satisfactory agronomical performance.

2. Materials and Methods

2.1. Germplasm Collection

The collection of *C. caesia* germplasm was carried out in six different states of India, namely, Assam, Arunachal Pradesh, Mizoram, Nagaland, Manipur, and Madhya Pradesh, which led to the accretion of 135 accessions. The accessions were identified by breeder Dr Mohan Lal of the division of Agrotechnology and Rural Development, CSIR-NEIST, Jorhat, India. The germplasm was also submitted to the repository of ICAR-NBPGR, New Delhi, and an Indigenous collection (IC) number was obtained. The herbarium for each genotype was prepared and added to the institute's departmental herbarium.

2.2. Setup of Experimental Trial

The collected accessions were planted in the experimental field with GPS coordinates: longitude 94°9′25.4628″ E, latitude 26°44′15.6948″ N and elevation 94 m over the average sea level. The germplasm was planted in randomised complete block design (RCBD) with spacing of 60 × 60 cm between plant to plant and row to row respectively. The plot size of the field was measured 4 m × 4 m length area. The NPK concentration available in the soil were Nitrogen (269 Kg/ha), P (49 Kg/ha), K (91 Kg/ha). The texture of the experimental soil was sandy loam, pH 5.2. The rhizomes were planted in the month of March 2016 and was harvested after every nine months for the three consecutive years for which the data were recorded.

2.3. Evaluation of Selection and Multilocation Trial

The three years (2016, 2017 and 2018) selection trial was performed on the agronomical traits, including number of tillers/plants (TPP), plant height (PH), length of leaf (LL), fresh rhizome yield (FRY), dry rhizome recovery (DRR) and essential oil yield (EO). These six agronomical traits were studied, out of which FRY, DRR and EO are the yield parameters preferred in crop improvement programme. Rhizome and essential oil are the most essential part of the crop due to the presence of bioactive components utilised in the aroma, food and pharmaceutical industries. The data were recorded using ten randomly picked plants of each genotype from each replication. Five high essential oil yielding strains were identified after a three-year selection trial which also yielded high rhizome. The multilocation trial of the selected strains along with one check variety were then conducted at five different locations in northeast India, i.e., Jorhat (Assam), Shillong (Meghalaya), Pasighat (Arunachal Pradesh), Imphal (Manipur) and Lakhamijan (Assam), during the years 2019 and 2020. The check variety under consideration was one of the high rhizome yielding local genotype of Pasighat, Arunachal Pradesh, India, mostly used for commercial cultivation. The environments E1, E2, E3, E4, E5 and E6, E7, E8, E9, E10 depict the locations of Jorhat, Lakhamijan, Imphal, Pasighat and Shillong for the first year and second year, respectively.

The trial continued for two more years for the agronomical and quality traits that were considered for the selection trial evaluation.

2.4. Isolation and GC-MS Analysis of the Essential Oil

The isolation of essential oil was performed according to the protocol suggested by [38] and modified by [19]. The Clevenger apparatus was used, in which 300 g of fresh rhizome was boiled in 3000 mL of distilled water for 6 h. The isolated essential oil was then collected in a glass vial and excess moisture content was removed by treatment with anhydrous sodium sulphate. The essential oil percentage (v/w) of the isolated essential oils was estimated using the following formula:

$$\text{Essential oil (FWB)} = \frac{volume\ of\ essential\ oil\ isolated\ (\text{mL})}{weight\ of\ rhizome\ used\ (\text{g})} \times 100$$

The TRACE ultra-gas chromatograph equipped with mass spectrophotometer (Thermo Fisher Scientific, Waltham, MA, USA) was used for the GC-MS analysis of the rhizome essential oil. The analysis through GC-MS was as per protocol modified by [19].

2.5. Statistical Analysis

Both the pooled and individual analysis of variance (ANOVA) was calculated for the multilocation trial data of the five selected genotypes along with the check variety based on two-year evaluation. The stability models such as Eberhart–Russell, AMMI (additive multiplicative mean interaction model) and GGE (genotype + genotype × environment) was measured to identify the better-performing stable genotypes. A total of ten environments (E1, E2, E3, E4, E5, E6, E7, E8, E9, E10) were studied to check the consistency of the genotype performance by metan R package developed by [39].

3. Results

3.1. Agronomical Data for MLT of the Five Identified Lines along with the Check

A total of 135 accessions were evaluated for three consecutive years which led to the identification of the five most high-yielding lines for rhizome and essential oil yield. The agronomical data were recorded for each year with five replications. The selected lines were compared with one of the local genotype of NE India denoted by the check variety. The evaluation of the multilocation trial in five different regions of northeast India for two years revealed a mean of 130.28 cm for plant height, tillers per plant (5.36), leaf length (46.41 cm), dry rhizome recovery (20.42%), fresh rhizome yield (19.85 tones/ha) and essential oil (0.61%). The coefficient of variation (CV) was highest for tillers per plant (21.76) followed by essential oil percentage (21.05), while the lowest CV was observed in the plant height (4.93). The maximum plant height (cm) was observed in the check variety (133.48) and minimum in KH-2 (125.68). Similarly, the range for tillers per plant varied from KH-120 (4.84) to KH-71 (5.78); leaf length (cm) from check (43.67) to KH-2 (48.58); dry rhizome recovery (%) from KH-58 (18.74) to KH-2 (25.92); fresh rhizome yield (tones/ha) from KH-100 (18.06) to KH-2 (26.23); and essential oil percentage from the check (0.42) to KH-2 (0.80). The highest and lowest environmental performances for the studied traits were depicted in Table 1.

The maximum environmental performances for the traits of plant height, fresh rhizome yield and essential oil yield were revealed in environment E9, while the best performance of dry rhizome recovery was exhibited in environment E4.

Table 1. Mean and range for all the studied traits of six high-yielding genotypes of *C. caesia* in MLT.

Traits	Mean	SE	SD	CV	MinENV	MaxENV	MinGEN	MaxGEN
PH	130.28	0.37	6.41	4.93	E6 (128.17)	E9 (133.53)	KH-2 (125.68)	Check (133.48)
TPP	5.36	0.07	1.17	21.76	E9 (5.13)	E3 (5.73)	KH-120 (4.84)	KH-71 (5.78)
LL	46.41	0.23	3.9	8.41	E5 (45.52)	E6 (48.78)	Check (43.67)	KH-2 (48.58)
DRR	20.42	0.19	3.32	16.29	E6 (19.13)	E4 (21.8)	KH-58 (18.74)	KH-2 (25.92)
FRY	19.85	0.19	3.36	16.98	E3 (19.23)	E9 (20.52)	KH-100 (18.06)	KH-2 (26.23)
EO	0.61	0.01	0.13	21.05	E7 (0.57)	E9 (0.66)	Check (0.42)	KH-2 (0.8)

PH: plant height; TPP: tillers per plant; LL: leaf length; DRR: dry rhizome recovery; FRY: fresh rhizome yield; EO: essential oil yield, SE: standard error; SD: standard deviation; CV: coefficient of variation; ENV: environment; GEN: genotype.

3.2. ANOVA of the Agronomical Traits for Six High-Yielding Accessions

The analysis of variance was performed for the MLT of six high-yielding accessions based on joint regression. This analysis clearly indicates that there were significant differences in the genotypes in terms of all the studied traits at $p > 0.05\%$. Further, the genotype by environment interaction was exhibited by tillers per plant, dry rhizome recovery, rhizome yield and essential oil yield (Table 2).

Table 2. ANOVA based on joint regression for the agronomical traits of six high-yielding genotypes of *C. caesia* in MLT.

Source	DF	PH	TPP	LL	DRR	FRY	EO
Total	59	68.87	2.12	35.28	39.30	46.89	0.08
GEN	5	342.63 ***	5.31 ***	157.91 ***	370.12 ***	495.60 ***	0.78 ***
ENV + (GEN × ENV)	54	43.52	1.82	23.93	8.66	5.34	0.01
ENV (linear)	1	776.14	9.16	242.96	150.14	54.60	0.20
GEN × ENV (linear)	5	45.50	4.87 **	37.88	16.68 **	8.74 *	0.01 *
Pooled deviation	48	28.05	1.35	17.91	4.88	3.97	0.01
Check	8	6.80	1.97	24.34 *	8.42	5.42 *	0.02 ***
KH-100	8	36.73	1.26	19.23	5.70	2.95	0.01 ***
KH-120	8	18.86	1.44	20.80	1.83	6.70 *	0.00
KH-2	8	22.61	1.65	9.46	1.04	1.92	0.00
KH-58	8	15.19	1.04	22.52 *	5.02	1.30	0.00 *
KH-71	8	68.11	0.75	11.12	7.27	5.51 *	0.00
Pooled error	200	35.51	1.14	10.72	4.29	2.63	0.00
Total	59	68.87	2.12	35.28	39.30	46.89	0.08

PH: plant height; TPP: tillers per plant; LL: leaf length; DRR: dry rhizome recovery; FRY: fresh rhizome yield; EO: essential oil yield, ENV: environment; GEN: genotype; DF: degree of freedom; *** significant at 0.5%; ** significant at 1%; * significant at 5%.

The analysis also indicated high significant variation in the check variety and KH-100 at $p > 0.05\%$ for the essential oil yield. Similarly, for rhizome yield, significant difference was revealed in the genotypes KH-120, KH-71 and the check variety. The traits like leaf length and plant height showed no significant environmental influence on the genotypes.

3.3. Stability Analysis through Joint Regression Method

The regression analysis was proposed by Eberhart and Russell, therefore, the model was named after them. In order to observe how the genotypes react differently to various environmental situations, stability parameters such as regression coefficient and deviation from regression were applied. The five identified lines and one control were subject to a stability analysis utilising the Eberhart–Russell (ER) model, which explains a genotype to be stable if it acquires high mean value (b_0), a regression coefficient (bi) equal to one and a regression from deviation (s^2di) equal to zero. The stability analysis was conducted for three yield traits, i.e., dry rhizome recovery, fresh rhizome yield and essential oil yield, since a significant genotype and environment interaction was observed in these traits.

The maximum DRR was expressed by genotype KH-2 followed by the check variety, but presence of significant regression coefficient in the check variety led to its rejection. The genotype KH-2 showed highest mean value (25.92%), bi = 1.01 and s^2di = −0.65 showing the values almost similar to that of stable line. Again, for FRY, the highest mean value was expressed in KH-2 followed by the check variety. The genotype KH-2 has a mean value of 26.33 tons/ha, bi = 0.76 and s^2di = −0.14, compared with the check variety demonstrating mean value 18.92 tons/ha, bi = 0.12 and s^2di = 0.56 (Table 3).

Table 3. Mean and stability parameters for the yield traits of six high-yielding genotypes of *C. caesia*.

GEN	DRR			FRY			EO		
	b_0	b_i	s^2di	b_0	b_i	s^2di	b_0	b_i	s^2di
KH-2	25.92	1.01	−0.65	26.33	0.76	−0.14	0.80	1.11	0.00
KH-58	18.74	0.10 *	0.15	18.83	2.56 ***	−0.27	0.57	0.20 ***	0.00 *
KH-71	19.56	1.49	0.60	18.25	0.85	0.57 *	0.67	1.22	0.00
KH-100	19.10	0.61	0.28	18.06	0.01	0.06	0.59	1.20	0.00 ***
KH-120	19.34	0.45	−0.49	18.79	1.71	0.81 *	0.62	1.54 *	0.00
Check	19.88	2.35 ***	0.83	18.92	0.12	0.56 *	0.42	0.72	0.00 ***

DRR: dry rhizome recovery%; FRY: fresh rhizome yield; EO: essential oil yield; GEN: genotype; *** significant at 0.5%; * significant at 5%; b_0: mean; b_i: regression coefficient; s^2di: regression from deviation.

The significant deviation from regression in the check variety makes it unsuitable for selection. The highest mean for essential oil yield was observed in KH-2 followed by KH-71 and KH-120 but the regression coefficient of KH-2 was nearer to the standard value of one compared with the other high-yielding genotypes. All these traits confirmed that KH-2 was the most suitable genotype to be considered as the superior line which can perform better in a wide range of environments. The nominal yield plots were also constructed for all the yield traits which clearly indicated that KH-2 was the winning genotype in the entire studied environment, with the highest yield in all the environments (Figure 1).

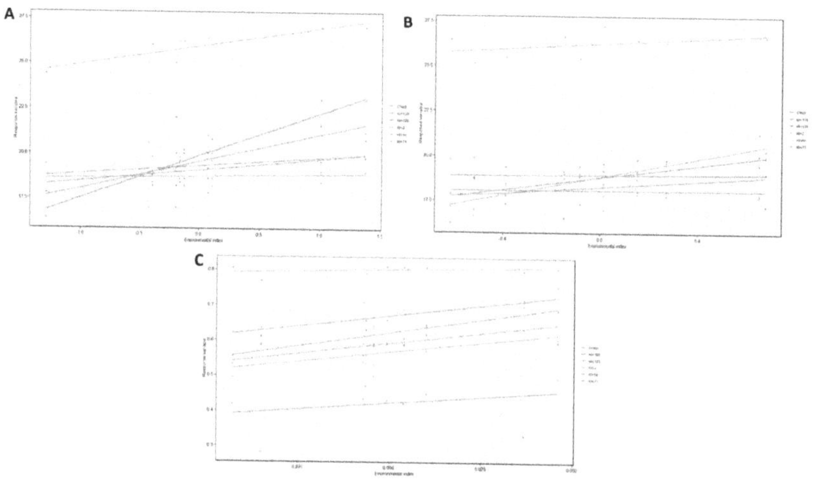

Figure 1. Nominal yield plot for the yield traits (**A**) DRR (**B**) FRY (**C**) EO yield.

3.4. *AMMI ANOVA for the Agronomical Traits for Six High-Yielding Accessions*

One of the most widely used AMMI model analyses was also performed to confirm the stability of the identified KH-2 genotype. AMMI ANOVA showed highly significant value for PH, LL, DRR, FRY and EO, indicating wide variation in the studied environment. This makes it necessary to study the consistency of the genotypes because environment can influence the phenotypic nature of the genotype. The genotypes also showed wide variation for all the traits, which was also predicted by the joint regression ANOVA. The environmental influence was observed in all the traits except plant height. The principal component analysis was divided into five principal components: PC1, PC2, PC3, PC4 and PC5. The first principal component correlates strongly with plant height, followed by LL, DRR, FRY, TPP and EO. Similarly, the second principal component corresponds with LL, followed by DRR, FRY, TPP and EO. Furthermore, the third principal component correlates with the traits FRY and EO (Table 4).

Table 4. AMMI analysis table for different agronomical traits of six high-yielding genotypes of *C. caesia* in the MLT.

Source	DF	PH	TPP	LL	DRR	FRY	EO
ENV	9	86.24 **	1.02	27.00 ***	16.68 ***	6.07 *	0.02 ***
REP(ENV)	40	28.90	1.38	8.28	3.33	2.52	0.00
GEN	5	342.63 ***	5.31 ***	157.91 ***	370.12 ***	495.60 ***	0.78 ***
GEN:ENV	45	34.98	1.98 **	23.31 ***	7.06 **	5.20 ***	0.01 ***
PC1	13	80.09 **	3.34 ***	33.48 ***	13.10 ***	7.05 ***	0.02 ***
PC2	11	21.57	2.47 *	29.51 ***	8.79 *	5.47 *	0.00 *
PC3	9	17.33	1.04	19.86	3.40	5.12 *	0.00 *
PC4	7	16.96	0.98	11.01	1.68	3.42	0.00
PC5	5	4.18	0.50	6.70	1.67	2.46	0.00
Residuals	200	35.51	1.14	10.72	4.29	2.63	0.00
Total	344	40.40	1.44	16.30	10.54	10.55	0.02

PH: plant height; TPP: tillers per plant; LL: leaf length; DRR: dry rhizome recovery; FRY: fry rhizome yield; EO: essential oil yield, ENV: environment; GEN: genotype; DF: degree of freedom; PC: principal component; *** significant at 0.5%; ** significant at 1%; * significant at 5%.

3.5. Stability Analysis through AMMI Model and GGE Biplot

AMMI biplot 1 and 2 were constructed for the yield traits (DRR, FRY and EO) which indicated KH-2 as the most stable genotype. AMMI1 biplot distinctly indicted KH-2 as the high dry rhizome recovery and stable genotype compared with the other five genotypes (Figure 2A).

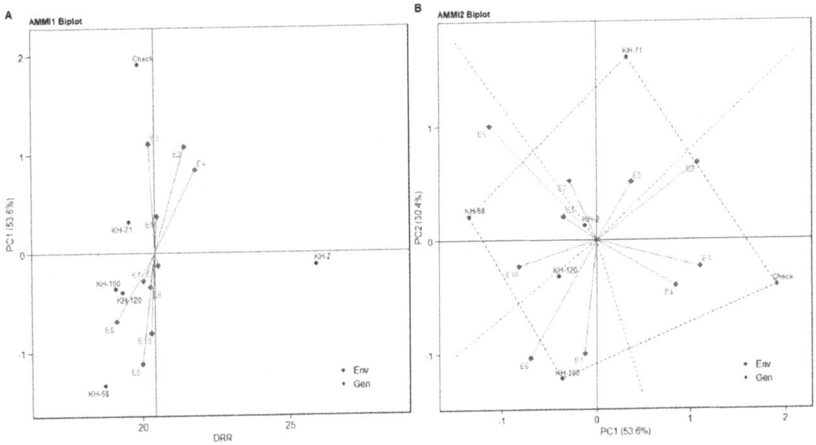

Figure 2. (**A**) AMMI1 biplot (**B**) AMMI2 biplot of DRR of five selected accessions and the check variety of *C. caesia* germplasm for eleven environments.

AMMI 2 biplot indicated KH-100, KH-58, KH-71 and the check variety as the high-yielding genotypes (Figure 2B). The genotype KH-100 was the most favourable genotype for the environments E1 and E6, while the genotype KH-58 showed better performance in E5, E8 and E10. The best suited genotype for the environments E7 and E9 is KH-71 while the check variety is most favoured in the E2, E3 and E4 environments. Similar results were predicted in GGE plot with KH-2 as the most consistent genotype, followed by KH-71, KH-100, KH-120, KH-58 and the check variety (Figure 3).

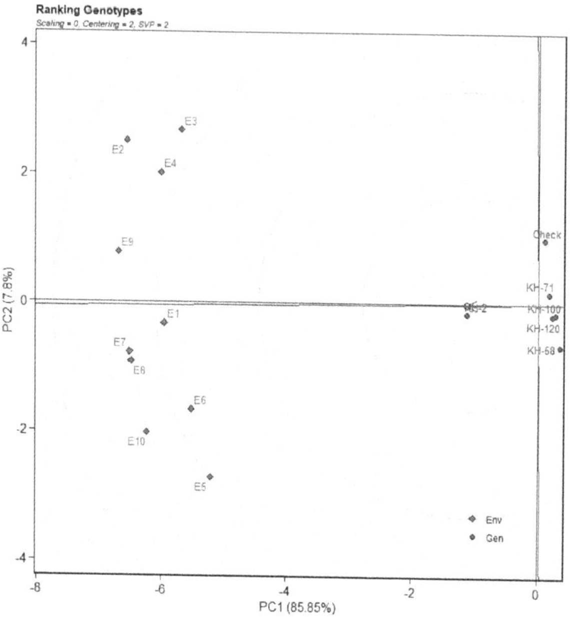

Figure 3. The genotype ranking through GGE plots for dry rhizome yield based on ideal genotype for eleven environments.

The ideal environment is the centre of the set of concentric lines that functions as a measure to gauge the separation between an environment and the ideal environment. It can be observed that E9 and E7 are the nearest to the ideal environment and therefore can be said to be the most desirable environments. E3 and E5 are the most undesirable studied environments (Figure 3). The trait FRY showed KH-100 as the most stable genotype but with low yield as compared to the other varieties. The genotype KH-2 is the high-yielding genotype for FRY with consistent performance as revealed in the biplot of AMMI1 for FRY (Figure 4A). The genotype KH-71, KH-58, KH-120 and the check variety are the highest yielders for fresh rhizome. The most favourable environments for the check variety are E3 and E4; and environments E6, E8 and E9 for the genotype KH-71. Similarly, the genotypes KH-58 and KH-120 were favourable for the environments E1, E2, E5, E7 and E10 (Figure 4B).

The ideal genotype identified for the trait FRY in the GGE biplot was KH-2. The other genotypes closest to the ideal genotypes were KH-58 and KH-100. The ranking of ideal environments for the trait FRY was E2 followed by E1, E9, E10. The least desirable environments were E3 and E8 (Figure 5).

Additionally, the AMMI1 biplot for trait EO yield revealed KH-71 and KH-2 as the stable and high essential oil yielding genotypes. However, KH-71 is more stable compared to KH-2 and KH-2 was high-yielding than KH-71. So, both the genotypes can be considered for EO yield (Figure 6A). In the AMMI2 biplot, it can be observed that genotype KH-71, KH-58, KH-100 and the check variety are high essential oil yielding genotypes in different environments (Figure 6B).

The most ideal genotype identified was KH-71, followed by KH-2, KH-120, KH-100 and the check variety, as depicted in the GGE biplot of EO. Environments E4, E2, E1 are the most favourable environments observed according to the GGE biplot of ranking environments for the trait EO (Figure 7).

All the analysis clearly depicts that genotype KH-2 is the most ideal genotype in terms of DRR, FRY and EO yield, maintaining its consistency in a wide range of environments. After the confirmation of the stability and high yield, this identified genotype, named as Jor Lab KH-2, was registered with ICAR-NBPGR, New Delhi, vide registration number INGR 21159.

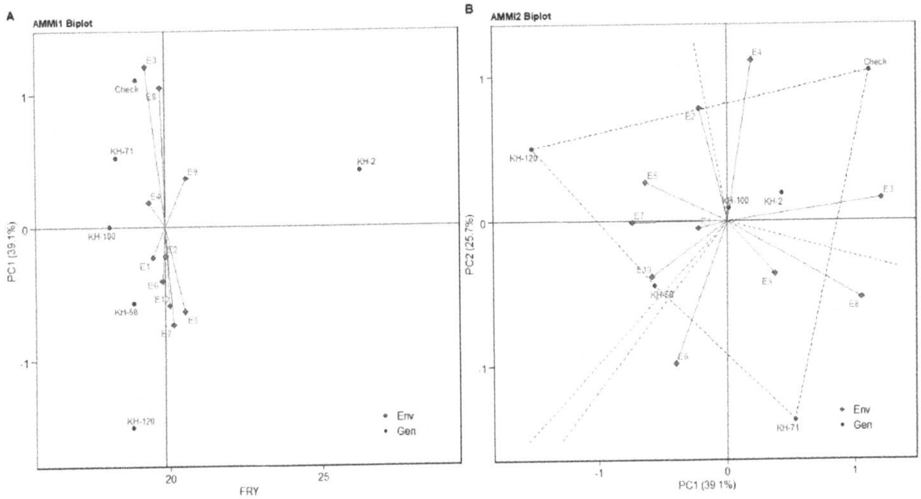

Figure 4. (**A**) AMMI1 biplot (**B**) AMMI2 biplot of FRY of five selected accessions and the check variety of *C. caesia* germplasm for ten environments.

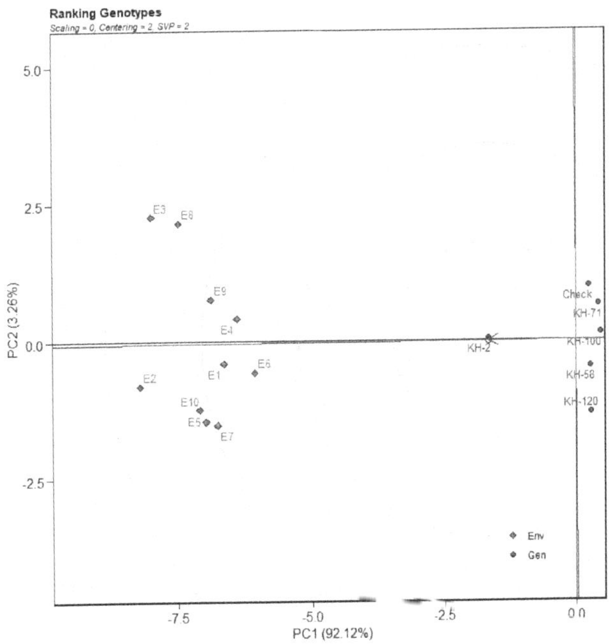

Figure 5. The genotype ranking through GGE plots for FRY based on ideal genotype for eleven environments.

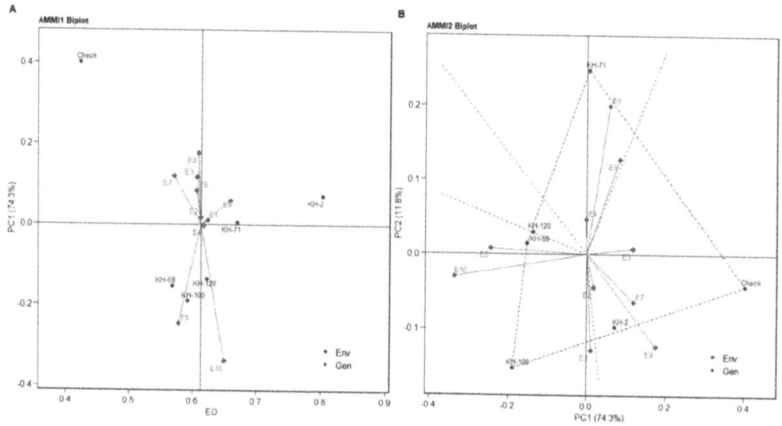

Figure 6. (**A**) AMMI1 biplot (**B**) AMMI2 biplot of EO yield of five selected accessions and the check variety of *C. caesia* germplasm for eleven environments.

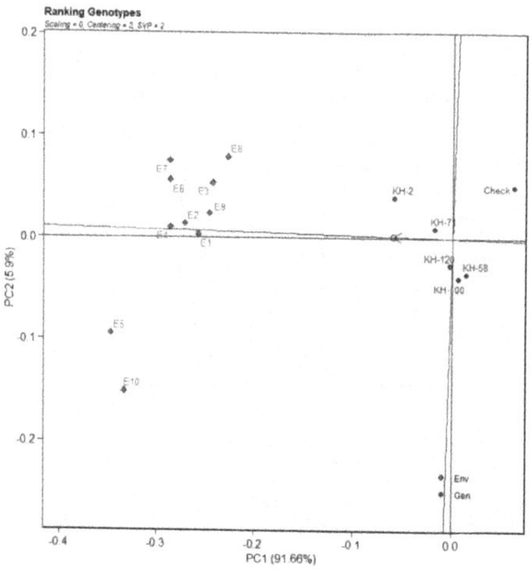

Figure 7. The genotype ranking through GGE plots for EO based on ideal genotype for eleven environments.

4. Discussion

The expression of superior traits with uniform result is highly desirable. However, the trait expression may differ due to various factors including environment, genotype, and the combined effect of genotype and environment [40–42]. This necessitates the need for the study of genotype–environment (G × E) interaction to evaluate the performance and stability for the identification of elite genotypes. Significant genotype–environment interaction was observed for the traits such as tillers per plant, dry rhizome recovery, rhizome yield and essential oil yield, therefore, the study of stability is essential for which a multivariate stability approach, which was used in the study. So far, no previous study has been reported in *C. caesia* for the identification of a stable genotype with superior desirable traits. Therefore, the results obtained in the present study were compared with the other species of Curcuma. The total curcuminoid content in 15 turmeric genotypes for three years was studied representing the major turmeric cultivating regions of India. The stability analysis was performed through the AMMI model, revealing IISR Prathiba and BSR 2 as stable genotypes for bisdemethoxycurcumin content. The genotypes SLP 389/1, Acc. 849 and CO 2 were found to be stable for demethoxycurcumin, while BSR 2, CO 2, IISR Prathiba and Duggirala Red were the genotypes reported to be stable for curcumin content [43]. The findings of the present study reported that genotype by environment interaction was exhibited by tillers per plant, dry rhizome recovery, rhizome yield and essential oil yield for the studied genotypes. A study was conducted on eleven cultivars of turmeric for fresh yield, curcumin yield, curing percentage, and dry yield at five different environments. For fresh yield, a large proportion of variation was attributed to environments (70.8%). For the traits such as curcumin content, curing percentage and dry yield, the variation due to genotype effect accounted for 17.7%, 31.2%, and 15.7%, respectively. The results from G × E interaction using the Eberhart and Russel model reported variation of 42.9% in curcumin content due to environment. For fresh yield with above average yield per plant across all environments, the genotype Mega Turmeric was found as most stable, while, for dry yield across all the environments, the genotypes IISR Prathiba, Mega Turmeric and IISR Kedaram were the ones found to be stable [38,44]. The genotypes by season interactions reported that 25 turmeric genotypes were found to be suitable for selection based on agro-morphological differences. GEIs contributed 34.88% variation in the rhizome yield, variation for weight per plant was 7.04%, 10.97% for rhizome width, the variation for petiole length was 11.95%, 47.25% variation was observed for lamina length and 25.44% variation was observed for lamina width [45]. Evaluation by GGE biplot of seven turmeric genotypes from Nepal was performed to identify traits helpful in turmeric selection. A total of 71.91% variation was observed from the two principal components (PC1 and PC2), among which 39.08% and 32.83% variation was exhibited by PC1 and PC2, respectively [46]. Another study in *Curcuma angustifolia* was carried out to identify a high-yielding stable genotype. The GGE biplot results exhibited the adaptation pattern of genotypes at various environments and the discrimination ability of the environments as well. The results revealed the genotype IGDMT-10-1(G5) as the most stable, and the most suitable environment for all the genotypes was revealed as E5 [47]. In a similar line, the present study revealed the environment E9 and E7 as the most desirable environments. In the environment E9, the traits of fresh rhizome yield, essential oil yield and plant height were found to be best performing, whereas in the environment E4, the dry rhizome recovery was found to be the highest performing trait. Another study was conducted on 17 ginger genotypes from Nigeria to identify stable genotypes and the best environment based on GGE biplot. The results from the GGE biplot model revealed UG2-9-01 as the most stable genotype with better yield for the traits of number of rhizome fingers per plant and rhizome length, while for the best environment, the Ikom environment was revealed as the best discriminating environment [48]. It was also reported that genotype by environment interactions was significant in some yield and growth traits of ginger [49]. The stability analysis for seventeen turmeric genotypes at two locations for two years using the Eberhart and Russel model revealed all the traits studied except oleoresin content and curcumin content to be highly significant for the mean square due to environment and linear. Similarly, all the traits studied except oleoresin content and curcumin content were found to be highly significant for the G × E interactions. Based on the results, stable genotypes with superior yield performance and stable genotypes with high curcumin yield under both favourable and unfavourable conditions were found [50]. A study on the three genotypes of *Curcuma zanthorrhiza* and twenty genotypes of *C. aeruginosa* at three different locations of Indonesia was carried out to evaluate their stability for total phenolic and antioxidant activity. The AMMI and GGE analysis revealed that G1, G6, G13, and G16 were the genotypes identified to be stable for phenolic antioxidant production. Among the identified stable genotypes, three *C. zanthorrhiza* genotypes were found to have higher total phenolic content and antioxidant activity compared to all the genotypes of *C. aeruginosa* in all the studied environments. Moreover, from the biplot analysis, 67.1% and 32.9% variation were found to be exerted by both the principal components (PC1 and PC2), respectively, for total phenolic content, while for the antioxidant activity parameters variation of 89.1% and 10.9% was found to be exerted by PC1 and PC2, respectively [51]. The GGE analysis in the present study confirmed the genotype Jor Lab KH-2 as the most stable genotype with superior yielding traits. A previous report [52,53] revealed that G × E interaction was found to be significant for the trait rhizome yield in turmeric. Another study on seventeen germplasm of turmeric was carried out to study their stability in three environments using the Eberhart and Russell model. The result of ANOVA revealed significant variation for the trait of rhizome yield and all the other thirteen traits studied in all the germplasm in all of the three environments, which accounted for the genetic variation among the studied germplasm. The G × E interaction was found to exhibit significant variation on the traits such as rhizome yield, rhizome thickness, rhizome fresh weight, rhizome dry weight and length of rhizome. CIMCH14130, CIMCH14208 and CIMCH14107 were the three identified stable genotypes with mean values, bi < 1 and s^2di 0 [54]. Similar results were obtained in our study for the genotype KH-2 with bi = 0.76 and s^2di = −0.14 while the check variety had bi = 0.12 and s^2di = 0.56 (Table 4). The result signifies the check variety significantly deviated from regression which highlights that the check variety was not selectable, and the genotype Jor Lab KH-2 was stable with high-yielding traits.

5. Conclusions

High crop productivity with consistent performance is the major goal and demand of breeders, researchers and growers. A multi-year study for three years was performed for 135 germplasms of *C. caesia* which led to the selection of five high rhizome yielding lines. These lines were evaluated in a multi-location trial and analysed using various multivariate analyses, such as Eberhart–Russell, AMMI and GGE biplot. The stability test was performed only on yield parameters because the yield parameters are the major goal of the breeder for the improvement of the crop. These analyses helped in the detection of the stable and superior genotype for fresh rhizome yield/plant and essential oil content. This genotype was named as Jor Lab KH-2 and was registered with ICAR-NBPGR, New Delhi. This selected genotype could be utilised for *C. caesia* varietal development programme and could be recommended for commercial cultivation. This superior genotype will also aid in conservation of this rare and endangered species which could be utilised for future prospects. This new variety will expand the possibilities for the commercial production of camphor which is the major component in the rhizome essential oil, and offer the pharmaceutical, flavour and fragrances industries new sources of raw materials which are cost-effective. This is the first report on the identification and registration of a superior variety of *C. caesia* confirmed through stability analysis showing consistent and satisfactory performance.

Author Contributions: Conceptualisation, M.L.; methodology, M.L.; software, S.M.; validation, M.L., S.M. and T.B.; formal analysis, M.P. and T.G.; investigation, S.M. and T.B.; resources, S.K.C. and H.L.; data curation, T.B. and M.P.; writing—original draft preparation, S.M., T.B. and T.G.; writing—review and editing, M.L. and S.M.; supervision, M.L. All authors have read and agreed to the published version of the manuscript.

Funding: This research was funded by NMPB, New Delhi, grant number R&D/AS/01/16-17.

Institutional Review Board Statement: Not Applicable.

Informed Consent Statement: Not Applicable.

Acknowledgments: The authors are thankful to the Director, CSIR-North East Institute of Science and Technology, Jorhat, Assam for providing necessary field and lab facilities to conduct the study.

Conflicts of Interest: The authors declare no conflict of interest.

References

1. Borah, A.; Kumar, D.; Paw, M.; Begum, T.; Lal, M. A review on ethnobotany and promising pharmacological aspects of an endangered medicinal plant, *Curcuma caesia* Roxb. *Turk. J. Bot.* **2020**, *44*, 205–213. [CrossRef]
2. Kress, W.J.; Prince, L.M.; Williams, K.J. The phylogeny and a new classification of the gingers (Zingiberaceae): Evidence from molecular data. *Am. J. Bot.* **2002**, *89*, 1682–1696. [CrossRef] [PubMed]
3. Sahu, R.; Saxena, J. A brief review on medicinal value of *Curcuma caesia*. *Int. J. Pharm. Life Sci.* **2013**, *4*, 2664–2666.

4. Sahu, B.; Kenwat, R.; Chandrakar, S. Medicinal value of *Curcuma cassia* Roxb: An overview. UK. *J. Pharm. Biosci.* **2016**, *4*, 69–74. [CrossRef]
5. Mahanta, B.P.; Sut, D.; Lal, M.; Haldar, S. Hydrodistillation alters the compositional originality in black turmeric (*Curcuma caesia* Roxb.) essential oil. *J. Essent. Oil Res.* **2021**, *33*, 240–246. [CrossRef]
6. Paliwal, P.; Pancholi, S.S.; Patel, R.K. Pharmacognostic parameters for evaluation of the rhizomes of *Curcuma caesia*. *J. Adv. Pharm. Technol. Res.* **2011**, *2*, 56–61. [CrossRef] [PubMed]
7. Paw, M.; Munda, S.; Borah, A.; Pandey, S.K.; Lal, M. Estimation of variability, genetic divergence, correlation studies of *Curcuma caesia* Roxb. *J. Appl. Res. Med. Aromat. Plants* **2020**, *17*, 100251. [CrossRef]
8. Available online: https://justagriculture.in/files/magazine/2021/october/002%20Black%20Turmeric.pdf (accessed on 15 September 2022).
9. Donipati, P.; Sreeramulu, S.H. Preliminary phytochemical screening of *Curcuma caesia*. *Int. J. Curr. Microbiol. App. Sci.* **2015**, *4*, 30–34.
10. Tag, H.; Das, A.K.; Loyi, H. Anti-inflammatory plant used by Khamti tribes of Lohit District in Arunachal Pradesh. *Nat. Prod. Rad.* **2007**, *4*, 340–343.
11. Israr, F.; Hassan, F.; Naqvi, B.S.; Azhar, I.; Jabeen, S.; Hssan, S.M.F. Studies on antibacterial activity of some traditional medicinal plants used in folk medicine *Pak. J. Pharm. Sci.* **2012**, *25*, 669–674.
12. Sasikumar, B. Genetic resource of *Curcuma*: Diversity, characterization and utilization. *Plant Genet. Res.* **2005**, *3*, 230–251. [CrossRef]
13. Rastogi, R.P.; Mehrotra, B.N. *Compendium of Indian Medicinal Plants*; CSIR: New Delhi, India, 1990; Volume 1, pp. 118–122.
14. Trivedi, P.C. *Ethnomedicinal Plant of India*; Aviskar Publisher: Jaipur, India, 2003; Volume 30, p. 46.
15. Mia, M.K.; Kadir, M.F.; Hossan, S.; Rahmatullah, M. Medicinal plants of the Garo tribe inhabiting the Madhupur forest region of Bangladesh. *Am. Eurasian J. Sustain. Agric.* **2009**, *3*, 165–171.
16. Idrisi, M.S.; Bordola, H.K.; Singh, R. Indigenous knowledge and medicinal use of plants by local communities in Rangit Valley, South Sikkim, India. *NeBIO* **2010**, *1*, 34–45.
17. Syamkumar, S.; Sasikumar, B. Molecular marker based genetic diversity analysis of Curcuma species from India. *Sci. Hortic.* **2007**, *112*, 235–241. [CrossRef]
18. Ravindran, P.N.; Nirmal Babu, K.; Sivaraman, K. *Turmeric: The Genus Curcuma*, 1st ed.; CRC Press: Boca Raton, FL, USA, 2007; p. 504.
19. Paw, M.; Gogoi, R.; Sarma, N.; Pandey, S.K.; Borah, A.; Begum, T.; Lal, M. Study of anti-oxidant, anti-inflammatory, genotoxicity, and antimicrobial activities and analysis of different constituents found in rhizome essential oil of *Curcuma caesia* Roxb., collected from North East India. *Curr. Pharm. Biotechnol.* **2020**, *21*, 403–413. [CrossRef]
20. Banerjee, A.; Nigam, S.S. Antifungal activity of the essential oil of *Curcuma caesia* Roxb. *Indian J. Med. Res.* **1976**, *64*, 1318–1321.
21. Arulmozhi, D.K.; Sridhar, N.; Veeranjaneyulu, A.; Arora, S.K. Preliminary mechanistic studies on the smooth muscle relaxant effect of hydroalcoholic extract of *Curcuma caesia*. *J. Herb. Pharmacother.* **2006**, *6*, 117–124. [CrossRef] [PubMed]
22. Mangla, M.; Shuaib, M.; Jain, J.; Kashyap, M. In-vitro evaluation of antioxidant activity of *Curcuma caesia* Roxb. *Int. J. Pharm. Sci. Res.* **2010**, *1*, 98–102.
23. Karmakar, I.; Saha, P.; Sarkar, N.; Bhattacharya, S.; Haldar, P.K. Neuropharmacological assessment of *Curcuma caesia* Roxb. rhizome in experimental animal models. *Orient. Pharm. Exp. Med.* **2011**, *11*, 251–255. [CrossRef]
24. Rajamma, A.G.; Bai, V.; Nambisan, B. Antioxidant and antibacterial activities of oleoresins isolated from nine *Curcuma* species. *Phytopharma* **2012**, *2*, 312–317.
25. Das, S.; Bordoloi, P.K.; Phukan, D.; Singh, S. Study of the anti-ulcerogenic activity of the ethanolic extracts of rhizome of *Curcuma caesia* against gastric ulcers in experimental animals. *Asian J. Pharm. Clin. Res.* **2012**, *5*, 200–203.
26. Sarangthem, K.; Haokip, M.J. Bioactive components in *Curcuma caesia* Roxb. growing in Manipur. *Bioscan* **2010**, *5*, 113–115.
27. Pandey, A.K.; Chowdhary, A.R. Volatile constituent of rhizome oil of *Curcuma caesia* Roxb. from central India. *Flavour Frag. J.* **2003**, *18*, 463. [CrossRef]
28. Song, E.K.; Cho, H.; Kim, J.S.; Kim, N.Y.; An, N.H.; Kim, J.A. Diarylheptanoids with free radical scavenging and hepatoprotective activity in vitro from *Curcuma longa*. *Planta Med.* **2001**, *67*, 876–877. [CrossRef] [PubMed]
29. Jayaprakasha, G.K.; Rao, L.J.; Sakariah, K.K. Antioxidant activities of curcumin, demethoxycurcumin and bisdemethoxy. *Food Chem.* **2006**, *98*, 720–724. [CrossRef]
30. Borah, A.; Paw, M.; Gogoi, R.; Loying, R.; Sarma, N.; Munda, S.; Pandey, S.K.; Lal, M. Chemical composition, antioxidant, anti-inflammatory, antimicrobial and in-vitro cytotoxic efficacy of essential of *Curcuma caesia* Roxb. leaves: An endangered medicinal plant of North East India. *Ind. Crop. Prod.* **2019**, *129*, 448–454. [CrossRef]
31. Bian, L.; Zhang, H.; Ge, Y.; Cepl, J.; Stejskal, J.; El-Kassaby, Y.A. Closing the gap between phenotyping and genotyping: Review of advanced, image-based phenotyping technologies in forestry. *Ann. For. Sci.* **2022**, *79*, 22. [CrossRef]
32. Anshori, M.F.; Purwoko, B.S.; Dewi, I.S.; Ardie, S.W.; Suwarno, W.B. Cluster heatmap for detection of good tolerance trait on doubled-haploid rice lines under hydroponic salinity screening. In Proceedings of the IOP Conference Series: Earth and Environmental Science, Makassar, Indonesia, 23–25 September 2019.
33. Fadli, M.; Farid, M.; Yassi, A.; Nasaruddin, N.; Anshori, M.F.; Nur, A.; Suratman, S. Evaluation of the advanced yield trial on tropical wheat (*Triticum aestivum*) mutant lines using selection index and multivariate analysis. *Biodiversitas* **2022**, *23*, 540–547. [CrossRef]

34. Singh, C.M.; Mishra, S.B.; Pandey, A.; Arya, M. Eberhart-Russell' and AMMI approaches of genotype by environment interaction (GEI) for yield and yield component traits in *Vigna radiata* L. Wilczek. *Int. J. Environ. Agric. Biotechnol.* **2014**, *7*, 277–292. [CrossRef]
35. Munda, S.; Sarma, N.; Lal, M. G × E interaction of 72 accessions with three year evaluation of *Cymbopogon winterianus* Jowitt. using regression coefficient and Additive Main effects and Multiplicative Interaction model (AMMI). *Ind. Crops Prod.* **2020**, *146*, 112169. [CrossRef]
36. Yan, W.; Kang, M.S.; Ma, B.; Woods, S.; Cornelius, P.L. GGE biplot vs. AMMI analysis of genotype-by-environment data. *Crop. Sci.* **2007**, *47*, 643–653. [CrossRef]
37. Lal, M.; Munda, S.; Dutta, S.; Pandey, S.K. Identification of a novel germplasm (Jor Lab L-9) of lemon grass (*Cymbopogon khasianus*) rich in methyl eugenol. *Crop. Breed. App. Biotechnol.* **2020**, *20*, e320720315. [CrossRef]
38. Clevenger, J.F. Apparatus for the determination of volatile oil. *J. Pharm. Sci.* **1928**, *17*, 345–349. [CrossRef]
39. Olivoto, T.; Lucio, A.D.C.; da Silva, J.A.G.; Marchioro, V.S.; de Souza, V.Q.; Jost, E. Mean performance and stability in multi-environment trials I: Combining features of AMMI and BLUP techniques. *Agron. J.* **2019**, *111*, 1–12. [CrossRef]
40. Ayaz, A.; Huang, H.; Zheng, M.; Zaman, W.; Li, D.; Saqib, S.; Zhao, H.; Lü, S. Molecular Cloning and Functional Analysis of *GmLACS2-3* Reveals Its Involvement in Cutin and Suberin Biosynthesis along with Abiotic Stress Tolerance. *Int. J. Mol. Sci.* **2021**, *22*, 9175. [CrossRef] [PubMed]
41. Naeem, M.; Muqarab, R.; Waseem, M. The *Solanum melongena* COP1 delays fruit ripening and influences ethylene signaling in tomato. *J. Plant Physiol.* **2019**, *240*, 152997. [CrossRef]
42. Ullah, F.; Gao, Y.; Sari, İ.; Jiao, R.-F.; Saqib, S.; Gao, X.-F. Macro-Morphological and Ecological Variation in Rosa sericea Complex. *Agronomy* **2022**, *12*, 1078. [CrossRef]
43. Aarthi, S.; Suresh, J.; Leela, N.K.; Prasath, D. Multi environment testing reveals genotype-environment interaction for curcuminoids in turmeric (*Curcuma longa* L.). *Ind. Crops Prod.* **2020**, *145*, 112090. [CrossRef]
44. Anandaraj, M.; Prasath, D.; Kandiannan, K.; Zachariah, T.J.; Srinivasan, V.; Jha, A.K.; Singh, B.K.; Singh, A.K.; Pandey, V.P.; Singh, S.P.; et al. Genotype by environment interaction effects on yield and curcumin in turmeric (*Curcuma longa* L.). *Ind. Crops Prod.* **2014**, *53*, 358–364. [CrossRef]
45. Aulia, R.; Maulana, H.; Filio, Y.L.; Shafira, N.A.; Anindita, P.A.; Suganda, T.; Concibido, V.; Karuniawan, A. Assessment of rhizome yield of local Indonesian turmeric (*Curcuma longa* L.) during two growing seasons. *Biodiversitas* **2022**, *23*, 2534–2543. [CrossRef]
46. Chapagain, T.R.; Timilsina, A.P.; Sharma, S.; Daha, K.M.; Ahamad, S. Evaluation of Turmeric (*Curcuma longa* L) genotypes for growth and yield attributes in plains of Eastern Nepal. *Asian J. Agric. Res.* **2021**, *8*, 57–64. [CrossRef]
47. Tiwari, V.; Shankar, D.; Singh, D.P. Comparison of tikhur genotype by using GGE Biplot. *J. Pharmacogn. Phytochem.* **2019**, *8*, 1470–1473.
48. Abua, M.N.; Iwo, G.A.; Ittah, M.A.; Obok, E.E.; Edugbo, R.E. GGE biplot analysis of multi-location yield trial of ginger (*Zingiber officinale* Rosc.) genotypes in South-Eastern, Nigeria. *Asian J. Agric. Rural Dev.* **2020**, *10*, 485–503.
49. Das, T.T.; Pradeepkumar, T.; Mayadevi, P.; Aiepe, K.C.; Kumaran, K. Stability analysis of ginger (*Zingiber officinale* Rosc). *J. Spices Aromat. Crops* **2000**, *9*, 165–167.
50. Singh, A.P.; Ojha, M.D.; Pandey, V.P.; Kumar, V. Genotypic x Environment interaction and stability analysis in Turmeric (*Curcuma longa* L.). *Int. J. Curr. Microbiol. App. Sci.* **2020**, *9*, 2656–2663. [CrossRef]
51. Suryani, S.; Al Anshory, A.C.; Marlin, M.; Artika, I.M.; Ambarsari, L.; Nurcholis, W. Variability total phenolic content and antioxidant activity of *Curcuma zanthorrhiza* and *C. aeruginosa* cultivated in three different locations in West Java, Indonesia. *Biodiversitas* **2022**, *23*, 1998–2003. [CrossRef]
52. Mehta, K.G.; Patel, R.H. Stability parameters for rhizome yield in *Curcuma longa*. *Ind. Cocoa, Arecanut Spices J.* **1983**, *6*, 77–80.
53. Singh, J.P.; Singh, M.K.; Singh, P.K.; Singh, R.D. Phenotypic stability in turmeric (*Curcuma longa* L.). *Ind. J. Arecanut Spices Med. Plants* **1995**, *19*, 40–42.
54. Mishra, R.; Gupta, A.K.; Lal, R.K. Genotype x environment interaction, stability analysis for yield and quality traits in turmeric (*Curcuma longa* L.). *Trends Phytochem. Res.* **2020**, *4*, 219–234.

genes

Article

Identification, Phylogeny, Divergence, Structure, and Expression Analysis of A20/AN1 Zinc Finger Domain Containing *Stress-Associated Proteins* (*SAPs*) Genes in *Jatropha curcas* L.

Abdul Jalal [1,2], Qurban Ali [2], Hakim Manghwar [3] and Daochen Zhu [1,*]

1 Biofuels Institute, School of the Environment and Safety Engineering, Jiangsu University, Zhenjiang 212013, China
2 Key Laboratory of Integrated Management of Crop Diseases and Pests, Ministry of Education, Department of Plant Pathology, College of Plant Protection, Nanjing Agricultural University, Nanjing 210095, China
3 Lushan Botanical Garden, Chinese Academy of Sciences, Jiujiang 332000, China
* Correspondence: dczhucn@ujs.edu.cn

Abstract: Jatropha is a small woody perennial biofuel-producing shrub. *Stress-associated proteins* (*SAPs*) are novel stress regulatory zinc-finger proteins and are mainly associated with tolerance against various environmental abiotic stresses in Jatropha. In the present study, the *JcSAP* gene family were analyzed comprehensively in *Jatropha curcas* and 11 *JcSAP* genes were identified. Phylogenetic analysis classified the *JcSAP* genes into four groups based on sequence similarity, similar gene structure features, conserved A20 and/or AN1 domains, and their responsive motifs. Moreover, the divergence analysis further evaluated the evolutionary aspects of the *JcSAP* genes with the predicted time of divergence from 9.1 to 40 MYA. Furthermore, a diverse range of cis-elements including light-responsive elements, hormone-responsive elements, and stress-responsive elements were detected in the promoter region of *JcSAP* genes while the *miRNA* target sites predicted the regulation of *JcSAP* genes via a candid *miRNA* mediated post-transcriptional regulatory network. In addition, the expression profiles of *JcSAP* genes in different tissues under stress treatment indicated that many *JcSAP* genes play functional developmental roles in different tissues, and exhibit significant differential expression under stress treatment. These results collectively laid a foundation for the functional diversification of *JcSAP* genes.

Keywords: abiotic stress; divergence; expression profile; jatropha; phylogeny; *SAP* genes

Citation: Jalal, A.; Ali, Q.; Manghwar, H.; Zhu, D. Identification, Phylogeny, Divergence, Structure, and Expression Analysis of A20/AN1 Zinc Finger Domain Containing *Stress-Associated Proteins* (*SAPs*) Genes in *Jatropha curcas* L. *Genes* **2022**, *13*, 1766. https://doi.org/10.3390/genes13101766

Academic Editor: Jean Molinier

Received: 12 September 2022
Accepted: 27 September 2022
Published: 30 September 2022

1. Introduction

Plant growth and productivity is severely affected by various biotic and abiotic stresses because of their sessile nature. These environmental stresses down- or upregulate a large pool of genes. To eliminate or reduce the harsh environmental effect, plants have evolved complex internal molecular mechanisms to modulate stress-responsive or regulatory genes [1,2]. Regulatory genes code for sensors that perceive stress signals, kinases that transmit the signals, and transcription factors that are down- or upregulated as a result of perceived stress. Thus, defence mechanisms start with the perception of stress, followed by signal transduction, synthesis of transcription factors, and finally down- or upregulation of genes that produce protective proteins and metabolites [2]. Hitherto, a large number of genes have been characterized that play an important role during different stresses and are involved at different levels of regulation, such as perception, signalling, transcription, and production of protective biomolecules [3–7]. The characterisation of genes belonging to the signal transduction and transcription factor categories is of great importance because of their effect on a wide range of stress-related genes. Zinc finger proteins are associated with this category and are considered the important proteins that contribute to protection against environmental stress [8–11].

Stress-associated proteins (SAP) are the newly identified stress regulatory zinc-finger proteins in plants with the A20 domain at N-terminal and AN1 domain at C-terminal, and are mainly associated with tolerance against various environmental abiotic stresses [1,12–15]. *SAP* genes in plants were first identified in *Oryza sativa*, and *OsSAP1* was declared to be largely involved in abscisic acid, healing wounds, heavy metals, salt, drought, and cold [9]. Similarly, other *OsSAP* genes have also been expressed under several environmental stresses, indicating their role in stress-response [5]. Recently, the response of *SAP* genes has been identified in several plant species and has been studied against various environmental stresses. The evidence from the previous literature has shown the presence of the *SAP* gene family in many species, like 18 *SAP* gene members in rice [5], 14 in Arabidopsis [5], 17 in barley [13], 27 in soybean [12], 37 in cotton [16], 13 in tomato [2], 12 in cucumber [17], 21 in eggplant [1], and 9 in almond [18]. However, the identification of *SAP* genes in Jatropha genome is not systematically deliberated.

Jatropha (*Jatropha curcas* L.) is a small woody perennial shrub that belongs to the subfamily Crotonoideae of Euphorbiaceae. Jatropha is a biofuel-producing non-food crop mostly grown in subtropical as well as tropical regions, with 40 to 50% of oil content in its seeds. Particularly important is the fact that Jatropha can grow on degraded soils, making it an attractive crop for biodiesel feedstock as it can be widely planted on marginal land considered inappropriate for food crops [19]. In addition, Jatropha is adaptable to grow in a wide range of agro-climatic conditions and possesses several properties like easy propagation, short gestation period, rapid growth, high oil content, low seed cost, and tolerance to salt and drought, making it suitable for biodiesel production [19–21]. With the increasing costs, gradual depletion of fossil energy resources, and having a strong potential for biofuel production, Jatropha is recently attracting the widespread attention of researchers. The identification of the *SAP* gene family in Jatropha might encompass a novel understanding of the flow and appearance of the mechanism of stress regulatory genes. In the current study, we identified the *SAP* gene family genome-wide in Jatropha and comprehensively analysed their phylogeny, domain, motif, gene structure, chromosome location, gene duplication, cis-acting elements, *miRNA* target sites, and the expression analysis in various tissues against stress, which laid a foundation for further studies on the biological function of *SAP* genes in Jatropha.

2. Materials and Methods

2.1. Identification, Sequence Alignment, and Phylogenetic Analysis of SAP Genes in Jatropha Curcas

To identify *SAP* genes in Jatropha, Pfam server (http://pfam.xfam.org/) (accessed on 31 August 2022) was searched for the A20 domain (PF01754) and AN1 domain (PF01428), and then the HMMER (https://www.ebi.ac.uk/Tools/hmmer/search/hmmsearch) (accessed on 31 August 2022) Hidden Markov model was used as a probe to screen all the candidate proteins. To ensure the reliability of the sequences and to remove redundant sequences, the search results of all candidate *SAP* protein sequences were further searched for the presence of A20/AN1 domains using the Pfam database (http://pfam.janelia.org/) (accessed on 31 August 2022), MOTIF search (https://www.genome.jp/tools/motif/) (accessed on 31 August 2022), NCBI conserved domain database (http://www.ncbi.nlm.nih.gov/Structure/cdd/wrpsb.cgi) (accessed on 31 August 2022), SMART database (http://smart.embl-heidelberg.de/) (accessed on 31 August 2022), and Inter ProScan program (http://www.ebi.ac.uk/Tools/pfa/ iprscan5/) (accessed on 31 August 2022). The protein sequences of the *SAP* gene family of rice and Arabidopsis (Supplementary Table S1) were retrieved from the previously published report [5]. Multiple sequence alignments of all the sequences of Jatropha, Arabidopsis, and rice were performed using Muscle. Subsequently, to create a phylogenetic tree, the alignments were imported to MEGA7 software (https://www.megasoftware.net/home) (accessed on 1 September 2022) using the neighbor-joining (NJ) method with a bootstrap option of 1000 replications. The phylogenetic tree was further visualized and edited via MEGA7 software (https://www.megasoftware.net/home) (accessed on 1 September 2022) [22–26].

2.2. *JcSAP* Protein Physicochemical Characteristics

To find the physicochemical properties of *JcSAP* genes, the NCBI database (https://www.ncbi.nlm.nih.gov/) (accessed on 1 September 2022) was searched to find the amino acid (bp), CDS (bp) and location on Chromosome. The ProtParam tool of ExPASy server (http://web.expasy.org/protparam) (accessed on 1 September 2022) was searched to predict the isoelectric points (pI), molecular weight (MW), grand average of hydropathicity (GRAVY), and Formula of each *SAP* gene in *Jatropha curcas*. Softberry server for plant protein location (http://www.softberry.com/berry.phtml?topic=protcomppl&group=programs&subgroup=proloc) (accessed on 1 September 2022) was employed to predict the subcellular localization each *SAP* protein in *Jatropha curcas* [22].

2.3. Gene Duplication Events, Homology and Synteny Analysis of *JcSAP* Genes

For Gene conservation, duplication events, homology, and Synteny analysis, a comparative Synteny analysis was performed by using circoletto Tool (https://www.tools.bat.infspire.org/circoletto/) (accessed on 2 September 2022) to visualize genome conservation. Protein sequences of Arabidopsis *SAP* genes were used against the identified *JcSAP* sequences and were finally exhibited by circus by running the Circoletto tool [27].

2.4. Conserved Domains, Motifs, and Gene Structure Organization of *JcSAP* Genes

The identified *JcSAP* protein sequences were subjected to NCBI CDD online software (https://www.ncbi.nlm.nih.gov/Structure/cdd/wrpsb.cgi) (accessed on 2 September 2022) for domain analysis, and the obtained results were visualized via the TBtools software (https://github.com/CJ-Chen/TBtools) (accessed on 2 September 2022). Similarly, to analyze the *JcSAP* proteins for the conserved motifs, the protein sequences of *JcSAP* were submitted to MEME suite software 5.4.1 (https://meme-suite.org/meme/tools/meme) (accessed on 2 September 2022). Consistently, to display the gene structure organization of *JcSAP* genes, the gene structure display server (http://gsds.gao-lab.org/) (accessed on 2 September 2022) was used by submitting the CDS and genomic sequences of *JcSAP* genes [22,24].

2.5. Divergence Analysis

For divergence analysis of *JcSAP* genes, the server *Ka*/*Ks* calculation tool (http://services.cbu.uib.no/tools/kaks) (accessed on 2 September 2022) was used and the non-synonymous substitution per nonsynonymous site (*Ka*) and synonymous substitution per synonymous site (*Ks*) was determined by inputting the DNA sequences of *JcSAP* genes using default parameters. The divergence time was calculated by the given formula [24,28].

$$Time\ of\ divergence\ (\mathrm{T}) = \frac{Synonymous\ substitution\ rate\ (dS \text{ or } Ks)}{2 \times Divergence\ rate\ (6.56 \times 10^{-9})} \times TMY\ (10^{-6})$$

2.6. Protein Structure Analysis of *JcSAP* Genes

To identify the structural composition of 11 *JcSAP* genes, an online tool for the prediction of secondary structure SOPMA (https://npsa-prabi.ibcp.fr/cgi-bin/secpredsopma.pl) (accessed on 2 September 2022) were used. The tertiary structure of *JcSAP* proteins were visualized via uniprot (https://www.uniprot.org) (accessed on 2 September 2022) to further support the secondary structure of *JcSAP* proteins [29].

2.7. Cis-Elements Analysis and Predicted miRNA Target Sites

To analyse the cis-regulatory element, the upstream region of 1500 bp of each genomic sequence of the *JcSAP* genes was submitted to PlantCARE server (https://bioinformatics.

psb.ugent.be/webtools/plantcare/html/) (accessed on 5 September 2022) and was searched for the presence of cis-regulatory elements. The results were then visualized using TBtools software (https://github.com/CJ-Chen/TBtools) (accessed on 6 September 2022) [24].

For the prediction *miRNA* target sites for *JcSAP* genes, the published miRNAs of *Jatropha curcas* were downloaded from the plant *miRNA* encyclopaedia (https://pmiren.com/download) (accessed on 6 September 2022) and were then subjected to the psRNATarget database (https://www.zhaolab.org/psRNATarget/analysis?function=3) (accessed on 6 September 2022) along with the CDS of *JcSAP* genes to predict the *miRNA* target sites and interaction with *JcSAP* genes. Finally, the results were visualized using Cytoscape software (Cytoscape Consortium, San Francisco, CA, USA) (http://apps.cytoscape.org/apps/stringapp) (accessed on 7 September 2022) [24,30].

2.8. Gene Expression Profiling of JcSAP Genes

For *JcSAP* gene expression patterns in different plant tissues under abiotic stress, the raw expression data of leaf and root tissues subjected to drought stress was retrieved from the public database of NCBI (https://www.ncbi.nlm.nih.gov/geo/query/acc.cgi?acc=GSE61109) (accessed on 11 September 2022) under the GEO accession number GSE61109. The expression profiles of all *JcSAP* genes were exhibited as reads per kilobase per million (RPKM) values and illustrated via heat map using TBtools [30,31].

3. Results

3.1. Identification and Phylogenetic Analysis of JcSAP Genes

To gain insight into the identification and evolution of *JcSAP* genes, the Hidden Markov model profiles of the A20 domain and AN1 domain were used as a probe to screen all the candidate proteins and were then used with the protein sequences of *Arabidopsis thaliana* and *Oryza sativa* to construct a phylogenetic tree (see materials and methods). Results demonstrated that *A. thaliana*, having a genome size of 135 mb and 2n chromosome number of 10, had a total of 14 *SAP* genes following both the A20 and AN1 domain. *Oryza sativa*, having a genome size of 372 mb and 2n chromosome number of 24, had a total of 18 *SAP* genes in which 12 members were following both the A20 and AN1 domain, 5 members were following only the AN1 domain, while 1 member was following only the A20 domain. *Jatropha curcas*, having a genome size of 416 mb and 2n chromosome number of 22, had a total of 11 *SAP* genes in which 8 members were following both the A20 and AN1 domain, 3 members were following only the AN1 domain, while there were no single gene members following the A20 domain only (Figure 1B, Table S1). To further investigate the classification and the evolutionary characteristics of the *JcSAP* proteins, an unrooted phylogenetic tree was constructed based on the *SAP* protein sequences of *Arabidopsis thaliana*, *Oryza sativa*, and Jatropha (Figure 1A). All available 34 sequences, including 14 *AtSAP*, 18 *OsSAP*, and 11 *JcSAP*, were mainly clustered into four groups. Group A contained four *JcSAP* members (*JcSAP810a*, *JcSAP810b*, *JcSAP111314*, and *JcSAP12*), clustered to *AtSAP8*, *AtSAP10*, *AtSAP11*, *AtSAP13*, and *AtSAP14*. Group B consisted of three JcSAP members (*JcSAP179*, *JcSAP3*, *JcSAP46*), clustered to *AtSAP1*, *AtSA3*, *AtSAP4*, *AtSAP6*, *AtSAP7*, and *AtSAP9*. Group C consisted of three *JcSAP* members (*JcSAP2a*, *JcSAP2b*, and *JcSAP2c*), clustered to *AtSAP2*. Group D included only one *JcSAP* member (*JcSAP5*) clustered to *AtSAP5*. These results of the comparative phylogenetic relationship predicted that the SAP members clustered together or with other species may share similar biological functions against stresses.

(A)

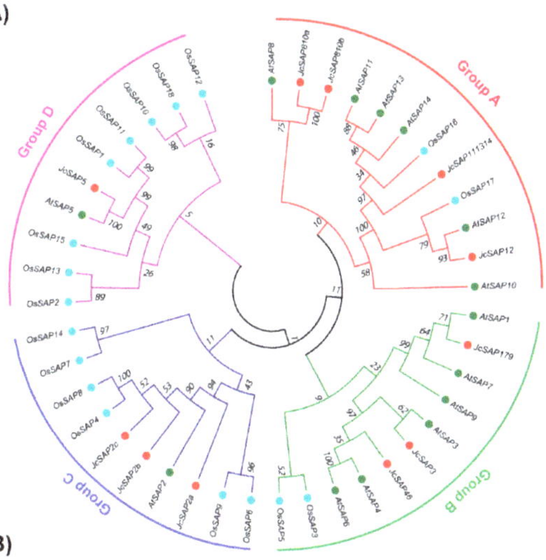

(B)

Species	Abbr.	All	A20 Only	AN1 Only	A20/AN1	Genome size (mb)	Chromosome
Arabidopsis Thaliana	At	14	0	0	14	135	2n=10
Oryza Sativa	Os	18	1	5	12	372	2n=24
Jatropha curcas	Jc	11	0	3	8	416	2n=22

Figure 1. Identification and phylogenetic Analysis of *SAP* members of Jatropha against Arabidopsis and rice. (**A**) Phylogenetic tree of 14 arabidopsis, 18 rice, and 11 Jatropha *SAP* members, (**B**) Identified *SAP* members of Arabisopsis, rice, and Jatropha with followed domains, genome size, and chromosome numbers.

3.2. Physicochemical Characteristics of JcSAP Genes

To further get into the physicochemical characteristic of the total 11 identified *SAP* genes in Jatropha in a comprehensive manner (Figure 1), the physicochemical parameters of *JsSAP* genes including gene code, location on chromosome, amino acids and CDS length, molecular weight (MW/kDa), isoelectric point (PI), GRAVY, Formula, and Predicted subcellular localization were investigated insilico and exhibited in Tables 1 and S1. Results demonstrated that the amino acid length *JcSAP* genes were varied from 133 (*JcSAP2a*) to 288 (*JcSAP111314*). CDS was ranged from 402 (*JcSAP2a*) to (867) (*JcSAP111314*). Molecular weight (MW/kDa) *JcSAP* genes varied from 14686.65 kDA (*JcSAP2a*) to 31998.28 (*JcSAP111314*), while the isoelectric point (PI) ranged from 8 PI (*JcSAP179*) to 9.4 PI (*JcSAP5*). Moreover, the predicted subcellular localization revealed that all JcSAP genes are cytoplasmic expect *JcSAP111314* and *JcSAP12*, which appeared to be extracellular. The presence of each *JcSAP* gene member on a specific chromosome was not predicted in silico, however the GRAVY and formula were given in Table 1. These results provide information about the basic known parameters of the *JcSAP* genes.

Table 1. Physicochemical characteristics of *JcSAP* genes.

Gene Name	Gene ID (Uniprot)	Chr	AA	CDS	MW/kDa	PI	GRAVY	Formula	Predicted Subcellular Localization
JcSAP179	A0A067JZZ9_JATCU	Un	178	537	18834.23	8	−0.453	$C_{809}H_{1279}N_{231}O_{261}S_{13}$	Cytoplasmic
JcSAP2a	A0A067KTX6_JATCU	Un	133	402	14686.65	9.1	−0.612	$C_{630}H_{990}N_{188}O_{194}S_{12}$	Cytoplasmic
JcSAP2b	A0A067JE11_JATCU	Un	171	516	18507.08	8.5	−0.442	$C_{785}H_{1266}N_{236}O_{249}S_{16}$	Cytoplasmic
JcSAP2c	A0A067L4U9_JATCU	Un	170	513	18087.44	8.8	−0.304	$C_{761}H_{1233}N_{229}O_{251}S_{15}$	Cytoplasmic
JcSAP3	A0A067KRF9_JATCU	Un	170	513	18661.18	8.8	−0.492	$C_{793}H_{1276}N_{236}O_{254}S_{15}$	Cytoplasmic
JcSAP46	A0A067L6B6_JATCU	Un	159	480	17559.01	8.6	−0.521	$C_{754}H_{1197}N_{225}O_{231}S_{14}$	Cytoplasmic
JcSAP5	A0A067L4Q5_JATCU	Un	185	558	20311.78	9.4	−0.665	$C_{849}H_{1395}N_{271}O_{283}S_{12}$	Cytoplasmic
JcSAP810a	A0A067JER8_JATCU	Un	135	408	14879.04	9.3	−0.49	$C_{643}H_{1038}N_{186}O_{199}S_{10}$	Cytoplasmic
JcSAP810b	A0A067JQU7_JATCU	Un	204	615	22711.95	9.3	−0.775	$C_{972}H_{1602}N_{282}O_{314}S_{14}$	Cytoplasmic
JcSAP111314	A0A067K2L3_JATCU	Un	288	867	31998.28	8.6	−0.665	$C_{1367}H_{2182}N_{416}O_{426}S_{23}$	Extracellular
JcSAP12	A0A067LAI2_JATCU	Un	197	594	21773.79	9	−0.639	$C_{924}H_{1487}N_{285}O_{288}S_{18}$	Extracellular

3.3. Synteny Analysis of *JcSAP* Genes

To further support the identification and homologous relationship among *SAP* genes, Synteny analysis of all the identified *SAP* genes of Jatropha and Arabidopsis was subjected to the circoletto tool to make a map of comparative synteny circos (see materials and methods). The Synteny map illustrated the relationship among the *SAP* genes of Jatropha and Arabidopsis species regarding their function, expression, duplication events, and evolution (Figure 2). The sequences were placed clockwise around a circle, starting at 12 o'clock, and the ideograms were placed in order to maximally untangle the ribbons; the queries and database entries were intertwined. Ribbon colors of the map diagram represent the alignment length, visualizing the sequence similarity and identity level, i.e., blue ≤ 0.25, green ≤ 0.50, orange ≤ 0.75, red > 0.75, providing an essential first glimpse at sequence relationships. The obtained results revealed that Arabidopsis *AtSAP1*, *AtSAP7*, and *AtSAP9* showed Synteny with *JcSAP179* of Jatropha. Similarly, *AtSAP2* showed Synteny with *JcSAP2a*, *JcSAP2b*, and *JcSAP2c*. *AtSAP3* showed Synteny with *JcSAP3*. *AtSAP4* and *AtSAP6* showed Synteny with *JcSAP46*. *AtSAP5* showed Synteny with *JcSAP5*. AtSAP10 showed Synteny with *JcSAP810a*, *JcSAP810b*. *AtSAP11*, *AtSAP13*, and *AtSAP14* showed Synteny with *JcSAP111314*. *AtSAP12* showed Synteny with *JcSAP12*. Moreover, based on the color intensity, the inward and outward tangling of ribbons showed conservation and duplication events, respectively, suggesting that SAP genes were conserved in Jatropha during evolution.

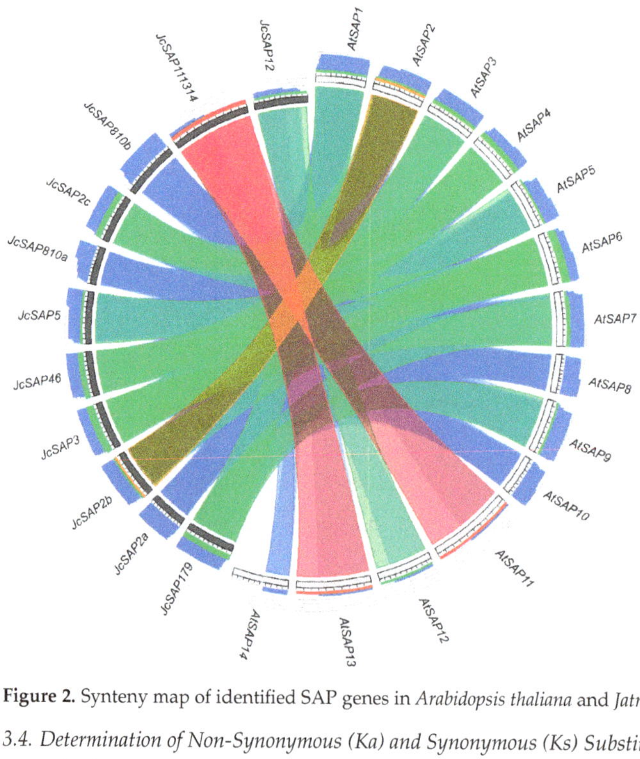

Figure 2. Synteny map of identified SAP genes in *Arabidopsis thaliana* and *Jatropha curcas*.

3.4. Determination of Non-Synonymous (Ka) and Synonymous (Ks) Substitution Rate

The divergence analysis was performed to gain insight into the evolutionary aspects of *JcSAP* genes by determining the non-synonymous substitution per non-synonymous site (*Ka*) and synonymous substitution per synonymous site (*Ks*) for each pair of paralogous *JcSAP* genes according to the phylogenetic tree (Figure S1) generated by *Ka/Ks* calculation server, in order to indicate the evolutionary discretion among *JcSAP* genes (Table 2). Results with a *Ka/Ks* value of <1 each pair of paralogous genes indicated the purifying selection pressure during the evolution. The divergence time for each pair of *JcSAP* genes ranged from 9.1 to 40 million years ago (MYA) (Table 3).

Table 2. Secondary Structure of 11 *JcSAP* proteins.

	H (%)	T (%)	E (%)	RC (%)
JcSAP11/13/14	22.22	1.04	11.11	65.62
JcSAP12	13.2	2.54	13.71	70.56
JcSAP1/7/9	21.35	2.81	11.8	64.04
JcSAP3	27.65	3.53	11.76	57.06
JcSAP2b	25.15	4.68	10.53	59.65
JcSAP4/6	30.19	4.4	12.58	52.83
JcSAP2a	26.32	4.51	13.53	55.64
JcSAP5	37.84	2.7	10.27	49.19
JcSAP8/10a	14.07	3.7	13.33	68.89
JcSAP2c	28.82	4.12	11.76	55.29
JcSAP8/10b	21.08	2.94	13.24	62.75

Table 3. Non-synonymous (*Ka*) and synonymous (*Ks*) substitution rate and divergence time of *JcSAP* genes.

Paralogous Genes		*Ka*	*Ks*	*Ka/Ks*	T (MYA)
JcSAP111314	*JcSAP12*	0.2614	0.5275	0.495545	40.2
JcSAP810b	*JcSAP810a*	0.04609	0.11945	0.385851	9.1
JcSAP2b	*JcSAP2c*	0.12225	0.33925	0.360354	25.8
JcSAP4/6	*JcSAP3*	0.1221	0.4072	0.299853	31.0
JcSAP5	*JcSAP179*	0.26015	0.46745	0.55653	35.6

3.5. Gene Structure, Domain and Motif Analysis of JcSAP Genes

To further explore the critical fundamental function of the identified *JcSAP* genes, the *JcSAP* were further investigated for conserved domains, gene structure organization, and motifs. Eight *JcSAP* genes like *JcSAP179*, *JcSAP2a*, *JcSAP2b*, *JcSAP2c*, *JcSAP3*, *JcSAP46*, *JcSAP5*, and *JcSAP810b* followed both the A20 domain at the N terminal and AN1 domain at C terminal, except *JcSAP111314*, *JcSAP12*, and *JcSAP810a*, which followed only the AN1 domain and had no A20 domain (Figure 3A). Similarly, the motif analysis revealed that all the *JcSAP* genes followed five conserved motifs corresponding to the A20 and AN1 domains (Figure 4A). The logos of the five identified motifs are present in Figure 4B. The demonstration of the gene structure organization further revealed the coding region (CDS) distribution (Figure 3B), indicating that *JcSAP179*, *JcSAP2a*, *JcSAP5*, *JcSAP810a*, and *JcSAP810b* have one exon, 2 *JcSAP3*, *JcSAP46*, *JcSAP111314*, and *JcSAP12* have two exons, while *JcSAP2b* and *JcSAP2c* have three exons. All these results collectively further support the conserved domain, motif, and structure and thus suggested to share similar biological functions in response to environmental abiotic stresses.

3.6. Protein Structure Analysis of JcSAP Genes

To delve further into the structure of the *JcSAP* genes, the secondary and tertiary structures of the *JcSAP* proteins were visualized (Table 2, Figure 5). The secondary structure of the *JcSAP* proteins explored the way these proteins fold and coil. The secondary structure of the JcSAP proteins consisted of four main elements including the α-helix (H%), β-turn (T%), extended chain (E%), and random coil (RC%). In the secondary structure of *JcSAP* proteins, the random coil (RC%) had the highest value ranging from 49.19% (*JcSAP5*) to 68.89% (*JcSAP8*/10a), followed by the α-helix (H%) ranging from 13.2% (*JcSAP12*) to 37.84% (*JcSAP5*), again followed by the extended chain (E%) ranging from 10.27% (*JcSAP5*) to 13.71% (*JcSAP12*), and β-turn (T%) ranging from 1.04% (*JcSAP11/13/14*) to 4.68% (*JcSAP2b*). Similarly, the tertiary structures of the *JcSAP* proteins visualized via uniprot (https://www.uniprot.org) (accessed on 2 September 2022) further supported the secondary structure of *JcSAP* proteins.

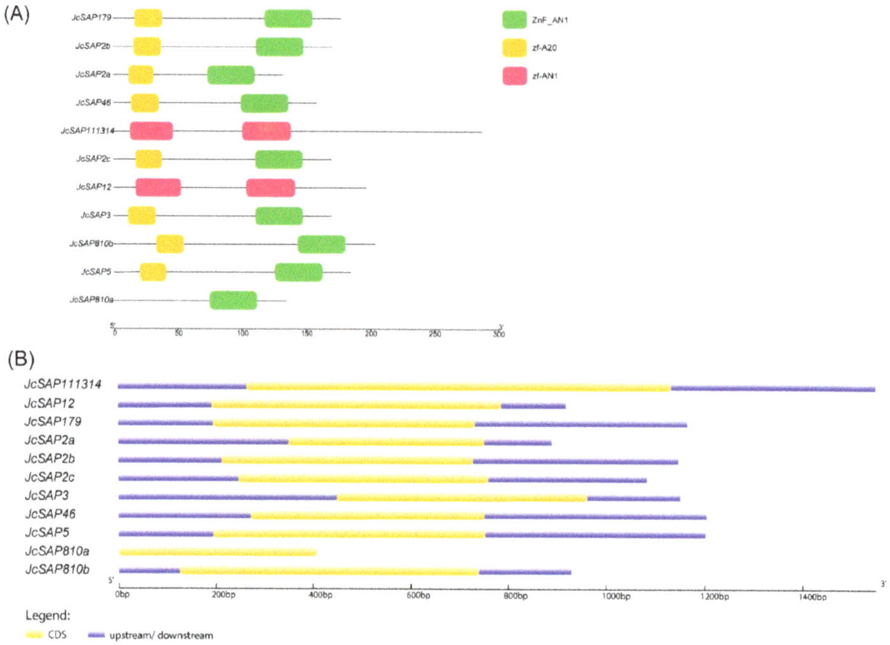

Figure 3. Domain (**A**) and Gene structure organization (**B**) of *JcSAP* genes.

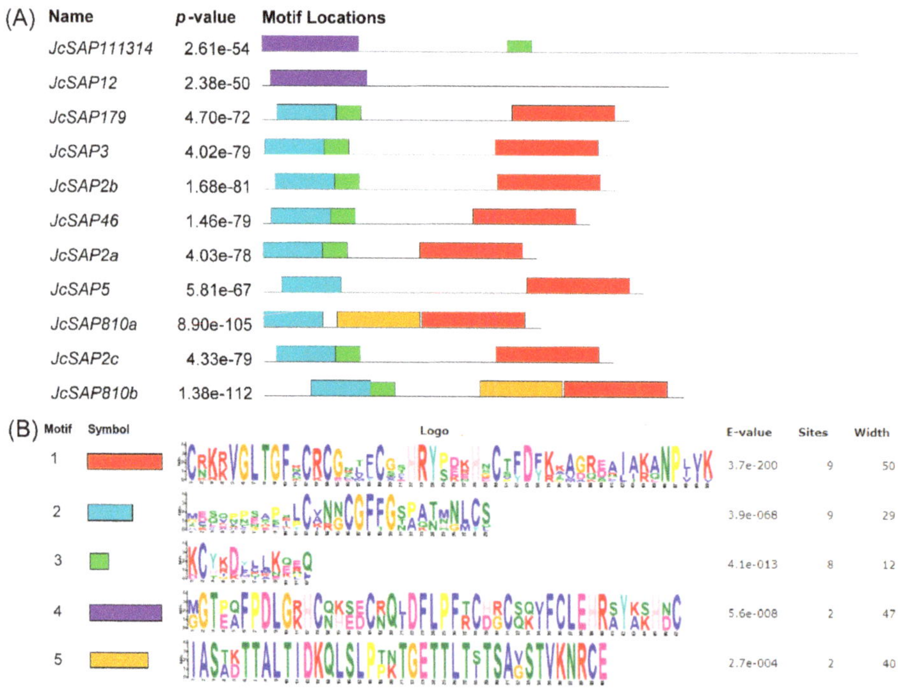

Figure 4. Conserved motifs of *JcSAP* genes in *Jatropha curcas*. (**A**) Identified motifs; (**B**) logos of the identified motifs.

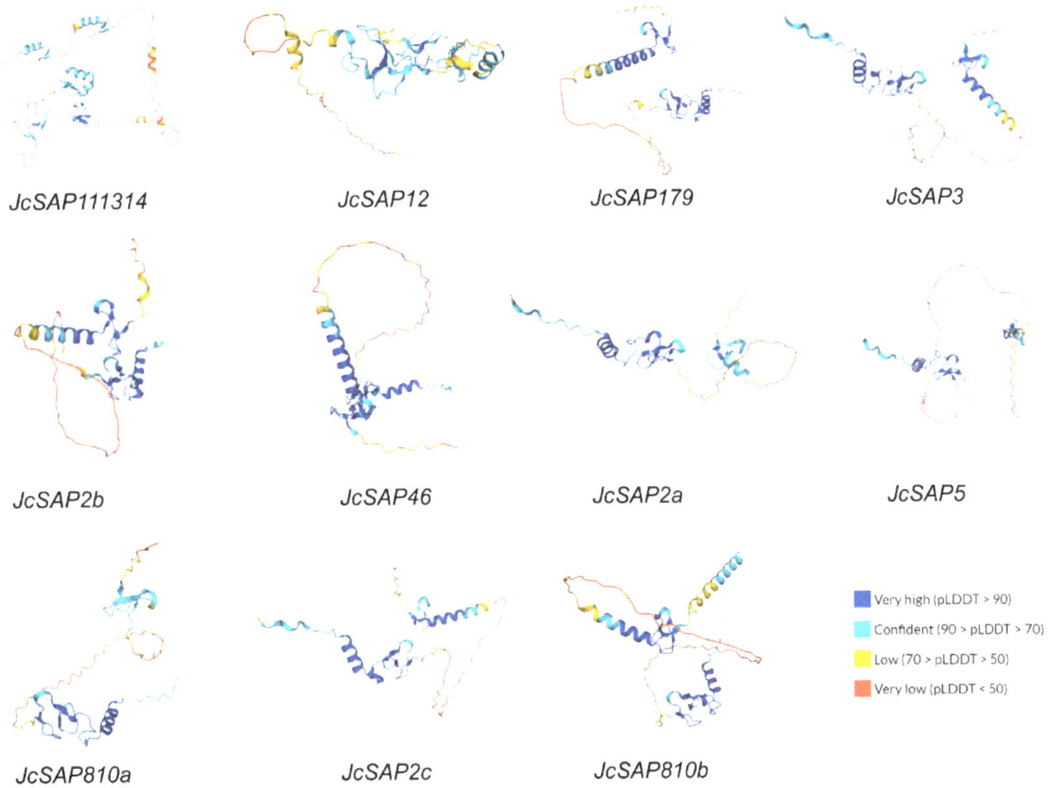

Figure 5. Illustration of the predicted tertiary structure of 11 *JcSAP* proteins. The protein structures all have the same domain colour schemes, revealing the degree of homology. The structures reveal a high degree of structural homology in most gene members.

3.7. Promoter Region Analysis and Prediction of miRNA Target Sites of JcSAP Genes

The promoter region analysis of the *JcSAP* genes resulted in the presence of a diverse range of cis-elements. The cis-elements identified during the promoter analysis of *JcSAP* genes were classified into three major groups/categories including light-responsive elements, hormone-responsive elements, and stress-responsive elements. The results revealed that there were 20 light responsive elements naming GT1-motif, chs-CMA1a, chs-CMA2a, AE-box, G-Box, Sp1, Box 4, MRE, I-box, GTGGC-motif, GATA-motif, TCT-motif, ATC-motif, TCT-motif, TCCC-motif, LAMP-element, 3-AF1 binding site, ACE, AT1-motif, GA-motif, 10 hormone responsive elements naming GARE-motif, TCA-element, TGACG-motif, CGTCA-motif, ABRE, TATC-box, TGA-element, P-box, AuxRR-core, AuxRE, and 12 Stress Responsive Elements naming TC-rich repeats, MBS, LTR, ARE, GCN4_motif, RY-element, HD-Zip 1, O2-site, WUN-motif, CAT-box, NON-box, CAT-box (Figure 6A). Moreover, the mean values of light-responsive elements ranged from 1.20 (*JcSAP810b*) to 3.50 (*JcSAP2a*), the mean values of hormone responsive elements ranged from 1.13 (*JcSAP111314*) to 4.00 (*JcSAP2b*), and the mean values of stress responsive elements ranged from 1.00 (*JcSAP2a*) to 2.33 (*JcSAP810a* and *JcSAP2c*) (Figure 6B).

Similarly, to predict the *miRNA* target sites of the *JcSAP* genes, the coding sequences of the identified *JcSAP* genes were used against the *miRNAs* of Jatropha (see material and methods). The results revealed that only three *miRNAs* (*Jcu-miR393*, *Jcu-miR1628*, and *Jcu-miR2111*) showed interaction with only four *JcSAP* genes (*JcSAP2a*, *JcSAP2b*, *JcSAP810a*, and *JcSAP12*) (Figure 6C). Moreover, *miRNA Jcu-miR393* showed interaction with two *JcSAP* genes, *JcSAP2a* and *JcSAP2b*, *miRNA Jcu-miR1628* showed interaction with *JcSAP12*, while *miRNA Jcu-miR2111* showed interaction with *JcSAP810a*. The Excel spreadsheet containing targeting sites predicted *miRNAs* ID, and the alignment with *JcSAP* gens are given in the Supplementary Table S2.

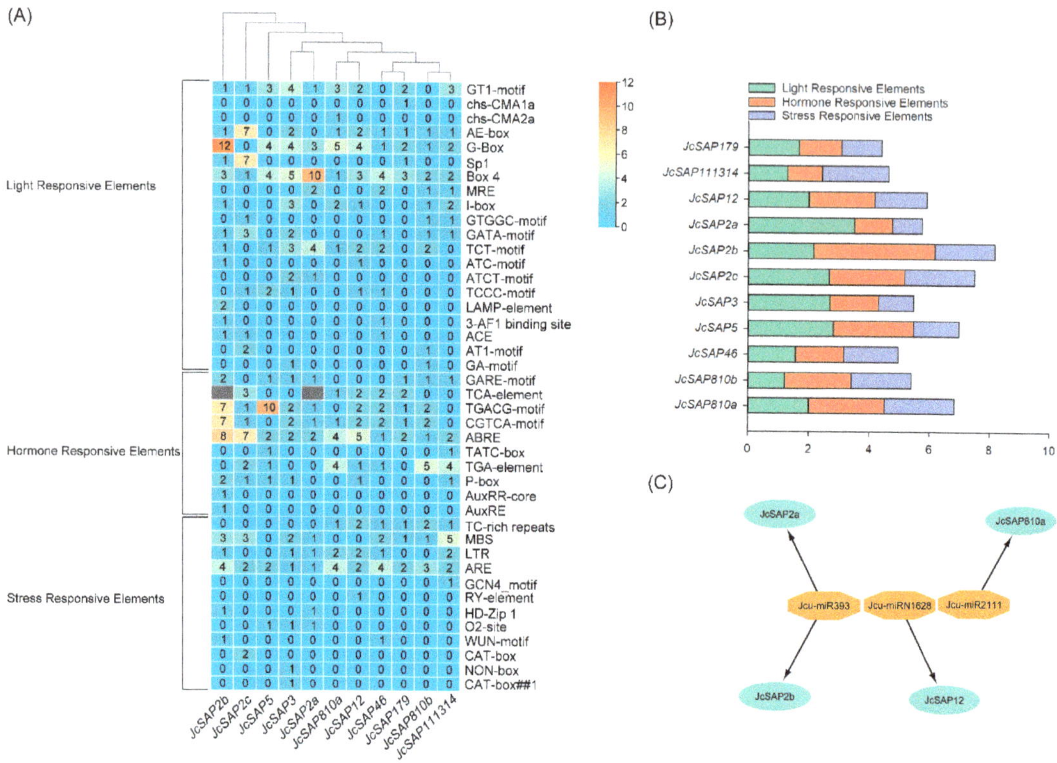

Figure 6. Cis regulatory elements and prediction of *miRNA* target sites of *JcSAP* genes. (**A**) The heat map of the Cis elements and their classification, present in the promoter region of *JcSAP* genes, (**B**) Graphical visualization of the groups of Cis elements in the promoter region of *JcSAP* genes, (**C**) Predicted *miRNA* target sites of *JcSAP* genes.

3.8. Gene Expression Profiling of *JcSAP* Genes

To investigate the possible function of *JcSAP* genes during environmental abiotic stress in Jatropha, the expression data were analysed from leaf and root tissues of Jatropha plants subjected to drought stress [32] (Figure 7A). The heat map showed the expression level of 11 *JcSAP* genes in different tissues of leaf and root during drought stress and after recovery (Figure 7B). The results further revealed that *JcSAP179*, *JcSAP3*, and *JcSAP2b* were highly expressed in leaf and in root, followed by *JcSAP111314*, *JcSAP810a*, and *JcSAP810b* as compared with the other *JcSAP* genes. Contrarily, *JcSAP2a*, *JcSAP5*, *JcSAP2c* were highly expressed in roots only. These results provide the possible involvement of *JcSAP* genes in abiotic stress tolerance in Jatropha.

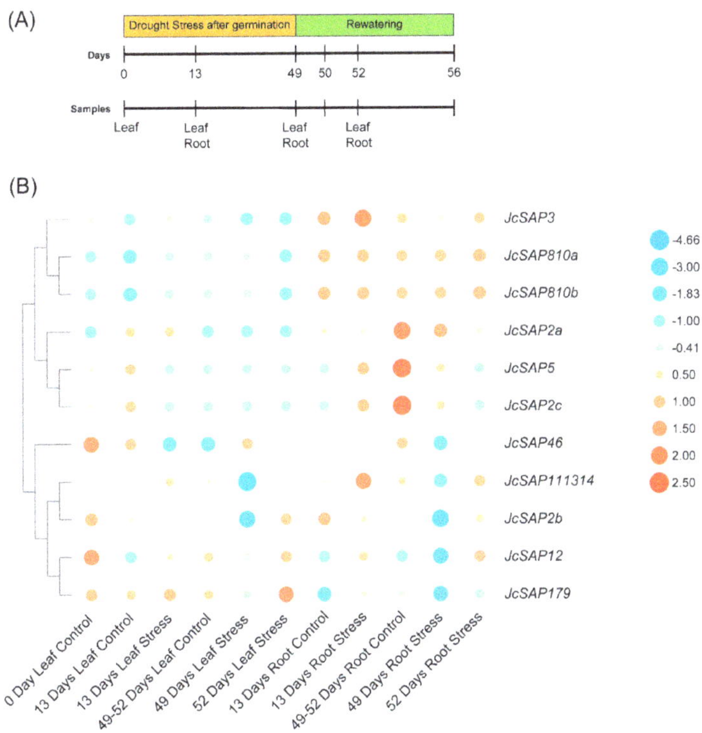

Figure 7. Expression levels of *JcSAP* genes in leaf and root tissues under abiotic stress. (**A**) Drought stress subjection and samples collected at different intervals during and after recovery of heat stress, (**B**) The expression levels the *JcSAP* genes at different intervals of leaf and root with their respective controls.

4. Discussion

Stress-associated proteins (SAP) are novel stress regulatory zinc-finger proteins and are strongly associated with tolerance against various abiotic stresses [1,12–15]. *SAP* gene family has previously identified and comprehensively studied in many plant species including *Oryza sativa* [5], *Arabidopsis thaliana* [5], *Hordeum vulgare* [13], *Glycine max* [12], *Gossypium hirsutum* [16], *Solanum lycopersicum* [2], *Cucumis sativus* [17], *Solanum melongena* [1], and *Prunus dulcis* [18], etc. However, there is no systematic study of *SAP* genes in the most important biofuel-producing shrub, Jatropha. In present study, 11 *SAP* genes were identified genome-wide in Jetropha and the phylogenetic divided the *JcSAP* genes into four groups (Figure 1); Synteny analysis showed that Jatropha *SAP* genes had a high homology with the Arabidopsis *SAP* genes (Figure 2). These results are highly in consistence with the previously reported results of [1,2], suggesting very little variation in *SAP* gene family members.

To obtain further insight into the similar features and biological functions of *JcSAP* genes, a search for the conserved domain, motif, and structure was conducted (Figures 3 and 4), resulting in the presence of A20 or AN1 domains and their respective motifs, thus suggesting that they share a similar biological function in response to stresses. Zinc-finger A20 or AN1 domains are highly conserved in all plant species and these results are in agreement with the previous reports [15,33]. Moreover, the domain organization revealed some domain-wise grouping, illustrating the presence of solely the AN1 domain in three *JcSAP* genes, with the others having both the A20 and AN1 domains (Figure 3). Except for Arabidopsis and tomato, similar reports were also exhibited in some other plant species like *Amborella trichopoda*, soybean, rice, and eggplant [1]. Such results may be due to the existence of a homology structure beyond the domain sequences [2]. These cases may also indicate the ancient origin of the specific AN1 domain with respect to its characteristics as compared with the A20 domain, or may also be due to the loss of one domain during evolution, as such cases occur in prologue genes (Table 2) and were also in line with the previous report on *Brassica napus* [14]. Moreover, the presence of corresponding motifs of the A20 and AN1 domain and the diversity of exon in the gene structure organization further strengthen the understanding of the evolutionary mechanisms in the *JcSAP* gene members [34,35]. In the present study, no intron has been found in *JcSAP* genes (Figure 3B), and this intron-free characteristic of *SAP* genes is usually found in other plant species like eggplant, rice, and apple [1,5,12,36]. It may be attributed to the fact that intron-free gene families can reduce posttranscriptional processing and rapidly adjust transcript expression [37].

The differential expression pattern of *JcSAP* genes in leaf and root tissues against drought stress revealed a potential role of these genes in stress response (Figure 7). Various studies have declared the role of *SAP* genes in different biotic and abiotic stresses [12,13]. Previous studies indicated that *SAP* genes play a great role in mediating abiotic stresses including cold, salt, and drought [1,2]. Our results are also in line with the previous studies on other species [12,36]. Altogether, the present study provides a baseline for understanding the molecular role of *JcSAP* genes and for further study of these genes against different abiotic stresses.

5. Conclusions

In conclusion, a total 11 *SAP* genes were identified during this study of Jatropha and were divided into four groups based on the phylogenetic analysis and amino acid sequence similarity; they may share similar biological functions against stresses. The physicochemical properties of *JcSAP* genes uncovered the basic gene parameters like amino acids and CDS length, molecular weight (MW/kDa), isoelectric point (PI), GRAVY, and molecular formula, and further revealed that most of the *JcSAP* genes are cytoplasmic, however, no detailed information was found regarding the chromosomal localization of *JcSAP* genes. The Synteny analysis showed that most of the *JcSAP* proteins were highly homologous to the Arabidopsis SAP proteins, indicating that *SAP* genes are conserved in Jatropha during evolution. Further domains and motifs analysis revealed that the A20 and AN1 domains are conserved in *JcSAP* genes and their similar gene structure features may be due to the duplication events during evolution. The divergence analysis further provided insight into the evolutionary aspects of *JcSAP* genes revealing the time of divergence from 9.1 to 40 MYA. The promoter region analysis of *JcSAP* genes resulted in a diverse range of cis-elements including light-responsive elements, hormone-responsive elements, and stress-responsive elements. The predicted *miRNA* target sites revealed that *JcSAP* genes may be regulated by a complicated *miRNA* mediated posttranscriptional regulatory network. In addition, the expression profiles of *JcSAP* genes in different tissues and stress treatments indicated that many *JcSAP* genes play functional developmental roles in different tissues, and exhibit significant differential expression under different stress treatments. All these results collectively suggested that *JcSAP* genes share similar biological functions in response to stresses and provide valuable clues for further investigation of *JcSAP* genes' function and diversity.

Supplementary Materials: The following supporting information can be downloaded at: https://www.mdpi.com/article/10.3390/genes13101766/s1, Figure S1: Phylogenetic tree for *Ka/Ks* paralogous genes; Table S1: Basic gene information; Table S2: *miRNA* target sites.

Author Contributions: Conceptualization, A.J.; methodology, A.J.; software, A.J.; validation, A.J. and Q.A. and H.M.; formal analysis, A.J.; investigation, A.J.; resources, A.J.; data curation, Q.A. and H.M.; writing—original draft preparation, A.J.; writing—review and editing, D.Z.; visualization, A.J.; supervision, D.Z.; project administration, D.Z.; funding acquisition, D.Z. All authors have read and agreed to the published version of the manuscript.

Funding: This work was supported by the Key Research and Development Program of Jiangsu Province (Grant No. BE2021691), and the Priority Academic Program Development of Jiangsu Higher Education Institutions.

Institutional Review Board Statement: Not applicable.

Informed Consent Statement: Not applicable.

Data Availability Statement: The data supporting the results are already mentioned in the main text and in supplementary files.

Conflicts of Interest: The authors declare no conflict of interest.

References

1. Wan, F.; Xu, Y.; Wang, S.; Gao, J.; Lu, D.; Zhou, C.; Liao, Y.; Ma, Y.; Zheng, Y. Identification and Expression Analysis of Zinc Finger A20/AN1 Stress-Associated Genes SmSAP Responding to Abiotic Stress in Eggplant. *Horticulturae* **2022**, *8*, 108. [CrossRef]
2. Solanke, A.U.; Sharma, M.K.; Tyagi, A.K.; Sharma, A.K. Characterization and Phylogenetic Analysis of Environmental Stress-Responsive SAP Gene Family Encoding A20/AN1 Zinc Finger Proteins in Tomato. *Mol. Genet. Genom.* **2009**, *282*, 153–164. [CrossRef] [PubMed]
3. Yamaguchi-Shinozaki, K.; Shinozaki, K. Transcriptional Regulatory Networks in Cellular Responses and Tolerance to Dehydration and Cold Stresses. *Annu. Rev. Plant Biol.* **2006**, *57*, 781–803. [CrossRef] [PubMed]
4. Chinnusamy, V.; Zhu, J.; Zhu, J.-K. Cold Stress Regulation of Gene Expression in Plants. *Trends Plant Sci.* **2007**, *12*, 444–451. [CrossRef]
5. Vij, S.; Tyagi, A.K. Emerging Trends in the Functional Genomics of the Abiotic Stress Response in Crop Plants. *Plant Biotechnol. J.* **2007**, *5*, 361–380. [CrossRef]
6. Bhatnagar-Mathur, P.; Vadez, V.; Sharma, K.K. Transgenic Approaches for Abiotic Stress Tolerance in Plants: Retrospect and Prospects. *Plant Cell Rep.* **2008**, *27*, 411–424. [CrossRef]
7. Solanke, A.U.; Sharma, A.K. Signal Transduction during Cold Stress in Plants. *Physiol. Mol. Biol. Plants* **2008**, *14*, 69–79. [CrossRef]
8. Davletova, S.; Schlauch, K.; Coutu, J.; Mittler, R. The Zinc-Finger Protein Zat12 Plays a Central Role in Reactive Oxygen and Abiotic Stress Signaling in Arabidopsis. *Plant Physiol.* **2005**, *139*, 847–856. [CrossRef]
9. Mukhopadhyay, A.; Vij, S.; Tyagi, A.K. Overexpression of a Zinc-Finger Protein Gene from Rice Confers Tolerance to Cold, Dehydration, and Salt Stress in Transgenic Tobacco. *Proc. Natl. Acad. Sci. USA* **2004**, *101*, 6309–6314. [CrossRef]
10. Ciftci-Yilmaz, S.; Mittler, R. The Zinc Finger Network of Plants. *Cell. Mol. Life Sci.* **2008**, *65*, 1150–1160. [CrossRef]
11. Xu, D.-Q.; Huang, J.; Guo, S.-Q.; Yang, X.; Bao, Y.-M.; Tang, H.-J.; Zhang, H.-S. Overexpression of a TFIIIA-Type Zinc Finger Protein Gene ZFP252 Enhances Drought and Salt Tolerance in Rice (*Oryza Sativa* L.). *FEBS Lett.* **2008**, *582*, 1037–1043. [CrossRef] [PubMed]
12. Zhang, X.-Z.; Zheng, W.-J.; Cao, X.-Y.; Cui, X.-Y.; Zhao, S.-P.; Yu, T.-F.; Chen, J.; Zhou, Y.-B.; Chen, M.; Chai, S.-C. Genomic Analysis of Stress Associated Proteins in Soybean and the Role of GmSAP16 in Abiotic Stress Responses in Arabidopsis and Soybean. *Front. Plant Sci.* **2019**, *10*, 1453. [CrossRef]
13. Baidyussen, A.; Aldammas, M.; Kurishbayev, A.; Myrzabaeva, M.; Zhubatkanov, A.; Sereda, G.; Porkhun, R.; Sereda, S.; Jatayev, S.; Langridge, P. Identification, Gene Expression and Genetic Polymorphism of Zinc Finger A20/AN1 Stress-Associated Genes, HvSAP, in Salt Stressed Barley from Kazakhstan. *BMC Plant Biol.* **2020**, *20*, 156. [CrossRef]
14. He, X.; Xie, S.; Xie, P.; Yao, M.; Liu, W.; Qin, L.; Liu, Z.; Zheng, M.; Liu, H.; Guan, M. Genome-Wide Identification of Stress-Associated Proteins (SAP) with A20/AN1 Zinc Finger Domains Associated with Abiotic Stresses Responses in Brassica Napus. *Environ. Exp. Bot.* **2019**, *165*, 108–119. [CrossRef]
15. Giri, J.; Dansana, P.K.; Kothari, K.S.; Sharma, G.; Vij, S.; Tyagi, A.K. SAPs as Novel Regulators of Abiotic Stress Response in Plants. *Bioessays* **2013**, *35*, 639–648. [CrossRef] [PubMed]
16. Gao, W.; Long, L.; Tian, X.; Jin, J.; Liu, H.; Zhang, H.; Xu, F.; Song, C. Genome-Wide Identification and Expression Analysis of Stress-Associated Proteins (SAPs) Containing A20/AN1 Zinc Finger in Cotton. *Mol. Genet. Genom.* **2016**, *291*, 2199–2213. [CrossRef]
17. Lai, W.; Zhou, Y.; Pan, R.; Liao, L.; He, J.; Liu, H.; Yang, Y.; Liu, S. Identification and Expression Analysis of Stress-Associated Proteins (SAPs) Containing A20/AN1 Zinc Finger in Cucumber. *Plants* **2020**, *9*, 400. [CrossRef] [PubMed]
18. Fatima, S.; Zafar, Z.; Gul, A.; Bhatti, M.F. Genome-Wide Identification of Stress-Associated Proteins (SAPs) Encoding A20/AN1 Zinc Finger in Almond (Prunus Dulcis) and Their Differential Expression during Fruit Development. *Plants* **2021**, *11*, 117. [CrossRef]
19. Fairless, D. The Little Shrub That Could–Maybe: India, like Many Countries, Has High Hopes for Jatropha as a Biofuel Source, but Little Is Known about How to Make It a Successful Crop. Daemon Fairless Digs for the Roots of a New Enthusiasm. *Nature* **2007**, *449*, 652–656. [CrossRef]
20. Bhasanutra, R.; Sutiponpeibun, S. Jatropha Curcas Oil as a Substitute for Diesel Engine Oil. *Int. Energy J.* **2017**, *4*, 56–70.
21. Openshaw, K. A Review of Jatropha Curcas: An Oil Plant of Unfulfilled Promise. *Biomass Bioenergy* **2000**, *19*, 1–15. [CrossRef]

22. Dong, Y.; Lu, J.; Liu, J.; Jalal, A.; Wang, C. Genome-Wide Identification and Functional Analysis of JmjC Domain-Containing Genes in Flower Development of Rosa Chinensis. *Plant Mol. Biol.* **2020**, *102*, 417–430. [CrossRef] [PubMed]
23. Liu, J.; Wu, S.; Sun, J.; Sun, J.; Wang, H.; Cao, X.; Lu, J.; Jalal, A.; Wang, C. Genome-Wide Analysis Reveals Widespread Roles for RcREM Genes in Floral Organ Development in Rosa Chinensis. *Genomics* **2021**, *113*, 3881–3894. [CrossRef] [PubMed]
24. Jalal, A.; Sun, J.; Chen, Y.; Fan, C.; Liu, J.; Wang, C. Evolutionary Analysis and Functional Identification of Clock-Associated PSEUDO-RESPONSE REGULATOR (PRRs) Genes in the Flowering Regulation of Roses. *Int. J. Mol. Sci.* **2022**, *23*, 7335. [CrossRef] [PubMed]
25. Liu, J.; Ren, M.; Chen, H.; Wu, S.; Yan, H.; Jalal, A.; Wang, C. Evolution of SHORT VEGETATIVE PHASE (SVP) Genes in Rosaceae: Implications of Lineage-Specific Gene Duplication Events and Function Diversifications with Respect to Their Roles in Processes Other than Bud Dormancy. *Plant Genome* **2020**, *13*, e20053. [CrossRef]
26. Ayaz, A.; Saqib, S.; Huang, H.; Zaman, W.; Lü, S.; Zhao, H. Genome-Wide Comparative Analysis of Long-Chain Acyl-CoA Synthetases (LACSs) Gene Family: A Focus on Identification, Evolution and Expression Profiling Related to Lipid Synthesis. *Plant Physiol. Biochem.* **2021**, *161*, 1–11. [CrossRef]
27. Errum, A.; Rehman, N.; Khan, M.R.; Ali, G.M. Genome-Wide Characterization and Expression Analysis of Pseudo-Response Regulator Gene Family in Wheat. *Mol. Biol. Rep.* **2021**, *48*, 2411–2427. [CrossRef]
28. Gaut, B.S.; Morton, B.R.; McCaig, B.C.; Clegg, M.T. Substitution Rate Comparisons between Grasses and Palms: Synonymous Rate Differences at the Nuclear Gene Adh Parallel Rate Differences at the Plastid Gene RbcL. *Proc. Natl. Acad. Sci. USA* **1996**, *93*, 10274–10279. [CrossRef]
29. Ge, M.; Zhong, R.; Sadeghnezhad, E.; Hakeem, A.; Xiao, X.; Wang, P.; Fang, J. Genome-Wide Identification and Expression Analysis of Magnesium Transporter Gene Family in Grape (Vitis Vinifera). *BMC Plant Biol.* **2022**, *22*, 217. [CrossRef]
30. Wang, J.; Jiang, X.; Bai, H.; Liu, C. Genome-Wide Identification, Classification and Expression Analysis of the JmjC Domain-Containing Histone Demethylase Gene Family in *Jatropha Curcas* L. *Sci. Rep.* **2022**, *12*, 6543. [CrossRef]
31. Ali, Q.; Ayaz, M.; Mu, G.; Hussain, A.; Yuanyuan, Q.; Yu, C.; Xu, Y.; Manghwar, H.; Gu, Q.; Wu, H. Revealing Plant Growth-Promoting Mechanisms of Bacillus Strains in Elevating Rice Growth and Its Interaction with Salt Stress. *Front. Plant Sci.* **2022**, *13*, 3190. [CrossRef] [PubMed]
32. Sapeta, H.; Lourenço, T.; Lorenz, S.; Grumaz, C.; Kirstahler, P.; Barros, P.M.; Costa, J.M.; Sohn, K.; Oliveira, M.M. Transcriptomics and Physiological Analyses Reveal Co-Ordinated Alteration of Metabolic Pathways in Jatropha Curcas Drought Tolerance. *J. Exp. Bot.* **2016**, *67*, 845–860. [CrossRef] [PubMed]
33. Wei, Q.; Wang, J.; Wang, W.; Hu, T.; Hu, H.; Bao, C. A High-Quality Chromosome-Level Genome Assembly Reveals Genetics for Important Traits in Eggplant. *Hortic. Res.* **2020**, *7*, 153. [CrossRef] [PubMed]
34. Rogozin, I.B.; Sverdlov, A.V.; Babenko, V.N.; Koonin, E.V. Analysis of Evolution of Exon-Intron Structure of Eukaryotic Genes. *Brief. Bioinform.* **2005**, *6*, 118–134. [CrossRef]
35. Rose, A.B. Intron-Mediated Regulation of Gene Expression. *Nucl. pre-mRNA Process. Plants* **2008**, *326*, 277–290. [CrossRef]
36. Dong, Q.; Duan, D.; Zhao, S.; Xu, B.; Luo, J.; Wang, Q.; Huang, D.; Liu, C.; Li, C.; Gong, X.; et al. Genome-Wide Analysis and Cloning of the Apple Stress-Associated Protein Gene Family Reveals MdSAP15, Which Confers Tolerance to Drought and Osmotic Stresses in Transgenic Arabidopsis. *Int. J. Mol. Sci.* **2018**, *19*, 2478. [CrossRef]
37. Grzybowska, E.A. Human Intronless Genes: Functional Groups, Associated Diseases, Evolution, and MRNA Processing in Absence of Splicing. *Biochem. Biophys. Res. Commun.* **2012**, *424*, 1–6. [CrossRef]

genes

MDPI

Article

Molecular Phylogeny, DNA Barcoding, and *ITS2* Secondary Structure Predictions in the Medicinally Important *Eryngium* Genotypes of East Coast Region of India

Gobinda Chandra Acharya [1,†], Sansuta Mohanty [1,†], Madhumita Dasgupta [2], Supriya Sahu [1,3], Satyapriya Singh [1], Ayyagari V. V. Koundinya [1], Meenu Kumari [4], Ponnam Naresh [5] and Manas Ranjan Sahoo [1,*]

1 Central Horticultural Experiment Station, ICAR–Indian Institute of Horticultural Research, Bhubaneswar 751019, Odisha, India
2 ICAR Research Complex for Northeastern Hill Region, Manipur Centre, Imphal 795004, Manipur, India
3 All India Institute of Medical Sciences, Bhubaneswar 751019, Odisha, India
4 ICAR Research Complex for Eastern Region, Research Centre, Ranchi 834010, Jharkhand, India
5 ICAR–Indian Institute of Horticultural Research, Bengaluru 560089, Karnataka, India
* Correspondence: manas.sahoo@icar.gov.in; Tel.: +91-674-2471867; Fax: +91-674-2471712
† These authors contributed equally to this work.

Citation: Acharya, G.C.; Mohanty, S.; Dasgupta, M.; Sahu, S.; Singh, S.; Koundinya, A.V.V.; Kumari, M.; Naresh, P.; Sahoo, M.R. Molecular Phylogeny, DNA Barcoding, and *ITS2* Secondary Structure Predictions in the Medicinally Important *Eryngium* Genotypes of East Coast Region of India. *Genes* **2022**, *13*, 1678. https://doi.org/10.3390/genes13091678

Academic Editors: Wajid Zaman and Hakim Manghwar

Received: 7 September 2022
Accepted: 15 September 2022
Published: 19 September 2022

Abstract: Commercial interest in the culinary herb, *Eryngium foetidum* L., has increased worldwide due to its typical pungency, similar to coriander or cilantro, with immense pharmaceutical components. The molecular delimitation and taxonomic classification of this lesser-known medicinal plant are restricted to conventional phenotyping and DNA-based marker evaluation, which hinders accurate identification, genetic conservation, and safe utilization. This study focused on species discrimination using DNA sequencing with chloroplast–plastid genes (*matK*, *Kim matK*, and *rbcL*) and the nuclear *ITS2* gene in two *Eryngium* genotypes collected from the east coast region of India. The results revealed that *matK* discriminated between two genotypes, however, *Kim matK*, *rbcL*, and *ITS2* identified these genotypes as *E. foetidum*. The ribosomal nuclear *ITS2* region exhibited significant inter- and intra-specific divergence, depicted in the DNA barcodes and the secondary structures derived based on the minimum free energy. Although the efficiency of *matK* genes is better in species discrimination, *ITS2* demonstrated polyphyletic phylogeny, and could be used as a reliable marker for genetic divergence studies understanding the mechanisms of RNA molecules. The results of this study provide insights into the scientific basis of species identification, genetic conservation, and safe utilization of this important medicinal plant species.

Keywords: DNA barcoding; *ITS2* secondary structure; medicinal herbs; species discrimination; spiny coriander

1. Introduction

Spiny coriander (*Eryngium* spp.), one of the largest taxonomically complex genera in the order, *Apiales*, and family, *Apiaceae*, is a perennial herb used as food and medicine in the world's tropics. The taxonomical evolution of the genus, *Eryngium*, is coupled with rapid radiation, hybridization, and long–distance dispersal. *Eryngium* comprises about 250 species with a cosmopolitan distribution worldwide, and South America is believed to be the center of diversity [1]. *Eryngium foetidum* L. is known as spiny or false coriander or culantro, and is used as a culinary herb and traditional medicine in Latin America and South-East Asia, having a similar taste and odor to coriander or cilantro. *E. foetidum* is widely cultivated in South-East Asia, tropical Africa, the Pacific Islands, and the southern tropics of Europe [2]. This perennial herb, *E. foetidum*, hailed from tropical America and the Caribbean Islands, and was further dispersed to South-East Asian countries in the 18th century [3]. *E. foetidum* is cultivated widely in the eastern and northeastern states, including the east coast and the Andaman and Nicobar Islands of India [4]. Despite its wide adaptability, the domestication and evolution of *Eryngium* are still debatable.

The leaves and pseudo stems of *E. foetidum* have multiple pharmaceutical uses, such as anti-diabetic [5], anti-inflammatory [6], anti-cancerous [4], and anti-clastogenic [7] properties. The phytochemical constituent eryngial in *Eryngium* leaves, the reason for the typical pungency, is also used for treating skin disease, arthritis, parasites, and other physiological diseases [4,8]. A mixture of the shoot and root portion is used as a folk medicine worldwide, such as an appetizer, antitussive, treating diarrhea and gastrointestinal disorders, hypoglycemic, and anti-venom agent [8]. Apart from treating several ailments, *Eryngium* is a potent spice for food garnishing and a flavoring agent in the coastal region of India. A recent study depicted that the component, 10-undecenal, is the reason for a strong aroma similar to the seasonal coriander in the *Eryngium* spp. collected from the east coast of India [9]. This perennial lesser-known herb possesses high chemo diversity importance for the pharmaceutical and aroma industries.

The use of traditional herbal medicine is increasing day by day. The correct nomenclature and accurate species identification need the utmost attention for the safe use and commercialization of medicinally important plants. *Eryngium* is a hemicryptophyte perennial herb with spiny leaf margins, light blue inflorescence, and a tap root system [9]. Reports on molecular characterization for species discrimination in *Eryngium* are scanty. The subgenera and species of *Eryngium* have been categorized based on their morphological features, such as leaf margin, color, odor, and chemical constituents [9,10]. A few reports are available on the infrageneric relationship using chloroplast DNA (*trnQ–trnK*) and nuclear internal transcribed spacer (*ITS*) sequencing [11]. Considering morphological variability, species discrimination in closely related genera is quite difficult. Molecular characterization using random DNA markers showed a compressive genetic variability among the genotypes. However, DNA barcoding is an advanced tool to investigate variations at the most conserved region to specify accurate species delimitation [12].

DNA barcoding is a robust emerging technique that confirms species boundaries from a small plant tissue using a short DNA section from a specific gene or genes [13], recommended by the Consortium for the Barcode of Life (CBOL) plant group [14]. The gene region includes a chloroplast–plastid region (*matK*, *rbcL*, *ycf*, *psbA–trnH*, etc.) and a nuclear *ITS* region with less intraspecific barcode gaps [15]. Maturase K (*matK* (XF/5R)) and *Kim matK* (3F_KIM/1R_KIM), the plant plastidial gene in the chloroplast region that splices group II introns in the conserved domain X of the reverse transcriptase domain can be used as a universal DNA barcode primer for plants [16]. The chloroplast DNA sequence, ribulose–bisphosphate carboxylase (*rbcL*), is one of the universal barcode genes used in species discrimination studies due to high amplification and low mutation rate [17]. *Ycf1* is the second largest gene in the plastid genome, encodes about 1800 amino acids, and is a reliable candidate for DNA barcoding [18]. The *psbA–trnH* gene is a non-coding fragment between *psbA* and *trnH* in the chloroplast genome, which delimitates faster evolution than *matK* and slower than *ITS*. The *ITS* region is a DNA spacer, localized between the small and large subunits of ribosomal RNA (rRNA) in the chromosomal or corresponding polycistronic transcript region, most commonly used for species discrimination studies [19]. The CBOL–Plant working group, suggested a combination of plastid (*matk*/*Kim matK*/*rbcL*) and nuclear region (ITS) as the efficient barcode tool to investigate plant species discrimination [20].

RNA secondary structure predictions at the conserved ITS rRNA region is a key ribosomal structure that predicts the function of rRNAs and tRNAs [21]. Computationally-predicted RNA structures represent the native RNA folding status of an organism, shedding light on novel RNA regulatory mechanisms [22]. RNA secondary structure prediction is an advanced tool for species discrimination, as it restricts sequencing error and eliminates pseudogene footprints [23].

The present study aimed to discriminate two important *Eryngium* genotypes grown in two vulnerable coastal agroclimatic regions of India using *matK*, *Kim matK*, *rbcL*, and *ITS* DNA barcode genes. We have also evaluated the potential of different barcode primers at chloroplast–plastid and nuclear regions discriminating the *Eryngium* spp. The phylogenetic relationship among these two *Eryngium* genotypes was established concerning the barcode regions. To elucidate the variation among the two contrast genotypes, *ITS2* secondary structure prediction was performed to validate the signature barcodes. The result of this study provides insights into the scientific scope for species identification in this lesser-known crop, using DNA barcoding as a robust tool for understanding genetic relationship studies for future crop improvement strategies.

2. Materials and Methods

2.1. Plant Materials

The two most popularly grown *Eryngium* genotypes were collected from the east and southeastern coastal plain zone of Odisha, Delanga, India (voucher specimen: Eryngium_CHB, and biological reference number: IC 0629514), and from the ICAR–Central Island Agricultural Research Institute (CIARI), the Island region of Car Nicobar, Port Blair, India (voucher number: Eryngium_AND), and maintained at the Central Horticultural Experiment Station (CHES), ICAR–Indian Institute of Horticultural Research (ICAR–IIHR), Bhubaneswar, India, and were used as the source materials for the present study. The study area is located at a latitude of 20°15′ N, a longitude of 85°52′ E, and an altitude of 35 m above mean sea level. The geographical location of the center of the collection of the two *Eryngium* samples is presented in Figure 1.

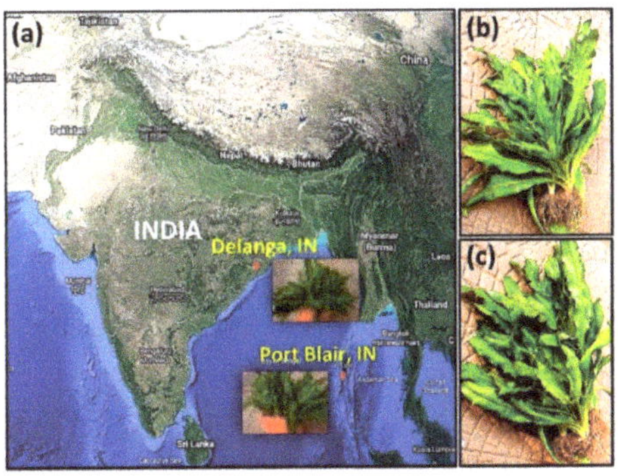

Figure 1. The geographical location (**a**) of the *Eryngium* genotypes collected from Delanga (20°03′ N, 85°76′ E, and 0 masl), Odisha, India (**b**), and Port Blair (11°62′ N, 92°72′ E, and 16 masl), Andaman and Nicobar Island, India (**c**) (source: http://earth.google.com/web, assessed on 4 September 2022).

2.2. Plant Sample Preparation

The first fully opened juvenile leaf samples were collected from the two *Eryngium* genotypes grown under a naturally ventilated poly house. The herbarium for the voucher specimen numbers, Eryngium_CHB and Eryngium_AND, was prepared following the standard procedures and submitted to the repository of ICAR–IIHR–CHES. A set of fresh leaf samples from the respective genotypes were used for DNA barcoding and RNA secondary structure prediction studies.

2.3. Total Genomic DNA Isolation

Total genomic DNA (gDNA) was isolated from the fresh leaf tissues of the two *Eryngium* genotypes using a plant gDNA extraction kit (GSure® Plant Mini Kit with WLN Buffer, GCC Biotech Pvt Ltd., Kolkata, India) following the manufacturer's instructions. The gDNAs were quantified using a nanodrop (Eppendorf, Hamburg, Germany) and checked on 0.8% agarose gel electrophoresis (Tarson, Kolkata, India). The gDNA concentration was adjusted to 50 ng μL^{-1}, and was used for PCR amplification with the DNA barcode primers, *matK*, *Kim matK*, *rbcL*, and *ITS*, as recommended by the Consortium of

Barcode of Life–Plant Group [14]. The details of the barcode primers are presented in Table 1.

Table 1. List of the barcoding primers used for PCR amplification of *Eryngium* spp.

Sl. No.	Region	Primer Name	Sequences (5′ to 3′)	Amplicon Size
1	*matK*	*matK XF*	TAATTTACGATCAATTCATTC	1500 bp
		matK 5R	GTTCTAGCACAAGAAAGTCG	
2	*Kim matK*	*3F_Kim matK*	CGTACAGTACTTTTGTGTTTACGAG	1500–1580 bp
		1R_Kim matK	ACCCAGTCCATCTGGAAATCTTGGTTC	
3	*rbcL*	*rbcLa F*	ATGTCACCACAAACAGAGACTAAAGC	600–800 bp
		rbcLa R	GTAAAATCAAGTCCACCRCG	
4	*ITS2*	*ITS2 S2F*	ATGCGATA CTTGGTGTGAATTATAGAAT	400–800 bp
		ITS2 S3R	GACGCTTCTCCAGACTACAAT	

2.4. PCR Amplification

The DNA barcode primers, *matK*, *Kim matK*, *rbcL*, and *ITS*, were synthesized at M/S Bioserve Biotechnologies India Pvt. Ltd., Hyderabad, India. PCR amplification for each primer was performed in a volume of 25 μL, containing 50 ng of gDNA (1 μL) as a template, 12.5 μL 2 × PCR master mix (GCC Biotech Pvt Ltd., Kolkata, India), primers (10 pM, 1 μL each of forward and reverse primers), and 9.5 μL Milli–Q water following Ayaz et al. (2020) [24] with partial modification. The PCR amplifications were carried out in a thermal cycler (Eppendorf, Hamburg, Germany). The PCR conditions follow a denaturation of 5 min at 95 °C, 40 cycles of 1 min at 95 °C, 1 min at 55 °C of annealing, 1 min at 72 °C, and a final extension of 10 min at 72 °C. Purification of the PCR products was performed using a PCR purification kit (GCC Biotech Pvt. Ltd., Kolkata, India) by following the manufacturer's instructions. The purified PCR products were visualized in 1.5% agarose TAE gels in an E–Box gel documentation system (Vilber, Eberhardzell, Germany). Good-quality PCR products (50 ng) were considered for further sequencing.

2.5. DNA Sequencing

The purified PCR products were sequenced at M/S Bioserve Biotechnologies India Pvt. Ltd., Hyderabad, India, using Sanger sequencing (ABI Genetic Analyzer 3730, 48 capillaries, 50 cm, Thermo Fisher, Waltham, MA, USA), and viewed in FinchTV v1.4.0 (Geospiza, Denver, CO, USA).

2.6. Bioinformatics Analysis

The forward and the reverse sequences obtained from all the PCR products amplified with *matK*, *Kim matK*, *rbcL*, and *ITS* primers were trimmed using SnapGene v 5.3 (https://www.snapgene.com; accessed on 4 September 2022). The contig of the forward and the reverse sequences was submitted in the Basic Local Alignment Search Tool (BLAST) of the National Centre for Biotechnology Information (NCBI) for homology search. The barcode gaps were manually edited in pairwise alignment view using BLAST, and species identification was performed following nucleotide blast with the maximum similarity score and lowest E value. The gene bank accession number was obtained using the barcode sequences in the bankIt submission portal of NCBI (https://submit.ncbi.nlm.nih.gov/subs/genbank; accessed on 21 July 2022). Multiple sequence alignment was performed using ClustalW v10.1.8 (https://www.genome.jp/tools-bin/clustalw; accessed on 4 September 2022) [25] with all obtained sequences in "muscle algorithm" using the neighbor-joining cluster method in MEGA11 software (Molecular Evolutionary Genetic Analysis; https://www.megasoftware.net; accessed on 4 September 2022). The phylogenetic analysis was performed following the neighbor-joining tree and minimum evolution method with the 1000 "*Bootstrap phylogeny*" test method and "*kimura–2–parameter*" substitution model (*d–transitions*) in MEGA software, considering the transitional and transversional nucleotide substitution [26].

2.7. DNA Barcoding and ITS2 Secondary Structure Predictions

DNA barcodes of the two *Eryngium* genotypes were generated using the Bio-Rad DNA barcode generator (http://biorad-ads.com/DNABarcodeWeb; accessed on 4 September 2022), considering the aligned DNA nucleotide sequences of *ITS2* primers. Similarly, the *RNA* secondary structure predictions were performed using the nucleotide sequences from *ITS2* primers using the rRNA database of *RNAfold* WebServer v2.4.18 (http://rna.tbi.univie.ac.at/cgi-bin/RNAWebSuite/RNAfold.cgi; accessed on 4 September 2022) [27].

3. Results

3.1. Amplification, Sequencing, Multiple Sequence Alignment, and Species Identification

The DNA barcode primers, *matK*, *Kim matK*, *rbcL*, and *ITS*, produced amplicons of 1500 bp, 1500–1580 bp, 600–800 bp, and 400–800 bp, respectively. The success rate of PCR amplification was higher (>90%) in *matK*, *Kim matK*, and *rbcL*, whereas *ITS2* showed lower reaction efficiency (80%). The standardization of template DNA concentration and specific primer at different ITS regions are required to validate the amplification success rate. Analyzing the sequences, *matK* exhibited the largest sequence length (534–823 bp), followed by *Kim matK* (763–780 bp), *rbcL* (476–493 bp), and *ITS* (300–357 bp). All the sequences were submitted to GenBank, and the accession numbers were obtained. The details of voucher specimens, species, and accession numbers are presented in Table 2. Using the BLASTn tool, both species were identified as *E. foetidum* at different barcode regions of *Kim matK*, *rbcL*, and *ITS2*. However, the *matK* gene successfully discriminates the species among the two genotypes. The specimen, Eryngium_AND, was showing similarities with *E. vaseyi*, whereas Eryngium_CHB was identical with *E. foetidum*, with 99% similarity at the conserved *matK* region.

Table 2. Molecular identification of *Eryngium* sp. using *matK*, *Kim matK*, *rbcL*, and *ITS2* barcode genes.

Barcode Genes	Voucher Specimen	Scientific Name	Accession Number	E Value	Query Coverage	Percent Identity
matK	Eryngium_AND	*E. vaseyi*	OP086070	0.0	100%	99.76%
	Eryngium_CHB	*E. foetidum*	OP086071	0.0	100%	99.81%
Kim matK	Eryngium_AND	*E. foetidum*	OP086074	0.0	98%	99.74%
	Eryngium_CHB	*E. foetidum*	OP086075	0.0	100%	99.87%
rbcL	Eryngium_AND	*E. foetidum*	OP086072	0.0	100%	100%
	Eryngium_CHB	*E. foetidum*	OP086073	0.0	100%	100%
ITS2	Eryngium_AND	*E. foetidum*	ON797393	2×10^{-153}	100%	100%
	Eryngium_CHB	*E. foetidum*	ON797394	2×10^{-153}	100%	100%

Figure 2 depicts the multiple sequence alignments obtained from all sequences at each chloroplast–plastid and nuclear region. Among the tested barcode primers, *matK* resulted in the largest alignments (583 bp), followed by *Kim matK* (780 bp), *rbcL* (493 bp), and *ITS* (357 bp). Alignments obtained from *ITS* showed the highest similarities (100%) among the two genotypes, whereas *matK* discriminated two genotypes as *E. vaseyi* (Eryngium_AND; ON797393) with 99.76% identity and *E. foetidum* (Eryngium_CHB; ON797394) with 99.81% identity (Figure 2).

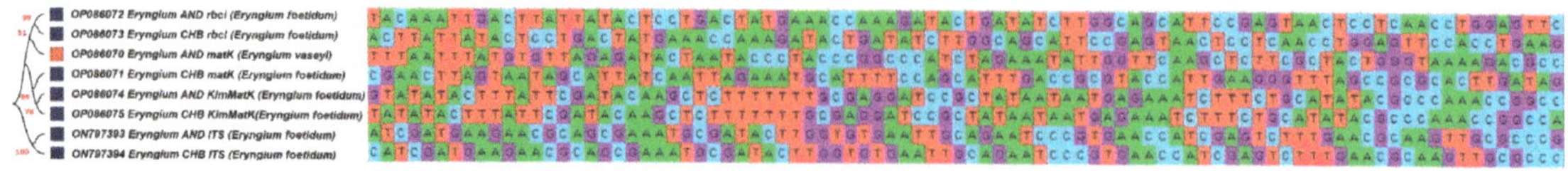

Figure 2. Multiple sequence alignment of *matK*, *Kim matK*, *rbcL*, and *ITS2* barcode primers and their phylogenetic relationship among two *Eryngium* genotypes.

3.2. Phylogenetic Studies

Phylogenetic analysis using a maximum likelihood tree (MLT) in a *K2P* model with bootstrap-1000 of *Eryngium* species showed a high similarity in the BLAST search to the available sequences in the NCBI database (Figure 3). Phylogenetic analysis revealed that Eryngium_AND_*matK* (823 bp) had >99.76% similarity with *E. vaseyi* (OP086070), which belongs to the class of *Magnoliopsida*. Other closest relatives of this species include *E. armatum* (OL689984.1), *E. yuccifolium* (MK520075.1), with 99.88% and 99.60% similarity, respectively, as per the BLAST search results. Similarly, the specimen, Eryngium_CHB_*matK* (534 bp), showed the closest similarity, >99.81% to *E. foetidum*, as its closest phylogenetic relative. The other closest matches are *E. baldwinii* OSBAR (MH551983.1), *E. ebracteatum* BioBot01719 (JQ586532.1), *E. cuneifolium* voucher FLAS: Judd 5564 (KY607232.1), with an average similarity of 99% in all exceeding species. The voucher specimens, Eryngium_CHB_*Kim matK* (780 bp), Eryngium_AND_*Kim matK* (763 bp), Eryngium_AND_*rbcL* (493 bp), Eryngium_CHB_*rbcL* (476 bp), Eryngium_AND_CHB (357 bp), and Eryngium_AND_*ITS* (300 bp), belong to *E. foetidum*, with the closest similarity of 99.74%, 99.87%, 100.00%, 100.00%, 100.00%, and 100.00%, respectively, belong to the class of *Magnoliopsida* and the family of *Apiaceae*. The other closest relatives of Eryngium_CHB_*Kim matK* are *Sanicula canadensis* isolate AD3HK81 maturase K (*matK*) gene (MF350053). Eryngium_AND_*rbcL* showed similarity with *Sanicula graveolens* voucher HU WAP 14409 (KX371924), having a similarity of 98.58%. Eryngium_CHB_*ITS* shows the highest similarity (98.88%) with *E. balansae ITS1*, 5.8S ribosomal RNA gene (EU070608). The phylogenetic analysis indicates that the plant genotypes collected from two locations showed species variation with different percentage indexes (Figure 3).

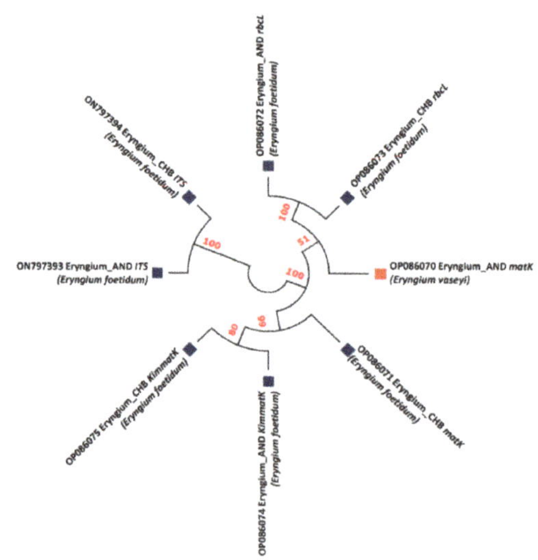

Figure 3. Maximum likelihood tree of *matK*, *Kim matK*, *rbcL*, and *ITS2* barcode primers depicting the phylogenetic relationship among two *Eryngium* genotypes. The bootstrap scores (1000 replicates) are shown (≥50%) for each branch.

3.3. DNA Barcoding and ITS2 Secondary Structure Predictions

Figure 4 represents the DNA barcodes and *ITS2* secondary structure predictions based on the minimum free energy (MFE). The highest MFE for Eryngium_AND (ON797393) was observed at 145–150 bp (Figure 4a; Table 3), which was recorded highest at 275–280 bp in Eryngium_CHB (ON797394) [Figure 4b]. The partition function and centroids followed a similar pattern as the MFE in both genotypes.

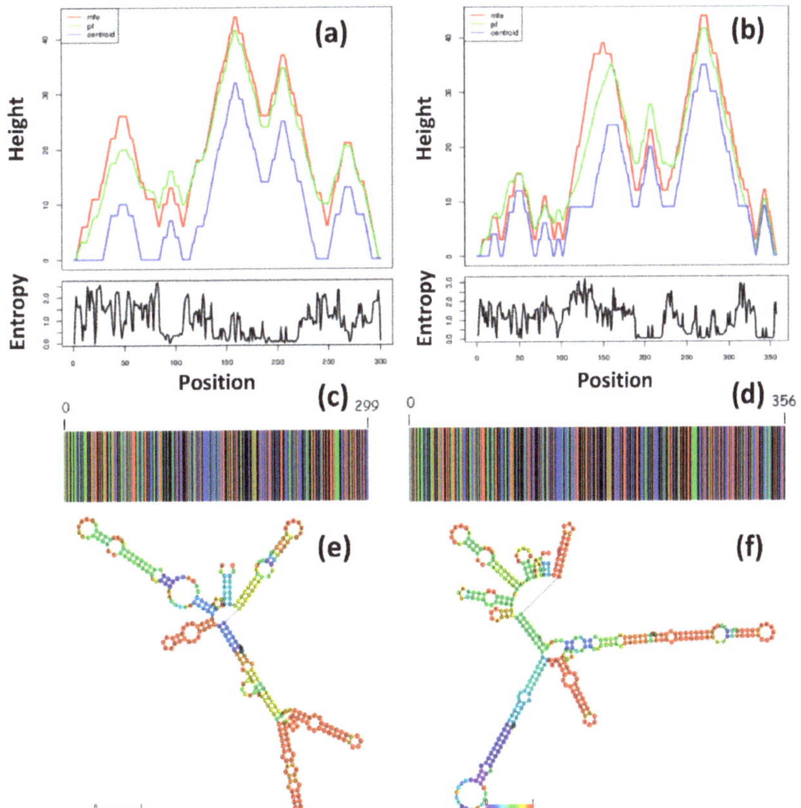

Figure 4. The variations observed in the predicted minimum free energy (MFE) (**a**,**b**), DNA barcodes (**c**,**d**), and secondary structures of *ITS2* region (consensus structure; **e**,**f**) for the *Eryngium* genotypes.

DNA barcodes derived from the *ITS2* sequences showed variations among the two tested genotypes (Figure 4c,d). Eryngium_CHB (ON797394) exhibited comparatively higher barcode length (356 bp) compared to Eryngium_AND (ON797393) [299 bp]. Similarly, *ITS2* secondary structure predictions discriminate both the tested genotypes representing a central ring with different helical orientations (Figure 4e,f). The loop number, position, size, and angle from the centroid are distinguishable in both genotypes. Eryngium_CHB (ON797394) represented a more complex structure than Eryngium_AND (ON797393), with varied loop numbers and angles from the spiral. The unique genetic structure at the conserved nuclear region would also be useful to develop species-specific primers to identify the lesser-known species at a shorter pace. The secondary structure predictions guide comparative sequence analysis and design of species–specific RNA molecules.

Table 3. Sequence characteristics, and the intraspecific and interspecific genetic divergence of candidate barcodes.

Voucher Specimen	Eryngium_AND				Eryngium_CHB			
Markers	*matK*	*Kim matK*	*rbcL*	*ITS2*	*matK*	*Kim matK*	*rbcL*	*ITS2*
Sequence length	823	763	493	300	534	780	476	357
Alignment length	823	763	492	300	533	773	419	357
Maximum score	1509	1399	909	555	979	1424	774	660
GC content (%)	37.18	36.70	43.00	62.33	35.96	36.28	43.28	63.31
CDS region	2	1	1	–	2	1	1	–
Amino acid	274	254	164	–	177	260	159	–

4. Discussion

Perennial *Eryngium* is an important food and medicinal herb gaining significant industrial and economical values in pharmaceutical industries. The medicinal use of traditional herbal plants needs accurate identification for their safe use in alimenting human diseases [28]. DNA barcoding, at the chloroplast–plastid and nuclear regions or in a combination of both regions, is a newer tool receiving increasing attention over the conventional phenotyping-based taxonomy. Molecular signatures at the conserved barcode region provide insights into the genetic importance and possible scope for crop improvement.

Our study aimed at species discrimination in the lesser-known *Eryngium* genotypes collected from two different coastal microclimatic zones of India by deploying DNA barcode genes. This study is the first report to identify *Eryngium* in this region using chloroplast–plastid and nuclear regions. A few reports [29,30] suggest the plastid barcode gene significantly creates variation at the species level. However, reports on species discrimination in *Eryngium* at both the plastid and nuclear region is scanty. Our study revealed significant variations at the plastid and nuclear regions with 99–100% similarities. Discrimination efficiency was higher in *matK* than *Kim matK*, *rbcL*, and Nuclear *ITS2*. Of the barcode genes tested, only *matK* could perform species delimitation (Figure 3; Table 2), which could be further used for large-scale species discrimination studies. The results of the study would be helpful for a reference database for accurate species identification in *Eryngium*.

DNA barcoding technology is an advanced method of identifying plant species with conservational and consumer marketable value [31]. The abundant relatives of the plant species fulfil the demand for applied herbal medicines obtained from medical plant genotypes. Likewise, amplified demand for medicinal products may lead to the over-harvesting and extinction of the non-threatened biological species. Therefore, identifying the plant samples is urgently needed for the conservation and prevention of perceived biodiversity loss. Thus, appropriate and prominent measures could provide a reliable genetic method for systematically identifying medicinal plant valuables for species preservation and genetic improvement.

Generating the importance of DNA barcoding onto the well-defined gene from genetic regions in a medicinal plant sample and its appearance of species in a specific manner are being studied in this research. To increase the prominence selection in monitoring, the discrimination based on DNA-based identity is found authentic for any medicinal plant specifically related to drug delivery. Specific universal DNA markers/primers for plants recommends genes from the chloroplast region, such as *Kim matK* maturase K (*matK*) gene; maturase K (*matK*) gene; ribulose 1,5-bisphosphate carboxylase/oxygenase large subunit (*rbcL*) gene marker, belonging to the enzyme (RuBisCo); and internal transcribed spacer (*ITS*) primer are selected as a core marker for the barcoding of DNA for medicinal plants [32]. However, this marker does not show 100% discrimination of the plant species, but *matK* is recommended as the marginal selection of the variation found in our collected *Eryngium* species. Ayaz et al. [24] performed genetic diversity studies in the *Lamiaceae* family using *rps 14* genes following molecular phylogeny. DNA-based markers exhibited significant variation among the species; however, barcode markers recommended by CBOL confirm the discriminations at conserved barcode regions.

The main goal of DNA barcoding is to assemble the reference library at a shorter pace, which includes DNA extraction, amplification, and sequencing. We have investigated the discrimination efficiency of four barcode regions (*matK, Kim matK, rbcL*, and *ITS*) to observe species discrimination, which varied between 80–90%. *matK* resulted in a higher sequence length to discriminate the voucher specimen into two different species, *E. vaseyi* and *E. foetidum*. We have estimated the amplicon universality and the success rate of amplification of various chloroplast–plastid (<90%) and nuclear genes (80%) in the rare taxa, *Eryngium*. Previous reports showed a low amplification rate of 60% in *matK* for species discrimination in other medicinal plants [12]. Molecular phylogeny analyzes the genetic heredity based on the DNA sequences to understand the evolutionary interrelationship, deriving the genetic diversity among the species. Molecular phylogenetics following the maximum likelihood tree depicts the molecular systematics of taxonomy evolution from different geo-coordinating regions [33]. Figure 3 distinguishes the molecular phylogeny based on the maximum likelihood neighbor-joining tree using the minimum evolution method with the bootstrap-phylogeny-1000 method and d-transitions substitution model. Various genetic divergence studies emphasize the phylogenetic exploration of medicinally important plants of the *Lamiaceae* family [24,28]. Among the conserved barcode genes, *ITS2*-anchored evolution, structure, and function establish a better understanding of the species through sequence, structure, and phylogeny based on the nucleotide sequences and independent DNA evolutionary models [34]. In our study, *ITS2*-based phylogeny distinguished the species into a different clade with 100% similarities (Figure 3).

According to the diversity-based discrimination on a species-specific level, the major explanation is that barcodes are improved candidates for divergence in a reliable conserved DNA sequence [35]. The specific approach in the level of discrimination and universality for medicinal plants is found to be more precise with the ribosomal RNA maturase K (*matK*) marker, which is considered a core DNA barcode spacer which would be exclusive in identifying the plant genotypes at the species level. Moreover, the mitochondrial genome arrangement of plants is genetically altered quickly, evading the existence of universal markers amplifying the coding genes (CDS region). Researchers [36,37] investigated the improved discrimination efficiency of different plant species using the combination of *rbcL* and *matK*, and *rbcL* and *ITS2* markers for more suitable variation. Supporting the intensified possibility of the *matK* DNA barcode marker, genetic data infested the quality of divergence [38].

The RNA secondary structure determines the understanding between the structural and functional relationship in designing the therapeutics and diagnostics of the target RNA. Minimum free-energy-anchored RNA secondary structure prediction indicated that site-specific mutagenesis occurred due to tertiary protein interaction. The ribosomal RNA structure has the advantage of understanding RNA chemistry compared to the nuclease mapping [39]. In our study, both genotypes represented diverse secondary structures, with distinguishable loop numbers, positions, and elevation from the centroid. The unique nuclear segments can be further used to develop species-specific markers for identifying the genotypes. *ITS2* could not show different variations at the species level, but could predict the differential secondary structure, indicating variations among the RNA molecules in both genotypes. The study would encourage RNA molecule studies for genetic evolution analysis of plant taxa.

In general, molecular approaches appear to be an ideal advancement suited to categorizing indistinguishable plant species. In supplementing the DNA sequencing database to the gene bank, reliably identified species exist for the barcode reference for the chosen plant genotypes for the possible implication of informatics in future medicinal analysis. The DNA database can target a broad range of conservational and ethnobiological traditional potentiality of this underutilized medicinal crop. For the participatory improvement of the crop, the medicinal germplasm is systematically maintained in a sequence database for future quality yield with the genetic variation. The suitable application of the traditional products concerning genetic complexity and identified plant samples can address a good companion integrated with conserving this medicinal plant.

5. Conclusions

The DNA barcoding results at the chloroplast–plastid and nuclear region discriminate the two *Eryngium* species. The plastid gene, *matK*, could be a potent barcode gene to discriminate between two genotypes, which may be further used for a large-scale divergence study. However, the ribosomal nuclear *ITS2* region could be a reliable nuclear region to distinguish inter- and intra–specific divergence, DNA barcoding, predictions of the secondary structure, and understanding and re-designing RNA molecules. The results of this study provide insights into the scientific basis of species identification, genetic conservation, and safe utilization of this important medicinal plant species. The DNA barcode tools could be used for species delimitation in medicinally and commercially important plants, and could also be helpful in the detection of adulteration in food and pharmaceutical industries. The results also encourage understanding genetic relationships for future crop improvement strategies for food, nutrition, and therapeutics.

Author Contributions: Conceptualization, M.R.S. and G.C.A.; data curation, M.R.S., S.M. and M.D.; formal analysis, M.R.S., S.M., M.D. and S.S. (Supriya Sahu); investigation, M.R.S., S.M. and M.D.; methodology, S.M., M.D. and M.R.S.; resources, M.R.S. and G.C.A.; software, S.M., M.R.S. and M.D.; writing—original draft, M.R.S., S.M., M.D. and G.C.A.; writing—review and editing, M.R.S., S.S. (Satyapriya Singh), A.V.V.K., M.K., P.N. and G.C.A. All authors have read and agreed to the published version of the manuscript.

Funding: This research received no external funding.

Institutional Review Board Statement: Not applicable.

Informed Consent Statement: Not applicable.

Data Availability Statement: Not applicable.

Acknowledgments: The infrastructure facility provided by the Director, ICAR Indian Institute of Horticultural Research (ICAR–IIHR), Bengaluru, India, and the Head, Central Horticultural Experiment Station (ICAR–IIHR–CHES), Bhubaneswar, India, are gratefully acknowledged. The financial assistance of Mission for Integrated Development in Horticulture (MIDH), Govt. Of Odisha, India, for the outsourcing of Sanger sequencing at M/S BioServe Biotechnologies, Hyderabad, is duly acknowledged.

Conflicts of Interest: The authors declare no conflict of interest.

References

1. Wang, P.; Su, Z.; Yuan, W.; Deng, G.; Li, S. Phytochemical constituents and pharmacological activities of *Eryngium* L. (Apiaceae). *Pharm. Crops* **2012**, *3*, 99–120. [CrossRef]
2. Le Claire, E.; Schwaiger, S.; Banaigs, B.; Stuppner, H.; Gafner, F. Distribution of a new rosmarinic acid derivative in *Eryngium alpinum* L. and other Apiaceae. *J. Agric. Food Chem.* **2005**, *53*, 4367–4372. [CrossRef] [PubMed]
3. Ramcharan, C. Culantro: A much utilized, little–understood herb. In *Perspectives on New Crops and New Uses*; Janick, J., Ed.; ASHS Press: Alexandria, VA, USA, 1999; pp. 506–509.
4. Singh, B.K.; Ramakrishna, Y.; Ngachan, S.V. Spiny coriander (*Eryngium foetidum* L.): A commonly used, neglected spicing–culinary herb of Mizoram, India. *Genet. Resour. Crop Evol.* **2014**, *61*, 1085–1090. [CrossRef]
5. Perez-Munoz, E.P.; Antunes-Ricardo, M.; Martínez-Ávila, M.; Guajardo-Flores, D. *Eryngium* species as a potential ally for treating metabolic syndrome and diabetes. *Front. Nutr.* **2022**, *9*, 878306. [CrossRef] [PubMed]
6. Mekhora, C.; Muangnoi, C.; Chingsuwanrote, P.; Dawilai, S.; Svasti, S.; Chasri, K.; Tuntipopipat, S. *Eryngium foetidum* suppresses inflammatory mediators produced by macrophages. *Asian Pac. J. Cancer Prev.* **2012**, *13*, 723–734. [CrossRef]
7. Promkum, C.; Butryee, C.; Tuntipopipat, S.; Kupradinun, P. Anticlastogenic effect of *Eryngium foetidum* L. Assessed by erythrocyte micronucleus assay. *Asian Pac. J. Cancer Prev.* **2012**, *13*, 3343–3347. [CrossRef]

8. Vukic, M.D.; Vukovic, N.L.; Djelic, G.T.; Obradovic, A.; Kacaniova, M.M.; Markovic, S.; Popović, S.; Baskić, D. Phytochemical analysis, antioxidant, antibacterial and cytotoxic activity of different plant organs of *Eryngium serbicum* L. *Ind. Crops Prod.* **2018**, *115*, 88–97. [CrossRef]
9. Acharya, G.C.; Ponnam, N.; Kumari, M.; Roy, T.K.; Shivashankara, K.S.; Sahoo, M.R. Phytochemical profiling of spiny coriander (*Eryngium foetidum* L.)—A potential perennial spicing–culinary herb of eastern India. *Acta Chromatogr.* **2021**, *34*, 197–202. [CrossRef]
10. Worz, A. A new subgeneric classification of the genus *Eryngium* L. (Apiaceae, Saniculoideae). *Bot. Jahrb. Syst.* **2005**, *126*, 253–259. [CrossRef]
11. Calvino, C.I.; Martínez, S.G.; Downie, S.R. The evolutionary history of *Eryngium* (Apiaceae, Saniculoideae): Rapid radiations, long distance dispersals, and hybridizations. *Mol. Phylogenet. Evol.* **2007**, *46*, 1129–1150. [CrossRef]
12. Jamdade, R.; Mosa, K.A.; El-Keblawy, A.; Al Shaer, K.; Al Harthi, E.; Al Sallani, M.; Al Jasmi, M.; Gairola, S.; Shabana, H.; Mahmoud, T. DNA barcodes for accurate identification of selected medicinal plants (Caryophyllales): Toward barcoding flowering plants of the United Arab Emirates. *Diversity* **2022**, *14*, 262. [CrossRef]
13. Hartvig, I.; Czako, M.; Kjaer, E.D.; Nielsen, L.R.; Theilade, I. The Use of DNA Barcoding in Identification and Conservation of Rosewood (*Dalbergia* spp.). *PLoS ONE* **2015**, *10*, e0138231. [CrossRef]
14. CBOL Plant Working Group. A DNA barcode for land plants. *Proc. Natl. Acad. Sci. USA* **2009**, *106*, 12794–12797. [CrossRef] [PubMed]
15. Paulay, G.; Meyer, C.P. DNA barcoding: Error rates based on comprehensive sampling. *PLoS Biol.* **2005**, *3*, e422. [CrossRef]
16. Yu, J.; Xue, J.-H.; Zhou, S.-L. New universal *matK* primers for DNA barcoding angiosperms. *J. Syst. Evol.* **2011**, *49*, 176–181. [CrossRef]
17. Nurhasanah, S.; Papuangan, N. Amplification and analysis of *rbcl* gene (ribulose-1,5-bisphosphate carboxylase) of clove in ternate Island. *IOP Conf. Ser. Earth Environ. Sci.* **2019**, *276*, 012061. [CrossRef]
18. Dong, W.; Xu, C.; Li, C.; Sun, J.; Zuo, Y.; Shi, S.; Cheng, T.; Guo, J.; Zhou, S. *ycf1*, the most promising plastid DNA barcode of land plants. *Sci. Rep.* **2015**, *5*, 8348. [CrossRef]
19. Bolson, M.; de Camargo Smid, E.; Brotto, M.L.; Silva-Pereira, V. ITS and *trnH-psbA* as Efficient DNA Barcodes to Identify Threatened Commercial Woody Angiosperms from Southern Brazilian Atlantic Rainforests. *PLoS ONE* **2015**, *10*, e0143049. [CrossRef]
20. Han, S.; Sebastin, R.; Wang, X.; Lee, K.J.; Cho, G.-T.; Hyun, D.Y.; Chung, J.W. Identification of Vicia Species Native to South Korea Using Molecular and Morphological Characteristics. *Front. Plant Sci.* **2021**, *12*, 608559. [CrossRef]
21. Gawronski, P.; Palac, A.; Scharff, L.B. Secondary Structure of Chloroplast mRNAs In Vivo and In Vitro. *Plants* **2020**, *9*, 323. [CrossRef]
22. Yang, X.; Yang, M.; Deng, H.; Ding, Y. New Era of Studying RNA Secondary Structure and Its Influence on Gene Regulation in Plants. *Front. Plant Sci.* **2018**, *9*, 671. [CrossRef] [PubMed]
23. Rampersad, S.N. ITS1, 5.8S and ITS2 secondary structure modelling for intra–specific differentiation among species of the *Colletotrichum gloeosporioides sensu lato* species complex. *SpringerPlus* **2014**, *3*, 684. [CrossRef] [PubMed]
24. Ayaz, A.; Zaman, W.; Saqib, S.; Ullah, F.; Mahmood, T. Phylogeny and diversity of *Lamiaceae* based on rps14 gene in Pakistan. *Genetika* **2020**, *52*, 435–452. [CrossRef]
25. Al-Juhani, W.S.; Khalik, K.N.A. Identification and phylogenetics study of *Arabis alpina* L. from the Kingdom of Saudi Arabia. *Pak. J. Bot.* **2021**, *53*, 1057–1064. [CrossRef]
26. Kumar, S.; Stecher, G.; Li, M.; Knyaz, C.; Tamura, K. MEGA X: Molecular Evolutionary Genetics Analysis across Computing Platforms. *Mol. Biol. Evol.* **2018**, *35*, 1547–1549. [CrossRef]
27. Lorenz, R.; Bernhart, S.H.; HönerzuSiederdissen, C.; Tafer, H.; Flamm, C.; Stadler, P.F.; Hofacker, I.L. ViennaRNA Package 2.0. *Algorithmic Mol. Biol.* **2011**, *6*, 26. [CrossRef]
28. Zaman, W.; Ye, J.; Ahmad, M.; Saqib, S.; Shinwari, Z.K.; Chen, Z. Phylogenetic exploration of traditional Chinese medicinal plants: A case study on *Lamiaceae*. *Pak. J. Bot.* **2022**, *54*, 1033–1040. [CrossRef]
29. Mathew, D. DNA barcoding and its application in horticultural crops. In *Biotechnology in Horticulture: Methods and Application*; Peter, K.K., Ed.; NIPA: New Delhi, India, 2013; pp. 25–50. [CrossRef]
30. Heise, W.; Barik, W.; Kubosz, D.; Kajtoch, L. A three–marker DNA barcoding approach for ecological studies of xerothermic plants and herbivorous insects from central Europe. *Bot. J. Lin. Soc.* **2015**, *177*, 576–592. [CrossRef]
31. Parveen, I.; Singh, H.K.; Malik, S.; Raghuvanshi, S.; Babbar, S.B. Evaluating five different loci (*rbc*L, *rpo*B, *rpo*C1, *matK*, and ITS) for DNA barcoding of Indian orchids. *Genome* **2017**, *60*, 665–671. [CrossRef]
32. Mathew, D.; Ramesh, G.A. A universal system for *matK* gene based diagnostic markers to identify the species in *Cucurbitaceae*. *Indian J. Hortic.* **2020**, *77*, 733–735. [CrossRef]
33. Felsenstein, J. (Ed.) *Inferring Phylogenies*; Sinauer Associates Inc.: Sunderland, MA, USA, 2004; p. 664.
34. Zhang, W.; Tian, W.; Gao, Z.; Wang, G.; Zhao, H. Phylogenetic Utility of rRNA *ITS2* Sequence-Structure under Functional Constraint. *Int. J. Mol. Sci.* **2020**, *21*, 6395. [CrossRef] [PubMed]
35. Kress, J.W.; Wurdack, K.J.; Zimmer, E.A.; Weigt, L.A.; Janzen, D.H. Use of DNA barcodes to identify flowering plants. *Proc. Natl. Acad. Sci. USA* **2005**, *102*, 8369–8374. [CrossRef]

36. Poovitha, S.; Stalin, N.; Balaji, R.; Parani, M.; Bank, M. Multi-locus DNA barcoding identifies *matK* as a suitable marker for species identification in *Hibiscus* L. *Genome* **2016**, *59*, 1150–1156. [CrossRef]
37. Ashfaq, M.; Asif, M.; Anjum, Z.; Zafar, Y. Evaluating the capacity of plant DNA barcodes to discriminate species of cotton (Gossypium: *Malvaceae*). *Mol. Ecol. Res.* **2013**, *13*, 573–582. [CrossRef]
38. Newmaster, S.; Grguric, M.; Shanmughanandhan, D.; Ramalingam, S.; Ragupathy, S. DNA barcoding detects contamination and substitution in North American herbal products. *BMC Med.* **2013**, *11*, 222. [CrossRef] [PubMed]
39. Mathews, D.H.; Disney, M.D.; Childs, J.L.; Schroeder, S.J.; Zuker, M.; Turner, D.H. Incorporating chemical modification constraints into a dynamic programming algorithm for prediction of RNA secondary structure. *Proc. Natl. Acad. Sci. USA* **2004**, *101*, 7287–7292. [CrossRef] [PubMed]

genes

MDPI

Article

Complete Chloroplast Genomes of 14 Subspecies of *D. glomerata*: Phylogenetic and Comparative Genomic Analyses

Yongjuan Jiao †, Guangyan Feng †, Linkai Huang, Gang Nie, Zhou Li, Yan Peng, Dandan Li, Yanli Xiong, Zhangyi Hu and Xinquan Zhang *

College of Grassland Science and Technology, Sichuan Agricultural University, Chengdu 611130, China
* Correspondence: zhangxq@sicau.edu.cn
† These authors contributed equally to this work.

Abstract: Orchardgrass (*Dactylis glomerata* L.) is a species in the Gramineae family that is highly important economically and valued for its role in ecology. However, the phylogeny and taxonomy of *D. glomerata* are still controversial based on current morphological and molecular evidence. The study of chloroplast (cp) genomes has developed into a powerful tool to develop molecular markers for related species and reveal the relationships between plant evolution and phylogenetics. In this study, we conducted comparative genomic analyses and phylogenetic inferences on 14 cp genomes of *D. glomerata* originating from the Mediterranean and Eurasia. The genome size ranged from 134,375 bp to 134,993 bp and exhibited synteny of gene organization and order. A total of 129–131 genes were identified, including 85–87 protein coding genes, 38 tRNA genes and 8 rRNA genes. The cp sequences were highly conserved, and key sequence variations were detected at the junctions of inverted repeats (IRs)/small single–copy (SSC) regions. Moreover, nine highly variable regions were identified among the subspecies based on a sequence divergence analysis. A total of 285 RNA editing sites were detected that were relevant to 52 genes, where *rpoB* exhibited the most abundant RNA editing sites. The phylogenetic analysis revealed that all *Dactylis* subspecies clustered into a monophyletic group and most branches provided a high support bootstrap. The main divergence time of *D. glomerata* was dated to the Miocene era, and this could have been due to changes in the climate. These findings will provide useful insights for further studies on phylogeny, the identification of subspecies and the development of hypotheses for the evolutionary history of the genus *Dactylis* and of the Gramineae family.

Keywords: chloroplast genome; comparative genomics; sequence divergence; RNA editing sites; phylogeny; divergence time; ploidy

Citation: Jiao, Y.; Feng, G.; Huang, L.; Nie, G.; Li, Z.; Peng, Y.; Li, D.; Xiong, Y.; Hu, Z.; Zhang, X. Complete Chloroplast Genomes of 14 Subspecies of *D. glomerata*: Phylogenetic and Comparative Genomic Analyses. *Genes* **2022**, *13*, 1621. https://doi.org/10.3390/genes13091621

Academic Editors: Wajid Zaman and Hakim Manghwar

Received: 17 August 2022
Accepted: 6 September 2022
Published: 9 September 2022

1. Introduction

Orchardgrass (*Dactylis glomerata* L.) is a member of the Gramineous family [1]. The genus contains one species and numerous subspecies that have different ploidy levels [2]. Orchardgrass is native to Eurasia and northern Africa, although it has been introduced to nearly every continent and utilized as excellent cool–season forage for the livestock industry that requires forage [3]. In China, orchardgrass has already become an elite forage crop for mixed pasture construction and rocky desertification improvement in southwestern regions. It is highly valued economically and has a strong potential for utilization, in addition to its substantial ecological importance [4]. Based on ploidy, there are primarily three categories of the genus *Dactylis*: diploid (2n = 2x = 14), tetraploid (2n = 2x = 28) and hexaploid (2n = 2x = 42) [5]. Tetraploid populations are widely used in forage production, while hexaploid populations are rarely reported [6]. For a long time, taxonomists focused on the evolutionary relationship between orchardgrass diploids and tetraploids. Nevertheless, the absence of taxonomically diagnostic characters, and the high morphological similarity among the species, has made it difficult to classify the genus [7]. Although previous research has attempted to classify the *Dactylis* genus at the cytological and genetic levels, a unified standard for the taxonomy of the subspecies is still lacking. *Dactylis* classification has been repeatedly revised based on herbaria and field studies [8], cytology and genetics [9]. Molecular genetic analysis of partial DNA sequencing recently [2,5] changed clustering from three (a diploid and two tetraploid groups) to two (comprising, respectively, 17 diploids and six tetraploids). In one sense, although these studies have greatly improved the understanding of the taxonomy and phylogeny of the *Dactylis* genus, controversy about the phylogenetic status and interspecific phylogenetic relationship of the *Dactylis* genus remains.

The chloroplast (cp) is a key organelle with substantial functions in photosynthesis, carbon fixation, translation, and transcription [10,11]. Typically, a cp has a highly conserved genome sequence that contains 100–130 genes with a range of sizes from 120–170 kb in most land plants [12,13]. A cp genome usually has a covalently closed circular molecular structure that contains two inverted repeats (IRs) separated by a large single–copy (LSC) region and a small single–copy (SSC) region [14]. Compared with nuclear genomes, cp genomes are characterized by a highly conserved genome structure, a moderate substitution rate and uniparental inheritance, which have been widely used for plant phylogeny, and to identify species, estimate divergence and generate genetic markers [15,16]. In recent years, rapid advances in next–generation sequencing (NGS) technologies have enabled many studies with high–quality genomes with raw reads—such genome sequencing generates genome sequences much more rapidly and economically than using traditional Sanger sequencing [17,18]. Therefore, we have witnessed a greatly increased number of complete plant cp genomes in recent years. To date, more than 6500 complete cp genome sequences have become available in the National Center for Biotechnology Information (NCBI) since the first cp genome, that of tobacco (*Nicotiana tabacum*), was sequenced in 1986 [19].

However, there have been few studies on the comparative analysis of cp genomes in the *Dactylis* genus to date. In this study, we sequenced and assembled complete cp genomes of 14 subspecies of *D. glomerata* using next–generation sequencing. Subsequently, we analyzed the structural characteristics of the genome and identified the variant regions and RNA editing sites in the cp genomes. In addition, the divergence date was estimated, and phylogenetic relationships were reconstructed to assess the taxonomic positions of *Dactylis* species. This study aims to provide a reference for the taxonomy, phylogeny, and population genetics of *D. glomerata*. The research should facilitate the exploration and utilization of forage resources.

2. Materials and Methods

2.1. Plant Material and DNA Extraction

Fresh leaves of 14 subspecies of *D. glomerata* were collected from the Sichuan Agricultural University experimental greenhouse (30°42′ N, 103°51′ E; Chengdu, China). The lighting was set to 14 h/10 h (day/night) with a temperature of 22 °C/15 °C (day/night) (Table S1). The total genomic DNA was extracted from 100 mg of fresh leaves using a Plant Genomic DNA Kit (Tiangen, Beijing, China).

2.2. Chloroplast Genome Sequencing, Assembly and Annotation

The fragmented DNA was prepared for libraries using a VAHTS Universal DNA Library Prep Kit for Illumina V3 (Jisi Huiyuan Biotechnologies Co., Ltd., Nanjing, China); the sequencing read length was PE150. The libraries that passed quality inspection were sequenced on an Illumina Nova Seq 6000 Platform (San Diego, CA, USA). The raw reads were filtered with a threshold of average quality <Q5 and a number of N > 5 using the fastp tool (version 0.20.0, https://github.com/OpenGene/fastp, accessed on 1 April 2022) [20]. The cp genomes of 14 *Dactylis* subspecies were assembled using the SPAdes pipeline (v3.10.1) [21]. To annotate the cp genome coding sequences (CDS), rRNA and tRNA were acquired using prodigal v2.6.3 (http://www.github.com/hyattpd/Prodigal, accessed on 1 April 2022) [22], hmmer v3.1b2 (http://www.hummer.org/, accessed on 1 April 2022) [23] and ARAGORN v1.2.38 (http://130.235.244.92/ARAGORN accessed on 1 April 2022) [24]. In addition, cp genomic data were extracted from NCBI using BLAST v2.6 (http://blast.ncbi.nlm.nih.gov/Blast.cqi, accessed on 1 April 2022) to align the assembled sequences [25]. The annotation results were then manually checked, and incorrect annotations were removed, along with redundant annotations. The boundaries of multiple exons were determined. Finally, circle maps of all the cp genomes were drawn using the program OGDRAW v1.1 [26].

2.3. Genome Comparison

To compare the structure of the cp genomes of all the *D. glomerata* subspecies, the borders between the IR and SC regions were analyzed. The program mVISTA was used to compare the cp genomes of 14 *Dactylis* subspecies, with the annotation of *D. aschersoniana* used as the reference [27]. Mauve (v2.3.1) software was used to analyze the homology and collinearity to align the genome [28].

2.4. Repeat Element Analysis and RNA Editing Identification

MISA (Microsatellite Identification Tool, v1.0) software was used to analyze the cpSSR using parameters 1–8 (single base repeat 8 times or more) [29]. The software Vmatch combined with a perl script was used to identify repeat sequences [30]. The parameters were as follows: minimum length = 30 bp, Hamming distance = 3, and the four identification forms were forward, palindromic, reverse, and complement. The editing sites of genes among the 14 subspecies of *D. glomerata* were predicted using PREP–CP with a cutoff value of 0.8 (http://prep.unl.edu/cgi-bin/cp-input.pl, accessed on 1 April 2022) [31].

2.5. Phylogenetic Analysis and the Estimation of Divergence Time

The cp genome sequences were used to construct a phylogenetic tree. The interspecies sequences were aligned by MAFFT software (v7.427–auto mode), and trimAl (v1.4. rev 15) was used to trim the well–aligned data [32]. Using RAxML software (v8.2.10) [33], a GTAGAMMA model with bootstrap analysis executed with 1000 replicates was used to construct the maximum likelihood evolutionary tree. Based on the phylogenetic tree obtained, mcmctree software in the paml (v4.9) package was used to construct the phylogenetic tree of divergence time, and the fossil time of species was searched from the timetree website (http://www.timetree.org/, accessed on 1 April 2022) [34]. Time calibration was based on fossil records of *Oryza sativa*, with a confidence range of 42–52 Mya. *Trapa* and *Corynocarpus* served as outgroups.

3. Results

3.1. Features of the Cp Genomes of the 14 Subspecies of D. glomerata

All the cp genomes of *D. glomerata* subspecies contained a typical quadripartite structure, comprising two IR (IRA and IRB) regions separated by a long single copy (LSC) and a small single copy (SSC) region. The length of the cp genomes of 14 subspecies of *D. glomerata* ranged from 134,375 bp (*D. glomerata* subsp. *lobata*) to 134,993 bp (*D. glomerata* subsp. *lusitanica*) (Figure 1). The lengths of LSC, SSC and IRs ranged from 79,753 to 79,773 bp, 12,246 to 12,276 bp and from 21,236 to 21,479 bp, respectively. The size of the LSC, SSC and IRs, as well as those of the whole cp genomes, were shorter than those of others in the two cp genomes of the subspecies *D. glomerata* subsp. *lobata* and *D. glomerata* subsp. *hispanica.1* (Table S2). The total GC content was 38–38.44%, while the average GC contents of the LSC, SSC and IR regions were 36.32%, 32.86% and 43.94%, respectively (Table S2). Moreover, a total of 129 genes were identified in the *D. glomerata* subspecies, excluding 131 genes identified in *D. glomerata* subsp. *hispanica.1*. The number of genes encoding the rRNA and tRNA was highly conserved, with eight genes encoding rRNA, and 38 genes encoding tRNA. The cp genomes of *D. glomerata* contained 83 protein–coding genes, while 85 protein–coding genes were identified in the cp genomes of *D. glomerata* subsp. *hispanica.1* (Table S2). Eight tRNA genes (*trnH–GUG, trnI–CAU, trnL–CAA, trnV–*

GAC, trnI–GAU, trnA–UGC, trnR–ACG and *trnN–GUU*), four rRNA (*rrn4.5, rrn5, rrn16* and *rrn23*), and seven protein–coding genes (*rps19, rpl2, rpl23, ndhB, rps7, rps12* and *rps15*) were duplicated in the IRs, excluding *D glomerata* subsp. *hispanica.1* (Figure 1). A putative gene of unknown function, *ycf15*, was present only in IRs in *D. glomerata* subsp. *hispanica.1* among 14 subspecies of *D. glomerata* (Table S3). Interestingly, we found that a unique gene, *infA*, was distributed in all *Dactylis* subspecies. A total of 21 intron–containing genes were found in the cp genomes of all the subspecies, including 13 protein–coding genes and eight tRNA genes (Table S4). Twelve protein–coding genes and eight tRNA genes contained one intron, and two genes (*ycf3* and *rps12*) contained two introns. Slight differences were found in the sizes of introns between different types of *D. glomerata* subspecies, as shown in Table S4. In all the cp genomes of *D. glomerata* subspecies, the *trnK–UUU* intron, including *matK*, contained the longest introns (2497–2523 bp).

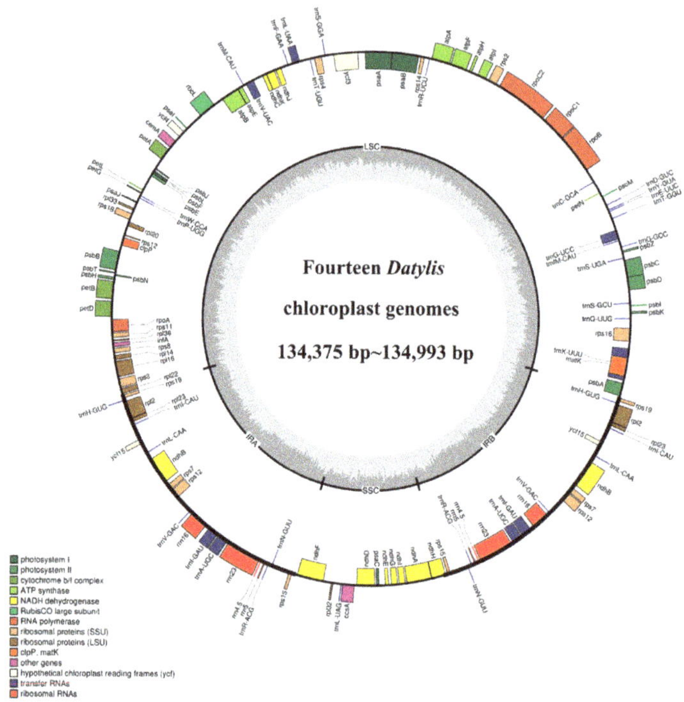

Figure 1. Representative cp genome of *Dactylis* subspecies. Genes drawn inside and outside of the circle are transcribed in clockwise and counter–clockwise directions, respectively. The colored bar indicates chloroplast gene groups. The dark gray bar graphs inner circle shows the GC content, and the light gray bar graphs show the AT content.

3.2. Comparative Genome Analysis

Fourteen cp genomes of the *D. glomerata* subspecies had highly similar sequences (Figure 2). The IR regions were more conserved than the SC regions. Among these cp genomes, the sequences in coding regions were almost identical or nearly identical. On the other hand, sequences in the noncoding regions were highly variable relative to the sequences in coding regions. The highly divergent regions were found in the intergenic spacers and introns, including regions of *rps19–psbA, psbM–petN, trnG–UCC–trnT–GGU, psaA–ycf3, rbcL–psaI, psbE–petL, rps12–trnV–GAC, trnV–GAC–rps12* and *rpl2–rpl23*. These highly divergent regions have the potential to be used for discrimination or phylogeny investigations of the 14 subspecies. Furthermore, we checked the possible rearrangement events, which indicated that the genome structures and gene sequences were basically identical, and no gene rearrangement had occurred (Figure S1).

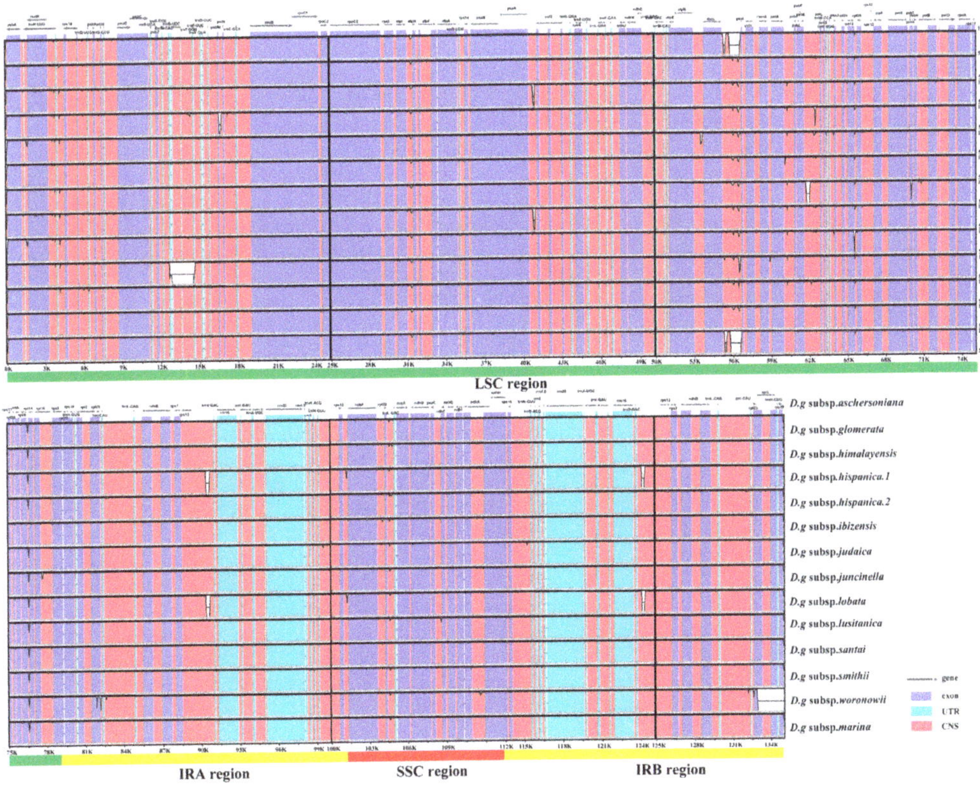

Figure 2. Sequence identity plot for cp genomes of *Dactylis* subspecies with *D.g* subsp. *aschersoniana*, as reference visualized by mVISTA. The gray arrows indicate the orientations of genes, the red blocks represent the intergenic region, the purple blocks represent exons, the blue blocks represent untranslated regions (UTRs), and the Y–axis represents the percent identity within 50–100%.

3.3. IR Contraction and Expansion

The contraction and expansion of the chloroplast IR regions appear to have a substantial role in the process of plant evolution, which is regarded as the main reason for the variation in genome size and gene quantity in different plants. Among the 14 subspecies of *D. glomerata*, *D. glomerata* subsp. *lusitanica* showed the longest cp genome (134,993 bp), with an IR of 21,472 bp, while the shortest cp genome sizes (134,375–134,376 bp) and IRs (21,236 bp) were detected in two subspecies, *D. glomerata* subsp. *lobata* and *D. glomerata* subsp. *hispanica.1* (Figure 3). We compared the IR/SC junctions of the 14 subspecies of *D. glomerata* and found that the IR/SSC junctions varied slightly. The location of *ndhF* was 109–110 downstream of the IRb–SSC junction.

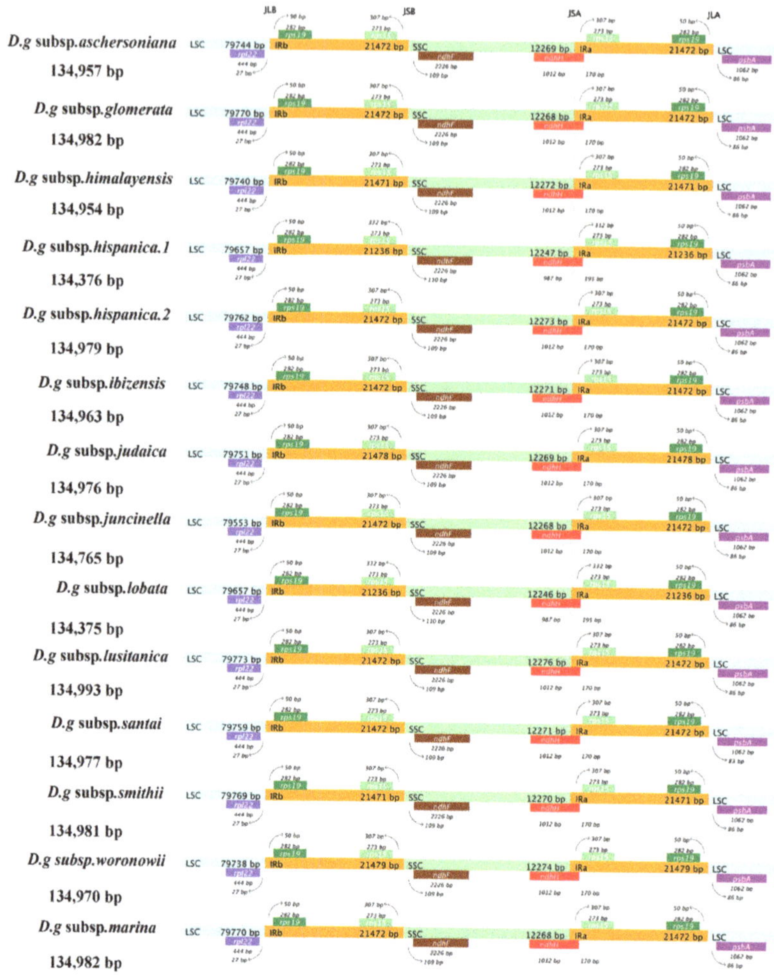

Figure 3. Comparison of the boundaries of LSC, SSC, and IR regions within cp genomes of 14 *D. glomerata* subspecies. Genes are denoted by bars, the gaps between the genes and the boundaries are indicated by the base lengths (bp). Extensions of the genes are indicated above the bars.

The SSC–IRa junction was located within the coding region of *ndhH*, and the partial sequence of *ndhH* was 1012 bp within the SSC region, while only 987 bp of *ndhH* was found within the SSC region in the subspecies *D. glomerata* subsp. *lobata* and *D. glomerata* subsp. *hispanica.1*. The sequence length of *ndhH*, which was located in IRa, was 170 bp, except in *D. glomerata* subsp. *lobata* and *D. glomerata* subsp. *hispanica.1*. Correspondingly, *rps15* was located 332 bp upstream of the IRa/SSC junction in *D. glomerata* subsp. *lobata* and *D. glomerata* subsp. *hispanica.1*, which had a 25 bp difference from the other 12 subspecies.

3.4. Characterization of SSRs and Repeat Sequences

Six categories of SSRs were detected in the cp genomes of 14 subspecies of *D. glomerata* (Figure 4). The number of cpSSRs varied in all the subspecies and ranged from 181 (*D. glomerata* subsp. *ibizensis*) to 185 (*D. glomerata* subsp. *judaica*). Most of the SSRs were mononucleotide SSRs, which comprised 66.65% of the total SSRs, followed by trinucleotide (22.97%) and tetranucleotide repeats (6.60%) (Figure 4B). Moreover, the number of tetranucleotide repeats was higher than the number of pentanucleotide and dinucleotide repeats. Pentanucleotide repeats were found in the cp genomes of 12 subspecies of *D. glomerata* except for *D. glomerata* subsp. *hispanica.2* and *D. glomerata* subsp. *ibizensis*. Notably, the hexanucleotide repeats were exceedingly rare across the cp genomes of these subspecies, and only one was identified in the cp genome of *D. glomerata* subsp. *santai* (Figure 4A). Most of the SSRs were distributed in the LSC regions, followed by the IR and SSC regions (Figure 4C). In contrast to the exon and intron regions, most of the SSRs were detected in the intergenic regions (Figure 4D). Moreover, most of the SSRs were comprised of the repeat type A/T, rather than the repeat type G/C (Table S5).

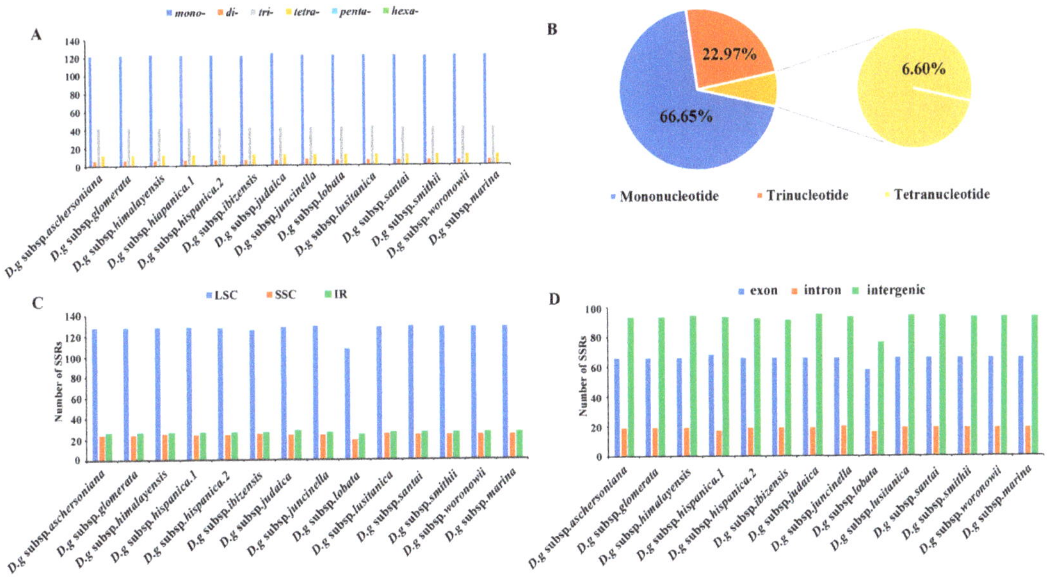

Figure 4. Analysis of simple sequence repeats (SSRs) in the cp genomes of 14 *D. glomerata* subspecies. (**A**) Number of different SSR types detected in cp genomes of *D. glomerata* subspecies; (**B**) The proportion of mononucleotide, trinucleotide and tetranucleotide SSRs; (**C**) Number of SSRs in LSC, SSC and IR regions; (**D**) Number of SSRs in exon, intron, and intergenic regions.

Interspersed repeated sequences are a type of repeat that differ from tandem repeats, which are distributed in a decentralized manner in the genome. Typically, interspersed repeats are classified into four types: forward, palindromic, reverse and complement repeats. In this study, we merely found palindromic repeats and forward repeats, which were similar to the number of interspersed repeats in all the subspecies (Figure 5A,B). The number of interspersed repeats varied from 38 to 44 in all the subspecies. *D. glomerata* subsp. *ibizensis* had the highest number of interspersed repeats, including 24 forward and 20 palindromic repeats. Although there were significant differences between the forward repeats (30–273 bp) and palindromic repeats (30–21,479 bp) in length, the size of most of the interspersed repeats was between 30–35 bp (Figure 5C,D).

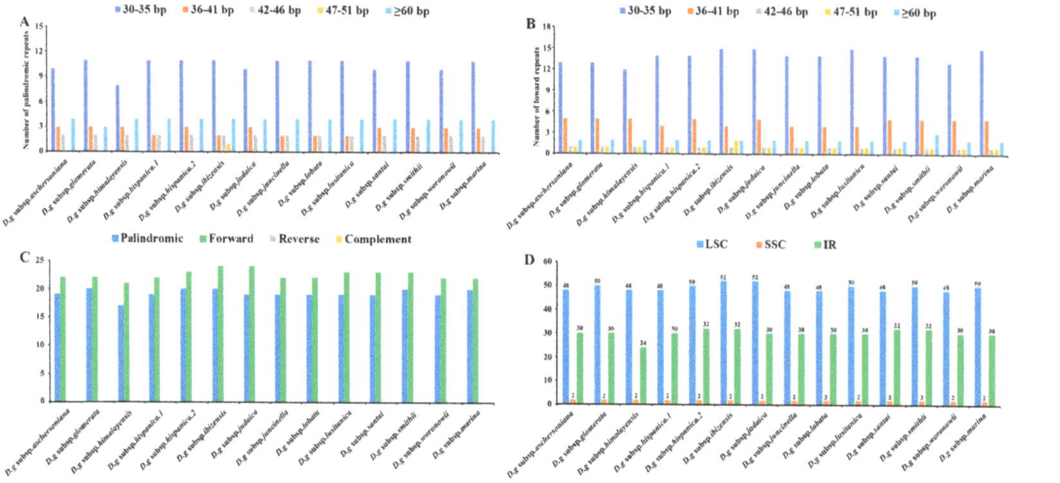

Figure 5. Analyses of repeated sequences in the cp genomes of 14 *D. glomerata* subspecies. (**A**) Number of palindromic repeats; (**B**) Number of forward repeats by length; (**C**) Number of different repeats; (**D**) Numbers of repeats in LSC, SSC and IR regions.

3.5. RNA Editing

We identified 285 RNA editing sites associated with 52 genes in the 14 cp genomes of *D. glomerata* subspecies (Table S6). Twelve types of RNA editing events were detected, including C to T, A to C, A to T, T to C, T to A, C to G, G to C, G to T, T to G, A to G, G to A, and C to A editing. In addition, all the editing events occurred in the region of proteincoding genes. In terms of the number of RNA editing sites of these genes, *rpoB* exhibited the most abundant RNA editing (15 sites), followed by *rpoC1* and *matk* (14 sites). *ndhF* had a value of 13. However, nine genes (*psbH, psbK, psbZ, rpl16, rpl22, rps8, ndhG, ndhI*, and *atpH*) had only one RNA editing site. The RNA editing of six genes (*atpA–1148, ndhF–62, ndhK–128, petB–611, rps8–3*, and *rpoA–527*) occurred in the cp genomes of all the subspecies. Although most of the editing events caused changes in the amino acids, there were 116 sites that did not alter any amino acids, including stop codons, among the subspecies. A total of 130 sites were edited in the third codon position, 79 sites were in the second codon position, and 74 sites were in the first codon position. Many of the editing events increased the hydrophobicity of amino acids, including conversions from hydrophilic amino acids to hydrophobic amino acids (57), from hydrophilic amino acids to hydrophilic amino acids (83), hydrophobic amino acids to hydrophilic amino acids (20), and hydrophobic amino acids to hydrophobic amino acids (107). Among the 169 edited amino acid sites, the largest proportion of changes was from Ser to Leu (13 sites), followed by Ser to Pro (8 sites), and Pro to Leu (5 sites). RNA editing in *D. glomerata* subsp. *glomerata* and *D. glomerata* subsp. *himalayensis* was completely consistent and contained 71 RNA editing sites that were involved in 27 genes. On the other hand, RNA editing of *D. glomerata* subsp. *ibizensis* and *D. glomerata* subsp. *hispanica.1* was different in many cases from those of the other subspecies, and 77 to 78 editing sites were identified in 34 genes of these two subspecies.

3.6. Phylogenetic Analyses

The clade composed of four *Triticum* species was the sister clade to the *D. glomerata* species and had BS of 100% (Figure 6). The *Dactylis* species was strongly supported as a monophyletic group (bootstrap support (BS) = 100%). These 14 subspecies of *D. glomerata* were divided into three groups. Group I contained three subspecies, group II contained 11 materials of 10 subspecies, and group III contained one subspecies. Although a few nodes still had lower BS values, most of the nodes of the ML tree had high BS values (> 80%). In Group II, it was clear that four *D. glomerata* subspecies—the diploid subspecies *aschersoniana*, the tetraploid subspecies *glomerata*, the tetraploid subspecies *woronowii*, and the diploid subspecies *judaica*–were the sister lineages of the diploid subspecies *himalayensis* and demonstrated higher BS (100%). The diploid subspecies *aschersoniana* showed the closest relationship with the tetraploid subspecies *glomerata* followed by the diploid subspecies *judaica* and the tetraploid subspecies *woronowii*. The diploid subspecies *judaica* was closely related to the tetraploid subspecies *woronowii*. The tetraploid subspecies *marina* showed a closer relationship with the evolutionary branches composed of the diploid subspecies *lusitanica* and *ibizensis*. *The* tetraploid subspecies *lobata* and *hispanica.1* are sister species, with BS of 100%.

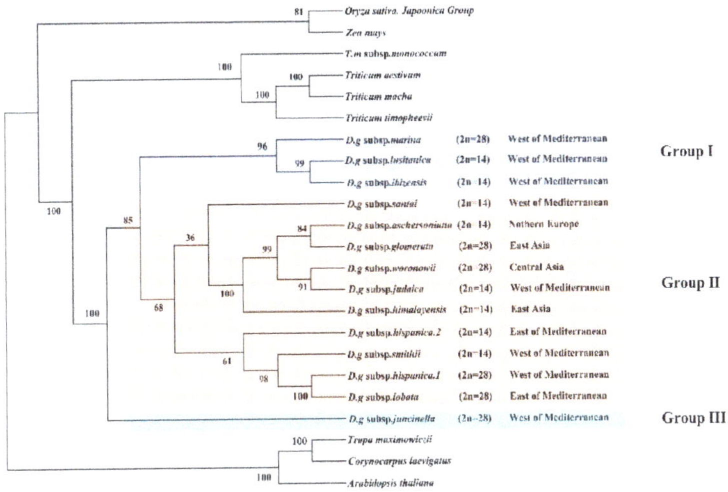

Figure 6. Phylogeny of *Dactylis* species inferred from complete chloroplast genome dataset. Phylogenetic tree constructed from *D. glomerata* subspecies and nine crop complete cp genome sequences using maximum likelihood (ML) method. Bootstrap values are marked above the branches. *Arabidopsis, Corynocarpus* and *Trapa* are used as the outgroups.

3.7. Divergence Time

Serial divergence times from the crop plants to *D. glomerata* were estimated at 53.0985 to 29.5353 Mya (Figure 7). The divergence of the four *Triticum* species (*T. aestivum, T. macha, T. timopheevii*, and *T. monococcum* subsp. *monococcum*) lagged far behind that of *Dactylis glomerata*, which shared a common ancestor at 28.5353 Mya. Our study revealed climate change might have been a driver for the divergence of *D. glomerata*. Based on our results, *D. glomerata* originated in the early Oligocene era (28.5383 Mya) and gradually began to diversify after the mid–Miocene era (15.3985–7.4043 Mya). The divergence times of the tetraploid subspecies *marina* and the diploid subspecies *santi, lusitanica* and *ibizensis* (15.3985 to 13.2162 Mya), all of which originated from the west of the Mediterranean, were much earlier than those of the other remaining subspecies. Moreover, the divergence event of the diploid subspecies *himalayensis*, the diploid subspecies *aschersoniana*, the diploid subspecies *judaica*, and the tetraploid subspecies *glomerata* and *woronowii* occurred in 11.7017 Mya.

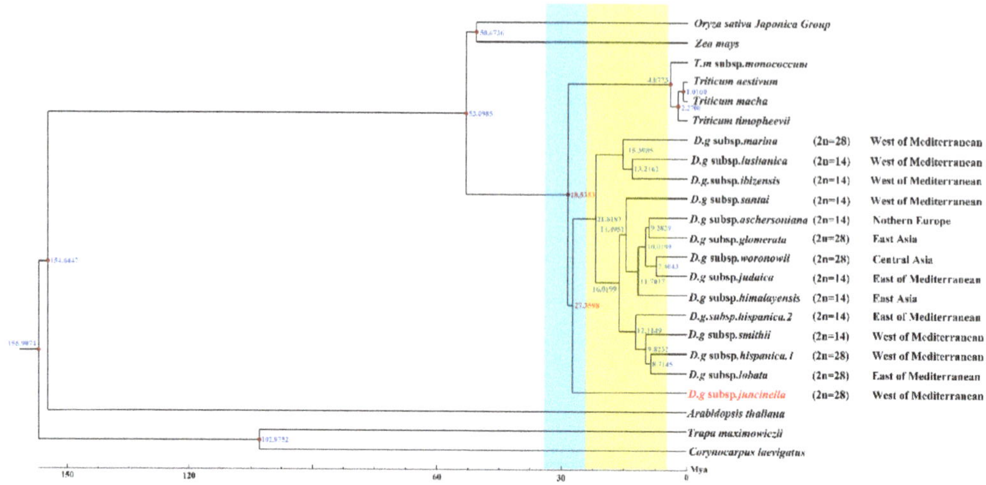

Figure 7. Divergence time of *D. glomerata* subspecies. Molecular dating of 14 *Dactylis* subspecies based on the cp genome sequences in cp genomes. Blue bar represents Oligocene period and yellow bar represents Miocene period.

4. Discussion

4.1. *The Features of Dactylis glomerata Subspecies Cp Genomes*

In this study, we assembled and compared the characteristics of the cp genomes of 14 subspecies of *D. glomerata*, with sizes within the range of angiosperms [35]. In agreement with many higher plants, the cp genomes of all the *D. glomerata* subspecies were highly conserved and displayed a typical quadripartite molecular structure [36,37]. Upon comparing the cp genomes from 14 subspecies of *D. glomerata*, we found that the genome structures were relatively conserved, with no rearrangement occurring in gene organization, but there were some very significant differences in terms of size among the cp genomes of the analyzed *D. glomerata* subspecies. The overall genome, LSC, SSC, and IR regions of the *D. glomerata* subsp. *hispanica.1* and the *D. glomerata* subsp. *lobata* were shorter than those of the others, which might have been caused by the contraction and expansion of the IR and SSC boundary regions [37,38]. As such, the existing differences may provide insights into the unique differences defining Gramineae species and subspecies [39,40]. Though there were some differences in cp, the genome size among the 14 subspecies, the total GC content and the number of genes were similar, partially reflecting the conservation of the cp genome in angiosperms. In terms of the GC content of the cp genomes of the 14 subspecies of *D. glomerata*, there was higher GC content in the IR regions than in the LSC and SSC regions, and this could be the result of there being four rRNA genes in the IRs [41]. A high GC content has always been conducive to the stability of the genome structure, making mutation difficult [42]. Thus, the IR was the most conserved region in the cp genome of the 14 subspecies of *D. glomerata*.

The cp genomes of most angiosperms contain 74 protein–coding genes in their cp genomes; however, five other genes were found among certain species, with variation also being observed in the *D. glomerata* subspecies [43]. Interestingly, a translation initiation factor, *infA*, has been independently lost multiple times during the evolution of different land plants. However, it appears in all the subspecies in this study [43]. Thus, we deduced that the presence of *infA* is an ancestral condition in these subspecies. In addition, a gene with a high frequency of absence from the 14 *D. glomerata* subspecies was *ycf15*. The function of *ycf15* has been a concern in previous studies–its potential as a protein–coding gene in the cp genome of angiosperms has been questioned due to its high frequency as a pseudogene [44–46]. It is estimated that the *ycf15* gene has completely disappeared from about 29 terrestrial plant lineages [47–54]. However, comparative analysis revealed that this gene was found in the *D. glomerata* subsp. *hispanica.1* but that it was not present in other listed *Dactylis* subspecies. Based on these findings, parallel losses of particular genes have occurred over the course of *Dactylis* evolution. In this study, this finding increased our interest in studying the function and evolution of *ycf15* in angiosperms in more detail. Thus, the *ycf15* loss event may provide useful information for a more in–depth study of the evolutionary history of *Dactylis* or of Gramineae species.

4.2. Divergence of Active Regions

In order to explain the level of genome divergence, sequence identity among cp DNAs was detected. Though there was a high degree of similarity at the cp genomic scale, highly divergent regions were found, which included nine non–coding regions (*rps19–psbA*, *psbM–petN*, *trnG*–UCC–*trnT*–GGU, *psaA–ycf3*, *rbcL–psal*, *psbE–petL*, *rps12–trnV*–GAC, *trnV*–GAC–*rps12* and *rpl2–rpl23*), which may be highly useful for further studies of phylogenetic relationships, the identification of species, and population genetics. In general, the cp genomes of the 14 subspecies of *D. glomerata* showed lower divergence levels in their coding regions and in IRs than in their non–coding regions and SSC regions, and the IR regions were the most conserved. The higher GC content in IRs might partially explain the divergence of the conserved nature of the IR and SC regions [42].

IR is an important indicator for measuring the structural stability of cp genomes. Generally, a high number of rearrangements are detected in cp genomes lacking an IR, and these rearrangement events have also been reported in various terrestrial plant lineages [55]. However, the comparative analysis indicated that no rearrangement events took place in the cp genomes of the 14 subspecies of *D. glomerata* with IRs. The presence of an IR could prevent the occurrence of rearrangement events, resulting in a lower species rearrangement rate, which might account for the absence of rearrangement events detected in the 14 subspecies of *D. glomerata* [56]. In addition, large and complex repeat sequences in cp genomes are associated with genome rearrangement and stabilization; they provide important information for understanding the evolutionary history of plant species [57–59]. In this study, about 38 to 44 repeats in the cp genomes of the 14 subspecies of *D. glomerata* were detected, a relatively small number [60]. The size of the repeats was studied in the cp genomes that were sequenced and compared–we found that most of the repeats ranged in size from 30 to 35 bp. Although almost all the repeats were not large repeats (>100 bp), the repeats in *Dactylis* were more abundant than those in other angiosperm cp genomes, ranging between 7 and 13 [61–63]. Given the correlation between repeat sequences and rearrangements, high rearrangement rates are likely to be observed in cp genomes with a high frequency of large repeat sequences (>100 bp) [64]. However, no rearrangement events were found to have occurred in our 14 subspecies of *D. glomerata*, which might be due to a lack of large repeat sequences. It is, therefore, not surprising that rearrangement events were not observed in the 14 *D. glomerata* subspecies, which explains the relatively higher degree of stability and conservation among the different *Dactylis* subspecies.

4.3. IR Contraction and Expansion

Although the IR is the most conserved region in the cp genome, the border contraction and expansion of IRs are regarded as common evolutionary events in plants [65]. Overall, there was a close association between the IR length and genome size. For instance, the *D. glomerata* subsp. *lusitanica* possessed the largest cp genome (134,993 bp) and had a relatively long IR (21,472 bp), whereas the *D. glomerata* subsp. *lobata* and *D. glomerata* subsp. *hispanica.1* had the smallest cp genomes (134,375–134,376 bp) and the shortest IRs (21,236 bp). Analysis of the contraction and expansion of IRs from 14 subspecies of *D. glomerata* revealed that the gene distribution at the boundary of the four regions of the cp genome followed a similar rule. However, relatively independent characteristics were observed in the microstructure, such as the locations of *ndhH*, *ndhF* and *rps15*. In Gramineae, *ndhH* exists near both ends of the SSC region and can extend to the IRs, and it plays an important role in the stability of the IR/SSC junction [66]. As a typical cp genome structure, *ndhH* extends into the IRs in *Dactylis*, which may represent an ancestral symplesiomorphy in Gramineae [67]. Longer extension of *ndhH* into IRa unique to the monophyletic *D glomerata* subsp. *lobata* and subsp. *hispanica.1* cluster among other subspecies indicates that the genome size variation of the 14 subspecies of *D. glomerata* was mainly due to border shifts and variation in IRs [68]. Thus, we speculated that *ndhH* might play an important role in the cp evolution within *Dactylis*. These variations broaden our knowledge of the evolution and genomic structure of the cp genome of *Dactylis*.

4.4. Characterization of SSRs

SSRs are considered the results of slipped strand mispairing during DNA replication, which is usually observed in plant cp genomes [69]. The SSRs of plant cp genomes have been widely used for molecular markers for studying species genetic variations due to a high polymorphism rate at the species level [70]. In total, six types of SSRs were identified, and similar numbers of SSRs were detected in the 14 subspecies of *D. glomerata*. These cpSSRs appeared more frequently in the LSC region than in the SSC and IR regions, consistent with results of other Gramineae species [71]. Among the repeat units that were identified, the mononucleotide repeat (A/T) unit was the most abundant repeat unit, of which A or T repeats accounted for the majority. Our results support the concept that the SSRs in cp genomes are typically composed of short polyadenine (polyA) or polythymine (polyT) repeats and rarely contain tandem guanine (G) or cytosine (C) repeats [72]. Among these SSRs, hexanucleotide repeats showed different distribution patterns. Hexanucleotide repeats were the most common types in all the *Oryza* species and in certain *Phalaris arundinacea* [60,73], while hexanucleotide repeats were species–specific in the present study. An assessment of the SSR categories indicated that hexanucleotide repeats only existed in *D. glomerata* subsp. *santai*. Since the chloroplast is highly conserved in an–giosperms, cpSSRs can be transferred between species and genera [66]. So, the SSRs identified in the cp genome of *Dactylis* could be used as useful molecular markers to identify these *D. glomerata* subspecies and related species in future studies.

4.5. RNA Editing

As a crucial post–transcriptional RNA modification process, RNA editing appears in almost all land plants [74]. In the cp genomes of 14 subspecies of *D. glomerata*, most of the RNA editing events predicted were of C to T. In these 14 subspecies, the amino acid transformation from serine to leucine was the most common type. The dominance of hydrophilic serine to hydrophobic leucine substitution among non–synonymous RNA editing suggests the evolutionary conservation of RNA editing [75]. The RNA editing sites are often detected in the first or second base of codons, causing the conversion of hydrophilic amino acids to hydrophobic amino acids [76,77]. This transformation increases the hydrophobicity of proteins and enhances their stability [78]. In our present study, approximately 57 of the amino acid changes were from hydrophilicity to hydrophobicity, and the increase in amino acid hydrophobicity might facilitate the formation of core residues in proteins [79]. Thus, the structure formed by hydrophobic mutations in the protein kernel is more stable than that formed by hydrophilic mutations, which may ultimately affect the secondary structure and function of proteins and expand their genetic information [75,80].

However, abundant RNA editing sites were identified in the third codon among *D. glomerata* subspecies, which were consistent with some Gramineae plants [81,82]. The chloroplast RNA editing of *D. glomerata* subsp. *ibizensis* and *D. glomerata* subsp. *hispanica.1* varied more than those of the other *D. glomerata* subspecies, which may be due to changes in the environmental adaptability or species specificity during long–term evolution. Thus, some specific RNA editing sites were also identified in some of the subspecies, which provided useful information for the origin and evolution of the *Dactylis* genus based on in–sights from chloroplast RNA editing. To our knowledge, this is the first study to report RNA editing among *D. glomerata* subspecies, elaborating the evolution of 14 subspecies of

D. glomerata with a focus on RNA editing, and laying the foundation for further studies of RNA editing mechanisms in *D. glomerata* and other types of plants.

4.6. Phylogeny Analysis and Divergence Time

Although the phylogenetic relationships among *Dactylis* taxa have already been extensively studied using morphological, cytological, isozyme, phenolic flavonoid, and molecular techniques, the classification of the *Dactylis genus* remains controversial [2,9,83–85]. In the present study, the phylogenetic tree, based on the cp genome, contains subspecies with different ploidy levels, and the high bootstrap support shows the capacity of complete cp genomes to enhance the phylogenetic resolutions during the evolution of *Dactylis*, which might represent useful information on the origin and evolution of *Dactylis* [86,87]. *Dactylis* is strongly supported as a monophyletic group and as the sister group of clades consisting of four species of *Triticeae*. We estimated that *Dactylis* and four *Triticeae* species diverged at 28.5353 Mya, which reinforces previous findings using genomics data (17.5–29.6 Mya) [82]. Diversification events in the lineages of *Dactylis* are primarily predicted to have gradually begun after the mid–Miocene era (15.3985–7.4043 Mya) [88]. It is suggested that there were huge climate changes in the climate after the mid–Miocene era, with the late–Miocene era being colder than the early Oligocene era [88]. This global cooling led to the contraction of tropical forests to lower latitudes, resulting in many open habitats, which presented opportunities for Gramineae plants [89]. As a type of cool grass, the adaptive radiation and diversity of *D. glomerata* subspecies were probably driven by climate change [82,90]. The tetraploid subspecies *marina* and the diploid subspecies *santai, lusitanica* and *ibizensis* from the western Mediterranean were the species observed during the early successive divergence of *Dactylis*, suggesting that the western Mediterranean might have been a diverse phylogenetic center in the past.

This study showed that most phylogenetic clades underwent adaptive radiation of the diploid subspecies followed by the tetraploid species. Obviously, most diploids have a much earlier origin than tetraploids, which is also supported by previous conclusions from studies on flavonoid variability [83,84,87]. Notably, the tetraploid subspecies *marina* clustered with the diploid subspecies *lusitanica* and *ibizensis*, indicating that one of these diploid subspecies, or their common ancestor, could be a parental species of tetraploid subspecies. The morphology of these subspecies also supports this hypothesis since the tetraploid subspecies *marina* possesses characteristic papillose epidermal cells that are very similar to those found in the diploid subspecies *ibizensis*, indicating that they might be derived from a common ancestor [8]. It has previously been suggested that the tetraploid subspecies *glomerata* may have evolved from the hybridization of *aschersoniana* (2n = 14) and *woronowii* (2n = 48) [83]. We found that it formed a clade with the diploid subspecies *aschersoniana* in the cp genome tree, supporting this diploid subspecies as one of its parents. Moreover, our results revealed that the tetraploid subspecies *woronowii* and the diploid subspecies *judaica* were clustered into a clade, indicating a close relationship between the two. Based on morphological studies, mainly of phenotypic characteristics, Stebbins and Zohary considered the tetraploid subspecies *woronowii* to be more similar to the xeromorphic diploid subspecies *judaica*, which has somewhat adapted to a dry summer climate and is considered to be of more recent origin [83]. Thus, these two subspecies may originate from a recent common diploid ancestor. In the present study, the diploid subspecies *himalayensis* forms a well–supported clade with the diploid subspecies *aschersoniana*, tetraploid subspecies *glomerata* and *woronowii*, and the diploid subspecies *judaica*, indicating that the five subspecies are closely related. In addition, the tetraploid subspecies *woronowii* is closely associated with the ancient Eurasian temperate forest flora (including the diploid subspecies *himalayensis* and *aschersoniana*) [83]. As Stebbins and Zohary note, the diploid subspecies *aschersoniana* and the tetraploid subspecies *woronowii* have similar flavonoid components, whereas the diploid subspecies *himalayensis* and *aschersoniana* have more primitive compounds than the tetraploid subspecies *woronowii* [84]. Obviously, the diploid subspecies *himalayensis* and *aschersoniana* originated earlier than the tetraploid subspecies *woronowii*. In this study, the diploid subspecies *himalayensis* was found to share a common ancestor with four other subspecies with different ploidy levels at 11.7017 Mya. The diploid subspecies *himalayensis* revealed similar phylogenetic positions when tetraploids were included, suggesting that the diploid subspecies *himalayensis*, or its ancestral lineage, may have been a parent of these subspecies. Phylogenomic analysis of the cp genomes will help to uncover the mysteries and controversies surrounding the phylogeny of *Dactylis* species. However, the sequencing data of the *Dactylis* subspecies used in this study were limited. Therefore, we recommend the inclusion of cp genome sequences of other subspecies of *D. glomerata* in future studies to help elucidate the evolution from *Dactylis*.

5. Conclusions

This is the first report of the whole cp genome of *D. glomerata* subspecies. The 14 subspecies of *D. glomerata* showed synteny of gene order and contained similar IR boundary regions in their cp genomes. Moreover, we also obtained important genetic information, including SSRs, repeat sequences, divergent hotspot regions, RNA editing sites, and divergence times associated with the relationships between the *D. glomerata* species and other Gramineae. Our results may supply insights to resolve taxonomic discrepancies and phylogenetic relationships within the *D. glomerata* species, accelerating the identification and utilization of forage resources.

Supplementary Materials: The following supporting information can be downloaded at: https://www.mdpi.com/article/10.3390/genes13091621/s1. Table S1: The information of *Dactylis glomerata* subspecies. Table S2: Summary of complete cp genomes of *Dactylis* subspecies; Table S3: Gene composition within chloroplast genomes of *Dactylis* species. Note: Genes *: genes with one intron; Genes **: genes with two introns; Gene (2): Number of copies of multi–copy genes; Genes [a]: represents the *ycf15* is particular in *D.g* subsp. *hispanica.1*; Table S4: Location and length of intron–containing genes of *Dactylis* subspecies cp genomes; Table S5: SSRs of *Dactylis* subspecies cp genomes; Table S6: Comparison of the RNA editing sites in *Dactylis* subspecies cp genomes. Note: "a" indicates that it is completely edited; "b" represents unedited.

Author Contributions: X.Z. and L.H. conceived and designed the experiments; Y.J., G.F., Y.X. and Z.H. performed the experiments; Y.J. and G.F. analyzed the data; G.N. and Z.L. contributed reagents/materials/analysis tools; Y.J. and G.F. wrote the paper; and Y.P. and D.L. reviewed and edited the paper. All authors have read and agreed to the published version of the manuscript.

Funding: This research was funded by the National Natural Science Foundation of China, grant numbers NSFC 31872997 and NSFC 32101422, the Sichuan Province's Science Fund for Distinguished Young Scholars under Grant (2021JDJQ001) and the Sichuan Province's Science Fund for Intertional Cooperation (2022YFH0058). These funding sources contributed to the design of the study, data collection and analysis, and the writing of the manuscript.

Institutional Review Board Statement: No applicable.

Informed Consent Statement: No applicable.

Data Availability Statement: All the data pertaining to the present study has been included in the tables and/or figures in the manuscript. The authors are pleased to share analyzed/raw data and plant materials upon reasonable request. All the cp genome sequences of this study have been de–posited into the CNGB Sequence Archive (CNSA) of the China National Gene Bank Data Base (CNGBdb) with accession number CNP0002292, https://db.cngb.org/mycngbdb/submissions/pro--ject, accessed on 7 April 2022. The plant materials were provided by the Department of Forage Science, College of Grassland Science and Technology, Sichuan Agricultural University, Chengdu, China.

Acknowledgments: The authors would like to express their gratitude to the following: Yanli Xiong, Dandan Li, and Yi Xiong from Sichuan Agricultural University. The authors gratefully acknowledge grants from the National Natural Science Foundation of China (NSFC 31872997 and NSFC 32101422), the Sichuan Province Science Fund for Distinguished Young Scholars under Grant (2021JDJQ0001), and the Sichuan Province Science Fund for Intertional Cooperation (2022YFH0058) for this research.

Conflicts of Interest: The authors declare that they have no conflict of interests.

References

1. Casler, M.D.; Fales, S.L.; Undersander, D.J.; Mcelroy, A.R. Genetic Progress from 40 Years of Orchardgrass Breeding in North America Measured under Hay Management. *Can J. Plant Sci.* **2001**, *81*, 713–721. [CrossRef]
2. Yan, D.; Zhao, X.; Cheng, Y.; Ma, X.; Huang, L.; Zhang, X. Phylogenetic and Diversity Analysis of Dactylis Glomerata Subspecies Using SSR and IT-ISJ Markers. *Molecules* **2016**, *21*, 1459. [CrossRef] [PubMed]
3. Lindner, R.; Garcia, A. Genetic differences between natural populations of diploid and tetraploid *Dactylis Glomerata* ssp. izcoi. *Grass Forage Sci.* **2008**, *52*, 291–297. [CrossRef]
4. Feng, G.; Xu, X.; Xu, L.; Yang, Z.; Nie, G.; Ma, X.; Huang, L.; Zhang, X. Comparative Transcript Profiling Suggests Distinct Flowering Response of Early- and Late-Flowering Phenotypes in Forage Grass *Dactylis glomerata* L. *J. Plant Growth Regul.* **2021**, *40*, 2124–2138. [CrossRef]
5. Stewart, A.V.; Ellison, N.W. A molecular phylogenetic framework for cocksfoot (*Dactylis glomerata* L.) improvement. *Crop. Pasture Sci.* **2014**, *65*, 780–786. [CrossRef]
6. Jones, K.; Borrill, M. Chromosomal status, gene exchange and evolution in *Dactylis*. *Genetica* **1962**, *32*, 296–322. [CrossRef]
7. Lumaret, R.; Guillerm, J.L.; Delay, J.; Loutfi, A.A.L.; Izco, J.; Jay, M. Polyploidy and habitat differentiation in *Dactylis glomerata* L. from Galicia (Spain). *Oecologia* **1987**, *73*, 436–446. [CrossRef]
8. Borrill, M. *Dactylis* marina Borrill, sp. nov., a natural group of related tetraploid forms. *J. Linn. Soc.* **1961**, *56*, 431–439. [CrossRef]
9. Lumaret, R.; Retired, M.B. sCytology, genetics, and evolution in the genus *Dactylis*. *Crit. Rev. Plant Sci.* **1988**, *7*, 55–91. [CrossRef]
10. Neuhaus, H.E.; Emes, M.J. Nonphotosyanthetic metabolism in plastids. *Annu. Rev. Plant Physiol. Plant Mol. Biol.* **2000**, *51*, 111–140. [CrossRef]
11. Arab, M.M.; Brown, P.J.; Abdollahi-Arpanahi, R.; Sohrabi, S.S.; Askari, H.; Aliniaeifard, S.; Mokhtassi-Bidgoli, A.; Mesgaran, M.B.; Leslie, C.A.; Marrano, A.; et al. Genome-Wide Association Analysis and Pathway Enrichment Provide Insights into the Genetic Basis of Photosynthetic Responses to Drought Stress in Persian Walnut. *Hortic. Res.* **2022**, *9*, uhac124. [CrossRef] [PubMed]
12. Bina, H.; Yousefzadeh, H.; Ali, S.S.; Esmailpour, M. Phylogenetic relationships, molecular taxonomy, biogeography of Betula, with emphasis on phylogenetic position of Iranian populations. *Tree Genet. Genomes* **2016**, *12*, 84. [CrossRef]
13. Jansen, R.K.; Wojciechowski, M.F.; Sanniyasi, E.; Lee, S.-B.; Daniell, H. Complete Plastid Genome Sequence of the Chickpea (Cicer Arietinum) and the Phylogenetic Distribution of Rps12 and ClpP Intron Losses among Legumes (Leguminosae). *Mol. Phylogenet. Evol.* **2008**, *48*, 1204–1217. [CrossRef] [PubMed]
14. Wicke, S.; Schneeweiss, G.M.; dePamphilis, C.W.; Müller, K.F.; Quandt, D. The Evolution of the Plastid Chromosome in Land Plants: Gene Content, Gene Order, Gene Function. *Plant Mol. Biol.* **2011**, *76*, 273–297. [CrossRef] [PubMed]
15. Drouin, G.; Daoud, H.; Xia, J. Relative Rates of Synonymous Substitutions in the Mitochondrial, Chloroplast and Nuclear Genomes of Seed Plants. *Mol. Phylogenet. Evol.* **2008**, *49*, 827–831. [CrossRef]
16. Krak, K.; Vít, P.; Belyayev, A.; Douda, J.; Hreusová, L.; Mandák, B. Allopolyploid Origin of Chenopodium Album s. Str. (Chenopodiaceae): A Molecular and Cytogenetic Insight. *PLoS ONE* **2016**, *11*, e0161063. [CrossRef] [PubMed]
17. Sadat-Hosseini, M.; Bakhtiarizadeh, M.R.; Boroomand, N.; Tohidfar, M.; Vahdati, K. Combining Independent de Novo Assemblies to Optimize Leaf Transcriptome of Persian Walnut. *PLoS ONE* **2020**, *15*, e0232005. [CrossRef]
18. Alkan, C.; Sajjadian, S.; Eichler, E.E. Limitations of next-generation genome sequence assembly. *Nat. Methods* **2010**, *8*, 61–65. [CrossRef]
19. Singh, B.P.; Kumar, A.; Kaur, H.; Singh, H.; Nagpal, A.K. CpGDB: A Comprehensive Database of Chloroplast Genomes. *Bioinformation* **2020**, *16*, 171–175. [CrossRef]
20. Chen, S.; Zhou, Y.; Chen, Y.; Gu, J. Fastp: An Ultra-Fast All-in-One FASTQ Preprocessor. *Bioinformatics* **2018**, *34*, i884–i890. [CrossRef]
21. Bankevich, A.; Nurk, S.; Antipov, D.; Gurevich, A.A.; Dvorkin, M.; Kulikov, A.S.; Lesin, V.M.; Nikolenko, S.I.; Pham, S.; Prjibelski, A.D.; et al. SPAdes: A New Genome Assembly Algorithm and Its Applications to Single-Cell Sequencing. *J. Comput. Biol.* **2012**, *19*, 455–477. [CrossRef]
22. Hyatt, D.; Chen, G.-L.; LoCascio, P.F.; Land, M.L.; Larimer, F.W.; Hauser, L.J. Prodigal: Prokaryotic Gene Recognition and Translation Initiation Site Identification. *BMC Bioinform.* **2010**, *11*, 119. [CrossRef]
23. Mistry, J.; Finn, R.D.; Eddy, S.R.; Bateman, A.; Punta, M. Challenges in Homology Search: HMMER3 and Convergent Evolution of Coiled-Coil Regions. *Nucleic Acids Res.* **2013**, *41*, e121. [CrossRef]
24. Laslett, D. ARAGORN, a Program to Detect TRNA Genes and TmRNA Genes in Nucleotide Sequences. *Nucleic Acids Res.* **2004**, *32*, 11–16. [CrossRef]
25. Altschul, S.F.; Gish, W.; Miller, W.; Myers, E.W.; Lipman, D.J. Basic Local Alignment Search Tool. *J. Mol. Biol.* **1990**, *215*, 403–410. [CrossRef]
26. Greiner, S.; Lehwark, P.; Bock, R. OrganellarGenomeDRAW (OGDRAW) Version 1.3.1: Expanded Toolkit for the Graphical Visualization of Organellar Genomes. *Nucleic Acids Res.* **2019**, *47*, W59–W64. [CrossRef]
27. Frazer, K.A.; Pachter, L.; Poliakov, A.; Rubin, E.M.; Dubchak, I. VISTA: Computational Tools for Comparative Genomics. *Nucleic Acids Res.* **2004**, *32*, W273–W279. [CrossRef]
28. Darling, A.C.E.; Mau, B.; Blattner, F.R.; Perna, A.N.T. Mauve: Multiple alignment of conserved genomic sequence with rearrangements. *Genome Res.* **2004**, *14*, 1394–1403. [CrossRef]

29. Thiel, T.; Michalek, W.; Varshney, R.; Graner, A. Exploiting EST Databases for the Development and Characterization of Gene-Derived SSR-Markers in Barley (*Hordeum vulgare* L.). *Theor. Appl. Genet.* **2003**, *106*, 411–422. [CrossRef]
30. Kurtz, S. The Vmatch large scale sequence analysis software. *Cent. Bioinform.* **2011**, *170*, 391–392.
31. Mower, J.P. PREP-Mt: Predictive RNA Editor for Plant Mitochondrial Genes. *BMC Bioinform.* **2005**, *6*, 96. [CrossRef] [PubMed]
32. Capella-Gutierrez, S.; Silla-Martinez, J.M.; Gabaldon, T. TrimAl: A Tool for Automated Alignment Trimming in Large-Scale Phylogenetic Analyses. *Bioinformatics* **2009**, *25*, 1972–1973. [CrossRef] [PubMed]
33. Silvestro, D.; Michalak, I. RaxmlGUI: A Graphical Front-End for RAxML. *Org. Divers. Evol.* **2012**, *12*, 335–337. [CrossRef]
34. Kumar, S.; Stecher, G.; Suleski, M.; Hedges, S.B. TimeTree: A Resource for Timelines, Timetrees, and Divergence Times. *Mol. Biol. Evol.* **2017**, *34*, 1812–1819. [CrossRef]
35. Raubeson, L.A.; Peery, R.; Chumley, T.W.; Dziubek, C.; Fourcade, H.M.; Boore, J.L.; Jansen, R.K. Comparative Chloroplast Genomics: Analyses Including New Sequences from the Angiosperms Nuphar Advena and Ranunculus Macranthus. *BMC Genom.* **2007**, *8*, 174. [CrossRef]
36. Li, X.; Zuo, Y.; Zhu, X.; Liao, S.; Ma, J. Complete Chloroplast Genomes and Comparative Analysis of Sequences Evolution among Seven Aristolochia (Aristolochiaceae) Medicinal Species. *IJMS* **2019**, *20*, 1045. [CrossRef]
37. Yao, X.H.; Tang, P.; Li, Z.Z.; Li, D.W.; Liu, Y.F.; Huang, H.W. The First Complete Chloroplast Genome Sequences in Actinidiaceae: Genome Structure and Comparative Analysis. *PLoS ONE* **2015**, *10*, e0129347.
38. 38. Asaf, S.; Khan, A.L.; Khan, A.; Khan, G.; Lee, I.J.; Al-Harrasi, A. Expanded Inverted Repeat Region with Large Scale Inversion in the First Complete Plastid Genome Sequence of Plantago Ovata. *Sci. Rep.* **2020**, *10*, 3881. [CrossRef]
39. Miller, J.T.; Bayer, R.J. Molecular phylogenetics of *Acacia* (Fabaceae: Mimosoideae) based on the chloroplast *MATK* coding sequence and flanking *TRNK* intron spacer regions. *Am. J. Bot.* **2001**, *88*, 697–705. [CrossRef]
40. Scharaschkin, T.; Doyle, J.A. Phylogeny and historical biogeography of Anaxagorea (Annonaceae) using morphology and non-coding chloroplast sequence data. *Syst. Bot.* **2005**, *30*, 712–735. [CrossRef]
41. Qian, J.; Song, J.; Gao, H.; Zhu, Y.; Xu, J.; Pang, X.; Yao, H.; Sun, C.; Li, X.; Li, C.; et al. The Complete Chloroplast Genome Sequence of the Medicinal Plant Salvia Miltiorrhiza. *PLoS ONE* **2013**, *8*, e57607.
42. Terakami, S.; Matsumura, Y.; Kurita, K.; Kanamori, H.; Katayose, Y.; Yamamoto, T.; Katayama, H. Complete Sequence of the Chloroplast Genome from Pear (Pyrus Pyrifolia): Genome Structure and Comparative Analysis. *Tree Genet. Genomes* **2012**, *8*, 841–854. [CrossRef]
43. Millen, R.S.; Olmstead, R.G.; Adams, K.L.; Palmer, J.D.; Lao, N.T.; Heggie, L.; Kavanagh, T.A.; Hibberd, J.M.; Gray, J.C.; Morden, C.W.; et al. Many Parallel Losses of InfA from Chloroplast DNA during Angiosperm Evolution with Multiple Independent Transfers to the Nucleus. *Cochrane Database Syst. Rev.* **2001**, *13*, 645–658.
44. Goremykin, V.; Hirsch-Ernst, K.I.; Wolfl, S.; Hellwig, F.H. The Chloroplast Genome of the "Basal" Angiosperm Calycanthus Fertilis-Structural and Phylogenetic Analyses. *Plant Syst. Evol.* **2003**, *242*, 119–135. [CrossRef]
45. Steane, D.A. Complete Nucleotide Sequence of the Chloroplast Genome from the Tasmanian Blue Gum, Eucalyptus Globulus (Myrtaceae). *DNA Res.* **2005**, *12*, 215–220. [CrossRef]
46. Schmitz-Linneweber, C.; Maier, R.M.; Alcaraz, J.-P.; Cottet, A.; Herrmann, R.G.; Mache, R. The Plastid Chromosome of Spinach (Spinacia Oleracea): Complete Nucleotide Sequence and Gene Organization. *Plant Mol. Biol.* **2001**, *45*, 307–315. [CrossRef]
47. Kim, H.T.; Chung, M.G.; Kim, K.-J. Chloroplast Genome Evolution in Early Diverged Leptosporangiate Ferns. *Mol. Cells* **2014**, *37*, 372–382. [CrossRef]
48. Wakasugi, T.; Tsudzuki, J.; Ito, S.; Nakashima, K.; Tsudzuki, T.; Sugiura, M. Loss of All Ndh Genes as Determined by Sequencing the Entire Chloroplast Genome of the Black Pine Pinus Thunbergii. *Proc. Natl. Acad. Sci. USA* **1994**, *91*, 9794–9798. [CrossRef]
49. Wu, C.S.; Lai, Y.T.; Lin, C.P.; Wang, Y.N.; Chaw, S.M. Evolution of reduced and compact chloroplast genomes (cp DNAs) in gnetophytes: Selection toward a lower-cost strategy. *Mol. Phylogenet. Evol.* **2009**, *52*, 115–124. [CrossRef]
50. Tsudzuki, J.; Ito, S.; Tsudzuki, T.; Wakasugi, T.; Sugiura, M. A new gene encoding tRNA (GGG) is present in the chloroplast genome of black pine: A compilation of 32 tRNA genes from black pine chloroplasts. *Curr. Genet.* **1994**, *26*, 153–158. [CrossRef]
51. Li, Z.H.; Qian, Z.H.; Liu, Z.L.; Deng, T.T.; Zu, Y.M.; Zhao, P.; Zhao, G.F. The complete chloroplast genome of Armand pine Pinus armandii, an endemic conifer tree species to China. *Mitochondrial. DNA A* **2016**, *27*, 2635–2636. [CrossRef] [PubMed]
52. Morton, B.R.; Clegg, M.T. A Chloroplast DNA Mutational Hotspot and Gene Conversion in a Noncoding Region near RbcL in the Grass Family (Poaceae). *Curr. Genet.* **1993**, *24*, 357–365. [CrossRef]
53. Ogihara, Y.; Isono, K.; Kojima, T.; Endo, A.; Hanaoka, M.; Shiina, T.; Terachi, T.; Utsugi, S.; Murata, M.; Mori, N.; et al. Structural Features of a Wheat Plastome as Revealed by Complete Sequencing of Chloroplast DNA. *Mol. Gene Genom.* **2002**, *266*, 740–746. [CrossRef] [PubMed]
54. McNeal, J.R.; Kuehl, J.V.; Boore, J.L.; de Pamphilis, C.W. Complete Plastid Genome Sequences Suggest Strong Selection for Retention of Photosynthetic Genes in the Parasitic Plant Genus Cuscuta. *BMC Plant Biol.* **2007**, *7*, 57. [CrossRef]
55. Sveinsson, S.; Cronk, Q. Evolutionary Origin of Highly Repetitive Plastid Genomes within the Clover Genus (Trifolium). *BMC Evol. Biol.* **2014**, *14*, 228. [CrossRef]
56. Palmer, J.D.; Osorio, B.; Aldrich, J.; Thompson, W.F. Chloroplast DNA Evolution among Legumes: Loss of a Large Inverted Repeat Occurred Prior to Other Sequence Rearrangements. *Curr. Genet.* **1987**, *11*, 275–286. [CrossRef]
57. Cavalier-Smith, T. Chloroplast Evolution: Secondary Symbiogenesis and Multiple Losses. *Curr. Biol.* **2002**, *12*, R62–R64. [CrossRef]

58. Milligan, B.G.; Hampton, J.N.; Palmer, J.D. Dispersed Repeats and Structural Reorganization in Subclover Chloroplast DNA. *Mol. Biol. Evol.* **1989**, *6*, 355–368.
59. Lin, W.-H.; Kussell, E. Evolutionary Pressures on Simple Sequence Repeats in Prokaryotic Coding Regions. *Nucleic Acids Res.* **2012**, *40*, 2399–2413. [CrossRef]
60. Xiong, Y.; Xiong, Y.; Jia, S.; Ma, X. The Complete Chloroplast Genome Sequencing and Comparative Analysis of Reed Canary Grass (*Phalaris arundinacea*) and Hardinggrass (*P. aquatica*). *Plants* **2020**, *9*, 748. [CrossRef]
61. Yang, M.; Zhang, X.; Liu, G.; Yin, Y.; Chen, K.; Yun, Q.; Zhao, D.; Al-Mssallem, I.S.; Yu, J. The Complete Chloroplast Genome Sequence of Date Palm (*Phoenix Dactylifera* L.). *PLoS ONE* **2010**, *5*, e12762. [CrossRef]
62. Uthaipaisanwong, P.; Chanprasert, J.; Shearman, J.R.; Sangsrakru, D.; Yoocha, T.; Jomchai, N.; Jantasuriyarat, C.; Tragoonrung, S.; Tangphatsornruang, S. Characterization of the Chloroplast Genome Sequence of Oil Palm (Elaeis Guineensis Jacq.). *Gene* **2012**, *500*, 172–180. [CrossRef]
63. Huang, Y.-Y.; Matzke, A.J.M.; Matzke, M. Complete Sequence and Comparative Analysis of the Chloroplast Genome of Coconut Palm (Cocos Nucifera). *PLoS ONE* **2013**, *8*, e74736. [CrossRef]
64. Guisinger, M.M.; Kuehl, J.V.; Boore, J.L.; Jansen, R.K. Extreme Reconfiguration of Plastid Genomes in the Angiosperm Family Geraniaceae: Rearrangements, Repeats, and Codon Usage. *Mol. Biol. Evol.* **2011**, *28*, 583–600. [CrossRef] [PubMed]
65. He, L.; Qian, J.; Li, X.; Sun, Z.; Xu, X.; Chen, S. Complete Chloroplast Genome of Medicinal Plant Lonicera Japonica: Genome Rearrangement, Intron Gain and Loss, and Implications for Phylogenetic Studies. *Molecules* **2017**, *22*, 249. [CrossRef]
66. Davis, J.I.; Soreng, R.J. Migration of Endpoints of Two Genes Relative to Boundaries between Regions of the Plastid Genome in the Grass Family (Poaceae). *Am. J. Bot.* **2010**, *97*, 874–892. [CrossRef]
67. Ma, P.-F.; Liu, Y.-L.; Jin, G.-H.; Liu, J.-X.; Wu, H.; He, J.; Guo, Z.-H.; Li, D.-Z. The *Pharus Latifolius* Genome Bridges the Gap of Early Grass Evolution. *Plant Cell* **2021**, *33*, 846–864. [CrossRef]
68. Li, W.; Zhang, C.; Guo, X.; Liu, Q.; Wang, K. Complete Chloroplast Genome of Camellia Japonica Genome Structures, Comparative and Phylogenetic Analysis. *PLoS ONE* **2019**, *14*, e0216645. [CrossRef]
69. Levinson, G.; Gutman, G.A. Slipped-Strand Mispairing: A Major Mechanism for DNA Sequence Evolution. *Mol. Biol. Evol.* **1987**, *4*, 203–221.
70. Pandey, S.; Yadav, P.S.; Ansari, W.A.; Pandey, M.; Yang, L.; Singh, B.; Dubey, R.K.; Singh, P.M.; Singh, J. Development of High Conserved Cross-Species Microsatellite Markers from Cucumber Genome and Their Applicability in Genetic Diversity and Comparative Mapping. *Sci. Hortic.* **2021**, *288*, 110408. [CrossRef]
71. Kuang, D.Y.; Wu, H.; Wang, Y.L.; Gao, L.M.; Zhang, S.Z.; Lu, L. Complete chloroplast genome sequence of Magnolia kwangsiensis (Magnoliaceae): Implication for DNA barcoding and population genetics. *Genome* **2011**, *54*, 663–673. [CrossRef] [PubMed]
72. Tripathy, K.; Singh, B.; Misra, G.; Singh, N.K. Identification, distribution and comparative analysis of microsatellites in the chloroplast genome of Oryza species. *Indian J. Genet. Pl. Br.* **2019**, *79*, 536–544. [CrossRef]
73. Hiesel, R.; Combettes, B.; Brennicke, A. Evidence for RNA editing in mitochondria of all major groups of land plants except the Bryophyta. *Proc. Natl. Acad. Sci. USA* **1994**, *91*, 629–633. [CrossRef]
74. He, P.; Huang, S.; Xiao, G.; Zhang, Y.; Yu, J. Abundant RNA Editing Sites of Chloroplast Protein-Coding Genes in Ginkgo Biloba and an Evolutionary Pattern Analysis. *BMC Plant Biol.* **2016**, *16*, 257. [CrossRef]
75. Brenner, W.G.; Mader, M.; Müller, N.A.; Hoenicka, H.; Schroeder, H.; Zorn, I.; Fladung, M.; Kersten, B. High Level of Conservation of Mitochondrial RNA Editing Sites Among Four Populus Species. *G3-Genes Genom. Genet.* **2019**, *9*, 709–717. [CrossRef]
76. Odintsova, M.S.; Yurina, N.P. RNA editing in plant chloroplasts and mitochondria. *Russ. J. Plant Phys1.* **2000**, *47*, 274–284.
77. Koo, H.J.; Yang, T.-J. RNA Editing May Stabilize Membrane-Embedded Proteins by Increasing Phydrophobicity: A Study of Zanthoxylum Piperitum and Z. Schinifolium Chloroplast NdhG. *Gene* **2020**, *746*, 144638. [CrossRef]
78. Chen, T.C.; Su, Y.Y.; Wu, C.H.; Liu, Y.C.; Huang, C.H.; Chang, C.C. Analysis of mitochondrial genomics and transcriptomics reveal abundant RNA edits and differential editing status in moth orchid, *Phalaenopsis aphrodite* subsp. formosana. *Sci. Hortic.* **2020**, *267*, 109304. [CrossRef]
79. Halder, R.; Jana, B. Exploring the role of hydrophilic amino acids in unfolding of protein in aqueous ethanol solution. *Proteins* **2021**, *89*, 116–125. [CrossRef]
80. Shi, W.Q.; Deng, P.C.; Li, B.L.; Niu, S.C.; Nie, X.J.; Wang, L.; Baloch, A.W.; Song, W.N. Prediction and Identification of RNA Editing Sites in Chloroplast Transcripts of *Brachypodium distachyon*. *Turk. J. Field Crops* **2012**, *32*, 28–35.
81. Chen, H.Y.; Deng, L.K.; Jiang, Y.; Lu, P.; Yu, J.N. RNA Editing Sites Exist in Protein-coding genes in the Chloroplast Genome of Cycas taitungensis. *J. Integr. Plant Biol.* **2011**, *53*, 961–970. [CrossRef]
82. Huang, L.K.; Feng, G.Y.; Yan, H.D.; Zhang, Z.R.; Bushman, B.S.; Wang, J.P.; Bombarely, A.; Li, M.Z.; Yang, Z.F.; Nie, G.; et al. Genome assembly provides insights into the genome evolution and flowering regulation of orchardgrass. *Plant Biotechnol. J.* **2020**, *18*, 373–388. [CrossRef]
83. Stebbins, G.L.; Zohary, D. Cytogenetic and evolutionary studies in the genus *Dactylis*: I: Morphology, distribution and interrelationships of the diploid subspecies. *Univ. Calif Berkeley Publ. Bot.* **1959**, *31*, 1.
84. Fiasson, J.-L.; Ardouin, P.; Jay, M. A Phylogenetic Groundplan of the Specific Complex Dactylis Glomerata. *Biochem. Syst. Ecol.* **1987**, *15*, 225–229. [CrossRef]
85. Moore, M.J.; Bell, C.D.; Soltis, P.S.; Soltis, D.E. Using Plastid Genome-Scale Data to Resolve Enigmatic Relationships among Basal Angiosperms. *Proc. Natl. Acad. Sci. USA* **2007**, *104*, 19363–19368. [CrossRef]

86. Moore, M.J.; Soltis, P.S.; Bell, C.D.; Burleigh, J.G.; Soltis, D.E. Phylogenetic Analysis of 83 Plastid Genes Further Resolves the Early Diversification of Eudicots. *Proc. Natl. Acad. Sci. USA* **2010**, *107*, 4623–4628. [CrossRef]
87. Stebbins, G.L. Cytogenetics and evolution in the grass family. *Am. J. Bot.* **1956**, *43*, 890. [CrossRef]
88. Beerling, D.J.; Royer, D.L. Convergent Cenozoic CO_2 history. *Nat. Geosci.* **2011**, *4*, 418–420. [CrossRef]
89. Prothero, D.R. The late Eocene-Oligocene extinctions. *Annu. Rev. Earth Planet Sci.* **1994**, *22*, 145–165. [CrossRef]
90. Zhang, L.; Zhu, X.; Zhao, Y.; Guo, J.; Zhang, T.; Huang, W.; Huang, J.; Hu, Y.; Huang, C.-H.; Ma, H. Phylotranscriptomics Resolves the Phylogeny of Pooideae and Uncovers Factors for Their Adaptive Evolution. *Mol. Biol. Evol.* **2022**, *39*, msac026. [CrossRef]

MDPI
St. Alban-Anlage 66
4052 Basel
Switzerland
www.mdpi.com

Genes Editorial Office
E-mail: genes@mdpi.com
www.mdpi.com/journal/genes

www.ingramcontent.com/pod-product-compliance
Lightning Source LLC
Chambersburg PA
CBHW061500220926
51850CB00004B/136
* 9 7 8 3 7 2 5 8 0 4 6 4 1 *